Africa 68-69

70

74

76

72

78

Europe

84

82

94

98

108

106

88

86

90

100

110

92

96

104

102

ESSENTIAL
WORLD
ATLAS

FOR THE NINTH EDITION

SENIOR CARTOGRAPHIC EDITOR Simon Mumford
PRODUCTION CONTROLLER Vivienne Yong   PRODUCER, PRE-PRODUCTION Luca Frassinetti
JACKET DESIGN DEVELOPMENT Sophia MTT
PUBLISHER Jonathan Metcalf   ASSOCIATE PUBLISHER Liz Wheeler
ART DIRECTOR Karen Self

FOR PREVIOUS EDITIONS

DORLING KINDERSLEY CARTOGRAPHY

PROJECT CARTOGRAPHY AND DESIGN
Julia Lunn, Julie Turner

CARTOGRAPHERS
James Anderson, Roger Bullen, Martin Darlison, Simon Mumford, John Plumer, Peter Winfield

DESIGN
Katy Wall

INDEX-GAZETTEER
Natalie Clarkson, Ruth Duxbury, Margaret Hynes, Margaret Stevenson

EDITORIAL DIRECTION
Andrew Heritage

ART DIRECTION
Chez Picthall

This American edition 2016. First American edition 1997. Second edition 1998. Third edition 2001.
Fourth edition 2003. Fifth edition 2005. Sixth edition 2008. Seventh edition 2011. Eighth edition 2013.
Published in the United States by DK Publishing,
345 Hudson Street, New York, New York 10014

A catalog record for this book is available from the Library of Congress.
ISBN: 978-1-4654-5069-2

DK books are available at special discounts when purchased
in bulk for sales promotions, premiums, fund-raising, or educational use. For details, contact:
DK Publishing Special Markets, 345 Hudson Street, New York, New York 10014

SpecialSales@dk.com

Printed in Malaysia

A WORLD OF IDEAS:
SEE ALL THERE IS TO KNOW

www.dk.com

Key to map symbols

Physical features

Elevation

6000m/19,686ft

4000m/13,124ft

3000m/9843ft

2000m/6562ft

1000m/3281ft

500m/1640ft

250m/820ft

0

Below sea level

△ Mountain

▽ Depression

⌅ Volcano

)(Pass/tunnel

Sandy desert

Drainage features

Major perennial river

Minor perennial river

- - - Seasonal river

Canal

| Waterfall

Perennial lake

Seasonal lake

Wetland

Ice features

Permanent ice cap/ice shelf

Winter limit of pack ice

Summer limit of pack ice

Borders

Full international border

- - - Disputed de facto border

· · · · Territorial claim border

×–×–× Cease-fire line

- - - Undefined boundary

Internal administrative boundary

Communications

Major road

Minor road

Railroad

✈ International airport

Settlements

◼ Above 500,000

◉ 100,000 to 500,000

○ 50,000 to 100,000

○ Below 50,000

● National capital

◉ Internal administrative capital

Miscellaneous features

+ Site of interest

⌇⌇⌇⌇ Ancient wall

Graticule features

Line of latitude/longitude/ Equator

- - - Tropic/Polar circle

25° Degrees of latitude/ longitude

Names

Physical features

Andes
Sahara   Landscape features
Ardennes

Land's End   Headland

Mont Blanc   Elevation/volcano/pass
4,807m

Blue Nile   River/canal/waterfall

Ross Ice Shelf   Ice feature

PACIFIC
OCEAN
Sulu Sea   Sea features
Palk Strait

Chile Rise   Undersea feature

Regions

FRANCE   Country

JERSEY   Dependent territory
(to UK)

KANSAS   Administrative region

Dordogne   Cultural region

Settlements

PARIS   Capital city

SAN JUAN   Dependent territory capital city

Chicago
Kettering   Other settlements
Burke

Inset map symbols

Urban area

City

Park

▫ Place of interest

▫ Suburb/district

Contents

The World Today

Flags of the World..................................6-7
The Political World...............................8-9
The Physical World..............................10-11
Standard Time Zones.............................12-13
Geology & Structure.............................14-15
World Climate...................................16-17
Ocean Currents..................................18-19
Life Zones......................................20-21
Population......................................22-23
Languages.......................................24-25
Religion..26-27
The Global Economy..............................28-29
Politics and Conflict...........................30-31

The World's Regions

North & Central America

North & Central America.........................34-35
Western Canada & Alaska.........................36-37
Eastern Canada..................................38-39
USA: The Northeast..............................40-41
USA: The Southeast..............................42-43
   Bermuda
USA: Central States.............................44-45
USA: The West...................................46-47
   Los Angeles & Hawaii
USA: The Southwest..............................48-49
Mexico..50-51
Central America.................................52-53
The Caribbean...................................54-55
   Jamaica, St. Lucia, & Barbados

South America

South America...................................56-57
Northern South America..........................58-59
Western South America...........................60-61
   Galápagos Islands
Brazil..62-63
Southern South America..........................64-65
The Atlantic Ocean..............................66-67

Africa

Africa..68-69
Northwest Africa................................70-71
Northeast Africa................................72-73
West Africa.....................................74-75
Central Africa..................................76-77
   Sao Tome & Principe
Southern Africa.................................78-79

Europe

Europe..80-81
The North Atlantic..............................82-83
Scandinavia & Finland...........................84-85
The Low Countries...............................86-87
The British Isles...............................88-89
   London

continued....

Europe *continued*

France, Andorra, & Monaco................90-91
  Paris, Andorra, & Monaco
Spain & Portugal................92-93
  Azores & Gibraltar
Germany & the Alpine States............94-95
  Liechtenstein
Italy................96-97
  San Marino & Vatican City
Central Europe................98-99
Southeast Europe................100-101
  Bosnia & Herzegovina
The Mediterranean................102-103
  Malta & Cyprus
Bulgaria & Greece................104-105
The Baltic States & Belarus............106-107
Ukraine, Moldova,
  & Romania................108-109
European Russia................110-111

North & West Asia

North & West Asia................112-113
Russia & Kazakhstan................114-115
Turkey & the Caucasus................116-117
The Near East................118-119
  West Bank
The Middle East................120-121
Central Asia................122-123

South & East Asia

South & East Asia................124-125
Western China & Mongolia............126-127
Eastern China & Korea................128-129
  Hong Kong
Japan................130-131
  Tokyo & Nansei-Shoto
South India & Sri Lanka................132-133
North India, Pakistan,
  & Bangladesh................134-135
Mainland Southeast Asia................136-137
Maritime Southeast Asia................138-139
  Singapore
The Indian Ocean................140-141

Australasia & Oceania

Australasia & Oceania................142-143
The Southwest Pacific................144-145
Western Australia................146-147
Eastern Australia................148-149
  Sydney
New Zealand................150-151
The Pacific Ocean................152-153
Antarctica................154
The Arctic Ocean................155

Index– Gazetteer

Overseas Territories
  & Dependencies................156-159
Glossary of Geographical
  Terms................160-161
Continental Factfiles................162-163
Geographical Comparisons............164-166
Index-Gazetteer of Places
  & Countries................167-256

Flags of the World

NORTH & CENTRAL AMERICA

CANADA
PAGES 36-39

UNITED STATES
OF AMERICA
PAGES 40-49

MEXICO
PAGES 50-51

BELIZE
PAGES 52-53

COSTA RICA
PAGES 52-53

EL SALVADOR
PAGES 52-53

GUATEMALA
PAGES 52-53

HONDUR
PAGES 52-

SOUTH

GRENADA
PAGES 54-55

HAITI
PAGES 54-55

JAMAICA
PAGES 54-55

ST KITTS & NEVIS
PAGES 54-55

ST LUCIA
PAGES 54-55

ST VINCENT &
THE GRENADINES
PAGES 54-55

TRINIDAD &
TOBAGO
PAGES 54-55

COLOMB
PAGES 58-5

AFRICA

URUGUAY
PAGES 64-65

CHILE
PAGES 64-65

PARAGUAY
PAGES 64-65

ALGERIA
PAGES 70-71

LIBYA
PAGES 70-71

MOROCCO
PAGES 70-71

TUNISIA
PAGES 70-71

BURUND
PAGES 72-

SUDAN
PAGES 72-73

TANZANIA
PAGES 72-73

UGANDA
PAGES 74-75

BENIN
PAGES 74-75

BURKINA FASO
PAGES 74-75

CAPE VERDE
PAGES 74-75

CÔTE D'IVOIRE
(IVORY COAST)
PAGES 74-75

GAMBIA
PAGES 74-75

GHANA
PAGES 74-

SIERRA
LEONE
PAGES 74-75

TOGO
PAGES 74-75

CAMEROON
PAGES 76-77

CENTRAL AFRICAN
REPUBLIC
PAGES 76-77

CHAD
PAGES 76-77

CONGO
PAGES 76-77

DEM. REP.
CONGO
PAGES 76-77

EQUATOR
GUINEA
PAGES 76-

MAURITIUS
PAGES 78-79

MOZAMBIQUE
PAGES 78-79

NAMIBIA
PAGES 78-79

SEYCHELLES
PAGES 78-79

SOUTH
AFRICA
PAGES 78-79

SWAZILAND
PAGES 78-79

ZAMBIA
PAGES 78-79

ZIMBABV
PAGES 78-

UNITED
KINGDOM
PAGES 88-89

FRANCE
PAGES 90-91

MONACO
PAGES 90-91

ANDORRA
PAGES 90-91

PORTUGAL
PAGES 92-93

SPAIN
PAGES 92-93

AUSTRIA
PAGES 94-95

GERMAN
PAGES 94-

POLAND
PAGES 98-99

SLOVAKIA
PAGES 98-99

ALBANIA
PAGES 100-101

BOSNIA &
HERZEGOVINA
PAGES 100-101

CROATIA
PAGES 100-101

KOSOVO
PAGES 100-101

MACEDONIA
PAGES 100-101

MONTENE
PAGES 100-

ASIA

MOLDOVA
PAGES 108-109

ROMANIA
PAGES 108-109

UKRAINE
PAGES 108-109

RUSSIAN
FEDERATION
PAGES 110-115

KAZAKHSTAN
PAGES 114-115

ARMENIA
PAGES 116-117

AZERBAIJAN
PAGES 116-117

GEORGI
PAGES 116-

KUWAIT
PAGES 120-121

OMAN
PAGES 120-121

QATAR
PAGES 120-121

SAUDI ARABIA
PAGES 120-121

UNITED ARAB
EMIRATES
PAGES 120-121

YEMEN
PAGES 120-121

AFGHANISTAN
PAGES 122-123

KYRGYZST
PAGES 122-

JAPAN
PAGES 130-131

INDIA
PAGES 132-135

SRI LANKA
PAGES 132-133

MALDIVES
PAGES 132-133

PAKISTAN
PAGES 134-135

BANGLADESH
PAGES 134-135

BHUTAN
PAGES 134-135

NEPAL
PAGES 134-135

CAMBOD
PAGES 136-

AUSTRALASIA & OCEANIA

PHILIPPINES
PAGES 138-139

SINGAPORE
PAGES 138-139

FIJI
PAGES 144-145

KIRIBATI
PAGES 144-145

MARSHALL
ISLANDS
PAGES 144-145

MICRONESIA
PAGES 144-145

NAURU
PAGES 144-145

PALAU
PAGES 144-

NICARAGUA
PAGES 52-53

PANAMA
PAGES 52-53

ANTIGUA & BARBUDA
PAGES 54-55

BAHAMAS
PAGES 54-55

BARBADOS
PAGES 54-55

CUBA
PAGES 54-55

DOMINICA
PAGES 54-55

DOMINICAN REPUBLIC
PAGES 54-55

GUYANA
PAGES 58-59

SURINAME
PAGES 58-59

VENEZUELA
PAGES 58-59

BOLIVIA
PAGES 60-61

ECUADOR
PAGES 60-61

PERU
PAGES 60-61

BRAZIL
PAGES 62-63

ARGENTINA
PAGES 64-65

DJIBOUTI
PAGES 72-73

EGYPT
PAGES 72-73

ERITREA
PAGES 72-73

ETHIOPIA
PAGES 72-73

KENYA
PAGES 72-73

RWANDA
PAGES 72-73

SOMALIA
PAGES 72-73

SOUTH SUDAN
PAGES 72-73

GUINEA
PAGES 74-75

GUINEA–BISSAU
PAGES 74-75

LIBERIA
PAGES 74-75

MALI
PAGES 74-75

MAURITANIA
PAGES 74-75

NIGER
PAGES 74-75

NIGERIA
PAGES 74-75

SENEGAL
PAGES 74-75

GABON
PAGES 76-77

SAO TOME & PRINCIPE
PAGES 76-77

ANGOLA
PAGES 78-79

BOTSWANA
PAGES 78-79

COMOROS
PAGES 78-79

LESOTHO
PAGES 78-79

MADAGASCAR
PAGES 78-79

MALAWI
PAGES 78-79

EUROPE

ICELAND
PAGES 82-83

DENMARK
PAGES 84-85

FINLAND
PAGES 84-85

NORWAY
PAGES 84-85

SWEDEN
PAGES 84-85

BELGIUM
PAGES 86-87

LUXEMBOURG
PAGES 86-87

NETHERLANDS
PAGES 86-87

IRELAND
PAGES 88-89

LIECHTENSTEIN
PAGES 94-95

SLOVENIA
PAGES 94-95

SWITZERLAND
PAGES 94-95

ITALY
PAGES 96-97

MALTA
PAGES 96-97

SAN MARINO
PAGES 96-97

VATICAN CITY
PAGES 96-97

CZECH REPUBLIC
PAGES 98-99

HUNGARY
PAGES 98-99

SERBIA
PAGES 100-101

CYPRUS
PAGES 102-103

BULGARIA
PAGES 104-105

GREECE
PAGES 104-105

BELARUS
PAGES 106-107

ESTONIA
PAGES 106-107

LATVIA
PAGES 106-107

LITHUANIA
PAGES 106-107

TURKEY
PAGES 116-117

ISRAEL
PAGES 118-119

JORDAN
PAGES 118-119

LEBANON
PAGES 118-119

SYRIA
PAGES 118-119

BAHRAIN
PAGES 120-121

IRAN
PAGES 120-121

IRAQ
PAGES 120-121

TAJIKISTAN
PAGES 122-123

TURKMENISTAN
PAGES 122-123

UZBEKISTAN
PAGES 122-123

CHINA
PAGES 126-129

MONGOLIA
PAGES 126-127

NORTH KOREA
PAGES 128-129

SOUTH KOREA
PAGES 128-129

TAIWAN
PAGES 128-129

LAOS
PAGES 136-137

MYANMAR (BURMA)
PAGES 136-137

THAILAND
PAGES 136-137

VIETNAM
PAGES 136-137

BRUNEI
PAGES 138-139

EAST TIMOR
PAGES 138-139

INDONESIA
PAGES 138-139

MALAYSIA
PAGES 138-139

PAPUA NEW GUINEA
PAGES 144-145

SAMOA
PAGES 144-145

SOLOMON ISLANDS
PAGES 144-145

TONGA
PAGES 144-145

TUVALU
PAGES 144-145

VANUATU
PAGES 144-145

AUSTRALIA
PAGES 146-149

NEW ZEALAND
PAGES 150-151

The Political World

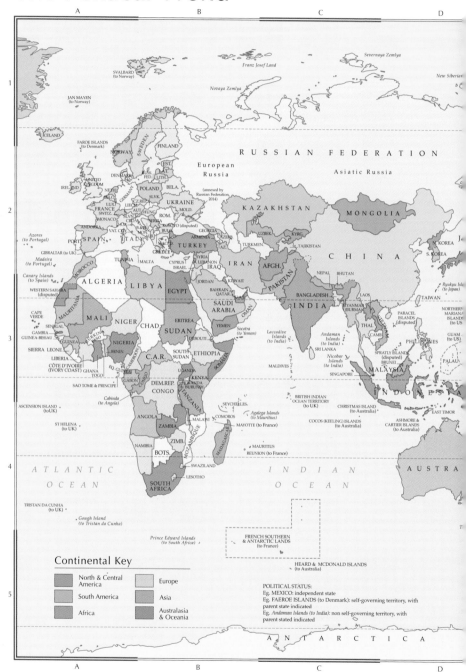

The Physical World

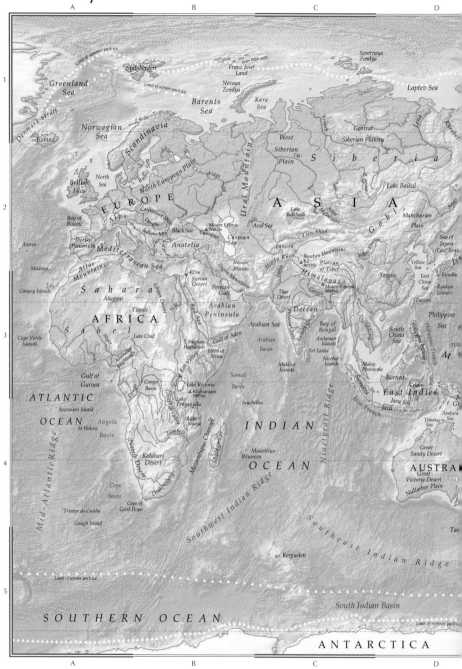

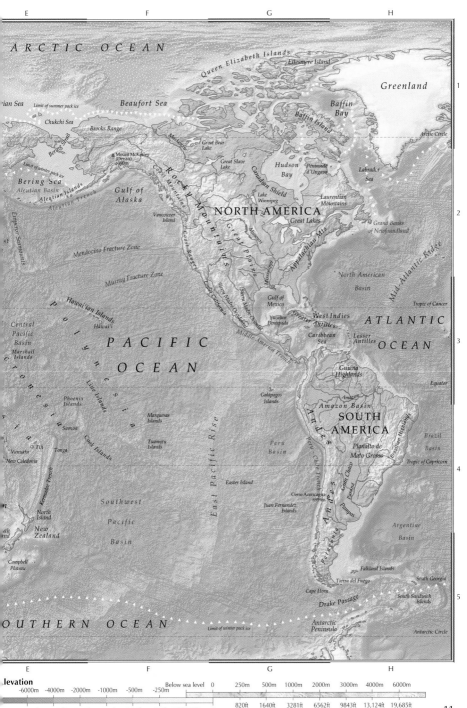

A R C T I C   O C E A N

Queen Elizabeth Islands
Ellesmere Island
Greenland

rian Sea
Limit of summer pack ice
Beaufort Sea
Baffin
Bay
Baffin Island

Chukchi Sea
Brooks Range
Mackenzie
Great Bear
Lake
Arctic Circle

Bering Strait
Mount McKinley
(Denali)
6194m
Great Slave
Lake
Hudson
Bay
Peninsula
d'Ungava
Labrador
Sea

Limit of winter pack ice
Bering Sea
Aleutian Basin
Aleutian Islands
Aleutian Trench
Gulf of
Alaska
Coast Mountains
Rocky Mountains
Canadian Shield
Lake
Winnipeg
Laurentian
Mountains

Emperor Seamounts
st
Vancouver
Island
NORTH AMERICA
Great Lakes
Grand Banks
of Newfoundland

Mendocino Fracture Zone
Coast Ranges
Great Plains
Missouri
Appalachian Mts.
North American
Basin
Mid-Atlantic Ridge

Murray Fracture Zone
Mississippi
Tropic of Cancer

Hawai'ian Islands
Hawai'i
Sierra Madre Occidental
Sierra Madre Oriental
Lower California
Gulf of
Mexico
Greater Antilles
West Indies
ATLANTIC

Central
Pacific
Basin
Marshall
Islands
P A C I F I C
Yucatan
Peninsula
Caribbean
Sea
Lesser
Antilles
O C E A N

Polynesia
O C E A N
Middle America Trench
Guiana
Highlands

Micronesia
Phoenix
Islands
Line Islands
Galápagos
Islands
Equator

Samoa
Marquesas
Islands
Amazon
Amazon Basin
SOUTH
AMERICA
Brazil
Basin

Vanuatu
Fiji
Tonga
Cook Islands
Tuamotu
Islands
Andes
Peru
Basin
Planalto de
Mato Grosso
Brazilian Highlands
Tropic of Capricorn

New Caledonia
Easter Island
Peru-Chile Trench
Gran Chaco
Paraná
Pampas
Argentine
Basin

North
Island
New
Zealand
Southwest
Pacific
Cerro Aconcagua
6959m
Juan Fernandez
Islands

th
nd
Kermadec Trench
Basin
Patagonia

Campbell
Plateau
Falkland Islands
South Georgia

Tierra del Fuego
South Sandwich
Islands

Cape Horn
Drake Passage

O U T H E R N   O C E A N
Limit of winter pack ice
Antarctic
Peninsula
Antarctic Circle

E   F   G   H

Elevation

| -6000m | -4000m | -2000m | -1000m | -500m | -250m | Below sea level 0 | 250m | 500m | 1000m | 2000m | 3000m | 4000m | 6000m |

| -19,658ft | -13,124ft | -6562ft | -3281ft | -1640ft | -820ft | -328ft/-100m | 0 | 820ft | 1640ft | 3281ft | 6562ft | 9843ft | 13,124ft | 19,685ft |

Standard Time Zones

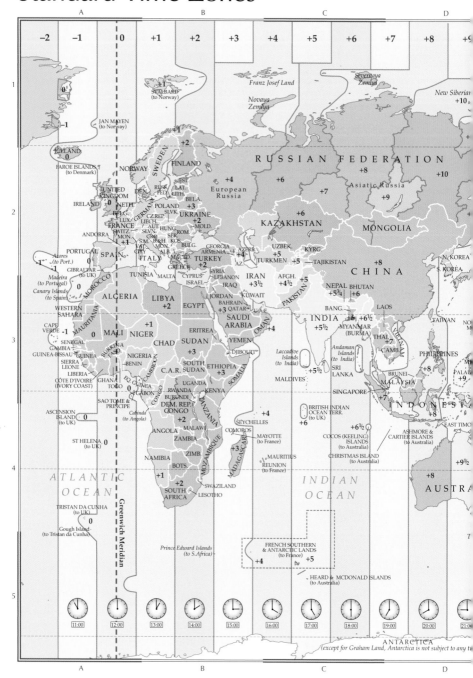

The numbers represented thus; +2/-2, indicate the number of hours each time zone is ahead or behind UCT (Coordinated Universal Tim...

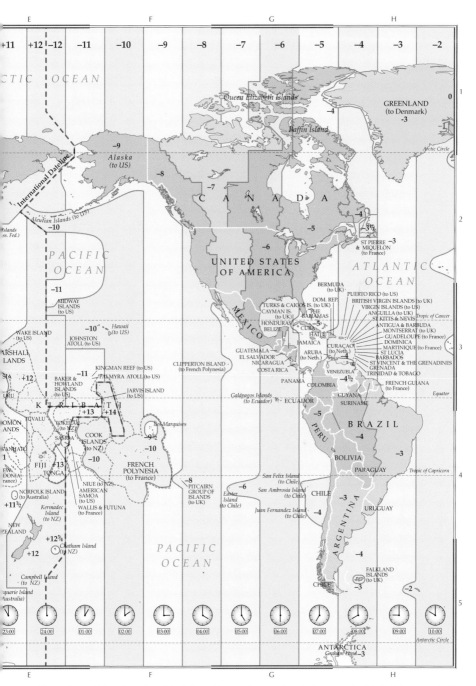

The clocks and 24-hour times given at the bottom of the map show time in each time zone when it is 12.00 hours noon UCT

Geology & Structure

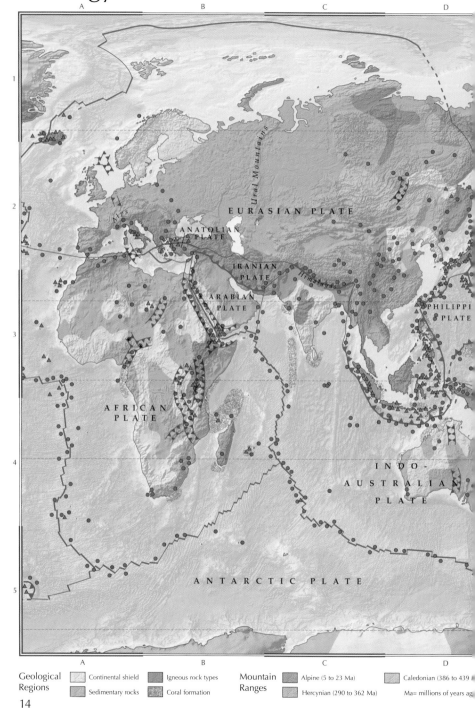

Ural Mountains

EURASIAN PLATE

Alps

ANATOLIAN PLATE

IRANIAN PLATE

Himalayas

ARABIAN PLATE

PHILIPPINE PLATE

AFRICAN PLATE

INDO-AUSTRALIAN PLATE

ANTARCTIC PLATE

Geological Regions				Mountain Ranges		
	Continental shield		Igneous rock types		Alpine (5 to 23 Ma)	Caledonian (386 to 439
	Sedimentary rocks		Coral formation		Hercynian (290 to 362 Ma)	Ma= millions of years ag

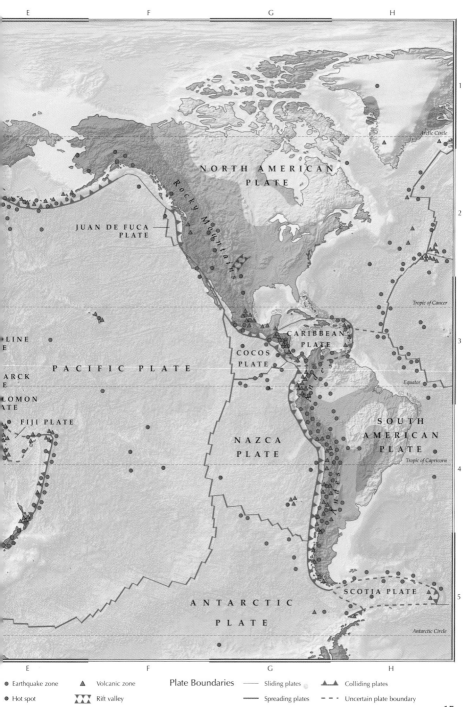

E   F   G   H

1

Arctic Circle

NORTH AMERICAN
PLATE

JUAN DE FUCA
PLATE

2

Rocky Mountains

Tropic of Cancer

CARIBBEAN
PLATE

3

·LINE
E

COCOS
PLATE

PACIFIC PLATE

Equator

·ARCK
E

·LOMON
·TE

FIJI PLATE

NAZCA
PLATE

SOUTH
AMERICAN
PLATE

Tropic of Capricorn

4

Andes

SCOTIA PLATE

5

ANTARCTIC

PLATE

Antarctic Circle

E   F   G   H

● Earthquake zone   ▲ Volcanic zone   Plate Boundaries   —— Sliding plates   ▲▲ Colliding plates
● Hot spot   ▼▼▼ Rift valley   —— Spreading plates   --- Uncertain plate boundary

15

World Climate

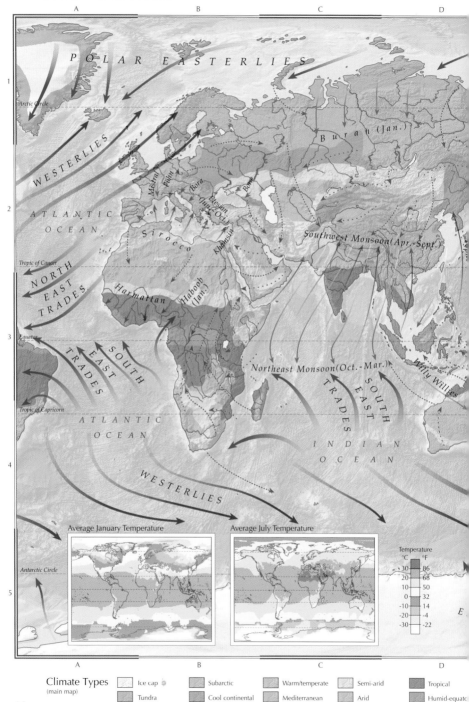

A   B   C   D

P O L A R   E A S T E R L I E S

Arctic Circle

W E S T E R L I E S

Mistral

Föhn

Bora

B u r a n (Jan.)

A T L A N T I C
O C E A N

Sirocco

Meltem
(Jun.–Oct.)

Bora

Khamsin

Southwest Monsoon(Apr.–Sept.)

Tropic of Cancer

N O R T H
E A S T
T R A D E S

Harmattan

Haboob
(Jan.)

Equator

S O U T H
E A S T
T R A D E S

Northeast Monsoon(Oct.–Mar.)

Willy Willies

T R A D E S

S O U T H
E A S T

Tropic of Capricorn

A T L A N T I C
O C E A N

I N D I A N
O C E A N

W E S T E R L I E S

Antarctic Circle

Average January Temperature

Average July Temperature

Temperature
°C   °F

°C	°F
30	86
20	68
10	50
0	32
-10	14
-20	-4
-30	-22

A   B   C   D

Climate Types
(main map)

- Ice cap
- Tundra
- Subarctic
- Cool continental
- Warm/temperate
- Mediterranean
- Semi-arid
- Arid
- Tropical
- Humid-equatorial

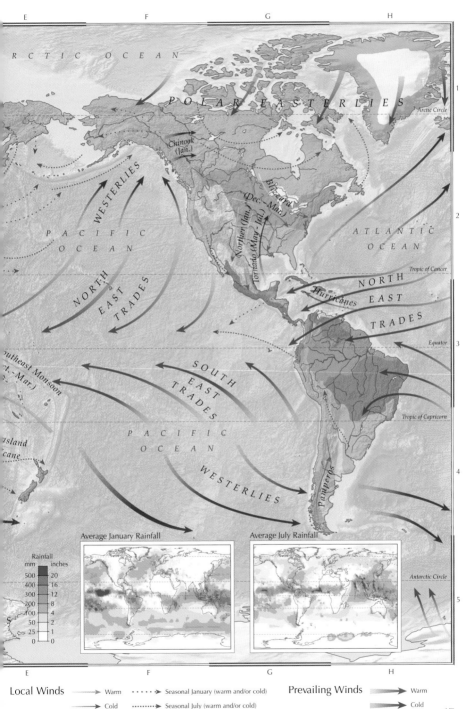

E   F   G   H

A R C T I C   O C E A N

1

Arctic Circle

P O L A R   E A S T E R L I E S

Chinook
(Jan.)

W E S T E R L I E S

Blizzard
(Dec. - Mar.)

Norther (Jan.)

Tornado (May - Jul.)

P A C I F I C
O C E A N

A T L A N T I C
O C E A N

2

Tropic of Cancer

N O R T H

E A S T   T R A D E S

N O R T H

E A S T

T R A D E S

Hurricanes

3

Equator

...outheast Monsoon
...ct. - Mar.)

S O U T H

E A S T

T R A D E S

...island
...cane

Tropic of Capricorn

P A C I F I C
O C E A N

4

W E S T E R L I E S

Pamperos

Average January Rainfall

Average July Rainfall

Rainfall
mm   inches
500   20
400   16
300   12
200   8
100   4
50   2
25   1
0   0

Antarctic Circle

5

E   F   G   H

Local Winds   → Warm   ····▸ Seasonal January (warm and/or cold)   Prevailing Winds   → Warm

→ Cold   ·····▸ Seasonal July (warm and/or cold)   → Cold

17

Ocean Currents

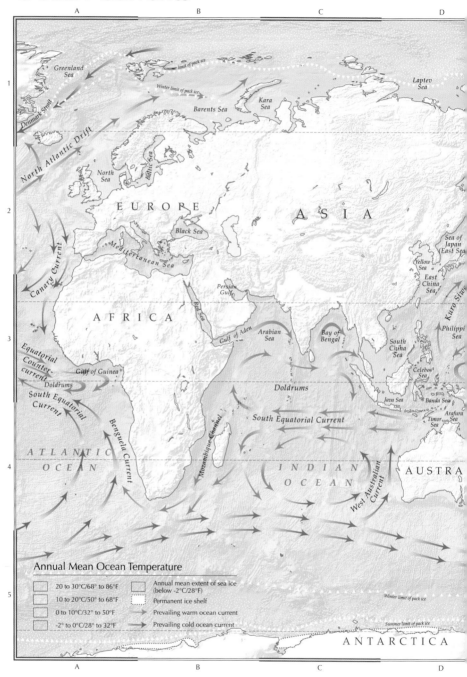

Greenland Sea

Summer limit of pack ice

Winter limit of pack ice

Laptev Sea

Denmark Strait

Barents Sea

Kara Sea

North Atlantic Drift

North Sea

Baltic Sea

EUROPE

ASIA

Black Sea

Sea of Japan (East Sea)

Mediterranean Sea

Canary Current

Yellow Sea

East China Sea

AFRICA

Red Sea

Persian Gulf

Gulf of Aden

Arabian Sea

Bay of Bengal

South China Sea

Philippine Sea

Kuro Siwo

Equatorial Counter-current

Gulf of Guinea

Doldrums

Doldrums

Celebes Sea

South Equatorial Current

South Equatorial Current

Java Sea

Banda Sea

Benguela Current

Mozambique Channel

Arafura Sea

Timor Sea

ATLANTIC OCEAN

INDIAN OCEAN

AUSTRA

West Australian Current

Annual Mean Ocean Temperature

Winter limit of pack ice

20 to 30°C/68° to 86°F	Annual mean extent of sea ice (below -2°C/28°F)
10 to 20°C/50° to 68°F	Permanent ice shelf
0 to 10°C/32° to 50°F	Prevailing warm ocean current
-2° to 0°C/28° to 32°F	Prevailing cold ocean current

Summer limit of pack ice

ANTARCTICA

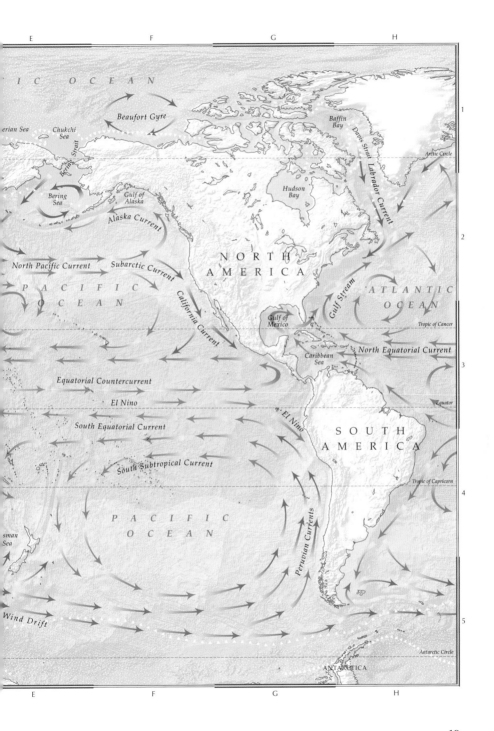

E F G H

1

2

3

4

5

IC OCEAN

Beaufort Gyre

Baffin Bay

erian Sea

Chukchi Sea

Bering Strait

Davis Strait

Arctic Circle

Bering Sea

Gulf of Alaska

Hudson Bay

Labrador Current

Alaska Current

North Pacific Current

Subarctic Current

NORTH AMERICA

PACIFIC OCEAN

California Current

Gulf of Mexico

Gulf Stream

ATLANTIC OCEAN

Tropic of Cancer

Caribbean Sea

North Equatorial Current

Equatorial Countercurrent

El Nino

Equator

South Equatorial Current

El Nino

SOUTH AMERICA

South Subtropical Current

Tropic of Capricorn

PACIFIC OCEAN

Peruvian Currents

sman Sea

Wind Drift

Antarctic Circle

ANTARCTICA

E F G H

Life Zones

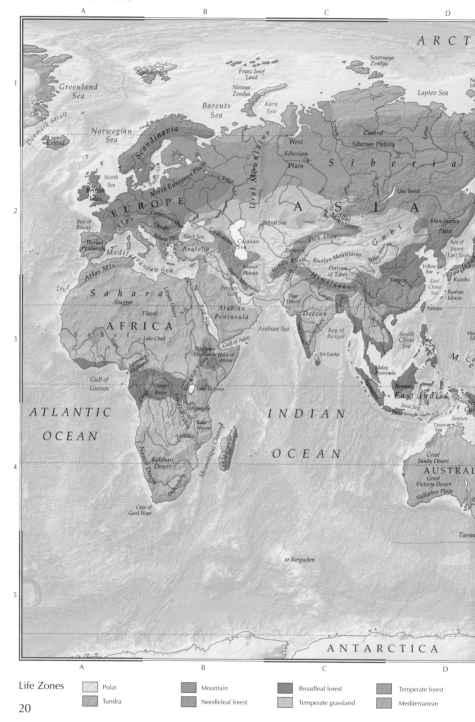

Life Zones

Polar	Mountain
Tundra	Needleleaf forest

Broadleaf forest	Temperate forest
Temperate grassland	Mediterranean

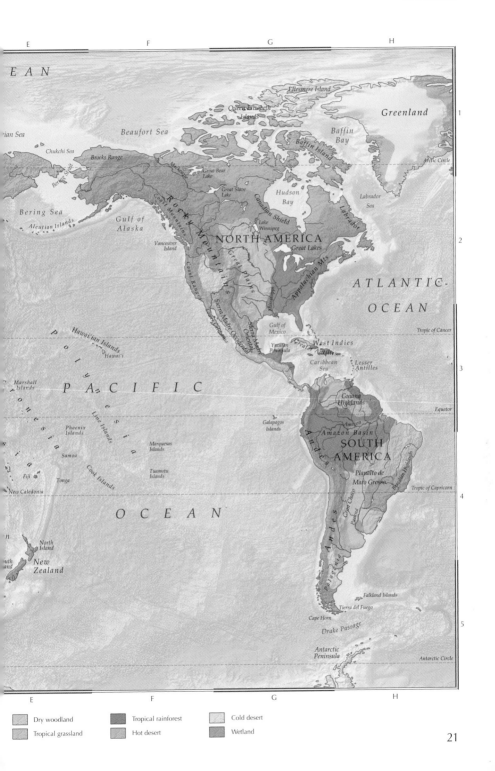

	E	F	G	H

Greenland

Ellesmere Island

Queen Elizabeth Islands

Beaufort Sea

Baffin Bay

Baffin Island

rian Sea

Chukchi Sea

Brooks Range

Arctic Circle

Mackenzie

Great Bear Lake

Bering Strait

Great Slave Lake

Hudson Bay

Labrador Sea

Bering Sea

Aleutian Islands

Gulf of Alaska

Vancouver Island

Canadian Shield

NORTH AMERICA

Rocky Mountains

Coast Mountains

Coast Ranges

Great Plains

Sierra Madre Occidental

Lake Winnipeg

Great Lakes

Appalachian Mts

ATLANTIC OCEAN

Sierra Madre Oriental

Gulf of Mexico

Yucatan Peninsula

Greater

West Indies

Antilles

Tropic of Cancer

Hawaiian Islands

Hawai'i

Caribbean Sea

Lesser Antilles

Marshall Islands

POLYNESIA

Guiana Highlands

Equator

cronesia

PACIFIC

Galapagos Islands

Amazon

Phoenix Islands

Line Islands

Marquesas Islands

Amazon Basin

SOUTH AMERICA

sia

Samoa

Tuamotu Islands

Brazilian Highlands

Fiji

Tonga

Cook Islands

Planalto de Mato Grosso

Tropic of Capricorn

New Caledonia

Gran Chaco

OCEAN

Pantanal

Andes

n

North Island

New Zealand

Patagonia

Falkland Islands

Tierra del Fuego

Cape Horn

Drake Passage

Antarctic Peninsula

Antarctic Circle

	E	F	G	H

☐ Dry woodland	☐ Tropical rainforest	☐ Cold desert	
☐ Tropical grassland	☐ Hot desert	☐ Wetland	

Population

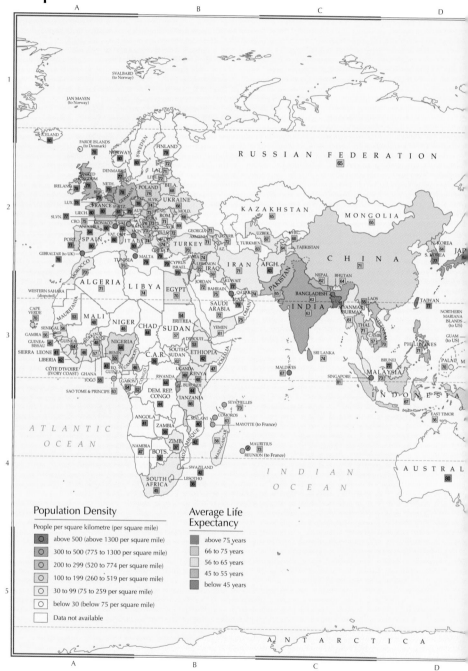

Population Density

People per square kilometre (per square mile)

- above 500 (above 1300 per square mile)
- 300 to 500 (775 to 1300 per square mile)
- 200 to 299 (520 to 774 per square mile)
- 100 to 199 (260 to 519 per square mile)
- 30 to 99 (75 to 259 per square mile)
- below 30 (below 75 per square mile)
- Data not available

Average Life Expectancy

- above 75 years
- 66 to 75 years
- 56 to 65 years
- 45 to 55 years
- below 45 years

Languages

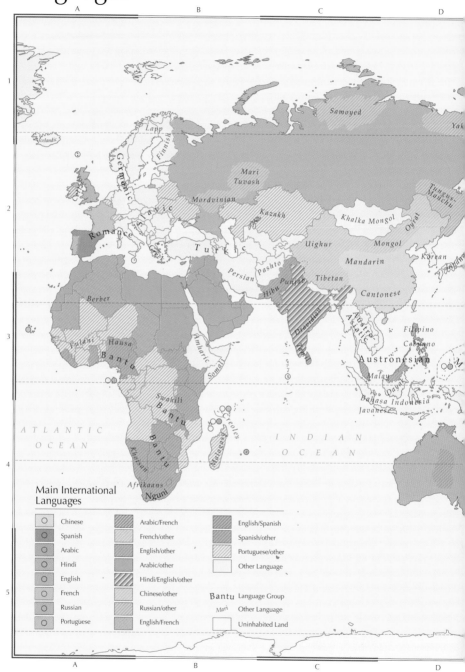

Main International Languages

○ Chinese	Arabic/French	English/Spanish
○ Spanish	French/other	Spanish/other
○ Arabic	English/other	Portuguese/other
○ Hindi	Arabic/other	Other Language
○ English	Hindi/English/other	
○ French	Chinese/other	**Bantu** Language Group
○ Russian	Russian/other	Mari Other Language
○ Portuguese	English/French	Uninhabited Land

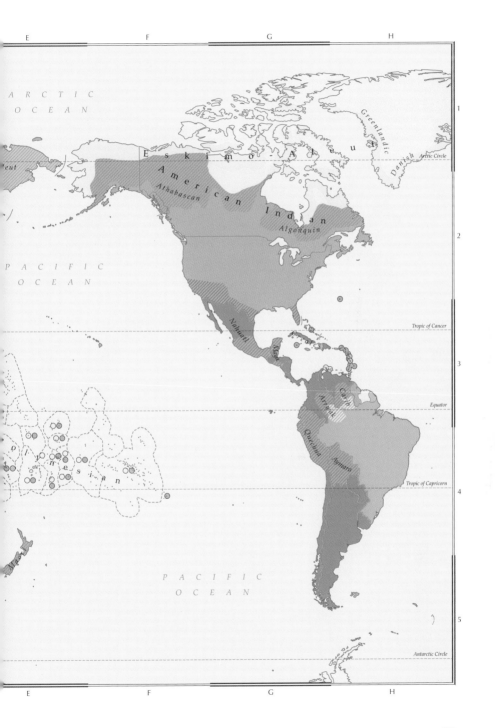

E   F   G   H

A R C T I C

O C E A N

eut

Greenlandic

Danish   Arctic Circle

E s k i m o - A l e u t

A m e r i c a n   I n d i a n

Athabascan

P A C I F I C

O C E A N

Algonquin

Nahuatl

Tropic of Cancer

Maya

Carib

Arawak

Carib

Equator

P o l y n e s i a n

Quechua

Aymara

Tropic of Capricorn

Maori

P A C I F I C

O C E A N

Antarctic Circle

E   F   G   H

Religion

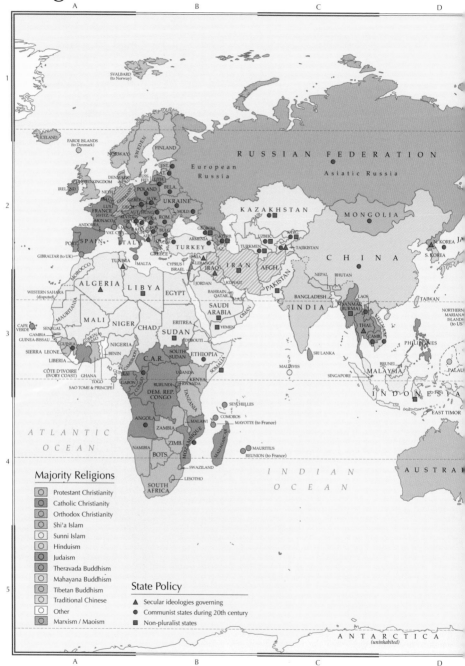

SVALBARD
(to Norway)

ICELAND

FAROE ISLANDS
(to Denmark)

NORWAY SWEDEN FINLAND

RUSSIAN FEDERATION

European
Russia

Asiatic Russia

DENMARK
UNITED KINGDOM

RUSS.
FED.

EST.
LAT.
LITH.

IRELAND

NETH.
POLAND
BELA.

KAZAKHSTAN

MONGOLIA

N. KOREA JA

FRANCE
LUX.
SWITZ.
SLVK.
AUT. HUNG.
UKRAINE
MOLD.
ROM.

S. KOREA

MONACO
ANDORRA

CZECH

SLVN. CRO.
BOS.

BULG.

GEORGIA

ARMENIA AZER.
UZBEK.

KYRG.

C H I N A

PORT. SPAIN
VAT. CITY
ITALY

MACED.
ALB.
GREECE

TURKEY

TURKMEN.
TAJIKISTAN

GIBRALTAR (to UK)

TUNISIA
MALTA

CYPRUS
ISRAEL
SYRIA
LEBANON
IRAQ

AZ.

I R A N
AFGH.

PAKISTAN

NEPAL BHUTAN

TAIWAN

MOROCCO

ALGERIA

LIBYA

EGYPT

JORDAN
KUWAIT

BANGLADESH

LAOS

NORTHERN
MARIANA
ISLANDS
(to US

WESTERN SAHARA
(disputed)

BAHRAIN
QATAR
U.A.E.

I N D I A

MYANMAR
(BURMA)

MAURITANIA

MALI

NIGER

CHAD

SUDAN

SAUDI
ARABIA

OMAN

THAI.

CAMB.

PHILIPPINES

CAPE
VERDE
SENEGAL
GAMBIA
GUINEA-BISSAU
GUINEA

ERITREA

YEMEN

SRI LANKA

BRUNEI

PALAU

SIERRA LEONE

BURKINA
FASO

NIGERIA

DJIBOUTI

SINGAPORE

MALAYSIA

LIBERIA

BENIN

CÔTE D'IVOIRE
(IVORY COAST)
GHANA

CAMEROON

C.A.R.

SOUTH
SUDAN

ETHIOPIA

SOMALIA

MALDIVES

TOGO

EQ. GUINEA

GABON
CONGO

UGANDA

KENYA

RWANDA

I N D O N E S I A

EAST TIMOR

SAO TOME & PRINCIPE

DEM. REP.
CONGO

BURUNDI

TANZANIA

SEYCHELLES

ANGOLA

ZAMBIA

MALAWI

COMOROS

MAYOTTE (to France)

A T L A N T I C

O C E A N

NAMIBIA

ZIMB.

MOZAMBIQUE

MADAGASCAR

MAURITIUS

REUNION (to France)

BOTS.

SWAZILAND

I N D I A N

O C E A N

AUSTRA

Majority Religions

◯	Protestant Christianity
◯	Catholic Christianity
◯	Orthodox Christianity
◯	Shi'a Islam
◯	Sunni Islam
◯	Hinduism
◯	Judaism
◯	Theravada Buddhism
◯	Mahayana Buddhism
◯	Tibetan Buddhism
◯	Traditional Chinese
◯	Other
◯	Marxism / Maoism

LESOTHO

SOUTH
AFRICA

State Policy

▲ Secular ideologies governing

● Communist states during 20th century

■ Non-pluralist states

A N T A R C T I C A
(uninhabited)

The Global Economy

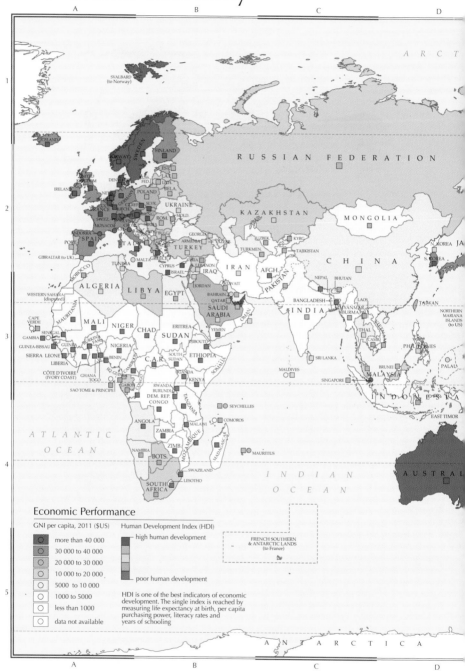

Economic Performance

GNI per capita, 2011 ($US)

- more than 40 000
- 30 000 to 40 000
- 20 000 to 30 000
- 10 000 to 20 000
- 5000 to 10 000
- 1000 to 5000
- less than 1000
- data not available

Human Development Index (HDI)

- high human development
- poor human development

HDI is one of the best indicators of economic development. The single index is reached by measuring life expectancy at birth, per capita purchasing power, literacy rates and years of schooling

E   F   G   H

1

GREENLAND
(to Denmark)

Arctic Circle

*Alaska
(to US)*

C A N A D A

2

P A C I F I C

O C E A N

UNITED STATES

OF AMERICA

A T L A N T I C

O C E A N

BERMUDA
(to UK)

PUERTO RICO
(to US)

DOM. REP.

ST KITTS & NEVIS

ANTIGUA & BAR.

Tropic of Cancer

M
E
X
I
C
O

TURKS & CAICOS ISLANDS (to UK)

CAYMAN ISLANDS
(to UK)

THE
BAHAMAS

HONDURAS

BELIZE

CUBA

JAMAICA

HAITI

CURAÇAO
(to Neth.)

GUADELOUPE (to France)

DOMINICA

MARTINIQUE (to France)

ST LUCIA

BARBADOS

*Hawai'i
(to US)*

GUATEMALA

EL SALVADOR

NICARAGUA

COSTA RICA

PANAMA

ARUBA
(to Neth.)

VENEZUELA

ST VINCENT &
THE GRENADINES

GRENADA

TRINIDAD & TOBAGO

3

RSHALL
ANDS

COLOMBIA

FRENCH GUIANA
(to France)

ECUADOR

GUYANA

SURINAME

Equator

NAURU

K I R I B A T I

P
E
R
U

B R A Z I L

TUVALU

TOKELAU
(to NZ)

SAMOA

COOK
ISLANDS
(to NZ)

BOLIVIA

PARAGUAY

Tropic of Capricorn

TONGA

FIJI

FRENCH POLYNESIA
(to France)

PITCAIRN, HENDERSON,
DUCIE & OENO ISLANDS
(to UK)

CHILE

A
R
G
E
N
T
I
N
A

URUGUAY

4

NEW
ALAND

P A C I F I C

O C E A N

FALKLAND ISLANDS
(to UK)

CHILE

5

Antarctic Circle

ANTARCTICA

E   F   G   H

29

Politics and Conflict

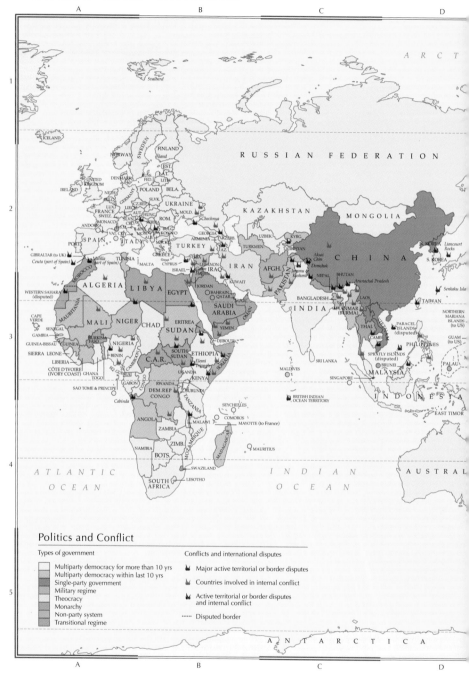

A   B   C   D

A R C T

1

Svalbard

ICELAND

R U S S I A N   F E D E R A T I O N

NORWAY   SWEDEN   FINLAND

Åland

EST.
RUS.
LAT.
FED.
DENMARK   LITH.

UNITED
KINGDOM

IRELAND   NETH.   POLAND   BELA.

GERMANY   CZECH
LUX.   SLVK   UKRAINE

FRANCE   LIECH.   AUT.   HUNG.   MOLD.

KAZAKHSTAN   MONGOLIA

2

SWITZ.   SLVN.   ROM.   Chechnya

ANDORRA   MONACO   CRO.   SERBIA

SPAIN   ITALY   VAT. CITY   MONT.   BOS.   MAC.
  ALB.   KOSOVO   BULG.

GEORGIA   UZBEK.
ARMENIA   AZERB.

KYRG.
TURKMEN.   TAJIKISTAN

N. KOREA   Liancourt
Rocks

PORT.   GREECE

GIBRALTAR (to UK)   TURKEY

Melilla
Ceuta (part of Spain)   TUNISIA   MALTA   CYPRUS   LEBANON   SYRIA

(part of Spain)   MOROCCO   ISRAEL   IRAQ

IRAN   AFGH.

C H I N A   S. KOREA

Aksai
Chin   Senkaku Isu.

WESTERN SAHARA
(disputed)   ALGERIA   LIBYA   EGYPT

JORDAN

PAKISTAN   Jammu
Kashmir   Demchok

NEPAL   BHUTAN

Arunachal Pradesh   TAIWAN

CAPE
VERDE

MAURITANIA   MALI   NIGER   CHAD

KUWAIT
BAHRAIN   QATAR
SAUDI
ARABIA   OMAN

ERITREA   YEMEN

BANGLADESH   I N D I A   MYANMAR
(BURMA)   LAOS

NORTHERN
MARIANA
ISLANDS
(to US)

SENEGAL   BURKINA
FASO

SUDAN   DJIBOUTI

THAI.   PARACEL
ISLANDS
(disputed)

GUAM
(to US)

3

GAMBIA
GUINEA-BISSAU   GUINEA
SIERRA LEONE   NIGERIA   BENIN   CAMEROON

LIBERIA   CÔTE D'IVOIRE
(IVORY COAST)   GHANA   EQ. GUINEA   TOGO

SOUTH
SUDAN   ETHIOPIA
C.A.R.   Ilemi
Triangle   SOMALIA

UGANDA   KENYA

CAMBODIA   PHILIPPINES

SRI LANKA   SPRATLY ISLANDS
(disputed)   BRUNEI   PALAU

MALDIVES   MALAYSIA

SAO TOME & PRINCIPE   GABON   CONGO   RWANDA   BURUNDI

SINGAPORE

Cabinda   DEM. REP.
CONGO   TANZANIA

I N D O N E S I A

SEYCHELLES

BRITISH INDIAN
OCEAN TERRITORY   EAST TIMOR

ANGOLA   MALAWI   COMOROS
MAYOTTE (to France)

ZAMBIA   MOZAMBIQUE

NAMIBIA   ZIMB.   MADAGASCAR   MAURITIUS

BOTS.

4

A T L A N T I C

O C E A N

SWAZILAND
SOUTH
AFRICA   LESOTHO

I N D I A N

O C E A N

A U S T R A L

Politics and Conflict

Types of government

	Multiparty democracy for more than 10 yrs
	Multiparty democracy within last 10 yrs
	Single-party government
	Military regime
	Theocracy
	Monarchy
	Non-party system
	Transitional regime

Conflicts and international disputes

    Major active territorial or border disputes

    Countries involved in internal conflict

    Active territorial or border disputes
    and internal conflict

·····   Disputed border

5

A N T A R C T I C A

A   B   C   D

The ——
WORLD'S
REGIONS

North & Central America

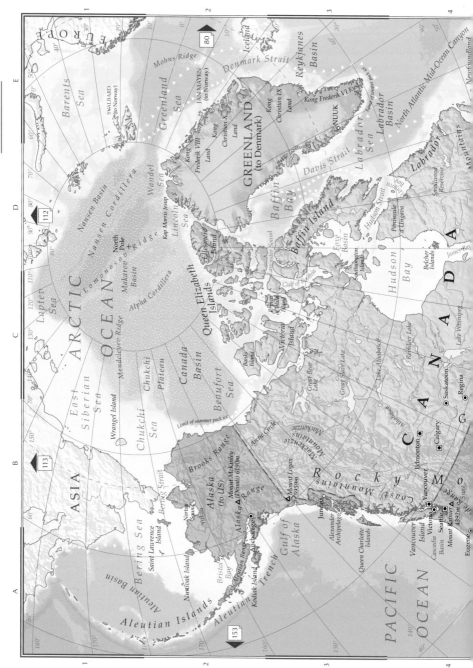

Population

- National capital
- o below 50,000
- o 50,000 to 100,000
- ⊙ 100,000 to 500,000
- ◼ above 500,000

0 km   1000

0 miles   1000

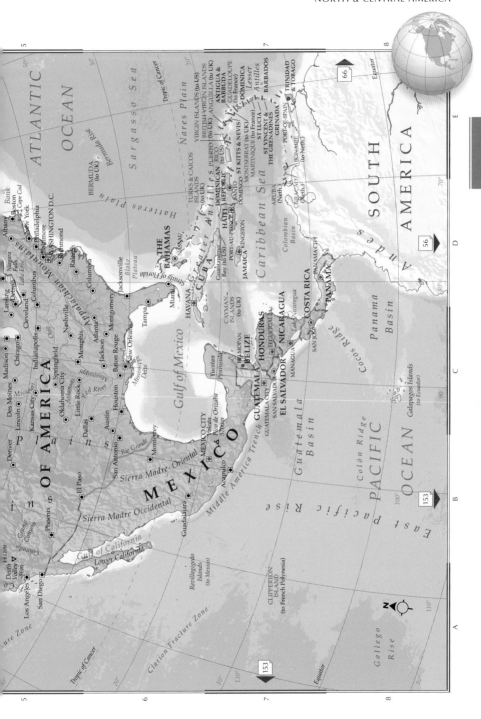

Western Canada & Alaska

Poluostrov Kamchatka

115

Arctic Circle

RUSSIAN
FEDERATION

Ostrov
Vrangelya

A R C T

Chukchi
Sea

Attu Island

Near
Islands

Wevok
Point Lay
Barrow

142

B e r i n g

Kivalina

S e a

Gambell

Wales

Saint Lawrence
Island

Bering Strait

Deering

Norton Sound

Rat
Islands

Amchitka
Island

Andreanof
Islands

Aleutian Islands

Atka

Alakanuk

Nunivak Island

Pribilof
Islands

Kwigillingok

Platinum

Grayling
Yukon River

Kokrines

Kuskokwim Mts

A L A S K A
(to US)

Fairbanks

Fort
Yukon

Alaska Range

McKinley
Park

Umnak Island

Dutch Harbor

Unalaska Island

Bristol
Bay

Unimak Island

Belkofski

Iliamna
Lake

Mount
McKinley
(Denali)
6190m

Susitna

Anchorage

Hope

Valdez

Gulkana

Chitina

Y U K

Alaska Peninsula

Shumagin
Islands

Kodiak

Cordova

Kodiak Island

Katalla

Yakutat

Mount Logan
△5959m

Whitehors

143

Gulf of

Alaska

Haines

Gustavus

Juneau

Kake

P A C I F I C

Alexander
Archipelago

Port
Alexander

Ketchikan

Prince Rupert

Kitimat

O C E A N

Queen Charlotte
Islands

Ocean Fall

Queen
Charlotte
Sound

Wadi

Port Hardy

Campbell

Vancouver Island

0 km   400

0 miles   400

Population

○ below 50,000   ○ 50,000 to 100,000   ◉ 100,000 to 500,000   ■ above 500,0

● Internal administrative capital

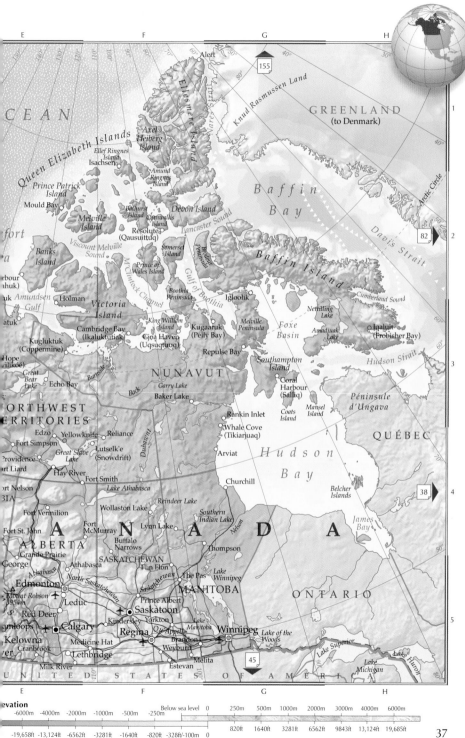

Alert

155

GREENLAND
(to Denmark)

Queen Elizabeth Islands

Ellesmere Island

Axel Heiberg Island

Ellef Ringnes Island Isachsen

Amund Ringnes Island

Knud Rasmussen Land

Baffin Bay

Arctic Circle

OCEAN

Prince Patrick Island

huk)

Mould Bay

Bathurst Island

Devon Island

Melville Island

Cornwallis Island

Resolute (Qausuittuq)

Lancaster Sound

Davis Strait

82

fort

Banks Island

Viscount Melville Sound

M'Clintock Channel

Somerset Island

Prince of Wales Island

Brodeur Peninsula

Baffin Island

a

rbour
ahuk)

Amundsen Gulf

Holman

Victoria Island

Boothia Peninsula

Igloolik

Cumberland Sound

uk

atuk

King William Island

Cambridge Bay (Ikaluktutiak)

Gjoa Haven (Uqsuqtuuq)

Kugaaruk (Pelly Bay)

Melville Peninsula

Foxe Basin

Nettilling Lake

Iqaluit (Frobisher Bay)

Hope
ılikóe)

Kugluktuk (Coppermine)

Great Bear Lake

Echo Bay

Burnside

Back

Repulse Bay

Amadjuak Lake

Hudson Strait

ORTHWEST
RRITORIES

Garry Lake

Baker Lake

NUNAVUT

Southampton Island

Coral Harbour (Salliq)

Coats Island

Mansel Island

Péninsule d'Ungava

Edzo

Yellowknife

Reliance

Fort Simpson

Great Slave Lake

Lutselk'e (Snowdrift)

Dubawnt

Rankin Inlet

Whale Cove (Tikiarjuaq)

QUÉBEC

Providence

rt Liard

Hay River

Fort Smith

Arviat

Hudson Bay

rt Nelson

Lake Athabasca

Churchill

3IA

Fort Vermilion

Wollaston Lake

Reindeer Lake

Southern Indian Lake

Belcher Islands

38

Fort St. John

Fort McMurray

A

Lynn Lake

N

Thompson

D

Nelson

A

James Bay

ALBERTA

George

Grande Prairie

Buffalo Narrows

SASKATCHEWAN

Athabasca

Flin Flon

Saskatchewan

The Pas

Lake Winnipeg

ONTARIO

Robson
954m

North Saskatchewan

Edmonton

Leduc

Prince Albert

MANITOBA

Red Deer

Saskatoon

amloops

Calgary

Kindersley

Yorkton

Regina

Lake Manitoba

Winnipeg

Lake of the Woods

Kelowna

Cranbrook

Medicine Hat

Qu'Appelle

Brandon

Weyburn

Lethbridge

Melita

Estevan

45

Lake Superior

Lake Huron

Lake Michigan

evation

-6000m	-4000m	-2000m	-1000m	-500m	-250m	Below sea level	0	250m	500m	1000m	2000m	3000m	4000m	6000m
-19,658ft	-13,124ft	-6562ft	-3281ft	-1640ft	-820ft	-328ft/-100m		820ft	1640ft	3281ft	6562ft	9843ft	13,124ft	19,685ft

Eastern Canada

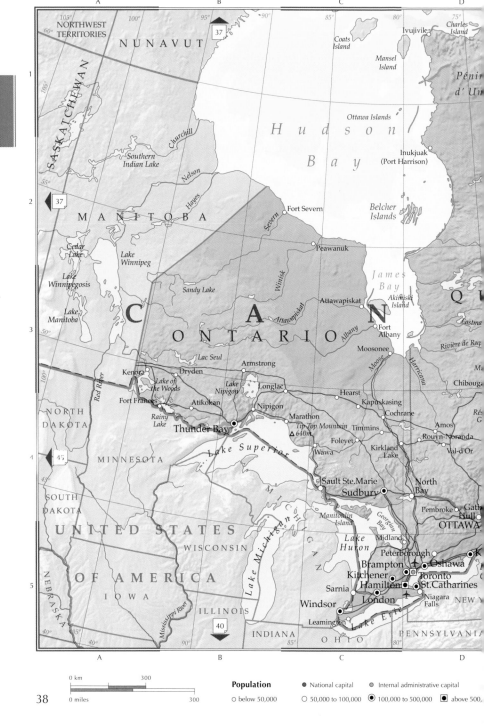

NORTHWEST TERRITORIES

SASKATCHEWAN

NUNAVUT

Coats Island

Mansel Island

Ivujivik

Charles Island

Péninsule d'Un

H u d s o n

Ottawa Islands

B a y

Churchill

Southern Indian Lake

Nelson

Inukjuak (Port Harrison)

M A N I T O B A

Hayes

Fort Severn

Belcher Islands

Cedar Lake

Lake Winnipeg

Lake Winnipegosis

Sandy Lake

Peawanuk

Severn

Winisk

J a m e s B a y

Akimiski Island

Q U

Lake Manitoba

C

A

Attawapiskat

N

Attawapiskat

Albany

Fort Albany

Eastma

O N T A R I O

Moosonee

Rivière de Ru

Lac Seul

Armstrong

Moose

Harricana

Chibouga

Kenora

Dryden

M

Lake of the Woods

Lake Nipigon

Longlac

Hearst

Kapuskasing

Red River

Fort Frances

Atikokan

Nipigon

Cochrane

Ré G

Rainy Lake

Marathon

Tip Top Mountain

Timmins

Amos

NORTH DAKOTA

Thunder Bay

△640m

Foleyet

Rouyn-Noranda

Val-d'Or

Lake Superior

Wawa

Kirkland Lake

MINNESOTA

Sault Ste.Marie

North Bay

Ma

SOUTH DAKOTA

L A K E   M I C H I G A N

Sudbury

Pembroke

Gati

UNITED STATES

Manitoulin Island

Georgian Bay

Hull

OTTAWA

NEBRASKA

WISCONSIN

Midland

Peterborough

K

O F   A M E R I C A

Lake Huron

Brampton

Oshawa

IOWA

Kitchener

Toronto

Hamilton

St.Catharines

Sarnia

London

NEW Y

Mississippi River

Windsor

Niagara Falls

ILLINOIS

Leamington

Lake Erie

INDIANA

O H I O

PENNSYLVANIA

0 km   300

0 miles   300

Population

● National capital   ◉ Internal administrative capital

○ below 50,000   ○ 50,000 to 100,000   ◉ 100,000 to 500,000   ■ above 500,

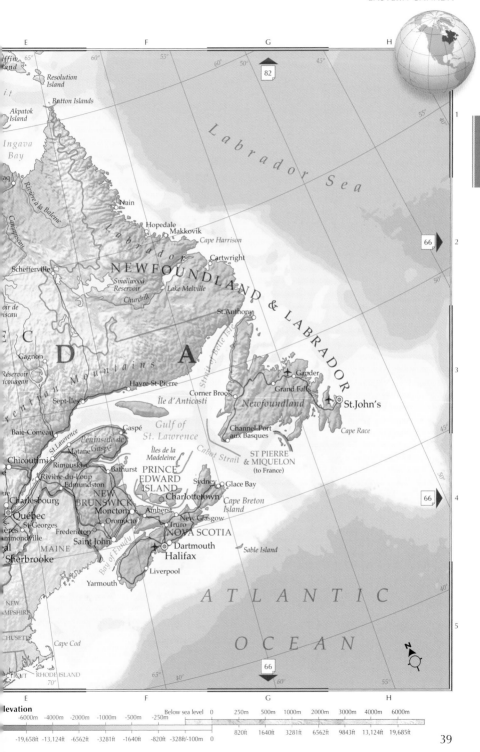

E   65°   60°   53°   F   60°   50°   G   45°   H

82

iffin
land

it

Akpatok
Island

Ingava
Bay

aq

Rivière à la Baleine

Canapiscau

Resolution
Island

Button Islands

40°

35°

1

Labrador Sea

Nain

Hopedale
Makkovik

Cape Harrison

Cartwright

66 ▶ 2

50°

Scheffervile

Smallwood
Reservoir

Churchill

Lake Melville

St.Anthony

N E W F O U N D L A N D & L A B R A D O R

oir de
iscau

C

D   **A**

Gagnon

Réservoir
icouagan

Sept-Îles

Havre-St-Pierre

Île d'Anticosti

Corner Brook

Gander

Grand Falls

Newfoundland

St.John's

45°

3

Strait of Belle Isle

Laurentian Mountains

Baie-Comeau

St.Lawrence

Péninsule de
Gaspé

Gaspé

Matane

Rimouski

Gulf of
St. Lawrence

Îles de la
Madeleine

Channel-Port
aux Basques

Cape Race

Chicoutimi

Rivière-du-Loup

Edmundston

Bathurst

**PRINCE
EDWARD
ISLAND**

Sydney

Glace Bay

Cabot Strait

**ST PIERRE
& MIQUELON**
(to France)

50°

Charlesbourg

**NEW
BRUNSWICK**

Moncton

Oromocto

Charlottetown

Cape Breton
Island

66 ▶ 4

Québec

St-Georges

Fredericton

Amherst

New Glasgow

ères

mmondville

Saint John

Truro

NOVA SCOTIA

al

MAINE

Bay of Fundy

Dartmouth

Sherbrooke

Halifax

Sable Island

40°

NEW
MPSHIRE

Liverpool

Yarmouth

A T L A N T I C

HUSETS

5

Cape Cod

O C E A N

35°

RHODE ISLAND
70°

66

N

E   65°   40°   F   G   60°   55°   H

Elevation

| -6000m | -4000m | -2000m | -1000m | -500m | -250m | Below sea level | 0 | 250m | 500m | 1000m | 2000m | 3000m | 4000m | 6000m |

| -19,658ft | -13,124ft | -6562ft | -3281ft | -1640ft | -820ft | -328ft/-100m | 0 | 820ft | 1640ft | 3281ft | 6562ft | 9843ft | 13,124ft | 19,685ft |

39

USA: The Northeast

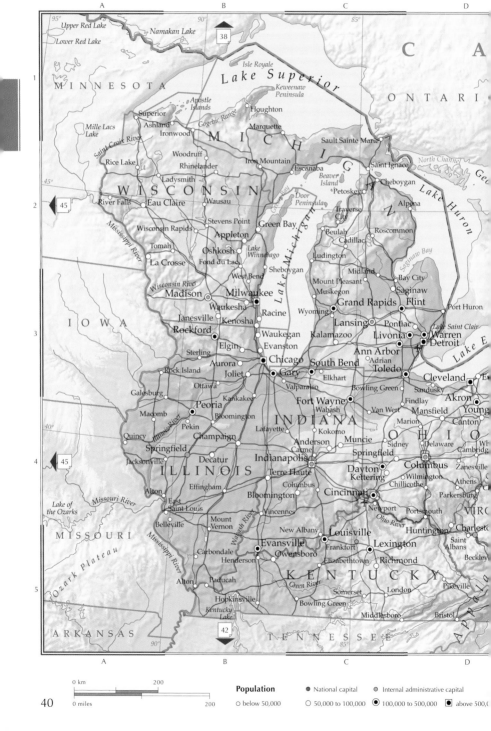

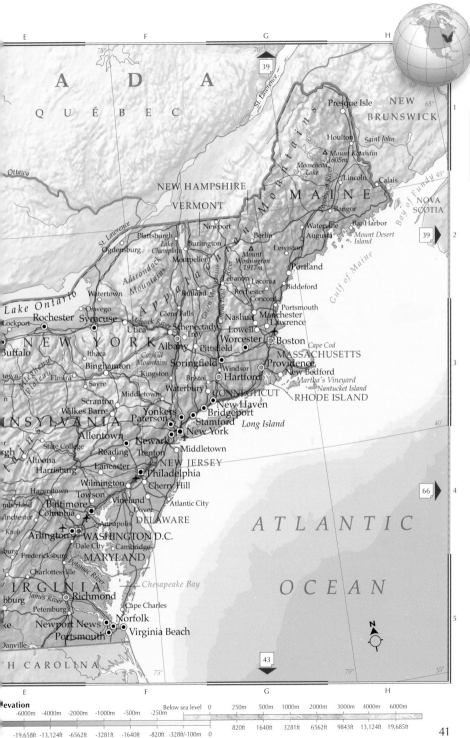

A   D   A

QUÉBEC

NEW
BRUNSWICK

Ottawa

NEW HAMPSHIRE

VERMONT

Presque Isle

Houlton   Saint John

Moosehead
Lake   △ Mount Katahdin
1605m

Lincoln   Calais

MAINE   NOVA
SCOTIA

St. Lawrence
Plattsburgh   Newport   Berlin   Waterville   Bar Harbor
Ogdensburg   Lake   Burlington   Augusta   Mount Desert
Champlain   Lewiston   Island

Montpelier   Mount
Washington   Portland
1917m

Lebanon   Laconia   Biddeford
Rutland   Rochester
Watertown   Concord

Lockport   Glens Falls   Portsmouth
Rochester   Oswego   Utica   Nashua   Portsmouth
Syracuse   Mohawk River   Schenectady   Manchester
Buffalo   Troy   Lowell   Lawrence

NEW YORK   Albany   Pittsfield   Worcester   Boston   Cape Cod
Ithaca
Binghamton   Catskill   Springfield   Providence
Mountains
town   Elmira   Kingston   Bristol   New Bedford
Sayre   Waterbury   Martha's Vineyard
Middletown   CONNECTICUT   Nantucket Island
Scranton   Hartford   New Haven   RHODE ISLAND
Wilkes Barre   Yonkers   Bridgeport
INSYLVANIA   Paterson   Stamford   Long Island

Allentown   Newark
rgh   State College   Reading   Trenton   New York
Altoona   Middletown
Harrisburg   Lancaster   NEW JERSEY

Hagerstown   Wilmington   Philadelphia
nberland   Towson   Cherry Hill
inchester   Baltimore   Vineland   Atlantic City
Columbia   Dover
Knob   Annapolis   DELAWARE

Arlington   WASHINGTON D.C.
burg   Fredericksburg   Dale City   Cambridge
MARYLAND
Charlottesville   Potomac River

VIRGINIA   Chesapeake Bay

hburg   James River   Richmond
Petersburg   Cape Charles

Newport News   Norfolk
Portsmouth   Virginia Beach
Danville

H CAROLINA

Lake Ontario
Allegheny Plateau
Adirondack Mountains
Green Mountains
Connecticut River
Appalachian Mountains
St. Lawrence R.
Penobscot River
Bay of Fundy
Gulf of Maine

ATLANTIC

OCEAN

N

75°   70°   65°   45°

40°

35°

E   F   G   H

1

2

3

4

5

evation

| -6000m | -4000m | -2000m | -1000m | -500m | -250m | Below sea level | 0 | 250m | 500m | 1000m | 2000m | 3000m | 4000m | 6000m |

| -19,658ft | -13,124ft | -6562ft | -3281ft | -1640ft | -820ft | -328ft/-100m | 0 | 820ft | 1640ft | 3281ft | 6562ft | 9843ft | 13,124ft | 19,685ft |

USA: The Southeast

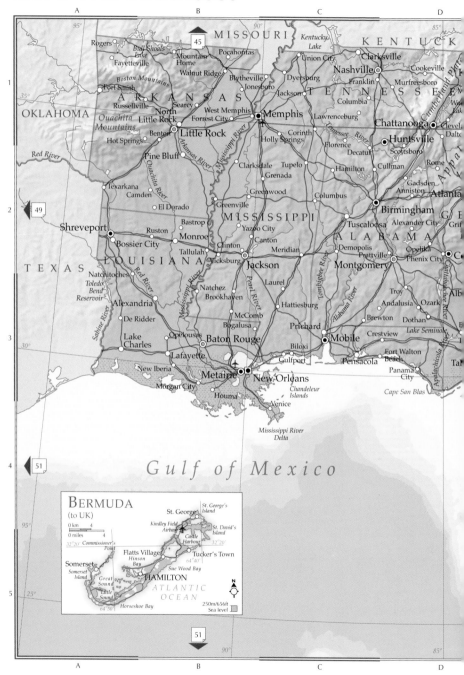

MISSOURI

Rogers
Bull Shoals Lake
Fayetteville
Mountain Home
Walnut Ridge
Pocahontas
Blytheville
Jonesboro
Dyersburg
Jackson
Boston Mountains
Fort Smith
A R K A N S A S
Russellville
Searcy
West Memphis
Forrest City
Corinth
Holly Springs
North Little Rock
Benton
Little Rock
Memphis
Lawrenceburg
Hot Springs
Pine Bluff
Clarksdale
Tupelo
Grenada
Hamilton
Cullman
Texarkana
Camden
El Dorado
Greenwood
Columbus
Shreveport
Ruston
Bastrop
Monroe
Yazoo City
Canton
Demopolis
Prattville
Opelika
Phenix City
Bossier City
Tallulah
Clinton
Meridian
Montgomery
L O U I S I A N A
Vicksburg
Jackson
Natchitoches
Laurel
Troy
Andalusia
Ozark
Natchez
Brookhaven
Hattiesburg
Brewton
Dothan
Alexandria
De Ridder
McComb
Bogalusa
Prichard
Crestview
Mobile
Lake Charles
Opelousas
Baton Rouge
Biloxi
Fort Walton Beach
Panama City
Lafayette
Gulfport
Pensacola
New Iberia
Metairie
New Orleans
Morgan City
Houma
Chandeleur Islands
Cape San Blas
Venice
Mississippi River Delta

Kentucky Lake
Union City
Clarksville
Cookeville
Nashville
Franklin
Murfreesboro
K E N T U C K
Columbia
T E N N E S S E E
Chattanooga
Huntsville
Florence
Decatur
Scottsboro
Rome
Gadsden
Anniston
Atlanta
Birmingham
Tuscaloosa
Alexander City
A L A B A M A

Gulf of Mexico

BERMUDA
(to UK)

0 km 4
0 miles 4

St. George
St. George's Island
Kindley Field Airbase
St. David's Island
Commissioner's Point
Castle Harbour
Flatts Village
Tucker's Town
Hinson Bay
Sue Wood Bay
Somerset
Somerset Island
HAMILTON
Great Sound
Little Sound
ATLANTIC OCEAN
Horseshoe Bay
250m/656ft
Sea level

Population

○ below 50,000
○ 50,000 to 100,000
◉ 100,000 to 500,000
◼ above 500,0

⊙ Internal administrative capital

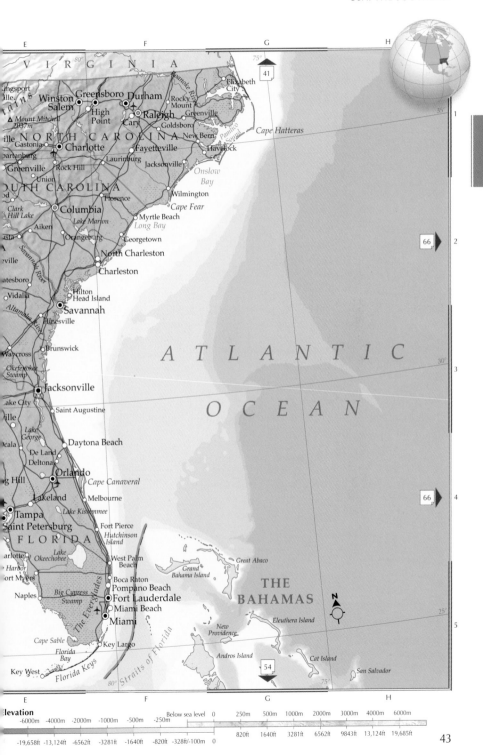

V I R G I N I A

Kingsport
ille

Winston Salem
Greensboro
Durham
High Point
Rocky Mount
Elizabeth City

Raleigh
Cary
Greenville
75°

Mount Mitchell
2037m
Cary
Goldsboro
New Bern
41

N O R T H   C A R O L I N A
Havelock
Pamlico Sound
Cape Hatteras
1

ille
Gastonia
Charlotte
Fayetteville

artanburg
Laurinburg
Jacksonville
Onslow Bay

Greenville
Rock Hill

OUTH CAROLINA
Florence
Wilmington

Union
Columbia
Cape Fear

Clark Hill Lake
Lake Marion
Myrtle Beach

ista
Aiken
Orangeburg
Georgetown
Long Bay

A T L A N T I C
66
2

eville
North Charleston

A
Charleston

atesboro

Vidalia
Hilton Head Island

Waycross
Savannah

Okefenokee Swamp
Hinesville

Brunswick
Altamaha River

O C E A N
30°
3

Lake City
Jacksonville

ille
Saint Augustine

Lake George
Ocala

De Land
Daytona Beach
Deltona

g Hill
Orlando
Cape Canaveral

Lakeland
Melbourne
66
4

Tampa
Lake Kissimmee

Saint Petersburg
Fort Pierce
Hutchinson Island

F L O R I D A
arlotte
Lake Okeechobee
West Palm Beach

Harbor
ort Myers
Boca Raton
Great Abaco

Grand Bahama Island

Pompano Beach
Naples
Big Cypress Swamp
Fort Lauderdale
THE BAHAMAS

The Everglades
Miami Beach
Miami
N

Cape Sable
Key Largo
Eleuthera Island
25°
5

Florida Bay
New Providence

Key West
Florida Keys
Straits of Florida
Andros Island
Cat Island

54
San Salvador
75°

levation
Below sea level 0   250m   500m 1000m 2000m 3000m 4000m 6000m

-6000m -4000m -2000m -1000m -500m -250m

-19,658ft -13,124ft -6562ft -3281ft -1640ft -820ft -328ft/-100m 0   820ft 1640ft 3281ft 6562ft 9843ft 13,124ft 19,685ft

43

USA: Central States

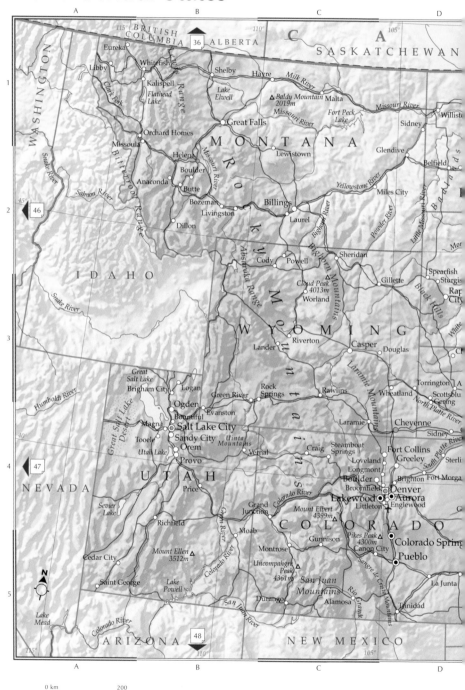

Population

○ below 50,000   ○ 50,000 to 100,000   ◉ 100,000 to 500,000   ■ above 500,0

◉ Internal administrative capital

0 km   200

0 miles   200

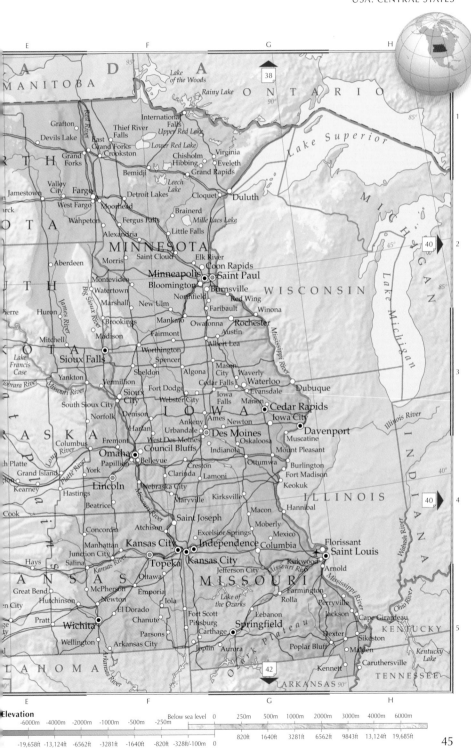

Elevation

-6000m	-4000m	-2000m	-1000m	-500m	-250m	Below sea level 0	250m	500m	1000m	2000m	3000m	4000m	6000m
-19,658ft	-13,124ft	-6562ft	-3281ft	-1640ft	-820ft	-328ft/-100m 0	820ft	1640ft	3281ft	6562ft	9843ft	13,124ft	19,685ft

USA: The West

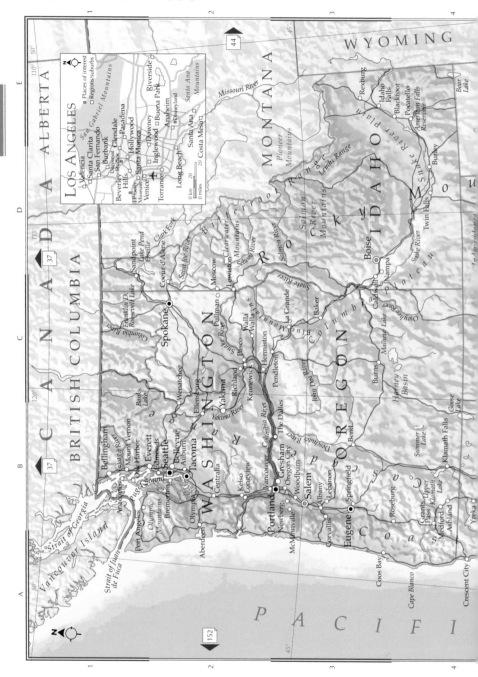

Los Angeles inset map:

Places of interest
Region/suburbs

Valencia
Santa Clarita
San Fernando
Burbank
Beverley Hills
Venice
Torrance
San Gabriel Mountains
Glendale
Hollywood
Universal Studios
Santa Monica
Pasadena
Riverside
Buena Park
Downey
Inglewood
Long Beach
Santa Ana
Disneyland
Anaheim
Santa Ana Mountains
Costa Mesa

1 Getty Museum
20 Costa Mesa

0 km   20
0 miles   20

WYOMING

MONTANA

IDAHO

ROCKY MOUNTAINS

Missouri River
Rexburg
Idaho Falls
Blackfoot
Pocatello
American Falls Reservoir
Bear Lake
Burley
Twin Falls
Snake River Plain
Snake River
Salmon River
Salmon River Mountains
Lemhi Range
Pioneer Mountains
Boise
Nampa
Caldwell
Owyhee River
Malheur Lake
Columbia Plateau

CANADA
ALBERTA
BRITISH COLUMBIA

Sandpoint
Lake Pend Oreille
Clark Fork
Coeur d'Alene
Franklin D. Roosevelt Lake
Columbia River
Spokane
Saint Joe River
Bitterroot Range
Clearwater Mountains
Moscow
Lewiston
Pullman
School River
Clearwater River
Snake River
La Grande
Baker
Burns
Harney Basin
Blue Mountains

WASHINGTON

Banks Lake
Wenatchee
Ellensburg
Yakima
Yakima River
Richland
Kennewick
Pasco
Walla Walla
Hermiston
Pendleton
Columbia River
John Day River
The Dalles
Deschutes River
Bend

Bellingham
Mount Vernon
Oak Harbor
Everett
Edmonds
Seattle
Bellevue
Auburn
Tacoma
Olympia
Bremerton
Port Angeles
Olympic Mountains
Puget Sound
Aberdeen
Centralia
Kelso
Longview
Vancouver
Portland
Oregon City
Gresham
Newberg
Woodburn
McMinnville
Salem
Albany
Corvallis
Lebanon
Springfield
Eugene
Roseburg
Grants Pass
Upper Klamath Lake
Medford
Ashland
Yreka
Klamath Falls
Summer Lake
Goose Lake

OREGON

Puget Sound
Strait of Georgia
Vancouver Island
Strait of Juan de Fuca

Coos Bay
Cape Blanco
Crescent City

PACIFIC

0 km   200
0 miles   200

Population

○ below 50,000
○ 50,000 to 100,000
◉ 100,000 to 500,000
■ above 500,0[...]
● Internal administrative capital

UTAH

NEVADA

Great Basin

Schell Creek Range

Ruby M.

Reese R.

Humboldt R.

Pyramid Lake

Honey Lake

Sparks
Reno
Carson City
Lake Tahoe
South Lake Tahoe
Carson Sink
Walker Lake
Mono Lake

Hawthorne
Tonopah

Ely
Alamo

Lake Powell

Grand Canyon
Colorado River

Lake Mead

Lake Mohave

Las Vegas
Henderson

ARIZONA

Gila River

Colorado River

Chocolate Mountains
Blythe
Bradley
Needles

MEXICO

CALIFORNIA

Sierra Nevada

Central Valley

San Joaquin Valley

Death Valley
-86m

Mount Whitney
△4418m

Ridgecrest
Tulare Lake Bed

Mojave Desert

Barstow
Victorville
Lancaster

Bakersfield
Delano
Porterville
Visalia
Hanford
Selma
Fresno
Madera
Atascadero

San Rafael Mountain

Santa Maria
Lompoc
Santa Barbara
Oxnard

Los Angeles
Pasadena
San Bernardino
Riverside
Santa Ana
Long Beach
Huntington Beach

Palm Springs

Escondido
Oceanside
Encinitas
Fallbrook

El Cajon
Lakeside
San Diego
Chula Vista

Salton Sea

Santa Catalina Island
Santa Rosa Island
San Clemente Island
Channel Islands

Santa Lucia Range

San Luis Obispo
Salinas
Gilroy
San Jose
Sunnyvale
Palo Alto
Santa Cruz
Monterey
Monterey Bay

San Francisco
Oakland
Berkeley
Vallejo
Napa
Fairfield
Stockton
Manteca
Modesto
Turlock

Sacramento
Citrus Heights
Woodland
Yuba City
Chico
Santa Rosa
Ukiah

Sacramento River

Sacramento Valley

OCEAN

Blythe

PACIFIC OCEAN

HAWAII

Kaua'i
Ni'ihau
Lihu'e
O'ahu
Kāne'ohe
Wahiawā
Honolulu
Waiāluā
Moloka'i
Maui
Wailuku
Mauna Kea
4205m
Hawai'i
Hilo

2000m/6562ft
1000m/3281ft
500m/1640ft
200m/656ft
Sea level

0 km 100
0 miles 100

Elevation

-6000m	-4000m	-2000m	-1000m	-500m	-250m	Below sea level	0	250m	500m	1000m	2000m	3000m	4000m	6000m
-19,658ft	-13,124ft	-6562ft	-3281ft	-1640ft	-820ft	-328ft/-100m	0	820ft	1640ft	3281ft	6562ft	9843ft	13,124ft	19,685ft

USA: The Southwest

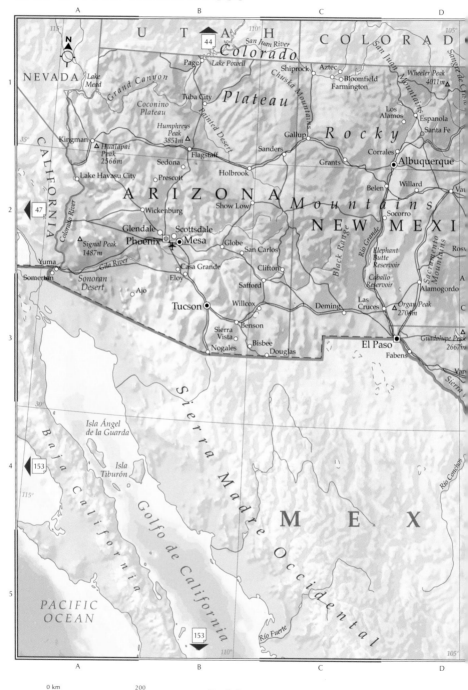

Population

○ below 50,000   ○ 50,000 to 100,000   ◉ 100,000 to 500,000   ◼ above 500,00

● Internal administrative capital

0 km   200

0 miles   200

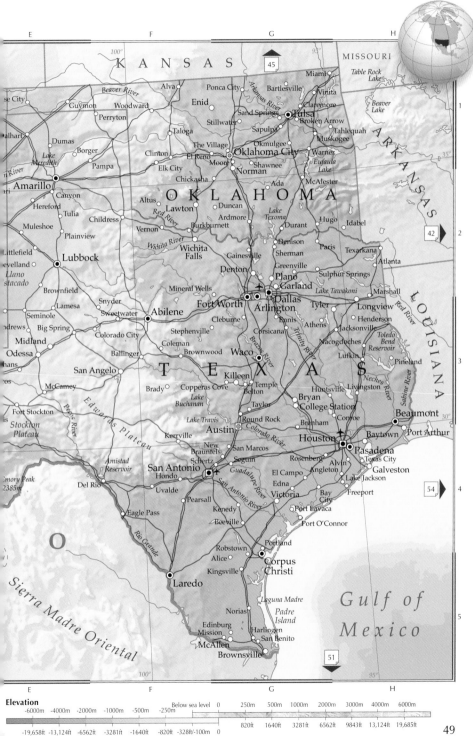

Elevation

-6000m	-4000m	-2000m	-1000m	-500m		Below sea level	0	250m	500m	1000m	2000m	3000m	4000m	6000m
					-250m									

| -19,658ft | -13,124ft | -6562ft | -3281ft | -1640ft | -820ft | -328ft/-100m | 0 | | 820ft | 1640ft | 3281ft | 6562ft | 9843ft | 13,124ft | 19,685ft |

Mexico

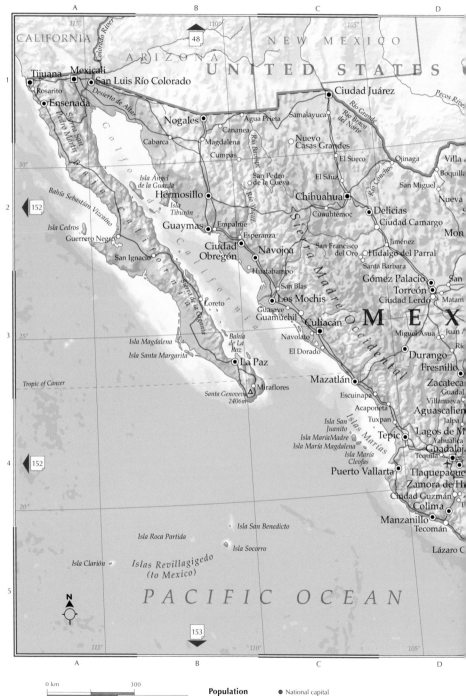

Population

National capital

○ below 50,000 ○ 50,000 to 100,000 ◉ 100,000 to 500,000 ◼ above 500,0

0 km   300

0 miles   300

E F G H

ALABAMA
FLORIDA

42

MISSISSIPPI

Brazos River
Red River
Sabine River
Mississippi River

A M E R I C A

LOUISIANA

T A S

Colorado River

Mississippi River
Delta

1

as Negras

Nuevo Laredo

66

Padre Island

2

Ciudad
Miguel Alemán

inas
algo

Reynosa

Río
Bravo

Matamoros

G u l f   o f

25°

85°

Monterrey

Laguna Madre

Tropic of Cancer

Montemorelos

M e x i c o

Linares

Yucatan Channel

3

Sierra

Ciudad Victoria

Cancún

Río Lagartos

Ciudad
Mante

Progreso

Tizimín

Isla
Cozumel

Ciudad Madero

Mérida

Motul

Madre

Pánuco

Tampico

Umán

20°

Ticul

Valladolid

Oriental

Ciudad Valles

Peto

Tekax

olores
Hidalgo

Laguna de Tamiahua

Oxkutzcab

Felipe Carrillo
Puerto

anajuato

Tamazunchale

Campeche

Yucatan

Querétaro

Tuxpán

Bahía de Campeche

Chetumal

ato Pachuca

Poza Rica

Champotón

Peninsula

52

4

MÉXICO

Papantla

Laguna de
Términos

MEXICO CITY

Tulancingo

Fransisco Escárcega

Toluca

Teziutlán

Xalapa

Frontera

uernavaca

Perote

Tlaxcala

Veracruz

Carmen

BELIZE

n

Popocatépetl
5452m

Puebla

Alvarado

Comalcalco

Villahermosa

Taxco

Córdoba

San

Coatzacoalcos

Macuspana

Zacatepec
Cuautla

Tehuacán

Andrés
Tuxtla

Minatitlán

Teapa

Iguala

Tuxtepec

Istmo de

Tuxtla

San Cristóbal
de Las Casas

Gulf of Honduras

Chilpancigo

Huajuapan

Tehuantepec

Ozocuautla

Chiapa de
Corzo

Comitán

Sierra

Oaxaca

Ixtepec

Matías Romero

15°

Madre   del   Sur

Arriaga

Presa de la
Angostura

Tecpan

Tehuantepec

Juchitán

capulco

Pinotepa
Nacional

Miahuatlán

Salina Cruz

Pijijiapan

GUATEMALA

HONDURAS

5

Puerto
Escondido

Puerto
Angel

Golfo de
Tehuantepec

Escuintla
Huixtla

Tapachula

Ciudad Hidalgo

EL SALVADOR

153

100°

95°

90°

E F G H

Elevation

-6000m	-4000m	-2000m	-1000m	-500m	-250m	Below sea level	0	250m	500m	1000m	2000m	3000m	4000m	6000m
-19,658ft	-13,124ft	-6562ft	-3281ft	-1640ft	-820ft	-328ft/-100m	0	820ft	1640ft	3281ft	6562ft	9843ft	13,124ft	19,685ft

51

Central America

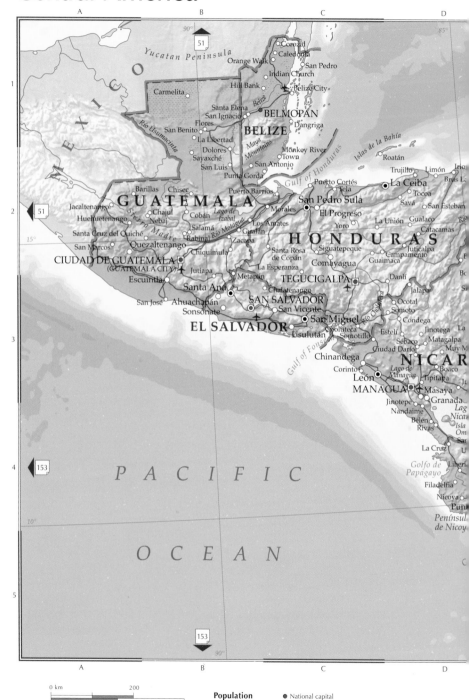

Population ● National capital

○ below 50,000   ○ 50,000 to 100,000   ◉ 100,000 to 500,000   ◼ above 500,0

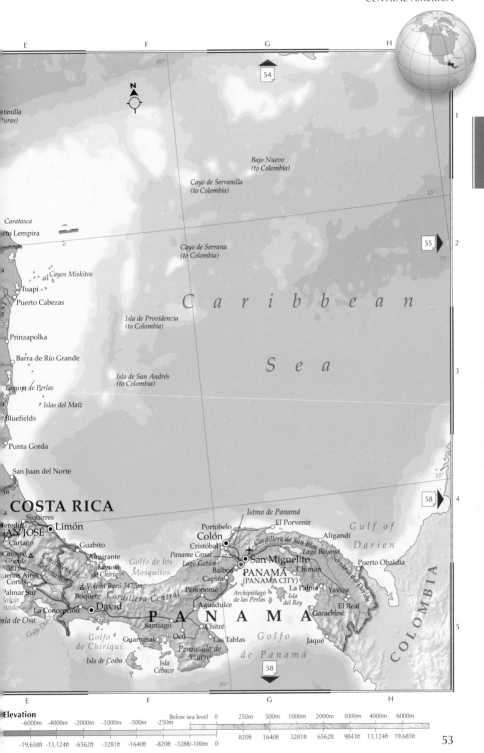

E   F   G   H

54

N

Bajo Nuevo
(to Colombia)

Cayo de Serranilla
(to Colombia)

80°

15°

Caratasca

rto Lempira

Cayo de Serrana
(to Colombia)

55

75°

a

Cayos Miskitos

Tuapi

Puerto Cabezas

C a r i b b e a n

Isla de Providencia
(to Colombia)

Prinzapolka

Barra de Río Grande

S e a

Lagun de Perlas

Isla de San Andrés
(to Colombia)

Islas del Maíz

Bluefields

Punta Gorda

San Juan del Norte

10°

58

m

Istmo de Panamá

COSTA RICA

Siquirres

El Porvenir

Gulf of
Darien

eredia

Limón

Portobelo

Aligandí

AN JOSÉ

Colón

Cristóbal

Cordillera de San Blas

Cartago

Guabito

Panama Canal

Lago Bayano

Puerto Obaldía

Chiripó

Grande
3819m

Almirante

Laguna
de Chiriquí

Golfo de los
Mosquitos

Lago Gatún

San Miguelito

Balboa

PANAMÁ
(PANAMA CITY)

Chimán

Serranía del Darién

enos Aires

Cortés

Volcán Barú 3475m

Capira

Penonomé

La Palma

Yaviza

almar Sur

Bahía

nado

Boquete

Cordillera Central

P A N A M A

Aguadulce

Archipiélago
de las Perlas

Isla
del Rey

El Real

Garachiné

La Concepción

David

Santiago

Chitré

Golfo

C O L O M B I A

la de Osa

Golfo Dulc

Golfo
de Chiriquí

Guarumal

Ocú

Las Tablas

de Panamá

Jaqué

Isla de Coiba

Península de
Azuero

Isla
Cébaco

58

80°

E   F   G   H

Elevation

-6000m	-4000m	-2000m	-1000m	-500m	-250m	Below sea level	0	250m	500m	1000m	2000m	3000m	4000m	6000m

| -19,658ft | -13,124ft | -6562ft | -3281ft | -1640ft | -820ft | -328ft/-100m | 0 | | 820ft | 1640ft | 3281ft | 6562ft | 9843ft | 13,124ft | 19,685ft |

53

The Caribbean

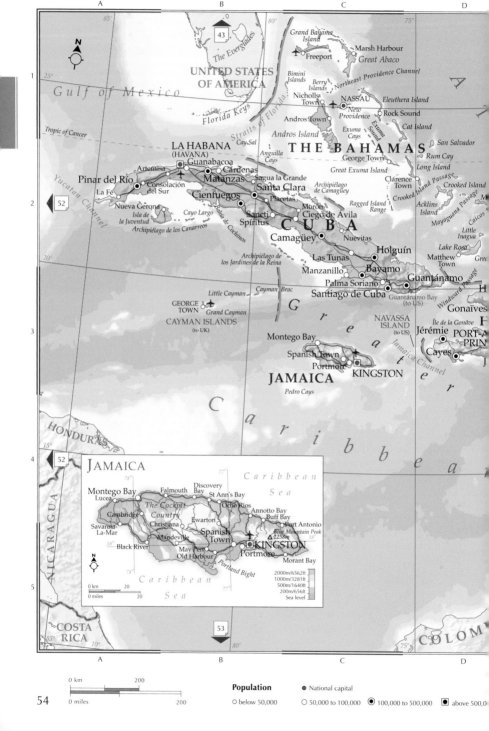

UNITED STATES OF AMERICA

Gulf of Mexico

The Everglades

Grand Bahama Island

Freeport

Marsh Harbour

Great Abaco

Bimini Islands

Berry Islands

Northeast Providence Channel

Nicholls Town

NASSAU

Eleuthera Island

Florida Keys

Andros Town

Rock Sound

New Providence

Cat Island

Straits of Florida

Cay Sal

Andros Island

Exuma Cays

Exuma Sound

Tropic of Cancer

LA HABANA (HAVANA)

Guanabacoa

THE BAHAMAS

San Salvador

Artemisa

Cárdenas

Anguilla Cays

George Town

Rum Cay

Pinar del Río

Matanzas

Sagua la Grande

Great Exuma Island

Long Island

Consolación del Sur

Santa Clara

Archipiélago de Camagüey

Clarence Town

Crooked Island

La Fé

Cienfuegos

Placetas

Morón

Crooked Island Passage

Acklins Island

Mayaguana Passage

Nueva Gerona

Cayo Largo

Sancti Spíritus

Ciego de Ávila

Ragged Island Range

Caicos

Isla de la Juventud

Archipiélago de los Canarreos

Bahía de Cochinos

CUBA

Camagüey

Nuevitas

Little Inagua

Lake Rosa

Yucatan Channel

Las Tunas

Holguín

Matthew Town

Archipiélago de los Jardines de la Reina

Manzanillo

Bayamo

Guantánamo

Palma Soriano

Little Cayman

Cayman Brac

Santiago de Cuba

Guantánamo Bay (to US)

Windward Passage

Gonaïves

GEORGE TOWN

Grand Cayman

Greater

NAVASSA ISLAND (to US)

Île de la Gonâve

Jérémie

PORT-A-PRIN

CAYMAN ISLANDS (to UK)

Montego Bay

Jamaica Channel

Cayes

Spanish Town

Portmore

KINGSTON

Antilles

JAMAICA

Pedro Cays

HONDUR

Caribbean

JAMAICA

Montego Bay

Lucea

Falmouth

Discovery Bay

St Ann's Bay

Caribbean Sea

Cambridge

The Cockpit Country

Ocho Rios

Annotto Bay

Savanna-La-Mar

Christiana

Ewarton

Buff Bay

Port Antonio

Mandeville

Spanish Town

Blue Mountain Peak △2258m

Black River

May Pen

Old Harbour

KINGSTON

Portmore

Morant Bay

Portland Bight

Caribbean Sea

2000m/6562ft
1000m/3281ft
500m/1640ft
200m/656ft
Sea level

0 km   20
0 miles   20

NICARAGUA

COSTA RICA

COLOM

Population

0 km   200
0 miles   200

● National capital

○ below 50,000   ○ 50,000 to 100,000   ◉ 100,000 to 500,000   ■ above 500,0

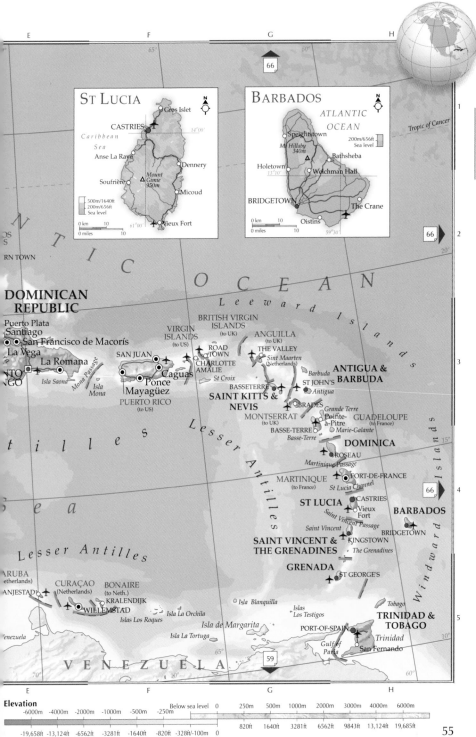

ST LUCIA

Caribbean Sea

Gros Islet
CASTRIES
14°00'
Anse La Raye
Dennery
Soufrière
△ Mount Gimie 950m
Micoud
61°00'
Vieux Fort

500m/1640ft
200m/656ft
Sea level

0 km   10
0 miles   10

BARBADOS

ATLANTIC OCEAN

Speightstown
Mt Hillaby 340m △
13°10'
Holetown
Bathsheba
Welchman Hall
BRIDGETOWN
Oistins
The Crane
59°30'

200m/656ft
Sea level

0 km   10
0 miles   10

Tropic of Cancer

ATLANTIC OCEAN

RN TOWN

DOMINICAN REPUBLIC

Puerto Plata
Santiago
San Francisco de Macorís
La Vega
La Romana
NTO
NGO
Isla Saona
Mona Passage
Isla Mona
SAN JUAN
Caguas
Ponce
Mayagüez
PUERTO RICO (to US)

Leeward Islands

VIRGIN ISLANDS (to US)
BRITISH VIRGIN ISLANDS (to UK)
ROAD TOWN
CHARLOTTE AMALIE
St Croix
ANGUILLA (to UK)
THE VALLEY
Sint Maarten (Netherlands)
Barbuda
ST JOHN'S
BASSETERRE
SAINT KITTS & NEVIS
Antigua
ANTIGUA & BARBUDA
MONTSERRAT (to UK)
BRADES
BASSE-TERRE
Basse-Terre
Grande Terre
Pointe-à-Pitre
GUADELOUPE (to France)
Marie-Galante
DOMINICA
ROSEAU
Martinique Passage
MARTINIQUE (to France)
FORT-DE-FRANCE
St Lucia Channel
ST LUCIA
CASTRIES
Vieux Fort
Saint Vincent Passage
BARBADOS
BRIDGETOWN
Saint Vincent
SAINT VINCENT & THE GRENADINES
KINGSTOWN
The Grenadines
GRENADA
ST GEORGE'S

Lesser Antilles

Windward Islands

Sea

ARUBA (Netherlands)
ANJESTAD
CURAÇAO (Netherlands)
BONAIRE (to Neth.)
KRALENDIJK
WILLEMSTAD
Islas Los Roques
Isla La Orchila
Isla Blanquilla
Islas Los Testigos
Tobago
TRINIDAD & TOBAGO
Isla de Margarita
PORT-OF-SPAIN
Trinidad
Gulf of Paria
San Fernando
Isla La Tortuga

VENEZUELA
enezuela

Lesser Antilles

Elevation

-6000m	-4000m	-2000m	-1000m	-500m	-250m	Below sea level 0	250m	500m	1000m	2000m	3000m	4000m	6000m
-19,658ft	-13,124ft	-6562ft	-3281ft	-1640ft	-820ft	-328ft/-100m 0	820ft	1640ft	3281ft	6562ft	9843ft	13,124ft	19,685ft

South America

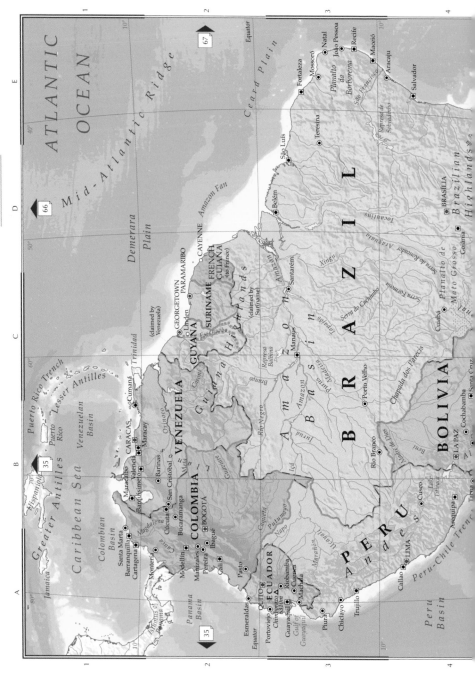

Population

● National capital

○ below 50,000   ○ 50,000 to 100,000   ◉ 100,000 to 500,000   ■ above 500,000

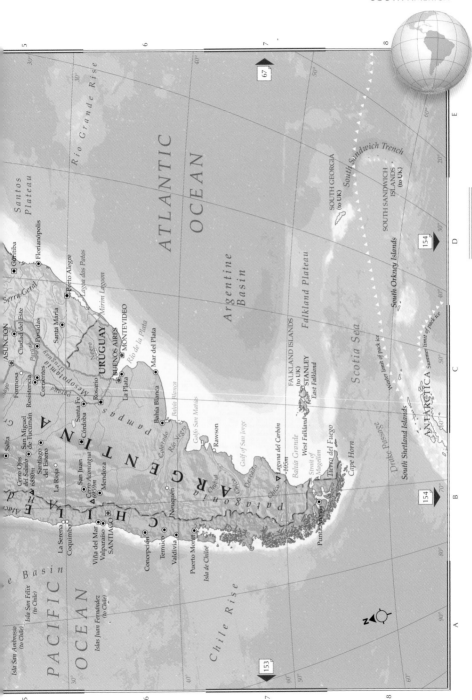

Northern South America

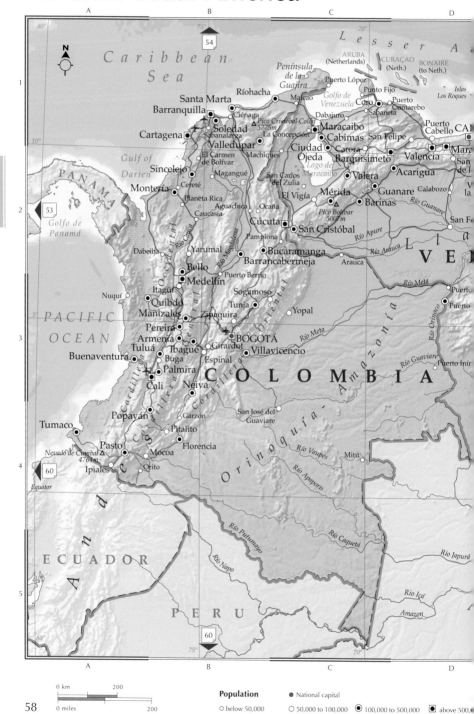

Population ● National capital

○ below 50,000   ○ 50,000 to 100,000   ◉ 100,000 to 500,000   ■ above 500,

0 km   200

0 miles   200

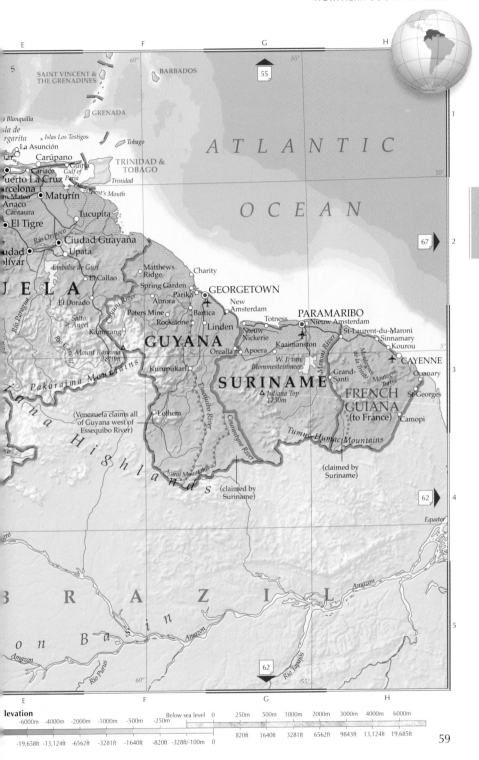

E   F   G   H

SAINT VINCENT &
THE GRENADINES

BARBADOS

60°

55°

55

a Blanquilla

Isla de
argarita

GRENADA

La Asunción

Islas Los Testigos

Tobago

Carúpano

nar

Cariaco

Güiria

TRINIDAD &
TOBAGO

Gulf of
Paria

uerto La Cruz

rcelona

Trinidad

10°

ATLANTIC

n Mateo

Anaco

Maturín

Serpent's Mouth

Cantaura

El Tigre

Tucupita

OCEAN

udad

olívar

Río Orinoco

Ciudad Guayana

Upata

67

2

JELA

Embalse de Guri

El Callao

Matthews
Ridge

Charity

Spring Garden

GEORGETOWN

El Dorado

Aurora

Parika

Salto
Ángel

Río Paragua

Cuyuni River

Peters Mine

Bartica

New
Amsterdam

Totness

PARAMARIBO

Nieuw Amsterdam

St-Laurent-du-Maroni

Sinnamary

Kamarang

Rockstone

Linden

Nieuw
Nickerie

Kaaimanston

Kourou

5°

Mount Roraima
2810m

GUYANA

Orealla

Apoera

W. J. van
Blommesteinmeer

Maroni River

Montagnes
de la Trinité

CAYENNE

Oyapock

Río Caroní

Pakaraima Mountains

Kurupukari

SURINAME

Juliana Top
1230m

Grand-
Santi

Montagne
Tortue

St-Georges

ana Highlands

(Venezuela claims all
of Guyana west of
Essequibo River)

Lethem

Essequibo River

Courantyne River

FRENCH
GUIANA
(to France)

Camopi

oco

Tumuc-Humac Mountains

(claimed by
Suriname)

62

4

Acarai Mountains

(claimed by
Suriname)

Equator

go

B   R   A   Z   I   L

Amazon

5

Río Purús

on   Ba   sin

Amazon

Amazon

Río Tapajós

Amazon

62

60°

55°

E   F   G   H

levation

-6000m   -4000m   -2000m   -1000m   -500m   -250m   Below sea level   0   250m   500m   1000m   2000m   3000m   4000m   6000m

-19,658ft   -13,124ft   -6562ft   -3281ft   -1640ft   -820ft   -328ft/-100m   0   820m   1640m   3281ft   6562ft   9843ft   13,124ft   19,685ft

Western South America

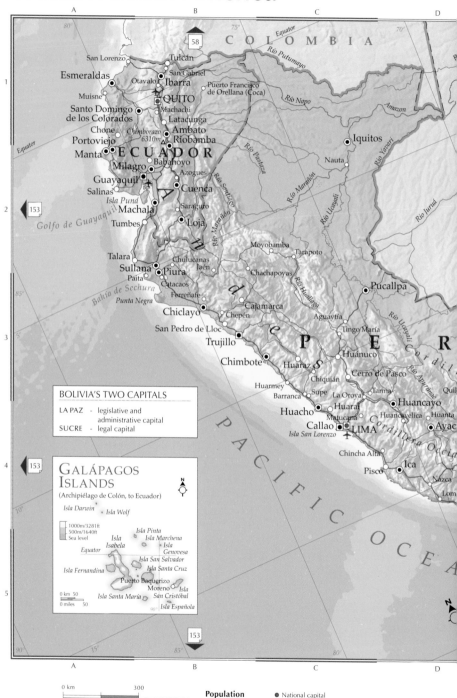

BOLIVIA'S TWO CAPITALS

LA PAZ - legislative and administrative capital

SUCRE - legal capital

GALÁPAGOS ISLANDS

(Archipiélago de Colón, to Ecuador)

1000m/3281ft
500m/1640ft
Sea level

Isla Darwin
Isla Wolf
Isla Pinta
Isla Marchena
Isla Isabela
Isla Genovesa
Isla San Salvador
Isla Fernandina
Isla Santa Cruz
Puerto Baquerizo Moreno
Isla San Cristóbal
Isla Santa María
Isla Española
Equator

0 km 50
0 miles 50

0 km 300
0 miles 300

Population
● National capital
○ below 50,000
○ 50,000 to 100,000
◉ 100,000 to 500,000
■ above 500,

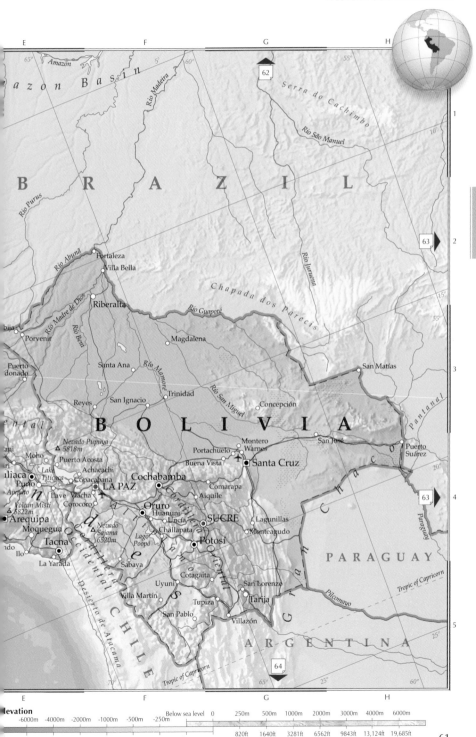

Elevation

						Below sea level	0	250m	500m	1000m	2000m	3000m	4000m	6000m
-6000m	-4000m	-2000m	-1000m	-500m	-250m									
-19,658ft	-13,124ft	-6562ft	-3281ft	-1640ft	-820ft	-328ft/-100m	0	820ft	1640ft	3281ft	6562ft	9843ft	13,124ft	19,685ft

Brazil

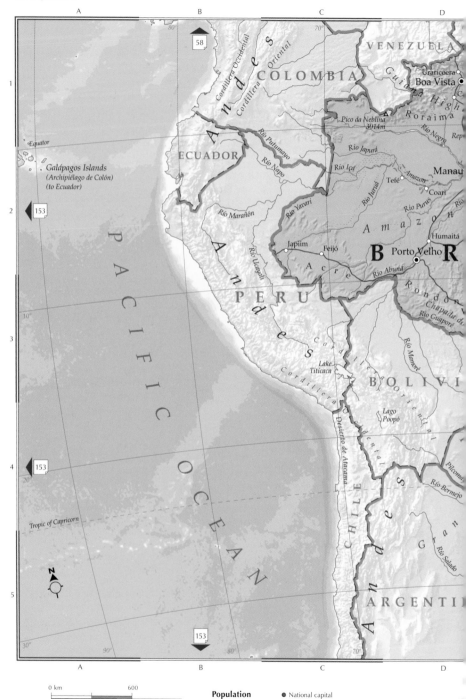

VENEZUELA

COLOMBIA

Guiana High

Uraricoera

Boa Vista

C

Roraima

Pico da Neblina
3014m

Rep

ECUADOR

Río Putumayo

Río Japurá

Río Napo

Río Içá

Manau

Amazon

Tefé

Coari

Equator

Galápagos Islands
(Archipiélago de Colón)
(to Ecuador)

Río Marañón

Río Yavari

Río Juruá

Río Purús

Río

153

Japiim

Feijó

B

Porto Velho

R

Río Ucayali

Acre

Humaitá

Rondôn

PERU

Río Abunã

Chapada do

Río Guaporé

Río Mamoré

Cordillera

Lake
Titicaca

BOLIVI

Cordillera

Desierto de Atacama

Occidental

Lago
Poopó

Oriental

Piícoma

PACIFIC OCEAN

153

Río Bermejo

Tropic of Capricorn

CHILE

A

n

G

Río Salado

N

153

ARGENTI

0 km   600

0 miles   600

Population

● National capital

○ below 50,000   ○ 50,000 to 100,000   ◉ 100,000 to 500,000   ◼ above 500,0

ATLANTIC OCEAN

FRENCH GUIANA
(to France)

nuc-Humac Mountains

Amapá

Macapá

Ilha Caviana de Fora

Mouths of the Amazon

Baía de Marajó

Ilha de Marajó

Belém

Baía de São Marcos

São Luís

Parnaíba

Camocim

Equator

Amazon

arém

Altamira

Represa de Tucuruí

Marabá

Imperatriz

Rio Xingu

tuba

P a r á

chinho

Bacabal

Piripiri

Teresina

Fortaleza

Maranhão

Ceará

Mossoró

Assu

Cabo de São Roque

Atol das Rocas

San Fernando de Noronha
(to Brazil)

Marabá

Carolina

Floriano

Picos

Rio Grande do Norte

Natal

Juazeiro do Norte

João Pessoa

Campina Grande

Balsas

Piauí

Pernambuco

Paraíba

Recife

Z I L

Represa de Sobradinho

Juazeiro

Alagoas

Maceió

Palmas do Tocantins

Rio São Francisco

Chapada Diamantina

Aracaju

Estância

Taguatinga

Tocantins

Serra dos Gradaús

Serra Formosa

Feira de Santana

Salvador

Baía de Todos os Santos

Rio Araguaia

Goiás

Bahia

Planalto Central

Itabuna

BRASÍLIA

Janaúba

Vitória da Conquista

Anápolis

Canavieiras

Montes Claros

Goiânia

Minas

Araçuaí

rosso

Jataí

iabá

Gerais

pólis

Araguari

Governador Valadares

Espírito Santo

to Grosso

Uberlândia

Uberaba

do Sul

Belo Horizonte

Campo Grande

Ribeirão Preto

Divinópolis

Vitória

anu

Marília

Juiz de Fora

Campos dos Goytacazes

te Prudente

Campinas

Londrina

Nova

Maringá

São Paulo

Iguaçu

Rio de Janeiro

Paraná

Santos

Tropic of Capricorn

Represa de Itaipu

Ponta Grossa

Curitiba

Saltos do Rio Iguaçu

Joinville

Iguaçu

Santa Catarina

Blumenau

Florianópolis

Grande

Passo Fundo

aria

Canoas

do Sul

Porto Alegre

Bagé

Lagoa dos Patos

Rio Grande

UAY

Mirim Lagoon

ATLANTIC OCEAN

Elevation

| -6000m | -4000m | -2000m | -1000m | -500m | | Below sea level | 0 | 250m | 500m | 1000m | 2000m | 3000m | 4000m | 6000m |
| | | | | | -250m | | | | | | | | | |

-19,658ft -13,124ft -6562ft -3281ft -1640ft -820ft -328ft/-100m 0   820ft 1640ft 3281ft 6562ft 9843ft 13,124ft 19,685ft

Southern South America

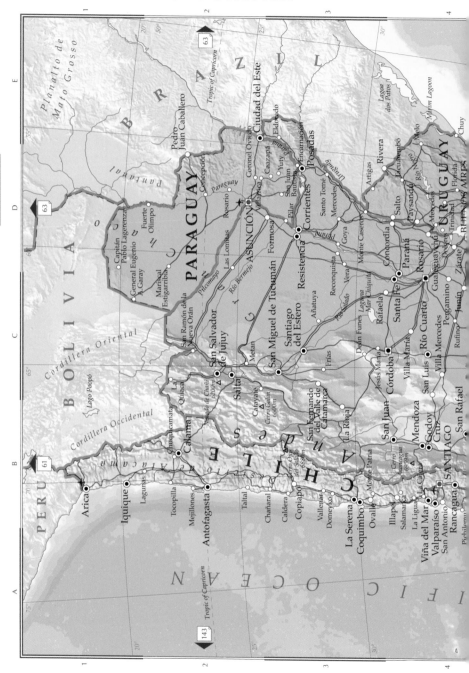

0 km   200

0 miles   200

Population   ● National capital

○ below 50,000   ○ 50,000 to 100,000   ◉ 100,000 to 500,000   ▣ above 500,

ATLANTIC

OCEAN

Mar del Plata

Necochea

Coronel
Dorrego

Bahía Blanca

Punta Alta

Bahía Blanca

Choele Choel

Cipolletti

Neuquén

Río Negro

Viedma

Golfo San Matías

Península
Valdés

Golfo Nuevo

San Antonio Oeste

Rawson

Trelew

Río Chubut

Comodoro Rivadavia

Golfo San Jorge

Caleta Olivia

Río Deseado

Puerto Deseado

Río Chico

Puerto
San Julián

Laguna del Carbón
-105m

Bahía
Grande

Río Gallegos

Strait of Magellan

FALKLAND ISLANDS
(to UK)

STANLEY

East
Falkland

Goose
Green

West
Falkland

Isla
de los Estados

Beagle Channel

Cabo de Hornos
(Cape Horn)

Drake Passage

Tierra del Fuego

Porvenir

Ushuaia

Punta Arenas

Puerto Natales

Cerro Fitzroy
3470m

El Calafate

Río Santa Cruz

Río Chico

Lago
Buenos Aires

Lago
Viedma

Cochrane

Perito
Moreno

Puerto
Aisén

San Martín

Puerto Chacabuco

Coyhaique

Cerro
Valdés Sur
4058m

Isla
Wellington

Río Cisnes

Sarmiento

Paso
de Indios

Esquel

Lago
Musters

Río Chubut

San Carlos de Bariloche

Lago
Nahuel Huapi

Temuco

Loncoche

Valdivia

Osorno

Puerto Varas

Puerto Montt

Ancud

Castro

Isla de Chiloé

Río Corcovado

Archipiélago
de los Chonos

Golfo de Penas

CHILE

Tapalqué

Zapala

d

Elevation

| -6000m | -4000m | -2000m | -1000m | -500m | -250m | Below sea level | 0 | 250m | 500m | 1000m | 2000m | 3000m | 4000m | 6000m |

| -19,658ft | -13,124ft | -6562ft | -3281ft | -1640ft | -820ft | -328ft/-100m | 0 | 820ft | 1640ft | 3281ft | 6562ft | 9843ft | 13,124ft | 19,685ft |

67

154

154

143

65

The Atlantic Ocean

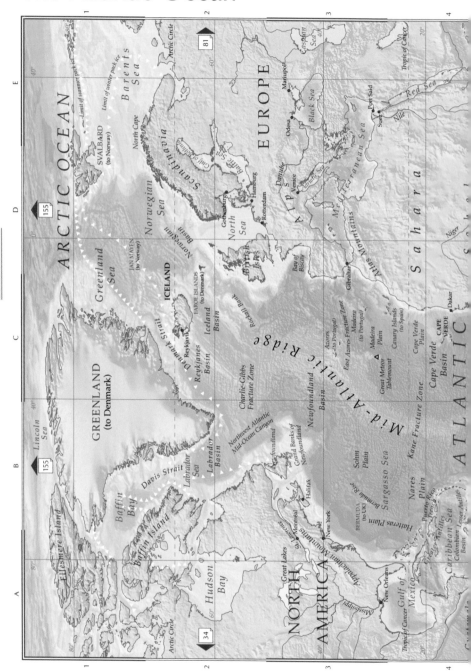

ARCTIC OCEAN

Limit of summer pack ice
Limit of winter pack ice

Barents Sea

EUROPE

Caspian Sea

Tropic of Cancer

Mariupol

Black Sea

Red Sea

Port Said

Odesa

Suez

Nile

North Cape

Svalbard
(to Norway)

Scandinavia

Gulf of Bothnia

Mediterranean Sea

Sahara

Danube

Venice

Adriatic Sea

Alps

Gothenburg

Hamburg

Rotterdam

North Sea

British Isles

Atlas Mountains

Niger

Sahel

Norwegian Sea

Norwegian Basin

Bay of Biscay

Gibraltar

Greenland Sea

JAN MAYEN
(to Norway)

ICELAND

Reykjavik

FAROE ISLANDS
(to Denmark)

Iceland Basin

Rockall Bank

Azores
(to Portugal)

East Azores Fracture Zone

Madeira
(to Portugal)

Madeira Plain

Canary Islands
(to Spain)

Dakar

CAPE VERDE

Cape Verde Plain

Denmark Strait

Reykjanes Basin

Charlie-Gibbs Fracture Zone

Mid-Atlantic Ridge

Great Meteor Tablemount

Cape Verde Basin

Lincoln Sea

GREENLAND
(to Denmark)

Labrador Basin

Labrador Sea

Northwest Atlantic Mid-Ocean Canyon

Newfoundland Basin

Newfoundland

Grand Banks of Newfoundland

Sohm Plain

Kane Fracture Zone

Nares Plain

ATLANTIC

Davis Strait

Baffin Bay

Baffin Island

Ellesmere Island

Halifax

St. Lawrence

Montreal

New York

Appalachian Mountains

BERMUDA
(to UK)

Bermuda Rise

Sargasso Sea

Hatteras Plain

Puerto Rico
Trench

Lesser Antilles

Caribbean Sea

Colombian Basin

Hudson Bay

Great Lakes

NORTH AMERICA

New Orleans

Gulf of Mexico

Mississippi

Tropic of Cancer

Arctic Circle

81

155

155

34

0 km   1000

0 miles   1000

● Major port

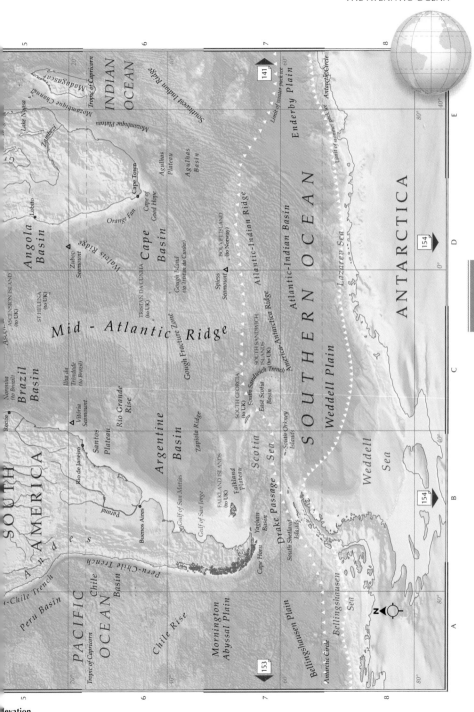

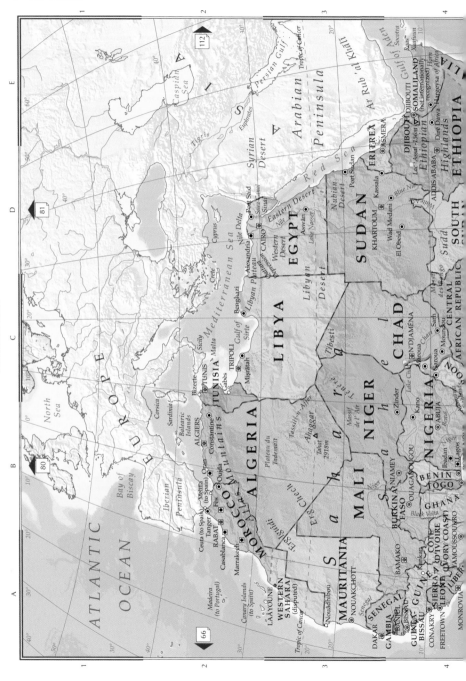

| 0 km | 1000 |
| 0 miles | 1000 |

Population   • National capital

o below 50,000   o 50,000 to 100,000   ◉ 100,000 to 500,000   ■ above 500,00

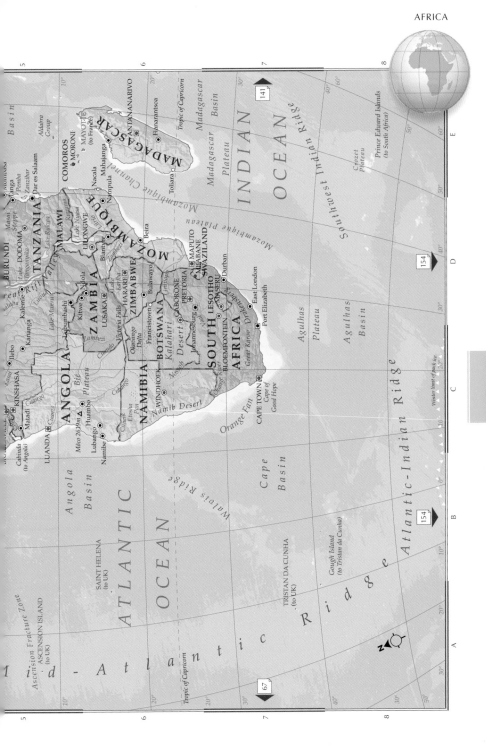

Northwest Africa

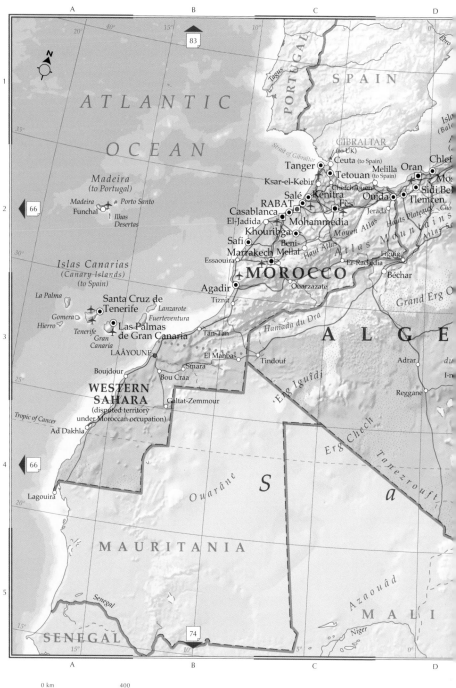

ATLANTIC

OCEAN

SPAIN

PORTUGAL

Tagus

Ebro

GIBRALTAR
(to UK)

Tanger
Ceuta (to Spain)
Strait of Gibraltar
Tetouan
Melilla
(to Spain)
Oran
Chlef
Mos
Ksar-el-Kebir
Chefchaouen
Sidi Bel
Salé
Kénitra
Ouijda
Tlemcen
RABAT
Fès
Jerada
Cho
Casablanca
Mohammedia
Moyen Atlas
Hauts Plateaux
El-Jadida
Khouribga
Mountains
Beni-
Haut Atlas
Atlas
Atlas
Safi
Mellal
Figuig
Marrakech
Er-Rachidia
Essaouira
Béchar
MOROCCO
Ouarzazate
Grand Erg
Agadir
Tiznit

Madeira
(to Portugal)
Madeira
Porto Santo
Funchal
Ilhas
Desertas

Islas Canarias
(Canary Islands)
(to Spain)

La Palma
Lanzarote
Santa Cruz de
Tenerife
Fuerteventura
Gomera
Hierro
Tenerife
Las Palmas
Gran
de Gran Canaria
Canaria
Tan-Tan
Hamada du Dra
ALGE

LAÂYOUNE
El Mahbas
Tindouf
Adrar
du
Smara
Boujdour
Bou Craa
Reggane
I-n-
WESTERN
Erg Iguîdi
SAHARA
(disputed territory
under Moroccan occupation)
Galtat-Zemmour
Erg Chech
Tropic of Cancer
Ad Dakhla
Tanezrouft
Lagouira
Ouarâne
S
a

MAURITANIA

Senegal

MALI

Azaouâd

Niger

SENEGAL

83

66

66

74

0 km   400

0 miles   400

70

Population ● National capital

○ below 50,000   ◯ 50,000 to 100,000   ◉ 100,000 to 500,000   ▣ above 500

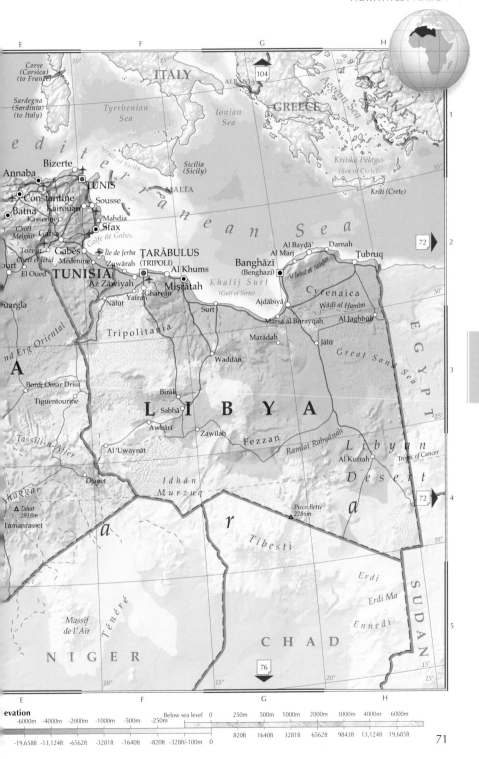

E   F   G   H

Corse (Corsica) (to France)

Tyrrhenian Sea

ITALY

ALBANIA

104

GREECE

Aegean Sea

Ionian Sea

Sardegna (Sardinia) (to Italy)

Strait of Sicily

Sicilia (Sicily)

Kritiko Pélagos (Sea of Crete)

Kríti (Crete)

MALTA

Bizerte

Annaba

TUNIS

Constantine

Sousse

Batna

Kairouan

Kasserine

Chott Melghir

Gafsa

Mahdia

Sfax

72

Tozeur

Gabès

Al Bayḍāʾ

Al Marj

Darnah

Ṭubruq

Chott el Jerid

Médenine

Île de Jerba

ṬARĀBULUS (TRIPOLI)

Banghāzī (Benghazi)

Golfe de Gabès

Zuwārah

Al Khums

El Oued

TUNISIA

Az Zāwiyah

Miṣrātah

Khalīj Surt (Gulf of Sirte)

Al Jabal al Akhḍar

Cyrenaica

uargla

Nālūt

Yafran

Gharyān

Surt

Ajdābiyā

Wādī al Ḥamīm

Marṣā al Burayqah

Al Jaghbūb

Tripolitania

Marādah

Jālū

Great Sand Sea

nd Erg Oriental

Waddān

A

Bordj Omar Driss

Birāk

L

I

B

Y

A

Tiguentourine

Sabhā

Al Kufrah

Tassili-n-Ajjer

Awbārī

Zawīlah

Fezzan

Ramlat Rabyānah

L i b y a n

Al 'Uwaynāt

Tropic of Cancer

72

Djanet

I d h ā n

M u r z u q

D e s e r t

△ **Tahat 2918m**

Picco Bette 2286m △

Tamanrasset

Tibesti

r

Haggar

Erdi

Erdi Ma

S U D A N

Ennedi

N I G E R

Massif de l'Aïr

Ténéré

C H A D

76

E   F   G   H

evation

| -6000m | -4000m | -2000m | -1000m | -500m | -250m | Below sea level | 0 | 250m | 500m | 1000m | 2000m | 3000m | 4000m | 6000m |

| -19,658ft | -13,124ft | -6562ft | -3281ft | -1640ft | -820ft | -328ft/-100m | 0 | 820ft | 1640ft | 3281ft | 6562ft | 9843ft | 13,124ft | 19,685ft |

Northeast Africa

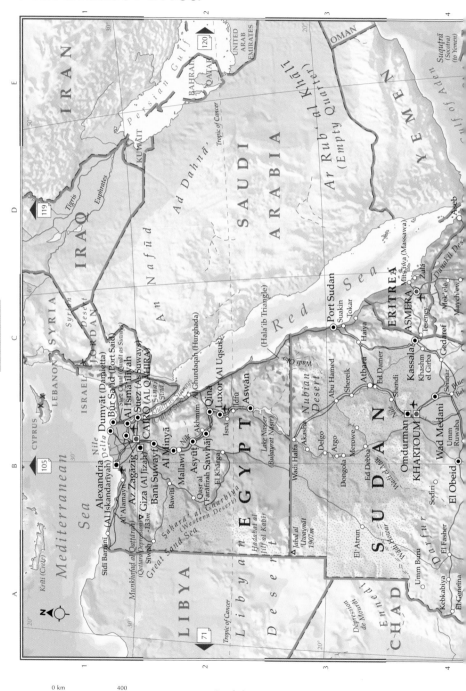

IRAN

Persian Gulf

BAHRAIN

QATAR

UNITED ARAB EMIRATES

OMAN

Suquţrā (Socotra) (to Yemen)

120

Tropic of Cancer

SAUDI ARABIA

Ad Dahnā

Ar Rub' al Khālī (Empty Quarter)

Y E M E N

Gulf of Aden

119

IRAQ

Tigris

Euphrates

An Nafūd

Aseb

SYRIA

Syrian Desert

JORDAN

Red Sea

Port Sudan

Suakin

Tokar

Wadi Oko

(Hala'ib Triangle)

ERITREA

Mitsiwa (Massawa)

Zula

ASMERA

Danakil Des.

Mek'elē

Maychew

LEBANON

105

ISRAEL

CYPRUS

Mediterranean Sea

Kríti (Crete)

Alexandria (Al Iskandariyah)

Sidi Barrani

Al 'Alamayn

Marsá Matrūḥ

Munkhafad al Qattāra (Qattara Depression) -133m

Siwah

Dumyāt (Damietta)

Būr Sa'īd (Port Said)

Al Ismā'īlīyah

Suez (As Suways)

Suez Canal (Qanāt as Suways)

CAIRO (AL QĀHIRA)

Az Zagāzīg

Giza (Al Jīzah)

Bani Suwayf

Al Minyā

Mallawi

Asyūt

Akhmīm

Qasr al Farāfirah

Bawītī

Sawhāj

El Khārga

Qina

Luxor

Al Ghurdaqah (Hurghada)

Isnā

Idfū

Aswān

Lake Nasser (Buhayrat Nāṣir)

Sinai (Sīnā')

Nubian Desert

Wadi Halfa

Akasha

Delgo

Argo

Abu Hamed

Shereik

Atbara

Haiya

Ed Damer

Shendi

Kassala

Khashm el Girba

Gedaref

Teseney

SUDAN

Omdurman

KHARTOUM

Wād Medani

Sennar

Blue Nile

El Obeid

Umm Ruwaba

Ed Debba

Dongola

Merowe

Delgo

LIBYA

Sahara' al Gharbiya (Western Desert)

Libyan

Desert

Great Sand Sea

Hadabat al Jilf al Kabīr

Jabal al 'Uwaynāt 1907m

El 'Atrun

Wadi al Malik

Wadi Howar

Sodiri

Ed Debba

D a r f u r

El Fasher

Umm Buru

Kebkabiya

El Geneina

C H A D

Ennedi

Depression de Mourdi

E G Y P T

N

72

Population

○ below 50,000

○ 50,000 to 100,000

● 100,000 to 500,000

■ above 500,000

● National capital

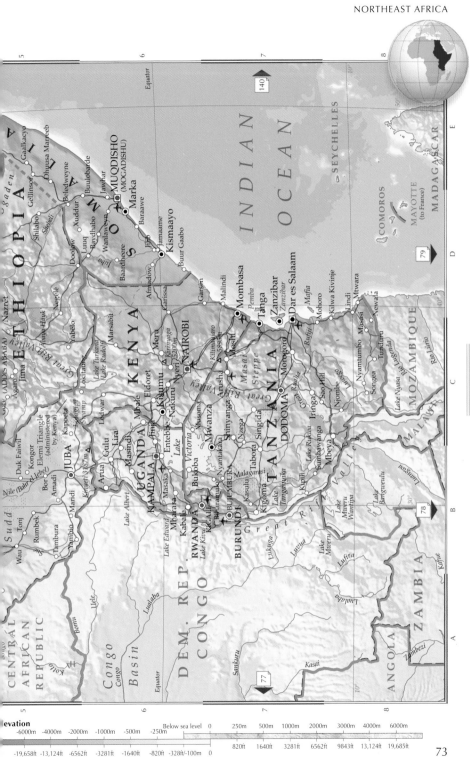

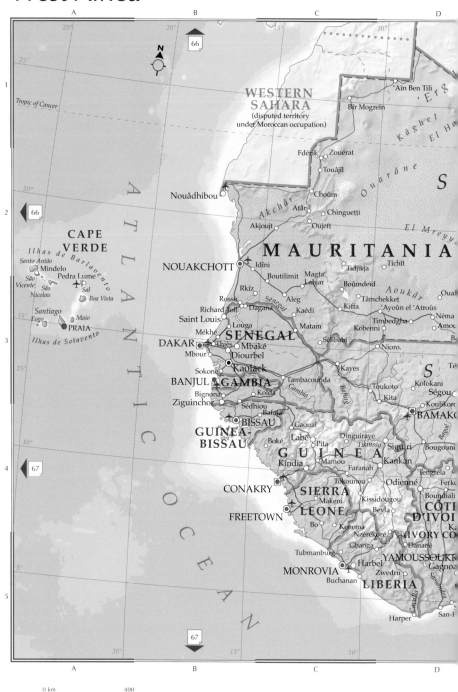

A B C D

66

N

Tropic of Cancer

WESTERN
SAHARA
(disputed territory
under Moroccan occupation)

Aïn Ben Tili
Bîr Mogreïn
Fdérik Zouérat
Touâjîl

Nouâdhibou
Choûm
Atâr Chinguetti
Akjoujt Oujeft

66
Idîni
NOUAKCHOTT
Boutilimit Magta' Tîdjikja Tîchît
Lahjar
Rkîz Boûmdeïd Aoukâr Oua
Rosso Aleg Tâmchekket
Richard Toll Dagana Kaédi Kiffa Ayoûn el 'Atroûs Néma
Saint Louis Matam Timbedgha Amou
Louga Kobenni B
Mékhé Schibabi Nioro
DAKAR Thiès Mbaké
Mbour Diourbel Nioro
SENEGAL

CAPE
VERDE
Ilhas de Barlavento
Santo Antão
Mindelo
São Pedra Lume
Vicente São Sal
Nicolau Boa Vista
Santiago Maio
Fogo
PRAIA
Ilhas de Sotavento

MAURITANIA

S

El Mreyye

Kolokani
Kayes S Té
Kaolack Kita Koulikoro
Sokone Toukoto Ségou
BANJUL GAMBIA Tambacounda BAMAKO
Bignona Kolda
Ziguinchor Sédhiou Gambia
Balaca Boké
BISSAU Gaoual Dinguiraye Siguiri Bougouni
GUINEA- Labé Pita Tikinsso
BISSAU Boké GUINEA Kankan
Kindia Mamou Faranah Odienné Tengréla
CONAKRY Tokounou CÔTE
SIERRA Kissidougou Boundiali D'IVOI
Makeni Beyla Ka
FREETOWN LEONE Konema IVORY CO
Bo Nzérékoré Danané
Gbanga YAMOUSSOUKF
Tubmanburg Harbel Gagnoa
MONROVIA Zwedru LIBERIA
Buchanan
Harper San-I

ATLANTIC OCEAN

67

67

74

0 km 400
0 miles 400

Population ● National capital

○ below 50,000 ○ 50,000 to 100,000 ◉ 100,000 to 500,000 ▣ above 500,

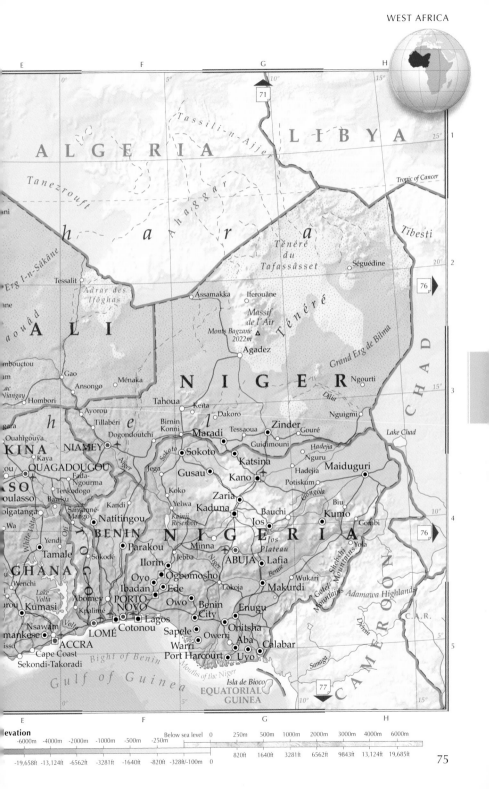

E   F   G   H

0°   5°   10°   15°

71

ALGERIA   LIBYA   1

Tanezrouft   *Ahaggar*   *Tibesti*

Tropic of Cancer

Tassili-n-Ajjer   25°

ni

Ténéré
du
Tafassâsset   Séguédine   20°   2

Assamakka   Iferouâne

Erg I-n-Sâkâne   Massif   *Ténéré*   76
Tessalit   de l'Aïr

Adrar des   Monts Bagzane   *Grand Erg de Bilna*
Ifôghas   2022m △

ane   Agadez

a o u â d   A L I   Ngourti

mbouctou   Gao   N I G E R   *Dilia*   15°   3
am   Ansongo   Ménaka

ac   Nguigmi
Viangay   Hombori   Tahoua   Keita   Dakoro   *Lake Chad*

gara   h   Ayorou   e   l   Tessaoua   Zinder   Gouré

ou   Ouahigouya   Tillabéri   Birnin   Maradi   Guidimouni   C
KINA   NIAMEY   Dogondoutchi   Konni   *Hadejia*   H

Kaya   Sokoto   Katsina   Nguru   A

QUAGADOUGOU   Fada-   Gusau   Kano   Hadejia   Maiduguri   D
oulasso   Ngourma   Potiskum   4

lgatanga   Bawku   Koko   Zaria   *Gongola*   Biu
Sansanné-   Yelwa   Kaduna   Bauchi   Kumo   10°
Wa   Mango   Kandi   *Kainji*   Jos   Gombi

Natitingou   *Reservoir*   *Jos*   G E R I A   76

Yendi   BENIN   N I   *Plateau*   Yola
Tamale   Sokodé   Parakou   Minna   Lafia
GHANA   Jebba   ABUJA   *Benue*   Wukari
irou   *Oti*   Ilorin   *Niger*   *Shebshi*   *Adamawa Highlands*
White Volta   Oyo   Ogbomosho   Lokoja   Makurdi   *Mountains*
Kumasi   Ede   *Gotel*   C.A.R.
Nsawam   Abomey   Ibadan   PORTO-   Owo   Enugu
mankese   Kpalimé   NOVO   Benin   Onitsha   *Djerem*
isso   LOMÉ   Cotonou   City   Aba   Calabar   5
ACCRA   Lagos   Sapele   Owerri   C A M E R O O N
Cape Coast   Warri   Uyo   *Samaga*
Sekondi-Takoradi   *Bight of Benin*   Port Harcourt
        Gulf of Guinea   *Mouths of the Niger*   77
                Isla de Bioco   10°
                EQUATORIAL   15°
                GUINEA

E   F   G   H

Central Africa

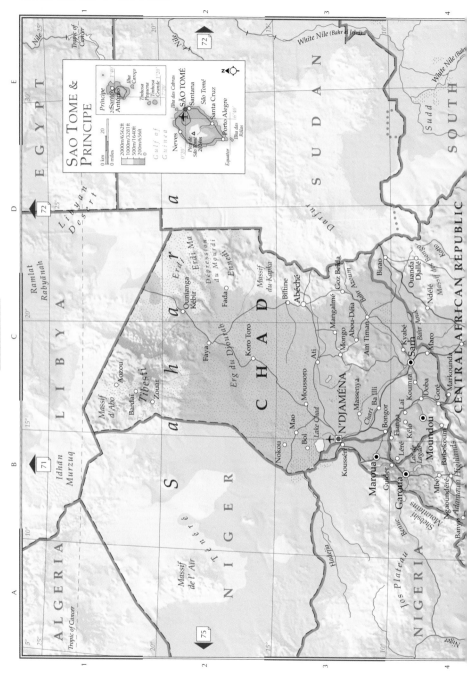

Population

○ below 50,000   ○ 50,000 to 100,000   ◉ 100,000 to 500,000   ▣ above 500,

● National capital

0 km   400
0 miles   400

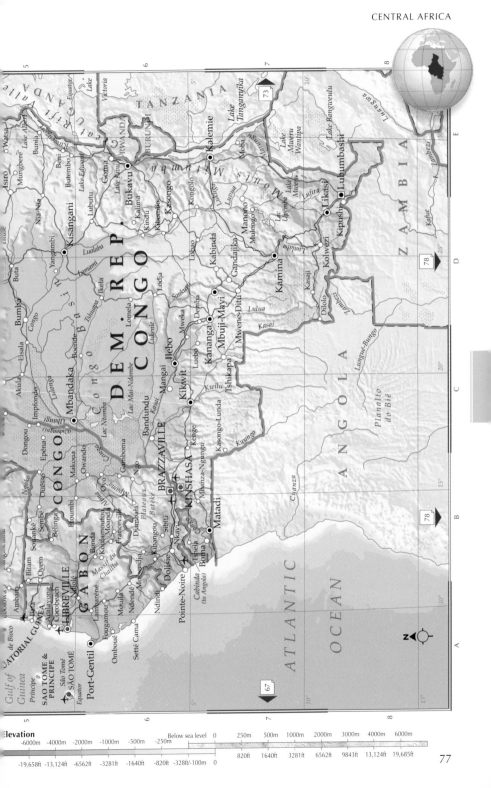

Southern Africa

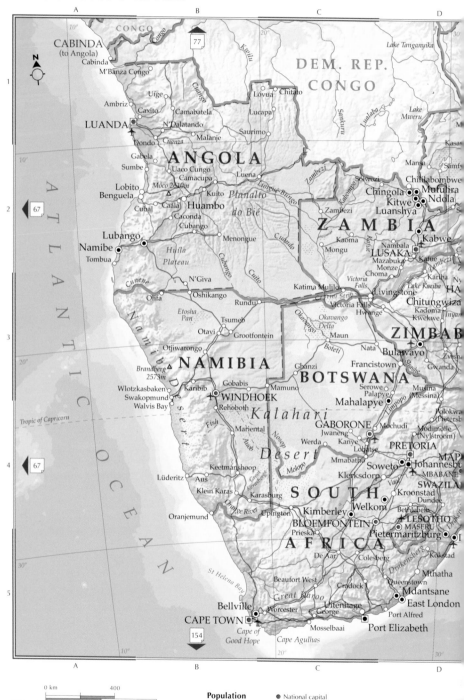

AFRICA

CABINDA
(to Angola)
Cabinda
M'Banza Congo

CONGO

77

DEM. REP.
CONGO

Lake Tanganyika

Uíge
Ambriz
Caxito
Camabatela
LUANDA
N'Dalatando
Dondo
Malanje

Lovua
Chitato
Lucapa
Saurimo

Lake
Mweru

Kasa

Gabela
Sumbe
Lobito
Benguela
Cubal
Caconda
Cubango

ANGOLA

Uaco Cungo
Camacupa
Móco 2610m
Caála
Huambo
Planalto
do Bié

Luena

Luanshya

Mansa
Samfy

Chililabombwe
Chingola
Mufulira
Kitwe
Ndola

Zambezi
Zambezi

Solwezi

Lubango
Namibe
Tombua

Menongue

Huíla
Plateau

Cubango

Cuando

ZAMBIA

Kaoma
Mongu

Nambala
LUSAKA
Mazabuka
Monze
Choma

Kabwe

Kafue

N'Giva

Cunene

Olifa
Oshikango

Rundu

Cuito

Katima Mulilo

Victoria
Falls
Victoria Falls
Caprivi Strip

Livingstone

Kariba
Lake Kariba

Chitungwiza
Kadoma
Kwekwe

HA
Ny

Etosha
Pan

Tsumeb

Okavango
Delta

Maun

Hwange

Inyan

Otavi
Grootfontein

Boteti

Nata

Bulawayo

ZIMBAB

Otjiwarongo

Brandberg
2573m

NAMIBIA

Ghanzi

Francistown

Gwanda

Zvish

Wlotzkasbaken
Swakopmund
Walvis Bay

Karibib

Gobabis
WINDHOEK
Rehoboth

Mamuno

BOTSWANA

Serowe
Palapye
Mahalapye

Musina
(Messina)

Tropic of Capricorn

Fish

Mariental

Kalahari

Nossob

GABORONE
Jwaneng
Kanye
Lobatse

Mochudi

Polokwa
(Pietersb

Modimolle
(Nylstroom)

PRETORIA

MAP

Lüderitz
Aus
Klein Karas
Karasburg

Keetmanshoop

Auob

Desert

Molopo

Werda

Mmabatho
Klerksdorp

Soweto
Johannesbu

SWAZIL

MBABANE

Oranjemund

Orange River

Upington

SOUTH

Kimberley

Prieska

Welkom
Kroonstad

Vaal

Dundee
Bethlehem

LESOTHO
MASERU

Bloemfontein

AFRICA

De Aar

Colesberg

Kokstad

Pietermaritzburg

I

St Helena Bay

Beaufort West

Cradock

Queenstown

Mthatha

Mdantsane
East London

Bellville
CAPE TOWN

154

Cape of
Good Hope

Worcester
George

Mosselbaai

Cape Agulhas

Uitenhage

Port Alfred

Port Elizabeth

Great Karoo

Population

○ below 50,000
○ 50,000 to 100,000
◉ 100,000 to 500,000
■ above 500,0

● National capital

0 km   400

0 miles   400

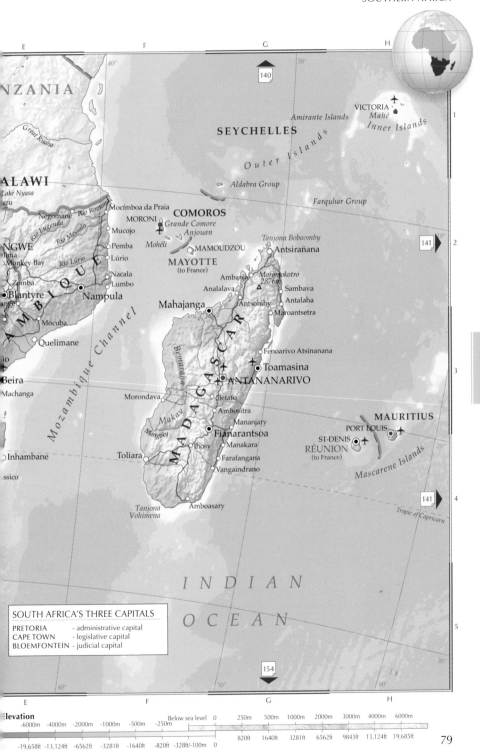

NZANIA

Great Ruaha

ALAWI
Lake Nyasa
zu

NGWE
ima
Monkey Bay
Zomba
nge
Blantyre

SEYCHELLES

Amirante Islands

VICTORIA
Mahé
Inner Islands

140

Outer Islands

Aldabra Group

Farquhar Group

141

Negomane
Rio Rovuma
Rio Lugenda
Rio Messalo
Rio Lúrio

Mocímboa da Praia
Mucojo
Pemba
Lúrio
Nacala
Lumbo
Nampula

COMOROS
MORONI
Grande Comore
Anjouan
Mohéli
MAMOUDZOU
MAYOTTE
(to France)

Ambanja
Analalava
Antsohihy

Tanjona Bobaomby
Antsiranana

Maromokotro
2876m
Sambava
Antalaha
Maroantsetra

Mocuba
Quelimane

Mahajanga

Beira
Machanga

Morondava

Fenoarivo Atsinanana
Toamasina
ANTANANARIVO
Betafo
Ambositra
Mananjary
Manakara
Farafangana
Vangaindrano

MADAGASCAR

Makay
Mangoky
Bemaraha

Ihosy
Fianarantsoa

MAURITIUS
PORT LOUIS
ST-DENIS
RÉUNION
(to France)

Inhambane
ssico

Toliara

Tanjona
Vohimena
Amboasary

Mascarene Islands

Tropic of Capricorn

141

I N D I A N

O C E A N

SOUTH AFRICA'S THREE CAPITALS	
PRETORIA	- administrative capital
CAPE TOWN	- legislative capital
BLOEMFONTEIN	- judicial capital

154

Elevation
-6000m -4000m -2000m -1000m -500m -250m   Below sea level 0   250m   500m   1000m   2000m   3000m   4000m   6000m

-19,658ft -13,124ft -6562ft -3281ft -1640ft -820ft -328ft/-100m 0   820ft   1640ft 3281ft 6562ft 9843ft 13,124ft 19,685ft

79

Europe

155

66

66

REYKJAVÍK
ICELAND
Vatnajökull

Arctic Circle
Limit of winter pack ice

Reykjanes Ridge

Iceland Basin

Norwegian Basin

FAROE ISLANDS
(to Denmark)

Norwegian Sea

Hatton Ridge

Faroe-Iceland Trough

Faroe-Shetland Trough
Shetland Islands

Trondheim

Rockall Bank
Rockall Trough

Outer Hebrides
Orkney Islands

Bergen

Stavanger

OSLO

British Isles

Glasgow
Edinburgh
Belfast
Ireland

IRELAND
DUBLIN
Isle of Man
UNITED KINGDOM
Liverpool
Manchester
Britain

North Sea

Gothenburg
Aalborg
Jutland
DENMARK
Odense

Celtic Sea
Cardiff
Birmingham
LONDON
Celtic Shelf
English Channel
Channel Islands

Hamburg
NETHERLANDS
THE HAGUE AMSTERDAM
Rotterdam
Hanover
Elbe
BELGIUM BRUSSELS
Düsseldorf
Bonn
GERMANY
Frankfurt am Main
LUXEMBOURG
LUXEMBOURG

Porcupine Plain

ATLANTIC OCEAN

Mid-Atlantic Ridge

Charlie-Gibbs Fracture Zone

Azores-Biscay Rise
Charcot Seamounts
Iberian Plain
Biscay Plain

le Havre
Rennes
PARIS
Nantes
Orléans
Strasbourg
Stuttgart

Zurich
Munich
BERN
SWITZERLAND
Innsbruck
AUS

Lyon
Milan
Turin
Venice
Trieste
Bologna
SAN MARINO

Bay of Biscay
A Coruña
Galicia Bank

Bordeaux
Bilbao
FRANCE
Mont Blanc 4807m
Massif Central

Porto

Cordillera Cantábrica
Duero
Pyrenees
Toulouse
ANDORRA
Nice
MONACO
Marseille
Pisa

PORTUGAL
Iberian Peninsula
Zaragoza
Ebro

Tagus Plain
LISBON
Tagus
MADRID

SPAIN
Peninsula
Barcelona
Valencia
Corsica
VATICAN CITY
ROME
ITALY
Apennines

Horseshoe Seamounts

Madeira
(to Portugal)

Seville
Guadalquivir
Palma
Balearic Islands
Sardinia
Naples

GIBRALTAR
(to UK)
Málaga
Ceuta
(to Spain)
Strait of Gibraltar
Algerian Basin
Cagliari
Tyrrhenian Sea

Cosenz

Palermo
Mount Etna 3340m
Sicily

Canary Islands
(to Spain)
Melilla
(to Spain)

N

Atlas Mountains

68

AFRICA

Mediterranean

MALTA
VALLETTA

0 km   500

0 miles   500

Population ● National capital

○ below 50,000   ○ 50,000 to 100,000   ◉ 100,000 to 500,000   ◼ above 500,00

The North Atlantic

A | B | C | D

EUROPE

Arctic Circle

37

Gulf of Boothia

Devon Island

Ellesmere Isl

Nares Strait

NUNAVUT

Hudson Bay

Southampton Island

Foxe Basin

38

CANADA

Baffin Island

Qaanaaq

Knud Rasmu

Innaanganeq

Savissivik

Qimusseriarsuaq

Baffin Bay

Kullorsuaq

Upernavik

Péninsule d'Ungava

QUÉBEC

Hudson Strait

Arnaud

Cumberland Sound

Frobisher Bay

Limit of summer pack ice

Uummannaq

Qeqertarsuaq

Qeqertarsuaq

Qeqertarsuup Tunua

Qasigiannguit

GREENLAN

(to Denma

Ungava Bay

George

Davis Strait

Sisimiut

Kong Frederik IX Land

Maniitsoq

NUUK

Kong Christian IX L

Gunn

Mont Forel 3360m

39

Paamiut

Ivittuut

Kong Frederik VI Kyst

Ammassalik

Den

NEWFOUNDLAND & LABRADOR

Labrador Sea

Qaqortoq

Nanortalik

Nunap Isua (Kap Farvel)

Limit of winter pack ice

Reykjanes Basin

ATLANTI

OCEAN

66

A | B | C | D

0 km 400

0 miles 400

Population

● National capital

○ below 50,000 ○ 50,000 to 100,000 ◉ 100,000 to 500,000 ▣ above 500,0

E 50° 40° 30° 20° 10° 0° 10° 20° 30° 40° 50° 60° F G H

ARCTIC OCEAN

155

Zemlya Frantsa-Iosifa

Kap Morris Jesup

Wandel Sea

1

Kvitøya

Novaya Zemlya

Independence Fjord

Nord

SVALBARD
(to Norway)

Nordaustlandet

Kong Karls Land

Kong Frederik VIII Land

Spitsbergen

Barentsøya

Barents Sea

Edgeøya

LONGYEARBYEN
Barentsburg

Storfjorden

110 2

Limit of winter pack ice

Greenland Sea

ristian X and

Bjørnøya
(to Norway)

Daneborg

Limit of summer pack ice

Nordkapp
(North Cape)

FINLAND

3

etermann Bjerg
940m

Kong Oscar Fjord

Mohns Ridge

Ittoqqortoormiit

Kangilinnguit

Kangikajik

JAN MAYEN
(to Norway)

Vestfjorden

Arctic Circle

84 4

ait

Norwegian Sea

S W E D E N

ICELAND

garvík

Siglufjörður Raufarhöfn

Norwegian Basin

Húsavík

Akureyri

ykkishólmur

Seyðisfjörður

Gulf of Bothnia

REYKJAVÍK

Neskaupstaður

Selfoss

Djúpivogur

Vatnajökull

shöfn

Hvannadalshnúkur
2119m

y Vestmannaeyjar

FAROE ISLANDS
(to Denmark)

N O R W A Y

5

N

TÓRSHAVN

Shetland Islands

85

Elevation

-6000m -4000m -2000m -1000m -500m -250m Below sea level 0 250m 500m 1000m 2000m 3000m 4000m 6000m

-19,658ft -13,124ft -6562ft -3281ft -1640ft -820ft -328ft/-100m 0 820ft 1640ft 3281ft 6562ft 9843ft 13,124ft 19,685ft

Scandinavia & Finland

0 km 200
0 miles 200

Population

○ below 50,000 ○ 50,000 to 100,000 ◉ 100,000 to 500,000 ■ above 500,00

● National capital

Elevation

-6000m	-4000m	-2000m	-1000m	-500m	-250m	Below sea level	0	250m	500m	1000m	2000m	3000m	4000m	6000m

| -19,658ft | -13,124ft | -6562ft | -3281ft | -1640ft | -820ft | -328ft/-100m | 0 | 820ft | 1640ft | 3281ft | 6562ft | 9843ft | 13,124ft | 19,685ft |

The Low Countries

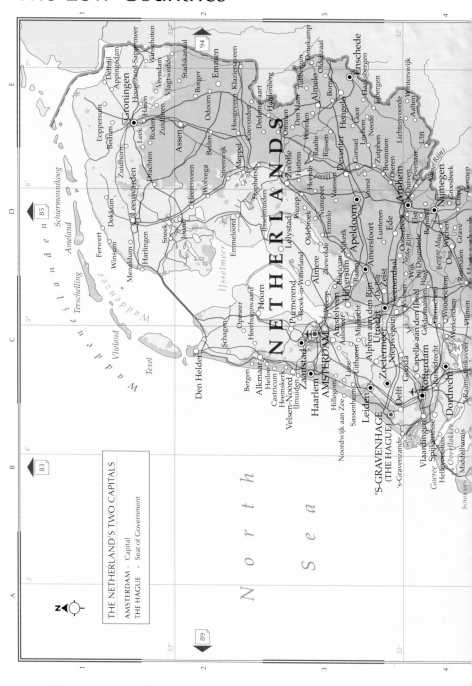

N

THE NETHERLAND'S TWO CAPITALS

AMSTERDAM - Capital
THE HAGUE - Seat of Government

Population

- ● National capital
- ○ below 50,000
- ○ 50,000 to 100,000
- ◉ 100,000 to 500,000
- ■ above 500,00

0 km   50
0 miles   50

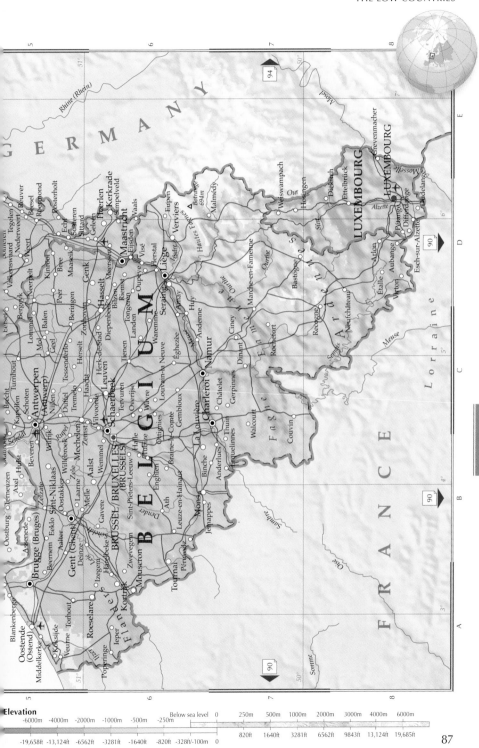

Elevation

-6000m	-4000m	-2000m	-1000m	-500m	-250m	Below sea level

0	250m	500m	1000m	2000m	3000m	4000m	6000m

| -19,658ft | -13,124ft | -6562ft | -3281ft | -1640ft | -820ft | -328ft/-100m | 0 |

820ft	1640ft	3281ft	6562ft	9843ft	13,124ft	19,685ft

87

The British Isles

North Sea

ATLANTIC OCEAN

Shetland Islands
Unst
Fetlar
Yell
Mainland
Lerwick

Fair Isle

Orkney Islands
Sanday
Kirkwall
Mainland
Hoy
John o'Groats

Thurso
Ben Hope 927m △
Ullapool
Inverness
Loch Ness
Aviemore
Moray Firth
Elgin
Findhorn
Dee
Mountains
Grampian
Fraserburgh
Peterhead
Aberdeen
Montrose
Arbroath
St Andrews
Forfar
Dundee
Perth
Tay
Firth of Forth
SCOTLAND
North West Highlands
Ben Nevis 1343m △
Fort William
Mallaig
Stornoway
Isle of Lewis
Harris
North Uist
South Uist
Barra
St Kilda
Outer Hebrides
The Little Minch
The Minch
Isle of Skye
Rhum
Eigg
Coll
Tiree
Isle of Mull
Oban
Firth of Lorn
Jura
Islay
Kintyre
Inner Hebrides
Loch Lomond
Loch Tay
Dunfermline
Stirling
Glasgow
Greenock
Paisley
East Kilbride
Hamilton
Kilmarnock
Prestwick
Clyde
Edinburgh
Berwick-upon-Tweed
Galashiels
Hawick
Peebles

△

0 km   100
0 miles   100

Population
National capital   Internal administrative capital
○ below 50,000   ○ 50,000 to 100,000   ◉ 100,000 to 500,000   ◼ above 500,0

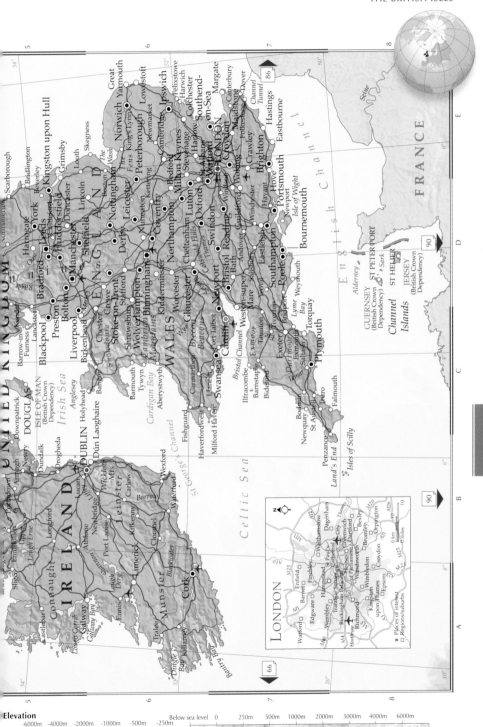

86

90

90

66

Elevation

							Below sea level	0	250m	500m	1000m	2000m	3000m	4000m	6000m
-6000m	-4000m	-2000m	-1000m	-500m	-250m										
-19,658ft	-13,124ft	-6562ft	-3281ft	-1640ft	-820ft	-328ft/-100m	0		820ft	1640ft	3281ft	6562ft	9843ft	13,124ft	19,685ft

89

France, Andorra & Monaco

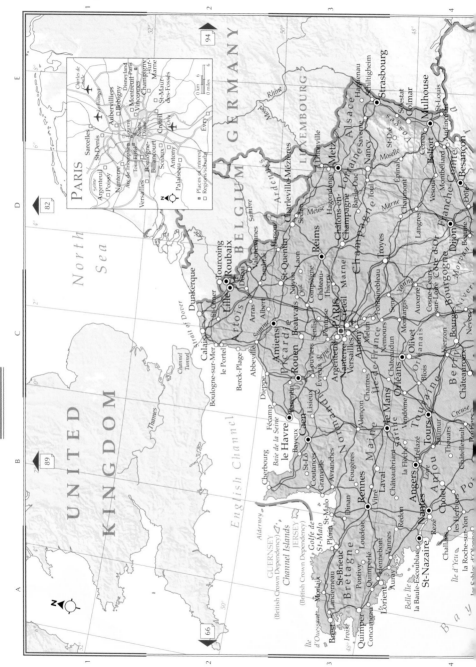

PARIS

□ Places of interest
□ Region's/suburb

N

UNITED

KINGDOM

North
Sea

Strait of Dover

Channel
Tunnel

English Channel

GUERNSEY
(British Crown Dependency)
Channel Islands
JERSEY
(British Crown Dependency)

GERMANY

BELGIUM

LUXEMBOURG

Rhine

Moselle

Meuse

ALSACE

Strasbourg
Haguenau
Schiltigheim
Sélestat
Colmar
Mulhouse
St-Louis
Cernay
Belfort
Audincourt
Montbéliard
Besançon
FRANCHE-COMTÉ
Dôle
Dijon
CÔTE D'OR
BOURGOGNE
Auxerre
Nevers
BERRY
Bourges
Vierzon
Châteauroux
Poitiers

Dunkerque
Tourcoing
Roubaix
Lille
St-Omer
Douai
Valenciennes
Cambrai
ARTOIS
Arras
Calais
Boulogne-sur-Mer
le Portel
Berck-Plage
Abbeville
PICARDIE
Amiens
Albert
Somme
St-Quentin
Laon
Oise
Compiègne
Beauvais
Senlis
Pontoise
Noyon
Château-Thierry
Reims
CHAMPAGNE
Châlons-en-Champagne
Épernay
Troyes
Marne
Sedan
Charleville-Mézières
ARDENNES
Thionville
Metz
LORRAINE
Nancy
Toul
Bar-le-Duc
Saverne
Épinal
Vesoul
Langres
Chaumont
Dieppe
Fécamp
le Havre
Baie de la Seine
Caen
Bayeux
Lisieux
Cherbourg
St-Lô
Coutances
Granville
Avranches
NORMANDIE
Louviers
Évreux
Rouen
Seine
Chartres
Mantes
ÎLE-DE-FRANCE
PARIS
Argenteuil
Nanterre
Créteil
Versailles
Melun
Fontainebleau
Nemours
Montargis
Sens
Yonne
Loing
Orléans
ORLÉANAIS
Olivet
Blois
Vendôme
Loire
Tours
TOURAINE
le Mans
MAINE
Sarthe
la Flèche
Alençon
Mortagne
Dinan
St-Malo
Golfe de St-Malo
St-Brieuc
Alderney
Plérin
Morlaix
Landerneau
Brest
Île d'Ouessant
Quimper
Concarneau
Lorient
Hennebont
Pontivy
Loudéac
BRETAGNE
Auray
Vannes
Redon
Rennes
Vitré
Laval
Châteaubriant
Fougères
Angers
Cholet
ANJOU
Saumur
Thouars
Châtellerault
Nantes
Rezé
St-Nazaire
la Baule-Escoublac
Challans
les Herbiers
la Roche-sur-Yon
Île d'Yeu
Belle Île
POITOU
Creuse
Indre

N

Population

● National capital

○ below 50,000   ○ 50,000 to 100,000   ◉ 100,000 to 500,000   ■ above 500,00

0 km   100

0 miles   100

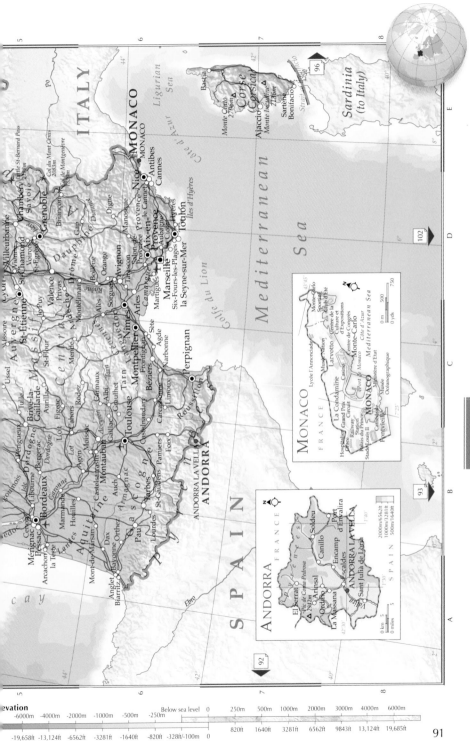

ITALY

Little St-Bernard Pass
Col du Mont Cenis
Col de Montgenèvre

Savoie
Chambéry
Grenoble
Voiron
Vienne
St-Chamond
St-Étienne
Lyon
Villeurbanne

Dauphiné
Gap
Briançon
Dôme

Alpes

Provence
Digne
Dr
Manosque

MONACO
MONACO
Antibes
Cannes
Nice
Provence
Aix-en-le Cannet

Côte d'Azur

Ligurian Sea

Corse
(Corsica)
Bastia
Monte Cinto
2706m
Monte Incudine
2136m
Ajaccio
Sartène
Bonifacio
Strait of Bonifacio

Sardinia
(to Italy)

Puy

Ardèche
Privas
Valence
Montélimar

Auvergne
Central
Massif
Aubrac
St-Flour
Aurillac

Orange
Avignon
Sorgues
Bollène
Tarascon

Camargue
Rhône
Nîmes
Arles
Salon-de-Provence
Martigues
Marseille
la Seyne-sur-Mer
Six-Fours-les-Plages
Toulon
Hyères
Îles d'Hyères

Mediterranean
Sea

Golfe du Lion

Ussel
Tulle
Brive-la-
Gaillarde
Figeac
Rodez
Cahors
Gaillac
Albi
Tarn
Castres
Mazamet
Béziers
Agde
Sète
Montpellier
Languedoc
Frontignan
Narbonne

Périgueux
Dordogne
Bergerac
Libourne
Lot
Moissac
Montauban
Agen
Castelsarrasin
Toulouse
Gaillac
Castelnaudary
Carcassonne
Limoux
Fox
Pamiers
Roussillon
Perpignan

Isle
Fronsac
Bordeaux
Pessac
Mérignac
Cenon
Arcachon
la Teste
Dordogne
Garonne
Marmande
Landes
Mont-de-Marsan
Dax
Bayonne
Anglet
Biarritz
Orthez
Pau
Tarbes
Lourdes
St-Gaudens
Auch
Armagnac
Gascogne
Aquitaine

ANDORRA LA VELLA
ANDORRA

Pyrénées

SPAIN

Ebro

[96]

[102]

[93]

[92]

[101]

MONACO

Monte-Carlo
Larvotto
La Condamine
Monte-Carlo
Port de Monaco
MONACO
FRANCE

Lycée l'Annonciade
Musée National
Grace
Centre de la
Culture et
d'Expositions
Centre de Congrès
Côte-d'Azur
Casino
Sporting
Club d'Été
Railway
Station
Palais du Prince
Cathédrale
Ministère d'État
Musée
Océanographique
Hospital
Grand Prix
Circuit
Stade Louis II
Fontvieille

Mediterranean Sea

0 m   500   750
0 yds   500   750

ANDORRA

FRANCE
Soldeu
El Serrat
Pic de Coma Pedrosa
2942m
Ordino
Arinsal
Camillo
Encamp
Port
d'Envalira
ANDORRA LA VELLA
La Massana
Escaldes
Sant Julià de Lòria

SPAIN

Pyrénées

2000m/6562ft
1000m/3281ft
500m/1640ft

0 km   5
0 miles 5

Elevation

-6000m	-4000m	-2000m	-1000m	-500m	-250m	Below sea level 0	250m	500m	1000m	2000m	3000m	4000m	6000m
-19,658ft	-13,124ft	-6562ft	-3281ft	-1640ft	-820ft	-328ft/-100m 0	820ft	1640ft	3281ft	6562ft	9843ft	13,124ft	19,685ft

Spain & Portugal

66

N

A Coruña
(La Coruña)
Ferrol   Luarca   Avilés   Gijon   Costa
                                (Xixón)
Laracha Betanzos   Pravia Villaviciosa Sa
Santa Catalina de Armada Vilálba   Tineo Oviedo Llane
Cabo Fisterra   Asturias   Mieres del Camín Torrelav
Oues   La Pola   Cabanaquinta Ca
Muros   Galicia Lugo   Cordillera Cantábrica Rein
Santiago de Compostela   Ponferrada León
Santa Uxía de Ribeira Lalín Chantada Monforte
Pontevedra O Carballiño de Lemos Astorga Castilla-León
Marín   Ourense (Orense)
Vigo   Ponteareas Xinzo de Limia Benavente Palencia
              Miño   Valladolid
Viana do Castelo Ponte da Barca Bragança Embalse de Zamora
              Braga Chaves Ricobayo Toro Due
Póvoa de Varzim Guimarães   Medina del Campo S
Vila do Conde Vila Real
Matosinhos Vila Real   Salamanca
Porto (Oporto) Lamego Embalse Ávila
Vila Nova de Gaia Douro de Almendra Sistema Ce
Ovar São João da Madeira   MA
Aveiro Albergaria-a-Velha Viseu   Sistema de Gredos
Ílhavo   Alto da Torre Guarda Ciudad-Rodrigo
            1993m▲ Serra da Estrêla Béjar Sierra de Gredos Talavera
Coimbra   Covilhã   de la Reina
Figueira da Foz   PORTUGAL Plasencia Tol
          Coria
OCEAN Leiria Castelo Branco Tagus Embalse de Embalse de
        Tomar Abrantes de Alcántara Cáceres Valdecañas
Entroncamento   Trujillo Herrera
Peniche Caldas da Rainha Portalegre   del Duque
Torres Vedras Santarém   Extremadura
        Coruche   Mérida Villanueva de la Serena
Sintra Estremoz Elvas   Ciudad
Cascais LISBOA (LISBON) Évora Badajoz Don Benito Puertolla
Almada Barreiro Serra d'Ossa Castuera
Setúbal   Almendralejo Villafranca de los Barros Pozobla
Alcácer do Sal Barragem Zafra Azuaga
Baía de Setúbal do Alqueva Jerez de los Caballeros
Sines Beja   Sierra Córdoba Morena Montoro
          Cortegana   Buj
Ourique Nerva La Algaba Palma del Río M
Valverde del Camino   Carmona Anda
Algarve Ayamonte Lepe Sevilla Écija Osuna E
Portimão   Isla   (Seville) Lu
Lagos Faro Cristina Huelva Dos Antequera
Cabo de Tavira   Las Cabezas de San Juan Hermanas Alora
São Vicente Olhão Golfo de Cádiz Lebrija Olvera Ronda
        Sanlúcar de Barrameda Úbrique C
        El Puerto de Santa María Jerez de la Frontera Marbella
        Cádiz San Fernando Estepona Fuen
Costa de la Luz Vejer de la Frontera Algeciras GIBRALTAR
        Barbate de Franco Ceuta (to Spain) (to UK)
Strait of Gibraltar
MOROCCO

ATLANTIC

AZORES (to Portugal)

N

Corvo
Flores São Graciosa
    Jorge Terceira
Faial Pico São Miguel
    Ponta Delgada
    Santa Maria

0 km   100
0 miles   100

200m/656ft
Sea level

70

0 km   100
0 miles   100

Population   ● National capital

○ below 50,000   ○ 50,000 to 100,000   ◉ 100,000 to 500,000   ◼ above 500

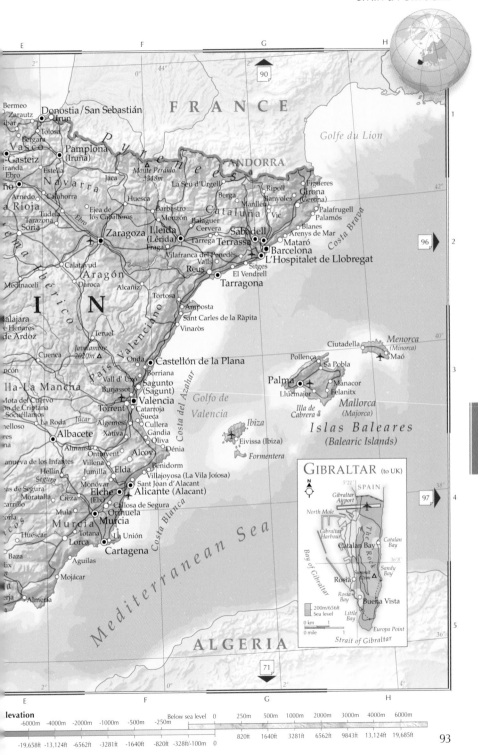

E F G H

2° 0° 44° 2° 4°

1

F R A N C E

Bermeo
Zarautz
ibat Donostia / San Sebastián *Golfe du Lion*
Irun
Tolosa
Bergara
Vasco
-Gasteiz Pamplona
iranda (Iruña)
Ebro Estella
ño Jaca A N D O R R A 42°
Rioja Monte Perdido Figueres
Arnedo N a v a r r a 3348m La Seu d'Urgell Ripoll Girona
Calahorra Huesca Berga Banyoles (Gerona)
Tarazona Ejea de Barbastro Manlleu Palafrugell
Soria los Caballeros Monzón Balaguer C a t a l u ñ a Vic Palamós
Tudela Lleida Cervera Blanes Costa Brava
Calatayud Zaragoza (Lérida) Tàrrega Sabadell Arenys de Mar 96 2
Medinaceli Fraga Terrassa Mataró
I N Daroca Aragón Vilafranca del Penedès Barcelona
Alcañiz Valls Sitges L'Hospitalet de Llobregat
alajara Tortosa Reus El Vendrell
e Henares Teruel Amposta Tarragona
de Ardoz Sant Carles de la Ràpita
Javalambre Menorca
Cuenca 2020m Vinaròs Ciutadella (Minorca) 40°
ncón Onda Castellón de la Plana Pollença Sa Pobla Maó 3
Ila-La Mancha Borriana País Valenciano Manacor
Mota del Cuervo Vall d'Uxó Sagunto Palma Felanitx
o de Criptana Burjassot (Sagunt) Costa del Azahar Illa de Mallorca
Socuéllamos Valencia Golfo de Cabrera (Majorca)
helloso Torrent Catarroja Valencia Islas Baleares
res La Roda Algemesí Sueca (Balearic Islands)
Albacete Xàtiva Cullera
anueva de los Infantes Almansa Gandia Ibiza
Hellín Ontinyent Oliva Dénia Eivissa (Ibiza)
as de Segura Villena Alcoy Formentera
Moratalla Jumilla Benidorm
carrillo Monóvar Elda Villajoyosa (La Vila Joíosa) 38°
cos Cieza Elche Sant Joan d'Alacant 97 4
Mula (Elx) Alicante (Alacant)
Huéscar Murcia Orihuela Callosa de Segura
Totana Murcia Costa Blanca
Baza Lorca La Unión
lix Cartagena
Aguilas
Mojácar M e d i t e r r a n e a n S e a
ría
Almería

A L G E R I A 36°

71

2° 0° 2° 4°

E F G H

levation
-6000m -4000m -2000m -1000m -500m -250m Below sea level 0 250m 500m 1000m 2000m 3000m 4000m 6000m

-19,658ft -13,124ft -6562ft -3281ft -1640ft -820ft -328ft/-100m 0 820ft 1640ft 3281ft 6562ft 9843ft 13,124ft 19,685ft

Germany & the Alpine States

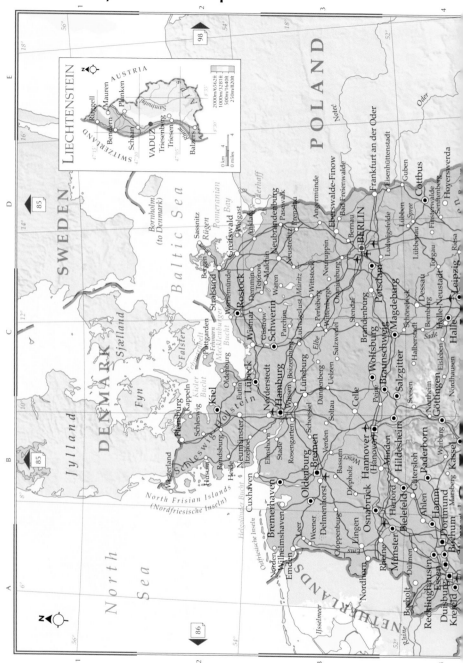

Population

○ below 50,000   ○ 50,000 to 100,000   ◉ 100,000 to 500,000   ■ above 500,0

● National capital

0 km   100

0 miles   100

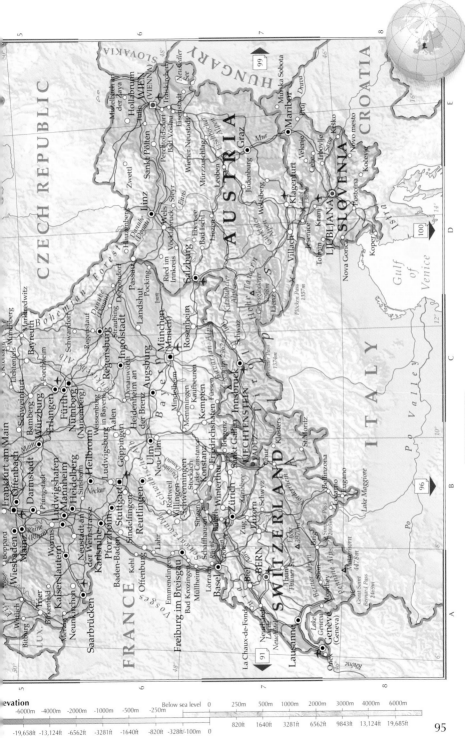

Elevation

-6000m	-4000m	-2000m	-1000m	-500m	-250m	Below sea level 0	250m	500m	1000m	2000m	3000m	4000m	6000m
-19,658ft	-13,124ft	-6562ft	-3281ft	-1640ft	-820ft	-328ft/-100m 0	820ft	1640ft	3281ft	6562ft	9843ft	13,124ft	19,685ft

0 km 100

0 miles 100

Population

- ● National capital
- ○ below 50,000
- ○ 50,000 to 100,000
- ◉ 100,000 to 500,000
- ■ above 500

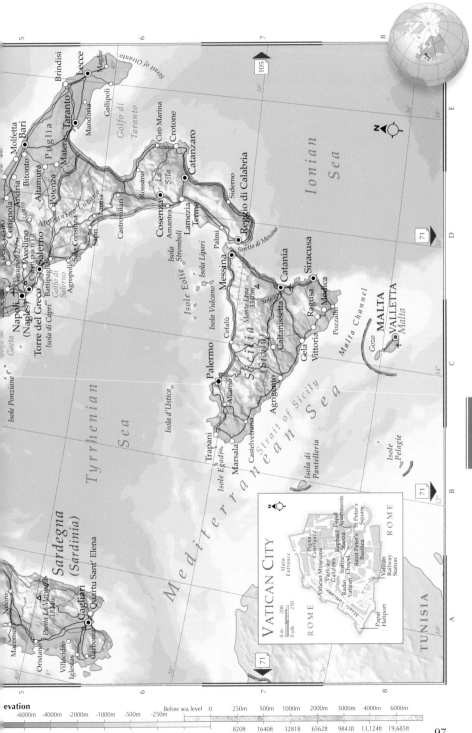

Brindisi
Lecce
Maglie
Gallipoli
Strait of Otranto
Golfo di Taranto
Taranto
Mandria
Molfetta
Bari
Bitonto
Puglia
Andria
Altamura
Cerignola
Matera
Avellino
Potenza
San Conischina
Campania
Salerno
Monte Vulture 1277m
Battipaglia
Campania
Napoli
(Naples)
Agropoli
Sapri
Torre del Greco
Isola di Capri
Golfo di
Salerno
Vesuvio 1277m
Gaeta
Isole Ponziane
Castrovillari
Cosenza
Amantea
Rossano
La
Sila
Appennino Lucano
Ciro Marina
Crotone
Catanzaro
Siderno
Reggio di Calabria
Lamezia
Terme
Palmi
Stretto di Messina
Isola
Stromboli
Isole Eolie
Isola Lipari
Isola Vulcano
Messina
Catania
Siracusa
Monte Etna 3340m
Simeto
Cefalù
Palermo
Caltanissetta
Caltagirone
Ragusa
Modica
Pozzallo
Gela
Vittoria
Alcamo
Sicilia
(Sicily)
Agrigento
Trapani
Isole Egadi
Marsala
Castelvetrano
Strait of Sicily
Isola di
Pantelleria
Isola d'Ustica
Ionian
Sea
Tyrrhenian
Sea
Mediterranean
Sea
Malta Channel
MALTA
VALLETTA
Malta
Gozo
Isole
Pelagie
105
71
71
71
71
5
6
7
8
E
D
C
B
A

Sardegna
(Sardinia)
Quartu Sant'Elena
Nuoro
Macomer
Oristano
Punta La Marmora 1834m
Cagliari
Carbonia
Villacidro
Iglesias
Gonnosfanadiga
Quartu Sant'Elena

TUNISIA

VATICAN CITY
ROME
ROME
Main
Entrance
Vatican Gardens
Radio
Vatican
Pigna
Courtyard
Vatican Museums
Sistine
Chapel
Raphael
Stanza
Papal
Apartments
St Peter's
Square
Saint Peter's
Basilica
Monte Vaticano
Vatican
Railway
Station
Papal
Heliport
0 m   200
0 yds 250

Elevation

| -6000m | -4000m | -2000m | -1000m | -500m | -250m | | Below sea level | 0 | 250m | 500m | 1000m | 2000m | 3000m | 4000m | 6000m |

-19,658ft -13,124ft -6562ft -3281ft -1640ft -820ft -328ft/-100m 0   820ft 1640ft 3281ft 6562ft 9843ft 13,124ft 19,685ft

Central Europe

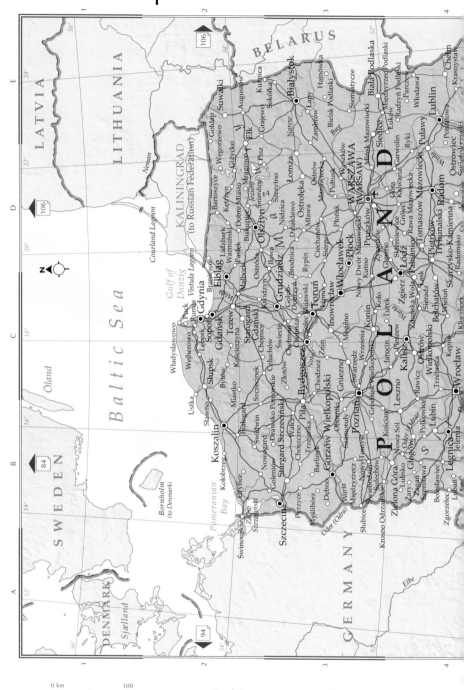

0 km   100

0 miles   100

Population   ● National capital

○ below 50,000   ○ 50,000 to 100,000   ◉ 100,000 to 500,000   ■ above 500

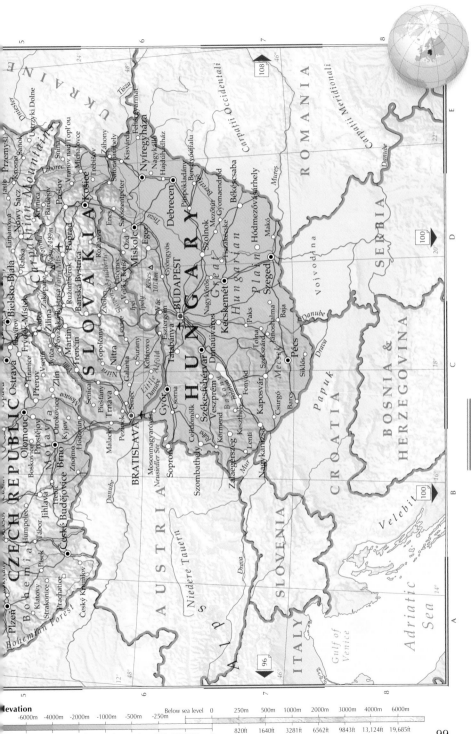

CZECH REPUBLIC

Plzeň
Strakonice
Klatovy
Prachatice
Český Krumlov
Písek
Tábor
Humpolec
Jihlava
České Budějovice
Třebíč
Znojmo
Boskovice
Brno
Kyjov
Prostějov
Otrokovice
Olomouc
Ostrava
Opava

B o h e m i a

M o r a v i a

H o h e m i a n F o r e s t

Morava

Danube

SLOVAKIA

Bielsko-Biała
Nowy Sącz
Přerov
Třinec
Frýdek-Místek
Rožnov
Vsetín
Zlín
Hranice
Přibor
Žilina
Čadca
Zákopane
Martin
Trenčín
Púchov
Považská Bystrica
Topoľčany
Nitra
Sered'
Galanta
Senec

Gerlachovský Štít 2655m

Bardejov
Stará Ľubovňa
Poprad
Banská Bystrica
Ružomberok
Levice
Zvolen
Lučenec

Laborec

Prešov
Košice
Michalovce
Snina
Vranov nad Topľ'ou
Trebišov
Sátoraljaújhely

UKRAINE

Carpathian Mountains

Dniester
Przemyśl
Sanok
Jasło
Krosno
Ustrzyki Dolne

BRATISLAVA
Malacky
Pezinok
Senica
Piešťany
Trnava

Mosonmagyaróvár
Neusiedler See
Sopron

Little Alföld
Győr
Csorna
Komárom
Tatabánya
Esztergom
Vác

BUDAPEST

Gyöngyös
Eger
Kékes 1014m
Miskolc
Ózd
Encs

Tisza
Tokaj
Nyíregyháza
Mátészalka
Fehérgyarmat
Kisvárda
Záhony

Hajdúnánás
Debrecen
Hajdúszoboszló
Berettyóújfalu

HUNGARY

Tisza

H u n g a r i a n   P l a i n

G r e a t

Szolnok
Szeged
Kecskemét
Nagykőrös
Tiszakécske
Mezőtúr
Békéscsaba
Gyomaendrőd
Hódmezővásárhely
Makó

ROMANIA

Carpaţii Occidentali
Carpaţii Meridionali

Mureş

Székesfehérvár
Dunaújváros
Veszprém
Paks
Kalocsa
Baja
Tolna
Szekszárd
Jánoshalma

Danube

Dunántúl
Dombóvár
Tata
Várpalota
Ajka
Várpalota

Bakony

Mecsek
Pécs
Siklós

Körmend
Szombathely
Keszthely
Zalaegerszeg
Nagykanizsa
Lenti
Csurgó
Barcs
Kaposvár
Fonyód

Papuk

Drava

CROATIA

SLOVENIA

ITALY

Gulf of Venice

Velebit

A d r i a t i c   S e a

BOSNIA & HERZEGOVINA

SERBIA

Vojvodina

Danube

Drava

AUSTRIA

Niedere Tauern

A l p s

Mur

108
100
100
96

Elevation

-6000m	-4000m	-2000m	-1000m	-500m	-250m	Below sea level	0	250m	500m	1000m	2000m	3000m	4000m	6000m
-19,658ft	-13,124ft	-6562ft	-3281ft	-1640ft	-820ft	-328ft/-100m	0	820ft	1640ft	3281ft	6562ft	9843ft	13,124ft	19,685ft

Southeast Europe

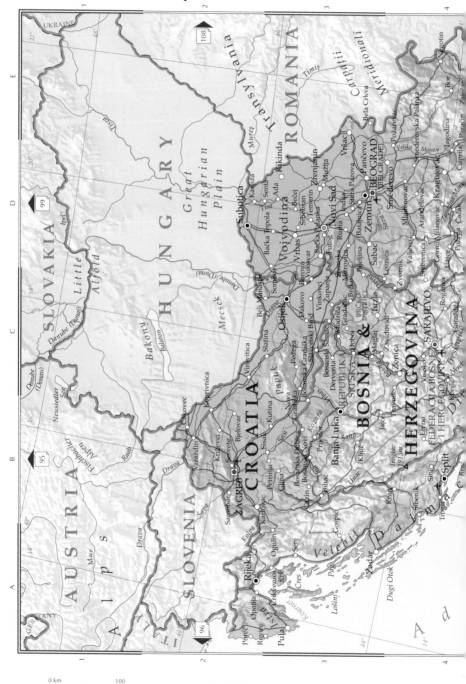

Population

0 km   100

0 miles   100

● National capital   ○ Internal administrative capital

○ below 50,000   ○ 50,000 to 100,000   ◉ 100,000 to 500,000   ◼ above 500,

BULGARIA

Strimónas

Thermaikós Kólpos

Aegean Sea

Évvoia
(Euboea)

MACEDONIA

Vardar

SKOPJE

Veles

Kumanovo
Kočani
Štip
Radoviš
Strumica

Kavadarci
Gevgelija

Prilep

BITOLA

Ohrid

Lake
Prespa

GREECE

Pindos
(Pindus
Mountains)

Patraikós

KOSOVO
(disputed)

PRISTINË

Peje / Peć

Vushtrri / Vučitrn

Podujevo

Gjilan / Gnjilane

Ferizaj

Urosevac

Prizren

Tetovo

Gostivar

Kičevo

Struga

Lake
Ohrid

Pogradec

Korçë

North
Albanian
Alps

Čakor
2698m

Debar

Peshkopi

Burrel

Elbasan

Lumi i Devollit

Bajram Curri

Gjakovë / Djakovica

Kukës

Lumi i Drinit

Black Drin

Berat

Pukë

Krujë

TIRANE
(TIRANA)

ALBANIA

Lumi i Osumit

Vlorë

Tepelenë

Sarandë

Gjirokastër

Konispol

Kérkyra
(Corfu)

Iónia Nisiá
(Ionian Islands)

PODGORICA

Lezhë

Durrës

Kavajë

Lushnjë

Fier

Ionian
Sea

Bar

Lake Scutari

Shkodër

Lac

Strait of Otranto

Golfo di

Adriatic
Sea

ITALY

Appennino Lucano

Kéfalloniá

104
105
103
97

In February 2008, Kosovo (a UN Protectorate within Serbia since 1999) declared independence. Although recognized by several countries, this decision has proved controversial with other states wary of setting a precedent for separatist groups within their own borders. It is therefore likely to be some time before Kosovo becomes universally recognized.

BOSNIA & HERZEGOVINA

CROATIA

SERBIA

Brčko

Tuzla

Drina

Bihać

Banja Luka

Sava

Prijedor

Zenica

Sarajevo

Goražde

Mostar

MONTENEGRO

Dubrovnik

Split

CROATIA

Adriatic Sea

Territorial extent
Republika Srpska
Federacija Bosne i Hercegovine
Brčko Distrikt

0   50 km
0   50 miles

Elevation

-6000m	-4000m	-2000m	-1000m	-500m	-250m	Below sea level 0	250m	500m	1000m	2000m	3000m	4000m	6000m
-19,658ft	-13,124ft	-6562ft	-3281ft	-1640ft	-820ft	-328ft/-100m 0	820ft	1640ft	3281ft	6562ft	9843ft	13,124ft	19,685ft

The Mediterranean

Population   ● National capital

○ below 50,000   ○ 50,000 to 100,000   ◉ 100,000 to 500,000   ■ above 500,0

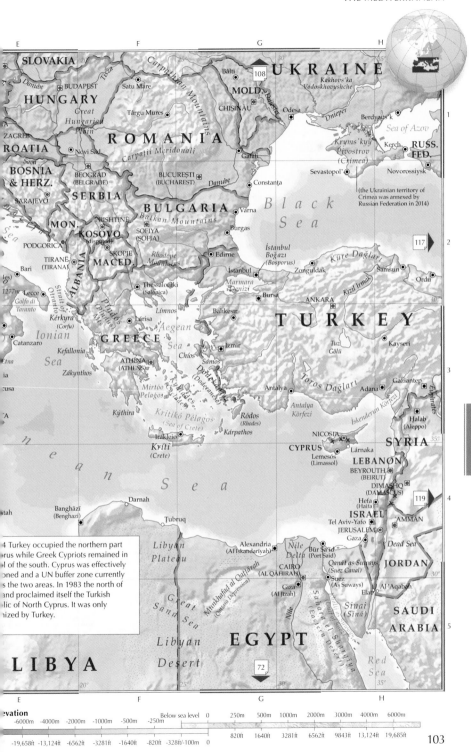

E   F   G   H

SLOVAKIA

Danube   BUDAPEST

HUNGARY

Great Hungarian Plain

ZAGREB

Tisza   Satu Mare

Carpathian Mountains

Bălţi   108

MOLD.

CHIŞINĂU

Kahovs'ka Vodoskhovyshche

U K R A I N E

Danube

Odesa

Dnieper

Berdyans'k

Târgu Mureş

R O M A N I A

Carpaţii Meridonali

Sea of Azov

ROATIA

Novi Sad

Galaţi

Kryms'kyy Pivostrov (Crimea)

Kerch

RUSS. FED.

Sevastopol'

Novorossiysk

BOSNIA & HERZ.

Sava

BEOGRAD (BELGRADE)

BUCUREŞTI (BUCHAREST)

Danube

Constanţa

B l a c k

SARAJEVO

S E R B I A

(the Ukrainian territory of Crimea was annexed by Russian Federation in 2014)

PRIŠTINË

B U L G A R I A

Varna

S e a

MON.

KOSOVO

Nišispand

SOFIYA (SOFIA)

Balkan Mountains

Burgas

PODGORICA

SKOPJE

Edirne

İstanbul Boğazı (Bosporus)

Zonguldak

Küre Dağları

117

Samsun

Bari

TIRANË (TIRANA)

MACED.

Rhodope Mountains

İstanbul

Marmara Denizi

Ordu

Straits of Otranto

Lecce

1277m

ALBANIA

Thessaloníki (Salónica)

Bursa

ANKARA

Kızıl Irmak

Golfo di Taranto

Kérkyra (Corfu)

Lárisa

Pindos Mountains

Límnos

Marmara Denizi

T U R K E Y

Catanzaro

Ionian

G R E E C E

Aegean Sea

Chíos

İzmir

Tuz Gölü

Kayseri

Etna

Sea

Zákynthos

Kefallonía

ATHÍNA (ATHENS)

Sámos

Gaziantep

cusa

ia

Kýthira

Mirtóo Pélagos

Kýthnos

Dodecanese

Antalya

Antalya Körfezi

Toros Dağları

Adana

İskenderun Körfezi

Halab (Aleppo)

Kritikó Pélagos (Sea of Crete)

Ródos (Rhodes)

Kárpathos

NICOSIA

CYPRUS

Lemesós (Limassol)

Lárnaka

SYRIA

Irakleio

Kríti (Crete)

BEYROUTH (BEIRUT)

LEBANON

DIMASHQ (DAMASCUS)

n e a n

Darnah

Hefa (Haifa)

119

atah

Banghāzī (Benghazi)

Ţubruq

S e a

Tel Aviv-Yafo

JERUSALEM

ISRAEL

'AMMĀN

Dead Sea

Gaza

4 Turkey occupied the northern part rus while Greek Cypriots remained in l of the south. Cyprus was effectively oned and a UN buffer zone currently s the two areas. In 1983 the north of and proclaimed itself the Turkish lic of North Cyprus. It was only nized by Turkey.

Alexandria (Al Iskandariyah)

Nile Delta

Būr Sa'īd (Port Said)

JORDAN

Libyan Plateau

CAIRO (AL QĀHIRAH)

Qanat as Suways (Suez Canal)

Suez (As Suways)

Al 'Aqabah

Munkhafad al Qaṭṭārah (Qattara Depression)

Giza (Al Jīzah)

Nile

Elat

Sinai (Sīnā)

SAUDI ARABIA

L I B Y A

Great Sand Sea

Libyan Desert

E G Y P T

72

Ṣaḥrāʾ ash Sharqīya (Eastern Desert)

Red Sea

E   F   G   H

evation

						Below sea level	0	250m	500m	1000m	2000m	3000m	4000m	6000m
-6000m	-4000m	-2000m	-1000m	-500m	-250m									
-19,658ft	-13,124ft	-6562ft	-3281ft	-1640ft	-820ft	-328ft/-100m	0		820ft	1640ft	3281ft	6562ft	9843ft	13,124ft	19,685ft

Bulgaria & Greece

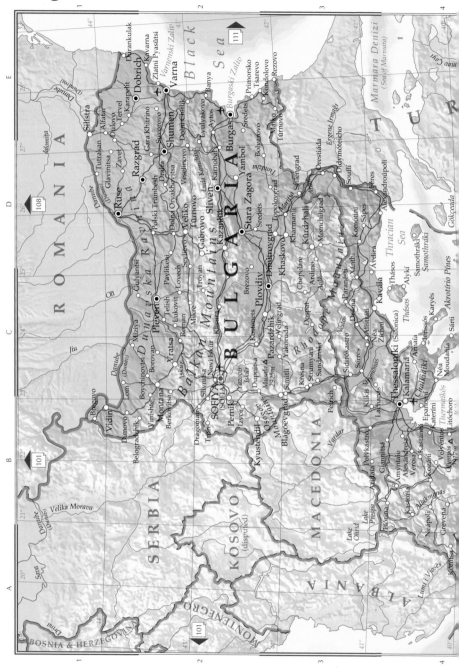

0 km   100

0 miles   100

Population

● National capital

○ below 50,000   ○ 50,000 to 100,000   ◉ 100,000 to 500,000   ◼ above 500

5   6   7   8

38°

37°

36°

35°

34°

Gediz Nehri

Büyükmenderes Nehri

Plomári

Psará

Antípsara

Chíos
Chíos

Sámos
Sámos

Ikaría

Foúrni

Arkí
Patmos
Lípsi
Léros
Agathonísi

Léros

Kálymnos

Kos
Kos

Nísyros

Tílos

Chálki

Sými

Saría

Kásos

116

Ródos
(Rhodes)

Ródos
Líndos

Kattaviá

Kárpathos

Kárpathos

Agía
Marína

Therma

Dodekánisa (Dodecanese)

Skýros

Vólos
Strofyliá

Malesína
Chalkída
Aliveri

Évvoia
(Euboea)

Kými

Kárystos

Marathónas

ATHÍNA
ATHENS
Kálamos

Peiraiás
(Piraeus)

Andros
Andros

Týnos
Tínos

Mýkonos
Mýkonos

Sýros
Sýros

Kéa

Tziá
Ioulís

Kýthnos
Kýthnos

Sérifos

Sífnos

Kimolos

Mílos

Folégandros

Kýklades (Cyclades)
- Kykládes -

Náxos
Náxos

Páros
Parikiá

Íos

Santoríni

Thíra
Thíra

Akrotírio Floída

Amorgós
Amorgós

Astypálaia

Antíti

Kéos

Kritikó Pélagos
(Sea of Crete)

Neápoli

Ágios Nikólaos

Sitía

Diní

Ierápetra

Mýrtos

Iráklio

Zarós

Tympáki

Spíli

Panormós

Lefká Or.

Chaniá

Kíssamos

Kántanos

Chóra Stakíon

Gávdos

Kríti (Crete)

Mediterranean Sea

28°   27°   26°   25°   24°   23°   22°   21°   20°

E

72

D

C

B

71

A

Kálamos

Ýdra

Ermióni

Aígina

Kýthira

Kýthira

Potamós

Antikýthira

Palaiá Epídavros

Póros

Leonídio

Monemvasía

Geráki

Daimoniá

Karavás

Gýtheio

Neápoli

Lakonikós Kólpos

Mírtoo Pelagos

Plomári

Aigínio

Lamía
Ámfissa

Áthrno

Lidoríki

Náfpaktos

Pátra

Ióni

Lefkáda
Lefkáda

Vasiliki

Amfilochía

Katoúna

Neochóri

Póros

Argostóli
Lixoúri

Kefalloniá

Níon Nisiá
(Ionian Islands)

Zákynthos

Alíartos

Agrínio

Mégos

Livanátes

Stylída

Korinthiakós
Kólpos

Káto Kástro

Xylókastro

Kórinthos
(Corinth)

Nemeá

Árgos

Trípoli

Sparti

Pelopónnisos
Peloponnese

Alfeiós

Lechainá

Gastoúni

Ker

Pýrgos

Zacháro

Kyparissía

Pýlos

Messíni

Kalamáta

Koróni

Areópoli

Gerolimenás

Lámpeia

Kató Achaïa

Lávrio
Keratéa

Agios
Ýliki

Thíva

Skýros

Aegean Sea

Sea

Ionian Sea

GREECE

N

Z

97

105

elevation

-6000m   -4000m   -2000m   -1000m   -500m   -250m   Below sea level   0   250m   500m   1000m   2000m   3000m   4000m   6000m

-19,658ft   -13,124ft   -6562ft   -3281ft   -1640ft   -820ft   -328ft/-100m   0   820ft   1640ft   3281ft   6562ft   9843ft   13,124ft   19,685ft

The Baltic States & Belarus

0 km	100
0 miles	100

Population   ● National capital

○ below 50,000   ○ 50,000 to 100,000   ◉ 100,000 to 500,000   ■ above 500

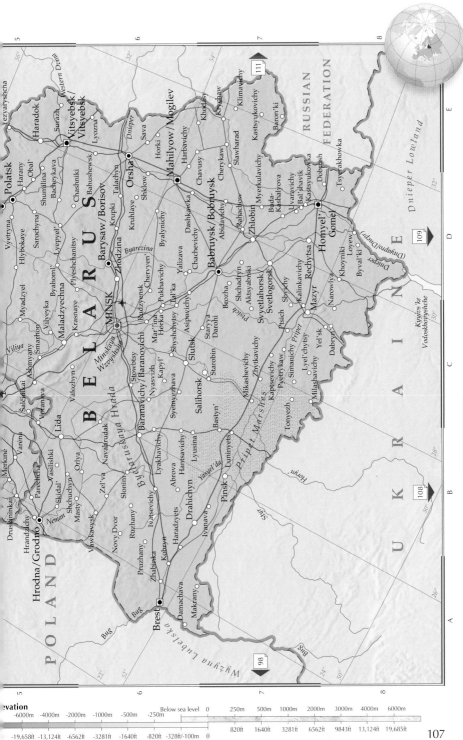

POLAND

BELARUS

RUSSIAN FEDERATION

UKRAINE

Polatsk

Hrodna/Grodno

Brest

MINSK

Vitsyebsk / Vitsyebsk

Mahilyow / Mogilev

Orsha

Barysaw / Borisov

Babruysk/Bobruysk

Homyel' / Gomel

Mazyr

Pinsk

Pripet Marshes

Dnieper Lowland

evation

| -6000m | -4000m | -2000m | -1000m | -500m | -250m | Below sea level 0 | 250m | 500m | 1000m | 2000m | 3000m | 4000m | 6000m |

| -19,658ft | -13,124ft | -6562ft | -3281ft | -1640ft | -820ft | -328ft/-100m 0 | 820ft | 1640ft | 3281ft | 6562ft | 9843ft | 13,124ft | 19,685ft |

107

Ukraine, Moldova & Romania

| 0 km | | 100 | |
| 0 miles | | | 100 |

Population   ● National capital

○ below 50,000   ◉ 50,000 to 100,000   ◉ 100,000 to 500,000   ◼ above 50

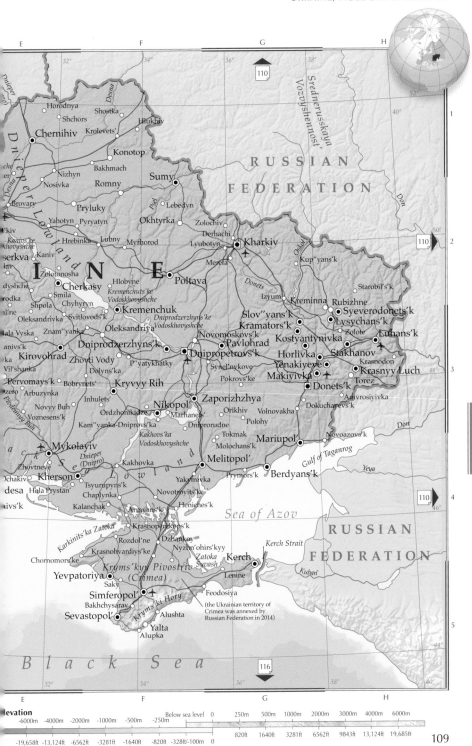

RUSSIAN

FEDERATION

Horodnya
Shchors
Shostka
Hlukhiv
Chernihiv
Krolevets'
Konotop
Nizhyn
Bakhmach
Nosivka
Romny
Sumy
Pryluky
Yahotyn
Pyryatyn
Lebedyn
Okhtyrka
Zolochiv
Derhachi
Hrebinka
Lubny
Myrhorod
Lyubotyn
Kharkiv
Kup"yans'k
Kaniv
Zolotonosha
Merefa
Cherkasy
Hlobyne
Poltava
Donets
Izyum
Starobil's'k
Smila
Kremenchuts'ke Vodoskhovyshche
Chyhyryn
Kreminna
Rubizhne
Svitlovods'k
Slov"yans'k
Syeverodonets'k
Oleksandrivka
Kramators'k
Lysychans'k
Znam"yanka
Oleksandriya
Dniprodzerzhyns'ke Vodoskhovyshche
Novomoskovs'k
Zolote
Luhans'k
Kirovohrad
Dniprodzerzhyns'k
Pavlohrad
Kostyantynivka
Stakhanov
Zhovti Vody
Dolyns'ka
Dnipropetrovs'k
Horlivka
Yenakiyeve
Krasnodon
Pervomays'k
Bobrynets'
P"yatykhatky
Synel'nykove
Makiyivka
Krasnyy Luch
Arbuzynka
Inhulets'
Kryvyy Rih
Pokrovs'ke
Donets'k
Torez
Novyy Buh
Ordzhonikidze
Zaporizhzhya
Amvrosiyivka
Voznesens'k
Kam"yanka-Dniprovs'ka
Orikhiv
Volnovakha
Dokuchayevs'k
Nikopol
Marhanets
Dniprorudne
Polohy
Kakhovs'ka Vodoskhovyshche
Tokmak
Novoazovs'k
Mykolayiv
Molochans'k
Mariupol'
Zhovtneve
Kakhovka
Melitopol'
Gulf of Taganrog
Kherson
Yakymivka
Prymors'k
Berdyans'k
Hola Prystan'
Chaplynka
Novotroyits'ke
Kalanchak
Armyans'k
Heniches'k
Sea of Azov
Rozdol'ne
Krasnoperekops'k
Dzhankoy
RUSSIAN
Krasnohvardiys'ke
Nyzhn'ohirs'kyy
Chornomors'ke
Zatoka Syvash
Kerch
FEDERATION
Yevpatoriya
Kryms'kyy Pivostriv (Crimea)
Saky
Feodosiya
Simferopol'
Lenine
Bakhchysaray
(the Ukrainian territory of Crimea was annexed by Russian Federation in 2014)
Sevastopol'
Alushta
Kryms'ki Hory
Yalta
Alupka

Black Sea

Elevation
-6000m -4000m -2000m -1000m -500m   Below sea level 0   250m   500m   1000m   2000m   3000m   4000m   6000m
-19,658ft -13,124ft -6562ft -3281ft -1640ft -820ft -328ft/-100m 0   820ft   1640ft   3281ft   6562ft   9843ft   13,124ft   19,685ft

European Russia

0 km   300

0 miles   300

Population   ● National capital

○ below 50,000   ○ 50,000 to 100,000   ◉ 100,000 to 500,000   ■ above 500,0

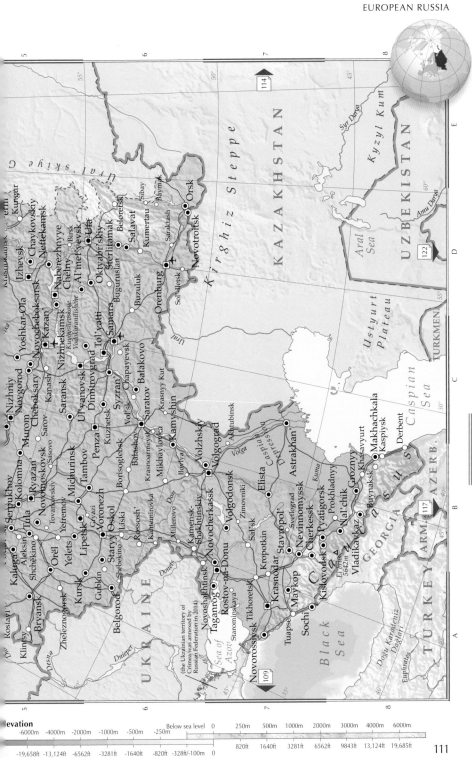

Elevation

				Below sea level	0	250m	500m	1000m	2000m	3000m	4000m	6000m
-6000m	-4000m	-2000m	-1000m	-500m	-250m							
-19,658ft	-13,124ft	-6562ft	-3281ft	-1640ft	-820ft	-328ft/-100m	0	820ft	1640ft	3281ft	6562ft	9843ft	13,124ft	19,685ft

North & West Asia

Population • National capital

○ below 50,000 ○ 50,000 to 100,000 ◉ 100,000 to 500,000 ◼ above 500,00

0 km   800
0 miles   800

E   120°   F   140°   G   160°   H   180°

155

OCEAN

80°

Chukchi   Chukchi
Plain   Plateau   1

pack ice   Ostrov Kotel'nyy   New Siberian Islands

Laptev Sea   Ostrov Kotel'nyy   East Siberian
Sea   Summer limit of pack ice

Yanskiy   Wrangel Island
Zaliv   Long Strait   Chukchi   70°
Sea
Indigirka   Ekiatapskiy Khrebet   Bering Strait
nd   Olenek   Lena   Kolyma   Arctic Circle
Khrebet Cherskogo   34   2
ERATION   Khrebet Verkhoyanskiy   Anadyr   Gulf of
i   Vilyuy   Aldan   Kolyma Range   Koryak Range   Anadyr   Bering
r   i   a   Yakutsk   Sea
Lena   Amga   Shelekhov   Bering
Gulf   Aleutian   Sea
Stanovoy Khrebet   Magadan   Basin   Winter limit of pack ice
Khrebet Dzhugdzhur   Sea of   Kamchatka   Aleutian Islands
y Khrebet   Okhotsk   Aleutian Trench   50°   3
IA   Amur   Zeya   Petropavlovsk-   Emperor Seamounts
Khabarovsk   Kamchatskiy
Sakhalin   Kuril Trench   Northwest Pacific   Chinook Trough   40°
Yuzhno-   Kuril Islands   Basin
Sakhalinsk   La Pérouse Strait
Vladivostok   (administered by   PACIFIC   34   4
Russian Federation,   Japan Trench
Sea of   claimed by Japan.)   30°
Japan   Sikhote-Alin   OCEAN
Yellow   (East Sea)   Shikoku Basin   Hawaiian Ridge
Sea   East   China   Ryukyu Trench   Tropic of Cancer   20°   Mid - Pacific Mountains   20°   5
Philippine   Sea
outh   Philippine Basin   Mariana Trench   10°
a   Sea   10°
Basin   120°   140°   160°   180°
143

E   F   G   H

Russia & Kazakhstan

0 km   600
0 miles   600

Population

● National capital

○ below 50,000   ○ 50,000 to 100,000   ◉ 100,000 to 500,000   ■ above 500,0

SVALBARD
(to Norway)

NORWAY

DENMARK

GERMAN

SWEDEN

FINLAND

KALININGRAD
(to Russ. Fed.)

Kaliningrad

POLAND   LITH.   LAT.   EST.

Sankt-Peterburg

Pskov

Velikiy Novgorod

BELARUS

Smolensk

Cherepovets

MOSKVA   Tver   Vologda
(MOSCOW)

Bryansk   Tula   Yaroslavl'   Kineshma

Belgorod   Ryazan'   Vladimir
Nizhniy Novgorod

Voronezh   Tambov   Kirov

Penza   Kazan'   Glazov   Solikamsk'

Mikhaylovka   Ul'yanovsk   Izheysk   Perm'   Serov

Rostov-na-   Saratov   Tol'yatti
Donu   Balakovo   Naberezhnyye   Yekaterinburg
Krasnodar   Volgograd   Chelny   Ufa
Sochi   Stavropol'   Samara   Tyumen'

Nal'chik   Ural'sk   Sterlitamak   Chelyabinsk   Tobol'sk
Astrakhan'   Orenburg
Vladikavkaz   Magnitogorsk   Ishim
Groznyy   Aktobe   Orsk   Petropavlovsk
Makhachkala   Atyrau   (Aktyubinsk)   Kostanay   Omsk
Alga   Rudnyy   Seversk
Fort-Shevchenko   Emba   Atbasar   Shchuchinsk   Novosibirsk
Aktau   Shalkar   ASTANA
Zhanaozen   KAZAKHSTAN   Pavlodar
Barnaul
Ustyurt   Aral   Aral'sk   Temirtau   Novokuznetsk
Plateau   Sea   Ayteke Bi   Saran'   Karagandy   Semey
Zhezkazgan   Ridder
Zhosaly   Zyryanovsk
Kyzylorda   Balkhash   Shar
Ust'-Kamenogorsk

Turkistan   Kentau   Ozero   Ayagoz
Karatau   Balkhash
Arys   Shu   Taldykorgan
IRAN   Shymkent   Taraz   Tekeli
Almaty
(Alma-Ata)

AFGHANISTAN   KYRGYZSTAN   CHINA

Murmansk

Barents
Sea

Nordkapp
(North Cape)

Kandalaksha

Severodvinsk   Ostrov
Arkhangel'sk   Kolguyev

Nar'yan-Mar

Vel'sk

Petrozavodsk

Kotlas   Ukhta   Vorkuta

Syktyvkar   Salekhard

Nadym

Nyagan'

Khanty-Mansiysk

Lesnoy

Surgut   Nizhnevarto

RUSSI

Tomsk

Kemerovo

Kras

Zapadno-
Sibirskaya
Ravnina

Black Sea

Caspian Sea

UKRAINE

MOLDOVA

(the Ukrainian
territory of
Crimea was
annexed by
Russia in 2014)

Sea of
Azov

GEORGIA

ARM.

AZERBAIJAN

TURKMENISTAN

UZBEKISTAN

TAJIKISTAN

ARCTIC

Novaya Zemlya

Karskoye More

Kara Sea

Arctic Circle

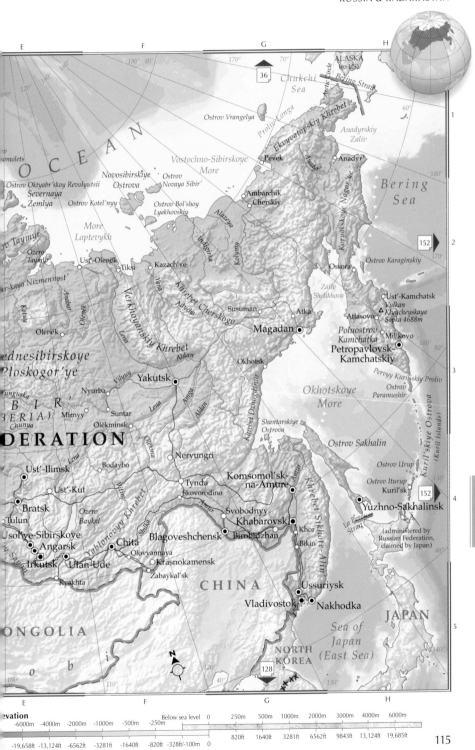

E F G H

180° 80° 170° 70°
36

ALASKA
(to US)
Arctic Circle
Chukchi
Sea

Bering Strait

60°

Ostrov Vrangelya

Proliv Longa

Ekvyeatapskiy Khrebet

Anadyrskiy
Zaliv

180°

O C E A N

omolets

Vostochno-Sibirskoye
More

Pevek

Anadyr'

Anadyr

B E R I N G
Sea

Ostrov Oktyabr'skoy Revolyutsii
Severnaya
Zemlya

Novosibirskiye
Ostrova

Ostrov
Novaya Sibir'

Ambarchik
Cherskiy

Koryakskoye Nagor'ye

170°

Ostrov Kotel'nyy

Ostrov Bol'shoy
Lyakhovskiy

Ossora

Ostrov Karaginskiy

152

v Taymyr

More
Laptevykh

Alazeya

Indigirka

Kolyma

Zaliv
Shelikhova

Ust'-Kamchatsk

170°

Ozero
Taymyr

Ust'-Olenëk

Tiksi

Kazach'ye

Yana

Khrebet Cherskogo

Susuman

Atka

Atlasovo

Vulkan
Klyucheyskaya
Sopka 4688m

rskaya Nizmennost'

Anabar

Olenëk

Lena

Adycha

Magadan

Okhotsk

Poluostrov
Kamchatka

Mil'kovo

160°

50°

Kotuy

Olenëk

Verkhoyanskiy Khrebet

Aldan

Petropavlovsk-
Kamchatskiy

ednesibirskoye
Ploskogor'ye

Nyurba

Vilyuy

Yakutsk

Aldan

Amga

Okhotskoye
More

Pervyy Kuril'skiy Proliv

Ostrov
Paramushir

S I B I R'
(S I B E R I A)

Mirnyy

Suntar

Lena

Olëkminsk

Shantarskiye
Ostrova

Ostrov Sakhalin

Kuril'skiye Ostrova
(Kuril Islands)

nguska

Chunya

Khrebet Dzhugdzhur

D E R A T I O N

Lena

Olëkma

Neryungri

Ostrov Urup

Ostrov Iturup

150°

Ust'-Ilimsk

Bodaybo

Vitim

Yablonovyy Khrebet

Tynda

Skovorodino

Komsomol'sk-
na-Amure

Amur

Khrebet Sikhote Alin'

Kuril'sk

152

Ust'-Kut

Amur

Svobodnyy

Khabarovsk

Birobidzhan

Khor

Yuzhno-Sakhalinsk

La Pérouse Strait

40°

Bratsk

Ozero
Baykal

Blagoveshchensk

Bikin

(administered by
Russian Federation,
claimed by Japan)

Tulun

Usol'ye-Sibirskoye

Angarsk

Irkutsk

Ulan-Ude

Chita

Shilka

Olovyannaya

Krasnokamensk

Zabaykal'sk

C H I N A

Ussuriysk

Kyakhta

Vladivostok

Nakhodka

J A P A N

ONGOLIA

o b

i

N

NORTH
KOREA

Sea of
Japan
(East Sea)

140°

128

120°

110°

40°

130°

E F G H

evation
-6000m -4000m -2000m -1000m -500m -250m Below sea level 0 250m 500m 1000m 2000m 3000m 4000m 6000m

-19,658ft -13,124ft -6562ft -3281ft -1640ft -820ft -328ft/-100m 0 820ft 1640ft 3281ft 6562ft 9843ft 13,124ft 19,685ft

Turkey & the Caucasus

ROMANIA
BULGARIA
UKRAINE
Kryms'kyy
Pivostriv
(Crimea)

(the Ukrainian territory of
Crimea was annexed by
Russian Federation in 2014)

Ticul Sinoie
Danube
Varnenski
Zaliv
Burgaski
Zaliv
Maritsa
Black Sea

108
104

Kırklareli
Edirne
Ergene Çayı
Çorlu
Tekirdağ
İstanbul
İzmit
Adapazarı
Çanakkale
Bandırma
Yalova
İznik Gölü
Bursa
Bilecik
Balıkesir
Bozüyük
Eskişehir
ANKARA
Edremit
Ayvalık
Simav Çayı
Kütahya
Polatlı
Kırıkkale
Kalecik
Akhisar
Simav
Gediz
Manisa
Uşak
Afyon
İzmir
Ödemiş
Alaşehir
Aydın
Nazilli
Dinar
Denizli
Burdur
Isparta
Söke
Büyükmenderes Nehri
Milas
Tavas
Burdur
Gölü
Muğla
Bodrum
Marmaris
Dalaman
Fethiye
Antalya
Kaş
Finike
Antalya
Körfezi
Manavgat
Alanya
Mut
Silifke
Anamur

Cide
İnebolu
Sinop
Bartın
Gerze
Küre Dağları
Zonguldak
Kastamonu
Bafra
Devrek
Karabük
Kargı
Çerkeş
Çankırı
Bolu
Gerede
Merzifon
Çorum
Alaca
Sorgun
Şarkışla
Hirfanlı
Barajı
Boğazlıyan
Tuz Gölü
Kulu
Bünyan
Cihanbeyli
Akşehir
İncesu
Nevşehir
Kayseri
Aksaray
Niğde
Göksun
Beyşehir
Gölü
Konya
Ereğli
Kahram
Şuğla Gölü
Karaman
Toros
Dağları
Ceyhan
Tarsus
Adana
Mersin (İçel)
İskenderun
Antakya

GREECE
Lésvos
Chíos
Sámos
Dodekánisa
(Dodecanese)
Ródos
(Rhodes)
Kárpathos
Mediterranean Sea

CYPRUS
TURKISH REPUBLIC OF
NORTHERN CYPRUS
(recognized only by Turkey)
LEBANON

T U R K E Y
Anatolia

Marmara Denizi
(Sea of Marmara)
İstanbul Boğazı
(Bosporus)
Çanakkale
Boğazı
(Dardanelles)

Sams
Tok
Yıld
Gü

105
72

0 km   200
0 miles   200

Population   ● National capital

○ below 50,000   ○ 50,000 to 100,000   ◉ 100,000 to 500,000   ◼ above 500,0

RUSSIAN
FEDERATION

C a u c a s u s

Gagra
Gudauta
Sokhumi
Ochamchire

Ap'khazet'i
Enguri
Mestia

Kazbek
5047m △

South
Ossetia

Caspian

Sea

Kutaisi

GEORGIA
Gori
Tsalka
TBILISI
Rustavi

Zaqatala
Şäki

Xaçmaz
Quba
Siyäzän

122

Samtredia
Poti
Kobuleti
Batumi
Hopa
Achara
Akhaltsikhe

Lesser
Caucasus
Kura

Mingäçevir
Märäzä

Sumqayıt
BAKI
(BAKU)

Pazar
Rize
Of
Çoruh Nehri
Artvin
Gyumri
Vanadzor
Gäncä
Yevlax

Greater Caucasus

Qäzimämmäd
Äli-Bayramı

ozon
ün
ül *Karadeniz Dağları*
ane

Kars
Artlk
Sarıkamış
ARMENIA
YEREVAN
Sevan
Sevana Lich

AZERBAIJAN
Nagorno-
Karabakh
Xankändi

Goris

İmişli

Biläsuvar

İspir
Askale
Pasinler
Horasan
Aras
Büyükağrı Dağı
(Mount Ararat) △
5137m
Artashat
Naxçıvan

AZERBAIJAN

Länkäran

Erzincan
Tercan
Y
Erzurum
Ağri
Doğubayazıt
Patnos
Erciş
Muradiye

Kemah
Kuban
Bingöl

Muş
Tatvan
Van
Gölü
Van
Gevaş

Daryācheh-ye
Orūmīyeh

Rashteh-ye Kūhhā-ye Alborz
(Elburz
Mountains)

IRAN

Elazığ
Silvan
Bitlis
Siirt

Keban
Baraji

Torosy

Diyarbakır
Batman
Şırnak

Silverek
irk
Viranşehir
Mardin
Nusaybin

anlıurfa
Ceylanpınar

Kurdistan

Tigris

Al Jazīrah
Euphrates
Jabal Bishrī

Kūhhā-ye Zagros
(Zagros
Mountains)

IRAN

RIA

IRAQ

Buhayrat
ath
Tharthār

120

The Near East

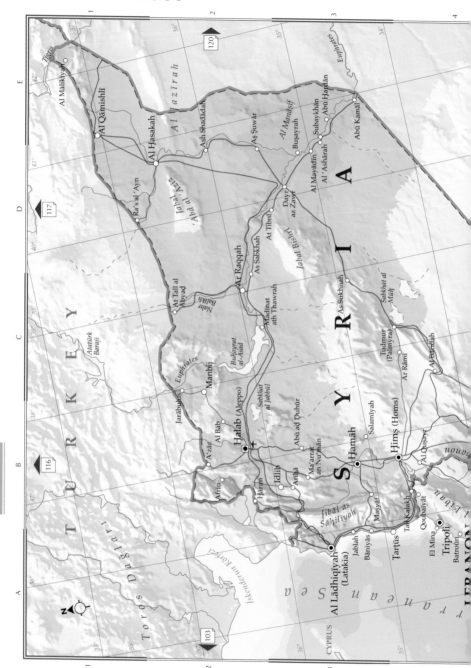

Population

● National capital

○ below 50,000   ○ 50,000 to 100,000   ◉ 100,000 to 500,000   ■ above 500,

0 km _____ 100

0 miles _____ 100

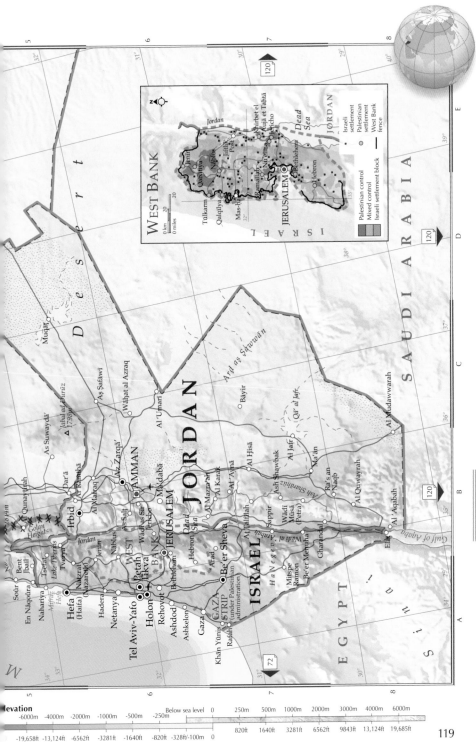

WEST BANK

Israeli settlement
Palestinian settlement
West Bank fence

Palestinian control
Mixed control
Israeli settlement block

Jordan

Jenin
Qabātiya
Nābius
Jilitit Post
Khirbet el
Auā et Tahtā
Jericho
Tulkarm
Qalqilya
Mas'ūdi
Ni'lin
Ramallah
JERUSALEM
Bethlehem
Hebron

Dead Sea

JORDAN

ISRAEL

0 km 20
0 miles 20

Elevation

-6000m	-4000m	-2000m	-1000m	-500m	-250m	Below sea level 0	250m	500m	1000m	2000m	3000m	4000m	6000m

-19,658ft -13,124ft -6562ft -3281ft -1640ft -820ft -328ft/-100m 0   820ft 1640ft 3281ft 6562ft 9843ft 13,124ft 19,685ft

The Middle East

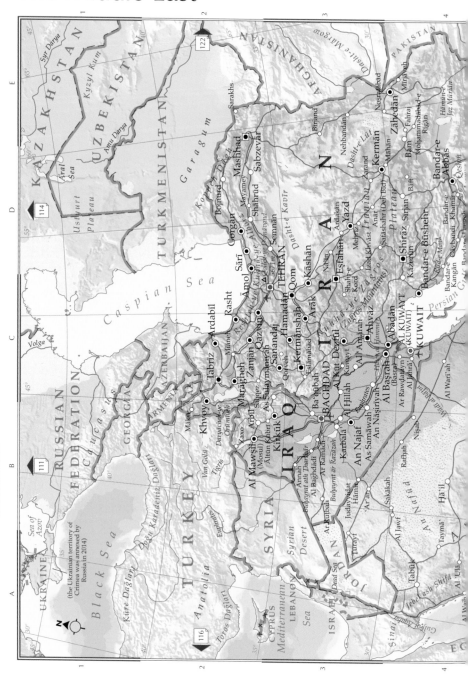

Population

● National capital

○ below 50,000
○ 50,000 to 100,000
◉ 100,000 to 500,000
◼ above 500,0

0 km   400
0 miles   400

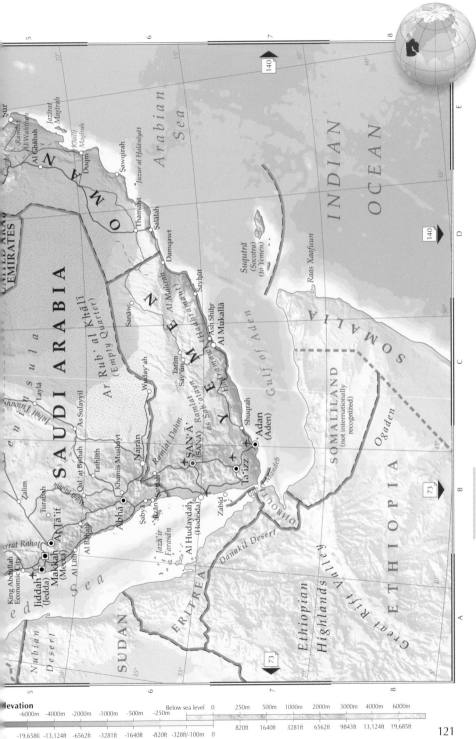

5   6   7   8

140

140

140

73

73

Sur
Ra's al Hadd
Ra's Mirbāt
Al Ghābah
Jazīrat Maṣīrah
Al Ashkharah
Khalīj Maṣīrah

Jabal Thanrāt
Duqm
Şawqirah
Al Wahībah

O M A N

Thamarīt
Şalālah
Jazīrat Ḩalāniyāt

A r a b i a n   S e a

I N D I A N

O C E A N

UNITED ARAB EMIRATES

S A U D I   A R A B I A

Damqawt

Sanāw

Al Mahrah

Y E M E N

Sayḩūt

Damqawt

Ar Rub' al Khālī
(Empty Quarter)

Suquṭrā
(Socotra)
(to Yemen)

Raas Xaafuun

Ar Rub' al Khālī

Tarīm
Sayʼūn
Ash Shiḩr

Hadramawt

Al Mukallā

Layla

As Sulayyil

Widayah

Ramlat as Sabʻatayn

Najrān

Ramlat Dahm

Khamis Mushayt

S O M A L I L A N D
(not internationally
recognized)

Shuqrah

Ogaden

Tathlīth

Qal'at Bīshah

Abhā

Al Bāḩah

Al Ta'if

SAN'Ā'
(SANA)

Ta'izz

Adan
(Aden)

Gulf of Aden

E T H I O P I A

Turabah

Zalim

Şabyā

Jīzān Şaʻdah

Zabid

Al Hudaydah
(Hodeida)

Jazā'ir
Farrasān

Danakil Desert

E R I T R E A

S U D A N

Makkah
(Mecca)

Jiddah
(Jedda)

King Abdullah
Economic City

Al Lithf

Ḩarrat Rahat

Nubian
Desert

Ethiopian
Highlands

Great Rift Valley

Bab al Mandeb

P e n i n s u l a

Jabal Tuwayq

Red Sea

Elevation

						Below sea level	0	250m	500m	1000m	2000m	3000m	4000m	6000m
-6000m	-4000m	-2000m	-1000m	-500m	-250m									

-19,658ft -13,124ft -6562ft -3281ft -1640ft -820ft -328ft/-100m 0   820ft 1640ft 3281ft 6562ft 9843ft 13,124ft 19,685ft

Central Asia

RUSSIAN
FEDERATION

GEORGIA

AZERBAIJAN

Caspian

Sea

Garabogaz
Aylagy

Ustyurt

Plateau

Aral
Sea

Mo'ynoq

Qoraqalpog'iston

Chimboy
Taxtako'pir

Köneürgenç   **Nukus**
Taxiatosh

Gurbansoltan Eje   Qubadag

Daşoguz   To'rtko'l

Xiva

UZBI

Kyz

Uch

Urganch

Gazojak   Lebap   Zar

Türkmenbaşy

Türkmenbaşy
Aylagy

Hazar

Balkanabat

Bereket

Türkmen
Aylagy

Ungüz   *Angyrsyndaky*   *Garagum*

Derweze

T u r a n   L o w l a n d

Amu Darya

Ganlygaýyr Platosy   Ûçtagan Gumy

TURKMENISTAN

Köpetdag Gershi

Serdar

Magtymguly   Baharly

Garagum

Seýdi

Galkynyş

Bu

Türkmenabat

Esenguly

Gökdepe
Gora Chapan
2889m

Abadan

AŞGABAT
(ASHGABAT)

Tejen

Kaka

Mary

Murgap

Bayramaly

Saý

Garag

Reşteh-ye Kūhhā-ye Alborz

Sarahs

Murgap

Garab
Belent
Ma

Bālā Murghāb

Serhetabat
Towraghoudī

Ghōriān   **Herāt**

Selseleh-ye Sefid Ki

Darya-ye

IRAN

Kūhhā-ye Zāgros

AFGHA

Shindand

Farah Rūd

Farah   Dilārām

Iranian

Plateau

Hāmūn-e
Şāberī

Dasht-e Khāsh

Lashkar Gāh
Chakhānsūr

Zaranj

Dasht-e Mārgow

Dīshū

Ků
Da

Darya-ye Helmand

Chāgai Hi

N

0 km   200

0 miles   200

Population

○ below 50,000

○ 50,000 to 100,000   ● 100,000 to 500,000   ■ above 500,0

● National capital

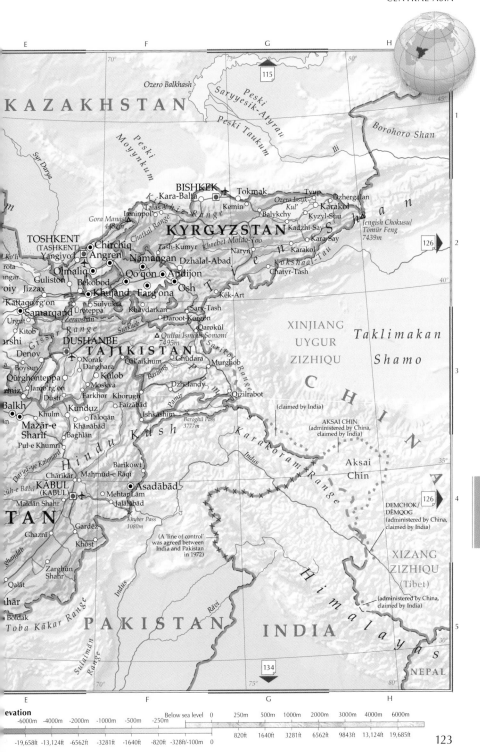

E   F   G   H

70°

115

Ozero Balkhash

Peski Saryyesik-Atyrau

Peski Taukum

KAZAKHSTAN

Peski Moyynkum

Borohoro Shan

Ili

45°

1

BISHKEK

Kara-Balta

Tokmak

Sur Darya

Talas

Kemin

Tyup

Dzhergalan

Ireninpol'

Chatkal Range

Ozero Issyk-Kul'

Karakol

Gora Manas

4482m

Balykchy

Kyzyl-Suu

Jengish Chokusu/

Tömür Feng

7439m

126

TOSHKENT

(TASHKENT)

Chirchiq

KYRGYZSTAN

Tash-Kumyr

Khrebet Moldo-Too

Kadzhi-Say

Kara-Say

Yangiyo'l

Angren

Namangan

Dzhalal-Abad

Naryn

Karakol

40°

Ko'li

rota

angar

oiy

Olmaliq

Qo'qon

Andijon

Kokshaal-Tau

2

Jizzax

Guliston

Bekobod

Osh

Chatyr-Tash

Khujand

Farg'ona

Kek-Art

Kattaqo'rg'on

Sulyukta

Khaydarkan

Sary-Tash

Samarqand

Uroteppa

Daroot-Korgon

Urgut

Zeravshan

Surkhob

Qarokül

Kitob

Range

△ Qullai Ismoili Somoní

7495m

XINJIANG

UYGUR

ZIZHIQU

Taklimakan

arshi

Gissar Range

DUSHANBE

Ghüdara

Murghob

Shamo

Denov

Norak

Qal'aikhum

Boysun

Danghara

Bartang

TAJIKISTAN

3

à

Qúrghònteppa

Kŭlob

Moskva

Pamir

Dzhelandy

Qizilrabot

C

Jarqo'rg'on

Dŭstí

Farkhor

Khorugh

(claimed by India)

Balkh

Khulm

Kunduz

Faizabad

Ishkashim

Panir

AKSAI CHIN

(administered by China,

claimed by India)

in

Tāloqān

35°

Mazār-e

Sharif

Khānābād

Baroghil Pass

3777m

Aksai

Chin

H

Pul-e Khumri

Baghlān

Hindu *Kush*

Indus

DEMCHOK/

DÉMQOG

(administered by China,

claimed by India)

126

Barikowt

Daryā-ye Kahmard

Chārikār

Mahmūd-e Rāqí

Asadābād

Karakoram

Range

I

N

A

4

uh-e Bābā

KABUL

(KABUL)

Mehtap Lām

Jalālābād

(A 'line of control'

was agreed between

India and Pakistan

in 1972)

XIZANG

ZIZHIQU

(Tibet)

Maïdān Shahr

Khyber Pass

1080m

(administered by China,

claimed by India)

TAN

Ghaznī

Gardēz

Khōst

Indus

Rāvi

H

i

m

a

l

a

y

a

5

Zarghūn

Shahr

Sulaiman Range

y

a

s

Qalāt

hār

PAKISTAN

INDIA

30°

Boldak

Toba Kākar Range

NEPAL

70°

75°

80°

134

E   F   G   H

evation

-6000m   -4000m   -2000m   -1000m   -500m   -250m   Below sea level   0   250m   500m   1000m   2000m   3000m   4000m   6000m

-19,658ft   -13,124ft   -6562ft   -3281ft   -1640ft   -820ft   -328ft/-100m   0   820ft   1640ft   3281ft   6562ft   9843ft   13,124ft   19,685ft

South & East Asia

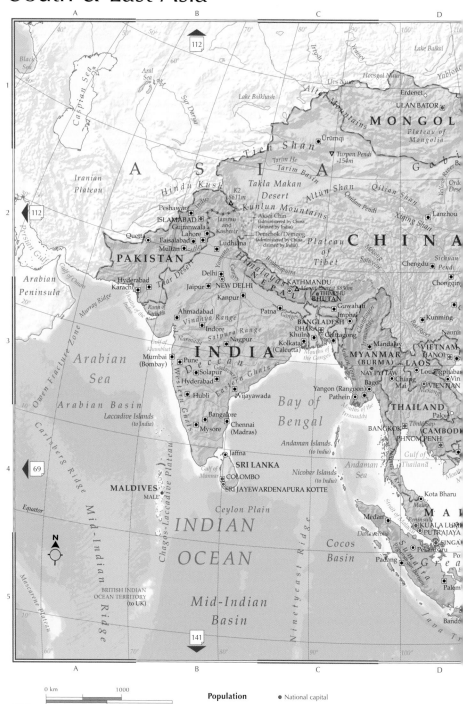

A   B   C   D

Black Sea
Caspian Sea
Aral Sea
Syr Darya
Lake Balkhash
Irtysh
Yenisey
Lake Baikal
Urs Nuur
Hosgol Nuur
Yablon...

112

Erdenet
ULAN BATOR

MONGOL
Plateau of Mongolia

Tien Shan
Ürümqi
Altai Mountains
▽ Turpan Pendi
-154m

Gobi
Orda Dese...

Iranian Plateau

A   S   I   A

Hindu Kush
Tarim He
Tarim Basin
K2
8611m
Takla Makan Desert
Altun Shan
Kunlun Mountains
Qaidam Pendi
Qilian Shan
Lanzhou
Xiqing Shan

112

Peshawar
ISLAMABAD
Gujranwala
Lahore
Faisalabad
Multan
Quetta

Jammu and Kashmir
Aksai Chin
(administered by China,
claimed by India)
Demchok/Demqog
(administered by China,
claimed by India)

Plateau of Tibet

CHINA

Chengdu
Sichuan Pendi
Chongqin...

PAKISTAN
Thar Desert
Indus
Ganges

Delhi
Jaipur
NEW DELHI
Ludhiana
Kanpur

Himalayas
Mount Everest 8850m
KATHMANDU
THIMPHU
BHUTAN
Guwahati
Imphal

NEPAL
Patna
Ganges
BANGLADESH
DHAKA
Khulna
Kolkata (Calcutta)
Chittagong

Hyderabad
Karachi
Rann of Kachchh

Ahmadabad
Vindhya Range
Indore
Satpura Range
Nagpur
Narmada
Godavari

Mumbai (Bombay)
Gulf of Khambhat

INDIA
Deccan
Pune
Solapur
Hyderabad
Hubli
Vijayawada

Arabian Peninsula
Persian Gulf
Gulf of Oman

Arabian Sea

Arabian Basin
Owen Fracture Zone
Murray Ridge

Laccadive Islands
(to India)

Bangalore
Mysore
Chennai (Madras)

Western Ghats
Eastern Ghats

Bay of Bengal

Mouths of the Ganges

MYANMAR (BURMA)
Mandalay
NAY PYI TAW
Bago
Yangon (Rangoon)
Pathein

VIETNAM
HANOI
LAOS
Louangphabang
Chiang Mai
VIENTIANE
Vin...

Mouths of the Irrawaddy

THAILAND
BANGKOK
Pakxe
Tonle Sap
CAMBOD...
PHNOM PENH
Gulf of Thailand

Jaffna
SRI LANKA
COLOMBO
SRI JAYEWARDENAPURA KOTTE
Gulf of Mannar

Andaman Islands
(to India)

Nicobar Islands
(to India)

Andaman Sea

Kota Bharu

Medan
Strait of Malacca
MAL...
KUALA LUMP...
PUTRAJAYA
SINGA...
Pekanbaru
Sumatra

MALDIVES
MALE

Ceylon Plain

INDIAN OCEAN

Chagos-Laccadive Plateau
Carlsberg Ridge
Mid-Indian Ridge
Mascarene Plateau

Equator

BRITISH INDIAN
OCEAN TERRITORY
(to UK)

Mid-Indian Basin

Cocos Basin

Padang
Palem...
Band...
Java Tr...

Ninetyeast Ridge

141

A   B   C   D

0 km   1000
0 miles   1000

69

Population   ● National capital

○ below 50,000   ◉ 50,000 to 100,000   ◉ 100,000 to 500,000   ■ above 500,00...

E   F   G   H

113

152

152

142

Qiqihar
Manchuria
Plain   Harbin   Lake Khanka
angchun
ao He
yang
Dandong   Dalian
ai

Sakhalin

Kuril Islands

Kuril Trench

Sapporo   Hokkaido

JAPAN

Sea of
Japan
(East Sea)

NORTH
KOREA
PYONGYANG

SOUTH
SEOUL   KOREA
SEJONG
CITY

Qingdao

Yellow
Sea

Honshu

Japan Trench

Sendai

Nagoya   Yokohama
Kyoto   TOKYO
Fujisan
3776m
Osaka
Hiroshima
Shikoku
Kitakyushu
Kyushu

Northwest
Pacific
Basin

Shatskiy Rise

Emperor Seamounts

Mapmaker Seamounts

East China
Shanghai   Sea
langzhou
achan

uzhou

TAIPEI
Gaoxiong
TAIWAN

Ryukyu Islands

Ryukyu Trench

Shikoku Basin

Kyushu

Mid - Pacific Mountains

PACIFIC
OCEAN

Philippine Sea

Philippine Basin

Palau Ridge

West
Mariana
Basin

Mariana Trench

East
Mariana
Basin

Melanesian
Basin

Luzon Strait

China
ea

Baguio
Luzon

MANILA

PHILIPPINES
Mindoro
Panay   Samar
Cebu
ISLANDS
(disputed)
Palawan   Bacolod
Negros
Sulu
Sea
Zamboanga

Mindanao
Davao

Yap Trench

Eauripik Rise

M i c r o n e s i a

Equator

Ontong
Java
Rise

M e l a n e s i a

ANDAR
RI BEGAWAN

Celebes
Sea
Manado

Halmahera

Jayapura

Bismarck Archipelago

Solomon
Islands

apan

Mollucas

Serahi
Ambon
Buru

Pegunungan Maoke

Solomon
Sea

DONESIA

Celebes
anjarmasin
Makassar

Makassar Strait

Flores

Lesser Sunda Islands
Flores
Bali

Banda Sea

New Guinea

EAST TIMOR

Timor
Sumba

Arafura
Sea

Timor Trough   Timor
Sea

AUSTRALIA

Coral
Sea

E   F   G   H

1

2

3

4

5

Western China & Mongolia

Eastern China & Korea

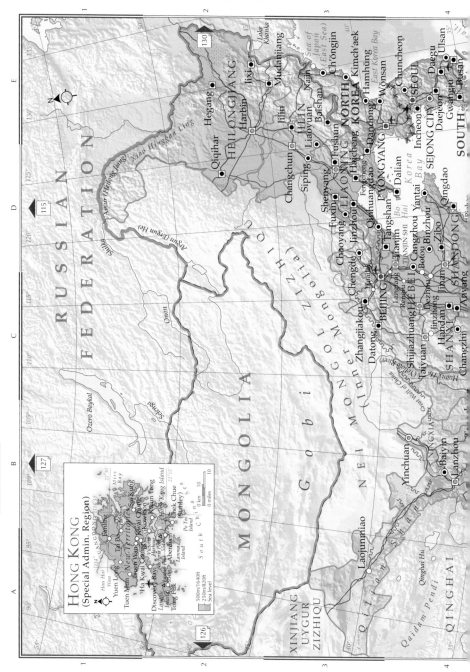

Population

National capital ● Internal administrative capital ●

○ below 50,000   ○ 50,000 to 100,000   ◉ 100,000 to 500,000   ■ above 500,0

0 km   400

0 miles   400

East China
Sea

Nansei Shoto
(Ryukyu Islands)

Okinawa

Tropic of Cancer

PACIFIC

OCEAN

152

TAIPEI (TAIBEI)

TAIWAN

Jilong

Taichung

Jiayi

Tainan

Gaoxiong

Luzon Strait

PHILIPPINES

139

(China and Taiwan claim
all of each other's territory)

SEOUL - Capital
SEJONG CITY - Administrative capital

SOUTH KOREA'S TWO CAPITALS

South China

Sea

PARACEL
ISLANDS
(disputed by China,
Taiwan and Vietnam)

Amphitrite Group

Crescent Group

Triton Island

SPRATLY ISLANDS
(disputed by China,
Malaysia, Philippines,
Taiwan and Vietnam)

Flat Island

Nanshan Island

Thitu
Island

Loaita Island

Namyit Island

Len Dao

Spratly Island

Hainan Dao

HAINAN

Xuwen

Danzhou

Dongfang

Beihai

Gulf of Tonkin

Red River

VIETNAM

CAMBODIA

136

LAOS

THAILAND

Gulf of Thailand

Mekong

Qinzhou

Nanning

Yulin

Zhanjiang

Maoming

Haikou

Suixi

Jiangmen

Zhaoqing

Macao
(Special Administrative Region)

Hong Kong
(Special Administrative Region)

Guangzhou

Dongguan Shantou

GUANGDONG

Shaoguan

Zhangzhou

Xiamen

Quanzhou

Yong'an

FUJIAN

Fuzhou

Longyan

Ganzhou

Nanping

JIANGXI

Shangrao

Jingdezhen

Nanchang

Jiujiang

Kangtian

Ji'an

Hengyang

Chuzhou

Yongzhou

HUNAN

Loudi

Changsha

Huaihua

ZHEJIANG

Ningbo

Wenzhou

Jinhua

Hangzhou

Jiaxing

Wuhu

Wuhan

HUBEI

Xizang

ANHUI

Anqing

Huangshi

Wuhan

Yichang

Lichuan

Wanzhou

CHONGQING SHI

Chongqing

Three Gorges
Reservoir

Dongting Hu

Yueyang

Zunyi

Neijiang

Zigong

Leshan

Yibin

GUIZHOU

Anshun

Guiyang

Guangan

Quanzhou

Liuzhou

GUANGXI
ZHUANGZU

Guilin

Changzhou

Yan'an

Chengdu

Sichuan
Pendi

Min Shan

Xichang

YUNNAN

Dali

Baoshan

Hengduan Shan

INDIA

MYANMAR
(BURMA)

Tropic of Cancer

Qiang Jiang

Wuliang Shan

Jinghong

Mekong

Salween

136

Nanhui Jia (Island)

Elevation

-6000m -4000m -2000m -1000m -500m -250m | Below sea level 0 | 250m 500m 1000m 2000m 3000m 4000m 6000m

-19,658ft -13,124ft -6562ft -3281ft -1640ft -820ft -328ft/-100m 0 | 820ft 1640ft 3281ft 6562ft 9843ft 13,124ft 19,685ft

129

Japan

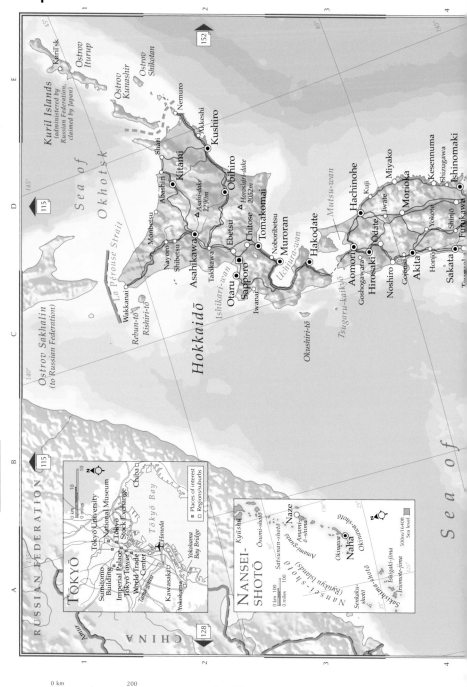

Kuril Islands
(administered by
Russian Federation,
claimed by Japan)

Kuril'sk

Ostrov
Iturup

Ostrov
Shikotan

Ostrov
Kunashir

Nemuro

Akkeshi

Kushiro

Shari

Kitami

Obihiro

Abashiri

△ Asahi-dake
2290m

△ Horoshiri-dake
2052m

Ebetsu

Tomakomai

Noboribetsu

Muroran

Hakodate

Mutsu-wan

Hachinohe

Kuji

Miyako

Kesennuma

Shizugawa

Ishinomaki

Monbetsu

Nayoro

Shibetsu

Asahikawa

Takikawa

Chitose

Udono-wan

Aomori

Goshogawara

Hirosaki

Odate

Iwate

Yokote

Morioka

Shinjō

Funakawa

Sea of
Okhotsk

Wakkanai

Rishiri-tō

Rebun-tō

Otaru

Sapporo

Iwanai

Ishikari-wan

Okushiri-tō

Tsugaru-kaikyō

Noshiro

Goyu

Honjō

Akita

Sakata

La Pérouse Strait

Ostrov Sakhalin
(to Russian Federation)

Hokkaidō

Sea of

RUSSIAN FEDERATION

Amur

CHINA

TOKYO

Chiba

Tōkyō University

National Museum

Tōkyō
Stock Exchange

Imperial Palace

Tōkyō Tower

World Trade
Center

Sumitomo
Building

Kawasaki

Tama-gawa

Yokohama

Haneda

Yokohama
Bay Bridge

Tōkyō Bay

■ Places of interest
□ Regions/suburbs

NANSEI-
SHOTŌ

Kyūshū

Ōsumi-shotō

Satsunan-shotō

Naze

Amami-
ō-shima

Amami-
guntō

Tokuno-shima

Okinawa

Naha

Okinawa-shotō

Ishigaki-jima

Iriomote-jima

Senkaku-
shotō

Sakishima-shotō

Nansei-shotō (Ryukyu Islands)

500m/1640ft
Sea level

Sea of

0 km   200

0 miles   200

Population

● National capital

○ below 50,000   ○ 50,000 to 100,000   ◉ 100,000 to 500,000   ■ above 500,

PACIFIC

OCEAN

Izu-shotō

Hachijō-jima

Miyake-jima
Mikura-jima

Nii-jima
O-shima

Kōzu-shima

Bōsō-hantō

Sagami-nada

Kashima-nada

Choshi

Hitachi

Utsunomiya
Mito
Oyama

Chiba
Yokohama

Kawagoe
Kawasaki

TOKYO

Mikuni-sanmyaku

Maebashi

Joetsu

Nagano
Toyama

Itoigawa

Matsumoto

Takaoka

Kanazawa

Komatsu

Hida-sanmyaku

Kōfu
Fuji-san
Fuji

Shizuoka
Toyota

Hamamatsu

Izu-hantō

Suruga-wan

Gifu
Nakatsugawa

Fukui

Tsuruga

Nagoya
Ōgaki

Okazaki
Ise

Ise-wan

Owase

Shingū
Wakayama

Tsu

Gobo
Tanabe

Kii-suidō

Shikoku

Tottori

Ōita

Bungo-suidō

Tosa-wan

Nakamura
Sukumo

Kyūshū

Nobeoka

Miyazaki

Miyakonojō

Shibushi-wan

Tanega-shima

Yaku-shima

Ōsumi-shotō

Ōsumi-kaikyō

Kagoshima

Satsuma-Sendai

Katsushiro

Kumamoto

Ōmuta

Kurume

Ōita

Ube

Hōfu

Iwakuni

Hiroshima

Kure

Matsuyama

Niihama

Kōchi

Takamatsu
Tokushima

Harima-nada

Awaji-shima

Akashi

Kobe
Ōsaka

Kyōto

Himeji

Okayama

Kurashiki

Chūgoku-sanchi

Masuda

Hamada

Gōtsu

Matsue

Yonago

Ōki-shotō
Dōgo
Dōzen

Liancourt Rocks
(under South
Korean control)

Toyama-wan

Wakasa-wan

Biwa-ko

Kōbō

SOUTH
KOREA

Tsushima

Kō-saki

Tsushima

Korea Straits

Iki

Kitakyūshū

Shimonoseki
Nagato

Yamaguchi

Fukuoka

Sasebo

Nagasaki

Gotō-rettō

Koshikijima-rettō

Satsuma-nada

Amakusa-nada

East

China Sea

Elevation

					Below sea level	0	250m	500m	1000m	2000m	3000m	4000m	6000m
-6000m	-4000m	-2000m	-1000m	-500m	-250m								
-19,658ft	-13,124ft	-6562ft	-3281ft	-1640ft	-820ft	-328ft/-100m	0		820ft	1640ft	3281ft	6562ft	9843ft	13,124ft	19,685ft

Southern India & Sri Lanka

A   B   C   D

134

Kalyān
Mumbai (Bombay)
Pune   Ahmadnagar   Nānded
Bārāmati   Nizāmābād   Telangana   Karīmnagar
I N D I A
Vizianaga
Solāpur   Secunderābad   Visāk
Sāngli
Kolhāpur   Gulbarga   Hyderābad   Rāj
Deccan
Belgaum   Karnātaka   Raichūr   Andhra   Vijayawā
Gadag   Kurnool   Krishna   Machil
Panaji   Hubli   Nandyāl   Pradesh   Chirāla
Ongole
Tungabhadra   Tādpatri   Kāvali
Dāvangere   Reservoir
Shimoga   Anantapur   Nellore
Bhadrāvati   Cuddapah
Udupi
Tumkūr
Mangalore   Bangalore   Vellore   Chennai (Madras)
Kāsaragod   Mandya   Kānchīpuram
Kannur (Cannanore)   Krishnagiri   Tiruppattūr
Mysore   Salem   Pondicherry
Kozhikode (Calicut)   Erode   Neyveli

Arabian

Sea

Amīndīvi
Islands'
Lakshadweep
(Laccadive Islands)
(to India)
Kavaratti
Island
Kalpeni
Island
Nine Degree
Channel
Minicoy Island
Eight Degree Channel

Coimbatore
Thrissur (Trichūr)   Tamil Nādu   Tiruchchirāppalli
Ernākulam   Dindigul   Madurai
Kochi (Cochin)   Palk Strait
Alappuzha (Alleppey)   Jaffna   SRI LA
Kollam (Quilon)   Rājapālaiyan   Mannar   Vavuniya
Thiruvananthapuram   Tuticorin   Trincon
(Trivandrum)   Puttalam   Anurādhapura
Nāgercoil   Gulf of   Matale   Bat
Mannar   Negombo   Kandy
COLOMBO
SRI JAYEWARDENAPURA   Kalutara   Ratnapura
KOTTE   Galle   Matara

Ihavandhippolhu
Atoll
MALDIVES
Faadhippolhu
Atoll
Horsburgh
Atoll
Male'Atoll
Ari Atoll   MALE'
Felidhu Atoll
Mulakatholhu
Kolhumadulu
Hadhdhunmathi Atoll

I N D I A

North Huvadhu Atoll
Equator
South Huvadhu
Atoll
Gan   140
Addu Atoll

SRI LANKA'S TWO CAPITALS	
COLOMBO	- Capital
SRI JAYEWARDENAPURA KOTTE	- Administrative capi

134   73   121

0 km   300
0 miles   300

132

Population
○ below 50,000
○ 50,000 to 100,000
◉ 100,000 to 500,000
■ above 500

● National capital

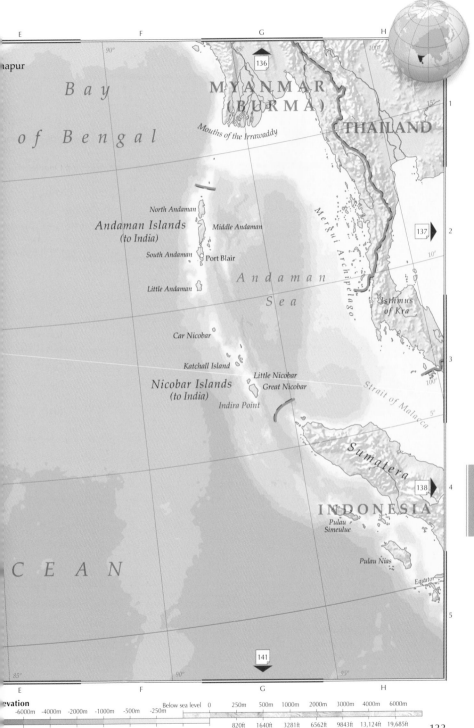

E   F   G   H

136

B a y

of B e n g a l

MYANMAR
(BURMA)

Mouths of the Irrawaddy

THAILAND

15°

1

North Andaman

Andaman Islands
(to India)

Middle Andaman

137

Mergui Archipelago

2

South Andaman

Port Blair

A n d a m a n

10°

Little Andaman

S e a

Isthmus
of Kra

Car Nicobar

Katchall Island

Nicobar Islands
(to India)

Little Nicobar
Great Nicobar

Strait of Malacca

100°

3

Indira Point

5°

Sumatera

138

INDONESIA

4

Pulau
Simeulue

C E A N

Pulau Nias

Equator

5

141

85°   90°   95°

E   F   G   H

evation

-6000m   -4000m   -2000m   -1000m   -500m   -250m   Below sea level   0   250m   500m   1000m   2000m   3000m   4000m   6000m

-19,658ft -13,124ft -6562ft -3281ft -1640ft   -820ft -328ft/-100m   0   820ft   1640ft   3281ft   6562ft   9843ft   13,124ft   19,685ft

Northern India, Pakistan & Bangladesh

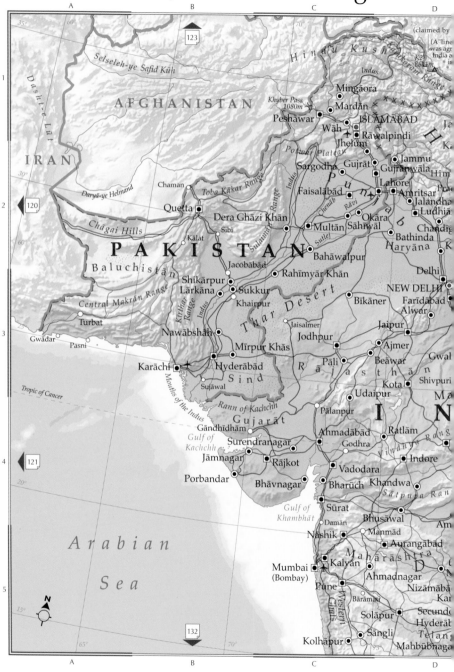

AFGHANISTAN

IRAN

Selseleh-ye Safīd Kūh

Dasht-e Lūt

Daryā-ye Helmand

Chāgai Hills

Chaman

Quetta

Kalat

Sibi

Toba Kākar Range

Central Makrān Range

Baluchistan

Turbat

Gwādar   Pasni

PAKISTAN

Jacobābād

Shikārpur

Lārkāna

Sukkur

Khairpur

Nawābshāh

Mīrpur Khās

Karāchī

Hyderābād

Sujāwal

Sind

Mouths of the Indus

Tropic of Cancer

Kirthar Range

Sulaimān Range

Indus

Hindu Kush

Mingāora

Mardān

Khyber Pass
1080m

Peshāwar

Wāh

Jhelum

ISLĀMĀBĀD

Rāwalpindi

Jammu

Sargodha

Gujrāt

Gujrānwāla

Faisalābād

Lahore

Amritsar

Jalandhar

Ludhiā

Multān

Sāhīwāl

Okara

Bathinda

Haryāna

Dera Ghāzi Khān

Bahāwalpur

Rahīmyār Khān

Bīkāner

Delhi

NEW DELHI

Farīdābād

Alwar

Jaisalmer

Jodhpur

Jaipur

Ajmer

Pāli

Beāwar

Kota

Shivpuri

Udaipur

Thar Desert

Rājasthān

Chenab

Rāvi

Sutlej

Potwar Plateau

Punjab

Chandīg

Gwal

Ma

Rann of Kachchh

Gujarāt

Gāndhīdhām

Surendranagar

Jāmnagar

Rājkot

Porbandar

Bhāvnagar

Bhāruch

Gulf of
Kachchh

Pālanpur

Ahmadābād

Godhra

Ratlām

Vadodara

Khandwa

Indore

Vindhya Range

Sātpura Ran

Gulf of
Khambhāt

Daman

Sūrat

Bhusāwal

Manmād

Am

Nashik

Aurangābād

Kalyān

Mumbai
(Bombay)

Pune

Bārāmati

Ahmadnagar

Nizāmābā

Mahārāshtra

D

Kar

Arabian

Sea

Western Ghats

Solāpur

Sāngli

Kolhāpur

Secunde

Hyderāb

Telan

Mahbūbnaga

N

Population

● National capital

○ below 50,000   ○ 50,000 to 100,000   ◉ 100,000 to 500,000   ◼ above 500

0 km   300

0 miles   300

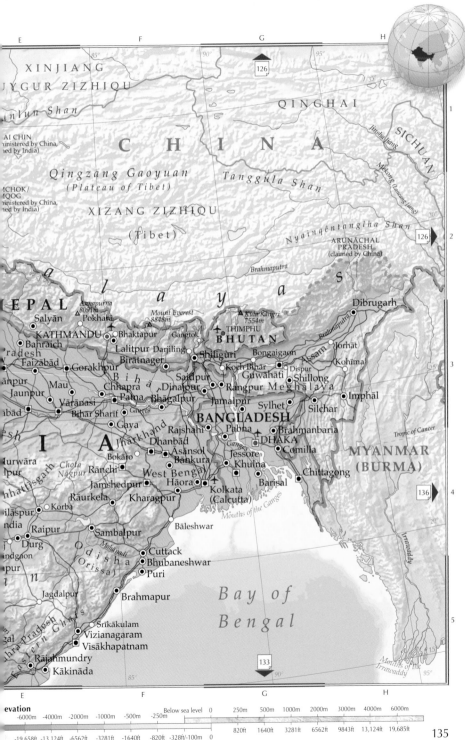

E   F   G   H

85°   90°   95°

126

XINJIANG
JYGUR ZIZHIQU

Nun Shan

QINGHAI

35

CHINA

1

AI CHIN
ministered by China,
ned by India)

Qingzang Gaoyuan
(Plateau of Tibet)

Tanggula Shan

Jinsha Jiang

SICHUAN

Mekong (Lancang Jiang)

CHOK/
IQOG
ninistered by China,
ned by India)

XIZANG ZIZHIQU
(Tibet)

Nyainqêntanglha Shan

30°

126

2

ARUNACHAL
PRADESH
(claimed by China)

Brahmaputra

a   l   a   y   a   s

EPAL

Annapurna
8091m
Mount Everest
8848m

Kula Kangri
7554m

Dibrugarh

Salyān   Pokhara
KATHMANDU
Bahrāich
Bhaktapur
Lalitpur Darjiling
THIMPHU
Gangtok
BHUTAN

Brahmaputra

Assam

Jorhat

Pradesh
Faizābād
Gorakhpur
Biratnagar
Shiliguri
Koch Bihār
Bongaigaon
Kohīma

ānpur
Jaunpur   Mau
Chhapra
Bihār
Saidpur
Dinajpur
Rangpur
Guwāhāti
Dispur
Shillong
MEGHĀLAYA

3

Vārānasi
Patna
Bhāgalpur
Jamalpur
Sylhet
Silchar
Imphāl

ābād
Bihar Sharif
Gaya
Ganges
BANGLADESH

I   A
Jharkhand
Rajshahi
Pabna
Brahmanbaria

urwāra
Chota
Nāgpur
Dhanbād
Asānsol
Bānkura
Ganges
DHAKA
Comilla

lpur
Bokāro
Ranchi
West Bengal
Jessore
Khulna

MYANMAR
(BURMA)

Tropic of Cancer

ilāspur
ndia
Jamshedpur
Hāora
Barisal
Chittagong

136

4

Raipur
Rāurkela
Kharagpur
Kolkata
(Calcutta)

Durg
Sambalpur
Bāleshwar
Mouths of the Ganges

20°

Irrawaddy

Jagdalpur
Cuttack
Bhubaneshwar
Puri

Bay of
Bengal

Brahmapur

15°

5

Srikākulam
Vizianagaram
Visākhapatnam

Mouths of the
Irrawaddy

133

Rajahmundry
Kākināda

85°   90°   95°

E   F   G   H

evation

-6000m   -4000m   -2000m   -1000m   -500m   -250m   Below sea level 0   250m   500m   1000m   2000m   3000m   4000m   6000m

-19,658ft   -13,124ft   -6562ft   -3281ft   -1640ft   -820ft   -328ft/-100m   0   820ft   1640ft   3281ft   6562ft   9843ft   13,124ft   19,685ft

Mainland Southeast Asia

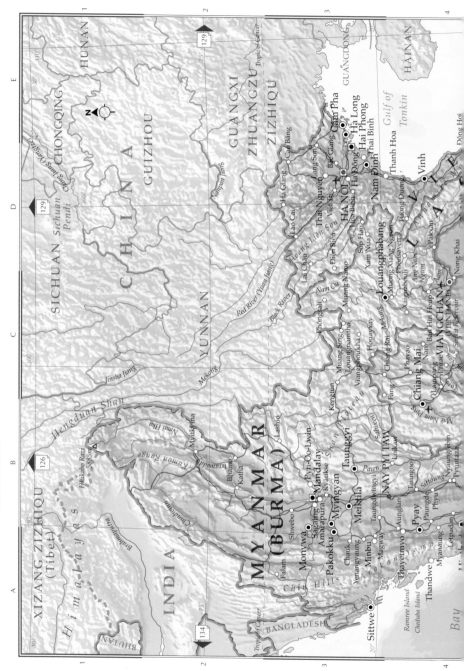

Population

● National capital

○ below 50,000　　○ 50,000 to 100,000　　◉ 100,000 to 500,000　　■ above 500,000

136

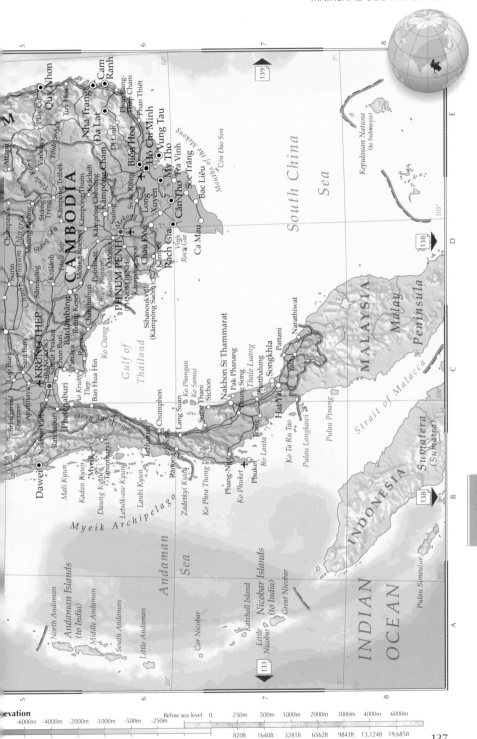

Quy Nhon
Plây Cu
Attapu
Tuy Hoa
Cam Ranh
Nha Trang
Da Lat
Di Linh
Phan Rang-Thap Cham
Phan Thiet
Biên Hoa
Hô Chi Minh
Vung Tau
My Tho
Soc Trăng
Tra Vinh
Bac Liêu
Ca Mau
Vinh
Rach Gia
Can Tho
Long Xuyên
Chau Doc
Kaiman
Kampong Spoe
Sihanoukville
(Kâmpông Saôm)

Mouths of the Mekong
Côn Dao Son

South China Sea

Kepulauan Natuna
(to Indonesia)

CAMBODIA
Stæng Trêng
Samráong
Kampong Trabek
Kâchéh
Kâmpông Thum
Kâmpông Cham
Kampong Chhnang
Svay Rieng
Kâmpông Chhnang
Suong
Pódôûhat
Kâmpông Chhnang
Prey Vêng
Kaôh Kong
Koun Pêam
PHNUM PENH
(PHNOM PENH)
Kâmpông
Kratié
Stæng Sên
Tônlé Sab
Môûng Roessei
Bâtdâmbâng
Chuor Phnum
Krâlanh
Kâmpông Trabek
Ba Mau

Phnum Dângrêk
Champasak
Kong
Veune
Virôchon
Tônlé Srepok
Kong

KRUNG THEP
(BANGKOK)
Samut Prakan
Chon Buri
Pattaya
Ayutthaya
Ratchaburi
Phetchaburi
Nakhon Pathom
Chachoengsao
Ban Hua Hin
Ao Krung Thep
Sra+ng
Ko Chang
Ko Krung
Chanthaburi
Rayong
Rêdat Kesei
Sâmrông
Sara Buri
Lop Buri
Surin
Chaiyaphum
Muang Khong
Khong
Prey Prakan
Srinagarind Reservoir

Gulf of
Thailand

MALAYSIA
Malay Peninsula

Nakhon Si Thammarat
Thung Song
Pak Phanang
Thale Luang
Phatthalung
Songkhla
Hat Yai
Yala
Pattani
Narathiwat

Chumphon
Lang Suan
Ko Phangan
Ko Samui
Surat Thani
Sichon
Phang-Nga
Phuket
Ko Phuket
Ko Lanta
Ko Ta Ru Tao
Krabi
Trang
Ko Phra Thong
Zadetkyi Kyun

Pulau Pinang
Pulau Langkawi

Strait of Malacca

Dawei
Mali Kyun
Kadan Kyun
Myeik
Tanintharyi
Daung Kyun
Letsôk-aw Kyun
Lanbi Kyun
Ranong
Kra Buri
Ko Chang

Myeik Archipelago

Andaman
Sea

Andaman Islands
(to India)
North Andaman
Middle Andaman
South Andaman
Little Andaman

Nicobar Islands
(to India)
Car Nicobar
Katchall Island
Little Nicobar
Great Nicobar

INDONESIA
Sumatera
(Sumatra)
Pulau Simeulue

INDIAN
OCEAN

139
138
138
133

evation
-6000m -4000m -2000m -1000m -500m -250m Below sea level 0 250m 500m 1000m 2000m 3000m 4000m 6000m
-19,658ft -13,124ft -6562ft -3281ft -1640ft -820ft -328ft/-100m 0 820ft 1640ft 3281ft 6562ft 9843ft 13,124ft 19,685ft

137

Maritime Southeast Asia

SINGAPORE

0 km   10
0 miles   10

MALAYSIA

Johore Strait

Causeway

Lim Chu
Kang
Choa Chu
Kang
Jurang
Industrial
Estate
Selat Pandan
Pulau Sudong
Pulau Pawai

Bukit Panjang New Town
Hougang
Bukit Timah 176m
Queenstown
City
Telok Blangah
Sentosa

Pulau
Ubin
Pulau
Tekong
Changi
Bedok
New Town

Urban areas
Open areas
Nature reserves

Strait of Singapore

MYANMAR
(BURMA)

LAOS

VIETNAM

THAILAND

Gulf of
Tonkin

Hainan Dao
(to China)

PARACEL ISLANDS
(disputed by China, Taiwan
and Vietnam)

South Ch

Sea

CAMBODIA

Mekong

Andaman
Sea

Nicobar Islands
(to India)

Isthmus of Kra

Gulf of
Thailand

Mouths of
the Mekong

SPRATLY ISLANDS
(disputed by China, Malaysia,
Philippines, Taiwan and Vietnam)

Banda Aceh   Sigli

Meulaboh

Langsa

Medan

Tebingtinggi

Pematangsiantar

Pulau Simeulue

Kepulauan
Banyak

Danau
Toba

Sibolga

Equator

Pulau Nias

Strait of Malacca

Pulau
Pinang

George
Town

Butterworth

Taiping

Ipoh

Klang

Kota Bharu

Kuala Terengganu

Dungun

Cukai

Kuantan

KUALA LUMPUR

PUTRAJAYA

Melaka

Muar

Batu Pahat

Keluang

Johor Bahru

SINGAPORE

Kepulauan
Natuna

Singkawang

Pontianak

Gunun

Kota Kinabal

BANDAR SERI
BEGAWAN

BRUNEI
Miri

Bintulu

Sarawak

Sibu

Sri Aman

Batang Rajang

Kuching

Sidas

Sungai Kapuas

Selat Serasan

M A L A Y S I A

Borneo

Sama

Balikpa

Banjarn

Pekanbaru

Solok

Rengat

Kualatungkal

Padang

Pulau Siberut

Kepulauan
Mentawai

Sungaipenuh

Jambi

Pangkalpinang

Bengkulu

Lahat

Palembang

Kepulauan
Lingga

Selat Karimata

Kalimantan

Sampit

Sungai Barito

Pegunungan Barisan

Batang Hari

Sumatera
(Sumatra)

Pulau
Belitung

Bangka

I   N   D

INDIAN

OCEAN

Kotabumi

Bandar Lampung

Serang

Sukabumi

Bogor

JAKARTA

Bandung

Cirebon

Tegal

Tasikmalaya

Cilacap

Magelang

Yogyakarta

Surakarta

Kediri

Pekalongan

Semarang

Kudus

Madiun

Malang

Java Sea

Pulau
Madura

Surabaya

Probolin

Jember

Bali

Selat Sunda

Jawa
(Java)

MALAYSIA'S TWO CAPITALS

KUALA LUMPUR   - Capital
PUTRAJAYA   - Administrative capital

0 km   200
0 miles   200

Population   ● National capital

○ below 50,000   ○ 50,000 to 100,000   ◉ 100,000 to 500,000   ◼ above 500

The Indian Ocean

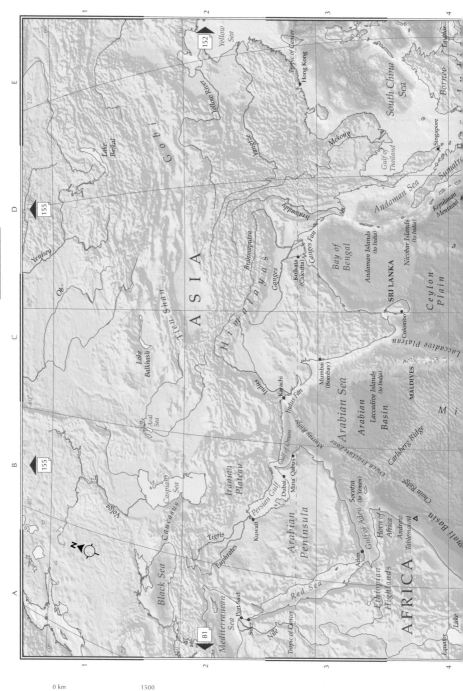

152

155

155

81

Yellow Sea

Yellow River

Tropic of Cancer

Hong Kong

Yangtze

Gobi

South China Sea

Mekong

Borneo

Equator

Lake Baikal

Gulf of Thailand

Singapore

Sumatra

Yenisey

Ob'

Tien Shan

A S I A

Brahmaputra

Irrawaddy

Andaman Sea

Kepulauan Mentawai

H i m a l a y a s

Ganges

Ganges Fan

Kolkata (Calcutta)

Bay of Bengal

Andaman Islands (to India)

Nicobar Islands (to India)

SRI LANKA

Ceylon Plain

Lake Balkhash

Indus

Karachi

Indus Fan

Mumbai (Bombay)

Colombo

Laccadive Plateau

Aral Sea

Arabian Sea

Laccadive Islands (to India)

MALDIVES

Arabian Basin

M i

Carlsberg Ridge

Owen Fracture Zone

Murray Ridge

Iranian Plateau

Caspian Sea

Volga

Persian Gulf

Dubai

Mina Qabus

Gulf of Oman

Socotra (to Yemen)

...ali Basin

Chain Ridge

Caucasus

Kuwait

Arabian Peninsula

Gulf of Aden

Horn of Africa

Andrew Tablemount

Tigris

Black Sea

Euphrates

Aden

Ethiopian Highlands

AFRICA

Mediterranean Sea

Port Said

Suez

Red Sea

Nile

Tropic of Cancer

Equator

Lake

N

0 km		1500	
0 miles		1500	

● Major port

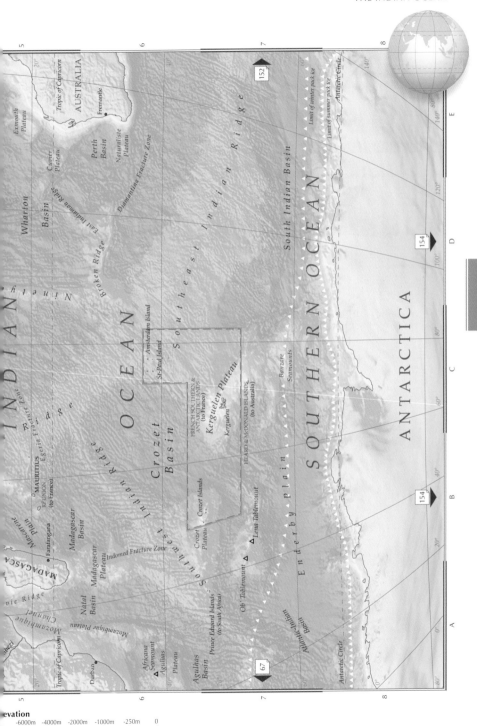

Elevation

-6000m	-4000m	-2000m	-1000m	-250m	0
-19,658ft	-13,124ft	-6562ft	-3281ft	-820ft	0

INDIAN OCEAN

SOUTHERN OCEAN

ANTARCTICA

AUSTRALIA

MADAGASCAR

Tropic of Capricorn

Exmouth Plateau

Fremantle

Perth Basin

Cuvier Plateau

Naturaliste Plateau

Diamantina Fracture Zone

East Indiaman Ridge

Wharton Basin

Broken Ridge

Ninetyeast Ridge

Southeast Indian Ridge

South Indian Basin

Limit of winter pack ice

Limit of summer pack ice

Antarctic Circle

Amsterdam Island

St.Paul Island

Egeria Fracture Zone

Rodrigues Fracture Zone

Central Indian Ridge

Crozet Basin

FRENCH SOUTHERN & ANTARCTIC LANDS (to France)

Kerguelen Plateau

Kerguelen

HEARD & McDONALD ISLANDS (to Australia)

Banzare Seamounts

Crozet Islands

Crozet Plateau

Southwest Indian Ridge

Lena Tablemount

Ob Tablemount

Enderby Plain

Atlantic-Indian Basin

Antarctic Circle

MAURITIUS

RÉUNION (to France)

Mascarene Plain

Farquhar

Madagascar Basin

Madagascar Plateau

Mozambique Plateau

Natal Basin

Prince Edward Islands (to South Africa)

Agulhas Plateau

Agulhas Basin

Africana Seamount

Agulhas

Durban

Tropic of Capricorn

Mozambique Channel

...vie Ridge

Zambezi

Indomed Fracture Zone

152

154

154

67

Australasia & Oceania

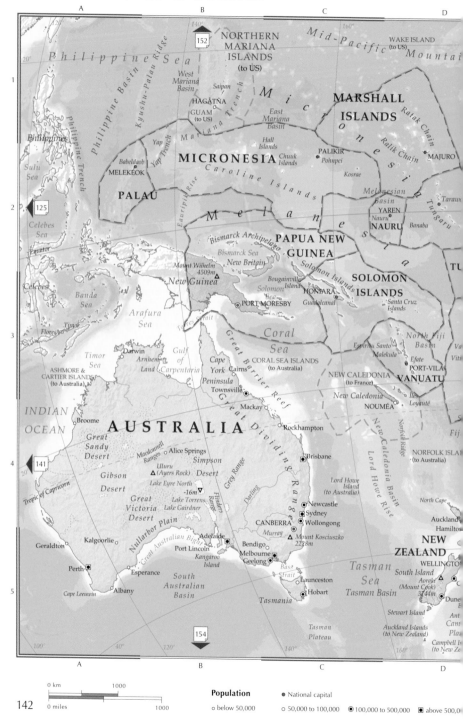

Population

● National capital

o below 50,000   o 50,000 to 100,000   ⊙ 100,000 to 500,000   ▣ above 500,0

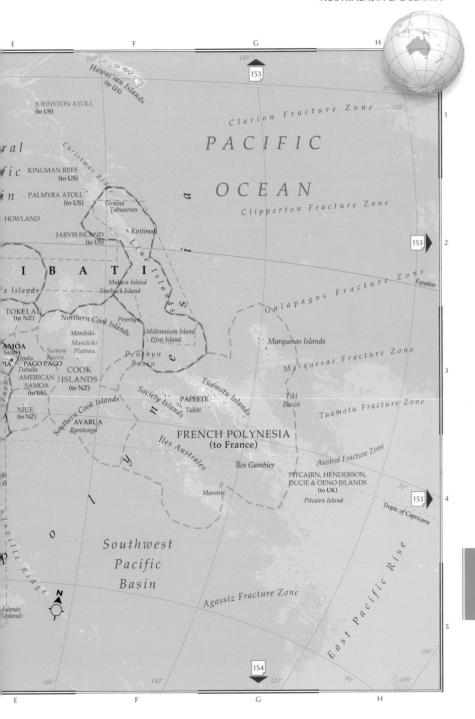

E   F   G   H

160°   140°   120°   20°

153

Hawai'ian Islands
(to US)

JOHNSTON ATOLL
(to US)

Clarion Fracture Zone

PACIFIC

ral

KINGMAN REEF
(to US)

fic

PALMYRA ATOLL
(to US)

in

HOWLAND

Teraina
Tabuaeran

JARVIS ISLAND
(to US)

Kiritimati

OCEAN

Clipperton Fracture Zone

153

2

KIRIBATI

x Islands

Malden Island
Starbuck Island

Equator

Galapagos Fracture Zone

TOKELAU
(to NZ)

Northern Cook Islands

Penrhyn

Millennium Island
Flint Island

Marquesas Islands

Marquesas Fracture Zone

3

Manihiki
Manihiki
Plateau

SAMOA
Savai'i
APIA   PAGO PAGO
Tutuila

Samoa
Basin

Upolu

Penrhyn
Basin

Tiki
Basin

AMERICAN
SAMOA
(to US)

COOK
ISLANDS
(to NZ)

Tuamotu Islands

Society Islands

PAPEETE
Tahiti

Tuamotu Fracture Zone

NIUE
(to NZ)

Southern Cook Islands

AVARUA
Rarotonga

FRENCH POLYNESIA
(to France)

Îles Australes

Austral Fracture Zone

20°

153

4

Îles Gambier

PITCAIRN, HENDERSON,
DUCIE & OENO ISLANDS
(to UK)

Marotiri

Pitcairn Island

Tropic of Capricorn

Southwest
Pacific
Basin

East Pacific Rise

Ridge

N

Islands
Zealand)

Agassiz Fracture Zone

5

100°

160°   140°   120°   40°   100°

154

E   F   G   H

The Southwest Pacific

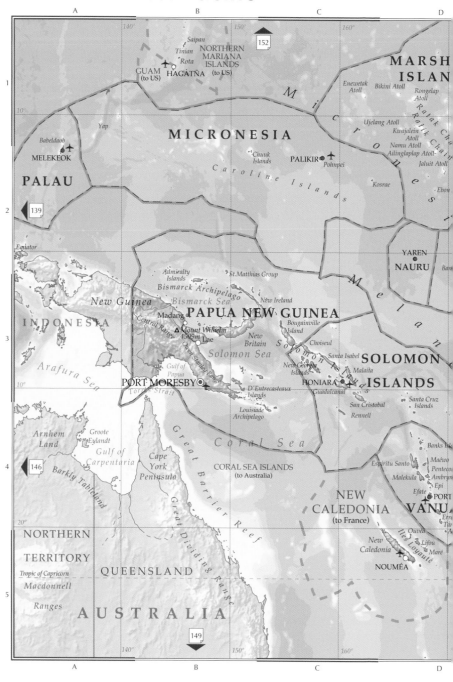

MARSH
ISLAN

Saipan
Tinian NORTHERN
Rota   MARIANA
        ISLANDS
GUAM   HAGATÑA (to US)
(to US)

152

Enewetak Bikini Atoll Rongelap
Atoll   Atoll

Ujelang Atoll
Kwajalein
Atoll
Namu Atoll
Ailinglaplap Atoll
Jaluit Atoll

Yap

Babeldaob
MELEKEOK

PALAU

MICRONESIA

Chuuk
Islands PALIKIR Pohnpei

Caroline Islands

Kosrae

Ebon

139

Equator

YAREN
NAURU

Admiralty
Islands   St.Matthias Group

Bismarck Archipelago

New Guinea   Bismarck Sea   New Ireland

Madang PAPUA NEW GUINEA

INDONESIA

Central Range △ Mount Wilhelm
                 4509m Lae

New
Britain   Solomon Sea

Bougainville
Island

Choiseul

Santa Isabel

SOLOMON

New Georgia
Islands   HONIARA

Malaita

ISLANDS

Arafura Sea

Gulf of
Papua

PORT MORESBY

Torres Strait

D'Entrecasteaux
Islands   Guadalcanal

Rennell

San Cristobal   Santa Cruz
                 Islands

Louisiade
Archipelago

Arnhem   Groote
Land   Eylandt

Gulf of
Carpentaria   Cape
              York
              Peninsula

Barkly Tableland

146

Coral Sea

Great Barrier Reef

CORAL SEA ISLANDS
(to Australia)

Banks Isl.

Espiritu Santo   Maéwo
                 Pentecos
Malekula   Ambryn
                 Epi
         Efate   PORT

NEW
CALEDONIA   VANU
(to France)

Etre
Tan

NORTHERN

Great Dividing Range

New
Caledonia   Ouvéa
             Lifou
Îles Loyauté Maré

NOUMÉA

TERRITORY

Tropic of Capricorn

Macdonnell

Ranges

QUEENSLAND

AUSTRALIA

149

0 km	750
0 miles	750

Population

○ below 50,000
○ 50,000 to 100,000
◉ 100,000 to 500,000
■ above 50

● National capital

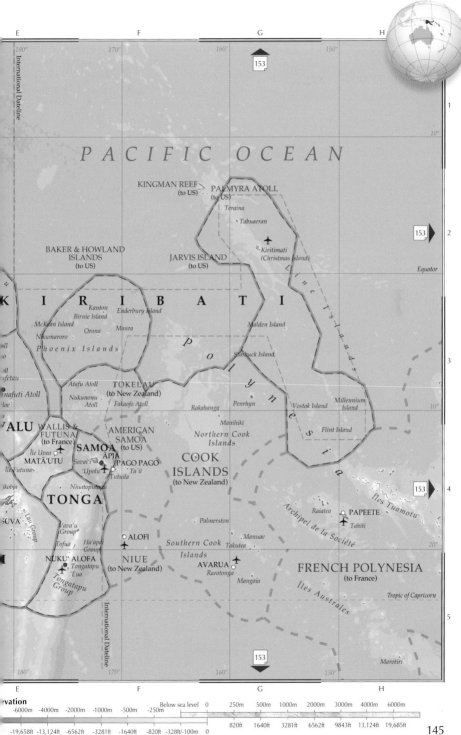

PACIFIC OCEAN

KINGMAN REEF (to US)   PALMYRA ATOLL (to US)
I. Teraina
· Tabuaeran

BAKER & HOWLAND
ISLANDS
(to US)

JARVIS ISLAND
(to US)

✈ Kiritimati
(Christmas Island)

Equator

K I R I B A T I

Kanton
Birnie Island   Enderbury Island
McKean Island
Nikumaroro   Orona   Manra   Malden Island

P h o e n i x   I s l a n d s

L i n e   I s l a n d s

P o l y n e s i a

Starbuck Island

Atafu Atoll   TOKELAU
Nukunonu   (to New Zealand)   Vostok Island   Millennium
Atoll   Fakaofo Atoll   Rakahanga   Penrhyn   Island

nafuti Atoll   Manihiki   Flint Island

ALU   WALLIS &   AMERICAN   Northern Cook
FUTUNA   SAMOA   Islands
(to France)   (to US)
Île Uvea   SAMOA
MATA'UTU   ✈APIA   ✈PAGO PAGO   COOK
Île Futuna   Savai'i   Ta'ū   ISLANDS
Upolu   Tutuila   (to New Zealand)

SUVA   Niuatoputapu   Îles Tuamotu

TONGA   Raiatea   PAPEETE
Vava'u   Palmerston   ✈ Tahiti
Group   Archipel de la Société
Tofua   · Manuae
Ha'apai   Takutea   Southern Cook
Group   Islands   Manae
ALOFI   AVARUA   FRENCH POLYNESIA
NUKU'ALOFA   Tongatapu   NIUE   Rarotonga   (to France)
'Eua   (to New Zealand)   Mangaia
Tongatapu   Îles Australes
Group   Tropic of Capricorn

Marotiri

Western Australia

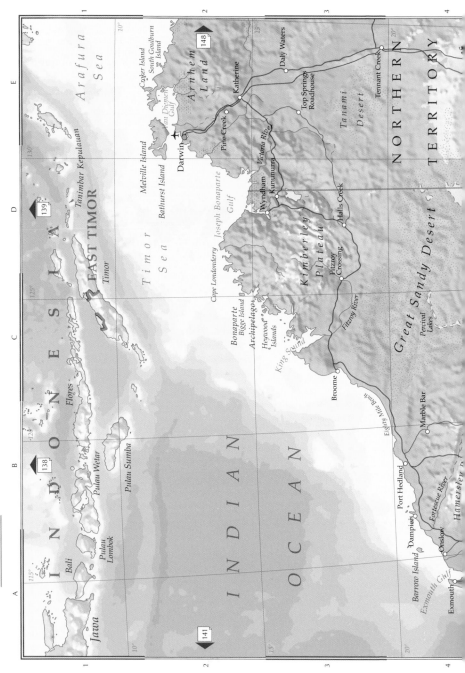

Arafura Sea

Coper Island
South Goulburn Island

148

Arnhem Land

Katherine

Daly Waters

Tennant Creek

Top Springs Roadhouse

Tanami Desert

NORTHERN   TERRITORY

Tanimbar Kepulauan

Van Diemen Gulf

Darwin

Pine Creek

Victoria River

Wyndham
Kununurra

Halls Creek

Kimberley Plateau

EAST TIMOR

139

Timor

Timor Sea

Melville Island

Bathurst Island

Joseph Bonaparte Gulf

Cape Londonderry

Fitzroy Crossing

Great Sandy Desert

Percival Lakes

INDONESIA

Flores

Bonaparte Archipelago
Bigge Island

Heywood Islands

King Sound

Fitzroy River

Fitzroy River

Pulau Wetar

138

Pulau Sumba

Broome

Marble Bar

Eighty Mile Beach

Port Hedland

Hamersley

INDIAN

Pulau Lombok

Bali

Fortescue River

OCEAN

Dampier
Onslow

Barrow Island

Jawa

141

Exmouth Gulf

Exmouth

0 km   300

146

0 miles   300

Population

○ below 50,000   ○ 50,000 to 100,000   ◉ 100,000 to 500,000   ■ above 500

● Internal administrative capital

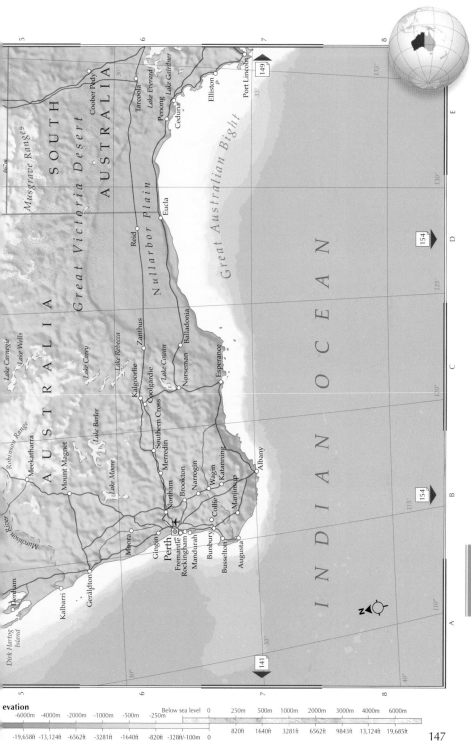

SOUTH AUSTRALIA

AUSTRALIA

Musgrave Ranges

Great Victoria Desert

Nullarbor Plain

Great Australian Bight

INDIAN OCEAN

Robinson Range

Murchison River

Dirk Hartog Island

Denham

Kalbarri

Geraldton

Lake Moore

Lake Barlee

Lake Wells

Lake Carnegie

Lake Carey

Lake Rebecca

Meekatharra

Mount Magnet

Moora

Gingin

Perth

Fremantle

Rockingham

Mandurah

Bunbury

Busselton

Augusta

Collie

Manjimup

Katanning

Wagin

Narrogin

Brookton

Northam

Merredin

Southern Cross

Kalgoorlie

Coolgardie

Lake Cowan

Zanthus

Norseman

Balladonia

Esperance

Albany

Eucla

Reid

Coober Pedy

Tarcoola

Penong

Ceduna

Elliston

Port Lincoln

Lake Gairdner

Lake Eterand

N

149

154

154

141

147

elevation

-6000m	-4000m	-2000m	-1000m	-500m	-250m	Below sea level	0	250m	500m	1000m	2000m	3000m	4000m	6000m
-19,658ft	-13,124ft	-6562ft	-3281ft	-1640ft	-820ft	-328ft/-100m	0	820ft	1640ft	3281ft	6562ft	9843ft	13,124ft	19,685ft

Eastern Australia

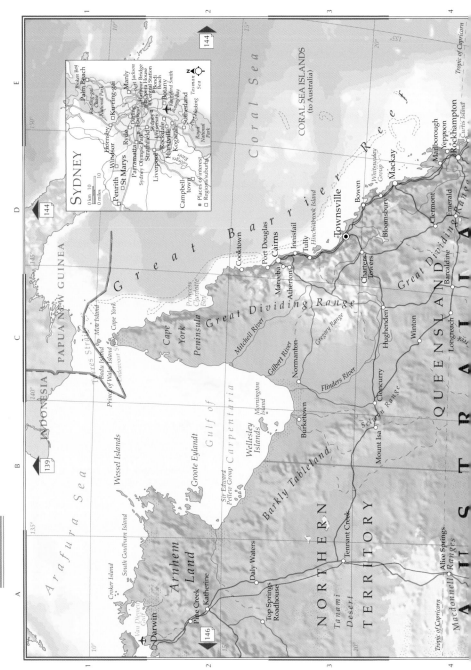

SYDNEY

Broken Bay
Palm Beach
Kings-gai
Chatswood
Manly
Hornsby
Darling Harbour Bridge
Ryde
Port Jackson
Windsor
Parramatta
Central Station
Penrith
Sydney Olympic Park
University
St Marys
Strathfield
Rockdale
Liverpool
Hurstville
Kogarah
Campbell
town
Bondi
Beach
Botany
Cronulla
Bondi Bay
Sutherland
Royal
National Park

Places of interest
Regions/suburbs

0 km   10
0 miles   10

Coral Sea

CORAL SEA ISLANDS
(to Australia)

Great Barrier Reef

PAPUA NEW GUINEA

INDONESIA

Arafura Sea

Wessel Islands

Groote Eylandt

Gulf of Carpentaria

Cape York Peninsula

Great Dividing Range

Cooktown
Port Douglas
Cairns
Innisfail
Mareeba
Atherton
Tully
Hinchinbrook Island

Townsville
Bowen
Whitsunday Group
Mackay

Bloomsbury
Clermont
Barcaldine
Emerald
Rockhampton
Yeppoon
Marlborough
Curtis Island

QUEENSLAND

Charters
Towers
Hughenden
Winton
Longreach

Mitchell River
Gilbert River
Flinders River
Normanton
Burketown

Mornington
Island
Wellesley
Islands

Barkly Tableland

Mount Isa
Cloncurry

Selwyn Range
Gregory Range

NORTHERN
TERRITORY

Tennant Creek
Daly Waters
Top Springs
Roadhouse
Katherine
Pine Creek
Darwin

Tanami Desert

Arnhem Land

Alice Springs
Macdonnell Ranges

Tropic of Capricorn

0 km   300
0 miles   300

Population

● National capital   ○ Internal administrative capital

○ below 50,000   ○ 50,000 to 100,000   ◉ 100,000 to 500,000   ◼ above 500,

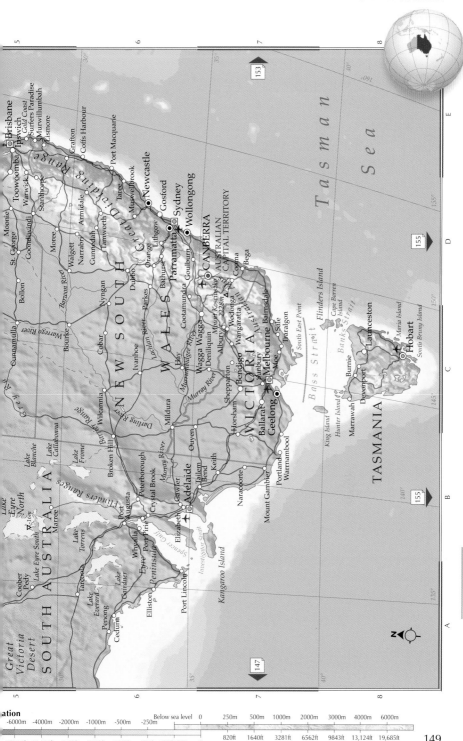

Eastern Australia map labels

Brisbane
Ipswich
Gold Coast
Surfers Paradise
Murwillumbah
Lismore
Toowoomba
Warwick
Stanthorpe
Grafton
Coffs Harbour
Port Macquarie
Great Dividing Range
Moonie
St. George
Goondiwindi
Moree
Walgett
Narrabri
Gunnedah
Armidale
Tamworth
Taree
Muswellbrook
Newcastle
Gosford
Sydney
Parramatta
Wollongong
Bollon
Cunnamulla
Bourke
Nyngan
Cobar
Ivanhoe
Dubbo
Orange
Lithgow
Bathurst
Parkes
Goulburn
CANBERRA
AUSTRALIAN CAPITAL TERRITORY
Cooma
Bega
South East Point
NEW SOUTH WALES
Warrego River
Barwon River
Darling River
Lachlan River
Murrumbidgee River
Hay
Cootamundra
Wagga Wagga
Deniliquin
Albury
Wodonga
Wangaratta
Mount Kosciuszko 2228m
Bairnsdale
Sale
Traralgon
Moe
VICTORIA
Grey Range
Bulloo River
Wilcannia
Broken Hill
Barrier Ranges
Mildura
Ouyen
Swan Hill
Shepparton
Bendigo
Sunbury
MELBOURNE
Geelong
Ballarat
Horsham
Naracoorte
Keith
Tailem Bend
Murray River
Portland
Warrnambool
Mount Gambier
SOUTH AUSTRALIA
Coober Pedy
Tarcoola
Ceduna
Penong
Elliston
Port Lincoln
Whyalla
Port Pirie
Port Augusta
Peterborough
Crystal Brook
Gawler
ADELAIDE
Elizabeth
Marree
Flinders Ranges
Lake Eyre North
Lake Eyre South
Lake Blanche
Lake Callabonna
Lake Frome
Lake Torrens
Lake Everard
Lake Gairdner
Eyre Peninsula
Spencer Gulf
Investigator Strait
Kangaroo Island
Great Victoria Desert
TASMANIA
Burnie
Devonport
Launceston
Hobart
Maria Island
South Bruny Island
Marrawah
Hunter Island
King Island
Flinders Island
Cape Barren Island
Banks Strait
Bass Strait
Tasman Sea

Elevation

						Below sea level	0	250m	500m	1000m	2000m	3000m	4000m	6000m
-6000m	-4000m	-2000m	-1000m	-500m	-250m									
-19,658ft	-13,124ft	-6562ft	-3281ft	-1640ft	-820ft	-328ft/-100m	0	820ft	1640ft	3281ft	6562ft	9843ft	13,124ft	19,685ft

New Zealand

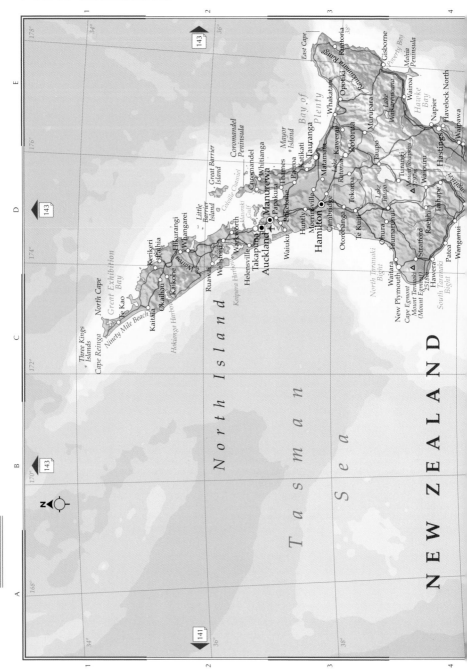

Population

- ● National capital
- ○ below 50,000
- ○ 50,000 to 100,000
- ◉ 100,000 to 500,000
- ■ above 500

The Pacific Ocean

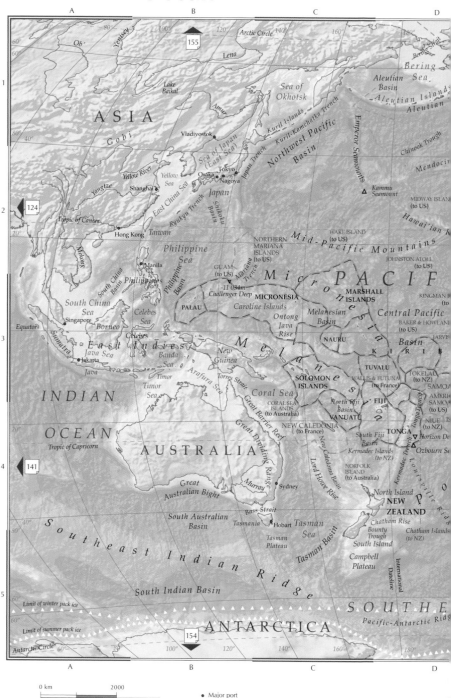

155

124

141

154

ASIA

Ob'
Yenisey
Lena
Arctic Circle
Bering Strait

Lake Baikal
Amur
Sea of Okhotsk
Bering Sea
Aleutian Basin
Aleutian Islands
Aleutian

Gobi
Vladivostok
Kuril Islands
Kuril-Kamchatka Trench
Kuril Trench
Northwest Pacific Basin
Emperor Seamounts
Chinook Trough
Mendocin

Sea of Japan (East Sea)
Tokyo
Osaka Nagoya
Japan Trench
△ Kammu Seamount
MIDWAY ISLAND (to US)

Yellow River
Yellow Sea
Shanghai
Japan
Ryukyu Trench
Shikoku Basin
Hawai'ian Ric

Yangtze
Tropic of Cancer
Hong Kong Taiwan
East China Sea
WAKE ISLAND (to US)
Mid-Pacific Mountains

Mekong
Philippine Sea
NORTHERN MARIANA ISLANDS (to US)
Micronesia
PACIFI
JOHNSTON ATOLL (to US)

Manila
GUAM (to US)
Mariana Trench
Central Pacific

South China Basin
Philippines
Philippine Basin
-11 034m
Challenger Deep
MICRONESIA
MARSHALL ISLANDS
KINGMAN R (to

South China Sea
Celebes Sea
PALAU
Caroline Islands
Melanesian Basin
BAKER & HOWLAN (to US)

Singapore
Borneo
Ontong Java Rise
NAURU
Central Pacific Basin
JARVIS

Equator
East Indies
Celebes
New Guinea
Melan
KIRIB

Sumatra
Java Sea
Banda Sea
TUVALU
TOKELA (to NZ)

Jakarta
Java
Timor
Arafura Sea
Torres Strait
SOLOMON ISLANDS
WALLIS & FUTUNA (to France)
AMERIC SAMO (to US)

INDIAN
Timor Sea
Coral Sea
e
s
FIJI
NIUE (to NZ)

CORAL SEA ISLANDS (to Australia)
North Fiji Basin
i
AMERIC SAMO

OCEAN
Great Barrier Reef
NEW CALEDONIA (to France)
VANUATU
a
TONGA
Horizon De

Tropic of Capricorn
Great Dividing Range
South Fiji Basin
New Caledonia Basin
Kermadec Islands (to NZ)
Kermadec Trench
Tonga Trench
Louisville Rid
Ozbourn S
P O

AUSTRALIA
NORFOLK ISLAND (to Australia)

Great Australian Bight
Murray
Sydney
Lord Howe Rise
North Island
NEW ZEALAND

South Australian Basin
Bass Strait
Tasmania Hobart
Tasman Sea
Chatham Rise
Bounty Trough
Chatham Islands (to NZ)

South
Tasman Plateau
Tasman Basin
South Island
International Dateline

east
Campbell Plateau

Indian
Ridge

South Indian Basin
SOUTHE

Limit of winter pack ice
Pacific-Antarctic Ridg

Limit of summer pack ice

ANTARCTICA

Antarctic Circle

0 km 2000
0 miles 2000
• Major port

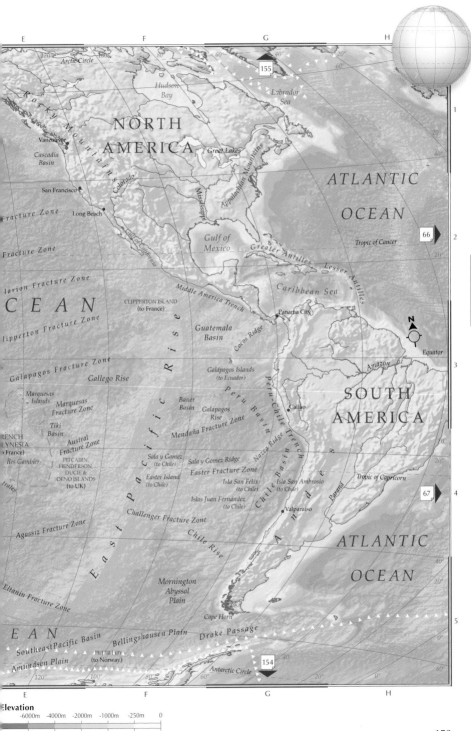

NORTH AMERICA

Arctic Circle

Hudson Bay

Labrador Sea

Rocky Mountains

Vancouver

Cascadia Basin

San Francisco

Colorado

Great Lakes

Appalachian Mountains

ATLANTIC OCEAN

Fracture Zone

Long Beach

Fracture Zone

Gulf of California

Mississippi

Gulf of Mexico

Greater Antilles

Tropic of Cancer

66

Lesser Antilles

larion Fracture Zone

OCEAN

lipperton Fracture Zone

CLIPPERTON ISLAND (to France)

Middle America Trench

Caribbean Sea

Guatemala Basin

Panama City

Cocos Ridge

N

Equator

Galapagos Fracture Zone

Gallego Rise

Galápagos Islands (to Ecuador)

Amazon

SOUTH AMERICA

Marquesas Islands

Marquesas Fracture Zone

Bauer Basin

Galapagos Rise

Callao

Peru Basin

Peru-Chile Trench

Tiki Basin

RENCH LYNESIA (France)

Austral Fracture Zone

Mendaña Fracture Zone

Nazca Ridge

Iles Gambier

PITCAIRN, HENDERSON, DUCIE & OENO ISLANDS (to UK)

Sala y Gomez (to Chile)

Sala y Gomez Ridge

Easter Island (to Chile)

Easter Fracture Zone

Isla San Félix (to Chile)

Isla San Ambrosio (to Chile)

Tropic of Capricorn

67

trales

Islas Juan Fernández (to Chile)

Valparaíso

Chile Basin

Andes

Paraná

East Pacific Rise

Challenger Fracture Zone

Agassiz Fracture Zone

Chile Rise

ATLANTIC OCEAN

Eltanin Fracture Zone

Mornington Abyssal Plain

Cape Horn

EAN

Southeast Pacific Basin

Bellingshausen Plain

Drake Passage

154

Amundsen Plain

PETER I ØY (to Norway)

Antarctic Circle

Elevation

-6000m	-4000m	-2000m	-1000m	-250m	0
-19,658ft	-13,124ft	-6562ft	-3281ft	-820ft	0

153

Antarctica

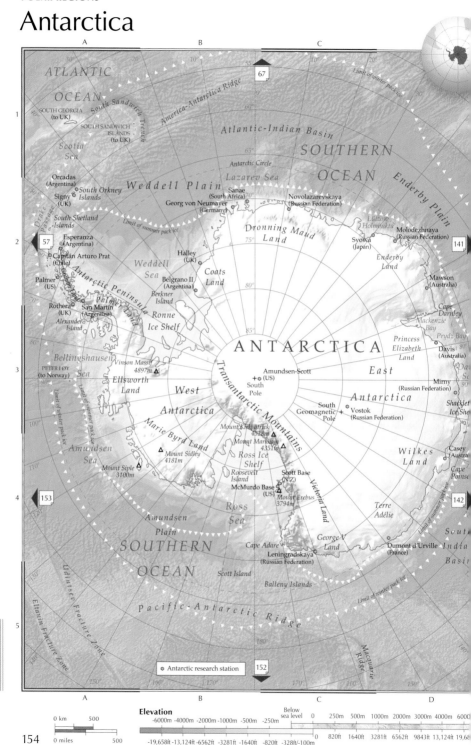

ATLANTIC
OCEAN

SOUTH GEORGIA
(to UK)

SOUTH SANDWICH
ISLANDS
(to UK)

*Scotia
Sea*

South Sandwich Trench

America-Antarctica Ridge

67

Limit of winter pack ice

Atlantic-Indian Basin

SOUTHERN

Antarctic Circle

OCEAN

Enderby Plain

Orcadas
(Argentina)

Lazarev Sea

Signy
(UK)

South Orkney
Islands

Weddell Plain

Sanae
(South Africa)

Novolazarevskaya
(Russian Federation)

Georg von Neumayer
(Germany)

Lützow
Holmbukta

Molodezhnaya
(Russian Federation)

South Shetland
Islands

Limit of summer pack ice

57

Esperanza
(Argentina)

Halley
(UK)

Dronning Maud
Land

Syowa
(Japan)

Enderby
Land

141

Capitán Arturo Prat
(Chile)

Weddell
Sea

Coats
Land

Mawson
(Australia)

Palmer
(US)

Belgrano II
(Argentina)

Cape
Darnley

Rothera
(UK)

San Martin
(Argentina)

Berkner
Island

Mackenzie
Bay

Prydz Bay

Alexander
Island

Ronne
Ice Shelf

ANTARCTICA

Princess
Elizabeth
Land

Davis
(Australia)

Bellingshausen

Vinson Massif
4897m

Amundsen-Scott
(US)

East

Sea

PETER I ØY
(to Norway)

57

Ellsworth
Land

South
Pole

Antarctica

Mirny
(Russian Federation)

West

Transantarctic Mountains

Antarctica

South
Geomagnetic
Pole

Vostok
(Russian Federation)

Shacklet
Ice She

Amundsen
Sea

Marie Byrd Land

Mount Kirkpatrick
4528m

Mount Markham
4351m

Wilkes
Land

Casey
Austra

Mount Sidley
4181m

Ross Ice
Shelf

Cape
Poinse

Mount Siple
3100m

Roosevelt
Island

Scott Base
(N.Z.)

Amundsen
Plain

153

McMurdo Base
(US)

Mount Erebus
3794m

Terre
Adélie

142

Ross
Sea

Victoria Land

SOUTHERN

Cape Adare

George V
Land

Dumont d'Urville
(France)

India

OCEAN

Leningradskaya
(Russian Federation)

Basir

Scott Island

Udintsev Fracture Zone

Balleny Islands

Limit of winter pack ice

Macquarie
Ridge

Eltanin Fracture Zone

P a c i f i c - A n t a r c t i c   R i d g e

○ Antarctic research station

152

0 km   500

0 miles   500

Elevation

-6000m -4000m -2000m -1000m -500m -250m

Below
sea level   0   250m 500m 1000m 2000m 3000m 4000m 6000

-19,658ft -13,124ft -6562ft -3281ft -1640ft -820ft -328ft/-100m

0   820ft 1640ft 3281ft 6562ft 9843ft 13,124ft 19,68

A   B   C   D

rctic Ocean

ALASKA
(to US)

Saint Lawrence
Island

Provideniya

Bering
Sea

Arctic Circle

RUSSIAN

152

Norton
Sound

Bering Strait

Chukchi
Sea

Ostrov
Vrangelya

East
Siberian
Sea

Limit of summer pack ice

uktoyaktuk

Limit of summer pack ice

Northwind
Plain

Chukchi
Plain

Chukchi
Plateau

Novosibirskiye
Ostrova

Limit of permanent ice cap

F E D E R A T I O N

113

Beaufort
Sea

NORTH AMERICA

Canada

Basin

Wrangel
Plain

Laptev
Sea

ictoria
sland

A R C T I C

Mendeleyev Ridge

Severnaya
Zemlya

CANADA

Queen
Elizabeth
Islands

North Geomagnetic Pole

Alpha Cordillera

Makarov
Basin

Lomonosov Ridge

+
North
Pole

Amundsen Basin

O C E A N

Gakkel Ridge

Nansen Basin

Svyataya Anna
Trough

Franz
Josef Land

Kara
Sea

Dikson

Ostrov
Belyy

36

38

affin
land

Baffin
Bay

Ellesmere Island

Lincoln
Sea

Kap Morris
Jesup

Knud Rasmussen
Land

Wandel
Sea

SVALBARD
(to Norway)

Novaya
Zemlya

East Novaya Zemlya Trough

112

GREENLAND
(to Denmark)

Kong Frederik VIII
Land

Greenland
Sea

Spitsbergen

Longyearbyen

Bjørnøya
(to Norway)

Limit of winter pack ice

Barents
Sea

Ostrov
Kotel'nyy

Chëshskaya Guba

JAN MAYEN
(to Norway)

Limit of summer pack ice

Mohns Ridge

North Cape

Murmansk

Kola
Peninsula

White Sea

Archangel

imit of winter pack ice

Denmark Strait

Iceland
Plateau

Norwegian
Sea

NORWAY

SWEDEN

FINLAND

66

E U R O P E

0 km   500

0 miles   500

● Major port

155

Overseas territories & dependencies

Despite the rapid process of global decolonization since the Second World War, around 8 million people in more than 50 territories around the world continue to live under the protection of France, Australia, the Netherlands, Denmark, Norway, New Zealand, the UK, or the USA. These remnants of former colonial empires may have persisted for economic, strategic or political reasons and are administered i a variety of ways.

AUSTRALIA

Australia's overseas territories have not been an issue since Papua New Guinea became independent in 1975. Consequently there is no overriding policy toward them. Norfolk Island is inhabited by descendants of the H.M.S Bounty mutineers and more recent Australian migrants.

Ashmore & Cartier Islands
Indian Ocean
Status: External territory
Claimed: 1931
Capital: Not applicable
Population: None
Area: 2 sq miles
(5.2 sq km)

Christmas Island
Indian Ocean
Status: External territory
Claimed: 1958
Capital: The Settlement
Population: 1530
Area: 52 sq miles
(135 sq km)

Cocos Islands
Indian Ocean
Status: External territory
Claimed: 1955
Capital: West Island
Population: 596
Area: 5.5 sq miles
(14 sq km)

Coral Sea Islands
South Pacific
Status: External territory
Claimed: 1969
Capital: None
Population: 8 (meteorologists)
Area: Less than 1.2 sq miles
(3 sq km)

Heard & McDonald Is.
Indian Ocean
Status: External territory
Claimed: 1947
Capital: Not applicable
Population: None
Area: 161 sq miles
(417 sq km)

Norfolk Island
South Pacific
Status: External territory
Claimed: 1774
Capital: Kingston
Population: 2210
Area: 13 sq miles
(34 sq km)

DENMARK

The Faroe Islands have been under Danish administration since Queen Margreth I of Denmark inherited Norway in 1380. The Home Rule Act of 1948 gave the Faroese control over all their internal affairs. Greenland first came under Danish rule in 1380. Today, Denmark is responsible for the island's foreign affairs and defense.

Faroe Islands
North Atlantic
Status: External territory
Claimed: 1380
Capital: Tórshavn
Population: 49,469
Area: 540 sq miles
(1399 sq km)

Greenland
North Atlantic
Status: External territory
Claimed: 1380
Capital: Nuuk
Population: 56,483
Area: 836,109 sq miles
(2,166,086 sq km)

FRANCE

France has developed economic ties with its *Territoires d'Outre-Mer* thereby stressing interdependence over independence. Overseas *départements*, officially part of France, have their own governments. Territorial *collectivités* and overseas *territoires* have varyin degrees of autonomy.

Clipperton Island
East Pacific
Status: Dependency of French Polynesia
Claimed: 1935
Capital: Not applicable
Population: None
Area: 2.7 sq miles
(7 sq km)

French Guiana
South America
Status: Overseas department
Claimed: 1817
Capital: Cayenne
Population: 250,109
Area: 35,135 sq miles
(91,000 sq km)

French Polynesia
South Pacific
Status: Overseas collectivity
Claimed: 1843
Capital: Papeete
Population: 276,831
Area: 1608 sq miles
(4165 sq km)

Guadeloupe
West Indies
Status: Overseas department
Claimed: 1635
Capital: Basse-Terre
Population: 405,739
Area: 629 sq miles
(1628 sq km)

Martinique
West Indies
Status: Overseas
department
Claimed: 1635
Capital: Fort-de-France
Population: 386,486
Area: 425 sq miles
(1100 sq km)

Mayotte
Indian Ocean
Status: Overseas
department
Claimed: 1843
Capital: Mamoudzou
Population: 212,645
Area: 144 sq miles
(374 sq km)

New Caledonia
South Pacific
Status: Special collectivity
Claimed: 1853
Capital: Nouméa
Population: 262,000
Area: 7347 sq miles
(19,100 sq km)

Réunion
Indian Ocean
Status: Overseas
department
Claimed: 1638
Capital: Saint-Denis
Population: 840,974
Area: 970 sq miles
(2500 sq km)

St. Pierre & Miquelon
North America
Status: Overseas collectivity
Claimed: 1604
Capital: Saint-Pierre
Population: 5716
Area: 93 sq miles
(242 sq km)

Wallis & Futuna
South Pacific
Status: Overseas collectivity
Claimed: 1842
Capital: Matá'Utu
Population: 15,561
Area: 106 sq miles
(274 sq km)

NETHERLANDS

The country's remaining overseas territories were formerly part of the Dutch West Indies. The Netherlands Antilles dissolved in 2010 leaving the constituent islands with varying degrees of autonomy, but the Netherlands remains responsible for their security.

Aruba
West Indies
Status: Autonomous
part of the Netherlands
Claimed: 1643
Capital: Oranjestad
Population: 102,911
Area: 75 sq miles (194 sq km)

Bonaire
West Indies
Status: Special municipality of the Netherlands
Claimed: 1816
Capital: Kralendijk
Population: 18,413
Area: 113 sq miles
(294 sq km)

Curaçao
West Indies
Status: Autonomous
part of the Netherlands
Claimed: 1816
Capital: Willemstad
Population: 153,500
Area: 171 sq miles
(444 sq km)

Sint Maarten
West Indies
Status: Autonomous
part of the Netherlands
Claimed: 1648
Capital: Philipsburg
Population: 39,689
Area: 13 sq miles (34 sq km)

NEW ZEALAND

New Zealand's government has no desire to retain any overseas territories. However, the economic weakness of its dependent territory Tokelau and its freely associated states, Niue and the Cook Islands, has forced New Zealand to remain responsible for their foreign policy and defense.

Cook Islands
South Pacific
Status: Associated territory
Claimed: 1901
Capital: Avarua
Population: 13,700
Area: 91 sq miles
(235 sq km)

Niue
South Pacific
Status: Associated territory
Claimed: 1901
Capital: Alofi
Population: 1190
Area: 102 sq miles
(264 sq km)

Tokelau
South Pacific
Status: Dependent territory
Claimed: 1926
Capital: Not applicable
Population: 1337
Area: 4 sq miles (10 sq km)

NORWAY

In 1920, 41 nations signed the Spits-bergen Treaty recognizing Norwegian sovereignty over Svalbard. There is a NATO base on Jan Mayen. Bouvet Island is a nature reserve.

Bouvet Island
South Atlantic
Status: Dependency
Claimed: 1928
Capital: Not applicable
Population: None
Area: 22 sq miles (58 sq km)

Jan Mayen
North Atlantic
Status: Dependency
Claimed: 1929
Capital: Not applicable
Population: 18 (meteorologists)
Area: 147 sq miles
(381 sq km)

Continued on page 158

Overseas territories & dependencies

Peter I. Island
Southern Ocean
Status: Dependency
Claimed: 1931
Capital: Not applicable
Population: None
Area: 69 sq miles (180 sq km)

Svalbard
Arctic Ocean
Status: Dependency
Claimed: 1920
Capital: Longyearbyen
Population: 1872
Area: 24,289 sq miles
(62,906 sq km)

UNITED KINGDOM

The UK still has the largest number of overseas territories. These are locally-governed by a mixture of elected representatives and appointed officials, and they all enjoy a large measure of internal self-government, but certain powers, such as foreign affairs and defense, are reserved for Governors of the British Crown.

Anguilla
West Indies
Status: Overseas territory
Claimed: 1650
Capital: The Valley
Population: 16,086
Area: 37 sq miles
(96 sq km)

Ascension Island
South Atlantic
Status: Overseas territory
Claimed: 1673
Capital: Georgetown
Population: 880
Area: 34 sq miles
(88 sq km)

Bermuda
North Atlantic
Status: Overseas territory
Claimed: 1612
Capital: Hamilton
Population: 65,024
Area: 20 sq miles (53 sq km)

British Indian Ocean Territory
Status: Overseas territory
Claimed: 1814
Capital: Diego Garcia
Population: 4000
Area: 23 sq miles
(60 sq km)

British Virgin Islands
West Indies
Status: Overseas territory
Claimed: 1672
Capital: Road Town
Population: 32,680
Area: 59 sq miles
(153 sq km)

Cayman Islands
West Indies
Status: Overseas territory
Claimed: 1670
Capital: George Town
Population: 58,435
Area: 100 sq miles (259 sq km)

Falkland Islands
South Atlantic
Status: Overseas territory
Claimed: 1832
Capital: Stanley
Population: 2840
Area: 4699 sq miles
(12,173 sq km)

Gibraltar
Southwest Europe
Status: Overseas territory
Claimed: 1713
Capital: Gibraltar
Population: 29,185
Area: 2.5 sq miles (6.5 sq km)

Guernsey
Channel Islands
Status: Crown Dependency
Claimed: 1066
Capital: St. Peter Port
Population: 65,849
Area: 25 sq miles (65 sq km)

Isle of Man
British Isles
Status: Crown Dependency
Claimed: 1765
Capital: Douglas
Population: 85,888
Area: 221 sq miles (572 sq km)

Jersey
Channel Islands
Status: Crown Dependency
Claimed: 1066
Capital: St. Helier
Population: 96,513
Area: 45 sq miles (116 sq km)

Montserrat
West Indies
Status: Overseas territory
Claimed: 1632
Capital: Plymouth *(de jure)*, Brades *(de facto)*
Population: 5215
Area: 40 sq miles (102 sq km)

Pitcairn Group of Islands
South Pacific
Status: Dependent territory
Claimed: 1887
Capital: Adamstown
Population: 48
Area: 18 sq miles (47 sq km)

St. Helena
South Atlantic
Status: Overseas territory
Claimed: 1673
Capital: Jamestown
Population: 7776
Area: 47 sq miles (122 sq km)

South Georgia & The South Sandwich Islands
South Atlantic
Status: Overseas territory
Claimed: 1775
Capital: Not applicable
Population: No permanent residents
Area: 1387 sq miles
(3592 sq km)

Tristan da Cunha
South Atlantic
Status: Overseas
territory
Claimed: 1612
Capital: Edinburgh
Population: 264
Area: 38 sq miles (98 sq km)

Turks & Caicos Islands
West Indies
Status: Overseas territory
Claimed: 1766
Capital: Cockburn Town
Population: 33,098
Area: 166 sq miles
(430 sq km)

UNITED STATES
OF AMERICA

America's overseas territories
have been seen as strategically
useful, if expensive, links with its
"backyards." The US has, in most
cases, given the local population a
say in deciding their own status.
A US Commonwealth territory, such
as Puerto Rico, has a greater level
of independence than that of a US
unincorporated territory.

American Samoa
South Pacific
Status: Unincorporated
territory
Claimed: 1900
Capital: Pago Pago
Population: 55,165
Area: 75 sq miles (195 sq km)

Baker &
Howland Islands
South Pacific
Status: Unincorporated
territory
Claimed: 1856
Capital: Not applicable
Population: None
Area: 0.5 sq miles (1.4 sq km)

Guam
West Pacific
Status: Unincorporated
territory
Claimed: 1898
Capital: Hagåtña
Population: 165,124
Area: 212 sq miles
(549 sq km)

Jarvis Island
South Pacific
Status: Unincorporated territory
Claimed: 1856
Capital: Not applicable
Population: None
Area: 1.7 sq miles (4.5 sq km)

Johnston Atoll
Central Pacific
Status: Unincorporated
territory
Claimed: 1858
Capital: Not applicable
Population: Not applicable
Area: 1 sq mile (2.8 sq km)

Kingman Reef
Central Pacific
Status: Unincorporated territory
Claimed: 1856
Capital: Not applicable
Population: None
Area: 0.4 sq mile
(1 sq km)

Midway Islands
Central Pacific
Status: Unincorporated
territory
Claimed: 1867
Capital: Not applicable
Population: 40
Area: 2 sq miles
(5.2 sq km)

Navassa Island
West Indies
Status: Unincorporated
territory
Claimed: 1856
Capital: Not applicable
Population: None
Area: 2 sq miles (5.2 sq km)

Northern
Mariana Islands
West Pacific
Status: Commonwealth
territory
Claimed: 1947
Capital: Saipan
Population: 53,855
Area: 177 sq miles (457 sq km)

Palmyra Atoll
Central Pacific
Status: Incorporated
territory
Claimed: 1898
Capital: Not applicable
Population: None
Area: 5 sq miles (12 sq km)

Puerto Rico
West Indies
Status: Commonwealth
territory
Claimed: 1898
Capital: San Juan
Population: 3.62 million
Area: 3515 sq miles
(9104 sq km)

Virgin Islands
West Indies
Status: Unincorporated
territory
Claimed: 1917
Capital: Charlotte Amalie
Population: 104,737
Area: 137 sq miles
(355 sq km)

Wake Island
Central Pacific
Status: Unincorporated
territory
Claimed: 1898
Capital: Not applicable
Population: 150 (US air base)
Area: 2.5 sq miles
(6.5 sq km)

Glossary of geographical terms

The following glossary lists all geographical terms occuring on the maps and in the main-entry names in the Index–Gazetteer. These terms may precede, follow or be run together with the proper elements of the name; where they precede it the term is reversed for indexing purposes – thus Poluostov Yamal is indexed as Yamal, Poluostrov.

A

Å *Danish, Norwegian*, River
Alpen *German*, Alps
Altiplanicie *Spanish*, Plateau
Älv(en) *Swedish*, River
Anse *French*, Bay
Archipiélago *Spanish*, Archipelago
Arcipelago *Italian*, Archipelago
Arquipélago *Portuguese*, Archipelago
Aukštuma *Lithuanian*, Upland

B

Bahía *Spanish*, Bay
Baía *Portuguese*, Bay
Baḥr *Arabic*, River
Baie *French*, Bay
Bandao *Chinese*, Peninsula
Banjaran *Malay*, Mountain range
Batang *Malay*, Stream
-berg *Afrikaans, Norwegian*, Mountain
Birket *Arabic*, Lake
Boğazı *Turkish*, Strait
Bucht *German*, Bay
Bugten *Danish*, Bay
Buḥayrat *Arabic*, Lake, reservoir
Buḥeiret *Arabic*, Lake
Bukit *Malay*, Mountain
-bukta *Norwegian*, Bay
bukten *Swedish*, Bay
Burnu *Turkish*, Cape, point
Buuraha *Somali*, Mountains

C

Cabo *Portuguese*, Cape
Cap *French*, Cape
Cascada *Portuguese*, Waterfall
Cerro *Spanish*, Hill
Chaîne *French*, Mountain range
Chau *Cantonese*, Island
Cháy *Turkish*, Stream
Chhâk *Cambodian*, Bay
Chhu *Tibetan*, River
-chôsuji *Korean*, Reservoir

Chott *Arabic*, Salt lake, depression
Ch'ün-tao *Chinese*, Island group
Cambodian, Mountains
Cordillera *Spanish*, Mountain range
Costa *Spanish*, Coast
Côte *French*, Coast
Cuchilla *Spanish*, Mountains

D

Dağı *Azerbaijani, Turkish*, Mountain
Dağları *Azerbaijani, Turkish*, Mountains
-dake *Japanese*, Peak
Danau *Indonesian*, Lake
Đao *Vietnamese*, Island
Daryá *Persian*, River
Daryácheh *Persian*, Lake
Dasht *Persian*, Plain, desert
Dawḥat *Arabic*, Bay
Dere *Turkish*, Stream
Dili *Azerbaijani*, Spit
-do *Korean*, Island
Dooxo *Somali*, Valley
Düzü *Azerbaijani*, Steppe
-dwíp *Bengali*, Island

E

Embalse *Spanish*, Reservoir
Erg *Arabic*, Dunes
Estany *Catalan*, Lake
Estrecho *Spanish*, Strait
-ey *Icelandic*, Island
Ezero *Bulgarian, Macedonian*, Lake

F

Fjord *Danish*, Fjord
-fjorden *Norwegian*, Fjord
-fjørdhur *Faeroese*, Fjord
Fleuve *French*, River
Fliegu *Maltese*, Channel
-fljór *Icelandic*, River

G

-gang *Korean*, River
Ganga *Nepali, Sinhala*, River
Gaoyuan *Chinese*, Plateau
-gawa *Japanese*, River

Gebel *Arabic*, Mountain
-gebirge *German*, Mountains
Ghubbat *Arabic*, Bay
Gjiri *Albanian*, Bay
Gol *Mongolian*, River
Golfe *French*, Gulf
Golfo *Italian, Spanish*, Gulf
Gora *Russian, Serbian*, Mountain
Gory *Russian*, Mountains
Guba *Russian*, Bay
Gunung *Malay*, Mountain

H

Ḥadd *Arabic*, Spit
-haehyôp *Korean*, Strait
Haff *German*, Lagoon
Hai *Chinese*, Sea, bay
Ḥammádat *Arabic*, Plateau
Hámún *Persian*, Lake
Hawr *Arabic*, Lake
Háyk' *Amharic*, Lake
He *Chinese*, River
Helodrano *Malagasy*, Bay
-hegység *Hungarian*, Mountain range
Hka *Burmese*, River
-ho *Korean*, Lake
Hô *Korean*, Reservoir
/olot *Hebrew*, Dunes
Hora *Belorussian*, Mountain
Hrada *Belorussian*, Mountains, ridge
Hsi *Chinese*, River
Hu *Chinese*, Lake

I

Île(s) *French*, Island(s)
Ilha(s) *Portuguese*, Island(s)
Ilhéu(s) *Portuguese*, Islet(s)
Irmak *Turkish*, River
Isla(s) *Spanish*, Island(s)
Isola (Isole) *Italian*, Island(s)

J

Jabal *Arabic*, Mountain
Jál *Arabic*, Ridge
-järvi *Finnish*, Lake
Jazírat *Arabic*, Island
Jazíreh *Persian*, Island

Jebel *Arabic*, Mountain
Jezero *Serbian/Croatian*, Lake
Jiang *Chinese*, River
-joki *Finnish*, River
-jökull *Icelandic*, Glacier
Juzur *Arabic*, Islands

K

Kaikyó *Japanese*, Strait
-kaise *Lappish*, Mountain
Kali *Nepali*, River
Kalnas *Lithuanian*, Mountain
Kalns *Latvian*, Mountain
Kang *Chinese*, Harbor
Kangri *Tibetan*, Mountain(s)
Kaôh *Cambodian*, Island
Kapp *Norwegian*, Cape
Kavír *Persian*, Desert
K'edi *Georgian*, Mountain range
Kediet *Arabic*, Mountain
Kepulauan *Indonesian, Malay*, Island group
Khalîg, Khalíj *Arabic*, Gulf
Khawr *Arabic*, Inlet
Khola *Nepali*, River
Khrebet *Russian*, Mountain range
Ko *Thai*, Island
Kolpos *Greek*, Bay
-kopf *German*, Peak
Körfäzi *Azerbaijani*, Bay
Körfezi *Turkish*, Bay
Kõrgustik *Estonian*, Upland
Koshi *Nepali*, River
Kowtal *Persian*, Pass
Kúh(há) *Persian*, Mountain(s)
-kundo *Korean*, Island group
-kysten *Norwegian*, Coast
Kyun *Burmese*, Island

L

Laaq *Somali*, Watercourse
Lac *French*, Lake
Lacul *Romanian*, Lake
Lago *Italian, Portuguese, Spanish*, Lake
Laguna *Spanish*,

Lagoon, Lake
Laht *Estonian*, Bay
Laut *Indonesian*, Sea
Lembalemba *Malagasy*,
 Plateau
Lerr *Armenian*,
 Mountain
Lerrnashght'a *Armenian*,
 Mountain range
Les *Czech*, Forest
Lich *Armenian*, Lake
Liqeni *Albanian*, Lake
Lumi *Albanian*, River
Lyman *Ukrainian*,
 Estuary

M

Mae Nam *Thai*, River
mägi *Estonian*, Hill
Maja *Albanian*, Mountain
man *Korean*, Bay
Marios *Lithuanian*, Lake
meer *Dutch*, Lake
Melkosopochnik
 Russian, Plain
meri *Estonian*, Sea
Mifraz *Hebrew*, Bay
Monkhafad *Arabic*,
 Depression
Mont(s) *French*,
 Mountain(s)
Monte *Italian*,
 Portuguese, Mountain
More *Russian*, Sea
Mörön *Mongolian*, River

N

Nagor'ye *Russian*,
 Upland
Najal *Hebrew*, River
Nahr *Arabic*, River
Nam *Laotian*, River
Nehri *Turkish*, River
Nevado *Spanish*,
 Mountain (snow-
 capped)
Nisoi *Greek*, Islands
Nizmennost' *Russian*,
 Lowland, plain
Nosy *Malagasy*, Island
Nur *Mongolian*, Lake
Nuruu *Mongolian*,
 Mountains
Nuur *Mongolian*, Lake
Nyzovyna *Ukrainian*,
 Lowland, plain

O

Ostrov(a) *Russian*,
 Island(s)
Oued *Arabic*,
 Watercourse
oy *Faeroese*, Island
øy(a) *Norwegian*,
 Island
Oya *Sinhala*, River
Ozero *Russian*,
 Ukrainian, Lake

P

Passo *Italian*, Pass
Pegunungan
 Indonesian, Malay,
 Mountain range
Pelagos *Greek*, Sea
Penisola *Italian*,
 Peninsula
Peski *Russian*, Sands
Phanom *Thai*, Mountain
Phou *Laotian*,
 Mountain
Pic *Catalan*, Peak
Pico *Portuguese*,
 Spanish, Peak
Pik *Russian*, Peak
Planalto *Portuguese*,
 Plateau
Planina, Planini
 Bulgarian, Macedonian,
 Serbian, Croatian,
 Mountain range
Ploskogor'ye *Russian*,
 Upland
Poluostrov *Russian*,
 Peninsula
Potamos *Greek*, River
Proliv *Russian*, Strait
Pulau *Indonesian*,
 Malay, Island
Pulu *Malay*, Island
Punta *Portuguese*,
 Spanish, Point

Q

Qá' *Arabic*, Depression
Qolleh *Persian*,
 Mountain

R

Raas *Somali*, Cape
-rags *Latvian*, Cape
Ramlat *Arabic*, Sands
Ra's *Arabic*, Cape,
 point, headland
Ravnina *Bulgarian*,
 Russian, Plain
Récif *French*, Reef
Represa (Rep.) *Spanish*,
 Portuguese, Reservoir
-rettō *Japanese*, Island
 chain
Riacho *Spanish*,
 Stream
Riban' *Malagasy*,
 Mountains
Rio *Portuguese*, River
Río *Spanish*, River
Riu *Catalan*, River
Rivier *Dutch*, River
Rivière *French*, River
Rowd *Pashtu*, River
Rúd *Persian*, River
Rudohorie *Slovak*,
 Mountains
Ruisseau *French*,
 Stream

S

Sabkhat *Arabic*, Salt
 marsh
Şaḥrá' *Arabic*, Desert
Samudra *Sinhala*,
 Reservoir
-san *Japanese*, Korean,
 Mountain
-sanchi *Japanese*,
 Mountains
-sanmaek *Korean*,
 Mountain
Sarír *Arabic*, Desert
Sebkha, Sebkhet *Arabic*,
 Salt marsh, depression
See *German*, Lake
Selat *Indonesian*, Strait
-selkä *Finnish*, Ridge
Selseleh *Persian*,
 Mountain range
Serra *Portuguese*,
 Mountain
Serranía *Spanish*,
 Mountain
Sha'íb *Arabic*,
 Watercourse
Shamo *Chinese*,
 Desert
Shan *Chinese*,
 Mountain(s)
Shan-mo *Chinese*,
 Mountain range
Shaṭṭ *Arabic*,
 Distributary
-shima *Japanese*, Island
Shui-tao *Chinese*,
 Channel
Sierra *Spanish*,
 Mountains
Sòn *Vietnamese*,
 Mountain
Sông *Vietnamese*, River
-spitze *German*, Peak
Štít *Slovak*, Peak
Stoeng *Cambodian*,
 River
Stretto *Italian*, Strait
Su Anbarı *Azerbaijani*,
 Reservoir
Sungai *Indonesian*,
 Malay, River
Suu *Turkish*, River

T

Tal *Mongolian*, Plain
Tandavan' *Malagasy*,
 Mountain range
Tangorombohitr'
 Malagasy, Mountain
 massif
Tao *Chinese*, Island
Tassili *Berber*, Plateau,
 mountain
Tau *Russian*,
 Mountain(s)
Taungdan *Burmese*,
 Mountain range

Teluk
Teluk *Indonesian*,
 Malay, Bay
Terara *Amharic*,
 Mountain
Tog *Somali*, Valley
Tônlé *Cambodian*,
 Lake
Top *Dutch*, Peak
-tunturi *Finnish*,
 Mountain
Tur'at *Arabic*,
 Channel

V

Väin *Estonian*, Strait
-vatn *Icelandic*, Lake
-vesi *Finnish*, Lake
Vinh *Vietnamese*, Bay
**Vodokhranilishche
 (Vdkhr.)** *Russian*,
 Reservoir
**Vodoskhovyshche
 (Vdskh.)** *Ukrainian*,
 Reservoir
Volcán *Spanish*,
 Volcano
Vozvyshennost'
 Russian, Upland,
 plateau
Vrh *Macedonian*,
 Peak
Vysochyna *Ukrainian*,
 Upland
Vysočina *Czech*,
 Upland

W

Waadi *Somali*,
 Watercourse
Wādí *Arabic*,
 Watercourse
Wâhat, Wâhat *Arabic*,
 Oasis
Wald *German*, Forest
Wan *Chinese*, Bay
Wyżyna *Polish*,
 Upland

X

Xé *Laotian*, River

Y

Yarımadası *Azerbaijani*,
 Peninsula
Yazovir *Bulgarian*,
 Reservoir
Yoma *Burmese*,
 Mountains
Yu *Chinese*, Islet

Z

Zaliv *Bulgarian*,
 Russian, Bay
Zatoka *Ukrainian*, Bay
Zemlya *Russian*, Land

Continental factfile

North & Central America

Total area:
9,400,000 sq miles
(24,346,000 sq km)

**Total number of
countries:** 23

Total population:
560 million

**Largest city with
population:** Mexico
City, Mexico 22.2 million

 **Country with highest
population density:** Barbados
1807 people per sq mile
(698 people per sq km)

Largest country:
Canada 3,855,171 sq miles
(9,984,670 sq km)

Smallest country:
St. Kitts & Nevis 101 sq miles
(261 sq km)

Largest lake: Lake Superior,
Canada/ USA 32,151 sq miles
(83,270 sq km)

 Longest river: Mississippi-
Missouri, USA 3710 miles
(5969 km)

Highest point: Mt. McKinley
(Denali), Alaska, USA 20,310 ft
(6190 m)

lowest point: Death Valley,
California, USA
282 ft (86 m) below sea level

South America

Total area:
6,880,000 sq miles
(17,819,000 sq km)

**Total number of
countries:** 12

Total population:
406 million

**Largest city with
population:** São Paulo,
Brazil 21.7 million

 **Country with highest
population density:** Ecuador
147 people per sq mile
(57 people per sq km)

Largest country:
Brazil 3,286,470 sq miles
(8,511,965 sq km)

Smallest country:
Suriname 63,039 sq miles
(163,270 sq km)

Largest lake: Lake Titicaca,
Bolivia/Peru 3220 sq miles
(8340 sq km)

 Longest river: Amazon,
Brazil 4049 miles
(6516 km)

Highest point: Cerro
Aconcagua, Argentina
22,831 ft (6959 m)

Lowest point: Laguna del
Carbón, Argentina
344 ft (105 m) below sea level

Africa

Total area:
11,677,250 sq miles
(30,244,050 sq km)

**Total number of
countries:** 54

Total population:
1109 million

**Largest city with
population:** Cairo,
Egypt 16.4 million

**Country with highest
population density:** Mauritius
1671 people per sq mile
(645 people per sq km)

Largest country:
Algeria 919,590 sq miles
(2,381,740 sq km)

Smallest country:
Seychelles 176 sq miles
(455 sq km)

Largest lake: Lake Victoria,
Uganda, Kenya, Tanzania
26,828 sq miles (69,484 sq km)

Longest river: Nile,
Uganda/Sudan/Egypt
4160 miles (6695 km)

Highest point: Kilimanjaro,
Tanzania 19,340 ft
(5895 m)

Lowest point: Lac',
Assal, Djibouti
512 ft (156 m) below sea level

Europe

Total area:
4,809,200 sq miles
(12,456,000 sq km)

**Total number of
countries:** 44

Total population:
721 million

**Largest city with
population:** Moscow,
Euro Russia 16.7 million

**Country with highest
population density:** Monaco
48,181 people per sq mile
(18,531 people per sq km)

Largest country: European
Russia 1,527,341 sq miles
(3,955,818 sq km)

Smallest country:
Vatican City, Italy 0.17 sq miles
(0.44 sq km)

Largest lake: Ladoga,
European Russia
7100 sq miles (18,390 sq km)

 Longest river: Volga,
European Russia
2290 miles (3688 km)

Highest point: El'brus,
Caucasus Mts, European
Russia 18,510 ft (5642 m)

Lowest point: Volga Delta,
Caspian Sea, European Russia
92 ft (28 m) below sea level

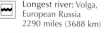

North & West Asia
 Total area: 9,585,500 sq miles (24,826,600 sq km)
 Total number of countries: 25
 Total population: 446 million
 Largest city with population: Tehran, Iran 13.4 million
 Country with highest population density: Bahrain 4762 people per sq mile (1841 people per sq km)
 Largest country: Asiatic Russia 5,065,471 sq miles (13,119,582 sq km)
 Smallest country: Bahrain 239 sq miles (620 sq km)
 Largest lake: Caspian Sea 142,243 sq miles (371,000 sq km)
 Longest river: Ob'-Irtysh, Asiatic Russia 3461 miles (5570 km)
 Highest point: Pik Pobedy, Kyrgyzstan/China 24,408 ft (7439 m)
 Lowest point: Dead Sea, Israel/Jordan 1401 ft (427 m) below sea level

South & East Asia
 Total area: 7,936,200 sq miles (20,554,700 sq km)
 Total number of countries: 24
 Total population: 3775 million
 Largest city with population: Tokyo, Japan 39.4 million
 Country with highest population density: Singapore 22,881 people per sq mile (8852 people per sq km)
 Largest country: China 3,705,386 sq miles (9,596,960 sq km)
 Smallest country: Maldives 116 sq miles (300 sq km)
 Largest lake: Tonle Sap, Cambodia 1000 sq miles (2850 sq km)
 Longest river: Chang Jiang (Yangtze) 3965 miles (6380 km)
 Highest point: Mount Everest, Nepal 29,029 ft (8848 m)
Lowest point: Turpan Hami, (Turfan basin), China 505 ft (154 m) below sea level

Australasia & Oceania
 Total area: 3,376,700 sq miles (8,745,750 sq km)
 Total number of countries: 14
 Total population: 37.5 million
 Largest city with population: Sydney, Australia 4.8 million
 Country with highest population density: Nauru 1165 people per sq mile (449 people per sq km)
 Largest country: Australia 2,967,893 sq miles (7,686,850 sq km)
 Smallest country: Nauru 8 sq miles (21 sq km)
 Largest lake: Lake Eyre, Australia 3700 sq miles (9583 sq km)
 Longest river: Murray-Darling, Australia 2330 miles (3750 km)
 Highest point: Mt. Wilhelm, Papua New Guinea 14,795 ft (4509 m)
 Lowest point: Lake Eyre, Australia 52 ft (16 m) below sea level

Antarctica
 Total area: 5,450,500 sq miles (14,000,000 sq km) of which approx. 324,300 sq miles (840,000 sq km) is ice-free.
 Total number of countries: The Antarctic Treaty has 30 participating nations and 14 with observer status. Claims by Australia, France, New Zealand, Norway, Argentina, Chile, and the UK are not recognized by other member states.
 Total Population: No indigenous population. 74 research stations, (42 are staffed all year-round). Population varies between about 1000 (winter) and 4000 (summer).
 Total volume of ice: 7,200,000 cu miles (30,000,000 cu km): contains 90% of Earth's fresh water
 Sea ice: 1,158,300 sq miles (3,000,000 sq km) in February. 7,722,000 sq miles (20,000,000 sq km) in October
 Lowest temperature: Vostok station -89.5°C (-129°F)
 Highest point: Vinson Massif 16,072 ft (4897 m)
 Lowest Point: Coastline 0ft/m

Geographical comparisons

Largest countries

Russian Fed. .... 6,592,735 sq miles....(17,075,200 sq km)
Canada ....3,855,171 sq miles .....(9,984,670 sq km)
USA....................3,717,792 sq miles .... (9,626,091 sq km)
China ....3,705,386 sq miles .....(9,596,960 sq km)
Brazil..................3,286,470 sq miles .....(8,511,965 sq km)
Australia ....2,967,893 sq miles .....(7,686,850 sq km)
India...................1,269,338 sq miles .....(3,287,590 sq km)
Argentina ....1,068,296 sq miles .....(2,766,890 sq km)
Kazakhstan ....1,049,150 sq miles .....(2,717,300 sq km)
Algeria .... 919,590 sq miles ....(2,381,740 sq km)

Smallest countries

Vatican City .... 0.17 sq miles ....(0.44 sq km)
Monaco................... 0.75 sq miles ....(1.95 sq km)
Nauru .... 8 sq miles ....(21 sq km)
Tuvalu .... 10 sq miles ....(26 sq km)
San Marino .... 24 sq miles ....(61 sq km)
Liechtenstein............. 62 sq miles ....(160 sq km)
Marshall Islands......... 70 sq miles ....(181 sq km)
St. Kitts & Nevis .... 101 sq miles ....(261 sq km)
Maldives................... 116 sq miles ....(300 sq km)
Malta....................... 122 sq miles ....(316 sq km)

Largest islands

Greenland..............840,000 sq miles (2,175,600 sq km)
New Guinea ....312,000 sq miles (808,000 sq km)
Borneo ....292,222 sq miles (757,050 sq km)
Madagascar ....226,656 sq miles (587,040 sq km)
Sumatra....................202,300 sq miles (524,000 sq km)
Baffin Island ....183,800 sq miles (476,000 sq km)
Honshu ....88,800 sq miles (230,000 sq km)
Britain.........................88,700 sq miles (229,800 sq km)
Victoria Island.............81,900 sq miles (212,000 sq km)
Ellesmere Island ....75,700 sq miles (196,000 sq km)

Richest countries   (GNI per capita, in US$)

Monaco 186,950
Liechtenstein.. 136,770
Norway... 102,610
Switzerland... 86,600
Qatar.. 85,550
Luxembourg.. 71,810
Australia 65,520
Denmark.. 61,160
Sweden... 59,240
Singapore 54,040

Poorest countries   (GNI per capita, in US$)

Malawi 270
Burundi 280
Somalia 288
Central African Republic....................... 320
Congo, Democratic Republic 400
Niger 410
Liberia 410
Madagascar 440
Guinea 460
Ethiopia 470

Most populous countries

China...1.386 billion
India ....1.252 billion
USA... 320 million
Indonesia... 250 million
Brazil.. 200 million
Pakistan.. 182 million
Nigeria.. 174 million
Bangladesh .... 157 million
Russian Federation............................. 143 million
Japan.. 127 million

Least populous countries

Vatican City 839
Nauru 9434
Tuvalu 10,698
Palau 21,108
San Marino 32,448
Monaco 36,136
Liechtenstein... 37,000
St. Kitts & Nevis 51,134
Marshall Islands.. 69,747
Dominica... 73,286

Most densely populated countries

Monaco.....48,181 people per sq mile (18,531 per sq km)
Singapore .....22,881 people per sq mile (8852 per sq km)
Vatican City ....4935 people per sq mile (1907 per sq km)
Bahrain ....4762 people per sq mile (1841 per sq km)
Malta...............3226 people per sq mile (1250 per sq km)
Bangladesh .....3029 people per sq mile (1169 per sq km)
Maldives ....2586 people per sq mile (1000 per sq km)
Taiwan ....1871 people per sq mile (722 per sq km)
Barbados...........1807 people per sq mile (698 per sq km)
Mauritius...........1671 people per sq mile (645 per sq km)

ost sparsely populated countries

ngolia......... 5 people per sq mile......... (2 per sq km)
mibia......... 7 people per sq mile......... (3 per sq km)
land......... 8 people per sq mile......... (3 per sq km)
iname......... 8 people per sq mile......... (3 per sq km)
stralia......... 8 people per sq mile......... (3 per sq km)
swana......... 9 people per sq mile......... (4 per sq km)
ya......... 9 people per sq mile......... (4 per sq km)
uritania..... 10 people per sq mile......... (4 per sq km)
nada......... 10 people per sq mile......... (4 per sq km)
yana......... 11 people per sq mile......... (4 per sq km)

ost widely spoken languages

Chinese (Mandarin)   6. Portuguese
Spanish   7. Bengali
English   8. Russian
Hindi   9. Japanese
Arabic   10. Javanese

rgest conurbations

kyo (Japan)..............................39,400,000
uangzhou (China)..........................32,600,000
anghai (China)............................29,600,000
carta (Indonesia)..........................27,000,000
elhi (India).............................25,300,000
oul (South Korea).........................24,200,000
rachi (Pakistan)..........................23,200,000
umbai (India).............................22,600,000
anila (Philippines)........................22,500,000
exico City (Mexico)........................22,200,000
ew York (USA)...............................21,800,000
o Paulo (Brazil)..........................21,700,000
eijing (China)............................19,900,000
saka (Japan)...............................17,800,000
s Angeles (USA)............................17,300,000
haka (Bangladesh)..........................16,700,000
oscow (Russian Federation)..............16,700,000
airo (Egypt)..............................16,400,000
olkata (India)............................15,800,000
uenos Aires (Argentina)...................15,700,000
angkok (Thailand).........................14,900,000
tanbul (Turkey)...........................14,000,000
ondon (UK)................................14,000,000
gos (Nigeria).............................13,500,000
hran (Iran)...............................13,400,000

Longest rivers

Nile (Northeast Africa)................4160 miles...... (6695 km)
Amazon (South America)...........4049 miles...... (6516 km)
Yangtze (China)........................3915 miles...... (6299 km)
Mississippi/Missouri (USA).........3710 miles........(5969 km)
Ob'-Irtysh (Russian Federation) 3461 miles...... (5570 km)
Yellow River (China)..................3395 miles...... (5464 km)
Congo (Central Africa)...............2900 miles...... (4667 km)
Mekong (Southeast Asia)........2749 miles...... (4425 km)
Lena (Russian Federation)........2734 miles...... (4400 km)
Mackenzie (Canada)...............2640 miles...... (4250 km)
Yenisey (Russian Federation)...2541 miles...... (4090 km)

Highest mountains (Height above sea level)

Everest..................... 29,029 ft....... (8848 m)
K2 .... 28,253 ft....... (8611 m)
Kanchenjunga I...................... 28,210 ft....... (8598 m)
Makalu I................................. 27,767 ft....... (8463 m)
Cho Oyu .... 26,907 ft....... (8201 m)
Dhaulagiri I........................... 26,796 ft....... (8167 m)
Manaslu I............................. 26,783 ft....... (8163 m)
Nanga Parbat I....................... 26,661 ft....... (8126 m)
Annapurna I .... 26,547 ft....... (8091 m)
Gasherbrum I......................... 26,471 ft....... (8068 m)

Largest bodies of inland water (Area & depth)

Caspian Sea
143,243 sq miles (371,000 sq km).......3215 ft (980 m)
Lake Superior
32,151 sq miles (83,270 sq km).......1289 ft (393 m)
Lake Victoria
26,560 sq miles (68,880 sq km).........328 ft (100 m)
Lake Huron
23,436 sq miles (60,700 sq km).........751 ft (229 m)
Lake Michigan
22,402 sq miles (58,020 sq km).........922 ft (281 m)
Lake Tanganyika
12,703 sq miles (32,900 sq km).... 4700 ft (1435 m)
Great Bear Lake
12,274 sq miles (31,790 sq km)...... 1047 ft (319 m)
Lake Baikal
11,776 sq miles (30,500 sq km).... 5712 ft (1741 m)
Great Slave Lake
10,981 sq miles (28,440 sq km).........459 ft (140 m)
Lake Erie
9915 sq miles (25,680 sq km)..........197 ft (60 m)

......continued on page 166

Geographical comparisons continued

Deepest ocean features

Challenger Deep, Mariana Trench (Pacific)
36,201 ft .... (11,034 m)
Vityaz III Depth, Tonga Trench (Pacific)
35,704 ft .... (10,882 m)
Vityaz Depth, Kurile-Kamchatka Trench (Pacific)
34,588 ft .... (10,542 m)
Cape Johnson Deep, Philippine Trench (Pacific)
34,441 ft .... (10,497 m)
Kermadec Trench (Pacific)
32,964 ft .... (10,047 m)
Ramapo Deep, Japan Trench (Pacific)
32,758 ft .... (9984 m)
Milwaukee Deep, Puerto Rico Trench (Atlantic)
30,185 ft .... (9200 m)
Argo Deep, Torres Trench (Pacific)
30,070 ft .... (9165 m)
Meteor Depth, South Sandwich Trench (Atlantic)
30,000 ft .... (9144 m)
Planet Deep, New Britain Trench (Pacific)
29,988 ft .... (9140 m)

Greatest waterfalls   (Mean flow of water)

Boyoma (D.R. Congo)..... 600,400 cu. ft/sec (17,000 cu.m/sec)
Khône (Laos/Cambodia) ... 410,000 cu. ft/sec (11,600 cu.m/sec)
Niagara (USA/Canada)......... 195,000 cu. ft/sec (5500 cu.m/sec)
Grande, Salto (Uruguay)..... 160,000 cu. ft/sec (4500 cu.m/sec)
Paulo Afonso (Brazil)........... 100,000 cu. ft/sec(2800 cu.m/sec)
Urubupungá (Brazil) ....97,000 cu. ft/sec (2750 cu.m/sec)
Iguaçu (Argentina/Brazil)........62,000 cu. ft/sec (1700 cu.m/sec)
Maribondo (Brazil)................53,000 cu. ft/sec (1500 cu.m/sec)
Victoria (Zimbabwe)..............39,000 cu. ft/sec (1100 cu.m/sec)
Murchison Falls (Uganda).....42,000 cu. ft/sec (1200 cu.m/sec)
Churchill (Canada)................35,000 cu. ft/sec (1000 cu.m/sec)
Kaveri Falls (India).................33,000 cu. ft/sec (900 cu.m/sec)

Highest waterfalls

Angel (Venezuela) ....3212 ft.............. (979 m)
Tugela (South Africa) ....3110 ft.............. (948 m)
Utigard (Norway)..........................2625 ft.............. (800 m)
Mongefossen (Norway) ....2539 ft.............. (774 m)
Mtarazi (Zimbabwe) ....2500 ft.............. (762 m)
Yosemite (USA) ....2425 ft.............. (739 m)
Ostre Mardola Foss (Norway) .....2156 ft.............. (657 m)
Tyssestrengane (Norway)...........2119 ft.............. (646 m)
*Cuquenan (Venezuela).............2001 ft.............. (610 m)
Sutherland (New Zealand)..........1903 ft.............. (580 m)
*Kjellfossen (Norway) ....1841 ft............(561 m)

indicates that the total height is a single leap

Largest deserts

Sahara...............3,450,000 sq miles (9,065,000 sq km)
Gobi.................... 500,000 sq miles (1,295,000 sq km)
Ar Rub al Khali ...... 289,600 sq miles (750,000 sq km)
Great Victorian ..... 249,800 sq miles (647,000 sq km)
Sonoran .... 120,000 sq miles (311,000 sq km)
Kalahari .... 120,000 sq miles (310,800 sq km)
Kara Kum............... 115,800 sq miles (300,000 sq km)
Takla Makan .... 100,400 sq miles (260,000 sq km)
Namib.....................52,100 sq miles (135,000 sq km)
Thar........................33,670 sq miles (130,000 sq km)

NB – Most of Antarctica is a polar desert, with only 2 inches (50 mm) of precipitation annually

Hottest inhabited places

Djibouti (Djibouti) ....86.0°F .... (30.0°C)
Tombouctou (Mali) ....84.7°F .... (29.3°C)
Tirunelveli (India) ....84.7°F .... (29.3°C)
Tuticorin (India)..........................84.7°F .... (29.3°C)
Nellore (India)............................84.5°F .... (29.2°C)
Santa Marta (Colombia) ....84.5°F .... (29.2°C)
Aden (Yemen)............................84.0°F .... (29.0°C)
Madurai (India)...........................84.0°F .... (29.0°C)
Niamey (Niger)............................84.0°F .... (29.0°C)

Driest inhabited places

Aswân (Egypt)..............................0.02 in .... (0.5 mm)
Luxor (Egypt).................................0.03 in .... (0.7 mm)
Arica (Chile).................................0.04 in .... (1.1 mm)
Ica (Peru)......................................0.10 in .... (2.3 mm)
Antofagasta (Chile)......................0.20 in .... (4.9 mm)
El Minya (Egypt) ....0.20 in .... (5.1 mm)
Asyut (Egypt)................................0.20 in .... (5.2 mm)
Callao (Peru).................................0.50 in ..... (12.0 mm)
Trujillo (Peru)................................0.55 in ..... (14.0 mm)
Al Fayyum (Egypt).........................0.80 in ..... (19.0 mm)

Wettest inhabited places

Mawsynram (India) ....467 in .. (11,862 mm)
Mt Waialeale (Hawaii, USA)....... 460 in .. (11,684 mm)
Cherrapunji (India)......................450 in .. (11,430 mm)
Cape Debundsha (Cameroon) ... 405 in .. (10,290 mm)
Quibdo (Colombia)......................354 in .... (8892 mm)
Buenaventura (Colombia) ....265 in .... (6743 mm)
Monrovia (Liberia) ....202 in .... (5131 mm)
Pago Pago (American Samoa).....196 in .... (4990 mm)
Moulmein (Myanmar) ....191 in .... (4852 mm)
Lae (Papua New Guinea) ....183 in .... (4645 mm)

GLOSSARY OF ABBREVIATIONS

This Glossary provides a comprehensive guide to the abbreviations used in this Atlas, and in the Index.

A
abbrev. abbreviated
Afr. Afrikaans
Alb. Albanian
Amh. Amharic
anc. ancient
Ar. Arabic
Arm. Armenian
Az. Azerbaijani

B
Basq. Basque
Bel. Belorussian
Ben. Bengali
Bibl. Biblical
Bret. Breton
Bul. Bulgarian
Bur. Burmese

C
Cam. Cambodian
Cant. Cantonese
Cast. Castilian
Cat. Catalan
Chin. Chinese
Cro. Croat
Cz. Czech

D
Dan. Danish
Dut. Dutch

E
Eng. English
Est. Estonian
est. estimated

F
Faer. Faeroese
Fij. Fijian
Fin. Finnish
Flem. Flemish
Fr. French
Fris. Frisian

G
Geor. Georgian
Ger. German
Gk. Greek
Guj. Gujarati

H
Haw. Hawaiian
Heb. Hebrew
Hind. Hindi
hist. historical
Hung. Hungarian

I
Icel. Icelandic
Ind. Indonesian
In. Inuit
Ir. Irish
It. Italian

J
Jap. Japanese

K
Kaz. Kazakh
Kir. Kirghiz
Kor. Korean
Kurd. Kurdish

L
Lao. Laotian
Lapp. Lappish
Lat. Latin
Latv. Latvian

Lith. Lithanian
Lus. Lusatian

M
Mac. Macedonian
Mal. Malay
Malg. Malagasy
Malt. Maltese
Mon. Montenegro
Mong. Mongolian

N
Nepali. Nepali
Nor. Norwegian

O
off. officially

P
Pash. Pashtu
Per. Persian
Pol. Polish
Port. Portuguese
prev. previously

R
Rmsch. Romansch
Roman. Romanian
Rus. Russian

S
SCr. Serbo - Croatian
Serb. Serbian
Slvk. Slovak
Slvn. Slovene
Som. Somali
Sp. Spanish
Swa. Swahili
Swe. Swedish

T
Taj. Tajik
Th. Thai
Tib. Tibetan
Turk. Turkish
Turkm. Turkmenistan

U
Uigh. Uighur
Ukr. Ukrainian
Uzb. Uzbek

V
var. variant
Vtn. Vietnamese

W
Wel. Welsh

X
Xh. Xhosa

Key to country factboxes within the Index:

Formation
Date of independence

Population
Total population / population density - based on total land area .

Calorie consumption
Average number of calories consumed daily per person.

A

Aa *see* Gauja
Aachen 94 A4 *Dut.* Aken, *Fr.* Aix-la-Chapelle; *anc.* Aquae Grani, Aquisgranum. Nordrhein-Westfalen, W Germany
Aaiún *see* Laâyoune
Aalborg 85 B7 *var.* Ålborg, Ålborg-Nørresundby; *anc.* Alburgum. Nordjylland, N Denmark
Aalen 95 B6 Baden-Württemberg, S Germany
Aalsmeer 86 C3 Noord-Holland, C Netherlands
Aalst 87 B6 Oost-Vlaanderen, C Belgium
Aalten 86 E4 Gelderland, E Netherlands
Aalter 87 B5 Oost-Vlaanderen, NW Belgium
Aanaarjävri *see* Inarijärvi
Äänekoski 85 D5 Länsi-Suomi, W Finland
Aar *see* Aare
Aare 95 A7 *var.* Aar. *river* W Switzerland
Aarhus *see* Århus
Aarlen *see* Arlon
Aat *see* Ath
Aba 77 E5 Orientale, NE Dem. Rep. Congo
Aba 73 G5 Abia, S Nigeria
Abá as Su'ûd *see* Najrân
Abaco Island *see* Great Abaco, N Bahamas
Ābādān 122 C4 Khūzestān, SW Iran
Abadan 122 C3 *prev.* Bezmein, Büzmeýin, *Rus.* Byuzmeyin. Ahal Welaýaty, C Turkmenistan
Abai *see* Blue Nile
Abakan 114 D4 Respublika Khakasiya, S Russian Federation
Abancay 60 D4 Apurímac, SE Peru
Abariringa *see* Kanton
Abashiri 130 D2 *var.* Abasiri. Hokkaidō, NE Japan
Abasiri *see* Abashiri
Ābay Wenz *see* Blue Nile
Abbaia *see* Åbaya Hâyk'
Abbatis Villa *see* Abbeville
Abbazia *see* Opatija
Abbeville 90 C2 *anc.* Abbatis Villa. Somme, N France
'Abd al 'Azīz, Jabal 118 D2 *mountain range* NE Syria
Abéché 76 C3 *var.* Abécher, Abeshr. Ouaddaï, SE Chad
Abécher *see* Abéché
Abela *see* Ávila
Abellinum *see* Avellino
Abemama 144 D2 *var.* Apamama; *prev.* Roger Simpson Island. *atoll* Tungaru, W Kiribati
Abengourou 75 E5 E Côte d'Ivoire
Aberbrothock *see* Arbroath
Abercorn *see* Mbala
Aberdeen 88 D3 *anc.* Devana. NE Scotland, United Kingdom
Aberdeen 45 E2 South Dakota, N USA
Aberdeen 46 B2 Washington, NW USA
Abergwaun *see* Fishguard
Abertawe *see* Swansea
Aberystwyth 89 C6 W Wales, United Kingdom
Abeshr *see* Abéché
Abhā 121 B6 'Asīr, SW Saudi Arabia
Abidavichy 107 D7 *Rus.* Obidovichi. Mahilyowskaya Voblasts', E Belarus
Abidjan 75 E5 S Côte d'Ivoire
Abilene 49 F3 Texas, SW USA
Abingdon *see* Pinta, Isla
Abkhazia *see* Apkhazeti
Åbo *see* Turku
Aboisso 75 E5 SE Côte d'Ivoire
Abo, Massif d' 76 B1 *mountain range* NW Chad
Abomey 75 F5 S Benin
Abou-Déïa 76 C3 Salamat, SE Chad
Aboudouhour *see* Abū aḏ Ḏuhūr
Abou Kémal *see* Abū Kamāl
Abrantes 92 B3 *var.* Abrántes. Santarém, C Portugal
Abrashlare *see* Brezovo
Abrolhos Bank 56 E4 *undersea bank* W Atlantic Ocean
Abrova 107 B6 *Rus.* Obrovo. Brestskaya Voblasts', SW Belarus
Abrud 108 B4 *Ger.* Gross-Schlatten, *Hung.* Abrudbánya. Alba, SW Romania
Abrudbánya *see* Abrud

Abruzzese, Appennino 96 C4 *mountain range* C Italy
Absaroka Range 44 B2 *mountain range* Montana/Wyoming, NW USA
Abū aḏ Ḏuhūr 118 B3 *Fr.* Aboudouhour. Idlib, NW Syria
Abu Dhabi *see* Abū Ẓabī
Abu Hamed 72 C3 River Nile, N Sudan
Abū Ḩardān 118 E3 *var.* Hajine. Dayr az Zawr, E Syria
Abuja 75 G4 *country capital* (Nigeria) Federal Capital District, C Nigeria
Abū Kamāl 118 E3 *Fr.* Abou Kémal. Dayr az Zawr, E Syria
Abula *see* Ávila
Abunã, Rio 62 C2 *var.* Río Abuná. *river* Bolivia/Brazil
Abut Head 151 B6 *headland* South Island, New Zealand
Abuye Meda 72 D4 *mountain* C Ethiopia
Abū Ẓabī 121 C5 *var.* Abū Ẓabī, *Eng.* Abu Dhabi. *country capital* (United Arab Emirates) Abū Ẓaby, C United Arab Emirates
Abū Ẓabī *see* Abū Ẓabī
Abyaḏ, Al Baḥr al *see* White Nile
Abyei Area 73 B5 *disputed region* Southern Kordofan, S Sudan
Abyla *see* Ávila
Abyssinia *see* Ethiopia
Acalayong 77 A5 SW Equatorial Guinea
Acaponeta 50 D4 Nayarit, C Mexico
Acapulco 51 E5 *var.* Acapulco de Juárez. Guerrero, S Mexico
Acapulco de Juárez *see* Acapulco
Acarai Mountains 59 F4 *Sp.* Serra Acaraí. *mountain range* Brazil/Guyana
Acaraí, Serra *see* Acarai Mountains
Acarigua 58 D2 Portuguesa, N Venezuela
Accra 75 E5 *country capital* (Ghana) SE Ghana
Achacachi 61 E4 La Paz, W Bolivia
Ach'ara 117 F2 *prev.* Achara, *var.* Ajaria. *autonomous republic* SW Georgia
Achara *see* Ach'ara
Acklins Island 54 C2 *island* SE Bahamas
Aconcagua, Cerro 64 B4 *mountain* W Argentina
Açores/Açores, Arquipélago dos/ Açores, Ilhas dos *see* Azores
A Coruña 92 B1 *Cast.* La Coruña, *Eng.* Corunna; *anc.* Caronium. Galicia, NW Spain
Acre 62 C2 *off.* Estado do Acre. *state* W Brazil
Acre 62 C2 *off.* Estado do Acre. *region* W Brazil
Açu *see* Assu
Acunum Acusio *see* Montélimar
Ada 100 D3 Vojvodina, N Serbia
Ada 49 G2 Oklahoma, C USA
Ada Bazar *see* Adapazarı
Adalia *see* Antalya
Adalia, Gulf of *see* Antalya Körfezi
Adama *see* Nazrēt
'Adan 121 B7 *Eng.* Aden. SW Yemen
Adana 116 D4 *var.* Seyhan. Adana, S Turkey
Adâncata *see* Horlivka
Adapazarı 116 B2 *prev.* Ada Bazar. Sakarya, NW Turkey
Adare, Cape 154 B4 *cape* Antarctica
Ad Dahna 120 C4 *desert* E Saudi Arabia
Ad Dakhla 70 A4 *var.* Dakhla. SW Western Sahara
Ad Dalanj *see* Dilling
Ad Damar *see* Ed Damer
Ad Damazīn *see* Ed Damazin
Ad Dāmir *see* Ed Damer
Ad Dammām 120 C4 *var.* Dammām. Ash Sharqīyah, NE Saudi Arabia
Ad Dawhah *see* Doha
Ad Dawḩah 120 C4 *Eng.* Doha. *country capital* (Qatar) C Qatar
Aḏ Ḏiffah *see* Libyan Plateau
Addis Ababa *see* Ādīs Ābeba
Addoo Atoll *see* Addu Atoll
Addu Atoll 132 A5 *var.* Addoo Atoll, Seenu Atoll. *atoll* S Maldives
Adelaide 149 B6 *state capital* South Australia
Adelsberg *see* Postojna
Aden *see* 'Adan
Aden, Gulf of 121 C7 *gulf* SW Arabian Sea
Adige 96 C2 *Ger.* Etsch. *river* N Italy
Adirondack Mountains 41 F2 *mountain range* New York, NE USA

Ādīs Ābeba 73 C5 Eng. Addis Ababa.
  country capital (Ethiopia) Ādīs Ābeba,
  C Ethiopia
Adıyaman 117 E4 Adıyaman,
  SE Turkey
Adjud 108 C4 Vrancea, E Romania
Admiralty Islands 144 B3 island group
  N Papua New Guinea
Adra 93 E5 Andalucía, S Spain
Adrar 70 D3 C Algeria
Adrian 40 C3 Michigan, N USA
Adrianople/Adrianopolis see Edirne
Adriatico, Mare see Adriatic Sea
Adriatic Sea 103 E2 Alb. Deti Adriatik,
  It. Mare Adriatico, SCr. Jadransko
  More, Slvn. Jadransko Morje. sea
  N Mediterranean Sea
Adriatik, Deti see Adriatic Sea
Adycha 115 F2 river NE Russian
  Federation
Aegean Sea 105 C5 Gk. Aigaíon Pelagos,
  Aigaío Pélagos, Turk. Ege Denizi. sea
  NE Mediterranean Sea
Aegviidu 106 D2 Ger. Charlottenhof.
  Harjumaa, NW Estonia
Aegyptus see Egypt
Aelana see Al 'Aqabah
Aelok see Ailuk Atoll
Aelönlaplap see Ailinglaplap Atoll
Aemona see Ljubljana
Aeolian Islands 97 C6 var. Isole Lipari,
  Eng. Aeolian Islands, Lipari Islands.
  island group S Italy
Aeolian Islands see Eolie, Isole
Æsernia see Isernia
Afar Depression see Danakil Desert
Afars et des Issas, Territoire Français
  des see Djibouti
Afghānestān, Dowlat-e Eslāmī-ye see
  Afghanistan
Afghanistan 122 C4 off. Islamic Republic
  of Afghanistan, Per. Dowlat-e Eslāmī-
  ye Afghānestān; prev. Republic of
  Afghanistan. country C Asia

AFGHANISTAN
Central Asia

Official name The Islamic Republic
  of Afghanistan
Formation 1919 / 1919
Capital Kabul
Population 30.6 million / 122 people per
  sq mile (47 people per sq km)
Total area 250,000 sq. miles
  (647,500 sq. km)
Languages Pashtu*, Tajik, Dari*, Farsi,
  Uzbek, Turkmen
Religions Sunni Muslim 80%,
  Shi'a Muslim 19%, Other 1%
Ethnic mix Pashtun 38%, Tajik 25%,
  Hazara 19%, Uzbek and Turkmen 15%,
  Other 3%
Government Nonparty system
Currency Afghani = 100 puls
Literacy rate 32%
Calorie consumption 2107 kilocalories

Afmadow 73 D6 Jubbada Hoose,
  S Somalia
Africa 68 continent
Africa, Horn of 68 E4 physical region
  Ethiopia/Somalia
Africana Seamount 141 A6 seamount
  SW Indian Ocean
'Afrin 118 B2 Ḥalab, N Syria
Afyon 116 B3 prev. Afyonkarahisar.
  Afyon, W Turkey
Agadès see Agadez
Agadez 75 G3 prev. Agadès. Agadez,
  C Niger
Agadir 70 B3 SW Morocco
Agaña/Agaña see Hagåtña
Ágaro 73 C5 Oromīya, C Ethiopia
Agassiz Fracture Zone 143 G5 fracture
  zone S Pacific Ocean
Agatha see Agde
Agathónisi 105 D6 island Dodekánisa,
  Greece, Aegean Sea
Agde 91 C6 anc. Agatha. Hérault,
  S France
Agedabia see Ajdābiyā
Agen 91 B5 anc. Aginnum. Lot-et-
  Garonne, SW France
Agendicum see Sens
Aghri Dagh see Büyükağrı Dağı
Agiá 104 B4 var. Ayiá. Thessalía,
  C Greece
Agialoúsa see Yenierenköy

Agía Marína 105 E6 Léros, Dodekánisa,
  Greece, Aegean Sea
Aginnum see Agen
Ágios Efstrátios 104 D4 var. Áyios
  Evstrátios, Hagios Evstrátios. island
  E Greece
Ágios Nikólaos 105 D8 var.
  Áyios Nikólaos. Kríti, Greece,
  E Mediterranean Sea
Āgra 134 D3 Uttar Pradesh, N India
Agra and Oudh, United Provinces of see
  Uttar Pradesh
Agram see Zagreb
Ağrı 117 F3 var. Karaköse; prev.
  Karakılısse. Ağrı, NE Turkey
Agri Dagi see Büyükağrı Dağı
Agrigento 97 C7 Gk. Akragas;
  prev. Girgenti. Sicilia, Italy,
  C Mediterranean Sea
Agriovótano 105 C5 Évvoia, C Greece
Agropoli 97 D5 Campania, S Italy
Aguachica 58 B2 Cesar, N Colombia
Aguadulce 53 F5 Coclé, S Panama
Agua Prieta 50 B1 Sonora,
  NW Mexico
Aguascalientes 50 D4 Aguascalientes,
  C Mexico
Aguaytía 60 C3 Ucayali, C Peru
Aguilas 93 E4 Murcia, SE Spain
Aguililla 50 D4 Michoacán,
  SW Mexico
Agulhas Basin 69 D8 undersea basin
  SW Indian Ocean
Agulhas, Cape 78 C5 headland
  SW South Africa
Agulhas Plateau 67 D6 undersea plateau
  SW Indian Ocean
Ahaggar 75 F2 high plateau region
  SE Algeria
Ahlen 94 B4 Nordrhein-Westfalen,
  W Germany
Ahmadābād 134 C4 var. Ahmedabad.
  Gujarāt, W India
Ahmadnagar 134 C5 var. Ahmednagar.
  Mahārāshtra, W India
Ahmedabad see Ahmadābād
Ahmednagar see Ahmadnagar
Ahuachapán 52 B3 Ahuachapán,
  W El Salvador
Ahvāz 120 C3 var. Ahwāz; prev. Nāsiri.
  Khūzestān, SW Iran
Ahvenanmaa see Åland
Ahwāz see Ahvāz
Aigaíon Pelagos/Aigaío Pélagos see
  Aegean Sea
Aígina 105 C6 var. Aíyina, Egina.
  Aígina, C Greece
Aígio 105 B5 var. Egio; prev. Aíyion.
  Dytikí Elláda, S Greece
Aiken 43 E2 South Carolina, SE USA
Ailinglaplap Atoll 144 D1 var.
  Aelönlaplap. atoll Ralik Chain,
  S Marshall Islands
Ailuk Atoll 144 D1 var. Aelok. atoll
  Ratak Chain, NE Marshall Islands
Aïnazi 106 D3 Est. Heinaste, Ger.
  Hainasch. N Latvia
'Aïn Ben Tili 74 D1 Tiris Zemmour,
  N Mauritania
Aintab see Gaziantep
Aïoun el Atrouss/Aïoun el Atroûss see
  'Ayoûn el 'Atroûs
Aiquile 61 F4 Cochabamba, C Bolivia
Aïr see Aïr, Massif de l'
Air du Azbine see Aïr, Massif de l'
Aïr, Massif de l' 75 G2 var. Aïr, Air
  du Azbine, Asben. mountain range
  NC Niger
Aiud 108 B4 Ger. Strassburg, Hung.
  Nagyenyed; prev. Engeten. Alba,
  SW Romania
Aix see Aix-en-Provence
Aix-en-Provence 91 D6 var. Aix; anc.
  Aquae Sextiae. Bouches-du-Rhône,
  SE France
Aix-la-Chapelle see Aachen
Aíyina see Aígina
Aíyion see Aígio
Aizkraukle 106 C4 S Latvia
Ajaccio 91 E7 Corse, France,
  C Mediterranean Sea
Ajaria see Ach'ara
Ajastan see Armenia
Aj Bogd Uul 126 D2 mountain
  SW Mongolia
Ajdābiyā 71 G2 var. Agedabia,
  Ajdābiyah. NE Libya
Ajdābiyah see Ajdābiyā

Ajmer 134 D3 var. Ajmere. Rājasthān,
  N India
Ajo 48 A3 Arizona, SW USA
Akaba see Al 'Aqabah
Akamagaseki see Shimonoseki
Akasha 72 B3 Northern, N Sudan
Akchâr 74 C2 desert W Mauritania
Aken see Aachen
Akermanceaster see Bath
Akhaltsikhe 117 F2 prev. Akhalts'ikhe.
  SW Georgia
Akhalts'ikhe see Akhaltsikhe
Akhisar 116 A3 Manisa, W Turkey
Akhmim 72 B2 var. Akhmîm; anc.
  Panopolis. C Egypt
Akhtubinsk 111 C7 Astrakhanskaya
  Oblast', SW Russian Federation
Akhtyrka see Okhtyrka
Akimiski Island 38 C3 island Nunavut,
  C Canada
Akinovka 109 F4 Zaporiz'ka Oblast',
  S Ukraine
Akita 130 D4 Akita, Honshū, C Japan
Akjoujt 74 C2 prev. Fort-Repoux.
  Inchiri, W Mauritania
Akkeshi 130 E2 Hokkaidō, NE Japan
Aklavik 36 D3 Northwest Territories,
  NW Canada
Akmola see Astana
Akmolinsk see Astana
Aknavásár see Târgu Ocna
Akpatok Island 39 E1 island Nunavut,
  E Canada
Akragas see Agrigento
Akron 40 D4 Ohio, N USA
Akrotiri see Akrotírion
Akrotírion 102 C5 var. Akrotiri. UK air
  base S Cyprus
Aksai Chin 124 D1 disputed region
  China/India
Aksaray 116 C4 Aksaray, C Turkey
Aksayqin see Aksai Chin
Akşehir 116 B4 Konya, W Turkey
Aktash see Oqtosh
Aktau 114 A4 Kaz. Aqtaū; prev.
  Shevchenko. Mangistau,
  W Kazakhstan
Aktjubinsk/Aktyubinsk see Aktobe
Aktobe 114 B4 Kaz. Aqtöbe; prev.
  Aktjubinsk, Aktyubinsk. Aktyubinsk,
  NW Kazakhstan
Aktsyabrski 107 C7 Rus. Oktyabr'skiy;
  prev. Karpilovka. Homyel'skaya
  Voblasts', SE Belarus
Aktyubinsk see Aktobe
Akula 77 C5 Équateur, NW Dem. Rep.
  Congo
Akureyri 83 E4 Nordhurland Eystra,
  N Iceland
Akyab see Sittwe
Alabama 42 C2 off. State of Alabama,
  also known as Camellia State,
  Heart of Dixie, The Cotton State,
  Yellowhammer State. state S USA
Alabama River 42 C3 river Alabama,
  S USA
Alaca 116 C3 Çorum, N Turkey
Alacant see Alicante
Alagoas 63 G2 off. Estado de Alagoas.
  region E Brazil
Alagoas 63 G2 off. Estado de Alagoas.
  state E Brazil
Alaïs see Alès
Alajuela 36 C2 Alajuela, C Costa Rica
Alakanuk 36 C2 Alaska, USA
Al 'Alamayn 72 B1 var. El 'Alamein.
  N Egypt
Al 'Amārah 120 C3 var. Amara. Maysān,
  E Iraq
Alamo 47 D6 Nevada, W USA
Alamogordo 48 D3 New Mexico,
  SW USA
Alamosa 44 C5 Colorado, C USA
Åland Islands 85 C6 var. Åland Islands,
  Fin. Ahvenanmaa. island group
  SW Finland
Åland Islands see Åland
Åland Islands see Åland Sea
Alandroal see Åland. sea. strait
  Baltic Sea/Gulf of Bothnia
Alands Hav see Åland Hav
Alanya 116 C4 Antalya, S Turkey
Alappuzha 132 C3 var. Alleppey. Kerala,
  SW India
Alaşehir 116 B4 Manisa, W Turkey
Al 'Ashārah 118 E3 var. Ashara.
  Zawr, E Syria
Alaska 36 C3 off. State of Alaska,
  known as Land of the Midnight Sun,
  The Last Frontier, Seward's Folly.
  Russian America. state NW US/
Alaska, Gulf of 36 C4 var. Golfo de
  Alasca. gulf Canada/USA
Alaska Peninsula 36 C3 peninsula
  Alaska, USA
Alaska Range 34 B2 mountain range
  Alaska, USA
Al-Asnam see Chlef
Alattio see Alta
Al Awaynāt see Al 'Uwaynāt
Alaykel'/Alay-Kuu see Kék-Art
Al 'Aynā 119 B7 Al Karak, W Jordan
Alazeya 115 G2 river NE Russian
  Federation
Al Bāb 118 B2 Ḥalab, N Syria
Albacete 93 E3 Castilla-La Mancha,
  C Spain
Al Baghdādī 120 B3 var. Khān al
  Baghdādī. Al Anbār, SW Iraq
Al Bāha see Al Bāḥah
Al Bāha 121 B5 var. Al Bāha. Al
  SW Saudi Arabia
Al Baḥrayn see Bahrain
Alba Iulia 108 B4 Ger. Weissenbur
  Hung. Gyulafehérvár; prev. Bālgra
  Karlsburg, Károly-Fehérvár. Alba
  W Romania
Albania 101 C7 off. Republic of
  Albania, Alb. Republika e Shqipë
  Shqipëria; prev. People's Socialist
  Republic of Albania. country SE I

ALBANIA
Southeast Europe

Official name Republic of Albania
Formation 1912 / 1921
Capital Tirana
Population 3.2 million / 302 people
  per sq mile (117 people per sq km)
Total area 11,100 sq. miles (28,748 sq
Languages Albanian*, Greek
Religions Sunni Muslim 70%, Alban
  Orthodox 20%, Roman Catholic 10%
Ethnic mix Albanian 98%, Greek 1%,
  Other 1%
Government Parliamentary system
Currency Lek = 100 qindarka (qintari
Literacy rate 97%
Calorie consumption 3023 kilocalorie

Albany see Aubagne
Albany 147 B7 Western Australia
Albany 42 D3 Georgia, SE USA
Albany 41 F3 state capital New York
  NE USA
Albany 46 B3 Oregon, NW USA
Albany 38 C3 river Ontario, S Cana
Alba Regia see Székesfehérvár
Al Bāridah 118 C4 var. Baridah. Ḥ
  S Syria
Al Başrah 120 C3 Eng. Basra, hist. I
  Bussora. Al Başrah, SE Iraq
Al Batrūn see Batroûn
Al Baydā' 71 G2 var. Beida. NE Lib
Albemarle Island see Isabela, Isla
Albemarle Sound 43 G1 inlet W At
  Ocean
Albergaria-a-Velha 92 B2 Aveiro,
  N Portugal
Albert 90 C3 Somme, N France
Alberta 37 E4 province SW Canada
Albert Edward Nyanza see
  Edward, Lake
Albert, Lake 73 B6 var. Albert Nyan
  Lac Mobutu Sese Seko. lake Ugano
  Dem. Rep. Congo
Albert Nyanza see Albert, Lake
Albertville see Kalemie
Albi 91 C6 anc. Albiga. Tarn, S Franc
Albiga see Albi
Alborg see Aalborg
Ålborg-Nørresundby see Aalborg
Albuquerque 48 D2 New Mexico,
  SW USA
Al Burayqah see Marsá al Burayqah
Alburgum see Aalborg
Albury 149 C7 New South Wales,
  SE Australia
Alcácer do Sal 92 B4 Setúbal, W Por
Alcalá de Henares 93 E3 Ar. Alkal'a
  Complutum. Madrid, C Spain

...no 97 C7 Sicilia, Italy,
Mediterranean Sea
...hiz 93 F2 Aragón, NE Spain
...ntara, Embalse de 92 C3 reservoir
Spain
...udete 92 D4 Andalucía, S Spain
...zar see Ksar-el-Kebir
...zarquivir see Ksar-el-Kebir
...i see Alcoy
...y 93 F4 Cat. Alcoi. Valenciana,
Spain
...bra Group 79 G2 island group
...V Seychelles
...n 115 F3 river NE Russian
...deration
...r al Baida see Rabat
...rney 90 A2 island Channel Islands
74 C3 Brakna, SW Mauritania
...sandriya see Oleksandriya
...sandropol' see Gyumri
...sandrovka see Oleksandrivka
...sandrovsk see Zaporizhzhya
...sin 111 B5 Tul'skaya Oblast',
Russian Federation
...sinac 100 E4 Serbia, SE Serbia
...çon 90 B3 Orne, N France
...quer 63 E2 Pará, NE Brazil
...J/Aleppo see Ḥalab
...t 37 F1 Ellesmere Island, Nunavut,
Canada
91 C6 prev. Alais. Gard, S France
...d 108 B3 Hung. Élesd. Bihor,
...W Romania
...sandria 96 B2 Fr. Alexandrie.
...emonte, N Italy
...und 85 A5 Møre og Romsdal,
Norway
...tian Basin 113 G3 undersea basin
...ering Sea
...tian Islands 36 A3 island group
...aska, USA
...tian Range 34 A2 mountain range
...aska, USA
...tian Trench 113 H3 trench
...Bering Sea
...ander Archipelago 36 D4 island
...oup Alaska, USA
...ander City 42 D2 Alabama, S USA
...ander Island 154 A3 island
...ntarctica
...ander Range see Kirghiz Range
...xandra 151 B7 Otago, South Island,
...ew Zealand
...xándreia 104 B4 var. Alexándria.
...entríki Makedonía, N Greece
...andretta see İskenderun
...andretta, Gulf of see İskenderun
...örfezi
...andria 72 B1 Ar. Al Iskandarīyah.
...Egypt
...andria 108 C5 Teleorman,
...Romania
...andria 42 B3 Louisiana, S USA
...andria 45 F2 Minnesota, N USA
...xandria see Alexándreia
...xandrie see Alessandria
...xandroúpoli 104 D4 var.
...lexandroúpolis, Turk. Dedeagaç,
...edeagach. Anatolikí Makedonía kai
...hráki, NE Greece
...androúpolis see Alexandroúpoli
...ashír see El Fasher
...tar 104 E1 Silistra, NE Bulgaria
...áös 105 B6 prev. Alfiós; anc.
...ipheus, Alpheus. river S Greece
...ós see Alfeiós
...ld see Great Hungarian Plain
...ıurát see Euphrates
...a 114 B4 Kaz. Algha. Aktyubinsk,
...W Kazakhstan
...arve 92 B4 cultural region
...Portugal
...eciras 92 C5 Andalucía, SW Spain
...emesí 93 F3 Valenciana, E Spain
...Genain see El Geneina
...er 71 E1 var. Algiers, El Djazaïr,
...l Jazair. country capital (Algeria)
...Algeria
...eria 70 C3 off. Democratic and
...opular Republic of Algeria. country
...Africa

ALGERIA
(continued)

Population 39.2 million / 43 people
per sq mile (16 people per sq km)
Total area 919,590 sq. miles
(2,381,740 sq. km)
Languages Arabic*, Tamazight (Kabyle,
Shawia, Tamashek), French
Religions Sunni Muslim 99%,
Christian and Jewish 1%
Ethnic mix Arab 75%, Berber 24%,
European and Jewish 1%
Government Presidential system
Currency Algerian dinar = 100 centimes
Literacy rate 73%
Calorie consumption 3220 kilocalories

Algeria, Democratic and Popular
Republic of see Algeria
Algerian Basin 80 C5 var. Balearic Plain.
undersea basin W Mediterranean Sea
Algha see Alga
Al Ghābah 121 E5 var. Ghaba. C Oman
Alghero 97 A5 Sardegna, Italy,
C Mediterranean Sea
Al Ghurdaqah 72 C2 var. Hurghada,
Ghurdaqah. E Egypt
Algiers see Alger
Al Golea see El Goléa
Algona 45 F3 Iowa, C USA
Al Hajar al Gharbi 121 D5 mountain
range N Oman
Al Hamad see Syrian Desert
Al Ḥasakah 118 D2 var. Al Hasijah, El
Haseke, Fr. Hassetché. Al Ḥasakah,
NE Syria
Al Hasijah see Al Ḥasakah
Al Ḥillah 120 B3 var. Hilla. Bābil, C Iraq
Al Ḥisā 119 B7 Aṭ Ṭafīlah, W Jordan
Al Ḥudaydah 121 B6 Eng. Hodeida.
W Yemen
Al Hufūf 120 C4 var. Hofuf. Ash
Sharqīyah, NE Saudi Arabia
Aliákmon see Aliákmonas
Aliákmonas 104 B4 prev. Aliákmon; anc.
Haliacmon. river N Greece
Aliartos 105 C5 Stereá Elláda, C Greece
Alicante 93 F4 Cat. Alacant, Lat.
Lucentum. Valenciana, SE Spain
Alice 49 G5 Texas, SW USA
Alice Springs 148 A4 Northern
Territory, C Australia
Alifu Atoll see Ari Atoll
Aligandi 53 G4 Kuna Yala, NE Panama
Alikí see Alykí
Alima 77 B6 river C Congo
Al Imārāt al 'Arabiyah Muttaḥidah
see United Arab Emirates
Alindao 76 C4 Basse-Kotto, S Central
African Republic
Aliquippa 40 D4 Pennsylvania,
NE USA
Al Iskandarīyah see Alexandria
Al Ismā'īliya 72 B1 var. Ismailia,
Ismâ'iliya. N Egypt
Alistráti 104 C3 Kentrikí Makedonía,
NE Greece
Alivéri 105 C5 var. Alivérion. Évvoia,
C Greece
Alivérion see Alivéri
Al Jabal al Akhḍar 71 G2 mountain
range NE Libya
Al Jafr 119 B7 Ma'ān, S Jordan
Al Jaghbūb 71 H3 NE Libya
Al Jahrā' 120 C4 var. Al Jahrah, Jahra.
C Kuwait
Al Jahrah see Al Jahrā'
Al Jamāhīriyah al 'Arabiyah al Lībiyah
ash Sha'biyah al Ishtirākiy see Libya
Al Jawf 120 B4 var. Jauf. Al Jawf,
NW Saudi Arabia
Al Jawlān see Golan Heights
Al Jazair see Alger
Al Jazīrah 118 E2 physical region
Iraq/Syria
Al Jīzah see Giza
Al Junaynah see El Geneina
Alkal'a see Alcalá de Henares
Al Karak 119 B7 var. El Kerak, Karak,
Kerak; anc. Kir Moab, Kir of Moab. Al
Karak, W Jordan
Al-Kasr al-Kebir see Ksar-el-Kebir
Al Khalil see Hebron
Al Khārijah 72 B2 var. El Khârga.
C Egypt
Al Khums 71 F2 var. Homs, Khoms,
Khums. NW Libya

Alkmaar 86 C2 Noord-Holland,
NW Netherlands
Al Kufrah 71 H4 SE Libya
Al Kūt 120 C3 var. Kūt al 'Amārah, Kut
al Imara. Wāsiṭ, E Iraq
Al-Kuwait see Al Kuwayt
Al Kuwayt 120 C4 var. Al-Kuwait, Eng.
Kuwait, Kuwait City; prev. Qurein.
country capital (Kuwait) E Kuwait
Al Lādhiqiyah 118 A3 Eng. Latakia, Fr.
Lattaquié; anc. Laodicea, Laodicea ad
Mare. Al Lādhiqīyah, W Syria
Allahābād 135 E3 Uttar Pradesh,
N India
Allanmyo see Aunglan
Allegheny Plateau 41 E3 mountain
range New York/Pennsylvania,
NE USA
Allenstein see Olsztyn
Allentown 41 F4 Pennsylvania, NE USA
Alleppey see Alappuzha
Alliance 44 D3 Nebraska, C USA
Al Lith 121 B5 Makkah, SW Saudi Arabia
Al Lubnān see Lebanon
Alma-Ata see Almaty
Almada 92 B4 Setúbal, W Portugal
Al Madīnah 121 A5 Eng. Medina. Al
Madīnah, W Saudi Arabia
Al Mafraq 119 B6 var. Mafraq. Al
Mafraq, N Jordan
Al Maḥdīyah see Mahdia
Al Mahrah 121 C6 mountain range
E Yemen
Al Majma'ah 120 B4 Ar Riyāḍ, C Saudi
Arabia
Al Mālikīyah 118 E1 var. Malkiye. Al
Ḥasakah, NE Syria
Almalyk see Olmaliq
Al Mamlakah see Morocco
Al Mamlaka al Urduniya al
Hashemiyah see Jordan
Al Manāmah 120 C4 Eng. Manama.
country capital (Bahrain) N Bahrain
Al Manāṣif 118 E3 mountain range
E Syria
Almansa 93 F4 Castilla-La Mancha,
C Spain
Al-Mariyya see Almería
Al Marj 71 G2 var. Barka, It. Barce.
NE Libya
Almaty 114 C5 var. Alma-Ata. Almaty,
SE Kazakhstan
Al Mawṣil 72 B2 Eng. Mosul. Nīnawá,
N Iraq
Al Mayādīn 118 D3 var. Mayadin, Fr.
Meyadine. Dayr az Zawr, E Syria
Al Mazra'ah 119 B6 var. Al Mazra'ah,
N Jordan
Al Mazra'ah see Al Mazra'ah
Almelo 86 E3 Overijssel, E Netherlands
Almendra, Embalse de 92 C2 reservoir
Castilla y León, NW Spain
Almendralejo 92 C4 Extremadura,
W Spain
Almere 86 C3 var. Almere-stad.
Flevoland, C Netherlands
Almere-stad see Almere
Almería 93 E5 Ar. Al-Mariyya; anc. Unci,
Lat. Portus Magnus. Andalucía, S Spain
Al Minyā 72 B2 var. El Minya, Minya.
N Egypt
Almirante 53 E4 Bocas del Toro,
NW Panama
Al Mudawwarah 119 B8 Ma'ān,
SW Jordan
Al Mukallā 121 C6 var. Mukalla.
SE Yemen
Alofi 145 F4 dependent territory capital
(Niue) W Niue
Aloha State see Hawaii
Aloja 106 D3 N Latvia
Alónnisos 105 C5 island Vóreies
Sporádes, Greece, Aegean Sea
Álora 92 D5 Andalucía, S Spain
Alor, Kepulauan 139 E5 island group
E Indonesia
Al Oued see El Oued
Alpen see Alps
Alpena 40 D2 Michigan, N USA
Alpes see Alps
Alpha Cordillera 155 B3 var. Alpha
Ridge. seamount range Arctic Ocean
Alpha Ridge see Alpha Cordillera
Alpheus see Alfeiós
Alphen see Alphen aan den Rijn

Alphen aan den Rijn 86 C3 var. Alphen.
Zuid-Holland, C Netherlands
Alpheus see Alfeiós
Alpi see Alps
Alpine 49 E4 Texas, SW USA
Alps 102 C1 Fr. Alpes, Ger. Alpen, It.
Alpi. mountain range C Europe
Al Qadārif see Gedaref
Al Qāhirah see Cairo
Al Qāmishli 118 E1 var. Kamishli,
Qāmishly. Al Ḥasakah, NE Syria
Al Qaṣrayn see Kasserine
Al Qayrawān see Kairouan
Al-Qsar al-Kbir see Ksar-el-Kebir
Al Qubayyāt see Qoubaïyât
Al Quds/Al Quds ash Sharif see
Jerusalem
Alqueva, Barragem do 92 C4 reservoir
Portugal/Spain
Al Qunayṭirah 119 B5 var. El Kuneitra,
El Quneitra, Kuneitra, Qunayṭra. Al
Qunayṭirah, SW Syria
Al Quṣayr 118 B4 var. El Quseir, Quṣayr,
Fr. Kousséir. Ḥimṣ, W Syria
Al Quwayrah 119 B8 var. El Quweira.
'Aqabah, SW Jordan
Alsace 90 E3 Ger. Elsass; anc. Alsatia.
cultural region NE France
Alsatia see Alsace
Alsdorf 94 A4 Nordrhein-Westfalen,
W Germany
Alt see Olt
Alta 84 D2 Fin. Alattio. Finnmark,
N Norway
Altai see Altai Mountains
Altai Mountains 126 C2 var. Altai, Chin.
Altay Shan, Rus. Altay. mountain range
Asia/Europe
Altamaha River 43 E3 river Georgia,
SE USA
Altamira 63 E2 Pará, NE Brazil
Altamura 97 E5 anc. Lupatia. Puglia,
SE Italy
Altar, Desierto de 50 A1 var. Sonoran
Desert. desert Mexico/USA
Altar, Desierto de see Sonoran Desert
Altay 126 D2 prev. Yösönbulag. Govĭ-
Altay, W Mongolia
Altay Altai Mountains, Asia/Europe
Altay Shan see Altai Mountains
Altbetsche see Bečej
Altenburg see Bucureşti, Romania
Altin Köprü see Altūn Kūbrī
Altiplano 61 F4 physical region W South
America
Altkanischa see Kanjiža
Alton 40 B5 Illinois, N USA
Alton 40 B4 Missouri, C USA
Altoona 41 E4 Pennsylvania, NE USA
Alto Paraná see Paraná
Altpasua see Stara Pazova
Alt-Schwanenburg see Gulbene
Altsohl see Zvolen
Altūn Kūbrī 120 B3 var. Altin Köprü.
At Ta'mim, N Iraq
Altun Shan 126 C4 mountain range
NW China
Altus 49 F2 Oklahoma, C USA
Altyn Tagh see Altun Shan
Al Ubayyid see El Obeid
Al Ubayyiḍ see El Obeid
Alūksne 106 D3 var. Marienburg.
NE Latvia
Al 'Ulā 120 A4 Al Madīnah, NW Saudi
Arabia
Al 'Umari 119 C6 'Ammān, E Jordan
Alupka 109 F5 Avtonomna Respublika
Krym, S Ukraine
Alushta 109 F5 Avtonomna Respublika
Krym, S Ukraine
Al 'Uwaynāt 71 F4 var. Al Awaynāt.
SW Libya
Alva 49 F1 Oklahoma, C USA
Alvarado 51 F4 Veracruz-Llave,
E Mexico
Alvin 49 H4 Texas, SW USA
Al Wāḥāt al Baḥrīyah see El Bawiti
Alwar 134 D3 Rājasthān, N India
Al Wari'ah 120 C4 Ash Sharqīyah,
N Saudi Arabia
Alytaus see Alytus
Alykí 104 C4 var. Alikí. Thásos,
N Greece
Alytus 107 B5 Pol. Olita. Alytus,
S Lithuania

Alzette *87 D8 river* S Luxembourg
Amadeus, Lake *147 D5 seasonal lake*
 Northern Territory, C Australia
Amadi *73 B5* Western Equatoria,
 SW South Sudan
Amadjuak Lake *37 G3 lake* Baffin Island,
 Nunavut, N Canada
Amakusa-nada *131 A7 gulf* SW Japan
Åmål *85 B6* Västra Götaland, S Sweden
Amami-gunto *130 A3 island group*
 SW Japan
Amami-o-shim *130 A3 island* S Japan
Amantea *97 D6* Calabria, SW Italy
Amapá *63 E1 off.* Estado de Amapá;
 prev. Território de Amapá. *region*
 NE Brazil
Amapá *63 E1 off.* Estado de Amapá;
 prev. Território de Amapá. *state*
 NE Brazil
Amapá, Estado de *see* Amapá
Amapá, Território de *see* Amapá
Amara *see* Al 'Amārah
Amarapura *136 B3* Mandalay,
 C Myanmar (Burma)
Amarillo *49 E2* Texas, SW USA
Amay *87 C6* Liège, E Belgium
Amazon *63 E1 Sp.* Amazonas. *river*
 Brazil/Peru
Amazonas *see* Amazon
Amazon Basin *62 D2 basin* N South
 America
Amazon, Mouths of the *63 F1 delta*
 NE Brazil
Ambam *77 B5* Sud, S Cameroon
Ambanja *79 G2* Antsiraňana,
 N Madagascar
Ambarchik *115 G2* Respublika Sakha
 (Yakutiya), NE Russian Federation
Ambato *60 B1* Tungurahua, C Ecuador
Ambérieu-en-Bugey *91 D5* Ain,
 E France
Ambianum *see* Amiens
Amboasary *79 F4* Toliara,
 S Madagascar
Amboina *see* Ambon
Ambon *139 F4 prev.* Amboina,
 Amboyna. Pulau Ambon, E Indonesia
Ambositra *79 G3* Fianarantsoa,
 SE Madagascar
Amboyna *see* Ambon
Ambracia *see* Árta
Ambre, Cap d' *see* Bobaomby, Tanjona
Ambrim *see* Ambrym
Ambriz *78 A1* Bengo, NW Angola
Ambrym *144 D4 var.* Ambrim. *island*
 C Vanuatu
Amchitka Island *36 A2 island* Aleutian
 Islands, Alaska, USA
Amdo *126 C5* Xizang Zizhiqu, W China
Ameland *86 D1 Fris.* It Amelân. *island*
 Waddeneilanden, N Netherlands
Amelân, It *see* Ameland
America *see* United States of America
America-Antarctica Ridge *67 C7*
 undersea ridge S Atlantic Ocean
America in Miniature *see* Maryland
American Falls Reservoir *46 E4*
 reservoir Idaho, NW USA
American Samoa *145 E4 US*
 unincorporated territory W Polynesia
Amersfoort *86 D3* Utrecht,
 C Netherlands
Ames *45 F3* Iowa, C USA
Amfilochía *105 A5 var.* Amfilokhía.
 Dytiki Elláda, C Greece
Amfilokhía *see* Amfilochía
Amga *115 F3 river* NE Russian
 Federation
Amherst *39 F4* Nova Scotia, SE Canada
Amherst *see* Kyaikkami
Amida *see* Diyarbakır
Amiens *90 C3 anc.* Ambianum,
 Samarobriva. Somme, N France
Amíndaion/Amindeo *see* Amýntaio
Amindivi Islands *132 A2 island group*
 Lakshadweep, India, N Indian Ocean
Amirante Islands *79 G1 var.* Amirantes
 Group. *island group* C Seychelles
Amirantes Group *see* Amirante Islands
Amistad, Presa de la *see* Amistad
 Reservoir
Amistad Reservoir *49 F4 var.* Presa de la
 Amistad. *reservoir* Mexico/USA
Amisus *see* Samsun
Ammaia *see* Portalegre
'Amman *119 B6 anc.* Philadelphia, *Bibl.*
 Rabbah Ammon, Rabbath Ammon.
 country capital (Jordan) 'Ammān,
 NW Jordan

Ammassalik *82 D4 var.* Angmagssalik.
 Tunu, S Greenland
Ammóchostos *see* Gazimağusa
Ammóchostos, Kólpos *see* Gazimağusa
 Körfezi
Amnok-kang *see* Yalu
Amoea *see* Portalegre
Amoentai *see* Amuntai
Āmol *120 D2 var.* Amul. Māzandarān,
 N Iran
Amorgós *105 D6* Amorgós, Kykládes,
 Greece, Aegean Sea
Amorgós *105 D6 island* Kykládes,
 Greece, Aegean Sea
Amos *38 D4* Québec, SE Canada
Amourj *74 D3* Hodh ech Chargui,
 SE Mauritania
Amoy *see* Xiamen
Ampato, Nevado *61 E4 mountain*
 S Peru
Amposta *93 F2* Cataluña, NE Spain
Amraoti *see* Amrāvati
Amrāvati *134 D4 prev.* Amraoti.
 Mahārāshtra, C India
Amritsar *134 D2* Punjab, N India
Amstelveen *86 C3* Noord-Holland,
 C Netherlands
Amsterdam *86 C3 country capital*
 (Netherlands) Noord-Holland,
 C Netherlands
Amsterdam Island *141 C6 island*
 NE French Southern and Antarctic
 Lands
Am Timan *76 C3* Salamat, SE Chad
Amu Darya *122 D2 Rus.* Amudar'ya,
 Taj. Dar''yoi Amu, *Turkm.* Amyderya,
 Uzb. Amudaryo; *anc.* Oxus. *river*
 C Asia
Amu-Dar'ya *see* Amyderýa
Amudar'ya/Amudaryo/Amu, Dar''yoi
 see Amu Darya
Amul *see* Āmol
Amund Ringnes Island *37 F2 Island*
 Nunavut, N Canada
Amundsen Basin *155 C3 prev.* Fram
 Basin. *undersea basin* Arctic Ocean
Amundsen Plain *154 A4 abyssal plain*
 S Pacific Ocean
Amundsen-Scott *154 B3 US research*
 station Antarctica
Amundsen Sea *154 A4 sea* S Pacific
 Ocean
Amuntai *138 D4 prev.* Amoentai.
 Borneo, C Indonesia
Amur *115 G4 Chin.* Heilong Jiang. *river*
 China/Russian Federation
Amvrosiyevka *see* Amvrosiyivka
Amvrosiyivka *109 H3 Rus.*
 Amvrosiyevka. Donets'ka Oblast',
 SE Ukraine
Amyderýa *123 E3 Rus.* Amu-Dar'ya.
 Lebap Welaýaty, NE Turkmenistan
Amyderya *see* Amu Darya
Amýntaio *104 B4 var.* Amindeo; *prev.*
 Amíndaion. Dytiki Makedonía,
 N Greece
Anabar *115 E2 river* NE Russian
 Federation
An Abhainn Mhór *see* Blackwater
Anaco *59 E2* Anzoátegui, NE Venezuela
Anaconda *44 B2* Montana, NW USA
Anacortes *46 B1* Washington,
 NW USA
Anadolu Dağları *see* Doğu Karadeniz
 Dağları
Anadyr' *115 H1* Chukotskiy
 Avtonomnyy Okrug, NE Russian
 Federation
Anadyr' *115 G1 river* NE Russian
 Federation
Anadyr, Gulf of *see* Anadyrskiy Zaliv
Anáfi *105 D7 anc.* Anaphe. *island*
 Kykládes, Greece, Aegean Sea
'Ānah *see* 'Annah
Anaheim *46 E2* California, W USA
Anaiza *see* 'Unayzah
Analalava *79 G2* Mahajanga,
 NW Madagascar
Anamur *116 C5* İçel, S Turkey
Anantapur *132 C2* Andhra Pradesh,
 S India
Anaphe *see* Anáfi
Anápolis *63 F3* Goiás, C Brazil
Anār *120 D3* Kermān, C Iran
Anatolia *116 C4 plateau* C Turkey
Anatom *see* Aneityum
Añatuya *64 C3* Santiago del Estero,
 N Argentina

An Bhearú *see* Barrow
Anchorage *36 C3* Alaska, USA
Ancona *96 C3* Marche, C Italy
Ancud *65 B6 prev.* San Carlos de Ancud.
 Los Lagos, S Chile
Ancyra *see* Ankara
Åndalsnes *85 A5* Møre og Romsdal,
 S Norway
Andalucía *92 D4 cultural region*
 S Spain
Andalusia *42 C3* Alabama, S USA
Andaman Islands *124 B4 island group*
 India, NE Indian Ocean
Andaman Sea *124 C4 sea* NE Indian
 Ocean
Andenne *87 C6* Namur, SE Belgium
Anderlues *87 B7* Hainaut, S Belgium
Anderson *40 C4* Indiana, N USA
Andes *64 B3 mountain range* W South
 America
Andhra Pradesh *135 E5 cultural region*
 E India
Andijon *123 F2 Rus.* Andizhan. Andijon
 Viloyati, E Uzbekistan
Andikíthira *see* Antikýthira
Andípaxi *see* Antípaxoi
Andípsara *see* Antípsara
Andíssa *see* Ántissa
Andizhan *see* Andijon
Andkhvöy *122 D3 prev.* Andkhvoy.
 Färyäb, N Afghanistan
Andorra *91 A7 off.* Principality of
 Andorra, *Cat.* Valls d'Andorra, *Fr.*
 Vallée d'Andorre. *country* SW Europe

ANDORRA
Southwest Europe

Official name Principality of Andorra
Formation 1278 / 1278
Capital Andorra la Vella
Population 85,293 / 474 people
 per sq mile (183 people per sq km)
Total area 181 sq. miles (468 sq. km)
Languages Spanish, Catalan*, French,
 Portuguese
Religions Roman Catholic 94%, Other 6%
Ethnic mix Spanish 46%, Andorran 28%,
 Other 18%, French 8%
Government Parliamentary system
Currency Euro = 100 cents
Literacy rate 99%
Calorie consumption Not available

Andorra *see* Andorra la Vella
Andorra la Vella *91 A8 var.* Andorra,
 Fr. Andorre la Vielle, *Sp.* Andorra
 la Vieja. *country capital* (Andorra)
 C Andorra
Andorra la Vieja *see* Andorra la Vella
Andorra, Principality of *see* Andorra
Andorra, Valls d'/Andorra, Vallée d'
 see Andorra
Andorre la Vielle *see* Andorra la Vella
Andover *89 D7* S England, United
 Kingdom
Andøya *84 C2 island* C Norway
Andreanof Islands *36 A3 island group*
 Aleutian Islands, Alaska, USA
Andria *97 D5* Puglia, SE Italy
Andropov *see* Rybinsk
Ándros *105 D6* Ándros, Kykládes,
 Greece, Aegean Sea
Ándros *105 C6 island* Kykládes, Greece,
 Aegean Sea
Andros Island *54 B2 island*
 NW Bahamas
Andros Town *54 C1* Andros Island,
 NW Bahamas
Andrug, Gora *see* Andrew Tablemount
Andrýu, Gora *see* Andrew Tablemount
Aneityum *144 D5 var.* Anatom; *prev.*
 Kéamu. *island* S Vanuatu
Anewetak *see* Enewetak Atoll
Angara *115 E4 river* S Russian
 Federation
Angarsk *115 E4* Irkutskaya Oblast',
 S Russian Federation
Ånge *85 C5* Västernorrland, C Sweden
Ángel de la Guarda, Isla *50 B2 island*
 NW Mexico
Angeles *139 E1 off.* Angeles City. Luzon,
 N Philippines
Angeles City *see* Angeles
Angel Falls *59 E3 Eng.* Angel Falls.
 waterfall E Venezuela

Angel Falls *see* Ángel, Salto
Angerburg *see* Węgorzewo
Ångermanälven *84 C4 river* N Swe
Angermünde *94 D3* Brandenburg,
 NE Germany
Angers *90 B4 anc.* Juliomagus. Ma
 Loire, NW France
Anglesey *89 C5 island* NW Wales,
 United Kingdom
Anglet *91 A6* Pyrénées-Atlantiques
 SW France
Angleton *49 H4* Texas, SW USA
Anglia *see* England
Anglo-Egyptian Sudan *see* Sudan
Angmagssalik *see* Ammassalik
Ang Nam Ngum *136 C4 lake* C Lac
Angola *78 B2 off.* Republic of Anga
 prev. People's Republic of Angola
 Portuguese West Africa. *country*
 SW Africa

ANGOLA
Southern Africa

Official name Republic of Angola
Formation 1975 / 1975
Capital Luanda
Population 21.5 million / 45 people
 per sq mile (17 people per sq km)
Total area 481,351 sq. miles
 (1,246,700 sq. km)
Languages Portuguese*, Umbundu,
 Kimbundu, Kikongo
Religions Roman Catholic 68%,
 Protestant 20%, Indigenous beliefs
 Ethnic mix Ovimbundu 37%, Other
 Kimbundu 25%, Bakongo 13%
Government Presidential system
Currency Readjusted kwanza = 100 t
Literacy rate 71%
Calorie consumption 2400 kilocalor

Angola Basin *69 B5 undersea basin*
 E Atlantic Ocean
Angola, People's Republic of *see*
 Angola
Angora *see* Ankara
Angostura *see* Ciudad Bolívar
Angostura, Presa de la *51 G5 reser*
 SE Mexico
Angouleme *91 B5 anc.* Iculisma.
 Charente, W France
Angoumois *91 B5 cultural region*
 W France
Angra Pequena *see* Lüderitz
Angren *123 F2* Toshkent Viloyati,
 E Uzbekistan
Anguilla *55 G3 UK dependent terri*
 E West Indies
Anguilla Cays *54 B2 islets* SW Baha
Anhui *128 C5 var.* Anhui Sheng,
 Anhwei, Wan. *province* E China
AnhuiSheng/Anhwei Wan *see* Anh
Anicium *see* le Puy
Anina *108 A4 Ger.* Steierdorf, Hung
 Stájerlakanina; *prev.* Staierdorf-An
 Steierdorf-Anina, Steyerlak-Anina
 Caraş-Severin, SW Romania
Anjou *90 B4 cultural region* NW Fra
Anjouan *see* Nzwani
Ankara *116 C3 prev.* Angora; *anc.*
 Ancyra. *country capital* (Turkey)
 Ankara, C Turkey
Ankeny *45 F3* Iowa, C USA
Anklam *94 D2* Mecklenburg-
 Vorpommern, NE Germany
An Mhuir Cheilteach *see* Celtic Sea
Annaba *71 E1 prev.* Bône. NE Alger
An Nafud *120 B4 desert* NW Saudi
 Arabia
An Najaf *120 B3 var.* 'Anah. Al Anbā
 NW Iraq
An Najaf *120 B3 var.* Najaf. An Naj
 S Iraq
Annamite Mountains *136 D4 Fr.*
 Annamitique, Chaîne.
 mountain range C Laos
Annamitique, Chaîne *see* Annamite
 Mountains
Annamses Cordillera *see* Annamite
 Mountains
Annapolis *41 F4 state capital* Maryla
 NE USA
Annapurna *135 E3 mountain* C N;
Ann Arbor *40 C3* Michigan, N USA
An Nāṣirīyah *120 C3 var.* Nasiriya.
 Qār, SE Iraq

iacum *see* Annecy
y 91 D5 *anc.* Anneciacum.
ate-Savoie, E France
al Abyaḍ *see* White Nile
al Azraq *see* Blue Nile
ton 42 D2 Alabama, S USA
to Bay 54 B4 C Jamaica
naigh *see* Omagh
g 128 D5 Anhui, E China
La Raye 55 F1 NW Saint Lucia
n 128 B6 Guizhou, S China
go 75 E3 Gao, E Mali
th Bán *see* Strabane
ya 116 D4 *anc.* Antioch,
iochia. Hatay, S Turkey
ha 79 G2 Antsiraňana,
Madagascar
ya 116 B4 *prev.* Adalia; *anc.*
leia, *Bibl.* Attalia. Antalya,
Turkey
ya, Gulf of 116 B4 *var.* Gulf of
lia, *Eng.* Gulf of Antalya. *gulf*
Turkey
ya, Gulf of *see* Antalya Körfezi
anarivo 79 G3 *prev.* Tananarive.
ntry capital (Madagascar)
tananarivo, C Madagascar
ctica 154 B3 *continent*
ctic Peninsula 154 A2 *peninsula*
arctica
a *see* Gaziantep
quera 92 D5 *anc.* Anticaria,
iquaria. Andalucía, S Spain
quera *see* Oaxaca
es 91 D6 *anc.* Antipolis. Alpes-
ritimes, SE France
aria *see* Antequera
osti, Île d' 39 F3 *Eng.* Anticosti
nd. *island* Québec, E Canada
osti Island *see* Anticosti, Île d'
ua 55 G3 *island* S Antigua and
rbuda, Leeward Islands
ua and Barbuda 55 G3 *country*
West Indies

IGUA & BARBUDA
West Indies

cial name Antigua and Barbuda
ation 1981 / 1981
tal St. John's
ulation 90,156 / 530 people
sq mile (205 people per sq km)
area 170 sq. miles (442 sq. km)
guages English*, English patois
gions Anglican 45%, Other Protestant
Roman Catholic 10%, Other 2%,
afarian 1%
ic mix Black African 95%, Other 5%
ernment Parliamentary system
ency Eastern Caribbean dollar
0 cents
acy rate 99%
rie consumption 2396 kilocalories

cythira 105 B7 *var.* Andikíthira.
and S Greece
Lebanon 118 B4 *var.* Jebel esh
arqi, *Ar.* Al Jabal ash Sharqī, *Fr.* Anti-
an. *mountain range* Lebanon/Syria
Liban *see* Anti-Lebanon
och *see* Antakya
chia *see* Antakya
paxoi 105 A5 *var.* Andipaxi.
and Iónia Nísiá, Greece,
Mediterranean Sea
odes Islands 142 D5 *island group*
New Zealand
olis *see* Antibes
sara 105 D5 *var.* Andípsara.
and E Greece
quaria *see* Antequera
sa 105 D5 *var.* Andíssa. Lésvos,
Greece
úr *see* Newry
rari *see* Bar
fagasta 64 B2 Antofagasta, N Chile
ny 90 F2 Hauts-de-Seine,
France
sionainn *see* Shannon
rañana 79 G2 *province*
Madagascar
ohihy 79 G2 Mahajanga,
N Madagascar
ung *see* Dandong
erpen 87 C5 *Eng.* Antwerp, *Fr.*
vers. Antwerpen, N Belgium
adhapura 132 D3 North Central
ovince, C Sri Lanka

Anvers *see* Antwerpen
Anyang 128 C4 Henan, C China
A'nyêmaqên Shan 126 D4 *mountain range* C China
Anykščiai 106 C4 Utena, E Lithuania
Anzio 97 C5 Lazio, C Italy
Ao Krung Thep 137 C5 *var.* Krung Thep Mahanakhon, *Eng.* Bangkok. *country capital* (Thailand) Bangkok, C Thailand
Aomori 130 D3 Aomori, Honshū, C Japan
Aóos *see* Vjosës, Lumi i
Aoraki 151 B6 *prev.* Aorangi, Mount Cook. *mountain* South Island, New Zealand
Aorangi *see* Aoraki
Aosta 96 A1 *anc.* Augusta Praetoria. Valle d'Aosta, NW Italy
Aoukâr 74 D3 *var.* Aouker. *plateau* C Mauritania
Aouk, Bahr 76 C4 *river* Central African Republic/Chad
Aouker *see* Aoukâr
Aozou 76 C1 Borkou-Ennedi-Tibesti, N Chad
Apalachee Bay 42 D3 *bay* Florida, SE USA
Apalachicola River 42 D3 *river* Florida, SE USA
Apamama *see* Abemama
Apaporis, Río 58 C4 *river* Brazil/Colombia
Apatity 110 C2 Murmanskaya Oblast', NW Russian Federation
Ape 106 D3 NE Latvia
Apeldoorn 86 D3 Gelderland, E Netherlands
Apennines 96 E2 *Eng.* Apennines. *mountain range* Italy/San Marino
Apennines *see* Appennino
Āpia 145 F4 *country capital* (Samoa) Upolu, SE Samoa
Apkhazeti 117 E1 *var.* Abkhazia; *prev.* Ap'khazet'i. *autonomous republic* NW Georgia
Ap'khazet'i *see* Apkhazeti
Apoera 59 G3 Sipaliwini, NW Suriname
Apostle Islands 40 B1 *island group* Wisconsin, N USA
Appalachian Mountains 35 D5 *mountain range* E USA
Appingedam 86 E1 Groningen, NE Netherlands
Appleton 40 B2 Wisconsin, N USA
Apulia *see* Puglia
Apure, Río 58 C2 *river* W Venezuela
Apurímac, Río 60 D3 *river* S Peru
Apuseni, Munții 108 A4 *mountain range* W Romania
Aqaba/'Aqaba *see* Al 'Aqabah
Aqaba, Gulf of 120 A4 *var.* Gulf of Elat, *Ar.* Khalīj al 'Aqabah; *anc.* Sinus Aelaniticus. *gulf* NE Red Sea
'Aqabah, Khalīj al *see* Aqaba, Gulf of
Āqchah 123 E3 *var.* Āqcheh. Jowzjān, N Afghanistan
Āqcheh *see* Āqchah
Aqmola *see* Astana
Aqtöbe *see* Aktobe
Aquae Augustae *see* Dax
Aquae Calidae *see* Bath
Aquae Flaviae *see* Chaves
Aquae Grani *see* Aachen
Aquae Sextiae *see* Aix-en-Provence
Aquae Solis *see* Bath
Aquae Tarbelicae *see* Dax
Aquidauana 63 E4 Mato Grosso do Sul, S Brazil
Aquila/Aquila degli Abruzzi *see* L'Aquila
Aquisgranum *see* Aachen
Aquitaine 91 B6 *cultural region* SW France
'Arabah, Wadi al 119 B7 *Heb.* Ha'Arava. *dry watercourse* Israel/Jordan
Arabian Basin 124 A4 *undersea basin* N Arabian Sea
Arabian Desert *see* Sahara el Sharqīya
Arabian Peninsula 121 B5 *peninsula* SW Asia
Arabian Sea 124 A3 *sea* NW Indian Ocean
Arabicus, Sinus *see* Red Sea
'Arabī, Khalīj al *see* Persian Gulf
'Arabīyah as Su'ūdīyah, Al Mamlakah al *see* Saudi Arabia
'Arabīyah Jumhūrīyah, Mişr al *see* Egypt

Arab Republic of Egypt *see* Egypt
Aracaju 63 G3 *state capital* Sergipe, E Brazil
Araçuaí 63 F3 Minas Gerais, SE Brazil
Arad 119 B7 Southern, S Israel
Arad 108 A4 Arad, W Romania
Arafura Sea 142 A3 *Ind.* Laut Arafuru. *sea* W Pacific Ocean
Arafuru, Laut *see* Arafura Sea
Aragón 93 E2 *autonomous community* E Spain
Araguaia, Río 63 E3 *var.* Araguaya. *river* C Brazil
Araguari 63 F3 Minas Gerais, SE Brazil
Araguaya *see* Araguaia, Río
Ara Jovis *see* Aranjuez
Arāk 123 C3 *prev.* Sultānābād. Markazī, W Iran
Arakan Yoma 136 A3 *mountain range* W Myanmar (Burma)
Araks/Arak's *see* Aras
Aral Sea 122 C2 *Kaz.* Aral Tengizi, *Rus.* Aral'skoye More, *Uzb.* Orol Dengizi. *inland sea* Kazakhstan/Uzbekistan
Aral'sk 114 B4 *Kaz.* Aral. Kzylorda, SW Kazakhstan
Aral'skoye More/Aral Tengizi *see* Aral Sea
Aranda de Duero 92 D2 Castilla y León, N Spain
Arandelovac 100 D4 *prev.* Arandjelovac. Serbia, C Serbia
Arandjelovac *see* Arandelovac
Aranjuez 92 D3 *anc.* Ara Jovis. C Spain
Araouane 75 E2 Tombouctou, N Mali
'Ar'ar 120 B3 Al Ḥudūd ash Shamālīyah, NW Saudi Arabia
Mount Ararat 117 F3 *var.* Aghri Dagh, Agri Dagi, Koh I Noh, Masis, *Eng.* Great Ararat, Mount Ararat. *mountain* E Turkey
Ararat, Mount *see* Büyükağrı Dağı
Aras 117 E3 *Arm.* Arak's, *Az.* Araz Nehri, *Per.* Rūd-e Aras, *Rus.* Araks; *prev.* Araxes. *river* SW Asia
Aras, Rūd-e *see* Aras
Arauca 58 C2 Arauca, NE Colombia
Arauca, Río 58 C2 *river* Colombia/Venezuela
Arauju *see* Orange
Araxes *see* Aras
Araz Nehri *see* Aras
Arbela *see* Arbīl
Arbīl 120 B2 *var.* Erbil, Irbīl, *Kurd.* Hawlêr; *anc.* Arbela. Arbīl, N Iraq
Arbroath 88 D3 *anc.* Aberbrothock. E Scotland, United Kingdom
Arbuzinka *see* Arbuzynka
Arbuzynka 109 E3 *Rus.* Arbuzinka. Mykolayivs'ka Oblast', S Ukraine
Arcachon 91 B5 Gironde, SW France
Arcae Remorum *see* Châlons-en-Champagne
Arcata 46 A4 California, W USA
Archangel *see* Arkhangel'sk
Archangel Bay *see* Chëshskaya Guba
Archidona 92 D5 Andalucía, S Spain
Arco 96 C2 Trentino-Alto Adige, N Italy
Arctic Mid Oceanic Ridge *see* Gakkel Ridge
Arctic Ocean 155 B3 *ocean*
Arda 104 C3 *var.* Ardhas, *Gk.* Ardas. *river* Bulgaria/Greece
Ardabīl 120 C2 *var.* Ardebil, Ardabīl, NW Iran
Ardākān 120 D3 Yazd, C Iran
Ardas/Ardhas *see* Arda
Ardebil *see* Ardabīl
Ardèche 91 C5 *cultural region* E France
Ardennes 87 C8 *physical region* Belgium/France
Ardhas *see* Arda/Ardas
Ardh aş Şawwān 119 C7 *var.* Ardh es Suwwān. *plain* S Jordan
Ardino *see* Arda
Ardmore 49 G2 Oklahoma, C USA
Arel *see* Arlon
Arelas/Arelate *see* Arles
Arendal 85 A6 Aust-Agder, S Norway
Arensburg *see* Kuressaare
Arenys de Mar 93 G2 Cataluña, NE Spain

Areópoli 105 B7 *prev.* Areópolis. Pelopónnisos, S Greece
Areópolis *see* Areópoli
Arequipa 61 E4 Arequipa, SE Peru
Arezzo 96 C3 *anc.* Arretium. Toscana, C Italy
Argalastí 105 C5 Thessalía, C Greece
Argenteuil 90 D1 Val-d'Oise, N France
Argentina 65 B5 *off.* Argentine Republic. *country* S South America

ARGENTINA
South America

Official name The Argentine Republic
Formation 1816 / 1816
Capital Buenos Aires
Population 41.4 million / 39 people per sq mile (15 people per sq km)
Total area 1,068,296 sq. miles (2,766,890 sq. km)
Languages Spanish*, Italian, Amerindian languages
Religions Roman Catholic 70%, Other 18%, Protestant 10%, Muslim 2%, Jewish 2%
Ethnic mix Indo-European 97%, Mestizo 2%, Amerindian 1%
Government Presidential system
Currency Argentine peso = 100 centavos
Literacy rate 98%
Calorie consumption 3155 kilocalories

Argentina Basin *see* Argentine Basin
Argentine Basin 57 C7 *var.* Argentina Basin. *undersea basin* SW Atlantic Ocean
Argentine Republic *see* Argentina
Argentine Rise *see* Falkland Plateau
Argentoratum *see* Strasbourg
Darya-ye Arghandab 123 E5 *river* SE Afghanistan
Argirocastro *see* Gjirokastër
Argo 72 B3 Northern, N Sudan
Argo Fracture Zone 141 C5 *tectonic Feature* C Indian Ocean
Árgos 105 B6 Pelopónnisos, S Greece
Argostóli 105 A5 *var.* Argostólion. Kefalloniá, Iónia Nísiá, Greece, C Mediterranean Sea
Argostólion *see* Argostóli
Argun 121 E1 *Chin.* Ergun He, *Rus.* Argun'. *river* China/Russian Federation
Argyrokastron *see* Gjirokastër
Århus 85 B7 *var.* Aarhus. Midtjylland, C Denmark
Aria *see* Herāt
Ari Atoll 132 A4 *var.* Alifu Atoll. *atoll* C Maldives
Arica 64 B1 *hist.* San Marcos de Arica. Arica y Parinacota, N Chile
Aridaía 104 B3 *var.* Aridea, Aridhaía. Dytikí Makedonía, N Greece
Aridea *see* Aridaía
Aridhaía *see* Aridaía
Arīhā 118 B3 Al Karak, W Jordan
Arīhā *see* Jericho
Ariminum *see* Rimini
Arinsal 91 A7 NW Andorra Europe
Arizona 48 A2 *off.* State of Arizona, also known as Copper State, Grand Canyon State. *state* SW USA
Arkansas 42 A1 *off.* State of Arkansas, also known as The Land of Opportunity. *state* S USA
Arkansas City 45 F5 Kansas, C USA
Arkansas River 49 G1 *river* C USA
Arkhangel'sk 114 B2 *Eng.* Archangel. Arkhangel'skaya Oblast', NW Russian Federation
Arkoi 105 E6 *island* Dodekánisa, Greece, Aegean Sea
Arles 91 D6 *var.* Arles-sur-Rhône; *anc.* Arelas, Arelate. Bouches-du-Rhône, SE France
Arles-sur-Rhône *see* Arles
Arlington 49 G3 Texas, SW USA
Arlington 41 E4 Virginia, NE USA
Arlon 87 D8 *Dut.* Aarlen, *Ger.* Arel, *Lat.* Orolaunum. Luxembourg, SE Belgium
Armagh 89 B5 *Ir.* Ard Mhacha. S Northern Ireland, United Kingdom
Armagnac 91 B6 *cultural region* SW France
Armenia 58 B3 Quindío, W Colombia

Armenia *117 F3 off.* Republic of Armenia, *var.* Ajastan, *Arm.* Hayastani Hanrapetut'yun; *prev.* Armenian Soviet Socialist Republic. *country* SW Asia

ARMENIA
Southwest Asia

Official name Republic of Armenia
Formation 1991 / 1991
Capital Yerevan
Population 3 million / 261 people per sq mile (101 people per sq km)
Total area 11,506 sq. miles (29,800 sq. km)
Languages Armenian*, Azeri, Russian
Religions Armenian Apostolic Church (Orthodox) 88%, Other 6%, Armenian Catholic Church 6%, Other 6%
Ethnic mix Armenian 98%, Other 1%, Yezidi 1%
Government Parliamentary system
Currency Dram = 100 luma
Literacy rate 99%
Calorie consumption 2809 kilocalories

Armenian Soviet Socialist Republic *see* Armenia
Armenia, Republic of *see* Armenia
Armidale *149 D6* New South Wales, SE Australia
Armstrong *38 B3* Ontario, S Canada
Armyans'k *109 F4 Rus.* Armyansk. Avtonomna Respublika Krym, S Ukraine
Arnaía *104 C4 Cont.* Arnea. Kentrikí Makedonía, N Greece
Arnaud *82 A3 river* Québec, E Canada
Arnea *see* Arnaía
Arnedo *93 E2* La Rioja, N Spain
Arnhem *86 D4* Gelderland, SE Netherlands
Arnhem Land *148 A2 physical region* Northern Territory, N Australia
Arno *96 B3 river* C Italy
Arnold *45 G4* Missouri, C USA
Arnswalde *see* Choszczno
Aroe Islands *see* Aru, Kepulauan
Arorae *145 E3 atoll* Tungaru, W Kiribati
Arrabona *see* Győr
Ar Rahad *see* Er Rahad
Ar Ramādī *120 B3 var.* Ramadi, Rumadiya. Al Anbār, SW Iraq
Ar Rāmī *118 C4* Ḩimş, C Syria
Ar Ramthā *119 B5 var.* Ramtha. Irbid, N Jordan
Arran, Isle of *88 C4 island* SW Scotland, United Kingdom
Ar Raqqah *118 C2 var.* Rakka; *anc.* Nicephorium. Ar Raqqah, N Syria
Arras *90 C2 anc.* Nemetocenna. Pas-de-Calais, N France
Ar Rawḑatayn *120 C4 var.* Raudhatain. N Kuwait
Arretium *see* Arezzo
Arriaca *see* Guadalajara
Arriaga *51 G5* Chiapas, SE Mexico
Ar Riyāḑ *121 C5 Eng.* Riyadh. *country capital* (Saudi Arabia) Ar Riyāḑ, C Saudi Arabia
Ar Rub 'al Khali *121 C6 var.* Empty Quarter, Great Sandy Desert. *desert* SW Asia
Ar Rustāq *121 E5 var.* Rostak, Rustaq. N Oman
Ar Ruţbah *120 B3 var.* Rutba. Al Anbār, SW Iraq
Árta *105 A5 anc.* Ambracia. Ípeiros, W Greece
Artashat *117 F3* S Armenia
Artemisa *54 B2* La Habana, W Cuba
Artesia *48 D3* New Mexico, SW USA
Arthur's Pass *151 C6 pass* South Island, New Zealand
Artigas *64 D3 prev.* San Eugenio, San Eugenio del Cuareim. Artigas, N Uruguay
Art'ik *117 F2* W Armenia
Artois *90 C2 cultural region* N France
Artsiz *see* Artsyz
Artsyz *108 D4 Rus.* Artsiz. Odes'ka Oblast', SW Ukraine
Artvin *117 F2* Artvin, NE Turkey
Arua *73 B6* NW Uganda
Aruba *58 C1 var.* Oruba. *Dutch autonomous region* S West Indies
Aru Islands *see* Aru, Kepulauan
Aru, Kepulauan *129 G4 Eng.* Aru Islands; *prev.* Aroe Islands. *island group* E Indonesia

Arunāchal Pradesh *135 G3 prev.* North East Frontier Agency, North East Frontier Agency of Assam. *cultural region* NE India
Arusha *73 C7* Arusha, N Tanzania
Arviat *37 G4 prev.* Eskimo Point. Nunavut, C Canada
Arvidsjaur *84 C4* Norrbotten, N Sweden
'Arys' *114 B5 prev.* Arys'. Yuzhnyy Kazakhstan, S Kazakhstan
Arys' *see* 'Arys'
Asadābād *123 F4 var.* Asaḑābād; *prev.* Chaghasarāy. Kunar, E Afghanistan
Asaḑābād *see* Asadābād
Asahi-dake *130 D2 mountain* Hokkaidō, N Japan
Asahikawa *130 D2* Hokkaidō, N Japan
Asamankese *75 E5* SE Ghana
Āsānsol *135 F4* West Bengal, NE India
Asben *see* Aïr, Massif de l'
Ascension Fracture Zone *69 A5 tectonic Feature* C Atlantic Ocean
Ascension Island *67 C5 UK dependent territory* C Atlantic Ocean
Ascoli Piceno *96 C4 anc.* Asculum Picenum. Marche, C Italy
Asculum Picenum *see* Ascoli Piceno
'Aseb *72 D4 var.* Assab, *Amh.* Āseb. SE Eritrea
Assen *86 E2* Drenthe, NE Netherlands
A Serra de Outes *92 B1* Galicia, NW Spain
Aşgabat *122 C3 prev.* Ashgabat, Ashkhabad, Poltoratsk. *country capital* (Turkmenistan) Ahal Welaýaty, C Turkmenistan
Ashara *see* Al 'Ashārah
Ashburton *151 C6* Canterbury, South Island, New Zealand
Ashburton River *146 A4 river* Western Australia
Ashdod *119 A6 anc.* Azotos, *Lat.* Azotus. Central, W Israel
Asheville *43 E1* North Carolina, SE USA
Ashgabat *see* Aşgabat
Ashkelon *119 A6 prev.* Ashqelon. Southern, C Israel
Ashkhabad *see* Aşgabat
Ashland *46 B4* Oregon, NW USA
Ashland *40 B1* Wisconsin, N USA
Ashmore and Cartier Islands *142 A3 Australian external territory* E Indian Ocean
Ashmyany *107 C5 Rus.* Oshmyany. Hrodzyenskaya Voblasts', W Belarus
Ashqelon *see* Ashkelon
Ash Shaddādah *118 D2 var.* Ash Shaddādah, Jisr ash Shadadi, Shaddādī, Shedadi, Tell Shedadi. Al Ḩasakah, NE Syria
Ash Shadādāh *see* Ash Shaddādah
Ash Sharah *119 B7 var.* Esh Sharā. *mountain range* W Jordan
Ash Shāriqah *120 D4 Eng.* Sharjah. Ash Shāriqah, NE United Arab Emirates
Ash Shawbak *119 B7* Ma'ān, W Jordan
Ash Shiḩr *121 C6* SE Yemen
Asia *112 continent*
Asinara *96 A4 island* W Italy
Asipovichy *107 D6 Rus.* Osipovichi. Mahilyowskaya Voblasts', C Belarus
Aşkale *117 E3* Erzurum, NE Turkey
Askersund *85 C6* Örebro, C Sweden
Asmara *see* Asmera
Asmera *72 C4 var.* Asmara. *country capital* (Eritrea) C Eritrea
Aspadana *see* Eşfahān
Asphaltites, Lacus *see* Dead Sea
Aspinwall *see* Colón
Assab *see* 'Aseb
As Sabkhah *118 D2 var.* Sabkha. Ar Raqqah, NE Syria
Assad, Lake *118 C2 Eng.* Lake Assad. *lake* N Syria
Assad, Lake *see* Assad, Buḩayrat al
Aş Şafāwī *119 C6* Al Mafraq, N Jordan
Aş Şaḩrā' ash Sharqīyah *see* Sahara el Sharqīya
As Salamīyah *see* Salamīyah
'Assal, Lac *68 E4* lake C Djibouti
As Salţ *119 B6 var.* Salt. Al Balqā', NW Jordan
Assamaka *see* Assamakka
Assamakka *75 F2 var.* Assamaka. Agadez, NW Niger
As Samāwah *120 B3 var.* Samawa. Al Muthanná, S Iraq
Assenede *87 B5* Oost-Vlaanderen, NW Belgium

Assiout *see* Asyūţ
Assiut *see* Asyūţ
Assling *see* Jesenice
Assouan *see* Aswān
Assu *63 G2 var.* Açu. Rio Grande do Norte, E Brazil
Assuan *see* Aswān
As Sukhnah *118 C3 var.* Sukhne, *Fr.* Soukhné. Ḩimş, C Syria
As Sulaymānīyah *120 C3 var.* Sulaimaniya, *Kurd.* Slēmānī. As Sulaymānīyah, NE Iraq
As Sulayyil *121 B5* Ar Riyāḑ, S Saudi Arabia
Aş Şuwār *118 D2 var.* Şuwār. Dayr az Zawr, E Syria
As Suwaydā' *119 B5 var.* Es Suweida, Suweida, *Fr.* Soueida. As Suwaydā', SW Syria
As Suways *see* Suez
Asta Colonia *see* Asti
Astacus *see* İzmit
Astana *114 C4 prev.* Akmola, Akmolinsk, Tselinograd, Aqmola. *country capital* (Kazakhstan) Akmola, N Kazakhstan
Asta Pompeia *see* Asti
Astarabad *see* Gorgān
Asterābād *see* Gorgān
Asti *96 A2 anc.* Asta Colonia, Asta Pompeia, Hasta Colonia, Hasta Pompeia. Piemonte, NW Italy
Astigi *see* Ecija
Astipálaia *see* Astypálaia
Astorga *92 C1 anc.* Asturica Augusta. Castilla y León, N Spain
Astrabad *see* Gorgān
Astrakhan' *111 C7* Astrakhanskaya Oblast', SW Russian Federation
Asturias *92 C1 autonomous community* NW Spain
Asturias *see* Oviedo
Asturica Augusta *see* Astorga
Astypálaia *105 D7 var.* Astipálaia, *It.* Stampalia. *island* Kykládes, Greece, Aegean Sea
Asunción *64 D2 country capital* (Paraguay) Central, S Paraguay
Aswān *72 B2 var.* Assouan, Assuan, Aswân; *anc.* Syene. SE Egypt
Aswân *see* Aswān
Asyūţ *72 B2 var.* Assiout, Assiut, Asyût, Siut; *anc.* Lycopolis. C Egypt
Asyût *see* Asyūţ
Atacama Desert *64 B2 Eng.* Atacama Desert. *desert* N Chile
Atacama Desert *see* Atacama Desert
Atafu Atoll *145 E3 island* NW Tokelau
Atamyrat *122 D3 prev.* Kerki. Lebap Welaýaty, E Turkmenistan
Aţâr *74 C2* Adrar, W Mauritania
Atas Bogd *126 D3 mountain* SW Mongolia
Atascadero *47 B7* California, W USA
Atatürk Baraji *117 E4 reservoir* S Turkey
Atbara *72 C3 var.* 'Aţbārah. River Nile, NE Sudan
'Aţbārah/'Aţbarah, Nahr *see* Atbara
Atbasar *114 C4* Akmola, N Kazakhstan
Atchison *45 F4* Kansas, C USA
Aternum *see* Pescara
Ath *87 B6 var.* Aat. Hainaut, SW Belgium
Athabasca *37 E5* Alberta, SW Canada
Athabasca *37 E5 var.* Athabaska. *river* Alberta, SW Canada
Athabasca, Lake *37 F4 lake* Alberta/Saskatchewan, SW Canada
Athabaska *see* Athabasca
Athenae *see* Athína
Athenae *see* Athína
Athens *43 E2* Georgia, SE USA
Athens *40 D4* Ohio, N USA
Athens *49 G3* Texas, SW USA
Athens *see* Athína
Athína *105 C6 Eng.* Athens, *prev.* Athínai; *anc.* Athenae. *country capital* (Greece) Attikí, C Greece
Athínai *see* Athína
Athlone *89 B5 Ir.* Baile Átha Luain. C Ireland
Ath Thawrah *see* Madīnat ath Thawrah
Ati *76 C3* Batha, C Chad
Atikokan *38 B4* Ontario, S Canada
Atka *115 G3* Magadanskaya Oblast', E Russian Federation
Atka *36 A3* Atka Island, Alaska, USA

Atlanta *42 D2 state capital* Georgi. SE USA
Atlanta *49 H2* Texas, SW USA
Atlantic City *41 F4* New Jersey, N
Atlantic-Indian Basin *67 D7 und basin* SW Indian Ocean
Atlantic-Indian Ridge *69 B8 und ridge* SW Indian Ocean
Atlantic Ocean *66 B4 ocean*
Atlas Mountains *70 C2 mountai.* NW Africa
Atlasovo *115 H3* Kamchatskiy Kr E Russian Federation
Atlas, Tell *102 C3 Eng.* Tell Atlas. *mountain range* N Algeria
Atlas, Tell *see* Atlas Tellien
Atlin *36 D4* British Columbia, W
Aţ Ţafīlah *119 B7 var.* Et Tafila, 'J Aţ Ţafīlah, W Jordan
Aţ Ţā'if *121 B5* Makkah, W Saudi
Attaleia/Attalia *see* Antalya
At Tall al Abyaḑ *118 C2 var.* Tall Abyaḑ, Tell Abyaḑ, *Fr.* Tell Abia Raqqah, N Syria
Aţ Ţanf *118 D4* Ḩimş, S Syria
Attapu *137 E5 var.* Samakhi, 'J Attapu, S Laos
Attawapiskat *38 C3* Ontario, C Ca
Attawapiskat *38 C3 river* Ontario S Canada
At Tibnī *118 D2 var.* Tibnī. Dayr Zawr, NE Syria
Attopeu *see* Attapu
Attu Island *36 A2 island* Aleutian: Islands, Alaska, USA
Atyrau *114 B4 prev.* Gur'yev. Aty W Kazakhstan
Aubagne *91 D6 anc.* Albania. Bou du-Rhône, SE France
Aubange *87 D8* Luxembourg, SE Belgium
Aubervilliers *90 E1* Seine-St-Den Île-de-France, N France Europe
Auburn *46 B2* Washington, NW U
Auch *91 B6 Lat.* Augusta Auscoru Elimberrum. Gers, S France
Auckland *150 D2* Auckland, Nort Island, New Zealand
Auckland Islands *142 C5 island g* S New Zealand
Audern *see* Audru
Audincourt *90 E4* Doubs, E Franc
Audru *106 D2 Ger.* Audern. Pärn SW Estonia
Augathella *149 D5* Queensland, E Australia
Augsbourg *see* Augsburg
Augsburg *95 C6 Fr.* Augsbourg; *a* Augusta Vindelicorum. Bayern, S Germany
Augusta *147 A7* Western Australi.
Augusta *43 E2* Georgia, SE USA
Augusta *41 G2 state capital* Maine NE USA
Augusta *see* London
Augusta Auscorum *see* Auch
Augusta Emerita *see* Mérida
Augusta Praetoria *see* Aosta
Augusta Trajana *see* Stara Zagora
Augusta Treverorum *see* Trier
Augusta Vangionum *see* Worms
Augusta Vindelicorum *see* Augsbu
Augustobona Tricassium *see* Troy
Augustodurum *see* Bayeux
Augustoritum Lemovicensium *se* Limoges
Augustów *98 E2 Rus.* Avgustov. Podlaskie, NE Poland
Aulie Ata/Auliye-Ata *see* Taraz
Aunglan *136 B4 var.* Allanmyo, M Magway, C Myanmar (Burma)
Auob *78 B4 var.* Oup. *river* Namibia South Africa
Aurangābād *134 D5* Mahārāshtra, C India
Auray *90 A3* Morbihan, NW Franc
Aurelia Aquensis *see* Baden-Baden
Aurelianum *see* Orléans
Aurès, Massif de l' *102 C4 mounta range* NE Algeria
Aurillac *91 C5* Cantal, C France
Aurium *see* Ourense
Aurora *59 F2* NW Guyana
Aurora *44 C4* Colorado, C USA
Aurora *40 B3* Illinois, N USA
Aurora *45 G5* Missouri, C USA
Aurora *see* Maéwo, Vanuatu
Aus *78 B4* Karas, SW Namibia
Ausa *see* Vic

sig *see* Ústí nad Labem
in *45 G3* Minnesota, N USA
in *49 G3 state capital* Texas, SW USA
rales, Archipel des *see* Australes, Îles
rales et Antarctiques Françaises,
erres *see* French Southern and
ntarctic Lands
rales, Îles *143 F4 var.* Archipel des
ustrales, Îles Tubuai, Tubuai Islands,
ng. Austral Islands. *island group*
W French Polynesia
tral Fracture Zone *143 H4 tectonic*
ature S Pacific Ocean
tralia *142 A4 off.* Commonwealth of
ustralia. *country*

STRALIA
Australasia & Oceania

icial name Commonwealth
Australia
mation 1901 / 1901
pital Canberra
pulation 23.3 million / 8 people
r sq mile (3 people per sq km)
al area 2,967,893 sq. miles
86,850 sq. km)
nguages English*, Italian, Cantonese,
eek, Arabic, Vietnamese,
original languages
igions Roman Catholic 26%,
nreligious 19%, Anglican 19%,
her 23%, Other Christian 13%,
ited Church 6%
nic mix European 90%, Asian 7%,
original 2%, Other 1%
vernment Parliamentary system
rrency Australian dollar = 100 cents
eracy rate 99%
lorie consumption 3265 kilocalories

tralia, Commonwealth of *see*
ustralia
tralian Alps *149 C7 mountain range*
E Australia
tralian Capital Territory *149 D7*
rev. Federal Capital Territory. *territory*
E Australia
tralie, Bassin Nord de l' *see* North
ustralian Basin
tral Islands *see* Australes, Îles
trava *see* Ostrov
tria *95 D7 off.* Republic of Austria,
er. Österreich. *country* C Europe

STRIA
Central Europe

icial name Republic of Austria
mation 1918 / 1919
pital Vienna
pulation 8.5 million / 266 people
r sq mile (103 people per sq km)
al area 32,378 sq. miles (83,858 sq. km)
nguages German*, Croatian, Slovenian,
ngarian (Magyar)
igions Roman Catholic 78%,
nreligious 9%, Other (including Jewish
d Muslim) 8%, Protestant 5%
nic mix Austrian 93%, Croat, Slovene,
d Hungarian 6%, Other 1%
vernment Parliamentary system
rrency Euro = 100 cents
eracy rate 99%
lorie consumption 3784 kilocalories

tria, Republic of *see* Austria
esiodorum *see* Auxerre
issiodorum *see* Auxerre
ricum *see* Chartres
ergne *91 C5 cultural region*
? France
erre *90 C4 anc.* Autesiodorum,
utissiodorum. Yonne, C France
aricum *see* Bourges
rua *145 G5 dependent territory*
apital (Cook Islands) Rarotonga,
 Cook Islands
sfelsófalu *see* Negreşti-Oaş
lles *104 C3* Anatolikí Makedonía kai
'hráki, NE Greece
eiro *92 B2 anc.* Talabriga. Aveiro,
la *see* Ávila
llino *97 D5 anc.* Abellinum.
ampania, S Italy
nio *see* Avignon
sta *85 C6* Dalarna, C Sweden
yron *91 C6 river* S France

Avezzano *96 C4* Abruzzo, C Italy
Avgustov *see* Augustów
Aviemore *88 C3* N Scotland, United
Kingdom
Avignon *91 D6 anc.* Avenio. Vaucluse,
SE France
Ávila *92 D3 var.* Avila; *anc.* Abela,
Abula, Abyla, Avela. Castilla y León,
C Spain
Avilés *92 C1* Asturias, NW Spain
Avranches *90 B3* Manche, N France
Avveel *see* Ivalo, Finland
Avvil *see* Ivalo
Awaji-shima *131 C6 island* SW Japan
Äwash *73 D5* Äfar, NE Ethiopia
Awbārī *71 F3* SW Libya
Ax *see* Dax
Axel *87 B5* Zeeland, SW Netherlands
Axel Heiberg Island *37 E1 var.* Axel
Heiburg. *island* Nunavut, N Canada
Axel Heiburg *see* Axel Heiberg Island
Axiós *see* Vardar
Ayacucho *60 D4* Ayacucho, S Peru
Ayagoz *114 C5 var.* Ayaguz, Kaz.
Ayakoz. *river* E Kazakhstan
Ayaguz *see* Ayagoz
Ayamonte *92 C4* Andalucía, S Spain
Ayaviri *61 E4* Puno, S Peru
Aydarko'l Ko'li *123 E2 Rus.* Ozero
Aydarkul'. *lake* C Uzbekistan
Aydarkul', Ozero *see* Aydarko'l Ko'li
Aydın *116 A4 var.* Aïdin; *anc.* Tralles
Aydin. Aydın, SW Turkey
Ayers Rock *see* Uluṟu
Ayeyarwady *see* Irrawaddy
Ayiá *see* Agiá
Áyios Evstrátios *see* Ágios Efstrátios
Áyios Nikólaos *see* Ágios Nikólaos
Ayorou *75 E3* Tillabéri, W Niger
'Ayoûn el 'Atroûs *74 D3 var.* Aïoun el
Atrous, Aïoun el Atroûss. Hodh el
Gharbi, SE Mauritania
Ayr *88 C4* W Scotland, United Kingdom
Ayteke Bi *114 B4 Kaz.* Zhangaqazaly;
prev. Novokazalinsk. Kzylorda,
SW Kazakhstan
Aytos *104 E2* Burgas, E Bulgaria
Ayutthaya *137 C5 var.* Phra Nakhon Si
Ayutthaya. Phra Nakhon Si Ayutthaya,
C Thailand
Ayvalık *116 A3* Balıkesir, W Turkey
Azahar, Costa del *93 F3 coastal region*
E Spain
Azaouâd *75 E3 desert* C Mali
Azärbaycan/Azärbaycan Respublikasi
see Azerbaijan
A'zäz *118 B2* Ḩalab, NW Syria
Azerbaijan *117 G2 off.* Azerbaijani
Republic, Az. Azärbaycan, Azärbaycan
Respublikasi; *prev.* Azerbaijan SSR.
country SE Asia

AZERBAIJAN
Southwest Asia

Official name Republic of Azerbaijan
Formation 1991 / 1991
Capital Baku
Population 9.4 million / 281 people
per sq mile (109 people per sq km)
Total area 33,436 sq. miles (86,600 sq. km)
Languages Azeri*, Russian
Religions Shi'a Muslim 68%, Sunni
Muslim 26%, Russian Orthodox 3%,
Armenian Apostolic Church
(Orthodox) 2%, Other 1%
Ethnic mix Azeri 91%, Other 3%, Lazs 2%,
Armenian 2%, Russian 2%
Government Presidential system
Currency New manat = 100 gopik
Literacy rate 99%
Calorie consumption 2952 kilocalories

Azerbaijani Republic *see* Azerbaijan
Azerbaijan SSR *see* Azerbaijan
Azimabad *see* Patna
Azizie *see* Telish
Azogues *60 B2* Cañar, S Ecuador
Azores *92 A4 var.* Açores, Ilhas dos
Açores, Port. Arquipélago dos Açores.
island group Portugal, NE Atlantic
Ocean
Azores-Biscay Rise *80 A3 undersea rise*
E Atlantic Ocean
Azotos/Azotus *see* Ashdod
Azoum, Bahr *76 C3 seasonal river*
SE Chad
Azov, Sea of *see* Dax
Azov, Sea of *109 H1 Rus.* Azovskoye
More, Ukr. Azovs'ke More. *sea*
NE Black Sea

Azov's'ke More/Azovskoye More *see*
Azov, Sea of
Azraq, Wāḩat al *119 C6 oasis*
N Jordan
Aztec *48 C1* New Mexico, SW USA
Azuaga *92 C4* Extremadura, W Spain
Azuero, Península de *53 F5 peninsula*
S Panama
Azul *65 D5* Buenos Aires, E Argentina
Azur, Côte d' *91 E6 coastal region*
SE France
'Azza *see* Gaza
Az Zaqāziq *72 B1 var.* Zagazig.
N Egypt
Az Zarqā' *119 B6 var.* Zarqa. Az Zarqā',
NW Jordan
Az Zarqā' *119 B6 var.* Zarqa. Zawia. N Libya
Az Zāwiyah *71 F2 var.* Zawia. Zawia. N Libya
Az Zilfī *120 B4* Ar Riyāḍ, N Saudi Arabia

B

Baalbek *118 B4 var.* Ba'labakk; *anc.*
Heliopolis. E Lebanon
Baardheere *73 D6 var.* Bardere, It.
Bardera. Gedo, SW Somalia
Baarle-Hertog *87 C5* Antwerpen,
N Belgium
Baarn *86 C3* Utrecht, C Netherlands
Babadag *108 D5* Tulcea, SE Romania
Babahoyo *60 B2 prev.* Bodegas. Los
Ríos, C Ecuador
Bābā, Kūh-e *123 E4 mountain range*
C Afghanistan
Babayevo *110 B4* Vologodskaya Oblast',
NW Russian Federation
Babeldaob *144 A1 var.* Babeldaop,
Babelthuap. N Palau
Babeldaop *see* Babeldaob
Bab el Mandeb *121 B7 strait* Gulf of
Aden/Red Sea
Babelthuap *see* Babeldaob
Babian Jiang *see* Black River
Babruysk *107 D7 Rus.* Bobruysk.
Mahilyowskaya Voblasts', E Belarus
Babuyan Channel *139 E1 channel*
N Philippines
Babuyan Islands *139 E1 island group*
N Philippines
Bacabal *63 F2* Maranhão, E Brazil
Bacău *108 C4 Hung.* Bákó. Bacău,
NE Romania
Bắc Bộ, Vịnh *see* Tonkin, Gulf of
Bắc Giang *136 D3* Ha Bắc, N Vietnam
Bacheykava *107 D5 Rus.* Bocheykovo.
Vitsyebskaya Voblasts', N Belarus
Back *37 F3 river* Nunavut, N Canada
Bačka Palanka *100 D3* Serbia,
NW Serbia
Bačka Topola *100 D3 Hung.* Topolya;
prev. Hung. Bácstopolya. Vojvodina,
N Serbia
Bac Liêu *137 D6 var.* Vinh Loi. Minh
Hai, S Vietnam
Bacolod *125 E4* Bacolod City.
Negros, C Philippines
Bacolod City *see* Bacolod
Bácsszenttamás *see* Srbobran
Bácstopolya *see* Bačka Topola
Bactra *see* Balkh
Badajoz *92 C4 anc.* Pax Augusta.
Extremadura, W Spain
Baden-Baden *95 B6 anc.* Aurelia
Aquensis. Baden-Württemberg,
SW Germany
Bad Freienwalde *94 D3* Brandenburg,
NE Germany
Badger State *see* Wisconsin
Bad Hersfeld *94 B4* Hessen, C Germany
Bad Homburg *see* Bad Homburg vor
der Höhe
Bad Homburg vor der Höhe *95 B5 var.*
Bad Homburg. Hessen, W Germany
Bad Ischl *95 D7* Oberösterreich,
N Austria
Bad Krozingen *95 A6* Baden-
Württemberg, SW Germany
Badlands *44 D2 physical region* North
Dakota/South Dakota, N USA
Badu Island *148 C1* island Queensland,
NE Australia
Bad Vöslau *95 E6* Niederösterreich,
NE Austria
Baeterrae/Baeterrae Septimanorum
see Béziers
Baetic Cordillera/Baetic Mountains *see*
Béticos, Sistemas
Bafatá *74 C4* C Guinea-Bissau
Baffin Bay *37 G2 bay* Canada/Greenland
NE Black Sea

Baffin Island *37 G2 island* Nunavut,
NE Canada
Bafing *74 C3 river* W Africa
Bafoussam *76 A4* Ouest, W Cameroon
Bafra *116 D2* Samsun, N Turkey
Bäft *120 D4* Kermān, S Iran
Bagaces *52 D4* Guanacaste, NW Costa
Rica
Bagdad *see* Baghdad
Bagé *63 E5* Rio Grande do Sul, S Brazil
Baghdad *120 B3 var.* Bagdad,
country capital (Iraq) Baghdād, C Iraq
Baghlān *123 E3* Baghlān, NE Afghanistan
Bago *136 B4 var.* Pegu. Bago,
SW Myanmar (Burma)
Bagoé *74 D4 river* Côte d'Ivoire/Mali
Bagrationovsk *106 A4 Ger.* Preussisch
Eylau. Kaliningradskaya Oblast',
W Russian Federation
Bagrax Hu *see* Bosten Hu
Baguio *139 E1 off.* Baguio City. Luzon,
N Philippines
Baguio City *see* Baguio
Bagzane, Monts *75 F3 mountain*
N Niger
Bahama Islands *see* Bahamas
Bahamas, The *54 C2 off.* Commonwealth
of the Bahamas. *country* N West Indies

BAHAMAS, THE
West Indies

Official name Commonwealth
of the Bahamas
Formation 1973 / 1973
Capital Nassau
Population 400,000 / 103 people
per sq mile (40 people per sq km)
Total area 5382 sq. miles (13,940 sq. km)
Languages English*, English Creole,
French Creole
Religions Baptist 32%, Anglican 20%,
Roman Catholic 19%, Other 17%,
Methodist 6%, Church of God 6%
Ethnic mix Black African 85%, Other 15%
Government Parliamentary system
Currency Bahamian dollar = 100 cents
Literacy rate 96%
Calorie consumption 2575 kilocalories

Bahamas *35 D6 var.* Bahama Islands.
island group N West Indies
Bahamas, Commonwealth of the *see*
Bahamas, The
Baharly *122 C3 var.* Bäherden, Rus.
Bakharden; *prev.* Bakherden. Ahal
Welaýaty, C Turkmenistan
Bahāwalpur *134 C2* Punjab, E Pakistan
Bäherden *see* Baharly
Bahia *63 F3 off.* Estado da Bahia. *state*
E Brazil
Bahia *63 F3 off.* Estado da Bahia. *state*
E Brazil
Bahía Blanca *65 C5* Buenos Aires,
E Argentina
Bahia, Estado da *see* Bahia
Bahir Dar *72 C4 var.* Bahr Dar, Bahrdar
Giyorgis. Āmara, N Ethiopia
Bahraich *135 E3* Uttar Pradesh, N India
Bahrain *120 C4 off.* State of Bahrain,
Dawlat al Bahrayn, Ar. Al Baḩrayn,
prev. Bahrein; *anc.* Tylos, Tyros.
country SW Asia

BAHRAIN
Southwest Asia

Official name Kingdom of Bahrain
Formation 1971 / 1971
Capital Manama
Population 1.3 million / 4762 people per
sq mile (1841 people per sq km)
Total area 239 sq. miles (620 sq. km)
Languages Arabic
Religions Muslim (mainly Shi'a) 99%,
Other 1%
Ethnic mix Bahraini 63%, Asian 19%,
Other Arab 10%, Iranian 8%
Government Monarchical /
parliamentary system
Currency Bahraini dinar = 1000 fils
Literacy rate 91%
Calorie consumption not available

Bahrain, State of *see* Bahrain
Bahrayn, Dawlat al *see* Bahrain
Bahr Dar/Bahrdar Giyorgis *see*
Bahir Dar
Bahrein *see* Bahrain

Bahr el, Azraq *see* Blue Nile
Bahr Tabariya, Sea of *see* Tiberias, Lake
Bahushewsk *107 E6 Rus.* Bogushëvsk.
  Vitsyebskaya Voblasts', NE Belarus
Baia Mare *108 B3 Ger.* Frauenbach,
  Hung. Nagybánya; *prev.* Neustadt.
  Maramureş, N Romania
Baia Sprie *108 B3 Ger.* Mittelstadt,
  Hung. Felsőbánya. Maramureş,
  NW Romania
Baïbokoum *76 B4* Logone-Oriental,
  SW Chad
Baidoa *see* Baydhabo
Baie-Comeau *39 E3* Québec, SE Canada
Baikal, Lake *115 E4 Eng.* Lake Baikal.
  lake S Russian Federation
Baikal, Lake *see* Baykal, Ozero
Baile Átha Cliath *see* Dublin
Baile Átha Luain *see* Athlone
Bailén *92 D4* Andalucía, S Spain
Baile na Mainistreach *see* Newtownabbey
Băileşti *108 B5* Dolj, SW Romania
Ba Illi *76 B3* Chari-Baguirmi, SW Chad
Bainbridge *42 D3* Georgia, SE USA
Bā'ir *see* Bāyir
Baireuth *see* Bayreuth
Bairiki *144 D2* Tarawa, NW Kiribati
Bairnsdale *149 C7* Victoria, SE Australia
Baishan *129 E3 prev.* Hunjiang. Jilin,
  NE China
Baiyin *128 B4* Gansu, C China
Baja *99 C7* Bács-Kiskun, S Hungary
Baja California Norte *50 B2 state*
  NW Mexico
Bajo Boquete *see* Boquete
Bajram Curri *101 D3* Kukës,
  N Albania
Bakala *76 C4* Ouaka, C Central African
  Republic
Bakan *see* Shimonoseki
Baker *46 C3* Oregon, NW USA
Baker and Howland Islands *145 E2 US*
  unincorporated territory W Polynesia
Baker Lake *37 F3 var.* Qamanittuaq.
  Nunavut, N Canada
Bakersfield *47 C7* California, W USA
Bakharden *see* Baharly
Bakhchisaray *see* Bakhchysaray
Bakhchysaray *109 F5 Rus.* Bakhchisaray.
  Avtonomna Respublika Krym,
  S Ukraine
Bakherden *see* Baharly
Bakhmach *109 F1* Chernihivs'ka Oblast',
  N Ukraine
Bākhtarān *see* Kermānshāh
Bakı *117 H2 Eng.* Baku. *country capital*
  (Azerbaijan) E Azerbaijan
Bákó *see* Bacău
Bakony *99 C7 Eng.* Bakony Mountains,
  Ger. Bakonywald. *mountain range*
  W Hungary
Bakony Mountains/Bakonywald *see*
  Bakony
Baku *see* Bakı
Bakwanga *see* Mbuji-Mayi
Balabac Island *129 C8 island*
  W Philippines
Balabac, Selat *see* Balabac Strait
Balabac Strait *138 D2 var.* Selat Balabac.
  strait Malaysia/Philippines
Ba'labakk *see* Baalbek
Balaguer *93 F2* Cataluña, NE Spain
Balakovo *111 C6* Saratovskaya Oblast',
  W Russian Federation
Bālā Morghāb *see* Bālā Murghāb
Bālā Murghāb *122 D4 prev.* Bālā
  Morghāb. Laghmān, NW Afghanistan
Balashov *111 B6* Saratovskaya Oblast',
  W Russian Federation
Balasore *see* Bāleshwar
Balaton *99 C7 var.* Lake Balaton,
  Ger. Plattensee. *lake* W Hungary
Balaton, Lake *see* Balaton
Balbina, Represa *62 D1 reservoir*
  NW Brazil
Balboa *53 G4* Panamá, C Panama
Balcarce *65 D5* Buenos Aires,
  E Argentina
Balclutha *151 B7* Otago, South Island,
  New Zealand
Baldy Mountain *44 C1 mountain*
  Montana, NW USA
Bâle *see* Basel
Balearic Plain *see* Algerian Basin
Baleares Major *see* Mallorca
Balearic Islands *93 G3 Eng.* Balearic
  Islands. *island group* Spain,
  W Mediterranean Sea
Balearic Islands *see* Baleares, Islas

Balearis Minor *see* Menorca
Baleine, Rivière à la *39 E2 river* Québec,
  E Canada
Balen *87 C5* Antwerpen, N Belgium
Bāleshwar *135 F4 prev.* Balasore.
  Odisha, E India
Bālgrad *see* Alba Iulia
Bali *138 D5 island* C Indonesia
Balıkesir *116 A3* Balıkesir, W Turkey
Balıkh, Nahr *118 C2 river* N Syria
Balikpapan *138 D4* Borneo,
  C Indonesia
Balkanabat *122 B2 Rus.* Nebitdag.
  Balkan Welaýaty, W Turkmenistan
Balkan Mountains *104 C2 Bul./*
  SCr. Stara Planina. *mountain range*
  Bulgaria/Serbia
Balkash *see* Balkhash
Balkash, Ozero *see* Balkhash, Ozero
Balkh *123 E3 anc.* Bactra. Balkh,
  N Afghanistan
Balkhash *114 C5 Kaz.* Balqash; *var.*
  Balkash. Karagandy, SE Kazakhstan
Balkhash, Lake *see* Balkhash, Ozero
Balkhash, Ozero *114 C5 Eng.* Lake
  Balkhash, *Kaz.* Balqash; *var.* Ozero
  Balkash. *lake* SE Kazakhstan
Balladonia *147 C6* Western Australia
Ballarat *149 C7* Victoria, SE Australia
Balleny Islands *154 B5 island group*
  Antarctica
Ballinger *49 F3* Texas, SW USA
Balochistan *134 B3 prev.* Baluchistan,
  Beluchistan. *province* SW Pakistan
Balqash *see* Balkhash/Balkhash, Ozero
Balş *108 B5* Olt, S Romania
Balsas *63 F2* Maranhão, E Brazil
Balsas, Río *51 E5 var.* Río Mexcala.
  river S Mexico
Bal'shavik *107 D7 Rus.* Bol'shevik.
  Homyel'skaya Voblasts', SE Belarus
Balta *108 D3* Odes'ka Oblast',
  SW Ukraine
Bălţi *108 D3 Rus.* Bel'tsy. N Moldova
Baltic Port *see* Paldiski
Baltic Sea *85 C7 Ger.* Ostee, *Rus.*
  Baltiskoye More. *sea* N Europe
Baltimore *41 F4* Maryland, NE USA
Baltischport/Baltiski *see* Paldiski
Baltiskoye More *see* Baltic Sea
Baltkrievija *see* Belarus
Baluchistan *see* Balochistan
Balvi *106 D4* NE Latvia
Balykchy *123 G2 Kir.* Ysyk-Köl; *prev.*
  Issyk-Kul', Rybach'ye. Issyk-Kul'skaya
  Oblast', NE Kyrgyzstan
Balzers *94 E2* S Liechtenstein
Bam *120 E4* Kermān, SE Iran
Bamako *74 D4 country capital* (Mali)
  Capital District, SW Mali
Bambari *76 C4* Ouaka, C Central African
  Republic
Bamberg *95 C5* Bayern, SE Germany
Bamenda *76 A4* Nord-Ouest,
  W Cameroon
Banaba *144 D2 var.* Ocean Island.
  island Tungaru, W Kiribati
Banaras *see* Vārānasi
Banda Aceh *138 A3 var.* Banda Atjeh;
  prev. Koetaradja, Kutaradja, Kutaraja.
  Sumatera, W Indonesia
Banda Atjeh *see* Banda Aceh
Banda, Laut *see* Banda Sea
Bandama *74 D5 var.* Bandama Fleuve.
  river S Côte d'Ivoire
Bandama Fleuve *see* Bandama
Bandar 'Abbās *see* Bandar-e 'Abbās
Bandarbeyla *73 E5 var.* Bender Beila,
  Bender Beyla. Bari, NE Somalia
Bandar-e 'Abbās *120 D4 var.* Bandar
  'Abbās; *prev.* Gombroon. Hormozgān,
  S Iran
Bandar-e Büshehr *120 C4 var.* Büshehr,
  Eng. Bushire. Büshehr, S Iran
Bandar-e Kangān *120 D4 var.* Kangān.
  Büshehr, S Iran
Bandar-e Khamīr *120 D4* Hormozgān,
  S Iran
Bandar-e Langeh *see* Bandar-e Lengeh
Bandar-e Lengeh *120 D4 var.* Bandar-e
  Langeh, Lingeh. Hormozgān, S Iran
Bandar Kassim *see* Boosaaso
Bandar Lampung *138 C4 var.*
  Bandarlampung, Tanjungkarang-
  Telukbetung; *prev.* Tandjoengkarang,
  Tanjungkarang, Teloekbetoeng,
  Telukbetung. Sumatera, W Indonesia
Bandarlampung *see* Bandar Lampung
Bandar Maharani *see* Muar

Bandar Masulipatnam *see*
  Machilipatnam
Bandar Penggaram *see* Batu Pahat
Bandar Seri Begawan *138 D3 prev.*
  Brunei Town. *country capital* (Brunei)
  N Brunei
Banda Sea *139 F5 var.* Laut Banda. *sea*
  E Indonesia
Bandiagara *75 E3* Mopti, C Mali
Bandırma *116 A3 var.* Penderma.
  Balıkesir, NW Turkey
Bandjarmasin *see* Banjarmasin
Bandoeng *see* Bandung
Bandundu *77 C6 prev.* Banningville.
  Bandundu, W Dem. Rep. Congo
Bandung *138 C5 prev.* Bandoeng. Jawa,
  C Indonesia
Bangalore *132 C2 var.* Bengalooru. *state*
  capital Karnātaka, S India
Bangassou *76 D4* Mbomou, SE Central
  African Republic
Banggai, Kepulauan *139 E4 island*
  group C Indonesia
Banghāzī *71 G2 Eng.* Bengazi, Benghazi,
  It. Bengasi. NE Libya
Bangka, Pulau *138 C4 island*
  W Indonesia
Bangkok *see* Ao Krung Thep
Bangkok, Bight of *see* Krung Thep, Ao
Bangladesh *135 G3 off.* People's
  Republic of Bangladesh; *prev.* East
  Pakistan. *country* S Asia

BANGLADESH
South Asia

Official name People's Republic
  of Bangladesh
Formation 1971 / 1971
Capital Dhaka
Population 157 million / 3029 people
  per sq mile (1169 people per sq km)
Total area 55,598 sq. miles
  (144,000 sq. km)
Languages Bengali*, Urdu, Chakma,
  Marma (Magh), Garo, Khasi, Santhali,
  Tripuri, Mro
Religions Muslim (mainly Sunni) 88%,
  Hindu 11%, Other 1%
Ethnic mix Bengali 98%, Other 2%
Government Parliamentary system
Currency Taka = 100 poisha
Literacy rate 59%
Calorie consumption 2430 kilocalories

Bangladesh, People's Republic of *see*
  Bangladesh
Ba Ngoi *see* Cam Ranh
Bangor *89 C6* NW Wales, United
  Kingdom
Bangor *89 B5 Ir.* Beannchar. E Northern
  Ireland, United Kingdom
Bangor *41 G2* Maine, NE USA
Bang Pla Soi *see* Chon Buri
Bangui *77 B5 country capital*
  (Central African Republic)
  Ombella-Mpoko,
  SW Central African Republic
Bangweulu, Lake *73 B8 var.* Lake
  Bangweolo. N Zambia
Ban Hat Yai *see* Hat Yai
Ban Hin Heup *136 C4* Viangchan,
  C Laos
Ban Houayxay/Ban Houei Sai *see*
  Houayxay
Ban Hua Hin *137 C6 var.* Hua Hin.
  Prachuap Khiri Khan, SW Thailand
Bani *74 D3 river* S Mali
Banias *see* Bāniyās
Banjska Palanka *see* Glina
Bani Suwayf *72 B2 var.* Beni Suef.
  N Egypt
Bāniyās *118 B3 var.* Banias, Baniyas,
  Paneas. Ṭarṭūs, W Syria
Banjak, Kepulauan *see* Banyak,
  Kepulauan
Banja Luka *100 B3* Republika Srpska,
  NW Bosnia and Herzegovina
Banjarmasin *138 D4 prev.*
  Bandjarmasin. Borneo, C Indonesia
Banjul *74 B3 prev.* Bathurst. *country*
  capital (Gambia) W Gambia
Banks, Îles *see* Banks Islands
Banks Island *37 E2 island* Northwest
  Territories, NW Canada
Banks Islands *144 D4 Fr.* Îles Banks.
  island group N Vanuatu
Banks Lake *46 B1 reservoir* Washington,
  NW USA

Banks Peninsula *151 C6 peninsula*
  South Island, New Zealand
Banks Strait *149 C8 strait*
  SW Tasman Sea
Bānkura *135 F4* West Bengal, NE Ir
Ban Mak Khaeng *see* Udon Thani
Banmo *see* Bhamo
Banningville *see* Bandundu
Bañolas *see* Banyoles
Ban Pak Phanang *see* Pak Phanang
Ban Sichon *see* Sichon
Banská Bystrica *99 C6 Ger.*
  Neusohl, *Hung.* Besztercebánya.
  Banskobystrický Kraj, C Slovakia
Bantry Bay *89 A7 Ir.* Bá Bheanntraí.
  bay SW Ireland
Banya *104 E2* Burgas, E Bulgaria
Banyak, Kepulauan *138 A3 prev.*
  Kepulauan Banjak. *island group*
  NW Indonesia
Banyo *76 B4* Adamaoua, NW Came
Banyoles *93 G2 var.* Bañolas. Cataluñ
  NE Spain
Banzare Seamounts *141 C7*
  seamount range S Indian Ocean
Banzart *see* Bizerte
Baoji *128 B4 var.* Pao-chi, Paoki.
  Shaanxi, C China
Baoro *76 B4* Nana-Mambéré,
  W Central African Republic
Baoshan *128 A6 var.* Pao-shan. Yur
  SW China
Baotou *127 F3 var.* Pao-t'ou, Paoto'
  Nei Mongol Zizhiqu, N China
Ba'qübah *120 B3 var.* Qubba. Diyāl
  C Iraq
Baquerizo Moreno *see*
  Puerto Baquerizo Moreno
Bar *101 C5 It.* Antivari. S Montenegr
Baraawe *73 D6 It.* Brava. Shabeellaht
  Hoose, S Somalia
Bārāmati *134 C5* Mahārāshtra, W Ir
Baranavichy *107 B6 Pol.* Baranowicz
  Rus. Baranovichi. Brestskaya Vobl
  SW Belarus
Baranovichi/Baranowicze *see*
  Baranavichy
Barbados *55 G1 country* SE West Ind

BARBADOS
West Indies

Official name Barbados
Formation 1966 / 1966
Capital Bridgetown
Population 300,000 / 1807 people
  per sq mile (698 people per sq km)
Total area 166 sq. miles (430 sq. km)
Languages Bajan (Barbadian English),
  English*
Religions Anglican 40%, Other 24%,
  Nonreligious 17%, Pentecostal 8%,
  Methodist 7%, Roman Catholic 4%
Ethnic mix Black African 92%, White
  Other 3%, Mixed race 2%
Government Parliamentary system
Currency Barbados dollar = 100 cents
Literacy rate 99%
Calorie consumption 3047 kilocaloric

Barbastro *93 F2* Aragón, NE Spain
Barbate de Franco *92 C5* Andalucía
  S Spain
Barbuda *55 G3 island* N Antigua and
  Barbuda
Barcaldine *148 C4* Queensland, Aus
Barcău *see* Berettyó
Barce *see* Al Marj
Barcelona *93 G2 anc.* Barcino,
  Barcinona. Cataluña, E Spain
Barcelona *59 E2* Anzoátegui,
  NE Venezuela
Barcino/Barcinona *see* Barcelona
Barcoo *see* Cooper Creek
Barcs *99 C7* Somogy, SW Hungary
Bardaï *76 C1* Borkou-Ennedi-Tibesti
  N Chad
Bardejov *99 D5 Ger.* Bartfeld, *Hung.*
  Bártfa. Prešovský Kraj, E Slovakia
Bardera/Bardere *see* Baardheere
Barduli *see* Barletta
Bareilly *135 E3 var.* Bareli. Uttar
  Pradesh, N India
Bareli *see* Bareilly
Barendrecht *86 C4* Zuid-Holland,
  SW Netherlands
Barentin *90 C3* Seine-Maritime,
  N France

tsburg *83 G2* Spitsbergen, Svalbard
tsevo More/Barents Havet *see* ents Sea
tsøya *83 G2 island* E Svalbard
ts Sea *110 C2 Nor.* Barents Havet, s. Barentsevo More. Arctic Ocean
arbor *41 H2* Mount Desert Island, ine, NE USA
97 E5 var. Bari delle Puglie; *anc.* ium. Puglia, SE Italy
ah *see* Al Bāridah
elle Puglie *see* Bari
ot *see* Barīkowṭ
owṭ *123 F4 var.* Barikot. Kunar, Afghanistan
as *52 A2 var.* Santa Cruz Barillas. ehuetenango, NW Guatemala
as *58 C2* Barinas, W Venezuela
al *135 G4* Barisal, S Bangladesh
an, Pegunungan *138 B4* *untain range* Sumatera, Indonesia
o, Sungai *138 D4 river* Borneo, ndonesia
m *see* Bari
a *see* Al Marj
y Tableland *148 B3 plateau* orthern Territory/Queensland, Australia
d *108 D4 prev.* Bîrlad. Vaslui, Romania
vento, Ilhas de *74 A2 var.* indward Islands. *island group* Cape Verde
e-Duc *90 D3 var.* Bar-sur-Ornain. euse, NE France
e, Lake *147 B6 lake* Western stralia
e Range *146 A4 mountain range* estern Australia
tta *97 D5 anc.* Barduli. Puglia, Italy
nek *98 B3 Ger.* Berlinchen. chodnio-pomorskie, NW Poland
nen-Elberfeld *see* Wuppertal
nouth *89 C6* NW Wales, United ngdom
aul *114 D4* Altayskiy Kray, Russian Federation
et *89 A7* United Kingdom
staple *89 C7* SW England, United ngdom
da *see* Vadodara
ghil Pass *123 F3 var.* Kowtal-e rowghīl. *pass* Afghanistan/Pakistan
n'ki *107 E7 Rus.* Boron'ki. ahilyowskaya Voblasts', E Belarus
wghīl, Kowtal-e *see* Baroghil Pass
uisimeto *58 C2* Lara, W Venezuela
a *88 B3 island* NW Scotland, United ingdom
a de Río Grande *53 E3* Región utónoma Atlántico Sur, E Nicaragua
anca *60 C3* Lima, W Peru
ancabermeja *58 B2* Santander, Colombia
anquilla *58 B1* Atlántico, Colombia
eiro *92 B4* Setúbal, W Portugal
ier Range *149 C6 hill range* ew South Wales, SE Australia
row *36 D2* Alaska, USA
ow *89 B6 Ir.* An Bhearú. *river* E Ireland
row-in-Furness *89 C5* NW England, nited Kingdom
row Island *146 A4 island* Western Australia
stow *47 C7* California, W USA
-sur-Ornain *see* Bar-le-Duc
tang *123 F3 river* SE Tajikistan
tenstein *see* Bartoszyce
tfa/Bartfeld *see* Bardejov
tica *59 F3* N Guyana
tın *116 C2* Bartın, NW Turkey
tlesville *49 G1* Oklahoma, C USA
toszyce *98 D2 Ger.* Bartenstein. Varmińsko-mazurskie, NE Poland
uun-Urt *127 F2* Sühbaatar, Mongolia
ú, Volcán *53 E5 var.* Volcán de hiriquí. *volcano* W Panama
won River *149 D5 river* New South Wales, SE Australia
ysaw *107 D6 Rus.* Borisov. Minskaya oblasts', NE Belarus
Basarabeasca *108 D4 Rus.* Bessarabka. SE Moldova
Basel *95 A7 Eng.* Basle, *Fr.* Bâle. Basel Stadt, NW Switzerland
Basilan *139 E3 island* Sulu Archipelago, SW Philippines
Basle *see* Basel
Basra *see* Al Baṣrah
Bassano del Grappa *96 C2* Veneto, NE Italy
Bassein *see* Pathein
Basseterre *55 G4 country capital* (Saint Kitts and Nevis) Saint Kitts, Saint Kitts and Nevis
Basse-Terre *55 G3 dependent territory capital* (Guadeloupe) Basse Terre, SW Guadeloupe
Basse Terre *55 G4 island* W Guadeloupe
Bassikounou *74 D3* Hodh ech Chargui, SE Mauritania
Bass, Îlots de *see* Marotiri
Bass Strait *149 C7 strait* SE Australia
Bassum *94 B3* Niedersachsen, NW Germany
Bastia *91 E7* Corse, France, C Mediterranean Sea
Bastogne *87 D7* Luxembourg, SE Belgium
Bastrop *42 B2* Louisiana, S USA
Bastyn' *107 B7 Rus.* Bostyn'. Brestskaya Voblasts', SW Belarus
Basuo *see* Dongfang
Basutoland *see* Lesotho
Bata *77 A5* NW Equatorial Guinea
Batae Coritanorum *see* Leicester
Batajnica *100 D3* Vojvodina, N Serbia
Batangas *139 E2 off.* Batangas City. Luzon, N Philippines
Batangas City *see* Batangas
Batavia *see* Jakarta
Bătdâmbâng *137 C5 prev.* Battambang. Bătdâmbâng, NW Cambodia
Batéké, Plateaux *77 B6 plateau* S Congo
Bath *89 D7 hist.* Akermanceaster; *anc.* Aquae Calidae, Aquae Solis. SW England, United Kingdom
Bathinda *134 D2* Punjab, NW India
Bathsheba *55 G1* E Barbados
Bathurst *149 D6* New South Wales, SE Australia
Bathurst *39 F4* New Brunswick, SE Canada
Bathurst *see* Banjul
Bathurst Island *146 D2 island* Northern Territory, N Australia
Bathurst Island *37 F2 island* Parry Islands, Nunavut, N Canada
Wadi al Batin *120 C4 dry watercourse* SW Asia
Batman *117 E4 var.* İluh. Batman, SE Turkey
Batna *71 E2* NE Algeria
Baton Rouge *42 B3 state capital* Louisiana, S USA
Batroûn *118 A4 var.* Al Batrūn. N Lebanon
Battambang *see* Bătdâmbâng
Batticaloa *132 D3* Eastern Province, E Sri Lanka
Battipaglia *97 D5* Campania, S Italy
Battle Born State *see* Nevada
Batumi *117 F2 prev.* Bat'umi. W Georgia
Bat'umi *see* Batumi
Batu Pahat *138 B3 prev.* Bandar Penggaram. Johor, Peninsular Malaysia
Bauchi *75 G4* Bauchi, NE Nigeria
Bauer Basin *153 F3 undersea basin* E Pacific Ocean
Bauska *106 C3 Ger.* Bauske. S Latvia
Bauske *see* Bauska
Bautzen *94 D4 Lus.* Budyšin. Sachsen, E Germany
Bauzanum *see* Bolzano
Bavaria *see* Bayern
Bavarian Alps *95 C7 Ger.* Bayrische Alpen. *mountain range* Austria/ Germany
Bavière *see* Bayern
Bavispe, Río *50 C2 river* NW Mexico
Bawīṭī *72 B2 var.* Bawîṭi. N Egypt
Bawîṭi *see* Bawīṭī
Bawku *75 E4* N Ghana
Bayamo *54 C3* Granma, E Cuba
Bayan Har Shan *126 D4 var.* Bayan Khar. *mountain range* C China
Bayanhongor *126 D2* Bayanhongor, C Mongolia
Bayan Khar *see* Bayan Har Shan
Bayano, Lago *53 G4 lake* E Panama
Bay City *40 C3* Michigan, N USA
Bay City *49 G4* Texas, SW USA
Baydhabo *73 D6 var.* Baydhowa, Isha Baydhabo, *It.* Baidoa. Bay, SW Somalia
Baydhowa *see* Baydhabo
Bayern *95 C6 Eng.* Bavaria, *Fr.* Bavière. *state* SE Germany
Bayeux *90 B3 anc.* Augustodurum. Calvados, N France
Bāyir *119 C7 var.* Bā'ir. Ma'ān, S Jordan
Bay Islands *52 C1 Eng.* Bay Islands. *island group* N Honduras
Bay Islands *see* Bahía, Islas de la
Baymak *111 D6* Respublika Bashkortostan, W Russian Federation
Bayonne *91 A6 anc.* Lapurdum. Pyrénées-Atlantiques, SW France
Bayou State *see* Mississippi
Bayram-Ali *see* Baýramaly
Baýramaly *122 D3 var.* Bayramaly; *prev.* Bayram-Ali. Mary Welaýaty, S Turkmenistan
Bayreuth *95 C5 var.* Baireuth. Bayern, SE Germany
Bayrische Alpen *see* Bavarian Alps
Bayrūt *see* Beyrouth
Bay State *see* Massachusetts
Baysun *see* Boysun
Bayt Laḥm *see* Bethlehem
Baytown *49 H4* Texas, SW USA
Baza *93 E4* Andalucía, S Spain
Bazargic *see* Dobrich
Bazin *see* Pezinok
Beagle Channel *65 C8 channel* Argentina/Chile
Béal Feirste *see* Belfast
Beannchar *see* Bangor, Northern Ireland, UK
Bear Island *see* Bjørnøya
Bear Lake *46 E4 lake* Idaho/Utah, NW USA
Beas de Segura *93 E4* Andalucía, S Spain
Beata, Isla *55 E3 island* SW Dominican Republic
Beatrice *45 F4* Nebraska, C USA
Beaufort Sea *36 D2 sea* Arctic Ocean
Beaufort-Wes *see* Beaufort West
Beaufort West *78 C5 Afr.* Beaufort-Wes. Western Cape, SW South Africa
Beaumont *49 H3* Texas, SW USA
Beaune *90 D4* Côte d'Or, C France
Beauvais *90 C3 anc.* Bellovacum, Caesaromagus. Oise, N France
Beaver Island *40 C2 island* Michigan, N USA
Beaver Lake *49 H1 reservoir* Arkansas, C USA
Beaver River *49 F1 river* Oklahoma, C USA
Beaver State *see* Oregon
Beāwar *134 C3* Rājasthān, N India
Bečej *100 D3 Ger.* Altbetsche, *Hung.* Óbecse, Rácz-Becse; *prev.* Magyar-Becse, Stari Bečej. Vojvodina, N Serbia
Béchar *70 D2 prev.* Colomb-Béchar. W Algeria
Beckley *40 D5* West Virginia, NE USA
Bécs *see* Wien
Bedford *89 D6* E England, United Kingdom
Bedum *86 E1* Groningen, NE Netherlands
Beehive State *see* Utah
Be'er Menuha *119 B7 prev.* Be'ér Menuḥa. Southern, S Israel
Be'ér Menuḥa *see* Be'er Menuha
Beernem *87 A5* West-Vlaanderen, NW Belgium
Beersheba *see* Be'er Sheva
Be'er Sheva *119 A7 var.* Beersheba, *Ar.* Bir es Saba; *prev.* Be'ér Sheva'. Southern, S Israel
Be'ér Sheva' *see* Be'er Sheva
Beesel *87 D5* Limburg, SE Netherlands
Beeville *49 G4* Texas, SW USA
Bega *149 D7* New South Wales, SE Australia
Begoml' *see* Byahoml'
Begovat *see* Bekobod
Behagle *see* Laï
Behar *see* Bihār
Beibu Wan *see* Tonkin, Gulf of
Beida *see* Al Bayḍā'
Beihai *128 B6* Guangxi Zhuangzu Zizhiqu, S China
Beijing *128 C3 var.* Pei-ching, Peking; *prev.* Pei-p'ing. *country capital* (China) Beijing Shi, E China
Beilen *86 E2* Drenthe, NE Netherlands
Beira *79 E3* Sofala, C Mozambique
Beirut *see* Beyrouth
Beit Lekhem *see* Bethlehem
Beiuș *108 B3 Hung.* Belényes. Bihor, NW Romania
Beja *92 B4 anc.* Pax Julia. Beja, SE Portugal
Béjar *92 C3* Castilla y León, N Spain
Bejraburi *see* Phetchaburi
Bekabad *see* Bekobod
Békás *see* Bicaz
Bek-Budi *see* Qarshi
Békéscsaba *99 D7 Rom.* Bichiș-Ciaba. Békés, SE Hungary
Bekobod *123 E2 Rus.* Bekabad; *prev.* Begovat. Toshkent Viloyati, E Uzbekistan
Bela Crkva *100 E3 Ger.* Weisskirchen, *Hung.* Fehértemplom. Vojvodina, W Serbia
Belarus *107 B6 off.* Republic of Belarus, *var.* Belorussia, *Latv.* Baltkrievija; *prev.* Belorussian SSR, *Rus.* Belorusskaya SSR. *country* E Europe

BELARUS
Eastern Europe

Official name Republic of Belarus
Formation 1991 / 1991
Capital Minsk
Population 9.4 million / 117 people per sq mile (45 people per sq km)
Total area 80,154 sq. miles (207,600 sq. km)
Languages Belarussian*, Russian*
Religions Orthodox Christian 80%, Roman Catholic 14%, Other 6%
Ethnic mix Belorussian 81%, Russian 11%, Polish 4%, Ukrainian 2%, Other 2%
Government Presidential system
Currency Belarussian rouble = 100 kopeks
Literacy rate 99%
Calorie consumption 3253 kilocalories

Belarus, Republic of *see* Belarus
Belau *see* Palau
Belaya Tserkov' *see* Bila Tserkva
Bełchatów *98 C4 var.* Belchatow. Łódzski, C Poland
Belchatow *see* Bełchatów
Belcher, Îles *see* Belcher Islands
Belcher Islands *38 C2 Fr.* Îles Belcher. *island group* Nunavut, SE Canada
Beledweyne *73 D5 var.* Belet Huen, *It.* Belet Uen. Hiiraan, C Somalia
Belém *63 F1 var.* Pará. *state capital* Pará, N Brazil
Belén *52 D4* Rivas, SW Nicaragua
Belen *48 D2* New Mexico, SW USA
Belényes *see* Beiuș
Belet Huen/Belet Uen *see* Beledweyne
Belfast *89 B5 Ir.* Béal Feirste. *national capital* E Northern Ireland, United Kingdom
Belfield *44 D2* North Dakota, N USA
Belfort *90 E4* Territoire-de-Belfort, E France
Belgard *see* Białogard
Belgaum *132 B1* Karnātaka, W India
Belgian Congo *see* Congo (Democratic Republic of)
België/Belgique *see* Belgium
Belgium *87 B6 off.* Kingdom of Belgium, *Dut.* België, *Fr.* Belgique. *country* NW Europe

BELGIUM
Northwest Europe

Official name Kingdom of Belgium
Formation 1830 / 1919
Capital Brussels
Population 11.1 million / 876 people per sq mile (338 people per sq km)
Total area 11,780 sq. miles (30,510 sq. km)
Languages Dutch*, French*, German*
Religions Roman Catholic 88%, Other 10%, Muslim 2%
Ethnic mix Fleming 58%, Walloon 33%, Other 6%, Italian 2%, Moroccan 1%
Government Parliamentary system
Currency Euro = 100 cents
Literacy rate 99%
Calorie consumption 3793 kilocalories

Belgium, Kingdom of *see* Belgium

Belgorod 111 A6 Belgorodskaya Oblast',
W Russian Federation
Belgrano II 154 A2 Argentinian research
station Antarctica
Belice see Belize/Belize City
Beligrad see Berat
Beli Manastir 100 C3 Hung.
Pélmonostor; prev. Monostor. Osijek-
Baranja, NE Croatia
Bélinga 77 B5 Ogooué-Ivindo, NE Gabon
Belitung, Pulau 138 C4 island
W Indonesia
Belize 52 B1 Sp. Belice; prev. British
Honduras, Colony of Belize. country
Central America

BELIZE
Central America

Official name Belize
Formation 1981 / 1981
Capital Belmopan
Population 294,385 / 33 people
per sq mile (13 people per sq km) / 48%
Total area 8867 sq. miles (22,966 sq. km)
Languages English*, English Creole,
Spanish, Mayan, Garifuna (Carib)
Religions Roman Catholic 62%,
Other 13%, Anglican 12%,
Methodist 6%, Mennonite 4%,
Seventh-day Adventist 3%
Ethnic mix Mestizo 49%, Creole 25%,
Maya 11%, Other 6%, Garifuna 6%,
Asian Indian 3%
Government Parliamentary system
Currency Belizean dollar = 100 cents
Literacy rate 75%
Calorie consumption 2757 kilocalories

Belize 52 B1 river Belize/Guatemala
Belize see Belize City
Belize City 52 C1 var. Belize, Sp. Belice.
Belize, NE Belize
Belize, Colony of see Belize
Beljak see Villach
Belkofski 36 B3 Alaska, USA
Belle Île 90 A4 island NW France
Belle Isle, Strait of 39 G3 strait
Newfoundland and Labrador, E Canada
Bellenz see Bellinzona
Belleville 40 B4 Illinois, N USA
Bellevue 45 F4 Iowa, C USA
Bellevue 46 B2 Washington, NW USA
Bellingham 46 B1 Washington, NW USA
Belling Hausen Mulde see Southeast
Pacific Basin
Bellingshausen Abyssal Plain see
Bellingshausen Plain
Bellingshausen Plain 153 F5 var.
Bellingshausen Abyssal Plain. abyssal
plain SE Pacific Ocean
Bellingshausen Sea 154 A3 sea
Antarctica
Bellinzona 95 B8 Ger. Bellenz. Ticino,
S Switzerland
Bello 58 B2 Antioquia, W Colombia
Bello Horizonte see Belo Horizonte
Bellovacum see Beauvais
Bellville 78 B5 Western Cape, SW South
Africa
Belmopan 52 C1 country capital (Belize)
Cayo, C Belize
Belogradchik 104 B1 Vidin,
NW Bulgaria
Belo Horizonte 63 F4 prev. Bello
Horizonte. state capital Minas Gerais,
SE Brazil
Belomorsk 110 B3 Respublika Kareliya,
NW Russian Federation
Beloretsk 111 D6 Respublika
Bashkortostan, W Russian Federation
Belorussia/Belorussian SSR
see Belarus
Belorusskaya Gryada see Byelaruskaya
Hrada
Belorusskaya SSR see Belarus
Beloshchel'ye see Nar'yan-Mar
Belostok see Białystok
Belovár see Bjelovar
Belozërsk 110 B4 Vologodskaya Oblast',
NW Russian Federation
Belton 49 G3 Texas, SW USA
Bel'tsy see Bălţi
Beluchistan see Balochistan
Belukha, Gora 114 D5 mountain
Kazakhstan/Russian Federation
Belynichi see Byalynichy
Belyy, Ostrov 114 D2 island N Russian
Federation

Bemaraha 79 F3 var. Plateau du
Bemaraha. mountain range
W Madagascar
Bemaraha, Plateau du see Bemaraha
Bemidji 45 F1 Minnesota, N USA
Bemmel 86 D4 Gelderland,
SE Netherlands
Benaco see Garda, Lago di
Benares see Vārānasi
Benavente 92 D2 Castilla y León,
N Spain
Bend 46 B3 Oregon, NW USA
Bender see Tighina
Bender Beila/Bender Beyla see
Bandarbeyla
Bender Cassim/Bender Qaasim see
Boosaaso
Bendern 94 E1 NW Liechtenstein Europe
Bendery see Tighina
Bendigo 149 C7 Victoria, SE Australia
Beneschau see Benešov
Beneški Zaliv see Venice, Gulf of
Benešov 99 B5 Ger. Beneschau.
Středočeský Kraj, W Czech Republic
Benevento 97 D5 anc. Beneventum,
Malventum. Campania, S Italy
Beneventum see Benevento
Bengal, Bay of 124 C4 bay N Indian
Ocean
Bengalooru see Bangalore
Bengasi see Banghāzī
Bengazi see Banghāzī
Bengbu 128 D5 var. Peng-pu. Anhui,
E China
Benghazi see Banghāzī
Bengkulu 138 B4 prev. Bengkoeloe,
Benkoelen, Benkulen. Sumatera,
W Indonesia
Benguela 78 A2 var. Benguella.
Benguela, W Angola
Benguella see Benguela
Bengweulu, Lake see Bangweulu, Lake
Ben Hope 88 B2 mountain N Scotland,
United Kingdom
Beni 77 E5 Nord-Kivu, NE Dem. Rep.
Congo
Benidorm 93 F4 Valenciana, SE Spain
Beni-Mellal 70 C2 C Morocco
Benin 74 F4 off. Republic of Benin; prev.
Dahomey. country W Africa

BENIN
West Africa

Official name Republic of Benin
Formation 1960 / 1960
Capital Porto-Novo
Population 10.3 million / 241 people
per sq mile (93 people per sq km)
Total area 43,483 sq. miles (112,620 sq. km)
Languages Fon, Bariba, Yoruba, Adja,
Houeda, Somba, French*
Religions Indigenous beliefs and
Voodoo 50%, Christian 30%, Muslim 20%
Ethnic mix Fon 41%, Other 21%, Adja 16%,
Yoruba 12%, Bariba 10%
Government Presidential system
Currency CFA franc = 100 centimes
Literacy rate 29%
Calorie consumption 2594 kilocalories

Benin, Bight of 75 F5 gulf W Africa
Benin City 75 F5 Edo, SW Nigeria
Benin, Republic of see Benin
Beni, Río 61 E3 river N Bolivia
Beni Suef see Banī Suwayf
Ben Nevis 88 C3 mountain N Scotland,
United Kingdom
Bénoué see Benue
Benson 48 B3 Arizona, SW USA
Bent Jbaïl 119 A5 var. Bint Jubayl.
S Lebanon
Benton 42 B1 Arkansas, C USA
Benue 76 B4 Fr. Bénoué. river
Cameroon/Nigeria
Beograd 100 D3 Eng. Belgrade. Serbia,
N Serbia
Berane 101 D5 prev. Ivangrad.
E Montenegro
Berat 101 C6 var. Berati, SCr. Beligrad.
Berat, C Albania
Berätäu see Berettyó
Berati see Berat
Berau, Teluk 139 G4 var. MacCluer
Gulf. bay Papua, E Indonesia
Berbera 72 D4 Woqooyi Galbeed,
NW Somalia
Berbérati 77 B5 Mambéré-Kadéï,
SW Central African Republic

Berck-Plage 90 C2 Pas-de-Calais,
N France
Berdichev see Berdychiv
Berdyans'k 109 G4 Rus. Berdyansk;
prev. Osipenko. Zaporiz'ka Oblast',
SE Ukraine
Berdychiv 108 D2 Rus. Berdichev.
Zhytomyrs'ka Oblast', N Ukraine
Beregovo/Beregszász see Berehove
Berehove 108 B3 Cz. Berehovo,
Hung. Beregszász, Rus. Beregovo.
Zakarpats'ka Oblast', W Ukraine
Berehovo see Berehove
Bereket 122 B2 prev. Rus. Gazandzhyk,
Kazandzhik, Turkm. Gazanjyk. Balkan
Welaýaty, W Turkmenistan
Beretäu see Berettyó
Berettyó 99 D7 Rom. Barcău; prev.
Berătău, Berettău. river Hungary/
Romania
Berettyóújfalu 99 D6 Hajdú-Bihar,
E Hungary
Berezhany 108 C2 Pol. Brzeżany.
Ternopil's'ka Oblast', W Ukraine
Berezina see Byarezina
Berezniki 111 D5 Permskiy Kray,
NW Russian Federation
Berga 93 G2 Cataluña, NE Spain
Bergamo 96 B2 anc. Bergomum.
Lombardia, N Italy
Bergara 93 E1 País Vasco, N Spain
Bergen 94 D2 Mecklenburg-
Vorpommern, NE Germany
Bergen 86 C2 Noord-Holland,
NW Netherlands
Bergen 85 A5 Hordaland, S Norway
Bergen see Mons
Bergerac 91 B5 Dordogne, SW France
Bergeyk 87 C5 Noord-Brabant,
S Netherlands
Bergomum see Bergamo
Bergse Maas 86 D4 river S Netherlands
Beringen 87 C5 Limburg, NE Belgium
Beringov Proliv see Bering Strait
Bering Sea 34 A2 sea N Pacific Ocean
Bering Strait 35 C2 Rus. Beringov Proliv.
strait Bering Sea/Chukchi Sea
Berja 93 E5 Andalucía, S Spain
Berkeley 47 B6 California, W USA
Berkner Island 154 A2 island Antarctica
Berkovitsa 104 C2 Montana,
NW Bulgaria
Berlin 94 D3 country capital (Germany)
Berlin, NE Germany
Berlin 41 G2 New Hampshire, NE USA
Berlinchen see Barlinek
Bermejo, Río 64 C2 river N Argentina
Bermeo 93 E1 País Vasco, N Spain
Bermuda 35 D6 var. Bermuda Islands,
Bermudas; prev. Somers Islands. UK
crown colony NW Atlantic Ocean
Bermuda Islands see Bermuda
Bermuda Rise 35 E6 undersea rise
C Sargasso Sea
Bermudas see Bermuda
Bern 95 A7 Fr. Berne. country capital
(Switzerland) Bern, W Switzerland
Bernau 94 D3 Brandenburg,
NE Germany
Bernburg 94 C4 Sachsen-Anhalt,
C Germany
Berne see Bern
Berner Alpen 95 A7 var. Berner
Oberland, Eng. Bernese Oberland.
mountain range SW Switzerland
Berner Oberland/Bernese Oberland see
Berner Alpen
Bernier Island 147 A5 island Western
Australia
Beroea see Ḥalab
Berry 90 C4 cultural region C France
Berry Islands 54 C1 island group
N Bahamas
Bertoua 77 B5 Est, E Cameroon
Beru 145 E2 var. Peru. atoll Tungaru,
W Kiribati
Berwick-upon-Tweed 88 D4 N England,
United Kingdom
Berytus see Beyrouth
Besançon 90 D4 anc. Besontium,
Vesontio. Doubs, E France
Beskra see Biskra
Besontium see Besançon
Bessarabeasca see Basarabeasca
Besztercze see Bistrita
Besztercebánya see Banská Bystrica
Betafo 79 G3 Antananarivo,
C Madagascar
Betanzos 92 B1 Galicia, NW Spain

Bethlehem 78 D4 Free State, C Sou...
Africa
Bethlehem 119 B6 var. Beit Lekher...
Ar. Bayt Laḥm, Heb. Bet Leḥem.
C West Bank
Béticos, Sistemas 92 D4 var. Sistema
Penibético, Eng. Baetic Cordillera
Baetic Mountains. mountain ran...
S Spain
Bet Leḥem see Bethlehem
Bétou 77 C5 Likouala, N Congo
Bette, Pic see Bette, Picco
Bette, Picco 71 G4 var. Bikkū Bittī,
Bette. mountain S Libya
Beulah 40 C2 Michigan, N USA
Beuthen see Bytom
Beveren 87 B5 Oost-Vlaanderen,
N Belgium
Beverley 89 D5 E England, United
Kingdom
Bexley 89 B8 Bexley, SE England, U...
Kingdom
Beyla 74 D4 SE Guinea
Beyrouth 118 A4 var. Bayrūt, Eng.
Beirut; anc. Berytus. country capi...
(Lebanon) W Lebanon
Beyşehir 116 B4 Konya, SW Turke...
Beyşehir Gölü 116 B4 lake C Turkey
Béziers 91 C6 anc. Baeterrae, Baet...
Septimanorum, Julia Beterrae. H...
S France
Bezmein see Abadan
Bezwada see Vijayawāda
Bhadrāvati 132 C2 Karnātaka, SW...
Bhāgalpur 135 F3 Bihār, NE India
Bhaktapur 135 F3 Central, C Nepa...
Bhamo 136 B2 var. Banmo. Kachir...
State, N Myanmar (Burma)
Bhārat see India
Bharūch 134 C4 Gujarāt, W India
Bhaunagar see Bhāvnagar
Bhāvnagar 134 C4 prev. Bhaunagar.
Gujarāt, W India
Bheanntraí, Bá see Bantry Bay
Bhopāl 134 D4 state capital Madhya
Pradesh, C India
Bhubaneshwar 135 F5 prev.
Bhubaneswar, Bhuvaneshwar. state
capital Odisha, E India
Bhubaneswar see Bhubaneshwar
Bhuket see Phuket
Bhusaval see Bhusāwal
Bhusāwal 134 D4 prev. Bhusaval.
Mahārāshtra, C India
Bhutan 135 G3 off. Kingdom of Bh...
var. Druk-yul. country S Asia

BHUTAN
South Asia

Official name Kingdom of Bhutan
Formation 1656 / 1865
Capital Thimphu
Population 800,000 / 44 people
per sq mile (17 people per sq km)
Total area 18,147 sq. miles (47,000 sq...
Languages Dzongkha*, Nepali, Assar
Religions Mahayana Buddhist 75%,
Hindu 25%
Ethnic mix Bhutanese 50%,
Nepalese 35%, Other 15%
Government Monarchical /
parliamentary system
Currency Ngultrum = 100 chetrum
Literacy rate 53%
Calorie consumption Not available

Bhutan, Kingdom of see Bhutan
Bhuvaneshwar see Bhubaneshwar
Biak, Pulau 139 G4 island E Indonesia
Biała Podlaska 98 E3 Lubelskie,
E Poland
Białogard 98 B2 Ger. Belgard.
Zachodnio-pomorskie, NW Poland
Białystok 98 E3 Rus. Belostok, Bieło
Podlaskie, NE Poland
Bianco, Monte see Blanc, Mont
Biarritz 91 A6 Pyrénées-Atlantiques...
SW France
Bicaz 108 C3 Hung. Békás. Neamţ,
NE Romania
Bichiş-Ciaba see Békéscsaba
Biddeford 41 G2 Maine, NE USA
Bideford 89 C7 SW England, United
Kingdom
Biel 95 A7 Fr. Bienne. Bern,
W Switzerland
Bielefeld 94 B4 Nordrhein-Westfale...
NW Germany

Bielitz-Biala see Bielsko-Biała
k see Białystok
-Biała 99 *C5 Ger.* Bielitz,
z-Biała. Śląskie, S Poland
Podlaski 98 *E3* Białystok,
and
en see Diên Biên Phu
Ong see South China Sea
oa 137 *E6* Đông Nai, S Vietnam
see Biel
e, Lac 38 *D2 lake* Québec,
nada
teau 78 *B2 var.* Bié Plateau.
au C Angola
teau see Bié, Planalto do
oress Swamp 43 *E5 wetland*
da, SE USA
sland 146 *C2 island* Western
*alia
a Mountains 44 *C2 mountain*
r Wyoming, C USA
n River 44 *C2 river* Montana/
ming, NW USA
a 74 *B3* SW Senegal
a see Tarbes
vo see Bihosava
aux River 45 *E2 river* Iowa/South
ota, N USA
ring 49 *E3* Texas, SW USA
400 B3 Federacija Bosne I
egovine, NW Bosnia and
egovina
**135 *F3 prev.* Behar. *cultural*
n N India
ee Bihar Sharif
mulo 73 *B7* Kagera, NW Tanzania
harif 135 *F3 var.* Bihār. Bihār,
dia
va 107 *D5 Rus.* Bigosovo.
**yebskaya Voblasts', NW Belarus
ia 100 *C3* Republika Srpska,
Bosnia and Herzegovina
Polje 101 *D5* E Montenegro
r 134 *C3* Rājasthān, NW India
**115 *G4* Khabarovskiy Kray,
ussian Federation
Atoll 144 *C1 var.* Pikinni. *atoll*
k Chain, NW Marshall Islands
Bitti see Bette, Picco
tr 135 *E4* Chhattīsgarh, C India
var 117 *H3 Rus.* Bilyasuvar; *prev.*
akino. SE Azerbaijan
erkva 109 *E2 Rus.* Belaya
kov'. Kyyivs'ka Oblast', N Ukraine
taung Range 137 *C6 var.*
nintari Taungdan. *mountain range*
nmar (Burma)/Thailand
**93 *E1 Basq.* Bilbo. País Vasco,
pain
see Bilbao
k 116 *B3* Bilecik, NW Turkey
s 44 *C2* Montana, NW USA
, Grand Erg de 75 *H3 desert*
Niger
a 148 *D4* Queensland, E Australia
42 C3 Mississippi, S USA
**76 *C3* Biltine, E Chad
ee Puerto Cabezas
uvar see Biläsuvar
**87 *D6* Limburg, NE Belgium
i Islands 54 *C1 island group*
ahamas
**87 *B7* Hainaut, S Belgium
be Island see Marchena, Isla
halfän, Jazā'ir see Ḩalāniyāt,
ur al
amton 41 *F3* New York, NE USA
**117 *E3* Bingöl, E Turkey
abayl see Bent Jbail
u 138 *D3* Sarawak, East Malaysia
ou 128 *D4* Shandong, E China
o, Río 65 *B5 river* C Chile
, Isla de 77 *A5 var.* Bioko, *Eng.*
ando Po, Sp. Fernando Póo;
. Macías Nguema Biyogo. *island*
V Equatorial Guinea
see Bioco, Isla de
**71 *F3 var.* Brak. C Libya
**76 *D3* Vakaga, N Central African
public
nagar 135 *F3* Eastern, SE Nepal
Saba see Ba'er Sheva
d 120 *E3* Khorāsān-e Janūbī,
an
nfeld 95 *A5* Rheinland-Pfalz,
Germany
nhead 89 *C5* NW England, United
gdom
1 see Bårland

Birmingham 89 *C6* C England, United Kingdom
Birmingham 42 *C2* Alabama, S USA
Bir Moghreïn see Bir Moghreïn
Bir Moghreïn 74 *C1 var.* Bir Moghreïn; *prev.* Fort-Trinquet. Tiris Zemmour, N Mauritania
Birnie Island 145 *E3 atoll* Phoenix Islands, C Kiribati
Birnin Konni 75 *F3 var.* Birni-Nkonni. Tahoua, SW Niger
Birni-Nkonni see Birnin Konni
Birobidzhan 115 *G4* Yevreyskaya Avtonomnaya Oblast', SE Russian Federation
Birsen see Biržai
Birsk 111 *D5* Respublika Bashkortostan, W Russian Federation
Biržai 106 *C4 Ger.* Birsen. Panevėžys, NE Lithuania
Birżebbuġa 102 *B5* SE Malta
Bisanthe see Tekirdağ
Bisbee 48 *B3* Arizona, SW USA
Biscaia, Baía de see Biscay, Bay of
Biscay, Bay of 80 *B4* Sp. Golfo de Vizcaya, *Port.* Baía de Biscaia. *bay* France/Spain
Biscay Plain 80 *B3 abyssal plain* SE Bay of Biscay
Bischofsburg see Biskupiec
Bishah, Wadi 121 *B5 dry watercourse* C Saudi Arabia
Bishkek 123 *G2 var.* Pishpek; *prev.* Frunze. *country capital* (Kyrgyzstan) Chuyskaya Oblast', N Kyrgyzstan
Bishop's Lynn see King's Lynn
Bishri, Jabal 118 *D3 mountain range* E Syria
Biskara see Biskra
Biskra 71 *E2 var.* Beskra, Biskara. NE Algeria
Biskupiec 98 *D2 Ger.* Bischofsburg. Warmińsko-Mazurskie, NE Poland
Bislig 139 *F2* Mindanao, S Philippines
Bismarck 45 *E2 state capital* North Dakota, N USA
Bismarck Archipelago 144 *B3 island group* NE Papua New Guinea
Bismarck Sea 144 *B3 sea* W Pacific Ocean
Bisnulok see Phitsanulok
Bissau 74 *B4 country capital* (Guinea-Bissau) W Guinea-Bissau
Bistrița 108 *B3 Ger.* Bistritz, *Hung.* Besztercze; *prev.* Nösen. Bistrița-Năsăud, N Romania
Bistritz see Bistrița
Bitam 77 *B5* Woleu-Ntem, N Gabon
Bitburg 95 *A5* Rheinland-Pfalz, SW Germany
Bitlis 117 *F3* Bitlis, SE Turkey
Bitoeng see Bitung
Bitola 101 *D6 Turk.* Monastir; *prev.* Bitolj. S FYR Macedonia
Bitolj see Bitola
Bitonto 97 *D5 anc.* Butuntum. Puglia, SE Italy
Bitterroot Range 46 *D2 mountain range* Idaho/Montana, NW USA
Bitung 139 *F3 prev.* Bitoeng. Sulawesi, C Indonesia
Biu 75 *H4* Borno, E Nigeria
Biwa-ko 131 *C6* lake Honshū, SW Japan
Bizerta see Bizerte
Bizerte 71 *E1 Ar.* Banzart, *Eng.* Bizerta. N Tunisia
Bjelovar 100 *B2 Hung.* Belovár. Bjelovar-Bilogora, N Croatia
Bjeshkët e Namuna see North Albanian Alps
Bjørneborg see Pori
Bjørnøya 83 *F3 Eng.* Bear Island. *island* N Norway
Blackall 148 *C4* Queensland, E Australia
Black Drin 101 *D6 Alb.* Lumi i Drinit të Zi, *SCr.* Crni Drim. *river* Albania/FYR Macedonia
Blackfoot 46 *E4* Idaho, NW USA
Black Forest 95 *B6 Eng.* Black Forest. *mountain range* SW Germany
Black Forest see Schwarzwald
Black Hills 44 *D3 mountain range* South Dakota/Wyoming, N USA
Blackpool 89 *C5* NW England, United Kingdom
Black Range 48 *C2 mountain range* New Mexico, SW USA
Black River 54 *A5* W Jamaica

Black River 136 *C3 Chin.* Babian Jiang, Lixian Jiang, *Fr.* Rivière Noire, *Vtn.* Sông Đa. *river* China/Vietnam
Black Rock Desert 47 *C5 desert* Nevada, W USA
Black Sand Desert see Garagum
Black Sea 116 *B1 var.* Euxine Sea, *Bul.* Cherno More, *Rom.* Marea Neagrã, *Rus.* Chernoye More, *Turk.* Karadeniz, *Ukr.* Chorne More. *sea* Asia/Europe
Black Sea Lowland 109 *E4 Ukr.* Prychornomor'ska Nyzovyna. *depression* SE Europe
Black Volta 75 *E4 var.* Borongo, Mouhoun, Moun Hou, *Fr.* Volta Noire. *river* W Africa
Blackwater 89 *A6 Ir.* An Abhainn Mhór. *river* S Ireland
Blackwater State see Nebraska
Blagoevgrad 104 *C3 prev.* Gorna Dzhumaya. Blagoevgrad, W Bulgaria
Blagoveshchensk 115 *G4* Amurskaya Oblast', SE Russian Federation
Blake Plateau 35 *D6 var.* Blake Terrace. *undersea plateau* W Atlantic Ocean
Blake Terrace see Blake Plateau
Blanca, Bahía 65 *C5 bay* E Argentina
Blanca, Costa 93 *F4 physical region* SE Spain
Blanche, Lake 149 *B5 lake* South Australia
Blanc, Mont 91 *D5 It.* Monte Bianco. *mountain* France/Italy
Blanco, Cape 46 *A4 headland* Oregon, NW USA
Blanes 93 *G2* Cataluña, NE Spain
Blankenberge 87 *A5* West-Vlaanderen, NW Belgium
Blankenheim 95 *A5* Nordrhein-Westfalen, W Germany
Blanquilla, Isla 59 *E1 var.* La Blanquilla. *island* N Venezuela
Blanquilla, La see Blanquilla, Isla
Blantyre 79 *E2 var.* Blantyre-Limbe. Southern, S Malawi
Blantyre-Limbe see Blantyre
Blaricum 86 *C3* Noord-Holland, C Netherlands
Blatnitsa see Durankulak
Blenheim 151 *C5* Marlborough, South Island, New Zealand
Blesae see Blois
Blida 70 *D2 var.* El Boulaida, El Boulaïda. N Algeria
Bloemfontein 78 *C4 var.* Mangaung. *country capital* (South Africa-judicial capital) Free State, C South Africa
Blois 90 *C4 anc.* Blesae. Loir-et-Cher, C France
Bloomfield 48 *C1* New Mexico, SW USA
Bloomington 40 *B4* Illinois, N USA
Bloomington 40 *C4* Indiana, N USA
Bloomington 45 *F2* Minnesota, N USA
Bloomsbury 148 *D3* Queensland, NE Australia
Bluefield 40 *D5* West Virginia, NE USA
Bluefields 53 *E3* Región Autónoma Atlántico Sur, SE Nicaragua
Bluegrass State see Kentucky
Blue Hen State see Delaware
Blue Law State see Connecticut
Blue Mountain Peak 54 *B5 mountain* E Jamaica
Blue Mountains 46 *C3 mountain range* Oregon/Washington, NW USA
Blue Nile 72 *C4 var.* Abai, Bahr el, Azraq, *Amh.* Ābay Wenz, *Ar.* An Nīl al Azraq, *Amh.* Bahr el Azraq. *river* Ethiopia/Sudan
Blumenau 63 *E5* Santa Catarina, S Brazil
Blythe 47 *D8* California, W USA
Blytheville 42 *C1* Arkansas, C USA
Bo 74 *C4* S Sierra Leone
Boaco 52 *D3* Boaco, S Nicaragua
Boa Vista 62 *D1 state capital* Roraima, N Brazil
Boa Vista 74 *A3 island* Ilhas de Barlavento, E Cape Verde
Bobaomby, Tanjona 79 *G2 Fr.* Cap d'Ambre. *headland* N Madagascar
Bobigny 90 *E1* Seine-St-Denis, N France
Bobo-Dioulasso 74 *D4* SW Burkina Faso
Bobrinets see Bobrynets'
Bobruysk see Babruysk
Bobrynets' 109 *E3 Rus.* Bobrinets. Kirovohrads'ka Oblast', C Ukraine
Boca Raton 43 *F5* Florida, SE USA
Bocay 52 *D2* Jinotega, N Nicaragua
Bocheykovo see Bacheykava

Bocholt 94 *A4* Nordrhein-Westfalen, W Germany
Bochum 94 *A4* Nordrhein-Westfalen, W Germany
Bocșa 108 *A4 Ger.* Bokschen, *Hung.* Boksánbánya. Caraș-Severin, SW Romania
Bodaybo 115 *F4* Irkutskaya Oblast', E Russian Federation
Bodegas see Babahoyo
Boden 84 *D4* Norrbotten, N Sweden
Bodensee see Constance, Lake, C Europe
Bodmin 89 *C7* SW England, United Kingdom
Bodø 84 *C3* Nordland, C Norway
Bodrum 116 *A4* Muğla, SW Turkey
Boeloekoemba see Bulukumba
Boende 77 *C5* Equateur, C Dem. Rep. Congo
Boeroe see Buru, Pulau
Boetoeng see Buton, Pulau
Bogale 136 *B4* Ayeyarwady, SW Myanmar (Burma)
Bogalusa 42 *B3* Louisiana, S USA
Bogatynia 98 *B4 Ger.* Reichenau. Dolnośląskie, SW Poland
Boğazlıyan 116 *D3* Yozgat, C Turkey
Bogendorf see Luków
Bogor 138 *C5 Dut.* Buitenzorg. Java, C Indonesia
Bogotá 58 *B3 prev.* Santa Fe, Santa Fe de Bogotá. *country capital* (Colombia) Cundinamarca, C Colombia
Boguchevsk see Bahushewsk
Boguslav see Bohuslav
Bo Hai 128 *D4 var.* Gulf of Chihli. *gulf* NE China
Bohemia 99 *A5 Cz.* Čechy, *Ger.* Böhmen. W Czech Republic
Bohemian Forest 95 *C5 Cz.* Český Les, Šumava, *Ger.* Böhmerwald. *mountain range* C Europe
Böhmen see Bohemia
Böhmerwald see Bohemian Forest
Böhmisch-Krumau see Český Krumlov
Bohol Sea 139 *E2 var.* Mindanao Sea. *sea* S Philippines
Bohuslav 109 *E2 Rus.* Boguslav. Kyyivs'ka Oblast', N Ukraine
Boise 46 *D3 var.* Boise City. *state capital* Idaho, NW USA
Boise City 49 *F1* Oklahoma, C USA
Boise City see Boise
Bois, Lac des see Woods, Lake of the
Bois-le-Duc see 's-Hertogenbosch
Boizenburg 94 *C3* Mecklenburg-Vorpommern, N Germany
Bojador see Boujdour
Bojnūrd 120 *D2 var.* Bujnurd. Khorāsān-e Shemālī, N Iran
Bokāro 135 *F4* Jhārkhand, N India
Boké 74 *C4* W Guinea
Bokhara see Buxoro
Boksánbánya/Bokschen see Bocșa
Bol 76 *B3* Lac, W Chad
Bolgatanga 75 *E4* N Ghana
Bolhrad 108 *D4 Rus.* Bolgrad. Odes'ka Oblast', SW Ukraine
Bolívar, Pico 58 *C2 mountain* W Venezuela
Bolivia 61 *F3 off.* Republic of Bolivia. *country* W South America

BOLIVIA
South America

Official name Plurinational State of Bolivia
Formation 1825 / 1938
Capital La Paz (administrative); Sucre (judicial)
Population 10.7 million / 26 people per sq mile (10 people per sq km)
Total area 424,162 sq. miles (1,098,580 sq. km)
Languages Aymara*, Quechua*, Spanish*
Religions Roman Catholic 93%, Other 7%
Ethnic mix Quechua 37%, Aymara 32%, Mixed race 13%, European 10%, Other 8%
Government Presidential system
Currency Boliviano = 100 centavos
Literacy rate 94%
Calorie consumption 2254 kilocalories

Bolivia, Republic of see Bolivia
Bollène 91 *D6* Vaucluse, SE France
Bollnäs 85 *C5* Gävleborg, C Sweden

177

Bollon *149 D5* Queensland, C Australia
Bologna *96 C3* Emilia-Romagna, N Italy
Bol'shevik *see* Bal'shavik
Bol'shevik, Ostrov *115 E2 island* Severnaya Zemlya, Russian Federation
Bol'shezemel'skaya Tundra *110 E3 physical region* NW Russian Federation
Bol'shoy Lyakhovskiy, Ostrov *115 F2 island* NE Russian Federation
Bolton *89 D5 prev.* Bolton-le-Moors. NW England, United Kingdom
Bolton-le-Moors *see* Bolton
Bolu *116 B3* Bolu, NW Turkey
Bolungarvík *83 E4* Vestfirdhir, NW Iceland
Bolyarovo *104 D3 prev.* Pashkeni. Yambol, E Bulgaria
Bolzano *96 C1 Ger.* Bozen; *anc.* Bauzanum. Trentino-Alto Adige, N Italy
Boma *77 B6* Bas-Congo, W Dem. Rep. Congo
Bombay *see* Mumbai
Bomu *76 D4 var.* Mbomou, Mbomu, M'Bomu. *river* Central African Republic/Dem. Rep. Congo
Bonaire *55 F5 Dutch special municipality* S West Indies
Bonanza *52 D2* Región Autónoma Atlántico Norte, NE Nicaragua
Bonaparte Archipelago *146 C2 island group* Western Australia
Bon, Cap *102 D3 headland* N Tunisia
Bonda *77 B6* Ogooué-Lolo, C Gabon
Bondoukou *75 E4* E Côte d'Ivoire
Bône *see* Annaba, Algeria
Bone *see* Watampone, Indonesia
Bone, Teluk *139 E4 bay* Sulawesi, C Indonesia
Bongaigaon *135 G3* Assam, NE India
Bongo, Massif des *76 D4 var.* Chaine des Mongos. *mountain range* NE Central African Republic
Bongor *76 B3* Mayo-Kébbi, SW Chad
Bonifacio *91 E7* Corse, France, C Mediterranean Sea
Bonifacio, Bocche de/Bonifacio, Bouches de *see* Bonifacio, Strait of
Bonifacio, Strait of *96 A4 Fr.* Bouches de Bonifacio, *It.* Bocche di Bonifacio. *strait* C Mediterranean Sea
Bonn *95 A5* Nordrhein-Westfalen, W Germany
Bononia *see* Vidin, Bulgaria
Bononia *see* Boulogne-sur-Mer, France
Boosaaso *72 E4 var.* Bandar Kassim, Bender Qaasim, Bosaso, *It.* Bender Cassim. Bari, N Somalia
Boothia Felix *see* Boothia Peninsula
Boothia, Gulf of *37 F2 gulf* Nunavut, NE Canada
Boothia Peninsula *37 F2 prev.* Boothia Felix. *peninsula* Nunavut, NE Canada
Boppard *95 A5* Rheinland-Pfalz, W Germany
Boquete *53 E5 var.* Bajo Boquete. Chiriquí, W Panama
Boquillas *50 D2 var.* Boquillas del Carmen. Coahuila, NE Mexico
Boquillas del Carmen *see* Boquillas
Bor *100 E4* Serbia, E Serbia
Bor *73 B5* Jonglei, E South Sudan
Borås *85 B7* Västra Götaland, S Sweden
Borbetomagus *see* Worms
Borborema, Planalto da *56 E3 plateau* NE Brazil
Bordeaux *91 B5 anc.* Burdigala. Gironde, SW France
Bordj Omar Driss *71 E3* E Algeria
Borgå *see* Porvoo
Børgefjell *84 C4 mountain range* C Norway
Borger *86 E2* Drenthe, NE Netherlands
Borger *49 E1* Texas, SW USA
Borgholm *85 C7* Kalmar, S Sweden
Borgo Maggiore *96 E1* NW San Marino
Borislav *see* Boryslav
Borisoglebsk *111 B6* Voronezhskaya Oblast', W Russian Federation
Borisov *see* Barysaw
Borlänge *85 C6* Dalarna, C Sweden
Borne *86 E3* Overijssel, E Netherlands
Borneo *138 C4 island* Brunei/Indonesia/Malaysia
Bornholm *85 B8 island* E Denmark
Borohoro Shan *126 B2 mountain range* NW China
Borongo *see* Black Volta
Boron'ki *see* Baron'ki
Borosjenő *see* Ineu

Borovan *104 C2* Vratsa, NW Bulgaria
Borovichi *110 B4* Novgorodskaya Oblast', W Russian Federation
Borovo *100 C3* Vukovar-Srijem, NE Croatia
Borriana *93 F3 var.* Burriana. País Valenciano, E Spain
Borşa *108 C3* Hung. Borsa. Maramureş, N Romania
Boryslav *108 B2 Pol.* Borysław, *Rus.* Borislav. L'vivs'ka Oblast', NW Ukraine
Borysław *see* Boryslav
Bosanska Dubica *100 B3 var.* Kozarska Dubica. Republika Srpska, NW Bosnia and Herzegovina
Bosanska Gradiška *100 B3 var.* Gradiška. Republika Srpska, N Bosnia and Herzegovina
Bosanski Novi *100 B3 var.* Novi Grad. Republika Srpska, NW Bosnia and Herzegovina
Bosanski Šamac *100 C3 var.* Šamac. Republika Srpska, N Bosnia and Herzegovina
Bosaso *see* Boosaaso
Bösing *see* Pezinok
Boskovice *99 B5 Ger.* Boskowitz. Jihomoravský Kraj, SE Czech Republic
Boskowitz *see* Boskovice
Bosna *100 C4 river* N Bosnia and Herzegovina
Bosne I Hercegovine, Federacija *100 B3 republic* Bosnia and Herzegovina
Bosnia and Herzegovina *100 B3 off.* Republic of Bosnia and Herzegovina. *country* SE Europe

BOSNIA & HERZEGOVINA
Southeast Europe

Official name Bosnia and Herzegovina
Formation 1992 / 1992
Capital Sarajevo
Population 3.8 million / 192 people per sq mile (74 people per sq km)
Total area 19,741 sq. miles (51,129 sq. km)
Languages Bosnian*, Serbian*, Croatian*
Religions Muslim (mainly Sunni) 40%, Orthodox Christian 31%, Roman Catholic 15%, Other 10%, Protestant 4%
Ethnic mix Bosniak 48%, Serb 34%, Croat 16%, Other 2%
Government Parliamentary system
Currency Marka = 100 pfeninga
Literacy rate 98%
Calorie consumption 3130 kilocalories

Bosnia and Herzegovina, Republic of *see* Bosnia and Herzegovina
Boso-hanto *131 D6 peninsula* Honshū, S Japan
Bosphorus/Bosporus *see* İstanbul Boğazı
Bosporus Cimmerius *see* Kerch Strait
Bosporus Thracius *see* İstanbul Boğazı
Bossangoa *76 C4* Ouham, C Central African Republic
Bossembélé *76 C4* Ombella-Mpoko, C Central African Republic
Bossier City *42 A2* Louisiana, S USA
Bosten Hu *126 C3 var.* Bagrax Hu. *lake* NW China
Boston *89 E6 prev.* St.Botolph's Town. E England, United Kingdom
Boston *41 G3 state capital* Massachusetts, NE USA
Boston Mountains *42 B1 mountain range* Arkansas, C USA
Bostyn' *see* Bastyn'
Botany *148 E2* New South Wales, E Australia
Botany Bay *148 E2 inlet* New South Wales, SE Australia
Boteti *78 C3 var.* Botletle. *river* N Botswana
Bothnia, Gulf of *85 D5 Fin.* Pohjanlahti, *Swe.* Bottniska Viken. *gulf* N Baltic Sea
Botletle *see* Boteti
Botoşani *108 C3 Hung.* Botosány. Botoşani, NE Romania
Botosány *see* Botoşani
Botou *128 C4 prev.* Bozhen. Hebei, E China
Botrange *87 D6 mountain* E Belgium
Botswana *78 C3 off.* Republic of Botswana. *country* S Africa

BOTSWANA
Southern Africa

Official name Republic of Botswana

BOTSWANA
(continued)

Formation 1966 / 1966
Capital Gaborone
Population 2 million / 9 people per sq mile (4 people per sq km)
Total area 231,803 sq. miles (600,370 sq. km)
Languages Setswana, English*, Shona, San, Khoikhoi, isiNdebele
Religions Christian 70%, Nonreligious 20%, Traditional beliefs 6%, Other 4%
Ethnic mix Tswana 79%, Kalanga 11%, Other 10%
Government Presidential system
Currency Pula = 100 thebe
Literacy rate 87%
Calorie consumption 2285 kilocalories

Botswana, Republic of *see* Botswana
Bottniska Viken *see* Bothnia, Gulf of
Bouar *76 B4* Nana-Mambéré, W Central African Republic
Bou Craa *70 B3 var.* Bu Craa. NW Western Sahara
Bougainville Island *142 B3 island* NE Papua New Guinea
Bougaroun, Cap *102 C3 headland* NE Algeria
Bougouni *74 D4* Sikasso, SW Mali
Boujdour *70 A3 var.* Bojador. W Western Sahara
Boulder *44 C4* Colorado, C USA
Boulder *44 B2* Montana, NW USA
Boulogne *see* Boulogne-sur-Mer
Boulogne-Billancourt *90 D1* Ile-de-France, N France Europe
Boulogne-sur-Mer *90 C2 var.* Boulogne; *anc.* Bononia, Gesoriacum, Gessoriacum. Pas-de-Calais, N France
Boûmdeïd *74 C3 var.* Boumdeït. Assaba, S Mauritania
Boumdeït *see* Boûmdeïd
Boundiali *74 D4* N Côte d'Ivoire
Bountiful *44 B4* Utah, W USA
Bounty Basin *see* Bounty Trough
Bounty Islands *142 D5 island group* S New Zealand
Bounty Trough *152 C5 var.* Bounty Basin. *trough* S Pacific Ocean
Bourbonnais *90 C4 cultural region* C France
Bourbon Vendée *see* la Roche-sur-Yon
Bourg *see* Bourg-en-Bresse
Bourgas *see* Burgas
Bourge-en-Bresse *see* Bourg-en-Bresse
Bourg-en-Bresse *91 D5 var.* Bourg, Bourge-en-Bresse. Ain, E France
Bourges *90 C4 anc.* Avaricum. Cher, C France
Bourgogne *90 C4 Eng.* Burgundy. *cultural region* E France
Bourke *149 C5* New South Wales, SE Australia
Bournemouth *89 D7* S England, United Kingdom
Boutilimit *74 C3* Trarza, SW Mauritania
Bouvet Island *67 D7 Norwegian dependency* S Atlantic Ocean
Bowen *148 D3* Queensland, NE Australia
Bowling Green *40 B5* Kentucky, S USA
Bowling Green *40 C3* Ohio, N USA
Boxmeer *86 D4* Noord-Brabant, SE Netherlands
Boyarka *109 E2* Kyyivs'ka Oblast', N Ukraine
Boychinovtsi *104 C2 prev.* Lekhchevo. Montana, NW Bulgaria
Boysun *123 E3 Rus.* Baysun. Surkhondaryo Viloyati, S Uzbekistan
Bozeman *44 B2* Montana, NW USA
Bozen *see* Bolzano
Bozhen *see* Botou
Bozüyük *116 B3* Bilecik, NW Turkey
Brač *100 B4 var.* Brach, *It.* Brazza; *anc.* Brattia. *island* S Croatia
Bracara Augusta *see* Braga
Brach *see* Brač
Brades *55 G3 de facto dependent territory capital, de jure capital ,* Plymouth, destroyed by volcano in 1995 (Montserrat) SW Montserrat
Bradford *89 D5* N England, United Kingdom
Brady *49 F3* Texas, SW USA
Braga *92 B2 anc.* Bracara Augusta. Braga, NW Portugal

Bragança *92 C2 Eng.* Braganza; [cut] Briga. Bragança, NE Portugal
Braganza *see* Bragança
Brahestad *see* Raahe
Brahmanbaria *135 G4* Chittagon E Bangladesh
Brahmapur *135 F5* Odisha, E Ind
Brahmaputra *135 H3 var.* Padm Tsangpo, *Ben.* Jamuna, *Chin.* Y Zangbo Jiang, *Ind.* Bramaputra Dihang, Siang. *river* S Asia
Bräila *108 D4* Bräila, E Romania
Braine-le-Comte *87 B6* Hainaut, SW Belgium
Brainerd *45 F2* Minnesota, N US
Brak *see* Birāk
Bramaputra *see* Brahmaputra
Brampton *38 D5* Ontario, S Cana
Branco, Rio *56 C3 river* N Brazil
Brandberg *78 A3 mountain* NW Namibia
Brandenburg *94 C3 var.* Brande an der Havel. Brandenburg, NE Germany
Brandenburg an der Havel *see* Brandenburg
Brandon *37 F5* Manitoba, S Cana
Braniewo *98 D2 Ger.* Braunsberg Warmińsko-mazurskie, N Polar
Brasil *see* Brazil
Brasília *63 F3 country capital* (Br Distrito Federal, C Brazil
Brasil, República Federativa do *see* Brazil
Braşov *108 C4 Ger.* Kronstadt, H Brassó; *prev.* Oraşul Stalin. Braş C Romania
Brassó *see* Braşov
Bratislava *99 C6 Ger.* Pressburg, Pozsony. *country capital* (Slova Bratislavský Kraj, W Slovakia
Bratsk *115 E4* Irkutskaya Oblast', C Russian Federation
Brattia *see* Brač
Braunsberg *see* Braniewo
Braunschweig *94 C4 Eng./Fr.* Brunswick. Niedersachsen, N G
Brava *see* Baraawe
Brava, Costa *93 H2 coastal region* NE Spain
Bravo del Norte, Río/Bravo, Río Grande, Rio
Bravo, Río *50 C1 river* Mexico/US North America
Brawley *47 D8* California, W USA
Brazil *62 C2 off.* Federative Repub Brazil, *Port.* República Federativa Brasil, *Sp.* Brasil; *prev.* United Sta Brazil. *country* South America

BRAZIL
South America

Official name Federative Republic of Brazil
Formation 1822 / 1828
Capital Brasília
Population 200 million / 61 people per sq mile (24 people per sq km)
Total area 3,286,470 sq. miles (8,511,965 sq. km)
Languages Portuguese*, German, Italian, Spanish, Polish, Japanese, Amerindian languages
Religions Roman Catholic 74%, Protestant 15%, Atheist 7%, Other Ethnic mix** White 54%, Mixed race Black 6%, Other 2%
Government Presidential system
Currency Real = 100 centavos
Literacy rate 91%
Calorie consumption 3287 kilocal

Brazil Basin *67 C5 var.* Brazilian E Brazil'skaya Kotlovina. *undersea* W Atlantic Ocean
Brazil, Federative Republic of *see* Brazil
Brazilian Basin *see* Brazil Basin
Brazilian Highlands *63 F3 var.* Brazil Highlands. *mountain range* E Brazil
Brazil'skaya Kotlovina *see* Brazil Basin
Brazil, United States of *see* Brazil
Brazos River *49 G3 river* Texas, S'
Brazza *see* Brač
Brazzaville *77 B6 country capital* (Congo) Capital District, S Cong
Brčko *100 C3* Brčko Distrikt, NE [cut] and Herzegovina

Distrikt 100 B3 *autonomous*
rict Bosnia and Herzegovina
t 87 C5 Antwerpen, N Belgium
n **Beacons** 89 C6 *mountain range*
ales, United Kingdom
86 C4 Noord-Brabant,
etherlands
87 D5 Limburg, NE Belgium
inica 101 E6 *river* E FYR
cedonia
nz 57 B7 *anc.* Brigantium.
arlberg, W Austria
vo 104 B1 Vidin, NW Bulgaria
en 94 B3 Fr. Brême. Bremen,
J Germany
erhaven 94 B3 Bremen,
J Germany
erton 46 B2 Washington, NW USA
am 49 G3 Texas, SW USA
ner, Col du/Brennero, Passo del
Brenner Pass
ner Pass 96 C1 *var.* Brenner Sattel,
Col du Brenner, Ger. Brennerpass,
Passo del Brennero. *pass* Austria/
y
nerpass *see* Brenner Pass
ner Sattel *see* Brenner Pass
ia 96 B2 *anc.* Brixia. Lombardia,
taly
au *see* Wrocław
anone 96 C1 Ger. Brixen.
ntino-Alto Adige, N Italy
107 A6 Pol. Brześć nad Bugiem,
s. Brest-Litovsk; *prev.* Brześć
ewski. Brestskaya Voblasts',
Belarus
90 A3 Finistère, NW France
Litovsk *see* Brest
gne 90 A3 Eng. Brittany, Lat.
tannia Minor. *cultural region*
V France
ster, Kap *see* Kangikajik
lton 42 C3 Alabama, S USA
nnev *see* Naberezhnyye Chelny
avo 104 D2 *prev.* Abrashlare.
wdiv, C Bulgaria
76 D4 Haute-Kotto, C Central
rican Republic
açon 91 D5 *anc.* Brigantio. Hautes-
pes, SE France
stow *see* Bristol
geport 41 F3 Connecticut, NE USA
getown 55 G2 *country capital*
arbados) SW Barbados
ington 89 D5 E England, United
ngdom
port 89 D7 S England, United
ngdom
g *see* Brzeg
95 A7 Fr. Brigue, It. Briga. Valais,
Switzerland
a *see* Brig
antio *see* Briançon
antium *see* Bregenz
ham City 44 B3 Utah, W USA
hton 89 E7 SE England, United
ngdom
hton 44 D4 Colorado, C USA
ae *see* Brig
disi 97 E5 *anc.* Brundisium,
undusium. Puglia, SE Italy
vera *see* St-Lô
bane 149 E5 *state capital*
ueensland, E Australia
tol 89 D7 *anc.* Bricgstow.
V England, United Kingdom
tol 41 F3 Connecticut, NE USA
tol 40 D5 Tennessee, S USA
tol Bay 36 B3 *bay* Alaska, USA
tol Channel 89 C7 *inlet* England/
ales, United Kingdom
ain 80 C3 *var.* Great Britain. *island*
ited Kingdom
annia Minor *see* Bretagne
sh Columbia 36 D4 Fr. Colombie-
ritannique. *province* SW Canada
sh Guiana *see* Guyana
sh Honduras *see* Belize
sh Indian Ocean Territory 141 B5
K *dependent territory* C Indian Ocean
sh Isles 89 *island group*
V Europe
sh North Borneo *see* Sabah
sh Solomon Islands Protectorate
e Solomon Islands
sh Virgin Islands 55 F3 *var.*
irgin Islands. UK *dependent territory*
West Indies
any *see* Bretagne
a Curretia *see* Brive-la-Gaillarde

Briva Isarae *see* Pontoise
Brive *see* Brive-la-Gaillarde
Brive-la-Gaillarde 91 C5 *prev.* Brive;
anc. Briva Curretia. Corrèze, C France
Brixen *see* Bressanone
Brixia *see* Brescia
Brno 99 B5 Ger. Brünn. Jihomoravský
Kraj, SE Czech Republic
Bročeni 106 B3 SW Latvia
Brod/Bród *see* Slavonski Brod
Brodeur Peninsula 37 F2 *peninsula*
Baffin Island, Nunavut, NE Canada
Brod na Savi *see* Slavonski Brod
Brodnica 98 C3 Ger. Buddenbrock.
Kujawski-pomorskie, C Poland
Broek-in-Waterland 86 C3 Noord-
Holland, C Netherlands
Broken Arrow 49 G1 Oklahoma, C USA
Broken Bay 148 E1 *bay* New South
Wales, SE Australia
Broken Hill 149 B6 New South Wales,
SE Australia
Broken Ridge 141 D6 *undersea plateau*
S Indian Ocean
Bromberg *see* Bydgoszcz
Bromley 89 B8 United Kingdom
Brookhaven 42 B3 Mississippi, S USA
Brookings 45 F3 South Dakota, N USA
Brooks Range 36 D2 *mountain range*
Alaska, USA
Brookton 147 B6 Western Australia
Broome 146 B3 Western Australia
Broomfield 44 D4 Colorado, C USA
Broucsella *see* Brussel/Bruxelles
Brovary 109 E2 Kyyivs'ka Oblast',
N Ukraine
Brownfield 49 E2 Texas, SW USA
Brownsville 49 G5 Texas, SW USA
Brownwood 49 F3 Texas, SW USA
Brozha 107 D7 Mahilyowskaya Voblasts',
E Belarus
Bruges *see* Brugge
Brugge 87 A5 Fr. Bruges. West-
Vlaanderen, NW Belgium
Brummen 86 D3 Gelderland,
E Netherlands
Brundisium/Brundusium *see* Brindisi
Brunei 138 D3 *off.* Brunei Darussalam,
Mal. Negara Brunei Darussalam.
country SE Asia

BRUNEI
Southeast Asia

Official name Brunei Darussalam
Formation 1984 / 1984
Capital Bandar Seri Begawan
Population 400,000 / 197 people
per sq mile (76 people per sq km)
Total area 2228 sq. miles (5770 sq. km)
Languages Malay*, English, Chinese
Religions Muslim (mainly Sunni) 66%,
Buddhist 14%, Christian 10%, Other 10%
Ethnic mix Malay 67%, Chinese 16%,
Other 11%, Indigenous 6%
Government Monarchy
Currency Brunei dollar = 100 cents
Literacy rate 95%
Calorie consumption 2949 kilocalories

Brunei Darussalam *see* Brunei
Brunei Town *see* Bandar Seri Begawan
Brünn *see* Brno
Brunner, Lake 151 C5 *lake* South Island,
New Zealand
Brunswick 43 E3 Georgia, SE USA
Brunswick *see* Braunschweig
Brusa *see* Bursa
Brus Laguna 52 D2 Gracias a Dios,
E Honduras
Brussa *see* Bursa
Brussel 87 C6 *var.* Brussels,
Fr. Bruxelles, Ger. Brüssel; *anc.*
Broucsella. *country capital*
(Belgium) Brussels, C Belgium
Brüssel/Brussels *see* Brussel/Bruxelles
Brüx *see* Most
Bruxelles *see* Brussel
Bryan 49 G3 Texas, SW USA
Bryansk 111 A5 Bryanskaya Oblast',
W Russian Federation
Brzeg 98 C4 Ger. Brieg; *anc.* Civitas Altae
Ripae. Opolskie, S Poland
Brześć Litewski/Brześć nad Bugiem
see Brest
Brzeżany *see* Berezhany
Bucaramanga 58 B2 Santander,
N Colombia
Buchanan 74 C5 *prev.* Grand Bassa.
SW Liberia

Buchanan, Lake 49 F3 *reservoir* Texas,
SW USA
Bucharest *see* București
Buckeye State *see* Ohio
Bu Craa *see* Bou Craa
București 108 C5 Eng. Bucharest, Ger.
Bukarest, *prev.* Altenburg; *anc.* Cetatea
Dâmboviței. *country capital* (Romania)
București, S Romania
Buda-Kashalyova 107 D7 Rus. Buda-
Koshelëvo. Homyel'skaya Voblasts',
SE Belarus
Buda-Koshelëvo *see* Buda-Kashalyova
Budapest 99 C6 *off.* Budapest Főváros,
SCr. Budimpešta. *country capital*
(Hungary) Pest, N Hungary
Budapest Főváros *see* Budapest
Budaun 134 D3 Uttar Pradesh, N India
Buddenbrock *see* Brodnica
Budimpešta *see* Budapest
Budweis *see* České Budějovice
Budyšin *see* Bautzen
Buena Park 46 E2 California, W USA
North America
Buenaventura 58 A3 Valle del Cauca,
W Colombia
Buena Vista 61 G4 Santa Cruz, C Bolivia
Buena Vista 93 H5 S Gibraltar Europe
Buena Vista 93 H5 S Gibraltar Europe
Buenavista 93 H5 Baja California Sur,
NW Mexico
Buenavista 93 H5 Sonora, NW Mexico
Buena Vista 93 H5 Cerro Largo,
Uruguay
Buena Vista 93 H5 Colorado, C USA
Buena Vista 93 H5 Georgia, SE USA
Buena Vista 93 H5 Virginia, NE USA
Buenos Aires 64 D4 *hist.* Santa Maria del
Buen Aire. *country capital* (Argentina)
Buenos Aires, E Argentina
Buenos Aires 53 E5 Puntarenas,
SE Costa Rica
Buenos Aires, Lago 65 B6 *var.* Lago
General Carrera. *lake* Argentina/Chile
Buffalo 41 E3 New York, NE USA
Buffalo Narrows 37 F4 Saskatchewan,
C Canada
Buff Bay 54 B5 E Jamaica
Buftea 108 C5 Ilfov, S Romania
Bug 81 E3 Bel. Zakhodni Buh, Eng.
Western Bug, Rus. Zapadnyy Bug, Ukr.
Zakhidnyy Buh. *river* E Europe
Buga 58 B3 Valle del Cauca,
W Colombia
Bughotu *see* Santa Isabel
Buguruslan 111 D6 Orenburgskaya
Oblast', W Russian Federation
Buitenzorg *see* Bogor
Bujalance 92 D4 Andalucía, S Spain
Bujanovac 101 D5 SE Serbia
Bujnurd *see* Bojnürd
Bujumbura 73 B7 *prev.* Usumbura.
country capital (Burundi) W Burundi
Bukarest *see* București
Bukavu 77 E6 *prev.* Costermansville.
Sud-Kivu, E Dem. Rep. Congo
Bukhara *see* Buxoro
Bukoba 73 B6 Kagera, NW Tanzania
Bülach 95 B7 Zürich, NW Switzerland
Bulawayo 78 D3 Matabeleland North,
SW Zimbabwe
Bulgan 127 E2 Bulgan, N Mongolia
Bulgaria 104 C2 *off.* Republic of Bulgaria,
Bul. Bŭlgariya; *prev.* People's Republic
of Bulgaria. *country* SE Europe

BULGARIA
Southeast Europe

Official name Republic of Bulgaria
Formation 1908 / 1947
Capital Sofia
Population 7.2 million / 169 people
per sq mile (65 people per sq km)
Total area 42,822 sq. miles (110,910 sq. km)
Languages Bulgarian*, Turkish, Romani
Religions Bulgarian Orthodox 83%,
Muslim 12%, Other 4%,
Roman Catholic 1%
Ethnic mix Bulgarian 84%, Turkish 9%,
Roma 5%, Other 2%
Government Parliamentary system
Currency Lev = 100 stotinki
Literacy rate 98%
Calorie consumption 2877 kilocalories

Bulgaria, People's Republic of *see*
Bulgaria
Bulgaria, Republic of *see* Bulgaria

Bŭlgariya *see* Bulgaria
Bullion State *see* Missouri
Bull Shoals Lake 42 B1 *reservoir*
Arkansas/Missouri, C USA
Bulukumba 139 E4 *prev.* Boeloekoemba.
Sulawesi, C Indonesia
Bumba 77 D5 Equateur, N Dem. Rep.
Congo
Bunbury 147 A7 Western Australia
Bundaberg 148 E4 Queensland,
E Australia
Bungo-suido 131 B7 *strait* SW Japan
Bunia 77 E5 Orientale, NE Dem. Rep.
Congo
Bünyan 116 D3 Kayseri, C Turkey
Buraida *see* Buraydah
Buraydah 120 B4 *var.* Buraida. Al
Qaşīm, N Saudi Arabia
Burdigala *see* Bordeaux
Burdur 116 B4 *var.* Buldur. Burdur,
SW Turkey
Burdur Gölü 116 B4 *salt lake* SW Turkey
Burê 72 C4 Āmara, N Ethiopia
Burgas 104 E2 *var.* Bourgas. Burgas,
E Bulgaria
Burgaski Zaliv 104 E2 *gulf* E Bulgaria
Burgos 92 D2 Castilla y León, N Spain
Burgundy *see* Bourgogne
Burhan Budai Shan 126 D3 *mountain*
range C China
Buriram 137 D5 *var.* Buri Ram,
Puriramya. Buri Ram, E Thailand
Buri Ram *see* Buriram
Burjassot 93 F3 Valenciana, E Spain
Burkburnett 49 F2 Texas, SW USA
Burketown 148 B3 Queensland,
NE Australia
Burkina *see* Burkina Faso
Burkina Faso 75 E4 *off.* Burkina Faso;
var. Burkina, *prev.* Upper Volta.
country W Africa

BURKINA FASO
West Africa

Official name Burkina Faso
Formation 1960 / 1960
Capital Ouagadougou
Population 16.9 million / 160 people
per sq mile (62 people per sq km)
Total area 105,869 sq. miles
(274,200 sq. km)
Languages Mossi, Fulani, French*,
Tuareg, Dyula, Songhai
Religions Muslim 55%, Christian 25%,
Traditional beliefs 20%
Ethnic mix Mossi 48%, Other 21%,
Peul 10%, Lobi 7%, Bobo 7%,
Mandé 7%
Government Transitional regime
Currency CFA franc = 100 centimes
Literacy rate 29%
Calorie consumption 2655 kilocalories

Burley 46 D4 Idaho, NW USA
Burlington 45 G4 Iowa, C USA
Burlington 41 F2 Vermont, NE USA
Burma 136 A3 *off.* Union of Myanmar.
country SE Asia. *See also* Myanmar
Burnie 149 C8 Tasmania, SE Australia
Burns 46 C3 Oregon, NW USA
Burnside 37 F3 *river* Nunavut,
NW Canada
Burnsville 45 F2 Minnesota, N USA
Burrel 101 D6 *var.* Burreli. Dibër,
C Albania
Burreli *see* Burrel
Burriana *see* Borriana
Bursa 116 B3 *var.* Brusa; *prev.* Brussa;
anc. Prusa. Bursa, NW Turkey
Bûr Sa'id 72 B1 *var.* Port Said. N Egypt
Burtnieks 106 C3 *var.* Burtnieks Ezers.
lake N Latvia
Burtnieks Ezers *see* Burtnieks
Burundi 73 B7 *off.* Republic of Burundi;
prev. Kingdom of Burundi, Urundi.
country C Africa

BURUNDI
Central Africa

Official name Republic of Burundi
Formation 1962 / 1962
Capital Bujumbura
Population 10.2 million / 1030 people
per sq mile (398 people per sq km)
Total area 10,745 sq. miles (27,830 sq. km)
Languages Kirundi*, French*, Kiswahili

BURUNDI
(continued)

Religions Roman Catholic 62%,
Traditional beliefs 23%, Muslim 10%,
Protestant 5%
Ethnic mix Hutu 85%, Tutsi 14%, Twa 1%
Government Presidential system
Currency Burundian franc = 100 centimes
Literacy rate 87%
Calorie consumption 1604 kilocalories

Burundi, Kingdom of see Burundi
Burundi, Republic of see Burundi
Buru, Pulau 139 F4 prev. Boeroe. island
E Indonesia
Busan 129 E4 off. Busan Gwang-yeoksi,
prev. Pusan, Jap. Fusan.
SE South Korea
Busan Gwang-yeoksi see Busan
Buşayrah 118 D3 Dayr az Zawr, E Syria
Büshehr/Bushire see Bandar-e Büshehr
Busra see Al Başrah, Iraq
Busselton 147 A7 Western Australia
Bussora see Al Başrah
Buta 77 D5 Orientale, N Dem. Rep.
Congo
Butembo 77 E5 Nord-Kivu, NE Dem.
Rep. Congo
Butler 41 E4 Pennsylvania, NE USA
Buton, Pulau 139 E4 var. Pulau Butung;
prev. Boetoeng. island C Indonesia
Bütow see Bytów
Butte 44 B2 Montana, NW USA
Butterworth 138 B3 Pinang, Peninsular
Malaysia
Button Islands 39 E1 island group
Nunavut, NE Canada
Butuan 139 F2 off. Butuan City.
Mindanao, S Philippines
Butuan City see Butuan
Butung, Pulau see Buton, Pulau
Butuntum see Bitonto
Buulobarde 73 D5 var. Buulo Berde.
Hiiraan, C Somalia
Buulo Berde see Buulobarde
Buur Gaabo 73 D6 Jubbada Hoose,
S Somalia
Buxoro 122 D2 var. Bokhara,
Rus. Bukhara. Buxoro Viloyati,
C Uzbekistan
Buynaksk 111 B8 Respublika Dagestan,
SW Russian Federation
Büyükmenderes Nehri 116 A4 river
SW Turkey
Buzău 108 C4 Buzău, SE Romania
Buzuluk 111 D6 Orenburgskaya Oblast',
W Russian Federation
Byahoml' 107 D5 Rus. Begoml'.
Vitsyebskaya Voblasts', N Belarus
Byalynichy 107 D6 Rus. Byelynichi.
Mahilyowskaya Voblasts', E Belarus
Byan Tumen see Choybalsan
Byarezina 107 D6 prev. Byerezino, Rus.
Berezina. river C Belarus
Bydgoszcz 98 C3 Ger. Bromberg.
Kujawski-pomorskie, C Poland
Byelaruskaya Hrada 107 B6 Rus.
Belorusskaya Gryada. ridge N Belarus
Byerezino see Byarezina
Byron Island see Nikunau
Bystrovka see Kemin
Bytča 99 C5 Zilínský Kraj, N Slovakia
Bytom 99 C5 Ger. Beuthen. Śląskie,
S Poland
Bytów 98 C2 Ger. Bütow. Pomorskie,
N Poland
Byuzmeyin see Abadan
Byval'ki 107 D8 Homyel'skaya Voblasts',
SE Belarus
Byzantium see İstanbul

C

Caála 78 B2 var. Kaala, Robert Williams,
Port. Vila Robert Williams. Huambo,
C Angola
Caazapá 64 D3 Caazapá, S Paraguay
Caballo Reservoir 48 C3 reservoir New
Mexico, SW USA
Cabanaquinta 92 D1 var. Cabañaquinta.
Asturias, N Spain
Cabañaquinta see Cabanaquinta
Cabanatuan 139 E1 off. Cabanatuan
City. Luzon, N Philippines
Cabanatuan City see Cabanatuan

Cabillonum see Chalon-sur-Saône
Cabimas 58 C1 Zulia, NW Venezuela
Cabinda 78 A1 var. Kabinda. Cabinda,
NW Angola
Cabinda 78 A1 var. Kabinda. province
NW Angola
Lake Cabora Bassa 78 D2 var.
Lake Cabora Bassa. reservoir
NW Mozambique
Cabora Bassa, Lake see Cahora Bassa,
Albufeira de
Caborca 50 B1 Sonora, NW Mexico
Cabot Strait 39 G4 strait E Canada
Cabo Verde, Ilhas do see Cape Verde
Cabras, Ilha das 76 E2 island
S Sao Tome and Principe, Africa,
E Atlantic Ocean
Cabrera, Illa de 93 G3 island E Spain
Cáceres 92 C3 Ar. Qazris. Extremadura,
W Spain
Cachimbo, Serra do 63 E2 mountain
range C Brazil
Caconda 78 B2 Huíla, C Angola
Čadca 99 C5 Hung. Csaca. Zilínský Kraj,
N Slovakia
Cadillac 40 C2 Michigan, N USA
Cadiz 139 E2 off. Cadiz City. Negros,
C Philippines
Cádiz 92 C5 anc. Gades, Gadier, Gadir,
Gadire. Andalucía, SW Spain
Cadiz City see Cadiz
Gulf of Cadiz 92 B5 Eng. Gulf of Cadiz.
gulf Portugal/Spain
Cádiz, Gulf of see Cádiz, Golfo de
Caducum see Cahors
Caen 90 B3 Calvados, N France
Caene/Caenepolis see Qinā
Caerdydd see Cardiff
Caer Glou see Gloucester
Caer Gybi see Holyhead
Caerleon see Chester
Caer Luel see Carlisle
Caesaraugusta see Zaragoza
Caesarea Mazaca see Kayseri
Caesarobriga see Talavera de la Reina
Caesarodunum see Tours
Caesaromagus see Beauvais
Caesena see Cesena
Cafayate 64 C2 Salta, N Argentina
Cagayan de Oro 139 E2 off. Cagayan de
Oro City. Mindanao, S Philippines
Cagayan de Oro City see Cagayan de Oro
Cagliari 97 A6 anc. Caralis. Sardegna,
Italy, C Mediterranean Sea
Caguas 55 F3 E Puerto Rico
Cahors 91 C5 anc. Cadurcum. Lot,
S France
Cahul 108 D4 Rus. Kagul. S Moldova
Caicos Passage 54 D2 strait Bahamas/
Turks and Caicos Islands
Caiffa see Hefa
Cailungo 96 E1 N San Marino
Caiphas see Hefa
Cairns 148 D3 Queensland, NE Australia
Cairo 72 B2 var. El Qâhira, Ar. Al
Qâhirah. country capital (Egypt)
N Egypt
Caisleán an Bharraigh see Castlebar
Cajamarca 60 B3 prev. Caxamarca.
Cajamarca, NW Peru
Čakovec 100 B2 Ger. Csakathurn, Hung.
Csáktornya; prev. Ger. Tschakathurn.
Medimurje, N Croatia
Calabar 75 G5 Cross River, S Nigeria
Calabozo 58 D2 Guárico, C Venezuela
Calafat 108 B5 Dolj, SW Romania
Calafate see El Calafate
Calahorra 93 E2 La Rioja, N Spain
Calais 90 C2 Pas-de-Calais, N France
Calais 41 H2 Maine, NE USA
Calais, Pas de see Dover, Strait of
Calama 64 B2 Antofagasta, N Chile
Călăraşi 108 D3 var. Călăras, Rus.
Kalarash. C Moldova
Călăraşi 108 C5 Călăraşi, SE Romania
Calatayud 93 E2 Aragón, NE Spain
Calbayog 139 E2 off. Calbayog City.
Samar, C Philippines
Calbayog City see Calbayog
Calcutta see Kolkata
Caldas da Rainha 92 B3 Leiria,
W Portugal
Caldera 64 B3 Atacama, N Chile
Caldwell 46 C3 Idaho, NW USA
Caledonia 52 C1 Corozal, N Belize
Caleta Olivia 65 B6 Santa Cruz,
SE Argentina
Calgary 37 E5 Alberta, SW Canada

Cali 58 B3 Valle del Cauca, W Colombia
Calicut see Kozhikode
California 47 B7 off. State of California,
also known as El Dorado, The Golden
State. state W USA
Gulf of California 50 B2 Eng. Gulf of
California; prev. Sea of Cortez. gulf
W Mexico
California, Gulf of see California,
Golfo de
Călimăneşti 108 B4 Vâlcea, SW Romania
Calisia see Kalisz
Callabonna, Lake 149 B5 lake South
Australia
Callao 60 C4 Callao, W Peru
Callatis see Mangalia
Callosa de Segura 93 F4 Valenciana,
E Spain
Calmar see Kalmar
Caloundra 149 E5 Queensland,
E Australia
Caltanissetta 97 C7 Sicilia, Italy,
C Mediterranean Sea
Caluula 72 E4 Bari, NE Somalia
Camabatela 78 B1 Cuanza Norte,
NW Angola
Camacupa 78 B2 var. General Machado,
Port. Vila General Machado. Bié,
C Angola
Camagüey 54 C2 prev. Puerto Príncipe.
Camagüey, C Cuba
Camagüey, Archipiélago de 54 C2
island group C Cuba
Camana 61 E4 var. Camaná. Arequipa,
SW Peru
Camargue 91 D6 physical region
SE France
Ca Mau 137 D6 var. Quan Long. Minh
Hai, S Vietnam
Cambay, Gulf of see Khambhāt, Gulf of
Camberia see Chambéry
Cambodia 137 D5 off. Kingdom
of Cambodia, var. Democratic
Kampuchea, Roat Kampuchea, Cam.
Kampuchea; prev. People's Democratic
Republic of Kampuchea. country
SE Asia

CAMBODIA
Southeast Asia

Official name Kingdom of Cambodia
Formation 1953 / 1953
Capital Phnom Penh
Population 14.3 million / 210 people
per sq mile (81 people per sq km)
Total area 69,900 sq. miles (181,040 sq. km)
Languages Khmer*, French, Chinese,
Vietnamese, Cham
Religions Buddhist 93%, Muslim 6%,
Christian 1%
Ethnic mix Khmer 90%, Other 5%,
Vietnamese 4%, Chinese 1%
Government Parliamentary system
Currency Riel = 100 sen
Literacy rate 79%
Calorie consumption 2411 kilocalories

Cambodia, Kingdom of see Cambodia
Cambrai 90 C2 Flem. Kambryk, prev.
Cambray; anc. Cameracum. Nord,
N France
Cambray see Cambrai
Cambrian Mountains 89 C6 mountain
range C Wales, United Kingdom
Cambridge 54 A4 W Jamaica
Cambridge 89 E6 Lat. Cantabrigia.
E England, United Kingdom
Cambridge 41 F4 Maryland, NE USA
Cambridge 40 D4 Ohio, NE USA
Cambridge Bay 37 F3 var. Ikaluktutiak.
Victoria Island, Nunavut, NW Canada
Camden 42 B2 Arkansas, C USA
Camellia State see Alabama
Cameracum see Cambrai
Cameroon 76 A4 off. Republic of
Cameroon, Fr. Cameroun. country
W Africa

CAMEROON
Central Africa

Official name Republic of Cameroon
Formation 1960 / 1961
Capital Yaoundé
Population 22.3 million / 124 people
per sq mile (48 people per sq km)

CAMEROON
(continued)

Total area 183,567 sq. miles
(475,400 sq. km)
Languages Bamileke, Fang, Fulani,
French*, English*
Religions Roman Catholic 35%,
Traditional beliefs 25%, Muslim 22%,
Protestant 18%
Ethnic mix Cameroon highlanders
Other 21%, Equatorial Bantu 19%,
Fulani 10%, Northwestern Bantu 8%
Government Presidential system
Currency CFA franc = 100 centimes
Literacy rate 71%
Calorie consumption 2586 kilocalc

Cameroon, Republic of see Camer
Cameroun see Cameroon
Camocim 63 F2 Ceará, E Brazil
Camopi 59 H3 E French Guiana
Campamento 52 C2 Olancho,
C Honduras
Campania 97 D5 Eng. Champagne
region S Italy
Campbell, Cape 151 D5 headland
Island, New Zealand
Campbell Island 142 D5 island S N
Zealand
Campbell Plateau 142 D5 undersea
plateau SW Pacific Ocean
Campbell River 36 D5 Vancouver
Island, British Columbia, SW Car
Campeche 51 G4 Campeche, SE M
Bay of Campeche 51 F4 Eng. Bay o
Campeche. bay E Mexico
Campeche, Bay of see Campeche,
Bahía de
Câm Pha 136 E3 Quang Ninh,
N Vietnam
Câmpina 108 C4 prev. Cimpina.
Prahova, SE Romania
Campina Grande 63 G2 Paraíba, E
Campinas 63 F4 São Paulo, S Brazil
Campobasso 97 D5 Molise, C Italy
Campo Criptana see Campo de Crip
Campo de Criptana 93 E3 var. Car
Criptana. Castilla-La Mancha, C !
Campo Grande 63 E4 state capital N
Grosso do Sul, SW Brazil
Campos dos Goytacazes 63 F4 Rio
Janeiro, SE Brazil
Câmpulung 108 B4 prev. Câmpulu
Muşcel, Cîmpulung. Argeş, S Ror
Câmpulung-Muşcel see Câmpulung
Campus Stellae see Santiago de
Compostela
Cam Ranh 137 E6 prev. Ba Ngoi. K
Hoa, S Vietnam
Canada 34 B4 country N North Am

CANADA
North America

Official name Canada
Formation 1867 / 1949
Capital Ottawa
Population 35.2 million / 10 people
per sq mile (4 people per sq km)
Total area 3,854,085 sq. miles
(9,984,670 sq. km)
Languages English*, French*, Chinese,
Italian, German, Ukrainian, Portuguese,
Inuktitut, Cree
Religions Roman Catholic 44%,
Protestant 29%, Other and
nonreligious 27%
Ethnic mix British, French, and other
European 87%, Asian 9%, Amerindia
Métis, and Inuit 4%
Government Parliamentary system
Currency Canadian dollar = 100 cent
Literacy rate 99%
Calorie consumption 3419 kilocalories

Canada Basin 34 C2 undersea basin
Arctic Ocean
Canadian River 49 E2 river SW USA
Çanakkale 116 A3 var. Dardanelli;
Chanak, Kale Sultanie. Çanakkal
W Turkey
Cananea 50 B1 Sonora, NW Mexico
Canarreos, Archipiélago de los 54 .
island group W Cuba
Canary Islands 70 A2 Eng. Canary
Islands. island group Spain,
NE Atlantic Ocean

ary Islands *see* Canarias, Islas
as *52 D4* Guanacaste, NW Costa Rica
averal, Cape *43 E4 headland* Florida,
E USA
avieiras *63 G3* Bahia, E Brazil
berra *142 C4 country capital*
(Australia) Australian Capital Territory,
E Australia
cún *51 H3* Quintana Roo, SE Mexico
dia *see* Irákleio
ea *see* Chaniá
gzhou *128 D4* Hebei, E China
iapíscau *39 E2 river* Québec,
Canada
iapíscau, Réservoir de *38 D3*
eservoir Québec, C Canada
ık Dağları *116 D2 mountain range*
l Turkey
illo *91 A7* Canillo, C Andorra Europe
kırı *116 C3 var.* Chankiri; *anc.*
angra, Germanicopolis. Çankın,
l Turkey
nanore *see* Kannur
nes *91 D6* Alpes-Maritimes,
E France
oas *63 E5* Rio Grande do Sul, S Brazil
on City *44 C5* Colorado, C USA
tabria *92 D1 autonomous*
ommunity N Spain
tábrica, Cordillera *92 C1 mountain*
ange N Spain
tabrigia *see* Cambridge
taura *59 E2* Anzoátegui,
JE Venezuela
terbury *89 E7 hist.* Cantwaraburh;
nc. Durovernum, *Lat.* Cantuaria.
JE England, United Kingdom
terbury Bight *151 C6 bight* South
**sland, New Zealand
terbury Plains *151 C6 plain* South
**sland, New Zealand
e Tho *137 E6* Cần Thơ, S Vietnam
ton *42 B2* Mississippi, S USA
ton *40 D4* Ohio, N USA
ton *see* Guangzhou
ton Island *see* Kanton
tuaria/Cantwaraburh *see*
**anterbury
yon *49 E2* Texas, SW USA
Bäng *136 D3 var.* Caobang. Cao
**ảng, N Vietnam
bang *see* Cao Bäng
-Breton, Ȋle du *see* Cape Breton
**sland
e Barren Island *149 C8 island*
**urneaux Group, Tasmania,
**E Australia
e Basin *69 B7 undersea basin*
**Atlantic Ocean
e Breton Island *39 G4 Fr.* Ȋle du
**Cap-Breton. *island* Nova Scotia,
**E Canada
e Charles *41 F5* Virginia, NE USA
e Coast *75 E5 prev.* Cape Coast
**astle. S Ghana
e Coast Castle *see* Cape Coast
e Girardeau *45 H5* Missouri, C USA
elle aan den IJssel *86 C4* Zuid-
**lolland, SW Netherlands
e Palmas *see* Harper
e Saint Jacques *see* Vung Tau
e Town *78 B5 var.* Ekapa, *Afr.*
**aapstad, Kapstad. *country capital*
**outh Africa-legislative capital)
**/estern Cape, SW South Africa
e Verde *74 A2 off.* Republic of Cape
**erde, *Port.* Cabo Verde, Ilhas do Cabo
**erde. *country* E Atlantic Ocean

•PE VERDE
Atlantic Ocean

•ficial name Republic of Cape Verde
•rmation 1975 / 1975
•pital Praia
•pulation 500,000 / 321 people
**•r sq mile (124 people per sq km)
•tal area 1557 sq. miles (4033 sq. km)
•nguages Portuguese Creole,
**ortuguese*
•igions Roman Catholic 97%, Other 2%,
**•otestant (Church of the Nazarene) 1%
•hnic mix Mestiço 71%, African 28%,
**uropean 1%
•vernment Presidential /
**arliamentary system
rrency Cape Verde escudo
**)0 centavos

CAPE VERDE
(continued)

Literacy rate 85%
Calorie consumption 2716 kilocalories

Cape Verde Basin *66 C4 undersea basin*
E Atlantic Ocean
Cape Verde Plain *66 C4 abyssal plain*
E Atlantic Ocean
Cape Verde, Republic of
see Cape Verde
Cape York Peninsula *148 C2 peninsula*
Queensland, N Australia
Cap-Haïtien *54 D3 var.* Le Cap.
N Haiti
Capira *53 G5* Panamá, C Panama
Capitán Arturo Prat *154 A2 Chilean*
research station South Shetland Islands,
Antarctica
Capitán Pablo Lagerenza *64 D1 var.*
Mayor Pablo Lagerenza. Chaco,
N Paraguay
Capodistria *see* Koper
Capri *97 C5 island* S Italy
Caprivi Concession *see* Caprivi Strip
Caprivi Strip *78 C3 Ger.* Caprivizipfel;
prev. Caprivi Concession.; *cultural*
region NE Namibia
Caprivizipfel *see* Caprivi Strip
Cap Saint-Jacques *see* Vung Tau
Caquetá, Río *58 C5 var.* Río Japurá,
Yapurá. *river* Brazil/Colombia
Caquetá, Río *see* Japurá, Rio
CAR *see* Central African Republic
Caracal *108 B5* Olt, S Romania
Caracaraí *62 D1* Rondônia, W Brazil
Caracas *58 D1 country capital*
(Venezuela) Distrito Federal,
N Venezuela
Caralis *see* Cagliari
Caratasca, Laguna de *53 E2 lagoon*
NE Honduras
Carballiño *see* O Carballiño
Carbón, Laguna del *64 B7 physical*
feature SE Argentina
Carbonale *40 B5* Illinois, N USA
Carbonia *97 A6 var.* Carbonia Centro.
Sardegna, Italy, C Mediterranean Sea
Carbonia Centro *see* Carbonia
Carcaso *see* Carcassonne
Carcassonne *91 C6 anc.* Carcaso. Aude,
S France
Cardamomes, Chaine des *see* Krâvanh,
Chuŏr Phnum
Cardamom Mountains *see* Krâvanh,
Chuŏr Phnum
Cárdenas *54 B2* Matanzas, W Cuba
Cardiff *89 C7 Wel.* Caerdydd. *national*
capital S Wales, United Kingdom
Cardigan Bay *89 C6 bay* W Wales,
United Kingdom
Carei *108 B3 Ger.* Gross-Karol, Karol,
Hung. Nagykároly; *prev.* Careii-Mari.
Satu Mare, NW Romania
Careii-Mari *see* Carei
Carey, Lake *147 B5 lake* Western
Australia
Cariaco *59 E1* Sucre, NE Venezuela
Caribbean Sea *54 C4 sea* W Atlantic
Ocean
Caribrod *see* Dimitrovgrad
Carlisle *88 C4 anc.* Caer Luel,
Luguvallium, Luguvallum.
NW England, United Kingdom
Carlow *89 B6 Ir.* Ceatharlach.
SE Ireland
Carlsbad *48 D3* New Mexico, SW USA
Carlsbad *see* Karlovy Vary
Carlsberg Ridge *140 B4 undersea ridge*
S Arabian Sea
Carlsruhe *see* Karlsruhe
Carmana/Carmania *see* Kermän
Carmarthen *89 C6* SW Wales, United
Kingdom
Carmaux *91 C6* Tarn, S France
Carmel *40 C4* Indiana, N USA
Carmelita *52 B1* Petén, N Guatemala
Carmen *51 G4 var.* Ciudad del Carmen.
Campeche, SE Mexico
Carmona *92 C4* Andalucía, S Spain
Carmona *see* Uíge
Carnaro *see* Kvarner
Carnarvon *147 A5* Western Australia
Carnegie, Lake *147 B5 salt lake* Western
Australia
Car Nicobar *133 F3 island* Nicobar
Islands, India, NE Indian Ocean

Caroço, Ilha *76 E1 island* N Sao Tome
and Principe, Africa, E Atlantic Ocean
Carolina *63 F2* Maranhão, E Brazil
Caroline Island *see* Millennium Island
Caroline Islands *144 B2 island group*
C Micronesia
Carolópolis *see* Châlons-en-Champagne
Caroní, Río *59 E3 river* E Venezuela
Caronium *see* A Coruña
Carora *58 C1* Lara, N Venezuela
Carpathian Mountains *81 E4 var.*
Carpathians, *Cz./Pol.* Karpaty, *Ger.*
Karpaten. *mountain range* E Europe
Carpathians *see* Carpathian Mountains
Carpatho/Carpathos *see* Kárpathos
Carpaţii Sudici
see Carpaţii Meridionali
Carpentaria, Gulf of *148 B2 gulf*
N Australia
Carpi *96 C2* Emilia-Romagna, N Italy
Carrara *96 B3* Toscana, C Italy
Carson City *47 C5 state capital* Nevada,
W USA
Carson Sink *47 C5 salt flat* Nevada,
W USA
Carstensz, Puntjak *see* Jaya, Puncak
Cartagena *58 B1 var.* Cartagena de los
Indes. Bolívar, NW Colombia
Cartagena *93 F4 anc.* Carthago Nova.
Murcia, SE Spain
Cartagena de los Indes *see* Cartagena
Cartago *53 F4* Cartago, C Costa Rica
Cartago *see* Carthage
Carthage *45 F5* Missouri, C USA
Carthago Nova *see* Cartagena
Cartwright *39 F2* Newfoundland and
Labrador, E Canada
Carúpano *59 E1* Sucre, NE Venezuela
Carusbur *see* Cherbourg
Caruthersville *45 H5* Missouri,
C USA
Cary *43 F1* North Carolina, SE USA
Casablanca *70 C2 Ar.* Dar-el-Beida.
NW Morocco
Casa Grande *48 B2* Arizona, SW USA
Cascade Range *46 B3 mountain range*
Oregon/Washington, NW USA
Cascadia Basin *34 A4 undersea basin*
NE Pacific Ocean
Cascais *92 B4* Lisboa, C Portugal
Caserta *97 D5* Campania, S Italy
Casey *154 D4 Australian research station*
Antarctica
Čáslav *99 B5 Ger.* Tschaslau. Střední
Čechy, C Czech Republic
Casper *44 C3* Wyoming, C USA
Caspian Depression *111 B7*
Kaz. Kaspiy Mangy Oypaty,
Rus. Prikaspiyskaya Nizmennost'.
depression Kazakhstan/Russian
Federation
Caspian Sea *114 A4 Az.* Xäzär Dänizi,
Kaz. Kaspiy Tengizi,
Per. Bahr-e Khazar, Daryā-ye Khazar,
Rus. Kaspiyskoye More.
inland sea Asia/Europe
Cassai *see* Kasai
Cassel *see* Kassel
Castamoni *see* Kastamonu
Casteggio *96 B2* Lombardia, N Italy
Castelló de la Plana *see* Castellón de
la Plana
Castellón de la Plana *93 F3 var.*
Castelló de la Plana. *Cat.* Castelló de la Plana.
Valenciana, E Spain
Castelnaudary *91 C6* Aude, S France
Castelo Branco *92 C3* Castelo Branco,
C Portugal
Castelsarrasin *91 B6* Tarn-et-Garonne,
S France
Castelvetrano *97 C7* Sicilia, Italy,
C Mediterranean Sea
Castilla-La Mancha *93 E3 autonomous*
community NE Spain
Castilla y León *92 C2 var.* Castillia Leon.
autonomous community
NW Spain
Castillia Leon *see* Castilla y León
Castlebar *89 A5 Ir.* Caisleán an
Bharraigh. W Ireland
Castleford *89 D5* N England,
United Kingdom
Castle Harbour *42 B5 inlet* Bermuda,
NW Atlantic Ocean
Castra Regina *see* Regensburg
Castricum *86 C3* Noord-Holland,
W Netherlands
Castries *55 F1 country capital*
(Saint Lucia) N Saint Lucia

Castro *65 B6* Los Lagos, W Chile
Castrovillari *97 D6* Calabria, SW Italy
Castuera *92 D4* Extremadura, W Spain
Caswell Sound *151 A7 sound* South
Island, New Zealand
Catacamas *52 D2* Olancho,
C Honduras
Catacaos *60 B3* Piura, NW Peru
Catalan Bay *93 H4 bay* E Gibraltar,
Mediterranean Sea
Cataluña *93 G2* N Spain
Catamarca *see* San Fernando del Valle
de Catamarca
Catania *97 D7* Sicilia, Italy,
C Mediterranean Sea
Catanzaro *97 D6* Calabria, SW Italy
Catarroja *93 F3* Valenciana, E Spain
Cat Island *54 C1 island* C Bahamas
Catskill Mountains *41 F3 mountain*
range New York, NE USA
Cattaro *see* Kotor
Cauca, Río *58 B2 river* N Colombia
Caucasia *58 B2* Antioquia,
NW Colombia
Caucasus *81 G4 Rus.* Kavkaz.
mountain range
Georgia/Russian Federation
Caura, Río *59 E3 river* C Venezuela
Cavaia *see* Kavajë
Cavalla *74 D5 var.* Cavally, Cavally
Fleuve. *river* Côte d'Ivoire/Liberia
Cavally/Cavally Fleuve *see* Cavalla
Caviana de Fora, Ilha *63 E1 var.* Ilha
Caviana. *island* N Brazil
Caviana, Ilha *see* Caviana de Fora, Ilha
Cawnpore *see* Känpur
Caxamarca *see* Cajamarca
Caxito *78 B1* Bengo, NW Angola
Cayenne *59 H3 dependent territory/*
arrondissement capital (French Guiana)
NE French Guiana
Cayes *see* Les Cayes. SW Haiti
Cayman Brac *54 B3 island* E Cayman
Islands
Cayman Islands *54 B3 UK dependent*
territory W West Indies
Cayo *see* San Ignacio
Cay Sal *54 B3 islet* SW Bahamas
Cazin *100 B3* Federacija Bosne
I Hercegovine, NW Bosnia and
Herzegovina
Cazorla *93 E4* Andalucía, S Spain
Ceadăr-Lunga *see* Ciadîr-Lunga
Ceará *63 F2 off.* Estado do Ceará.
region C Brazil
Ceará *63 F2 off.* Estado do Ceará.
state C Brazil
Ceará *see* Fortaleza
Ceara Abyssal Plain *see* Ceará Abyssal
Plain
Ceará, Estado do *see* Ceará
Ceará Abyssal Plain *66 D4* Ceara Abyssal
Plain. *abyssal plain* W Atlantic Ocean
Ceatharlach *see* Carlow
Cébaco, Isla *53 F5 island* SW Panama
Cebu *139 E2* Cebu City. Cebu,
C Philippines
Cebu City *see* Cebu
Čechy *see* Bohemia
Cecina *96 B3* Toscana, C Italy
Cedar City *44 A5* Utah, W USA
Cedar Falls *45 G3* Iowa, C USA
Cedar Lake *37 F5 var.* Cedar Lake.
C Canada
Cedar Rapids *45 G3* Iowa, C USA
Cedros, Isla *50 A2 island* W Mexico
Ceduna *149 A6* South Australia
Cefalù *97 C7 anc.* Cephaloedium. Sicilia,
Italy, C Mediterranean Sea
Celebes *see* Sulawesi
Celebes *see* Sulawesi
Celebes *139 E4 Ind.* Celebes. *island*
C Indonesia
Celebes Sea *139 E3 Ind.* Laut Sulawesi.
sea Indonesia/Philippines
Celje *95 E7 Ger.* Cilli. S Slovenia
Celldömölk *99 C6 Var.* N Hungary
Celle *94 B3 var.* Zelle. Niedersachsen,
N Germany
Celovec *see* Klagenfurt
Celtic Sea *89 B7 Ir.* An Mhuir Cheilteach.
sea SW British Isles
Celtic Shelf *80 B3 continental shelf*
E Atlantic Ocean
Cenderawasih, Teluk *139 E4*
var. Teluk Irian, Teluk Sarera.
bay W Pacific Ocean
Cenon *91 B5* Gironde, SW France
Centennial State *see* Colorado
Centrafricaine, République
see Central African Republic

Central African Republic 76 C4 var.
République Centrafricaine, abbrev.
CAR; prev. Ubangi-Shari, Oubangui-
Chari, Territoire de l'Oubangui-Chari.
country C Africa

CENTRAL AFRICAN REPUBLIC
Central Africa

Official name Central African Republic
Formation 1960 / 1960
Capital Bangui
Population 4.6 million / 19 people
per sq mile (7 people per sq km)
Total area 240,534 sq. miles
(622,984 sq. km)
Languages Sango, Banda, Gbaya, French*
Religions Traditional beliefs 35%, Roman
Catholic 25%, Protestant 25%, Muslim 15%
Ethnic mix Baya 33%, Banda 27%,
Other 17%, Mandjia 13%, Sara 10%,
Government Presidential system
Currency CFA franc = 100 centimes
Literacy rate 37%
Calorie consumption 2154 kilocalories

Central, Cordillera 58 B3 mountain
range W Colombia
Cordillera Central 55 E3 mountain
range C Dominican Republic
Cordillera Central 53 F5 mountain
range C Panama
Central, Cordillera 139 E1 mountain
range Luzon, N Philippines
Central Group see Inner Islands
Centralia 46 B2 Washington, NW USA
Central Indian Ridge see Mid-Indian
Ridge
Central Makran Range 134 A3
mountain range W Pakistan
Central Pacific Basin 142 D1 undersea
basin C Pacific Ocean
Central Provinces and Berar
see Madhya Pradesh
Central Range 144 B3 mountain range
NW Papua New Guinea
Central Russian Upland 109 G1 Eng.
Central Russian Upland. mountain
range W Russian Federation
Central Russian Upland see
Srednerusskaya Vozvyshennost'
Central Siberian Plateau 114 D3 var.
Central Siberian Uplands, Eng. Central
Siberian Plateau. mountain range
N Russian Federation
Central Siberian Plateau/Central
Siberian Uplands
see Srednesibirskoye Ploskogor'ye
Central, Sistema 92 D3 mountain range
C Spain
Central Valley 47 B6 valley California,
W USA
Centum Cellae see Civitavecchia
Ceos see Tziá
Cephaloedium see Cefalù
Ceram see Seram, Pulau
Ceram Sea see Laut Seram
Cerasus see Giresun
Cereté 58 B2 Córdoba, NW Colombia
Cergy-Pontoise see Pontoise
Cerignola 97 D5 Puglia, SE Italy
Çerkeş 116 C2 Çankin, N Turkey
Cernăuți see Chernivtsi
Cernay 90 E4 Haut-Rhin, NE France
Cerro de Pasco 60 C3 Pasco, C Peru
Cervera 93 F2 Cataluña, NE Spain
Cervino, Monte see Matterhorn
Cesena 93 C4 anc. Caesena.
Emilia-Romagna, N Italy
Cēsis 106 D3 Ger. Wenden. C Latvia
Česká Republika see Czech Republic
České Budějovice 99 B5 Ger. Budweis.
Jihočeský Kraj, S Czech Republic
Český Krumlov 99 A5 var. Böhmisch-
Krumau, Ger. Krummau. Jihočeský
Kraj, S Czech Republic
Český Les see Bohemian Forest
Cetatea Damboviței see București
Cetinje 101 C5 It. Cettigne.
S Montenegro
Cette see Sète
Cettigne see Cetinje
Ceuta 70 C2 enclave Spain, N Africa
Cévennes 91 C6 mountain range
S France
Ceyhan 116 D4 Adana, S Turkey
Ceylanpınar 117 E4 Şanlıurfa,
SE Turkey
Ceylon see Sri Lanka

Ceylon Plain 124 B4 abyssal plain
N Indian Ocean
Ceyre to the Caribs see Marie-Galante
Chachapoyas 60 B2 Amazonas, NW Peru
Chachevichy 107 D6 Rus. Chechevichi.
Mahilyowskaya Voblasts', E Belarus
Chaco see Gran Chaco
Chad 76 C3 off. Republic of Chad, Fr.
Tchad. country C Africa

CHAD
Central Africa

Official name Republic of Chad
Formation 1960 / 1960
Capital N'Djaména
Population 12.8 million / 26 people
per sq mile (10 people per sq km)
Total area 495,752 sq. miles
(1,284,000 sq. km)
Languages French*, Sara, Arabic*, Maba
Religions Muslim 51%, Christian 35%,
Animist 7%, Traditional beliefs 7%
Ethnic mix Other 30%, Sara 28%,
Mayo-Kebbi 12%, Arab 12%, Ouaddai 9%,
Kanem-Bornou 9%
Government Presidential system
Currency CFA franc = 100 centimes
Literacy rate 37%
Calorie consumption 2061 kilocalories

Chad, Lake 76 B3 Fr. Lac Tchad. lake
C Africa
Chad, Republic of see Chad
Chadron 44 D3 Nebraska, C USA
Chadyr-Lunga see Ciadir-Lunga
Chagai Hills 134 A2 var. Chāh Gay.
mountain range Afghanistan/Pakistan
Chaghasarāy see Asadābād
Chagos-Laccadive Plateau 124 B4
undersea plateau N Indian Ocean
Chagos Trench 141 C5 trench N Indian
Ocean
Chāh Gay see Chāgai Hills
Chaillu, Massif du 77 B6 mountain
range C Gabon
Chain Ridge 140 B4 undersea ridge
W Indian Ocean
Chajul 52 B2 Quiché, W Guatemala
Chakhānsūr 122 D5 Nīmrōz,
SW Afghanistan
Chala 60 D4 Arequipa, SW Peru
Chalatenango 52 C3 Chalatenango,
N El Salvador
Chalcidice see Chalkidikí
Chalcis see Chalkída
Chalki 105 E7 island Dodekánisa,
Greece, Aegean Sea
Chalkída 105 C5 var. Halkida,
prev. Khalkís; anc. Chalcis.
Evvoia, E Greece
Chalkidikí 104 C4 var. Khalkidhikí; anc.
Chalcidice. peninsula NE Greece
Challans 90 B4 Vendée, NW France
Challapata 61 F4 Oruro, SW Bolivia
Challenger Deep 152 B3 trench
W Pacific Ocean
Challenger Deep see Mariana Trench
Challenger Fracture Zone 153 F4
tectonic feature SE Pacific Ocean
Châlons-en-Champagne 90 D3 prev.
Châlons-sur-Marne, hist. Arcae
Remorum; anc. Carolopois. Marne,
NE France
Châlons-sur-Marne see
Châlons-en-Champagne
Chalon-sur-Saône 90 D4 anc.
Cabillonum. Saône-et-Loire, C France
Cha Mai see Thung Song
Chaman 134 B2 Baluchistān,
SW Pakistan
Chambéry 91 D5 anc. Cambéria. Savoie,
E France
Champagne 90 D3 cultural region
N France
Champagne see Campania
Champaign 40 B4 Illinois, N USA
Champasak 67 D5 Champasak, S Laos
Champlain, Lake 41 F2 lake Canada/
USA
Champotón 51 G4 Campeche,
SE Mexico
Chanak see Çanakkale
Chañaral 64 B3 Atacama, N Chile
Chan-chiang/Chanchiang see Zhanjiang
Chandeleur Islands 42 C3 island group
Louisiana, S USA
Chandigarh 134 D2 state capital Punjab,
N India

Chandrapur 135 E5 Mahārāshtra,
C India
Changan see Xi'an, Shaanxi, C China
Changane 79 E3 river S Mozambique
Changchun 128 D3 var. Ch'angch'un,
Ch'ang-ch'un; prev. Hsinking. province
capital Jilin, NE China
Ch'angch'un/Ch'ang-ch'un see
Changchun
Chang Jiang 128 A5 Eng. Yangtze; var.
Yangtze Kiang. river SW China
Changkiakow see Zhangjiakou
Chang, Ko 137 C6 island S Thailand
Changsha 128 C5 var. Ch'angsha,
Ch'ang-sha. province capital Hunan,
S China
Ch'angsha/Ch'ang-sha see Changsha
Changzhi 128 C4 Shanxi, C China
Chaniá 105 C7 var. Hania, Khaniá, Eng.
Canea; anc. Cydonia. Kriti, Greece,
E Mediterranean Sea
Chañi, Nevado de 64 B2 mountain
NW Argentina
Chankiri see Çankırı
Channel Islands 89 C8 Fr. Iles
Normandes. island group S English
Channel
Channel Islands 47 B8 island group
California, W USA
Channel-Port aux Basques 39
G4 Newfoundland and Labrador,
SE Canada
Channel, The see English Channel
Channel Tunnel 90 C2 tunnel France/
United Kingdom
Chantabun/Chantaburi see Chanthaburi
Chantada 92 C1 Galicia, NW Spain
Chanthaburi 137 C6 var. Chantabun,
Chantaburi. Chantaburi, S Thailand
Chanute 45 F5 Kansas, C USA
Chaouèn see Chefchaouen
Chaoyang 128 D3 Liaoning, NE China
Chapala, Lago de 50 D4 lake C Mexico
Chapan, Gora 122 B3 mountain
C Turkmenistan
Chapayevsk 111 C6 Samarskaya Oblast',
W Russian Federation
Chaplynka 109 F4 Khersons'ka Oblast',
S Ukraine
Chapra see Chhapra
Charcot Seamounts 80 B3 seamount
range E Atlantic Ocean
Chardzhev see Türkmenabat
Chardzhou/Chardzhui see Türkmenabat
Charente 91 B5 cultural region
W France
Charente 91 B5 river W France
Chari 76 B3 var. Shari. river Central
African Republic/Chad
Chārīkār 123 E4 Parwān,
NE Afghanistan
Charity 59 F2 NW Guyana
Charjew see Türkmenabat
Charkhlik/Charkhliq see Ruoqiang
Charleroi 87 C7 Hainaut, S Belgium
Charlesbourg 39 E4 Québec, SE Canada
Charles de Gaulle 90 E1 (Paris)
Seine-et-Marne, N France
Charles Island see Santa María, Isla
Charleston 43 F2 South Carolina,
SE USA
Charleston 40 D5 state capital West
Virginia, NE USA
Charleville 149 D5 Queensland,
E Australia
Charleville-Mézières 90 D3 Ardennes,
N France
Charlie-Gibbs Fracture Zone 66 C2
tectonic feature N Atlantic Ocean
Charlotte 43 E1 North Carolina, SE USA
Charlotte Amalie 55 F3 prev. Saint
Thomas. dependent territory capital
(Virgin Islands (US)) Saint Thomas,
N Virgin Islands (US)
Charlotte Harbor 43 E5 inlet Florida,
SE USA
Charlottenhof see Aegviidu
Charlottesville 41 E5 Virginia, NE USA
Charlottetown 39 F4 province capital
Prince Edward Island, Prince Edward
Island, SE Canada
Charlotte Town see Roseau, Dominica
Charsk see Shar
Charters Towers 148 D3 Queensland,
NE Australia
Chartres 90 C3 anc. Autricum, Civitas
Carnutum. Eure-et-Loir, C France

Chashniki 107 D5 Vitsyebskaya
Voblasts', N Belarus
Châteaubriant 90 B4 Loire-Atlantiq
NW France
Châteaudun 90 C3 Eure-et-Loir,
C France
Châteauroux 90 C4 prev. Indreville.
Indre, C France
Château-Thierry 90 C3 Aisne, N Fra
Châtelet 87 C7 Hainaut, S Belgium
Châtelherault see Châtellerault
Châtellerault 90 B4 var. Châtelherau
Vienne, W France
Chatham Island see San Cristóbal, Isl
Chatham Island Rise see Chatham R
Chatham Islands 143 E5 island grou
New Zealand, SW Pacific Ocean
Chatham Rise 142 D5 var. Chatham
Island Rise. undersea rise S Pacific
Ocean
Chatkal Range 123 F2 Rus. Chatkal'
Khrebet. mountain range Kyrgyzst
Uzbekistan
Chatkal'skiy Khrebet see Chatkal Ra
Chāttagām see Chittagong
Chattahoochee River 42 D3 river
SE USA
Chattanooga 42 D1 Tennessee, S US
Chatyr-Tash 123 G2 Narynskaya Ob
C Kyrgyzstan
Châu Đốc 137 D6 var. Chauphu, Chi
Phu. An Giang, S Vietnam
Chauk 136 A3 Magway, W Myanmar
(Burma)
Chaumont 90 D4 prev. Chaumont-e
Bassigny. Haute-Marne, N France
Chaumont-en-Bassigny see Chaumo
Chau Phu see Châu Đốc
Chausy see Chavusy
Chaves 92 C2 anc. Aquae Flaviae. V
Real, N Portugal
Chávez, Isla see Santa Cruz, Isla
Chavusy 107 E6 Rus. Chausy.
Mahilyowskaya Voblasts', E Belarus
Chaykovskiy 111 D5 Permskiy Kray
NW Russian Federation
Cheb 99 A5 Ger. Eger. Karlovarský K
W Czech Republic
Cheboksary 111 C5 Chuvashskaya
Respublika, W Russian Federation
Cheboygan 40 C2 Michigan, N USA
Chechaouèn see Chefchaouen
Chech, Erg 74 D1 desert Algeria/Mal
Chechevichi see Chachevichy
Che-chiang see Zhejiang
Cheduba Island 136 A4 island
W Myanmar (Burma)
Chefchaouen 70 C2 var. Chaouèn,
Chechaouèn, Sp. Xauen. N Morocc
Chefoo see Yantai
Cheju-do see Jeju-do
Cheju Strait see Jeju Strait
Chekiang see Zhejiang
Cheleken see Hazar
Chelkar see Shalkar
Chełm 98 E4 Rus. Kholm. Lubelskie
SE Poland
Chełmno 98 C3 Ger. Culm, Kulm.
Kujawski-pomorskie, C Poland
Chełmża 98 C3 Ger. Culmsee, Kulm.
Kujawski-pomorskie, C Poland
Cheltenham 89 D6 C England, Unite
Kingdom
Chelyabinsk 114 C3 Chelyabinskaya
Oblast', C Russian Federation
Chemnitz 94 D4 prev. Karl-Marx-St
Sachsen, E Germany
Chemulpo see Incheon
Chenāb 134 C2 river India/Pakistan
Chengchiatun see Liaoyuan
Cheng-chou/Chengchow see Zhengz
Chengde 128 D3 var. Jehol. Hebei,
E China
Chengdu 128 B5 var. Ch'eng-tu, Chengtu.
province capital Sichuan, C Chi
Chenghsien see Zhengzhou
Cheng-tu/Ch'eng-tu see Chengdu
Chennai 132 D2 prev. Madras. state
capital Tamil Nādu, S India
Chenstokhov see Częstochowa
Chen Xian/Chenxian/Chen Xiang x
Chenzhou
Chepelare 104 C3 Smolyan, S Bulgar
Chepén 60 B3 La Libertad, C Peru
Chepo 53 G5 Panamá, C Panama
Cher 90 C4 river C France
Cherbourg 90 B3 anc. Carusbur.
Manche, N France

epovets 110 B4 Vologodskaya
  last', NW Russian Federation
gui, Chott ech 70 D2 salt lake,
  W Algeria
ikov see Cherykaw
kassy see Cherkasy
kasy 109 E2 Rus. Cherkassy.
  erkas'ka Oblast', C Ukraine
kessk 111 B7 Karachayevo-
  erkesskaya Respublika, SW Russian
  deration
nigov see Chernihiv
nihiv 109 E1 Rus. Chernigov.
  ernihivs'ka Oblast', NE Ukraine
nivtsi 108 C3 Ger. Czernowitz,
  m. Cernăuţi, Rus. Chernovtsy.
  ernivets'ka Oblast', W Ukraine
no More see Black Sea
nomorskoye see Chornomors'ke
novtsy see Chernivtsi
noye More see Black Sea
nyakhovsk 106 A4 Ger. Insterburg.
  aliningradskaya Oblast', W Russian
  deration
ry Hill 41 F4 New Jersey, NE USA
ski Range see Cherskogo, Khrebet
skiy 115 G2 Respublika Sakha
  akutiya), NE Russian Federation
skogo, Khrebet 115 F2 var. Cherski
  nge. mountain range NE Russian
  deration
so see Cres
ven' see Chervyen'
vonograd see Chervonohrad
vonohrad 108 C2 Rus.
  ervonograd. L'vivs'ka Oblast',
  W Ukraine
wyen' 107 D6 Rus. Cherven'.
  nskaya Voblasts', C Belarus
ykaw 107 E7 Rus. Cherikov.
  hilyowskaya Voblasts', E Belarus
apeake Bay 41 F5 inlet NE USA
ha Bay see Chëshskaya Guba
hskaya Guba 155 D5 var.
  xchangel Bay, Chesha Bay, Dvina
  y. bay NW Russian Federation
ter 89 C6 Wel. Caerleon, hist.
  gaceaster, Lat. Deva, Devana Castra.
  England, United Kingdom
aintana Roo, SE Mexico
umal 51 H4 var. Payo Obispo.
  aintana Roo, SE Mexico
iot Hills 88 D4 hill range England/
  otland, United Kingdom
enne 44 D4 state capital Wyoming,
  USA
enne River 44 D3 river South
  kota/Wyoming, N USA
di-Oşorheiu see Târgu Secuiesc
pra 135 F3 prev. Chapra. Bihār,
  India
ttisgarh 135 E4 cultural region
  India
i see Jiayi
i see Jiayi
ng-hsi see Jiangxi
ng Mai 136 B4 var. Chiangmai,
  iengmai, Kiangmai. Chiang Mai,
  N Thailand
ngmai see Chiang Mai
ng Rai 136 C3 var. Chianpai,
  ienrai, Muang Chiang Rai. Chiang
  i, NW Thailand
ng-su see Jiangsu
nning/Chian-ning see Nanjing
npai see Chiang Rai
nti 96 C3 cultural region C Italy
pa see Chiapa de Corzo
pa de Corzo 51 G5 var. Chiapa.
  iapas, SE Mexico
vi see Jiayi
a 130 B1 var. Tiba. Chiba, Honshū,
  apan
ougamau 38 D3 Québec, SE Canada
ago 40 B3 Illinois, N USA
ch'i-ha-erh see Qiqihar
kasha 49 G2 Oklahoma, C USA
ayo 60 B3 Lambayeque, NW Peru
o 47 B5 California, W USA
o, Río 65 B7 river SE Argentina
o, Río 65 B6 river S Argentina
outimi 39 E4 Québec, SE Canada
agmai see Chiang Mai
arai see Chiang Rai
anuova 96 D2 SW San Marino
i 96 D4 var. Teate. Abruzzo, C Italy
ng 127 G2 var. Ulanhad. Nei
  ongol Zizhiqu, N China
rin see Chyhyryn
fu see Yantai

Chihli see Hebei
Chihli, Gulf of see Bo Hai
Chihuahua 50 C2 Chihuahua,
  NW Mexico
Childress 49 F2 Texas, SW USA
Chile 64 B3 off. Republic of Chile.
  country SW South America

CHILE
South America
Official name Republic of Chile
Formation 1818 / 1883
Capital Santiago
Population 17.6 million / 61 people
  per sq mile (24 people per sq km)
Total area 292,258 sq. miles (756,950
  sq. km)
Languages Spanish*,
  Amerindian languages
Religions Roman Catholic 89%,
  Other and nonreligious 11%
Ethnic mix Mestizo and European 90%,
  Other Amerindian 9%, Mapuche 1%
Government Presidential system
Currency Chilean peso = 100 centavos
Literacy rate 99%
Calorie consumption 2989 kilocalories

Chile Basin 57 A5 undersea basin
  E Pacific Ocean
Chile Chico 65 B6 Aisén, W Chile
Chile, Republic of see Chile
Chile Rise 57 A7 undersea rise SE Pacific
  Ocean
Chilia-Nouă see Kiliya
Chililabombwe 78 D2 Copperbelt,
  C Zambia
Chi-lin see Jilin
Chillán 65 B5 Bío Bío, C Chile
Chillicothe 40 D4 Ohio, N USA
Chill Mhantáin, Sléibhte see Wicklow
  Mountains
Chiloé, Isla de 65 A6 var. Isla Grande de
  Chiloé. island W Chile
Chilpancingo 51 E5 var. Chilpancingo
  de los Bravos. Guerrero, S Mexico
Chilpancingo de los Bravos see
  Chilpancingo
Chilung see Jilong
Chimán 53 G5 Panamá, E Panama
Chimbay see Chimboy
Chimborazo 60 A1 volcano C Ecuador
Chimbote 60 C3 Ancash, W Peru
Chimboy 122 D1 Rus. Chimbay.
  Qoraqalpog'iston Respublikasi,
  NW Uzbekistan
Chimoio 79 E3 Manica, C Mozambique
China 124 C2 off. People's Republic
  of China, Chin. Chung-hua Jen-min
  Kung-ho-kuo, Zhonghua Renmin
  Gongheguo; prev. Chinese Empire.
  country E Asia

CHINA
East Asia
Official name People's Republic of China
Formation 960 / 1999
Capital Beijing
Population 1.39 billion / 385 people
  per sq mile (149 people per sq km)
Total area 3,705,386 sq. miles
  (9,596,960 sq. km)
Languages Mandarin*, Wu, Cantonese,
  Hsiang, Min, Hakka, Kan
Religions Nonreligious 59%, Traditional
  beliefs 20%, Other 13%, Buddhist 6%,
  Muslim 2%
Ethnic mix Han 92%, Other 4%, Hui 1%,
  Miao 1%, Manchu 1%, Zhuang 1%
Government One-party state
Currency Renminbi (yuan)
  = 10 jiao = 100 fen
Literacy rate 95%
Calorie consumption 3074 kilocalories

Chi-nan/Chinan see Jinan
Chinandega 52 C3 Chinandega,
  NW Nicaragua
China, People's Republic of see China
China, Republic of see Taiwan
Chincha Alta 60 D4 Ica, SW Peru
Chin-chiang see Quanzhou
Chin-chou/Chinchow see Jinzhou
Chindwin see Chindwin
Chindwin 136 B2 var. Chindwin. river
  N Myanmar (Burma)

Chinese Empire see China
Chinghai see Qinghai
Ch'ing Hai see Qinghai Hu, China
Chingola 78 D2 Copperbelt, C Zambia
Ching-Tao/Ch'ing-tao see Qingdao
Chinguetti 74 C2 var. Chinguetti. Adrar,
  C Mauritania
Chin Hills 136 A3 mountain range
  W Myanmar (Burma)
Chinhsien see Jinzhou
Chinnereth see Tiberias, Lake
Chinook Trough 113 H4 trough
  N Pacific Ocean
Chioggia 96 C2 anc. Fossa Claudia.
  Veneto, NE Italy
Chíos 105 D5 var. Hios, Khíos, It. Scio,
  Turk. Sakiz-Adasi. Chíos, E Greece
Chíos 105 D5 var. Khíos. island E Greece
Chipata 78 D2 prev. Fort Jameson.
  Eastern, E Zambia
Chiquián 60 C3 Ancash, W Peru
Chiquimula 52 B2 Chiquimula,
  SE Guatemala
Chīrāla 132 D1 Andhra Pradesh, E India
Chirchik see Chirchiq
Chirchiq 123 E2 Rus. Chirchik.
  Toshkent Viloyati, E Uzbekistan
Chiriquí Gulf 53 E5 Eng. Chiriquí Gulf.
  gulf SW Panama
Chiriquí Gulf see Chiriquí, Golfo de
Chiriquí, Laguna de 53 E5 lagoon
  NW Panama
Chiriquí, Volcán de see Barú, Volcán
Chirripó, Cerro see Chirripó Grande,
  Cerro
Chirripó Grande, Cerro 52 D4 var.
  Cerro Chirripó. mountain
  SE Costa Rica
Chisec 52 B2 Alta Verapaz, C Guatemala
Chisholm 45 F1 Minnesota, N USA
Chisimaio/Chisimayu see Kismaayo
Chişinău 108 D4 Rus. Kishinev. country
  capital (Moldova) C Moldova
Chita 115 F4 Zabaykal'skiy Kray,
  S Russian Federation
Chitangwiza see Chitungwiza
Chitato 77 C4 Lunda Norte, NE Angola
Chitina 36 D3 Alaska, USA
Chitose 130 D2 var. Titose. Hokkaidō,
  NE Japan
Chitré 53 F5 Herrera, S Panama
Chittagong 135 G4 Ben. Chāttagām.
  Chittagong, SE Bangladesh
Chitungwiza 78 D3 prev. Chitangwiza.
  Mashonaland East, NE Zimbabwe
Chkalov see Orenburg
Chlef 90 D4 var. El Cheliff, Ech
  Cheliff; prev. Al-Asnam, El Asnam,
  Orléansville. NW Algeria
Chocolate Mountains 47 D8 mountain
  range California, W USA
Chodorów see Khodoriv
Chodzież 98 C3 Wielkopolskie,
  C Poland
Choele Choel 65 C5 Río Negro,
  C Argentina
Choiseul 144 C3 var. Lauru. island
  NW Solomon Islands
Chojnice 98 C2 Pomorskie, N Poland
Ch'ok'ē 72 C4 var. Choke Mountains.
  mountain range NW Ethiopia
Cholet 90 B4 Maine-et-Loire,
  NW France
Choluteca 52 C3 Choluteca, S Honduras
Choluteca, Río 52 C3 river
  SW Honduras
Choma 78 C2 Southern, S Zambia
Chomutov 98 A4 Ger. Komotau.
  NW Czech Republic
Chona 113 E2 river C Russian Federation
Chon Buri 137 C5 prev. Bang Pla Soi.
  Chon Buri, S Thailand
Chone 60 A1 Manabí, W Ecuador
Ch'ōngjin 129 E3 NE North Korea
Ch'ŏngju 129 E4 prev. Chŏngju.
  Chŏngju, W South Korea
Chongqing 129 B5 var. Ch'ung-ch'ing,
  Chungking, Pahsien,
  Tchongking, Yuzhou. Chongqing
  Shi, C China
Chongqing Shi 129 B5 province C China
Chonos, Archipiélago de los 65 A6
  island group S Chile
Chóra 105 D7 Kykládes, Greece,
  Aegean Sea
Chóra Sfakíon 105 C8 var. Sfákia. Kríti,
  Greece, E Mediterranean Sea
Chorne More see Black Sea

Chornomors'ke 109 E4 Rus.
  Chernomorskoye. Avtonomna
  Respublika Krym, S Ukraine
Chorokh/Chorokhi see Çoruh Nehri
Chortkiv 108 C2 Rus. Chortkov.
  Ternopil's'ka Oblast', W Ukraine
Chortkov see Chortkiv
Chorzów 99 C5 Ger. Königshütte; prev.
  Królewska Huta. Śląskie, S Poland
Chośebuz see Cottbus
Chōsen-kaikyō see Korea Strait
Chōshi 131 D5 var. Tyōsi. Chiba,
  Honshū, S Japan
Chosŏn-minjujuŭi-inmin-kanghwaguk
  see North Korea
Choszczno 98 B3 Ger. Arnswalde.
  Zachodnio-pomorskie, NW Poland
Chota Nagpur 135 E4 plateau N India
Choûm 74 C2 Adrar, C Mauritania
Choybalsan 127 F2 prev. Byan Tumen.
  Dornod, E Mongolia
Christchurch 151 C6 Canterbury, South
  Island, New Zealand
Christiana 54 B5 C Jamaica
Christiania see Oslo
Christiansand see Kristiansand
Christianshåb see Qasigiannguit
Christiansund see Kristiansund
Christmas Island 141 D5 Australian
  external territory E Indian Ocean
Christmas Island see Kiritimati
Christmas Ridge 143 E1 undersea ridge
  C Pacific Ocean
Chuan see Sichuan
Ch'uan-chou see Quanzhou
Chubek see Moskva
Chubut, Río 65 B6 river SE Argentina
Ch'u-chiang see Shaoguan
Chudskoye Ozero see Peipus, Lake
Chugoku-sanchi 131 B6 mountain range
  Honshū, SW Japan
Chuí see Chuy
Chukai see Cukai
Chukchi Plain 155 B2 abyssal plain
  Arctic Ocean
Chukchi Plateau 34 C2 undersea plateau
  Arctic Ocean
Chukchi Sea 34 C2 sea Arctic Ocean
  More. see Arctic Ocean
Chukchi Sea see Chukotskoye
Chula Vista 47 C8 California, W USA
Chulucanas 60 B2 Piura, NW Peru
Chulym 114 D4 river C Russian
  Federation
Chumphon 137 C6 var. Jumporn.
  SW Thailand
Chuncheon see Chuncheon
Ch'unch'ŏn 129 E4 Jap. Shunsen.
  N South Korea
Chungch'ŏng see Chuncheon
Chungking see Chongqing
Chunya 115 E3 river C Russian Federation
Chuquicamata 64 B2 Antofagasta,
  N Chile
Chuquisaca see Sucre
Chur 95 B7 Fr. Coire, It. Coira, Rmsch.
  Cuera, Curia Rhaetorum.
  Graubünden, E Switzerland
Churchill 37 G4 Manitoba, C Canada
Churchill 39 F2 river Manitoba/
  Saskatchewan, C Canada
Churchill 39 F2 river Newfoundland and
  Labrador, E Canada
Chuska Mountains 48 C1 mountain
  range Arizona/New Mexico, SW USA
Chusovoy 111 D5 Permskiy Kray,
  NW Russian Federation
Chust see Khust
Chuuk Islands 144 B2 var. Hogoley
  Islands. Truk Islands. island
  group Caroline Islands, C Micronesia
Chuy 65 E5 var. Chuí, Bra. Chuy.
  Rocha, E Uruguay
Chyhyryn 109 E2 Rus. Chigirin.
  Cherkas'ka Oblast', N Ukraine
Ciadır-Lunga 108 D4 var. Ceadâr-
  Lunga, Rus. Chadyr-Lunga. S Moldova
Cide 116 C2 Kastamonu, N Turkey
Ciechanów 98 D3 prev. Zichenau.
  Mazowieckie, C Poland
Ciego de Ávila 54 C2 Ciego de Ávila,
  C Cuba
Ciénaga 58 B1 Magdalena, N Colombia
Cienfuegos 54 B2 Cienfuegos, C Cuba
Cieza 93 E4 Murcia, SE Spain
Cihanbeyli 116 C3 Konya, C Turkey

Cikobia *145 E4 prev.* Thikombia.
island N Fiji
Cilacap *138 C5 prev.* Tjilatjap. Jawa,
C Indonesia
Cill Airne *see* Killarney
Cill Chainnigh *see* Kilkenny
Cilli *see* Celje
Cill Mhantáin *see* Wicklow
Cîmpina *see* Câmpina
Cîmpulung *see* Câmpulung
Cina Selatan, Laut *see* South China Sea
Cincinnati *40 C4* Ohio, N USA
Ciney *87 C7* Namur, SE Belgium
Cinto, Monte *91 E7 mountain* Corse,
France, C Mediterranean Sea
Cintra *see* Sintra
Cipolletti *65 B5* Río Negro, C Argentina
Cirebon *138 C4 prev.* Tjirebon. Jawa,
S Indonesia
Cirkvenica *see* Crikvenica
Cirò Marina *97 E6* Calabria, S Italy
Cirquenizza *see* Crikvenica
Cisnădie *108 B4 Ger.* Heltau, *Hung.*
Nagydisznód. Sibiu, SW Romania
Citharista *see* la Ciotat
Citlaltépetl *see* Orizaba, Volcán Pico de
Citrus Heights *47 B5* California, W USA
Ciudad Acuña *see* Villa Acuña
Ciudad Bolívar *59 E2 prev.* Angostura.
Bolívar, E Venezuela
Ciudad Camargo *50 D2* Chihuahua,
N Mexico
Ciudad Cortés *see* Cortés
Ciudad Darío *52 D3 var.* Dario.
Matagalpa, W Nicaragua
Ciudad de Dolores Hidalgo *see* Dolores
Hidalgo
Ciudad de Guatemala *52 B2 Eng.*
Guatemala City; *prev.* Santiago
de los Caballeros. *country capital*
(Guatemala) Guatemala, C Guatemala
Ciudad del Carmen *see* Carmen
Ciudad del Este *64 E2 prev.* Ciudad
Presidente Stroessner, Presidente
Stroessner, Puerto Presidente
Stroessner. Alto Paraná, SE Paraguay
Ciudad Delicias *see* Delicias
Ciudad de México *see* México
Ciudad de Panama *see* Panamá
Ciudad Guayana *59 E2 prev.* San Tomé
de Guayana, Santo Tomé de Guayana.
Bolívar, NE Venezuela
Ciudad Guzmán *50 D4* Jalisco,
SW Mexico
Ciudad Hidalgo *51 G5* Chiapas,
SE Mexico
Ciudad Juárez *50 C1* Chihuahua,
N Mexico
Ciudad Lerdo *50 D3* Durango, C Mexico
Ciudad Madero *51 E3 var.* Villa Cecilia.
Tamaulipas, C Mexico
Ciudad Mante *51 E3* Tamaulipas,
C Mexico
Ciudad Miguel Alemán *51 E2*
Tamaulipas, C Mexico
Ciudad Obregón *50 B2* Sonora,
NW Mexico
Ciudad Ojeda *58 C1* Zulia,
NW Venezuela
Ciudad Porfirio Díaz *see* Piedras Negras
Ciudad Presidente Stroessner *see*
Ciudad del Este
Ciudad Quesada *see* Quesada
Ciudad Real *92 D3* Castilla-La Mancha,
C Spain
Ciudad-Rodrigo *92 C3* Castilla y León,
N Spain
Ciudad Trujillo *see* Santo Domingo
Ciudad Valles *51 E3* San Luis Potosí,
C Mexico
Ciudad Victoria *51 E3* Tamaulipas,
C Mexico
Ciutadella *93 H3 var.* Ciutadella
de Menorca. Menorca, Spain,
W Mediterranean Sea
Ciutadella Ciutadella de Menorca *see*
Ciutadella
Civitanova Marche *96 D3* Marche,
C Italy
Civitas Altae Ripae *see* Brzeg
Civitas Carnutum *see* Chartres
Civitas Eburovicum *see* Évreux
Civitavecchia *96 C4 anc.* Centum Cellae,
Trajani Portus. Lazio, C Italy
Claremore *49 G1* Oklahoma, C USA
Clarence *151 C5* Canterbury, South
Island, New Zealand
Clarence *151 C5 river* South Island,
New Zealand

Clarence Town *54 D2* Long Island,
C Bahamas
Clarinda *45 F4* Iowa, C USA
Clarion Fracture Zone *153 E2*
tectonic feature NE Pacific Ocean
Clarión, Isla *50 A5 island* W Mexico
Clark Fork *44 A1 river* Idaho/Montana,
NW USA
Clark Hill Lake *43 E2 var.* J.Storm
Thurmond Reservoir. *reservoir*
Georgia/South Carolina, SE USA
Clarksburg *40 D4* West Virginia,
NE USA
Clarksdale *42 B2* Mississippi, S USA
Clarksville *42 C1* Tennessee, S USA
Clausentum *see* Southampton
Clayton *49 F1* New Mexico, SW USA
Clearwater *43 E4* Florida, SE USA
Clearwater Mountains *46 D2*
mountain range Idaho, NW USA
Cleburne *49 G3* Texas, SW USA
Clermont *148 D4* Queensland,
E Australia
Clermont-Ferrand *91 C5* Puy-de-Dôme,
C France
Cleveland *40 D3* Ohio, N USA
Cleveland *42 D1* Tennessee, S USA
Clifton *48 C2* Arizona, SW USA
Clinton *42 B2* Mississippi, S USA
Clinton *49 F1* Oklahoma, C USA
Clipperton Fracture Zone *153 E3*
tectonic feature E Pacific Ocean
Clipperton Island *35 A7 French*
dependency of French Polynesia
E Pacific Ocean
Cloncurry *148 B3* Queensland,
C Australia
Clonmel *89 B6 Ir.* Cluain Meala.
S Ireland
Cloppenburg *94 B3* Niedersachsen,
NW Germany
Cloquet *45 G2* Minnesota, N USA
Cloud Peak *44 C3 mountain* Wyoming,
C USA
Clovis *49 E2* New Mexico, SW USA
Cluain Meala *see* Clonmel
Cluj *see* Cluj-Napoca
Cluj-Napoca *108 B3 Ger.* Klausenburg,
Hung. Kolozsvár; *prev.* Cluj. Cluj,
NW Romania
Clutha *151 B7 river* South Island, New
Zealand
Clyde *88 C4 river* W Scotland, United
Kingdom
Coari *62 D2* Amazonas, N Brazil
Coast Mountains *36 D4 Fr.* Chaine
Côtière. *mountain range* Canada/USA
Coast Ranges *46 A4 mountain range*
W USA
Coats Island *37 G3 island* Nunavut,
NE Canada
Coats Land *154 B2 physical region*
Antarctica
Coatzacoalcos *51 G4 var.* Quetzalcoalco;
prev. Puerto México. Veracruz-Llave,
E Mexico
Cobán *52 B2* Alta Verapaz,
C Guatemala
Cobar *149 C6* New South Wales,
SE Australia
Cobija *61 E3* Pando, NW Bolivia
Coblence/Coblenz *see* Koblenz
Coburg *95 C5* Bayern, SE Germany
Coca *see* Puerto Francisco de Orellana
Cocanada *see* Kākināda
Cochabamba *61 G4 hist.* Oropeza.
Cochabamba, C Bolivia
Cochin *see* Kochi
Cochrane *38 C4* Ontario, S Canada
Cochrane *65 B7* Aisén, S Chile
Cocibolca *see* Nicaragua, Lago de
Cockade State *see* Maryland
Cockburn Town *55 E2* San Salvador,
E Bahamas
Cockpit Country, The *54 A4 physical*
region W Jamaica
Cocobeach *77 A5* Estuaire, NW Gabon
Coconino Plateau *48 B1 plain* Arizona,
SW USA
Coco, Rio *53 E2 var.* Río Wanki,
Segoviao Wangkí. *river* Honduras/
Nicaragua
Cocos Basin *124 C5 undersea basin*
E Indian Ocean
Cocos Island Ridge *see* Cocos Ridge
Cocos Islands *141 D5 island group*
E Indian Ocean
Cocos Ridge *35 C8 var.* Cocos Island
Ridge. *undersea ridge* E Pacific Ocean

Cod, Cape *41 G3 headland*
Massachusetts, NE USA
Codfish Island *151 A8 island* SW New
Zealand
Codlea *108 C4 Ger.* Zeiden, *Hung.*
Feketehalom. Brașov, C Romania
Cody *44 C2* Wyoming, C USA
Coeur d'Alene *46 C2* Idaho, NW USA
Coevorden *86 E2* Drenthe,
NE Netherlands
Coffs Harbour *149 E6* New South Wales,
SE Australia
Cognac *91 B5 anc.* Compniacum.
Charente, W France
Cohalm *see* Rupea
Coiba, Isla de *53 E5 island* SW Panama
Coihaique *65 B6 var.* Coyhaique. Aisén,
S Chile
Coimbatore *132 C3* Tamil Nādu, S India
Coimbra *92 B3 anc.* Conimbria,
Conímbriga. Coimbra, N Portugal
Coín *92 D5* Andalucía, S Spain
Coira/Coire *see* Chur
Coirib, Loch *see* Corrib, Lough
Colby *45 E4* Kansas, C USA
Colchester *89 E6* Connecticut, NE USA
Coleman *49 F3* Texas, SW USA
Coleraine *88 B4 Ir.* Cúil Raithin.
N Northern Ireland, United Kingdom
Colesberg *78 C5* Northern Cape,
C South Africa
Colima *50 D4* Colima, S Mexico
Coll *88 B3 island* W Scotland, United
Kingdom
College Station *49 G3* Texas, SW USA
Collie *147 A7* Western Australia
Collipo *see* Leiria
Colmar *90 E4 Ger.* Kolmar. Haut-Rhin,
NE France
Cöln *see* Köln
Cologne *see* Köln
Colomb-Béchar *see* Béchar
Colombia *58 B3 off.* Republic of
Colombia. *country* N South America

COLOMBIA
South America

Official name Republic of Colombia
Formation 1819 / 1903
Capital Bogotá
Population 48.3 million / 120 people
per sq mile (47 people per sq km) / 77%
Total area 439,733 sq. miles
(1,138,910 sq, km)
Languages Spanish*, Wayuu, Páez, and
other Amerindian languages
Religions Roman Catholic 95%, Other 5%
Ethnic mix Mestizo 58%, White 20%,
European–African 14%, African 4%,
African–Amerindian 3%, Amerindian 1%
Government Presidential system
Currency Colombian peso = 100 centavos
Literacy rate 94%
Calorie consumption 2593 kilocalories

Colombian Basin *56 A1 undersea basin*
SW Caribbean Sea
Colombia, Republic of *see* Colombia
Colombie-Britannique *see* British
Columbia
Colombo *132 C4 country capital* (Sri
Lanka) Western Province, W Sri Lanka
Colón *53 G4 prev.* Aspinwall. Colón,
C Panama
Colón, Archipiélago de *see* Galápagos
Islands
Colonia Agrippina *see* Köln
Colón Ridge *35 B8 undersea ridge*
E Pacific Ocean
Colorado *44 C4 off.* State of Colorado,
also known as Centennial State, Silver
State. *state* C USA
Colorado City *49 F3* Texas, SW USA
Colorado Plateau *48 B1 plateau*
W USA
Colorado, Río *65 C5 river* E Argentina
Colorado, Río *see* Colorado River
Colorado River *35 B5 var.* Río Colorado.
river Mexico/USA
Colorado River *49 G4 river* Texas,
SW USA
Colorado Springs *44 D5* Colorado,
C USA
Columbia *41 E4* Maryland, NE USA
Columbia *45 G4* Missouri, C USA
Columbia *43 E2 state capital* South
Carolina, SE USA
Columbia *42 C1* Tennessee, S USA

Columbia River *46 B3 river* Canad
USA
Columbia Plateau *46 C3 plateau* I
Oregon, NW USA
Columbus *42 D2* Georgia, SE USA
Columbus *40 C4* Indiana, N USA
Columbus *42 C2* Mississippi, S US
Columbus *45 F4* Nebraska, C USA
Columbus *40 D4 state capital* Ohio
N USA
Colville Channel *150 D2 channel* !
Island, New Zealand
Colville River *36 D2 river* Alaska, \
Comacchio *96 C3 var.* Commachic
Comactium. Emilia-Romagna, N
Comactium *see* Comacchio
Comalcalco *51 G4* Tabasco, SE Me
Coma Pedrosa, Pic de *91 A7 mour*
NW Andorra
Comarapa *61 F4* Santa Cruz, C Bol
Comayagua *52 C2* Comayagua,
W Honduras
Comer See *see* Como, Lago di
Comilla *135 G4 Ben.* Kumillā.
Chittagong, E Bangladesh
Comino *see* Kemmuna
Comitán *51 G5 var.* Comitán de
Domínguez. Chiapas, SE Mexico
Comitán de Domínguez *see* Comit
Commachio *see* Comacchio
Commissioner's Point *42 A5 head*
W Bermuda
Communism Peak *123 F3 prev.* Q
Kommunizm. *mountain* E Tajiki:
Como *96 B2 anc.* Comum. Lomba:
N Italy
Comodoro Rivadavia *65 B6* Chub
SE Argentina
Como, Lake *96 B2 var.* Lario, Eng.
Como, Ger. Comer See. *lake* N Ita
Como, Lake *see* Como, Lago di
**Comores, République Fédérale
Islamique des** *see* Comoros
Comoros *79 F2 off.* Federal Islamic
Republic of the Comoros, Fr.
République Fédérale Islamique de
Comores. *country* W Indian Ocea

COMOROS
Indian Ocean

Official name Union of the Comor
Formation 1975 / 1975
Capital Moroni
Population 800,000 / 929 people
per sq mile (359 people per sq km)
Total area 838 sq. miles (2170 sq, km
Languages Arabic*, Comoran*, Fren
Religions Muslim (mainly Sunni) 98%
Other 1%, Roman Catholic 1%
Ethnic mix Comoran 97%, Other 3%
Government Presidential system
Currency Comoros franc = 100 cent
Literacy rate 76%
Calorie consumption 2139 kilocalo

**Comoros, Federal Islamic Republ
the** *see* Comoros
Compiègne *90 C3* Oise, N France
Complutum *see* Alcalá de Henares
Compniacum *see* Cognac
Compostella *see* Santiago de Comp
Comrat *108 D4 Rus.* Komrat.
S Moldova
Comum *see* Como
Conakry *74 C4 country capital* (Gu
SW Guinea
Conca *see* Cuenca
Concarneau *90 A3* Finistère, NW F
Concepción *61 G3* Santa Cruz, E B
Concepción *65 B5* Bío Bío, C Chile
Concepción *64 D2 var.* Villa
Concepción. Concepción, C Parag
Concepción de la Vega *see* La Vega
Concepción de la Vega *see* La Vega
Conchos, Río *48 D4 river* NW Me:
Conchos, Río *50 B2 river* C Mexic
Concord *41 G3 state capital* New
Hampshire, NE USA
Concordia *45 E4* Kansas, C USA
Concordia *64 D4* Entre Ríos, E Arg
Côn Đao Sơn *137 E7 var.* Con Son
island S Vietnam
Condate *see* Rennes, Ille-et-Vilaine,
France
Condate *see* St-Claude, Jura, France
Condega *52 D3* Estelí, NW Nicarag
Condivincum *see* Nantes
Confluentes *see* Koblenz

g Hoa Xa Hôi Chu Nghia Viêt Nam
e Vietnam
go *77 D5 off.* Republic of the Congo,
: Moyen-Congo; *prev.* Middle Congo.
untry C Africa

NGO
Central Africa

ficial name Republic of the Congo
mation 1960 / 1960
pital Brazzaville
pulation 4.4 million / 33 people
sq mile (13 people per sq km)
al area 132,046 sq. miles
,000 sq. km)
nguages Kongo, Teke, Lingala, French*
igions Traditional beliefs 50%, Roman
holic 35%, Protestant 13%, Muslim 2%
nic mix Bakongo 51%, Teke 17%,
her 16%, Mbochi 11%, Mbédé 5%
vernment Presidential system
rrency CFA franc = 100 centimes
eracy rate 79%
orie consumption 2195 kilocalories

go, **Dem. Rep.** *77 C6 off.* Democratic
epublic of Congo; prev. Zaïre, Belgian
ongo, Congo (Kinshasa). *country*
Africa

NGO, DEM. REP.
Central Africa

ficial name Democratic Republic of
e Congo
rmation 1960 / 1960
pital Kinshasa
pulation 67.5 million / 77 people
sq mile (30 people per sq km)
al area 905,563 sq. miles
45,410 sq. km)
nguages Kiswahili, Tshiluba, Kikongo,
gala, French*
ligions Roman Catholic 50%,
otestant 20%, Traditional beliefs and
her 10%, Muslim 10%, Kimbanguist 10%
nic mix Other 55%, Mongo, Luba,
ngo, and Mangbetu-Azande 45%
vernment Presidential system
rrency Congolese franc = 100 centimes
eracy rate 61%
lorie consumption 1585 kilocalories

go *77 C6 var.* Kongo, Fr. Zaïre.
iver C Africa
go Basin *77 C6 drainage basin*
V Dem. Rep. Congo
go/Congo (Kinshasa) *see* Congo
Democratic Republic of
ai see Cuneo
aimbria/Conimbriga see Coimbra
ajeeveram see Kānchipuram
anacht see Connaught
nnaught *89 A5 var.* Connacht, *Ir.*
Chonnacht, Cúige. *province*
V Ireland
necticut *41 F3 off.* State of
Connecticut, also known as Blue Law
tate, Constitution State, Land of
teady Habits, Nutmeg State. *state*
NE USA
necticut *41 G3 river* Canada/USA
aroe *49 G3* Texas, SW USA
sentia see Cosenza
nsolación del Sur *54 A2* Pinar del
Río, W Cuba
a Son see Côn Đao Son
nstance see Konstanz
nstance, Lake *95 B7 Ger.* Bodensee.
ake C Europe
nstanța *108 D5 var.* Küstendje, *Eng.*
Constanza, Ger. Konstanza, *Turk.*
Küstence. Constanța, SE Romania
nstantia see Coutances
nstantia see Konstanz
nstantine *71 E2 var.* Qacentina, *Ar.*
Qoussantína. N Algeria
nstantinople see İstanbul
nstantiola see Oltenița
nstanz see Konstanz
nstanza see Constanța
nstitution State see Connecticut
s see Kos
ober Pedy *149 A5* South Australia
okeville *42 D1* Tennessee, S USA
ok Islands *145 F4 territory in free*
association with New Zealand S Pacific
Ocean

Cook, Mount *see* Aoraki
Cook Strait *151 D5 var.* Raukawa. *strait*
New Zealand
Cooktown *148 D2* Queensland,
NE Australia
Coolgardie *147 B6* Western Australia
Cooma *149 D7* New South Wales,
SE Australia
Coomassie *see* Kumasi
Coon Rapids *45 F2* Minnesota, N USA
Cooper Creek *148 C4 var.* Barcoo,
Cooper's Creek. *seasonal river*
Queensland/South Australia
Cooper's Creek *see* Cooper Creek
Coos Bay *46 A3* Oregon, NW USA
Cootamundra *149 D6* New South Wales,
SE Australia
Copacabana *61 E4* La Paz, W Bolivia
Copenhagen *see* København
Copiapó *64 B3* Atacama, N Chile
Copperas Cove *49 G3* Texas, SW USA
Coppermine *see* Kugluktuk
Copper State *see* Arizona
Coquilhatville *see* Mbandaka
Coquimbo *64 B3* Coquimbo, N Chile
Corabia *108 B5* Olt, S Romania
Coral Harbour *37 G3 var.* Salliq.
Southampton Island, Nunavut,
NE Canada
Coral Sea *142 B3 sea* SW Pacific Ocean
Coral Sea Islands *144 B4 Australian*
external territory SW Pacific Ocean
Corantijn River *see* Courantyne River
Corcovado, Golfo *65 B6 gulf* S Chile
Corcyra Nigra *see* Korčula
Cordele *42 D3* Georgia, SE USA
Córdoba *64 C3* Córdoba, C Argentina
Córdoba *51 F4* Veracruz-Llave, E Mexico
Córdoba *92 D4 var.* Cordoba, *Eng.*
Cordova; *anc.* Corduba. Andalucía,
SW Spain
Cordova *36 C3* Alaska, USA
Cordova/Cordoba *see* Córdoba
Corduba *see* Córdoba
Corentyne River *see* Courantyne River
Corfu *104 A4 var.* Kérkira, *Eng.*
Corfu. *island* Iónia Nisiá, Greece,
C Mediterranean Sea
Corfu *see* Kérkyra
Coria *92 C3* Extremadura, W Spain
Corinth *42 C1* Mississippi, S USA
Corinth *see* Kórinthos
Corinth, Gulf of *105 B5 Eng.* Gulf of
Corinth; *anc.* Corinthiacus Sinus. *gulf*
C Greece
Corinth, Gulf of/Corinthiacus Sinus *see*
Korinthiakós Kólpos
Corinthus *see* Kórinthos
Corinto *52 C3* Chinandega,
NW Nicaragua
Cork *89 A6 Ir.* Corcaigh. S Ireland
Çorlu *116 A2* Tekirdağ, NW Turkey
Corner Brook *39 G3* Newfoundland,
Newfoundland and Labrador, E Canada
Cornhusker State *see* Nebraska
Corn Islands *53 E3 var.* Corn Islands.
island group SE Nicaragua
Corn Islands *see* Maíz, Islas del
Cornwallis Island *37 F2 Island* Nunavut,
N Canada
Coro *58 C1* Falcón, N Venezuela
Coro *see* Santa Ana de Coro.
Falcón, N Venezuela
Corocoro *61 F4* La Paz, W Bolivia
Coromandel *150 D2* Waikato, North
Island, New Zealand
Coromandel Coast *132 D2 coast* E India
Coromandel Peninsula *150 D2*
peninsula North Island, New Zealand
Coronado, Bahía de *52 D5 bay*
S Costa Rica
Coronel Dorrego *65 C5* Buenos Aires,
E Argentina
Coronel Oviedo *64 D2* Caaguazú,
SE Paraguay
Corozal *52 C1* Corozal, N Belize
Corpus Christi *49 G4* Texas, SW USA
Corrales *48 D2* New Mexico, SW USA
Corrib, Lough *89 A5 Ir.* Loch Coirib.
lake W Ireland
Corrientes *64 D3* Corrientes,
NE Argentina
Corriza *see* Korçë
Corsica *91 E7 Eng.* Corsica. *island*
France, C Mediterranean Sea
Corsica *see* Corse
Corsicana *49 G3* Texas, SW USA
Cortegana *92 C4* Andalucía, S Spain
Cortés *53 E5 var.* Ciudad Cortés.
Puntarenas, SE Costa Rica

Cortez, Sea of *see* California, Golfo de
Cortina d'Ampezzo *96 C1* Veneto,
NE Italy
Coruche *92 B3* Santarém, C Portugal
Çoruh Nehri *117 E3 Geor.* Chorokh,
Rus. Chorokhi. *river* Georgia/Turkey
Çorum *116 D3 var.* Chorum. Çorum,
N Turkey
Corunna *see* A Coruña
Corvallis *46 B3* Oregon, NW USA
Corvo *92 A5 var.* Ilha do Corvo. *island*
Azores, Portugal, NE Atlantic Ocean
Corvo, Ilha do *see* Corvo
Cos *see* Kos
Cosenza *97 D6 anc.* Consentia. Calabria,
SW Italy
Cosne-Cours-sur-Loire *90 C4* Nièvre,
Bourgogne, C France Europe
Costa Mesa *46 D2* California, W USA
North America
Costa Rica *53 E4 off.* Republic of Costa
Rica. *country* Central America

COSTA RICA
Central America

Official name Republic of Costa Rica
Formation 1838 / 1838
Capital San José
Population 4.9 million / 249 people
per sq mile (96 people per sq km)
Total area 19,730 sq. miles (51,100 sq. km)
Languages Spanish*, English Creole,
Bribri, Cabecar
Religions Roman Catholic 71%,
Evangelical 14%, Nonreligious 11%,
Other 4%
Ethnic mix Mestizo and European
94%, Black 3%, Other 1%, Chinese 1%,
Amerindian 1%
Government Presidential system
Currency Costa Rican colón
= 100 céntimos
Literacy rate 97%
Calorie consumption 2898 kilocalories

Costa Rica, Republic of *see* Costa Rica
Costermansville *see* Bukavu
Cotagaita *61 F5* Potosí, S Bolivia
Côte d'Ivoire *74 D4 off.* Republic of
Côte d'Ivoire; *Fr.* République de la Côte
d'Ivoire, *Eng* Ivory Coast. *country*
W Africa

CÔTE D'IVOIRE (IVORY COAST)
West Africa

Official name Republic of Côte d'Ivoire
Formation 1960 / 1960
Capital Yamoussoukro
Population 20.3 million / 165 people
per sq mile (64 people per sq km)
Total area 124,502 sq. miles
(322,460 sq. km)
Languages Akan, French*, Krou. Voltaique
Religions Muslim 38%, Traditional beliefs
25%, Roman Catholic 25%, Other 6%,
Protestant 6%
Ethnic mix Akan 42%, Voltaique 18%,
Mandé du Nord 17%, Krou 11%, Mandé
du Sud 12%, Other 2%
Government Presidential system
Currency CFA franc = 100 centimes
Literacy rate 41%
Calorie consumption 2781 calories

Côte d'Ivoire, République de *see*
Côte d'Ivoire
Côte d'Or *90 D4 cultural region* C France
Côte Française des Somalis *see* Djibouti
Côtière, Chaine *see* Coast Mountains
Cotonou *75 F5 var.* Kotonu. S Benin
Cotrone *see* Crotone
Cotswold Hills *89 D6 var.* Cotswolds.
hill range S England, United Kingdom
Cotswolds *see* Cotswold Hills
Cottbus *94 D4 Lus.* Choébuz; *prev.*
Kottbus. Brandenburg, E Germany
Cottonou *see* Cotonou
Cotyora *see* Ordu
Couentrey *see* Coventry
Council Bluffs *45 F4* Iowa, C USA
Courantyne River *59 G4 var.* Corantijn
River, Corentyne River. *river* Guyana/
Suriname
Courland Lagoon *106 A4 Ger.* Kurisches
Haff, *Rus.* Kurskiy Zaliv. *lagoon*
Lithuania/Russian Federation
Courtrai *see* Kortrijk

Coutances *90 B3 anc.* Constantia.
Manche, N France
Couvin *87 C7* Namur, S Belgium
Coventry *89 D6 anc.* Couentrey.
C England, United Kingdom
Covilhã *92 C3* Castelo Branco, E Portugal
Cowan, Lake *147 B6 lake* Western
Australia
Coxen Hole *see* Roatán
Coxin Hole *see* Roatán
Coyhaique *see* Coihaique
Coyote State, The *see* South Dakota
Cozhê *126 C5* Xizang Zizhiqu, W China
Cozumel, Isla *51 H3 island* SE Mexico
Cracovia/Cracow *see* Kraków
Cradock *78 C5* Eastern Cape, S South
Africa
Craig *44 C4* Colorado, C USA
Craiova *108 B5* Dolj, SW Romania
Cranbrook *37 E5* British Columbia,
SW Canada
Crane *see* The Crane
Cranz *see* Zelenogradsk
Crawley *89 E7* SE England, United
Kingdom
Cremona *96 B2* Lombardia, N Italy
Creole State *see* Louisiana
Cres *100 A3 It.* Cherso; *anc.* Crexa.
island W Croatia
Crescent City *46 A4* California, W USA
Crescent Group *128 C7 island group*
C Paracel Islands
Creston *45 F4* Iowa, C USA
Crestview *42 D3* Florida, SE USA
Crete *105 C7 Eng.* Crete. *island* Greece,
Aegean Sea
Crétéil *90 E2* Val-de-Marne, N France
Sea of Crete *105 D7 var.* Kretikon
Delagos, *Eng.* Sea of Crete; *anc.* Mare
Creticum. *sea* Greece, Aegean Sea
Crete, Sea of/Creticum, Mare *see* Kritikó
Pélagos
Creuse *90 B4 river* C France
Crewe *89 D6* C England, United
Kingdom
Crexa *see* Cres
Crikvenica *100 A3 It.* Cirquenizza; *prev.*
Crikvenica, Crjkvenica. Primorje-
Gorski Kotar, NW Croatia
Crimea *see* Kryms'kyy Pivostriv
Cristóbal *53 G4* Colón, C Panama
Cristóbal Colón, Pico *58 B1 mountain*
N Colombia
Cristur/Cristuru Sácuiesc *see* Cristuru
Secuiesc
Cristuru Secuiesc *108 C4 prev.* Cristur,
Cristuru Sácuiesc; Sitaş Cristuru, *Ger.*
Kreutz, *Hung.* Székelykeresztúr, Szitás-
Keresztúr. Harghita, C Romania
Crjkvenica *see* Crikvenica
Crna Gora *see* Montenegro
Crna Reka *101 D6 river* S FYR
Macedonia
Crni Drim *see* Black Drin
Croatia *100 B3 off.* Republic of Croatia,
Ger. Kroatien, *SCr.* Hrvatska. *country*
SE Europe

CROATIA
Southeast Europe

Official name Republic of Croatia
Formation 1991 / 1991
Capital Zagreb
Population 4.3 million / 197 people
per sq mile (76 people per sq km)
Total area 21,831 sq. miles (56,542 sq. km)
Languages Croatian*
Religions Roman Catholic 88%,
Orthodox Christian 4%, Other 7%,
Muslim 1%
Ethnic mix Croat 90%, Other 5%,
Serb 5%
Government Parliamentary system
Currency Kuna = 100 lipa
Literacy rate 99%
Calorie consumption 3052 kilocalories

Croatia, Republic of *see* Croatia
Crocodile *see* Limpopo
Croia *see* Krujë
Croker Island *146 E2 island* Northern
Territory, N Australia
Cromwell *151 B7* Otago, South Island,
New Zealand
Crooked Island *54 D2 island*
SE Bahamas
Crooked Island Passage *54 D2 channel*
SE Bahamas

Crookston *45 F1* Minnesota, N USA
Crossen *see* Krosno Odrzańskie
Croton/Crotona *see* Crotone
Crotone *97 E6 var.* Cotrone; *anc.* Croton, Crotona. Calabria, SW Italy
Croydon *89 A8* SE England, United Kingdom
Crozet Basin *141 B6 undersea basin* S Indian Ocean
Crozet Islands *141 B7 island group* French Southern and Antarctic Lands
Crozet Plateau *141 B7 var.* Crozet Plateaus. *undersea plateau* SW Indian Ocean
Crozet Plateaus *see* Crozet Plateau
Crystal Brook *149 B6* South Australia
Csaca *see* Čadca
Csakathurn/Csáktornya *see* Čakovec
Csíkszereda *see* Miercurea-Ciuc
Csorna *99 C6* Győr-Moson-Sopron, NW Hungary
Csurgó *99 C7* Somogy, SW Hungary
Cuando *78 C2 var.* Kwando. *river* S Africa
Cuango *78 B1 var.* Kwango. *river* Angola/Dem. Rep. Congo
Cuango *see* Kwango
Cuanza *78 B1 var.* Kwanza. *river* C Angola
Cuauhtémoc *50 C2* Chihuahua, N Mexico
Cuautla *51 E4* Morelos, S Mexico
Cuba *54 B2 off.* Republic of Cuba. *country* W West Indies

CUBA
West Indies

Official name Republic of Cuba
Formation 1902 / 1902
Capital Havana
Population 11.3 million / 264 people per sq mile (102 people per sq km) / 76%
Total area 42,803 sq. miles (110,860 sq. km)
Languages Spanish
Religions Nonreligious 49%, Roman Catholic 40%, Atheist 6%, Other 4%, Protestant 1%
Ethnic mix Mulatto (mixed race) 51%, White 37%, Black 11%, Chinese 1%
Government One-party state
Currency Cuban peso = 100 centavos
Literacy rate 99%
Calorie consumption 3277 kilocalories

Cubal *78 B2* Benguela, W Angola
Cubango *78 B2 var.* Kuvango, *Port.* Vila Artur de Paiva, Vila da Ponte. Huíla, SW Angola
Cubango *78 B2 var.* Kavango, Kavengo, Kubango, Okavango, Okavanggo. *river* S Africa
Cuba, Republic of *see* Cuba
Cúcuta *58 C2 var.* San José de Cúcuta. Norte de Santander, N Colombia
Cuddapah *132 C2* Andhra Pradesh, S India
Cuenca *60 B2* Azuay, S Ecuador
Cuenca *93 E3 anc.* Conca. Castilla-La Mancha, C Spain
Cuera *see* Chur
Cuernavaca *51 E4* Morelos, S Mexico
Cuiabá *63 E3 prev.* Cuyabá. *state capital* Mato Grosso, SW Brazil
Cúige *see* Connaught
Cúige Laighean *see* Leinster
Cúige Mumhan *see* Munster
Cuijck *86 D4* Noord-Brabant, SE Netherlands
Cúil Raithin *see* Coleraine
Cuito *78 B2 var.* Kwito. *river* SE Angola
Cukai *138 B3 var.* Chukai, Kemaman. Terengganu, Peninsular Malaysia
Cularo *see* Grenoble
Culiacán *50 C3 var.* Culiacán Rosales, Culiacán-Rosales. Sinaloa, C Mexico
Culiacán-Rosales/Culiacán Rosales *see* Culiacán
Cullera *93 F3* Valenciana, E Spain
Cullman *42 C2* Alabama, S USA
Culm *see* Chełmno
Culmsee *see* Chełmża
Cumaná *59 E1* Sucre, NE Venezuela
Cumbal, Nevado de *58 A4 elevation* S Colombia
Cumberland *41 E4* Maryland, NE USA
Cumberland Plateau *42 D1 plateau* E USA

Cumberland Sound *37 H3 inlet* Baffin Island, Nunavut, NE Canada
Cumpas *50 B2* Sonora, NW Mexico
Cuneo *96 A2 Fr.* Coni. Piemonte, NW Italy
Cunnamulla *149 C5* Queensland, E Australia
Ćuprija *100 E4* Serbia, E Serbia
Curaçao *55 E5 Dutch island territory* S West Indies
Curaçao *55 E5 island* Lesser Antilles
Curia Rhaetorum *see* Chur
Curicó *64 B4* Maule, C Chile
Curieta *see* Krk
Curitiba *63 E4 prev.* Curytiba. *state capital* Paraná, S Brazil
Curtbunar *see* Tervel
Curtea de Argeş *108 C4 var.* Curtea-de-Arges. Argeş, S Romania
Curtea-de-Arges *see* Curtea de Argeş
Curtici *108 A4 Ger.* Kurtitsch, *Hung.* Kürtös. Arad, W Romania
Curtis Island *148 E4 island* Queensland, SE Australia
Curytiba *see* Curitiba
Curzola *see* Korčula
Cusco *61 E4 var.* Cuzco. Cusco, C Peru
Cusset *91 C5* Allier, C France
Cutch, Gulf of *see* Kachchh, Gulf of
Cuttack *135 F4* Odisha, E India
Cuvier Plateau *141 E6 undersea plateau* E Indian Ocean
Cuxhaven *94 B2* Niedersachsen, NW Germany
Cuyabá *see* Cuiabá
Cuyuni, Río *see* Cuyuni River
Cuyuni River *59 F3 var.* Río Cuyuni. *river* Guyana/Venezuela
Cuzco *see* Cusco
Cyclades *105 D6 var.* Kikládhes, *Eng.* Cyclades. *island group* SE Greece
Cyclades *see* Kykládes
Cydonia *see* Chaniá
Cymru *see* Wales
Cyprus *102 C4 off.* Republic of Cyprus, *Gk.* Kypros, *Turk.* Kıbrıs, Kıbrıs Cumhuriyeti. *country* E Mediterranean Sea

CYPRUS
Southeast Europe

Official name Republic of Cyprus
Formation 1960 / 1960
Capital Nicosia
Population 1.1 million / 308 people per sq mile (119 people per sq km)
Total area 3571 sq. miles (9250 sq. km)
Languages Greek*, Turkish*
Religions Orthodox Christian 78%, Muslim 18%, Other 4%
Ethnic mix Greek 81%, Turkish 11%, Other 8%
Government Presidential system
Currency Euro = 100 cents (In TRNC, new Turkish lira = 100 kurus)
Literacy rate 99%
Calorie consumption 2661 kilocalories

Cyprus, Republic of *see* Cyprus
Cythnos *see* Kýthnos
Czech Republic *99 A5 Cz.* Česká Republika. *country* C Europe

CZECH REPUBLIC
Central Europe

Official name Czech Republic
Formation 1993 / 1993
Capital Prague
Population 10.7 million / 351 people per sq mile (136 people per sq km)
Total area 30,450 sq. miles (78,866 sq. km)
Languages Czech*, Slovak, Hungarian (Magyar)
Religions Roman Catholic 39%, Atheist 38%, Other 18%, Protestant 3%, Hussite 2%
Ethnic mix Czech 90%, Other 4%, Moravian 4%, Slovak 2%
Government Parliamentary system
Currency Czech koruna = 100 haleru
Literacy rate 99%
Calorie consumption 3292 kilocalories

Czenstochau *see* Częstochowa
Czernowitz *see* Chernivtsi

Częstochowa *98 C4 Ger.* Czenstochau, Tschenstochau, *Rus.* Chenstokhov. Śląskie, S Poland
Człuchów *98 C3 Ger.* Schlochau. Pomorskie, NW Poland

D

Dabajuro *58 C1* Falcón, NW Venezuela
Dabeiba *58 B2* Antioquia, NW Colombia
Dąbrowa Tarnowska *99 D5* Małopolskie, S Poland
Dabryn' *107 C8 Rus.* Dobryn'. Homyel'skaya Voblasts', SE Belarus
Dacca *see* Dhaka
Daegu *107 E4 off.* Daegu Gwang-yeoksi, *prev.* Taegu, *Jap.* Taikyū. SE South Korea
Daegu Gwang-yeoksi *see* Daegu
Daejeon *107 E4 off.* Daejeon Gwang-yeoksi, *prev.* Taejŏn, *Jap.* Taiden. C South Korea
Daejeon Gwang-yeoksi *see* Daejeon
Dagana *74 B3* N Senegal
Dagda *106 D4* SE Latvia
Dagden *see* Hiiumaa
Dagenham *89 B8* United Kingdom
Dağlıq Quarabağ *see* Nagorno-Karabakh
Dagö *see* Hiiumaa
Dagupan *139 E1 off.* Dagupan City. Luzon, N Philippines
Dagupan City *see* Dagupan
Dahm, Ramlat *121 B6 desert* NW Yemen
Dahomey *see* Benin
Daihoku *see* Taipei
Daimiel *92 D3* Castilla-La Mancha, C Spain
Dainan *see* Tainan
Daingin, Bá an *see* Dingle Bay
Dairen *see* Dalian
Dakar *74 B3 country capital* (Senegal) W Senegal
Dakhla *see* Ad Dakhla
Dakoro *75 G3* Maradi, S Niger
Đakovica *see* Gjakovë
Đakovo *100 C3 var.* Djakovo, *Hung.* Diakovár. Osijek-Baranja, E Croatia
Dakshin *see* Deccan
Dalain Hob *126 D3 var.* Ejin Qi. Nei Mongol Zizhiqu, N China
Dalai Nor *see* Hulun Nur
Dalaman *116 A4* Muğla, SW Turkey
Dalandzadgad *127 E3* Ömnögovĭ, S Mongolia
Đa Lat *137 E6* Lâm Đông, S Vietnam
Dalby *149 D5* Queensland, E Australia
Dale City *41 E4* Virginia, NE USA
Dalhart *49 E1* Texas, SW USA
Dalian *128 D4 var.* Dairen, Dalien, Jay Dairen, Lüda, Ta-lien, *Rus.* Dalny. Liaoning, NE China
Dalien *see* Dalian
Dallas *49 G2* Texas, SW USA
Dalmacija *100 B4 Eng.* Dalmatia, *Ger.* Dalmatien, *It.* Dalmazia. *cultural region* S Croatia
Dalmatia/Dalmatien/Dalmazia *see* Dalmacija
Dalny *see* Dalian
Dalton *42 D1* Georgia, SE USA
Dálvvadis *see* Jokkmokk
Daly Waters *148 A2* Northern Territory, N Australia
Damachava *107 A6 var.* Damachova, *Pol.* Domaczewo, *Rus.* Domachevo. Brestskaya Voblasts', SW Belarus
Damachova *see* Damachava
Damān *134 C4* Damān and Diu, W India
Damara *76 C4* Ombella-Mpoko, S Central African Republic
Damas *see* Dimashq
Damasco *see* Dimashq
Damascus *see* Dimashq
Dämmäm *see* Ad Dammām
Damoûr *119 A5 var.* Ad Dāmūr. C Lebanon
Dampier *146 A4* Western Australia
Dampier, Selat *139 F4 strait* Papua, E Indonesia
Damqawt *121 D6 var.* Damqut. E Yemen

Damqut *see* Damqawt
Damxung *126 C5 var.* Gongtang. X Zizhiqu, W China
Danakil Desert *72 D4 var.* Afar Depression, Danakil Plain. *desert* E Africa
Danakil Plain *see* Danakil Desert
Danané *74 D5* W Côte d'Ivoire
Đà Nang *137 E5 prev.* Tourane. Qua Nam-Đa Nang, C Vietnam
Danborg *see* Daneborg
Dandong *128 D3 var.* Tan-tung; *prev* An-tung. Liaoning, NE China
Daneborg *83 E3 var.* Danborg. Tu... N Greenland
Dänew *see* Galkynyş
Dangara *see* Danghara
Dangerous Archipelago *see* Tuamot... Îles
Danghara *123 E3 Rus.* Dangara. SW Tajikistan
Danghe Nanshan *126 D3 mountain range* W China
Dang Raek, Phanom/Dangrek, Ch... des *see* Dângrêk, Chuôr Phnum
Chuor Phnum Dangrek *137 D5 va...* Phanom Dang Raek, Phanom Dor Rak, *Fr.* Chaîne des Dangrek. *mou range* Cambodia/Thailand
Dangriga *52 C1 prev.* Stann Creek. Stann Creek, E Belize
Danish West Indies *see* Virgin Island (US)
Danlí *52 D2* El Paraíso, S Honduras
Danmark *see* Denmark
Danmarksstraedet *see* Denmark Stra...
Dannenberg *94 C3* Niedersachsen, N Germany
Dannevirke *150 D4* Manawatu-Wanganui, North Island, New Zeal
Dantzig *see* Gdańsk
Danube *81 E4 Bul.* Dunav, *Cz.* Duna... *Ger.* Donau, *Hung.* Duna, *Rom.* Dunărea. *river* C Europe
Danubian Plain *104 C2 Eng.* Danub... Plain. *lowlands* N Bulgaria
Danubian Plain *see* Dunavska Ravni
Danum *see* Doncaster
Danville *41 E5* Virginia, NE USA
Danxian/Dan Xian *see* Danzhou
Danzhou *128 C7 prev.* Danxian, Da Xian, Nada. Hainan, S China
Danzig *see* Gdańsk
Danziger Bucht *see* Danzig, Gulf of
Danzig, Gulf of *98 C2 var.* Gulf of Gdańsk, *Ger.* Danziger Bucht, *Pol.* Zakota Gdańska, *Rus.* Gdan'skaya Bukhta. *gulf* N Poland
Daqm *see* Duqm
Dar'ā *119 B5 var.* Der'a, *Fr.* Déraa. Dar'ā, SW Syria
Darabani *108 C3* Botoşani, NW Romania
Daraut-Kurgan *see* Daroot-Korgon
Dardanelli *116 A2 Eng.* Dardanelle... *strait* NW Turkey
Dardanelles *see* Çanakkale Boğazı
Dardanelli *see* Dardanelli
Dar-el-Beida *see* Casablanca
Dar es Salaam *73 C7 var.* Dar es Salaam, E Tanzania
Darfield *151 C6* Canterbury, South Island, New Zealand
Darfur *72 A4 var.* Darfur Massif. *cultural region* W Sudan
Darfur Massif *see* Darfur
Darhan *127 E2* Darhan Uul, N Mong...
Darién, Golfo del *see* Darién, Gulf of
Darién, Gulf of *58 A2 Sp.* Golfo del Darién. *gulf* S Caribbean Sea
Darién, Serranía del *53 H5 mount...* *range* Colombia/Panama
Dario *see* Ciudad Darío
Dariorigum *see* Vannes
Därjiling *135 F3 prev.* Darjeeling. ... Bengal, NE India
Darling River *149 C6 river* New Sou Wales, SE Australia
Darmstadt *95 B5* Hessen, SW Germa...
Darnah *71 G2 var.* Derna. NE Libya
Darnley, Cape *154 D2 cape* Antarcti...
Daroca *93 E2* Aragón, NE Spain
Daroot-Korgon *123 F3 var.* Daraut-Kurgan. Oshskaya Oblast', SW Kyrgyzstan

ford 89 B8 SE England,
...ited Kingdom
...moor 89 C7 moorland SW England,
...ited Kingdom
...mouth 39 F4 Nova Scotia,
...Canada
...raza see Derweze, Turkmenistan
...win 146 D2 prev. Palmerston, Port
...arwin. *territory capital* Northern
...rritory, N Australia
...win, Isla 60 A4 *island* Galápagos
...lands, W Ecuador
...howuz see Daşoguz
...kawka 107 D6 Rus. Dashkovka.
...ahilyowskaya Voblasts', E Belarus
...kovka see Dashkawka
...guz 122 C2 Rus. Dashkhovuz,
...urkm. Dashhowuz; *prev.* Tashauz.
...soguz Welaýaty, N Turkmenistan
...šöng see Black River
...ong 128 C3 *var.* Tatung, Ta-t'ung.
...anxi, C China
...gava see Dagava
...gavpils 106 D4 Ger. Dünaburg; *prev.*
...us. Dvinsk. SE Latvia
...ng Kyun 137 B6 *island* S Myanmar
(...urma)
...phiné 91 D5 *cultural region* E France
...angere 132 C2 Karnātaka, W India
...ao 139 F3 *off.* Davao City.
...indanao, S Philippines
...ao City see Davao
...ao Gulf 139 F3 *gulf* Mindanao,
...Philippines
...enport 45 G3 Iowa, C USA
...d 53 E5 Chiriquí, W Panama
...ie Ridge 141 H5 *undersea ridge*
...Indian Ocean
...is 154 D3 *Australian research station*
...ntarctica
...as Sea 154 D3 *sea* Antarctica
...is Strait 82 B3 *strait* Baffin Bay/
...abrador Sea
...ei 137 B5 *var.* Tavoy, Htawei.
...anintharyi, S Myanmar (Burma)
...lat Qatar see Qatar
...91 B6 *var.* Ax; *anc.* Aquae Augustae,
...quae Tarbelicae. Landes, SW France
...z az Zawr 118 D3 *var.* Deir ez Zor.
...ayr az Zawr, E Syria
...on 40 C4 Ohio, N USA
...tona Beach 43 E4 Florida, SE USA
...ar 78 C5 Northern Cape, C South
...frica
...t Sea 119 B6 *var.* Bahret Lut, Lacus
...sphaltites, Ar. Al Baḥr al Mayyit,
...aḥrat Lūt, Heb. Yam HaMelaḥ. *salt*
...ke Israel/Jordan
...n Funes 64 C3 Córdoba, C Argentina
...h Valley 47 C7 *valley* California,
... USA
...howuz see Daşoguz
...r 101 D6 Ger. Dibra, Turk. Debre.
...FYR Macedonia
...ehagle see Laï
...ica 99 D5 Podkarpackie, SE Poland
...sildt see De Bilt
...ilt 86 C3 *var.* De Bildt. Utrecht,
...Netherlands
...no 98 B3 Zachodnio-pomorskie,
...W Poland
...re see Debar
...recen 99 D6 Ger. Debreczin, Rom.
...ebreţin; *prev.* Debreczen. Hajdú-
...har, E Hungary
...reczen/Debreczin see Debrecen
...reţin see Debrecen
...atur 42 C1 Alabama, S USA
...atur 40 B4 Illinois, N USA
...can 134 D5 Hind. Dakshin. *plateau*
...India
...in 98 B4 Ger. Tetschen. Ústecký Kraj,
...N Czech Republic
...eagaç/Dedeagach see Alexandroúpoli
...emsvaart 86 E3 Overijssel,
...Netherlands
...88 C3 *river* NE Scotland, United
...ingdom
...ring 36 C2 Alaska, USA
... see Dej
...gendorf 95 D6 Bayern, SE Germany
...rmenlik 102 C5 Gk. Kythréa.
...Cyprus
...Bid see Şafâshahr
...li see Delhi
...Shü see Dishü
...ze 87 B5 Oost-Vlaanderen,

Deir ez Zor see Dayr az Zawr
Deirgeirt, Loch see Derg, Lough
Dej 108 B3 Hung. Dés; *prev.* Deés. Cluj, NW Romania
De Jouwer see Joure
Dekéleia see Dhekélia
Dékoa 76 C4 Kémo, C Central African Republic
De Land 43 E4 Florida, SE USA
Delano 47 C7 California, W USA
Delārām see Dilārām
Delaware 40 D4 Ohio, N USA
Delaware 41 F4 *off.* State of Delaware, *also known as* Blue Hen State, Diamond State, First State. *state* NE USA
Delft 86 B4 Zuid-Holland, W Netherlands
Delfzijl 86 E1 Groningen, NE Netherlands
Delgo 72 B3 Northern, N Sudan
Delhi 134 D3 *var.* Dehli, Dilli, *hist.* Shahjahanabad. *union territory capital* Delhi, N India
Delicias 50 D2 *var.* Ciudad Delicias. Chihuahua, N Mexico
Déli-Kárpátok see Carpaţii Meridionalii
Delmenhorst 94 B3 Niedersachsen, NW Germany
Del Rio 49 F4 Texas, SW USA
Deltona 43 E4 Florida, SE USA
Demba 77 D6 Kasai-Occidental, C Dem. Rep. Congo
Dembia 76 D4 Mbomou, SE Central African Republic
Demchok 126 A5 *var.* Dêmqog. *disputed region* China/India
Demerara Plain 56 C2 *abyssal plain* W Atlantic Ocean
Deming 48 C3 New Mexico, SW USA
Demmin 94 C2 Mecklenburg-Vorpommern, NE Germany
Demopolis 42 C2 Alabama, S USA
Dêmqog 126 A5 *var.* Demchok. China/India
Denali see McKinley, Mount
Denau see Denov
Dender 87 B6 Fr. Dendre. *river* W Belgium
Dendre see Dender
Denekamp 86 E3 Overijssel, NE Netherlands
Den Haag see 's-Gravenhage
Denham 147 A5 Western Australia
Den Ham 86 E3 Overijssel, E Netherlands
Den Helder 86 C2 Noord-Holland, NW Netherlands
Dénia 93 F4 Valenciana, E Spain
Deniliquin 149 C7 New South Wales, SE Australia
Denison 45 F3 Iowa, C USA
Denison 49 G2 Texas, SW USA
Denizli 116 B4 Denizli, SW Turkey
Denmark 85 A7 *off.* Kingdom of Denmark, Dan. Danmark; *anc.* Hafnia. *country* N Europe

DENMARK
Northern Europe

Official name **Kingdom of Denmark**
Formation **950 / 1944**
Capital **Copenhagen**
Population **5.6 million / 342 people per sq mile (132 people per sq km)**
Total area **16,639 sq miles (43,094 sq km)**
Languages **Danish***
Religions **Evangelical Lutheran 95%, Roman Catholic 3%, Muslim 2%**
Ethnic mix **Danish 96%, Other (including Scandinavian and Turkish) 3%, Faroese and Inuit 1%**
Government **Parliamentary system**
Currency **Danish krone = 100 øre**
Literacy rate **99%**
Calorie consumption **3363 kilocalories**

Denmark, Kingdom of see Denmark
Denmark Strait 82 D4 *var.* Danmarksstraedet. *strait* Greenland/Iceland
Dennery 55 F1 E Saint Lucia
Denov 123 E3 Rus. Denau. Surkhondaryo Viloyati, S Uzbekistan
Denpasar 138 D5 *prev.* Paloe. Bali, C Indonesia
Denton 49 G2 Texas, SW USA
D'Entrecasteaux Islands 144 B3 *island group* SE Papua New Guinea

Denver 44 D4 *state capital* Colorado, C USA
Der'a/Derá/Déraa see Dar'ā
Dera Ghāzi Khān 134 C2 *var.* Dera Ghāzikhān. Punjab, C Pakistan
Dera Ghāzikhān see Dera Ghāzi Khān
Đeravica see Gjeravicë
Derbent 111 B8 Respublika Dagestan, SW Russian Federation
Derby 89 D6 C England, United Kingdom
Dereli see Gónnoi
Dergachi see Derhachi
Derg, Lough 89 A6 Ir. Loch Deirgeirt. *lake* W Ireland
Derhachi 109 G2 Rus. Dergachi. Kharkivs'ka Oblast', E Ukraine
De Ridder 42 A3 Louisiana, S USA
Dérna see Darnah
Derry see Londonderry
Dertosa see Tortosa
Derventa 100 B3 Republika Srpska, N Bosnia and Herzegovina
Derweze 122 C2 Rus. Darvaza. Ahal Welaýaty, C Turkmenistan
Dés see Dej
Deschutes River 46 B3 *river* Oregon, NW USA
Desē 71 C5 *var.* Desse, *It.* Dessie. Āmara, N Ethiopia
Deseado, Río 65 B7 *river* S Argentina
Desertas, Ilhas 70 A2 *island group* Madeira, Portugal, NE Atlantic Ocean
Des Moines 45 F3 *state capital* Iowa, C USA
Desna 109 E2 *river* Russian Federation/Ukraine
Dessau 94 C4 Sachsen-Anhalt, E Germany
Desse see Desē
Dessie see Desē
Dêstêro see Florianópolis
Detroit 40 D3 Michigan, N USA
Detroit Lakes 45 F2 Minnesota, N USA
Deurne 87 D5 Noord-Brabant, SE Netherlands
Deutschendorf see Poprad
Deutsch-Eylau see Iława
Deutsch Krone see Wałcz
Deutschland/Deutschland, Bundesrepublik see Germany
Deutsch-Südwestafrika see Namibia
Deva 108 B4 Ger. Diemrich, Hung. Déva. Hunedoara, W Romania
Deva see Aberdeen
Deva see Chester
Deventer 86 D3 Overijssel, E Netherlands
Devils Lake 45 E1 North Dakota, N USA
Devoll see Devollit, Lumi i
Devollit, Lumi i 101 D6 *var.* Devoll. *river* SE Albania
Devon Island. 37 F2 *prev.* North Devon Island. *island* Parry Islands, Nunavut, NE Canada
Devonport 149 C8 Tasmania, SE Australia
Devrek 116 C2 Zonguldak, N Turkey
Deynau see Galkynyş
Dezfūl 120 C3 *var.* Dizful. Khūzestān, SW Iran
Dezhou 128 D4 Shandong, E China
Dhaka 135 G4 *prev.* Dacca. *country capital* (Bangladesh) Dhaka, C Bangladesh
Dhanbād 135 F4 Jhārkhand, NE India
Dhekélia 102 C5 *Eng.* Dhekelia, *Gk.* Dekéleia. *UK air base* SE Cyprus
Dhidhimótikhon see Didymóteicho
Dhíkti Ori see Dhíkti Óri
Dhodhekánisos see Dodekánisa
Dhomokós see Domokós
Dhrepanon, Akrotírio see Drépano, Akrotírio
Dhún na nGall, Bá see Donegal Bay
Dhuusa Marreeb 73 E5 *var.* Dusa Marreb, *It.* Dusa Marreb. Galguduud, C Somalia
Diakovár see Đakovo
Diamantina, Chapada 63 F3 *mountain range* E Brazil
Diamantina Fracture Zone 141 E6 *tectonic feature* S Indian Ocean

Diamond State see Delaware
Diarbekr see Diyarbakır
Dibio see Dijon
Dibra see Debar
Dibrugarh 135 H3 Assam, NE India
Dickinson 44 D2 North Dakota, N USA
Dicle see Tigris
Didimotiho see Didymóteicho
Didymóteicho 104 D3 *var.* Dhidhimótikhon, Didimotiho. Anatolikí Makedonía kai Thráki, NE Greece
Diedenhofen see Thionville
Diekirch 87 D7 Diekirch, C Luxembourg
Diemrich see Deva
Điện Biên see Điện Biên Phu
Điện Biên Phu 136 D3 *var.* Bien Bien, Điện Biên. Lai Châu, N Vietnam
Diepenbeek 87 D6 Limburg, NE Belgium
Diepholz 94 B3 Niedersachsen, NW Germany
Dieppe 90 C2 Seine-Maritime, N France
Dieren 86 D4 Gelderland, E Netherlands
Differdange 87 D8 Luxembourg, SW Luxembourg
Digne 91 D6 *var.* Digne-les-Bains. Alpes-de-Haute-Provence, SE France
Digne-les-Bains see Digne
Digoel see Digul, Sungai
Digoin 90 C4 Saône-et-Loire, C France
Digul, Sungai 139 H5 *prev.* Digoel. *river* Papua, E Indonesia
Dihang see Brahmaputra
Dijlah see Tigris
Dijon 90 D4 *anc.* Dibio. Côte d'Or, C France
Dikhil 72 D4 SW Djibouti
Dikson 114 D2 Krasnoyarskiy Kray, N Russian Federation
Dikti 105 D8 *var.* Dhíkti Óri. *mountain range* Kríti, Greece, E Mediterranean Sea
Dilārām 122 D5 *prev.* Delārām. Nīmrūz, SW Afghanistan
Dili 139 F5 *var.* Dilli, Dilly. *country capital* (East Timor) N East Timor
Dilia 75 G3 *var.* Dillia. *river* SE Niger
Di Linh 137 E6 Lâm Đồng, S Vietnam
Dilli see Delhi, India
Dilli see Dili, East Timor
Dillia see Dilia
Dillon 44 B2 Montana, NW USA
Dilly see Dili
Dilolo 77 D8 Katanga, S Dem. Rep. Congo
Dimashq 119 B5 *var.* Ash Shām, Esh Sham, *Eng.* Damascus, *Fr.* Damas, *It.* Damasco. *country capital* (Syria) Rīf Dimashq, SW Syria
Dimitrovgrad 104 D3 Haskovo, S Bulgaria
Dimitrovgrad 111 C6 *prev.* Caribrod. Serbia, SE Serbia
Dimovo see Pernik
Dimovo 104 B1 Vidin, NW Bulgaria
Dinajpur 135 F3 Rajshahi, NW Bangladesh
Dinan 90 A3 Côte d'Armor, NW France
Dinant 87 C7 Namur, S Belgium
Dinar 116 B4 Afyon, SW Turkey
Dinara see Dinaric Alps
Dinaric Alps 100 C4 *var.* Dinara. Herzegovina/Croatia
Dindigul 132 C3 Tamil Nādu, SE India
Dingle Bay 89 A6 Ir. Bá an Daingin. *bay* SW Ireland
Dinguiraye 74 C4 N Guinea
Diourbel 74 B3 W Senegal
Diré Dawa 71 D5 E Ethiopia
Dirk Hartog Island 147 A5 *island* Western Australia
Dirschau see Tczew
Disappointment, Lake 146 C4 *salt lake* Western Australia
Discovery Bay 54 B4 Middlesex, Jamaica, Greater Kingston, C Jamaica
Dishū 123 A7 *var.* Deshū; *prev.* Deh Shū, Desho. Helmand, S Afghanistan
Disko Bugt see Qeqertarsuup Tunua
Dispur 135 H3 *state capital* Assam, NE India
Divinópolis 63 F4 Minas Gerais, SE Brazil
Divo 74 D5 S Côte d'Ivoire

Divodurum Mediomatricum *see* Metz
Diyarbakır *117 E4 var.* Diarbekr; *anc.*
Amida. Diyarbakır, SE Turkey
Dizful *see* Dezfūl
Djailolo *see* Halmahera, Pulau
Djajapura *see* Jayapura
Djakarta *see* Jakarta
Djakovo *see* Đakovo
Djambala *77 B6* Plateaux, C Congo
Djambi *see* Hari, Batang
Djambi *see* Jambi
Djanet *71 E4 prev.* Fort Charlet.
SE Algeria
Djéblé *see* Jablah
Djelfa *70 D2 var.* El Djelfa. N Algeria
Djéma *76 D4* Haut-Mbomou, E Central
African Republic
Djember *see* Jember
Djérablous *see* Jarābulus
Djerba *71 F2 var.* Djerba, Jazīrat Jarbah.
island E Tunisia
Djerba *see* Jerba, Île de
Djérem *76 B4 river* C Cameroon
Djevdjelija *see* Gevgelija
Djibouti *72 D4 var.* Jibuti. *country
capital* (Djibouti) E Djibouti
Djibouti *72 D4 off.* Republic of Djibouti,
var. Jibuti; *prev.* French Somaliland,
French Territory of the Afars and
Issas, *Fr.* Côte Française des Somalis,
Territoire Français des Afars et des
Issas. *country* E Africa

Djibouti, Republic of *see* Djibouti
Djokjakarta *see* Yogyakarta
Djourab, Erg du *76 C2 desert* N Chad
Djúpivogur *83 E5* Austurland,
SE Iceland
Dmitriyevsk *see* Makiyivka
Dnepr *see* Dnieper
Dneprodzerzhinsk *see* Romaniv
**Dneprodzerzhinskoye
Vodokhranilishche** *see*
Dniprodzerzhyns'ke Vodoskhovyshche
Dnepropetrovsk *see* Dnipropetrovs'k
Dneprorudnoye *see* Dniprorudne
Dnestr *see* Dniester
Dnieper *81 F4 Bel.* Dnyapro, *Rus.* Dnepr,
Ukr. Dnipro. *river* E Europe
Dnieper Lowland *109 E2 Bel.*
Prydnyaprowskaya Nizina, *Ukr.*
Prydniprovs'ka Nyzovyna. *lowlands*
Belarus/Ukraine
Dniester *81 E4 Rom.* Nistru, *Rus.*
Dnestr, *Ukr.* Dnister; *anc.* Tyras. *river*
Moldova/Ukraine
Dnipro *see* Dnieper
Dniprodzerzhyns'k *see* Romaniv
Dniprodzerzhyns'ke Vodoskhovyshche
109 F3 Rus. Dneprodzerzhinskoye
Vodokhranilishche. *reservoir*
C Ukraine
Dnipropetrovs'k *109 F3 Rus.*
Dnepropetrovsk; *prev.* Yekaterinoslav.
Dnipropetrovs'ka Oblast', E Ukraine
Dniprorudne *109 F3 Rus.*
Dneprorudnoye. Zaporiz'ka Oblast',
SE Ukraine
Dnister *see* Dniester
Dnyapro *see* Dnieper
Doba *76 C4* Logone-Oriental, S Chad
Döbeln *94 D4* Sachsen, E Germany
Doberai Peninsula *139 G4 Dut.*
Vogelkop. *peninsula* Papua,
E Indonesia
Doboj *100 C3* Republiks Srpska,
N Bosnia and Herzegovina
Dobre Miasto *98 D2 Ger.* Guttstadt.
Warmińsko-mazurskie, NE Poland
Dobrich *104 E1 Rom.* Bazargic; *prev.*
Tolbukhin. Dobrich, NE Bulgaria

Dobrush *107 D7* Homyel'skaya
Voblasts', SE Belarus
Dobryn' *see* Dabryn'
Dodecanese *105 D6 var.* Nóties
Sporádes, *Eng.* Dodecanese; *prev.*
Dhodhekánisos, Dodekanisos. *island
group* SE Greece
Dodecanese *see* Dodekánisa
Dodekanisos *see* Dodekánisa
Dodge City *45 E5* Kansas, C USA
Dodoma *69 D5 country capital*
(Tanzania) Dodoma, C Tanzania
Dogana *96 E1* NE San Marino Europe
Dogo *131 B6 island* Oki-shotō,
SW Japan
Dogondoutchi *75 F3* Dosso, SW Niger
Dogrular *see* Pravda
Doğubayazıt *117 F3* Ağrı, E Turkey
Doğu Karadeniz Dağları *117 E3 var.*
Anadolu Dağları. *mountain range*
NE Turkey
Doha *see* Ad Dawḥah
Doire *see* Londonderry
Dokdo *see* Liancourt Rocks
Dokkum *86 D1* Fryslân, N Netherlands
Dokuchayevs'k *109 G3 var.*
Dokuchayevsk. Donets'ka Oblast',
SE Ukraine
Dokuchayevsk *see* Dokuchayevs'k
Doldrums Fracture Zone *66 C4 fracture
zone* W Atlantic Ocean
Dôle *90 D4* Jura, E France
Dolina *see* Dolyna
Dolinskaya *see* Dolyns'ka
Dolisie *77 B6 prev.* Loubomo. Niari,
S Congo
Dolna Oryakhovitsa *104 D2 prev.*
Polikrayshte. Veliko Tŭrnovo,
N Bulgaria
Dolní Chlifik *104 E2 prev.* Rudnik.
Varna, E Bulgaria
Dolomites *96 C1 var.* Dolomiti, *Eng.*
Dolomites. *mountain range* NE Italy
Dolomites/Dolomiti *see* Dolomitiche,
Alpi
Dolores *64 D4* Buenos Aires, E Argentina
Dolores *52 B1* Petén, N Guatemala
Dolores *64 D4* Soriano, SW Uruguay
Dolores Hidalgo *51 E4 var.* Ciudad
de Dolores Hidalgo. Guanajuato,
C Mexico
Dolyna *108 B2 Rus.* Dolina. Ivano-
Frankivs'ka Oblast', W Ukraine
Dolyns'ka *109 F3 Rus.* Dolinskaya.
Kirovohrads'ka Oblast', S Ukraine
Domachëvo/Domaczewo *see* Damachava
Dombås *85 B5* Oppland, S Norway
Domel Island *see* Letsôk-aw Kyun
Domesnes, Cape *see* Kolkasrags
Domeyko *64 B3* Atacama, N Chile
Dominica *55 H4 off.* Commonwealth of
Dominica. *country* E West Indies

Dominica Channel *see* Martinique
Passage
Dominica, Commonwealth of *see*
Dominica
Dominican Republic *55 E2 country*
C West Indies

Domokós *105 B5 var.* Dhomokós. Stereá
Elláda, C Greece
Don *111 B6 var.* Duna, Tanais. *river*
SW Russian Federation
Donau *see* Danube
Donauwörth *95 C6* Bayern, S Germany
Don Benito *92 C3* Extremadura,
W Spain
Doncaster *89 D5 anc.* Danum.
N England, United Kingdom
Dondo *78 B1* Cuanza Norte, NW Angola
Donegal *89 B5 Ir.* Dún na nGall.
Donegal, NW Ireland
Donegal Bay *89 A5 Ir.* Bá Dhún na nGall.
bay NW Ireland
Donets *109 G2 river* Russian Federation/
Ukraine
Donets'k *109 G3 Rus.* Donetsk; *prev.*
Stalino. Donets'ka Oblast', E Ukraine
Dongfang *128 B7 var.* Basuo. Hainan,
S China
Dongguan *128 C6* Guangdong, S China
Đông Ha *136 E4* Quang Tri, C Vietnam
Dong Hai *see* East China Sea
Đông Hoi *136 D4* Quang Binh,
C Vietnam
Dongliao *see* Liaoyuan
Dongola *72 B3 var.* Donqola, Dunqulah.
Northern, N Sudan
Dongou *77 C5* Likouala, NE Congo
Dong Rak, Phanom *see* Dângrêk, Chuŏr
Phnum
Dongting Hu *128 C5 var.* Tung-t'ing
Hu. *lake* S China
Donostia *93 E1* País Vasco, N Spain *see
also* San Sebastián
Donqola *see* Dongola
Doolow *73 D5* Sumalē, E Ethiopia
Doornik *see* Tournai
Door Peninsula *40 C2 peninsula*
Wisconsin, N USA
Dooxo Nugaaleed *73 E5 var.* Nogal
Valley. *valley* E Somalia
Dordogne *91 B5 cultural region*
SW France
Dordogne *91 B5 river* W France
Dordrecht *86 C4 var.* Dordt, Dort.
Zuid-Holland, SW Netherlands
Dordt *see* Dordrecht
Dorohoi *108 C3* Botoşani, NE Romania
Dorotea *84 C4* Västerbotten, N Sweden
Dorpat *see* Tartu
Dorre Island *147 A5 island* Western
Australia
Dort *see* Dordrecht
Dortmund *94 A4* Nordrhein-Westfalen,
W Germany
Dos Hermanas *92 C4* Andalucía, S Spain
Dospad Dagh *see* Rhodope Mountains
Dospat *104 C3* Smolyan, S Bulgaria
Dothan *42 D3* Alabama, S USA
Dotnuva *106 B4* Kaunas, C Lithuania
Douai *90 C2 prev.* Douay; *anc.* Duacum.
Nord, N France
Douala *77 A5 var.* Duala. Littoral,
W Cameroon
Douglas *89 C5 dependent territory
capital* (Isle of Man) E Isle of Man
Douglas *43 C3* Arizona, SW USA
Douglas *44 D3* Wyoming, C USA
Douma *see* Dūmā
Douro *see* Duero
Douvres *see* Dover
Dover *89 E7 Fr.* Douvres, *Lat.* Dubris
Portus. SE England, United Kingdom
Dover *41 F4 state capital* Delaware,
NE USA
Dover, Strait of *90 C2 var.* Straits of
Dover, Fr. Pas de Calais. *strait* England,
United Kingdom/France
Dover, Straits of *see* Dover, Strait of
Dovrefjell *85 B5 plateau* S Norway

Downpatrick *89 B5 Ir.* Dún Pádraig.
SE Northern Ireland,
United Kingdom
Dozen *131 B6 island* Oki-shotō,
SW Japan
Dràa, Hammada du *see* Dra, Hamad
Drač/Draç *see* Durrës
Drachten *86 D2* Fryslân, N Netherla
Drăgăşani *108 B5* Vâlcea, SW Roma
Dragoman *104 B2* Sofiya, W Bulgari
Dra, Hamada du *70 C3 var.* Hamma
du Dràa, Haut Plateau du Dra. *pla*
W Algeria
Dra, Haut Plateau du *see* Dra,
Hamada du
Drahichyn *107 B6 Pol.* Drohiczyn
Poleski, *Rus.* Drogichin. Brestskay
Voblasts', SW Belarus
Drakensberg *78 D5 mountain range*
Lesotho/South Africa
Drake Passage *57 B8 passage* Atlant
Ocean/Pacific Ocean
Dralfa *104 D2* Tŭrgovishte, N Bulga
Dráma *104 C3 var.* Dhráma. Anatol
Makedonía kai Thráki, NE Greece
Dramburg *see* Drawsko Pomorskie
Drammen *85 B6* Buskerud, S Norwa
Drau *see* Drava
Drava *100 C3 var.* Drau, Eng. Drave
Hung. Dráva. *river* C Europe
Dráva/Drave *see* Drau/Drava
Drawsko Pomorskie *98 B3 Ger.*
Dramburg. Zachodnio-pomorskie
NW Poland
Drépano, Akrotírio *104 C4 var.*
Akrotírio Dhrepanon. *headland*
N Greece
Drepanum *see* Trapani
Dresden *94 D4* Sachsen, E Germany
Drin *see* Drinit, Lumi i
Drina *100 C3 river* Bosnia and
Herzegovina/Serbia
Drinit, Lumi i *101 D5 var.* Drin. *riv*
NW Albania
Drinit të Zi, Lumi i *see* Black Drin
Drissa *see* Drysa
Drobeta-Turnu Severin *108 B5
prev.* Turnu Severin. Mehedinţi,
SW Romania
Drogheda *89 B5 Ir.* Droichead Átha.
NE Ireland
Drogichin *see* Drahichyn
Drogobych *see* Drohobych
Drohiczyn Poleski *see* Drahichyn
Drohobych *108 B2 Pol.* Drohobycz,
Rus. Drogobych. L'vivs'ka Oblast',
NW Ukraine
Drohobycz *see* Drohobych
Droichead Átha *see* Drogheda
Drôme *91 D5 cultural region* E Franc
Dronning Maud Land *154 B2 physi*
region Antarctica
Drontheim *see* Trondheim
Drug *see* Durg
Druk-yul *see* Bhutan
Drummondville *39 E4* Québec,
SE Canada
Druskienniki *see* Druskininkai
Druskininkai *107 B5 Pol.* Druskienn
Alytus, S Lithuania
Dryden *38 B3* Ontario, C Canada
Drysa *107 D5 Rus.* Drissa. *river*
N Belarus
Duacum *see* Douai
Duala *see* Douala
Dubai *see* Dubayy
Dubăsari *108 D3 Rus.* Dubossary.
NE Moldova
Dubawnt *37 F4 river* Nunavut,
NW Canada
Dubayy *120 D4 Eng.* Dubai. Dubayy
NE United Arab Emirates
Dubbo *149 D6* New South Wales,
SE Australia
Dublin *89 B5 Ir.* Baile Átha Cliath; *an*
Eblana. *country capital* (Ireland)
Dublin, E Ireland
Dublin *43 E2* Georgia, SE USA
Dubno *108 C2* Rivnens'ka Oblast',
NW Ukraine
Dubossary *see* Dubăsari
Dubris Portus *see* Dover
Dubrovnik *101 B5 It.* Ragusa.
Dubrovnik-Neretva, SE Croatia
Dubuque *45 G3* Iowa, C USA
Dudelange *87 D8 var.* Forge du Sud,
Ger. Dudelingen. Luxembourg,
S Luxembourg
Dudelingen *see* Dudelange

o 92 D2 Port. Douro. *river*
tugal/Spain
eldorf *see* Düsseldorf
l 87 C5 Antwerpen, C Belgium
Otok 100 A4 var. Isola Grossa, It.
a Lunga. *island* W Croatia
ekerke *see* Dunkerque
urg 94 A4 prev. Duisburg-
nborn. Nordrhein-Westfalen,
Germany
urg-Hamborn *see* Duisburg
n 86 D4 Gelderland, E Netherlands
aiwil 73 B5 Jonglei, E South Sudan
a 126 D4 var. Qagan Us.
hina
, Golfo 53 E5 *gulf* S Costa Rica
, Golfo *see* Izabal, Lago de
en 94 A4 Nordrhein-Westfalen,
Germany
vo 104 E1 Silistra, NE Bulgaria
h 45 G2 Minnesota, N USA
i 119 B5 Fr. Douma. Rif Dimashq,
Syria
as 49 E1 Texas, SW USA
ries 88 C4 S Scotland, United
gdom
ont d'Urville 154 C4 French
arch station Antarctica
yât 72 B1 var. Dumyât, Eng.
mietta. N Egypt
see Danube, C Europe
see Western Dvina
see Don, Russian Federation
burg *see* Daugavpils
see Wien, Austria
see Danube, C Europe
pentele *see* Dunaújváros
rea *see* Danube
újváros 99 C7 prev. Dunapentele,
dinváros. Fejér, C Hungary
v *see* Danube
an 49 G2 Oklahoma, C USA
alk 89 B5 Ir. Dún Dealgan. Louth,
Ireland
Dealgan *see* Dundalk
lee 78 D4 KwaZulu/Natal, E South
rica
lee 88 C4 E Scotland, United
gdom
din 151 B7 Otago, South Island,
w Zealand
ermline 88 C4 C Scotland, United
gdom
u 77 E5 Orientale, NE Dem. Rep.
ngo
an 138 B3 var. Kuala Dungun.
engganu, Peninsular Malaysia
olme *see* Durham
erque 90 C2 Eng. Dunkirk, Flem.
inkerke; prev. Dunquerque. Nord,
rance
irk *see* Dunkerque
Laoghaire 89 B6 Eng. Dunleary;
v. Kingstown. E Ireland
eary *see* Dún Laoghaire
Pádraig *see* Downpatrick
uerque *see* Dunkerque
ulah *see* Dongola
itsa 104 C2 prev. Marek, Stanke
mitrov. Kyustendil, W Bulgaria
a 121 E5 var. Daqm. E Oman
nce 91 D6 river SE France
ngo 50 D3 var. Victoria de
rango. Durango, N Mexico
ngo 44 C5 Colorado, C USA
nkulak 104 E1 Rom. Răcari;
v. Blatnitsa, Duranulac. Dobrich,
Bulgaria
nt 49 G2 Oklahoma, C USA
nulac *see* Durankulak
zo *see* Durrës
an 78 D4 var. Port Natal. KwaZulu/
tal, E South Africa
e 106 B3 Ger. Durben. W Latvia
en *see* Durbe
135 E4 prev. Drug. Chhattisgarh,
ndia
am 89 D5 hist. Dunholme.
ngland, United Kingdom
am 43 F1 North Carolina,
USA
cortorum *see* Reims
storum *see* Silistra
vernum *see* Canterbury
s 101 C6 var. Durrësi, Dursi,
Durazzo, SCr. Drač, Turk. Draç.
rrës, W Albania
si *see* Durrës
see Durrës

Durüz, Jabal ad 119 C5 mountain
SW Syria
D'Urville Island 150 C4 island C New
Zealand
Dusa Mareb/Dusa Marreb *see* Dhuusa
Marreeb
Dushanbe 123 E3 var. Dyushambe;
prev. Stalinabad, Taj. Stalinobod.
country capital (Tajikistan)
W Tajikistan
Düsseldorf 94 A4 var. Duesseldorf.
Nordrhein-Westfalen, W Germany
Düstī 123 E3 Rus. Dusti.
Dutch East Indies *see* Indonesia
Dutch Guiana *see* Suriname
Dutch Harbor 36 B3 Unalaska Island,
Alaska, USA
Dutch New Guinea *see* Papua
Duzdab *see* Zāhedān
Dvina Bay *see* Chëshskaya Guba
Dvinsk *see* Daugavpils
Dyanev *see* Galkynyş
Dyersburg 42 C1 Tennessee, S USA
Dyushambe *see* Dushanbe
Dza Chu *see* Mekong
Dzaudzhikau *see* Vladikavkaz
Dzerzhinsk 111 C5 Nizhegorodskaya
Oblast', W Russian Federation
Dzerzhinskiy *see* Nar'yan-Mar
Dzhalal-Abad 123 F2 Kir. Jalal-
Abad. Dzhalal-Abadskaya Oblast',
W Kyrgyzstan
Dzhambul *see* Taraz
Dzhankoy 109 F4 Avtonomna
Respublika Krym, S Ukraine
Dzharkurgan *see* Jarqo'rg'on
Dzhelandy 123 F3 SE Tajikistan
Dzhergalan 123 G2 Kir. Jyrgalan. Issyk-
Kul'skaya Oblast', NE Kyrgyzstan
Dzhezkazgan *see* Zhezkazgan
Dzhizak *see* Jizzax
Dzhugdzhur, Khrebet 115 G3 mountain
range E Russian Federation
Dzhusaly *see* Zhosaly
Działdowo 98 D3 Warmińsko-
Mazurskie, C Poland
Dzuunmod 127 E2 Töv, C Mongolia
Dzüün Soyonï Nuruu *see* Vostochnyy
Sayan
Dzvina *see* Western Dvina

E

Eagle Pass 49 F4 Texas, SW USA
East Açores Fracture Zone *see* East
Azores Fracture Zone
East Antarctica 154 C3 prev. Greater
Antarctica. physical region Antarctica
East Australian Basin *see* Tasman Basin
East Azores Fracture Zone 66 C3 var.
East Açores Fracture Zone. tectonic
feature E Atlantic Ocean
Eastbourne 89 E7 SE England, United
Kingdom
East Cape 150 E3 headland North Island,
New Zealand
East China Sea 125 E2 Chin. Dong Hai.
sea W Pacific Ocean
Easter Fracture Zone 153 G4 tectonic
feature E Pacific Ocean
Easter Island 153 F4 var. Rapa Nui,
Easter Island. island E Pacific Ocean
Eastern Desert 68 B2 var. Aṣ Ṣaḥrā'
ash Sharqīyah, Eng. Arabian Desert,
Eastern Desert. desert E Egypt
Eastern Desert *see* Sahara el Sharqīya
Eastern Ghats 135 E3 mountain range
SE India
Eastern Sayans *see* Vostochnyy Sayan
Eastern Sierra Madre *see* Madre Oriental,
Sierra
East Falkland 65 D8 var. Isla Soledad.
island E Falkland Islands
East Frisian Islands 94 A3 Eng.
East Frisian Islands. island group
NW Germany
East Frisian Islands *see* Ostfriesische
Inseln
East Grand Forks 45 E1 Minnesota,
N USA
East Indiaman Ridge 141 D6 undersea
ridge E Indian Ocean
East Indies 152 A3 island group SE Asia
East Kilbride 88 C4 S Scotland, United
Kingdom
East Korea Bay 129 E3 bay N North
Korea

Eastleigh 89 D7 S England,
United Kingdom
East London 78 D5 Afr. Oos-Londen;
prev. Emonti, Port Rex. Eastern Cape,
S South Africa
Eastmain 38 D3 river Québec, C Canada
East Mariana Basin 142 B1 undersea
basin W Pacific Ocean
East Novaya Zemlya Trough 112 C1
var. Novaya Zemlya Trough. trough
W Kara Sea
East Pacific Rise 153 F4 undersea rise
E Pacific Ocean
East Pakistan *see* Bangladesh
East Saint Louis 40 B4 Illinois, N USA
East Scotia Basin 67 C7 undersea basin
SE Scotia Sea
East Sea 130 A4 var. Sea of Japan, Rus.
Yapanskoye More. Sea NW Pacific
Ocean
East Siberian Sea 115 F1 Eng. East
Siberian Sea. sea Arctic Ocean
East Siberian Sea *see* Vostochno-
Sibirskoye More
East Timor 139 F5 var. Loro Sae; prev.
Portuguese Timor, Timor Timur.
country S Indonesia

EAST TIMOR
Southeast Asia

Official name Democratic Republic of
Timor-Leste
Formation 2002 / 2002
Capital Dili
Population 1.1 million / 195 people
per sq mile (75 people per sq km)
Total area 5756 sq. miles (14,874 sq. km)
Languages Tetum* (Portuguese/
Austronesian), Bahasa Indonesia,
Portuguese*
Religions Roman Catholic 95%,
Other 5%
Ethnic mix Papuan groups approx. 85%,
Indonesian groups approx. 13%,
Chinese 2%
Government Parliamentary system
Currency US dollar = 100 cents
Literacy rate 58%
Calorie consumption 2083 kilocalories

Eau Claire 40 A2 Wisconsin, N USA
Eau Claire, Lac à L' *see* St. Clair, Lake
Eauripik Rise 142 B2 undersea rise
W Pacific Ocean
Ebensee 95 D6 Oberösterreich, N Austria
Eberswalde-Finow 94 D3 Brandenburg,
E Germany
Ebetsu 130 D2 var. Ebetu. Hokkaidō,
NE Japan
Ebetu *see* Ebetsu
Eblana *see* Dublin
Ebolowa 77 A5 Sud, S Cameroon
Ebon Atoll 144 D2 var. Epoon. atoll
Ralik Chain, S Marshall Islands
Ebora *see* Évora
Eboracum *see* York
Ebro 93 E2 river NE Spain
Ebusus *see* Eivissa
Ebusus *see* Eivissa
Ecbatana *see* Hamadān
Ech Cheliff/Ech Chleff *see* Chlef
Echo Bay 37 E3 Northwest Territories,
NW Canada
Echt 87 D5 Limburg, SE Netherlands
Ecija 93 E3 Rus. Dusti. Andalucía,
SW Spain
Ecuador 60 B1 off. Republic of Ecuador.
country NW South America

ECUADOR
South America

ECUADOR
(continued)

Government Presidential system
Currency US dollar = 100 cents
Literacy rate 93%
Calorie consumption 2477 kilocalories

Ecuador, Republic of *see* Ecuador
Ed Da'ein 72 A4 Southern Darfur,
W Sudan
Ed Damazin 72 C4 var. Ad Damazin.
Blue Nile, E Sudan
Ed Damer 72 C3 var. Ad Dāmir, Ad
Damar. River Nile, NE Sudan
Ed Debba 72 B3 Northern, N Sudan
Ede 86 D4 Gelderland, C Netherlands
Ede 75 F5 Osun, SW Nigeria
Edéa 77 A5 Littoral, SW Cameroon
Edessa *see* Şanlıurfa
Edfu *see* Idfū
Edgeøya 83 G2 island S Svalbard
Edgware 89 A7 Harrow, SE England,
United Kingdom
Edinburg 49 G5 Texas, SW USA
Edinburgh 88 C4 national capital
S Scotland, United Kingdom
Edingen *see* Enghien
Edirne 116 A2 Eng. Adrianople; anc.
Adrianopolis, Hadrianopolis. Edirne,
NW Turkey
Edmonds 46 B2 Washington,
NW USA
Edmonton 37 E5 province capital
Alberta, SW Canada
Edmundston 39 E4 New Brunswick,
SE Canada
Edna 49 G4 Texas, SW USA
Edolo 96 B1 Lombardia, N Italy
Edremit 116 A3 Balıkesir, NW Turkey
Edward, Lake 77 E5 var. Albert Edward
Nyanza, Edward Nyanza, Lac Idi
Amin, Lake Rutanzige. lake Uganda/
Dem. Rep. Congo
Edward Nyanza *see* Edward, Lake
Edwards Plateau 49 F3 plain Texas,
SW USA
Edzo 53 E4 prev. Rae-Edzo. Northwest
Territories, NW Canada
Eeklo 87 B5 var. Eekloo. Oost-
Vlaanderen, NW Belgium
Eekloo *see* Eeklo
Eersel 87 C5 Noord-Brabant,
S Netherlands
Eesti Vabariik *see* Estonia
Efate 144 D4 var. Efate, Fr. Vaté; prev.
Sandwich Island. island C Vanuatu
Efate *see* Efate
Effingham 40 B4 Illinois, N USA
Eforie-Sud 108 D5 Constanţa,
E Romania
Egadi Is. 97 B7 island group S Italy
Ege Denizi *see* Aegean Sea
Eger 99 D6 Ger. Erlau. Heves,
NE Hungary
Eger *see* Cheb, Czech Republic
Egeria Fracture Zone 141 C5 tectonic
feature W Indian Ocean
Éghezèe 87 C6 Namur, C Belgium
Egio *see* Aígio
Egmont, Cape 150 C4 headland North
Island, New Zealand
Egmont *see* Taranaki, Mount
Egoli *see* Johannesburg
Egypt 72 B2 off. Arab Republic of Egypt,
Ar. Jumhūrīyah Miṣr al 'Arabīyah, prev.
United Arab Republic; anc. Aegyptus.
country N Africa

EGYPT

Official name Arab Republic of Egypt
Formation 1936 / 1982
Capital Cairo
Population 82.1 million / 214 people
per sq mile (83 people per sq km)
Total area 386,660 sq. miles
(1,001,450 sq. km)
Languages Arabic*, French,
English, Berber
Religions Muslim (mainly Sunni) 90%,
Coptic Christian and other 10%
Ethnic mix Egyptian 99%, Nubian,
Armenian, Greek, and Berber 1%
Government Transitional regime

EGYPT
(continued)

Currency Egyptian pound = 100 piastres
Literacy rate 74%
Calorie consumption 3557 kilocalories

Eibar *93 E1* País Vasco, N Spain
Eibergen *86 E3* Gelderland,
  E Netherlands
Eidfjord *85 A5* Hordaland, S Norway
Eier-Berg *see* Suur Munamägi
Eifel *95 A5 plateau* W Germany
Eiger *95 B7 mountain* C Switzerland
Eigg *88 B3 island* W Scotland, United
  Kingdom
Eight Degree Channel *132 B3 channel*
  India/Maldives
Eighty Mile Beach *146 B4 beach*
  Western Australia
Eijsden *87 D6* Limburg, SE Netherlands
Eilat *see* Elat
Eindhoven *87 D5* Noord-Brabant,
  S Netherlands
Eipel *see* Ipel'
Éire *see* Ireland
Éireann, Muir *see* Irish Sea
Eisenhüttenstadt *94 D4* Brandenburg,
  E Germany
Eisenmarkt *see* Hunedoara
Eisenstadt *95 E6* Burgenland, E Austria
Eisleben *94 C4* Sachsen-Anhalt,
  C Germany
Eivissa *93 G3 var. Iviza, Cast.*
  Ibiza; *anc.* Ebusus. Ibiza, Spain,
  W Mediterranean Sea
Eivissa *93 G3 var. Iviza, Cast.*
  Ibiza; *anc.* Ebusus. Ibiza, Spain,
  W Mediterranean Sea
Eivissa *93 G3 var. Iviza, Cast.*
  Ibiza; *anc.* Ebusus. Ibiza, Spain,
  W Mediterranean Sea
Ejea de los Caballeros *93 E2* Aragón,
  NE Spain
Ejin Qi *see* Dalain Hob
Ekapa *see* Cape Town
Ekaterinodar *see* Krasnodar
Ekiatapskiy Khrebet *see* Ekvyvatapskiy
  Khrebet
Ekvyvatapskiy Khrebet *115 G1 prev.*
  Ekiatapskiy Khrebet. *mountain range*
  NE Russian Federation
El 'Alamein *see* Al 'Alamayn
El Asnam *see* Chlef
Elat *119 B8 var.* Eilat, Elath. Southern,
  S Israel
Elat, Gulf of *see* Aqaba, Gulf of
Elath *see* Elat, Israel
Elath *see* Al 'Aqabah, Jordan
El'Atrun *72 B3* Northern Darfur,
  NW Sudan
Elâziğ *117 E3 var.* Elâzig, Elâziz. Elâzığ,
  E Turkey
Elba *96 B4 island* Archipelago Toscano,
  C Italy
Elbasan *101 D6 var.* Elbasani. Elbasan,
  C Albania
Elbasani *see* Elbasan
Elbe *80 D3 Cz.* Labe. *river* Czech
  Republic/Germany
Elbert, Mount *144 C4 mountain*
  Colorado, C USA
Elbing *see* Elblag
Elblag *98 C2 var.* Elblag, *Ger.* Elbing.
  Warmińsko-Mazurskie, NE Poland
El Boulaïda/El Boulaïda *see* Blida
El'brus *111 A8 var.* Gora El'brus.
  mountain SW Russian Federation
El'brus, Gora *see* El'brus
El Burgo de Osma *93 E2* Castilla y
  León, C Spain
Elburz Mountains *120 C2 Eng.* Elburz
  Mountains. *mountain range* N Iran
Elburz Mountains *see* Alborz,
  Reshteh-ye Kühhā-ye
El Cajon *47 C8* California, W USA
El Calafate *65 B7 var.* Calafate. Santa
  Cruz, S Argentina
El Callao *59 E2* Bolívar, E Venezuela
El Campo *49 G4* Texas, SW USA
El Carmen de Bolívar *58 B2* Bolívar,
  NW Colombia
El Cayo *see* San Ignacio
El Centro *47 D8* California, W USA
Elche *93 F4 var.* Elx; *anc.* Ilici, *Lat.* Illicis.
  Valenciana, E Spain
Elda *93 F4* Valenciana, E Spain
El Djazaïr *see* Alger

El Djelfa *see* Djelfa
Eldorado *64 E3* Misiones, NE Argentina
El Dorado *50 C3* Sinaloa, C Mexico
El Dorado *42 B2* Arkansas, C USA
El Dorado *45 F5* Kansas, C USA
El Dorado *59 F2* Bolívar, E Venezuela
El Dorado *see* California
Eldoret *73 C6* Rift Valley, W Kenya
Elektrostal' *111 B5* Moskovskaya
  Oblast', W Russian Federation
Elemi Triangle *73 B5 disputed region*
  Kenya/Sudan
Elephant Butte Reservoir *48 C2*
  reservoir New Mexico, SW USA
Ėlesd *see* Aleşd
Eleuthera Island *54 C1 island*
  N Bahamas
El Fasher *72 A4 var.* Al Fāshir. Northern
  Darfur, W Sudan
El Ferrol/El Ferrol del Caudillo *see*
  Ferrol
El Gedaref *see* Gedaref
El Geneina *72 A4 var.* Ajjinena, Al-
  Genain, Al Junaynah. Western Darfur,
  W Sudan
Elgin *88 C3* NE Scotland, United
  Kingdom
Elgin *40 B3* Illinois, N USA
El Giza *see* Giza
El Goléa *70 D3 var.* Al Golea. C Algeria
El Ḥank *74 D1 cliff* N Mauritania
El Haseke *see* Al Ḥasakah
Elimberrum *see* Auch
Eliocroca *see* Lorca
Élisabethville *see* Lubumbashi
Elista *111 B7* Respublika Kalmykiya,
  SW Russian Federation
Elizabeth *149 B6* South Australia
Elizabeth City *43 G1* North Carolina,
  SE USA
Elizabethtown *40 C5* Kentucky, S USA
El-Jadida *70 C2 prev.* Mazagan.
  W Morocco
Elk *98 E2 Ger.* Lyck. Warmińsko-
  mazurskie, NE Poland
Elk City *49 F1* Oklahoma, C USA
El Khalil *see* Hebron
El Khârga *see* Al Khārijah
Elkhart *40 C3* Indiana, N USA
El Khartûm *see* Khartoum
Elk River *45 F2* Minnesota, N USA
El Kuneitra *see* Al Qunayţirah
Ellás *see* Greece
Ellef Ringnes Island *37 E1 island*
  Nunavut, N Canada
Ellen, Mount *44 B5 mountain* Utah,
  W USA
Ellensburg *46 B2* Washington, NW USA
Ellesmere Island *37 F1 island* Queen
  Elizabeth Islands, Nunavut, N Canada
Ellesmere, Lake *151 C6 lake* South
  Island, New Zealand
Ellice Islands *see* Tuvalu
Elliston *149 A6* South Australia
Ellsworth Land *154 A3 physical region*
  Antarctica
El Mahbas *70 B3 var.* Mahbés.
  SW Western Sahara
El Mina *118 B4 var.* Al Mīnā'.
  N Lebanon
El Minya *see* Al Minyā
Elmira *41 E3* New York, NE USA
El Mreyyé *74 D2 desert* E Mauritania
Elmshorn *94 B3* Schleswig-Holstein,
  N Germany
El Muglad *72 A4* Southern Kordofan,
  C Sudan
El Obeid *72 B4 var.* Al Obayyid, Al
  Ubayyiḍ. Northern Kordofan, C Sudan
El Ouâdi *see* El Oued
El Oued *71 E2 var.* Al Oued, El Ouâdi,
  El Wad. NE Algeria
Eloy *48 B2* Arizona, SW USA
El Paso *48 D3* Texas, SW USA
El Porvenir *53 G4* Kuna Yala,
  N Panama
El Progreso *52 C2* Yoro, NW Honduras
El Puerto de Santa María *92 C5*
  Andalucía, S Spain
El Qâhira *see* Cairo
El Quneitra *see* Al Qunayţirah
El Quseir *see* Al Quşayr
El Quweira *see* Al Quwayrah
El Rama *53 E3* Región Autónoma
  Atlántico Sur, SE Nicaragua
El Real *53 H5 var.* El Real de Santa María.
  Darién, SE Panama
El Real de Santa María *see* El Real
El Reno *49 F1* Oklahoma, C USA

El Salvador *52 B3 off.* Republica de El
  Salvador. *country* Central America

EL SALVADOR
Central America

Official name Republic of El Salvador
Formation 1841 / 1841
Capital San Salvador
Population 6.3 million / 788 people
  per sq mile (304 people per sq km)
Total area 8124 sq. miles (21,040 sq. km)
Languages Spanish*
Religions Roman Catholic 80%,
  Evangelical 18%, Other 2%
Ethnic mix Mestizo 90%,
  White 9%, Amerindian 1%
Government Presidential system
Currency Salvadoran colón =
  100 centavos; US dollar = 100 cents
Literacy rate 86%
Calorie consumption 2513 kilocalories

El Salvador, Republica de *see* El Salvador
Elsass *see* Alsace
El Sáuz *50 C2* Chihuahua, N Mexico
El Serrat *91 A7* N Andorra Europe
Elst *86 D4* Gelderland, E Netherlands
El Sueco *50 C2* Chihuahua, N Mexico
El Suweida *see* As Suwaydā'
El Suweis *see* Suez
Eltanin Fracture Zone *153 E5 tectonic*
  feature SE Pacific Ocean
El Tigre *59 E2* Anzoátegui, NE Venezuela
Elvas *92 C4* Portalegre, C Portugal
El Vendrell *93 G2* Cataluña, NE Spain
El Vígía *58 C2* Mérida, NW Venezuela
El Wad *see* El Oued
Elwell, Lake *44 B1 reservoir* Montana,
  NW USA
Elx *see* Elche
Ely *45 D5* Nevada, W USA
El Yopal *see* Yopal
Emajõgi *106 D3 Ger.* Embach. *river*
  SE Estonia
Emāmrūd *see* Shāhrūd
Emāmshahr *see* Shāhrūd
Emba *114 B4 Kaz.* Embi. Aktyubinsk,
  W Kazakhstan
Embach *see* Emajõgi
Embi *see* Emba
Emden *94 A3* Niedersachsen,
  NW Germany
Emerald *148 D4* Queensland, E Australia
Emerald Isle *see* Montserrat
Emesa *see* Ḥimş
Emmaste *106 C2* Hiiumaa, W Estonia
Emmeloord *86 D2* Flevoland,
  N Netherlands
Emmen *86 E2* Drenthe, NE Netherlands
Emmendingen *95 A6* Baden-
  Württemberg, SW Germany
Emona *see* Ljubljana
Emonti *see* East London
Emory Peak *49 E4 mountain* Texas,
  SW USA
Empalme *50 B2* Sonora, NW Mexico
Emperor Seamounts *113 G3 seamount*
  range NW Pacific Ocean
Empire State of the South *see* Georgia
Emporia *45 F5* Kansas, C USA
Empty Quarter *see* Ar Rub' al Khālī
Ems *94 A3 Dut.* Eems. *river*
  NW Germany
Enareträsk *see* Inarijärvi
Encamp *91 A8* Encamp, C Andorra
  Europe
Encarnación *64 D3* Itapúa, S Paraguay
Encinitas *47 C8* California, W USA
Encs *99 D6* Borsod-Abaúj-Zemplén,
  NE Hungary
Endeavour Strait *148 C1 strait*
  Queensland, NE Australia
Enderbury Island *145 F3 atoll* Phoenix
  Islands, C Kiribati
Enderby Land *154 C2 physical region*
  Antarctica
Enderby Plain *154 D2 abyssal plain*
  S Indian Ocean
Endersdorf *see* Jędrzejów
Enewetak Atoll *144 C1 var.* Ānewetak,
  Eniwetok. *atoll* Ralik Chain,
  W Marshall Islands
Enfield *89 A7* United Kingdom
Engeten *see* Aiud
England *89 D5 Lat.* Anglia. *cultural*
  region United Kingdom, United
Enghien *87 B6 Dut.* Edingen. Hainaut,
  SW Belgium

Englewood *44 D4* Colorado, C USA
English Channel *89 D8 var.* The
  Channel, *Fr.* la Manche. *channel*
  NW Europe
Engure *106 C3* W Latvia
Engures Ezers *106 B3 lake* NW Latvia
Enguri *117 F1 Rus.* Inguri. *river*
  NW Georgia
Enid *49 F1* Oklahoma, C USA
Enikale Strait *see* Kerch Strait
Eniwetok *see* Enewetak Atoll
En Nâqoûra *119 A5 var.* An Nāqūrā.
  SW Lebanon
En Nazira *see* Natzrat
Ennedi *76 D2 plateau* E Chad
Ennis *89 A6 Ir.* Inis. Clare, W Ireland
Ennis *49 G3* Texas, SW USA
Enniskillen *89 B5 var.* Inniskilling.
  Inis Ceithleann. SW Northern Ireland
  United Kingdom
Enns *95 D6 river* C Austria
Enschede *86 E3* Overijssel, E Netherlands
Ensenada *50 A1* Baja California Norte
  NW Mexico
Entebbe *73 B6* S Uganda
Entroncamento *92 B3* Santarém,
  C Portugal
Enugu *75 G5* Enugu, S Nigeria
Epanomi *104 B4* Kentrikí Makedonía
  N Greece
Epéna *77 B5* Likouala, NE Congo
Eperies/Eperjes *see* Prešov
Epi *144 D4 var.* Épi. *island* C Vanuatu
Épi *see* Epi
Épinal *90 D4* Vosges, NE France
Epiphania *see* Ḥamāh
Epoon *see* Ebon Atoll
Epsom *89 A8* United Kingdom
Equality State *see* Wyoming
Equatorial Guinea *77 A5 off.* Equatorial
  Guinea, Republic of. *country* C Africa

EQUATORIAL GUINEA
Central Africa

Official name Republic of
  Equatorial Guinea
Formation 1968 / 1968
Capital Malabo
Population 800,000 / 74 people
  per sq mile (29 people per sq km)
Total area 10,830 sq. miles (28,051 sq
Languages Spanish*, Fang, Bubi, Fren
Religions Roman Catholic 90%,
  Other 10%
Ethnic mix Fang 85%, Other 11%, Bubi
Government Presidential system
Currency CFA franc = 100 centimes
Literacy rate 94%
Calorie consumption Not available

Equatorial Guinea, Republic of *see*
  Equatorial Guinea
Erautini *see* Johannesburg
Erbil *see* Arbīl
Erciş *117 F3* Van, E Turkey
Erdély *see* Transylvania
Erdélyi-Havasok *see* Carpaţii
  Meridionalii
Erdenet *127 E2* Orhon, N Mongolia
Erdi *76 C2 plateau* NE Chad
Erdi Ma *76 D2 desert* NE Chad
Erebus, Mount *154 B4 volcano* Ross
  Island, Antarctica
Ereğli *116 C4* Konya, S Turkey
Erenhot *127 F2 var.* Erlian. Nei Mongol
  Zizhiqu, NE China
Erfurt *94 C4* Thüringen, C Germany
Ergene Çayı *see* Ergene Irmaği
Ergene Irmaği *116 A2 var.* Ergene
  river NW Turkey
Ergun *see* Labudalin;
  prev. Ergun Youqi.
  Nei Mongol Zizhiqu, N China
Ergun He *see* Argun
Ergun Youqi *see* Ergun
Erie *40 D3* Pennsylvania, NE USA
Érié, Lac *see* Erie, Lake
Erie, Lake *40 D3 Fr.* Lac Érié. *lake*
  Canada/USA
Eritrea *72 C4 off.* State of Eritrea, f
  country E Africa

ERITREA
East Africa

Official name State of Eritrea
Formation 1993 / 2002
Capital Asmara

ITREA
(continued)

pulation 6.3 million / 139 people
r sq mile (54 people per sq km)
tal area 46,842 sq. miles (121,320 sq. km)
nguages Tigrinya*, English*, Tigre,
ar, Arabic*, Saho, Bilen, Kunama,
ra, Hadareb
ligions Christian 50%, Muslim 48%,
her 2%
nic mix Tigray 50%, Tigre 31%, Other
, Afar 5%, Saho 5%
vernment Presidential /
rliamentary system
rrency Nakfa = 100 cents
eracy rate 70%
lorie consumption 1640 kilocalories

rea, State of *see* Eritrea
an *see* Yerevan
angen 95 C5 Bayern, S Germany
au *see* Eger
an *see* Erenhot
elo 86 D3 Gelderland,
Netherlands
iíóni 105 C6 Pelopónnisos, S Greece
oúpoli 105 D6 *var.* Hermoupolis;
rev. Ermoúpolis. Sýyros, Kykládes,
reece, Aegean Sea
oúpolis *see* Ermoúpoli
akulam 132 C3 Kerala, SW India
de 132 C2 Tamil Nādu, SE India
uelinnes 87 B7 Hainaut, S Belgium
Rachidia 70 C2 *var.* Ksar al Soule.
Morocco
ahad 72 B4 *var.* Ar Rahad. Northern
ordofan, C Sudan
omango 144 D4 *island* S Vanuatu
s *see* Irtysh, C Asia
a *see* Eritrea
rum *see* Erzurum
ncan 117 E3 *var.* Erzinjan. Erzincan,
Turkey
njan *see* Erzincan
arum 117 E3 *prev.* Erzerum.
rzurum, NE Turkey
erg 85 A7 Syddtjylland, W Denmark
o *see* Espoo
aldes 91 A8 Escaldes Engordany,
Andorra Europe
anaba 40 C2 Michigan, N USA
aut *see* Scheldt
n-sur-Alzette 87 D8 Luxembourg,
Luxembourg
inda, Grand Lac des *see* Great
ave Lake
ondido 47 C8 California, W USA
iinapa 50 D3 *var.* Escuinapa de
lidalgo. Sinaloa, C Mexico
iinapa de Hidalgo *see* Escuinapa
iintla 52 B2 Escuintla, S Guatemala
iintla 51 G5 Chiapas, SE Mexico
aguly 122 B3 *Rus.* Gasan-Kuli.
alkan Welaýaty, W Turkmenistan
hān 120 C3 *Eng.* Isfahan; *anc.*
spadana. Eşfahān, C Iran
Sharā *see* Ash Sharāh
see Ishim,
azakhstan/Russian Federation
mo Point *see* Arviat
işehir 116 B3 *var.* Eskishehr.
skişehir, W Turkey
ishehr *see* Eskişehir
mābād *see* Eslāmābād-e Gharb
mābād-e Gharb 120 C3 *var.*
slāmābād; *prev.* Harunabad,
ahābād. Kermānshāhān, W Iran
eraldas 60 A1 Esmeraldas,
Ecuador
a *see* Isnā
aña *see* Spain
anola 48 D1 New Mexico, SW USA
erance 147 B7 Western Australia
eranza 50 B2 Sonora, NW Mexico
eranza 154 A2 *Argentinian research
ation* Antarctica
inal 58 B3 Tolima, C Colombia
nhaço, Serra do 56 D4 *mountain
ange* SE Brazil
rito Santo 63 F4 *off.* Estado do
spírito Santo. *state* E Brazil
rito Santo 63 F4 *off.* Estado do
spírito Santo. *region* E Brazil
ento
rito Santo, Estado do *see* Espírito
anto
ritu Santo 144 C4 *var.* Santo. *island*
r Vanuatu

Espoo 85 D6 *Swe.* Esbo. Etelä-Suomi,
S Finland
Esquel 65 B6 Chubut, SW Argentina
Essaouira 70 B2 *prev.* Mogador.
W Morocco
Esseg *see* Osijek
Es Semara *see* Smara
Essen 87 C5 Antwerpen, N Belgium
Essen 94 A4 *var.* Essen an der Ruhr.
Nordrhein-Westfalen, W Germany
Essen an der Ruhr *see* Essen
Essequíbo River 59 F3 *river* C Guyana
Es Suweida *see* As Suwaydā'
Estacado, Llano 49 E2 *plain* New
Mexico/Texas, SW USA
Estados, Isla de los 65 C8 *prev. Eng.*
Staten Island. *island* S Argentina
Estância 63 G3 Sergipe, E Brazil
Esteli 52 D3 Esteli, NW Nicaragua
Estella 93 E1 *Bas.* Lizarra. Navarra,
N Spain
Estepona 92 D5 Andalucía, S Spain
Estevan 37 F5 Saskatchewan, S Canada
Estland *see* Estonia
Estonia 106 D2 *off.* Republic of Estonia,
Est. Eesti Vabariik, *Ger.* Estland, *Latv.*
Igaunija; *prev.* Estonian SSR, *Rus.*
Estonskaya SSR. *country* NE Europe

ESTONIA
Northeast Europe

Official name Republic of Estonia
Formation 1991 / 1991
Capital Tallinn
Population 1.3 million / 75 people
per sq mile (29 people per sq km)
Total area 17,462 sq. miles (45,226 sq. km)
Languages Estonian*, Russian
Religions Evangelical Lutheran 14%,
Orthodox Christian 25%, Other 19%
Ethnic mix Estonian 69%, Russian 25%,
Other 4%, Ukrainian 2%
Government Parliamentary system
Currency Euro = 100 cents
Literacy rate 99%
Calorie consumption 3214 kilocalories

Estonian SSR *see* Estonia
Estonia, Republic of *see* Estonia
Estonskaya SSR *see* Estonia
Estrela, Serra de 92 C3 *mountain range*
C Portugal
Estremadura *see* Extremadura
Estremoz 92 C4 Évora, S Portugal
Eszék *see* Osijek
Esztergom 99 C6 *Ger.* Gran; *anc.*
Strigonium. Komárom-Esztergom,
N Hungary
Étalle 87 D8 Luxembourg, SE Belgium
Etāwah 134 D3 Uttar Pradesh, N India
Ethiopia 73 C5 *off.* Federal Democratic
Republic of Ethiopia; *prev.* Abyssinia,
People's Democratic Republic of
Ethiopia. *country* E Africa

ETHIOPIA
East Africa

Official name Federal Democratic
Republic of Ethiopia
Formation 1896 / 2002
Capital Addis Ababa
Population 94.1 million / 220 people
per sq mile (85 people per sq km)
Total area 435,184 sq. miles
(1,127,127 sq. km)
Languages Amharic*, Tigrinya, Galla,
Sidamo, Somali, English, Arabic
Religions Orthodox Christian 40%,
Muslim 40%, Traditional beliefs 15%,
Other 5%
Ethnic mix Oromo 40%, Amhara 25%,
Other 13%, Sidama 9%, Tigray 7%,
Somali 6%
Government Parliamentary system
Currency Ethiopian birr = 100 cents
Literacy rate 39%
Calorie consumption 2105 kilocalories

Ethiopia, Federal Democratic Republic
of *see* Ethiopia
Ethiopian Highlands 73 C5
var. Ethiopian Plateau. *plateau*
N Ethiopia
Ethiopian Plateau *see* Ethiopian
Highlands
Ethiopia, People's Democratic Republic
of *see* Ethiopia

Mount Etna 97 C7 *Eng.* Mount
Etna. *volcano* Sicilia, Italy,
C Mediterranean Sea
Etna, Mount *see* Etna, Monte
Etosha Pan 78 B3 *salt lake* N Namibia
Etoumbi 77 B5 Cuvette Ouest,
NW Congo
Etsch *see* Adige
Et Tafila *see* Aţ Ţafilah
Ettelbrück 87 D8 Diekirch,
C Luxembourg
'Eua 145 E5 *prev.* Middleburg Island.
island Tongatapu Group, SE Tonga
Euboea 105 C5 *Lat.* Euboea. *island*
C Greece
Euboea *see* Évvoia
Eucla 147 D6 Western Australia
Euclid 40 D3 Ohio, N USA
Eufaula Lake 49 G1 *var.* Eufaula
Reservoir. *reservoir* Oklahoma, C USA
Eufaula Reservoir *see* Eufaula Lake
Eugene 46 B3 Oregon, NW USA
Eumolpias *see* Plovdiv
Eupen 87 D6 Liège, E Belgium
Euphrates 112 B4 *var.* Al-Furāt, *Turk.*
Firat Nehri. *river* SW Asia
Eureka 47 A5 California, W USA
Eureka 44 A1 Montana, NW USA
Europa Point 93 H5 *headland*
S Gibraltar
Europe 80 *continent*
Eutin 94 C2 Schleswig-Holstein,
N Germany
Euxine Sea *see* Black Sea
Evanston 45 G3 Iowa, C USA
Evanston 44 B4 Wyoming, C USA
Evansville 40 B5 Indiana, N USA
Eveleth 45 G1 Minnesota, N USA
Everard, Lake 149 A6 *salt lake* South
Australia
Everest, Mount 126 B5 *Chin.*
Qomolangma Feng, *Nep.* Sagarmāthā.
mountain China/Nepal
Everett 46 B2 Washington, NW USA
Everglades, The 43 F5 *wetland* Florida,
SE USA
Evje 85 A6 Aust-Agder, S Norway
Evmolpia *see* Plovdiv
Évora 92 B4 *anc.* Ebora, *Lat.* Liberalitas
Eure, N France
Évreux 90 C3 *anc.* Civitas Eburovicum.
Eure, N France
Évros *see* Maritsa
Évry 90 E2 Essonne, N France
Ewarton 54 B5 C Jamaica
Excelsior Springs 45 F4 Missouri,
C USA
Exe 89 C7 *river* SW England, United
Kingdom
Exeter 89 C7 *anc.* Isca Damnoniorum.
SW England, United Kingdom
Exmoor 89 C7 *moorland* SW England,
United Kingdom
Exmouth 146 A4 Western Australia
Exmouth 89 C7 SW England, United
Kingdom
Exmouth Gulf 146 A4 *gulf* Western
Australia
Exmouth Plateau 141 E5 *undersea
plateau* N Indian Ocean
Extremadura 92 C3 *var.* Estremadura.
autonomous community W Spain
Exuma Cays 54 C1 *islets* C Bahamas
Exuma Sound 54 C1 *sound* C Bahamas
Eyre Mountains 151 A7 *mountain range*
South Island, New Zealand
Eyre North, Lake 149 A5 *salt lake* South
Australia
Eyre Peninsula 149 A6 *peninsula* South
Australia
Eyre South, Lake 149 A5 *salt lake* South
Australia
Ezo *see* Hokkaidō

Færoerne *see* Faroe Islands
Faetano 96 E2 E San Marino
Făgăraş 108 C4 *Ger.* Fogarasch, *Hung.*
Fogaras. Braşov, C Romania
Fagibina, Lake *see* Faguibine, Lac
Fagne 87 C7 *hill range* S Belgium
Faguibine, Lac 75 E3 *var.* Lake Fagibina.
lake NW Mali
Fahlun *see* Falun
Fahraj 120 E4 Kermān, SE Iran
Faial 92 A5 *var.* Ilha do Faial. *island*
Azores, Portugal, NE Atlantic Ocean
Faial, Ilha do *see* Faial
Faifo *see* Hôi An
Fairbanks 36 D3 Alaska, USA
Fairfield 47 B6 California, W USA
Fair Isle 88 D2 *island* NE Scotland,
United Kingdom
Fairlie 151 B6 Canterbury, South Island,
New Zealand
Fairmont 45 F3 Minnesota, N USA
Faisalābād 134 C2 *prev.* Lyallpur.
Punjab, NE Pakistan
Faizābād 135 E3 Uttar Pradesh, N India
Faizābād 123 F3 *var.* Feïzābād,
Feyzābād, Fyzabad; *prev.* Feyzābād.
Badakhshān, NE Afghanistan
Fakaofo Atoll 145 F3 *island* SE Tokelau
Falam 136 A3 Chin State, W Myanmar
(Burma)
Falconara Marittima 96 C3 Marche,
C Italy
Falkenau an der Eger *see* Sokolov
Falkland Islands 65 D7 *var.* Falklands,
Islas Malvinas. *UK dependent territory*
SW Atlantic Ocean
Falkland Plateau 57 D7 *var.* Argentine
Rise. *undersea feature* SW Atlantic
Ocean
Falklands *see* Falkland Islands
Falknov nad Ohří *see* Sokolov
Fallbrook 47 C8 California, W USA
Falmouth 54 A4 W Jamaica
Falmouth 89 C7 SW England, United
Kingdom
Falster 85 B8 *island* SE Denmark
Fălticeni 108 C3 *Hung.* Falticsén.
Suceava, NE Romania
Falticsén *see* Fălticeni
Falun 85 C6 *var.* Fahlun. Kopparberg,
C Sweden
Famagusta *see* Gazimağusa
Famagusta Bay 102 C5 *var.* Famagusta
Bay, *Gk.* Kólpos Ammóchostos. *bay*
E Cyprus
Famagusta Bay *see* Gazimağusa Körfezi
Famenne 87 C7 *physical region*
SE Belgium
Fang 136 C3 Chiang Mai, NW Thailand
Fanning Island *see* Tabuaeran
Fanø 96 G3 *island* W Denmark
Farafangana 79 G4 Fianarantsoa,
SE Madagascar
Farāh 122 D4 *var.* Farah, Fararud.
Farāh, W Afghanistan
Farah Rud 122 D4 *river* W Afghanistan
Faranah 74 C4 Haute-Guinée, S Guinea
Fararud *see* Farāh
Farasan, Jaza'ir 121 A6 *island group*
SW Saudi Arabia
Farewell, Cape 150 C4 *headland* South
Island, New Zealand
Farewell, Cape *see* Nunap Isua
Fargo 45 F2 North Dakota, N USA
Farg'ona 123 F3 *var.* Fergana; *prev.*
Novyy Margilan. Farg'ona Viloyati,
E Uzbekistan
Faribault 45 F2 Minnesota, C USA
Farīdābād 134 D3 Haryāna, N India
Farkhor 123 E3 *Rus.* Parkhar.
SW Tajikistan
Farmington 45 G5 Missouri, C USA
Farmington 48 C1 New Mexico,
SW USA
Faro 92 B5 Faro, S Portugal
Faroe-Iceland Ridge 80 C1 *undersea
ridge* N Norwegian Sea
Faroe Islands 83 E5 *var.* Faroes,
Dan. Færoerne, *Faer.* Føroyar. *Danish
external territory* N Atlantic Ocean
Faroe-Shetland Trough 80 C2 *trough*
NE Atlantic Ocean
Farquhar Group 79 G2 *island group*
S Seychelles
Faros, Khalīj-e *see* Persian Gulf
Farvel, Kap *see* Nunap Isua
Fastiv 109 E2 *Rus.* Fastov. Kyyivs'ka
Oblast', NW Ukraine
Fastov *see* Fastiv

F

Faadhippolhu Atoll 132 B4 *var.*
Fadiffolu, Lhaviyani Atoll. *atoll*
N Maldives
Fabens 48 D3 Texas, SW USA
Fada 76 C2 Borkou-Ennedi-Tibesti,
E Chad
Fada-Ngourma 75 E4 E Burkina Faso
Fadiffolu *see* Faadhippolhu Atoll
Faenza 96 C3 *anc.* Faventia. Emilia-
Romagna, N Italy
Faeroe Islands *see* Faroe Islands

Fauske *84 C3* Nordland, C Norway
Faventia *see* Faenza
Faxa Bay *see* Faxaflói
Faxaflói *82 D5 Eng.* Faxa Bay. *bay* W Iceland
Faya *76 C2 prev.* Faya-Largeau, Largeau. Borkou-Ennedi-Tibesti, N Chad
Faya-Largeau *see* Faya
Fayetteville *42 A1* Arkansas, C USA
Fayetteville *43 F1* North Carolina, SE USA
Fdérick *see* Fdérik
Fdérik *74 C2 var.* Fdérick, *Fr.* Fort Gouraud. Tiris Zemmour, NW Mauritania
Fear, Cape *43 F2 headland* Bald Head Island, North Carolina, SE USA
Fécamp *90 B3* Seine-Maritime, N France
Fédala *see* Mohammedia
Federal Capital Territory *see* Australian Capital Territory
Fehérgyarmat *99 E6* Szabolcs-Szatmár-Bereg, E Hungary
Fehértemplom *see* Bela Crkva
Fehmarn *94 C2 island* N Germany
Fehmarn Belt *94 C2 Dan.* Femern Bælt, *Ger.* Fehmarnbelt. *strait* Denmark / Germany
Fehmarnbelt Fehmarn Belt/Femer Bælt
Feijó *62 C2* Acre, W Brazil
Feilding *150 D4* Manawatu-Wanganui, North Island, New Zealand
Feira *see* Feira de Santana
Feira de Santana *63 G3 var.* Feira. Bahia, E Brazil
Feizābād *see* Faizābād
Feketehalom *see* Codlea
Felanitx *93 G3* Mallorca, Spain, W Mediterranean Sea
Felicitas Julia *see* Lisboa
Felidhu Atoll *132 B4 atoll* C Maldives
Felipe Carrillo Puerto *51 H4* Quintana Roo, SE Mexico
Felixstowe *89 E6* E England, United Kingdom
Fellin *see* Viljandi
Felsőbánya *see* Baia Sprie
Felsőmuzslya *see* Mužlja
Femunden *85 B5 lake* S Norway
Fénérive *see* Fenoarivo Atsinanana
Fengcheng *128 D3 var.* Feng-cheng, Fenghwangcheng. Liaoning, NE China
Feng-cheng *see* Fengcheng
Fenghwangcheng *see* Fengcheng
Fengtien *see* Liaoning, China
Fengtien *see* Shenyang, China
Fenoarivo Atsinanana *79 G3 prev./Fr.* Fénérive. Toamasina, E Madagascar
Fens, The *89 E6 wetland* E England, United Kingdom
Feodosiya *109 F5 var.* Kefe, *It.* Kaffa; *anc.* Theodosia. Avtonomna Respublika Krym, S Ukraine
Ferdinand *see* Montana, Bulgaria
Ferdinandsberg *see* Oţelu Roşu
Féres *104 D3* Anatolikí Makedonía kai Thráki, NE Greece
Fergana *see* Farg'ona
Fergus Falls *45 F2* Minnesota, N USA
Ferizaj *101 D5 Serb.* Uroševac. C Kosovo
Ferkessédougou *74 D4* N Côte d'Ivoire
Fermo *96 C4 anc.* Firmum Picenum. Marche, C Italy
Fernandina, Isla *60 A5 var.* Narborough Island. *island* Galápagos Islands, Ecuador, E Pacific Ocean
Fernando de Noronha *63 H2 island* E Brazil
Fernando Po/Fernando Póo *see* Bioco, Isla de
Ferrara *96 C2 anc.* Forum Alieni. Emilia-Romagna, N Italy
Ferreñafe *60 B3* Lambayeque, W Peru
Ferro *see* Hierro
Ferrol *92 B1 var.* El Ferrol; *prev.* El Ferrol del Caudillo. Galicia, NW Spain
Fertő *see* Neusiedler See
Ferwerd *see* Ferwert
Ferwert *86 D1 Dutch.* Ferwerd. Friesland, N Netherlands
Fès *70 C2 Eng.* Fez. N Morocco
Feteşti *108 D5* Ialomiţa, SE Romania
Fethiye *116 B4* Muğla, SW Turkey
Fetlar *88 D1 island* NE Scotland, United Kingdom
Feuilles, Rivière aux *16 D2 river* Québec, E Canada
Feyzābād *see* Faizābād
Feyẓābād *see* Faizābād

Fez *see* Fès
Fianarantsoa *79 F3* Fianarantsoa, C Madagascar
Fianga *76 B4* Mayo-Kébbi, SW Chad
Fier *101 C6 var.* Fieri. Fier, SW Albania
Fieri *see* Fier
Figeac *91 C5* Lot, S France
Figig *see* Figuig
Figueira da Foz *92 B3* Coimbra, W Portugal
Figueres *93 G2* Cataluña, E Spain
Figuig *70 D2 var.* Figig. E Morocco
Fiji *145 E5 off.* Sovereign Democratic Republic of Fiji, *Fij.* Viti. *country* SW Pacific Ocean

FIJI
Australasia & Oceania

Official name	Republic of the Fiji Islands
Formation	1970 / 1970
Capital	Suva
Population	900,000 / 128 people per sq mile (49 people per sq km)
Total area	7054 sq. miles (18,270 sq. km)
Languages	Fijian, English*, Hindi, Urdu, Tamil, Telugu
Religions	Hindu 38%, Methodist 37%, Roman Catholic 9%, Muslim 8%, Other 8%
Ethnic mix	Melanesian 51%, Indian 44%, Other 5%
Government	Parliamentary system
Currency	Fiji dollar = 100 cents
Literacy rate	94%
Calorie consumption	2930 kilocalories

Fiji, Sovereign Democratic Republic of *see* Fiji
Filadelfia *52 D4* Guanacaste, NW Costa Rica
Filiaşi *108 B5* Dolj, SW Romania
Filipstad *85 B6* Värmland, C Sweden
Finale Ligure *96 A3* Liguria, NW Italy
Finchley *89 A7* United Kingdom
Findlay *40 C4* Ohio, N USA
Finike *116 B4* Antalya, SW Turkey
Finland *84 D4 off.* Republic of Finland, *Fin.* Suomen Tasavalta, Suomi. *country* N Europe

FINLAND
Northern Europe

Official name	Republic of Finland
Formation	1917 / 1947
Capital	Helsinki
Population	5.4 million / 46 people per sq mile (18 people per sq km)
Total area	130,127 sq. miles (337,030 sq. km)
Languages	Finnish*, Swedish*, Sámi
Religions	Evangelical Lutheran 83%, Other 15%, Orthodox Christian 1%, Roman Catholic 1%
Ethnic mix	Finnish 93%, Other (including Sámi) 7%
Government	Parliamentary system
Currency	Euro = 100 cents
Literacy rate	99%
Calorie consumption	3285 kilocalories

Finland, Gulf of *85 D6 Est.* Soome Laht, *Fin.* Suomenlahti, *Ger.* Finnischer Meerbusen, *Rus.* Finskiy Zaliv, *Swe.* Finska Viken. *gulf* E Baltic Sea
Finland, Republic of *see* Finland
Finnischer Meerbusen *see* Finland, Gulf of
Finnmarksvidda *84 D2 physical region* N Norway
Finska Viken/Finskiy Zaliv *see* Finland, Gulf of
Finsterwalde *94 D4* Brandenburg, E Germany
Fiordland *151 A7 physical region* South Island, New Zealand
Fiorina *96 E1* NE San Marino
Firat Nehri *see* Euphrates
Firenze *96 C3 Eng.* Florence; *anc.* Florentia. Toscana, C Italy
Firmum Picenum *see* Fermo
First State *see* Delaware
Fischbacher Alpen *95 E7 mountain range* E Austria
Fischhausen *see* Primorsk
Fish *78 B4 var.* Vis. *river* S Namibia
Fishguard *89 C6 Wel.* Abergwaun. SW Wales, United Kingdom

Fisterra, Cabo *92 B1 headland* NW Spain
Fitzroy Crossing *146 C3* Western Australia
Fitzroy River *146 C3 river* Western Australia
Fiume *see* Rijeka
Flagstaff *48 B2* Arizona, SW USA
Flanders *see* Vlaanderen
Flandre *see* Vlaanderen
Flathead Lake *44 B1 lake* Montana, NW USA
Flat Island *128 C8 island* NE Spratly Islands
Flatts Village *42 B5 var.* The Flatts Village. C Bermuda
Flensburg *94 B2* Schleswig-Holstein, N Germany
Flessingue *see* Vlissingen
Flickertail State *see* North Dakota
Flinders Island *149 C8 island* Furneaux Group, Tasmania, SE Australia
Flinders Ranges *149 B6 mountain range* South Australia
Flinders River *148 C3 river* Queensland, NE Australia
Flin Flon *37 F5* Manitoba, C Canada
Flint *40 C3* Michigan, N USA
Flint Island *145 G4 island* Line Islands, E Kiribati
Floreana, Isla *see* Santa María, Isla
Florence *42 C1* Alabama, S USA
Florence *43 F2* South Carolina, SE USA
Florence *see* Firenze
Florencia *58 B4* Caquetá, S Colombia
Florentia *see* Firenze
Flores *52 B1* Petén, N Guatemala
Flores *139 E5 island* Nusa Tenggara, C Indonesia
Flores *92 A5 island* Azores, Portugal, NE Atlantic Ocean
Flores, Laut *see* Flores Sea
Flores Sea *138 D5 Ind.* Laut Flores. *sea* C Indonesia
Floriano *63 F2* Piauí, E Brazil
Florianópolis *63 F5 prev.* Destêrro. *state capital* Santa Catarina, S Brazil
Florida *64 D4* Florida, S Uruguay
Florida *43 E4 off.* State of Florida, *also known as* Peninsular State, Sunshine State. *state* SE USA
Florida Bay *43 E5 bay* Florida, SE USA
Florida Keys *43 E5 island group* Florida, SE USA
Florida, Straits of *54 B1 strait* Atlantic Ocean/Gulf of Mexico
Flórina *104 B4 var.* Phlórina. Dytikí Makedonía, N Greece
Florissant *45 G4* Missouri, C USA
Floúda, Akrotírio *105 D7 headland* Astypálaia, Kykládes, Greece, Aegean Sea
Flushing *see* Vlissingen
Flylân *see* Vlieland
Foča *100 C4 var.* Srbinje. SE Bosnia and Herzegovina
Foçşani *108 C4* Vrancea, E Romania
Fogaras/Fogarasch *see* Făgăraş
Foggia *97 D5* Puglia, SE Italy
Fogo *74 A3 island* Ilhas de Sotavento, SW Cape Verde
Foix *91 B6* Ariège, S France
Folégandros *105 C7 island* Kykládes, Greece, Aegean Sea
Foleyet *38 C4* Ontario, S Canada
Foligno *96 C4* Umbria, C Italy
Folkestone *89 E7* SE England, United Kingdom
Fond du Lac *40 B2* Wisconsin, N USA
Fongafale *145 E3 var.* Funafuti. Funafuti Atoll, SE Tuvalu
Fonseca, Golfo de *see* Fonseca, Gulf of
Fonseca, Gulf of *52 C3 Sp.* Golfo de Fonseca. *gulf* C Central America
Fontainebleau *90 C3* Seine-et-Marne, N France
Fontenay-le-Comte *90 B4* Vendée, NW France
Fontvieille *91 B8 SW* Monaco Europe
Fonyód *99 C7* Somogy, W Hungary
Foochow *see* Fuzhou
Foochow *see* Fuzhou
Forchheim *95 C5* Bayern, SE Germany
Foo-hsin *see* Fuxin
Forel, Mont *82 D4 mountain* SE Greenland
Forfar *88 C3* E Scotland, United Kingdom
Forge du Sud *see* Dudelange

Forlì *96 C3 anc.* Forum Livii. Emilia Romagna, N Italy
Formentera *93 G4 anc.* Ophiusa, *La* Frumentum. *island* Islas Baleares, Spain, W Mediterranean Sea
Formosa *64 D2* Formosa, NE Argen
Formosa/Formo'sa *see* Taiwan
Formosa, Serra *63 E3 mountain rar* C Brazil
Formosa Strait *see* Taiwan Strait
Føroyar *see* Faroe Islands
Forrest City *42 B1* Arkansas, C USA
Fort Albany *38 C3* Ontario, C Canac
Fortaleza *61 F2* Pando, N Bolivia
Fortaleza *63 G2 prev.* Ceará. *state ca* Ceará, NE Brazil
Fort-Archambault *see* Sarh
Fort-Bayard *see* Zhanjiang
Fort-Cappolani *see* Tidjikja
Fort Charlet *see* Djanet
Fort-Chimo *see* Kuujjuaq
Fort Collins *44 D4* Colorado, C USA
Fort-Crampel *see* Kaga Bandoro
Fort Davis *49 E3* Texas, SW USA
Fort-de-France *55 H4 prev.* Fort-Ro dependent territory capital (Martin W Martinique
Fort Dodge *45 F3* Iowa, C USA
Fortescue River *146 A4 river* Weste Australia
Fort-Foureau *see* Kousséri
Fort Frances *38 B4* Ontario, S Canac
Fort Good Hope *37 E3 var.* Rãdeyíl Northwest Territories, NW Canada
Fort Gouraud *see* Fdérik
Forth *88 C4 river* C Scotland, United Kingdom
Forth, Firth of *88 C4 estuary* E Scotl United Kingdom
Fortín General Eugenio Garay *see* General Eugenio A. Garay
Fort Jameson *see* Chipata
Fort-Lamy *see* Ndjamena
Fort Lauderdale *43 F5* Florida, SE U
Fort Liard *37 E4 var.* Liard. Northw Territories, W Canada
Fort Madison *45 G4* Iowa, C USA
Fort McMurray *37 A4* Alberta, C Ca
Fort McPherson *36 D3 var.* McPher Northwest Territories, NW Canada
Fort Morgan *44 D4* Colorado, C US
Fort Myers *43 E5* Florida, SE USA
Fort Nelson *37 E4* British Columbia W Canada
Fort Peck Lake *44 C1 reservoir* Mon NW USA
Fort Pierce *43 F4* Florida, SE USA
Fort Providence *37 E4 var.* Provider Northwest Territories, W Canada
Fort-Repoux *see* Akjoujt
Fort Rosebery *see* Mansa
Fort Rousset *see* Owando
Fort-Royal *see* Fort-de-France
Fort St. John *37 E4* British Columbia W Canada
Fort Scott *45 F5* Kansas, C USA
Fort Severn *38 C2* Ontario, C Canad
Fort-Shevchenko *114 A4* Mangistau W Kazakhstan
Fort-Sibut *see* Sibut
Fort Simpson *37 E4 var.* Simpson. Northwest Territories, W Canada
Fort Smith *42 B1* Arkansas, C USA
Fort Stockton *49 E3* Texas, SW USA
Fort-Trinquet *see* Bïr Mogreïn
Fort Vermilion *37 E3* Alberta, W Canada
Fort Victoria *see* Masvingo
Fort Walton Beach *42 C3* Florida, SE USA
Fort Wayne *40 C3* Indiana, N USA
Fort William *88 C3* N Scotland, Unit Kingdom
Fort Worth *49 G2* Texas, SW USA
Fort Yukon *36 D3* Alaska, USA
Forum Alieni *see* Ferrara
Forum Livii *see* Forlì
Fossa Claudia *see* Chioggia
Fougamou *77 A6* Ngounié, C Gabon
Fougères *90 B3* Ille-et-Vilaine, NW France
Foulwind, Cape *151 B5 headland* So Island, New Zealand
Foumban *76 A4* Ouest, NW Cameroon
Fou-shan *see* Fushun

eaux Strait *151 A8 strait* S New
Zealand
se Basin *37 G3 sea* Nunavut,
N Canada
x Glacier *151 B6* West Coast, South
sland, New Zealand
ga *93 F2* Aragón, NE Spain
m Basin *see* Amundsen Basin
nce *90 B4 off.* French Republic, *It./Sp.*
rancia; *prev.* Gaul, Gaule, *Lat.* Gallia.
country W Europe

ANCE
Western Europe

fficial name French Republic
rmation 987 / 1919
apital Paris
pulation 64.3 million / 303 people
er sq mile (117 people per sq km)
tal area 211,208 sq. miles
(547,030 sq. km)
nguages French*, Provençal, German,
eton, Catalan, Basque
ligions Roman Catholic 88%,
uslim 8%, Protestant 2%,
addhist 1%, Jewish 1%
hnic mix French 90%, North African
ainly Algerian) 6%, German (Alsace) 2%,
eton 1%, Other (including Corsicans) 1%
overnment Presidential /
arliamentary system
urrency Euro = 100 cents
teracy rate 99%
alorie consumption 3524 kilocalories

anceville *77 B6 var.* Massoukou,
Masuku. Haut-Ogooué, E Gabon
ncfort *see* Frankfurt am Main
nche-Comté *90 D3 cultural region*
E France
ancia *see* France
ancis Case, Lake *45 E3 reservoir*
South Dakota, N USA
ncisco Escárcega *51 G4* Campeche,
SE Mexico
ancistown *78 D3* North East,
NE Botswana
anconian Jura *see* Fränkische Alb
ankenalb *see* Fränkische Alb
ankenstein/Frankenstein in
Schlesien *see* Ząbkowice Śląskie
ankfort *40 C5 state capital* Kentucky,
S USA
ankfort on the Main *see* Frankfurt
am Main
ankfurt *see* Frankfurt am Main,
Germany
ankfurt *see* Słubice, Poland
ankfurt am Main *95 B5 var.*
Frankfurt, *Fr.* Francfort; *prev. Eng.*
Frankfort on the Main. Hessen,
SW Germany
ankfurt an der Oder *94 D3*
Brandenburg, E Germany
änkische Alb *95 C6*
var. Frankenalb, *Eng.* Franconian Jura.
mountain range S Germany
anklin *42 C1* Tennessee, S USA
anklin D. Roosevelt Lake *46 C1
reservoir* Washington, NW USA
anz Josef Land *114 D1 Eng.* Franz
Josef Land. *island group* N Russian
Federation
anz Josef Land *see* Frantsa-Iosifa,
Zemlya
aserburgh *88 D3* NE Scotland,
United Kingdom
aser Island *148 E4
var.* Great Sandy Island.
island Queensland, E Australia
auenbach *see* Baia Mare
auenburg *see* Saldus, Latvia
edericksburg *41 E5* Virginia,
NE USA
edericton *39 F4 province capital* New
Brunswick, SE Canada
ederikshåb *see* Paamiut
edrikshald *see* Halden
edrikstad *85 B6* Østfold, S Norway
eeport *54 C1* Grand Bahama Island,
N Bahamas
eeport *49 H4* Texas, SW USA
ee State *see* Maryland
eetown *74 C4 country capital* (Sierra
Leone) W Sierra Leone
eiburg *see* Freiburg im Breisgau,
Germany

Freiburg im Breisgau *95 A6 var.*
Freiburg, *Fr.* Fribourg-en-Brisgau.
Baden-Württemberg, SW Germany
Freiburg in Schlesien *see* Świebodzice
Fremantle *147 A6* Western Australia
Fremont *45 F4* Nebraska, C USA
French Guiana *59 H3 var.* Guiana,
Guyane. *French overseas department*
N South America
French Guinea *see* Guinea
French Polynesia *143 F4 French
overseas territory* S Pacific Ocean
French Republic *see* France
French Somaliland *see* Djibouti
French Southern and Antarctic
Lands *141 B7 Fr.* Terres Australes
et Antarctiques Françaises. *French
overseas territory* S Indian Ocean
French Sudan *see* Mali
French Territory of the Afars and Issas
see Djibouti
French Togoland *see* Togo
Fresnillo *50 D3 var.* Fresnillo de
González Echeverría. Zacatecas,
C Mexico
Fresnillo de González Echeverría *see*
Fresnillo
Fresno *47 C6* California, W USA
Frías *64 C3* Catamarca, N Argentina
Fribourg-en-Brisgau *see* Freiburg im
Breisgau
Friedek-Mistek *see* Frýdek-Místek
Friedrichshafen *95 B7* Baden-
Württemberg, S Germany
Friendly Islands *see* Tonga
Frisches Haff *see* Vistula Lagoon
Frobisher Bay *82 B3 inlet* Baffin Island,
Nunavut, NE Canada
Frobisher Bay *see* Iqaluit
Frohavet *84 B4 sound* C Norway
Frome, Lake *149 B6 salt lake* South
Australia
Frontera *51 G4* Tabasco, SE Mexico
Frontignan *91 C6* Hérault, S France
Frostviken *see* Kvarnbergsvattnet
Frøya *84 A4 island* W Norway
Frumentum *see* Formentera
Frunze *see* Bishkek
Frýdek-Místek *99 C5 Ger.* Friedek-
Mistek. Moravskoslezský Kraj, E Czech
Republic
Fu-chien *see* Fujian
Fu-chou *see* Fuzhou
Fu-chou *see* Fuzhou
Fuengirola *92 D5* Andalucía, S Spain
Fuerte Olimpo *64 D2 var.* Olimpo. Alto
Paraguay, NE Paraguay
Fuerte, Río *48 C5 river* C Mexico
Fuerteventura *70 B3 island* Islas
Canarias, Spain, NE Atlantic Ocean
Fuhkien *see* Fujian
Fu-hsin *see* Fuxin
Fuji *131 D6 var.* Huzi. Shizuoka,
Honshū, S Japan
Fujian *128 D6 var.* Fu-chien, Fuhkien,
Fukien, Min, Fujian Sheng. *province*
SE China
Fujian Sheng *see* Fujian
Mount Fuji *131 C6 var.* Fujiyama,
Eng. Mount Fuji. *mountain* Honshū,
SE Japan
Fuji, Mount/Fujiyama *see* Fuji-san
Fukang *126 C2* Xinjiang Uygur Zizhiqu,
W China
Fukien *see* Fujian
Fukui *131 C6 var.* Hukui. Fukui,
Honshū, SW Japan
Fukuoka *131 A7 var.* Hukuoka, *hist.*
Najima. Fukuoka, Kyūshū, SW Japan
Fukushima *130 D4 var.* Hukusima.
Fukushima, Honshū, C Japan
Fulda *95 B5* Hessen, C Germany
Funafuti *see* Fongafale
Funafuti Atoll *145 E3 atoll* C Tuvalu
Funchal *70 A2* Madeira, Portugal,
NE Atlantic Ocean
Fundy, Bay of *39 F5 bay* Canada/USA
Fünen *see* Fyn
Fünfkirchen *see* Pécs
Furnes *see* Veurne
Fürth *95 C5* Bayern, S Germany
Furukawa *130 D4 var.* Hurukawa,
Ōsaki. Miyagi, Honshū, C Japan
Fusan *see* Busan
Fushë Kosovë *101 D5 Serb.* Kosovo
Polje. C Kosovo
Fushun *128 D3 var.* Fou-shan, Fu-shun.
Liaoning, NE China
Fu-shun *see* Fushun
Fusin *see* Fuxin
Füssen *95 C7* Bayern, S Germany

Futog *100 D3* Vojvodina, NW Serbia
Futuna, Île *145 E4 island* S Wallis and
Futuna
Fuxin *128 D3 var.* Fou-hsin, Fu-hsin,
Fusin. Liaoning, NE China
Fuzhou *128 D5 var.* Foochow, Fu-chou.
province capital Fujian, SE China
Fuzhou *128 D6 var.* Foochow, Fu-chou.
province capital Fujian, SE China
Fyn *85 B8 Ger.* Fünen. *island* C Denmark
FYR Macedonia/FYROM *see* Macedonia,
FYR
Fyzabad *see* Feizābād

G

Gaafu Alifu Atoll *see* North Huvadhu
Atoll
Gaalkacyo *73 E5 var.* Galka'yo, *It.*
Galcaio. Mudug, C Somalia
Gabela *78 B2* Cuanza Sul, W Angola
Gaberones *see* Gaborone
Gabès *71 E2 var.* Qābis. E Tunisia
Gabès, Golfe de *71 F2 Ar.* Khalīj Qābis.
gulf E Tunisia
Gabon *77 B6 off.* Gabonese Republic.
country C Africa

GABON
Central Africa

Official name Gabonese Republic
Formation 1960 / 1960
Capital Libreville
Population 1.7 million / 17 people
per sq mile (7 people per sq km)
Total area 103,346 sq. miles
(267,667 sq. km)
Languages Fang, French*, Punu, Sira,
Nzebi, Mpongwe
Religions Christian (mainly Roman
Catholic) 55%, Traditional beliefs 40%,
Other 4%, Muslim 1%
Ethnic mix Fang 26%, Shira-punu 24%,
Other 16%, Foreign residents 15%,
Nzabi-duma 11%, Mbede-Teke 8%
Government Presidential system
Currency CFA franc = 100 centimes
Literacy rate 89%
Calorie consumption 2781 kilocalories

Gabonese Republic *see* Gabon
Gaborone *78 C4 prev.* Gaberones.
country capital (Botswana) South East,
SE Botswana
Gabrovo *104 D2* Gabrovo, N Bulgaria
Gadag *132 C1* Karnātaka, W India
Gades/Gadier/Gadir/Gadire *see* Cádiz
Gadsden *42 D2* Alabama, S USA
Gaeta *97 C5* Lazio, C Italy
Gaeta, Gulf of *97 C5 var.* Gulf of Gaeta.
gulf C Italy
Gaeta, Gulf of *see* Gaeta, Golfo di
Gäfle *see* Gävle
Gafsa *71 E2 var.* Qafşah. W Tunisia
Gagnoa *74 D5* C Côte d'Ivoire
Gagra *121 F1* NW Georgia
Gaillac *91 C6 var.* Gaillac-sur-Tarn.
Tarn, S France
Gaillac-sur-Tarn *see* Gaillac
Gaillimh *see* Galway
Gaillimhe, Cuan na *see* Galway Bay
Gainesville *43 E3* Florida, SE USA
Gainesville *42 D2* Georgia, SE USA
Gainesville *49 G2* Texas, SW USA
Lake Gairdner *149 A6 salt lake* South
Australia
Gaizina Kalns *see* Gaiziņkalns
Gaiziņkalns *106 C3 var.* Gaizina Kalns.
mountain E Latvia
Gakkel Ridge *155 B3 var.* Arctic Mid
Oceanic Ridge, Nansen Cordillera.
undersea feature Arctic Ocean
Galán, Cerro *63 B3 mountain*
NW Argentina
Galanta *99 C6 Hung.* Galánta. Trnavský
Kraj, W Slovakia
Galapagos Fracture Zone *153 E3
tectonic feature* E Pacific Ocean
Galápagos Islands *see*
Galápagos Islands, Sp. Archipiélago
de Colón, Islas de los Galápagos,
Tortoise Islands. *island group* Ecuador,
E Pacific Ocean
Galápagos, Islas de los *see* Galápagos
Islands

Galapagos Rise *153 F3 undersea rise*
E Pacific Ocean
Galashiels *88 C4* SE Scotland, United
Kingdom
Galați *108 D4 Ger.* Galatz. *Rom.* Galaţi,
E Romania
Galatz *see* Galaţi
Galcaio *see* Gaalkacyo
Galesburg *40 B3* Illinois, N USA
Galicia *92 B1 var.* Galiza.
autonomous community NW Spain
Galicia Bank *80 B4 undersea bank*
E Atlantic Ocean
Galilee, Sea of *see* Tiberias, Lake
Galka'yo *see* Gaalkacyo
Galkynyş *122 D3 prev. Rus.* Deynau,
Dyanev, *Turkm.* Dänew. Lebap
Welaýaty, NE Turkmenistan
Gallaecia *see* Galicia
Galle *132 D4 prev.* Point de Galle.
Southern Province, SW Sri Lanka
Gallego Rise *153 F3 undersea rise*
E Pacific Ocean
Gallegos *see* Río Gallegos
Gallia *see* France
Gallipoli *97 E6* Puglia, SE Italy
Gällivare *84 C3 Lapp.* Váhtjer.
Norrbotten, N Sweden
Gallup *48 C1* New Mexico, SW USA
Galtat-Zemmour *70 B3* C Western
Sahara
Galveston *49 H4* Texas, SW USA
Galway *89 A5 Ir.* Gaillimh. W Ireland
Galway Bay *89 A6 Ir.* Cuan na Gaillimhe.
bay W Ireland
Gámas *see* Kaamanen
Gambell *36 C2* Saint Lawrence Island,
Alaska, USA
Gambia *74 B3 off.* Republic of The
Gambia, The Gambia. *country*
W Africa

GAMBIA
West Africa

Official name Republic of the Gambia
Formation 1965 / 1965
Capital Banjul
Population 1.8 million / 466 people
per sq mile (180 people per sq km)
Total area 4363 sq. miles (11,300 sq. km)
Languages Mandinka, Fulani, Wolof, Jola,
Soninke, English*
Religions Sunni Muslim 90%,
Christian 8%, Traditional beliefs 2%
Ethnic mix Mandinka 42%, Fulani
18%,Wolof 16%, Jola 10%, Serahuli 9%,
Other 5%
Government Presidential system
Currency Dalasi = 100 butut
Literacy rate 52%
Calorie consumption 2849 kilocalories

Gambia *74 C3 Fr.* Gambie. *river*
W Africa
Gambia, Republic of The *see* Gambia
Gambia, The *see* Gambia
Gambier, Îles *143 G4 island group*
E French Polynesia
Gamboma *77 B6* Plateaux, E Congo
Ganaklakerleby *see* Kokkola
Gan *132 B5* Addu Atoll, C Maldives
Gananagar *see* Sri Ganganagar
Ganaanee *see* Juba
Gäncä *117 G2 Rus.* Gyandzha; *prev.*
Kirovabad, Yelisavetpol. W Azerbaijan
Gand *see* Gent
Gandajika *77 D7* Kasai-Oriental, S Dem.
Rep. Congo
Gändhidhām *134 C4* Gujarāt, W India
Gandia *93 F3* Valenciano, E Spain
País Valenciano, E Spain
Gandía *see* Gandia
Ganges *135 F3 Ben.* Padma.
Bangladesh/India
Ganges Cone *see* Ganges Fan
Ganges Fan *140 D3 var.* Ganges Cone.
undersea fan N Bay of Bengal
Ganges, Mouths of the *135 G4 delta*
Bangladesh/India
Gangtok *135 F3 state capital* Sikkim,
N India
Gansu *128 B4 var.* Gan, Gansu Sheng,
Kansu. *province* N China
Gansu Sheng *see* Gansu

Gantsevichi *see* Hantsavichy
Ganzhou *128 D6* Jiangxi, S China
Gao *75 E3* Gao, E Mali
Gaocheng *see* Litang
Gaoual *74 C4* N Guinea
Gaoxiong *128 D6 var.* Kaohsiung, *Jap.* Takao, Takow. S Taiwan
Gap *91 D5 anc.* Vapincum. Hautes-Alpes, SE France
Gaplañgyr Platosy *122 C2 Rus.* Plato Kaplangky. *ridge* Turkmenistan/ Uzbekistan
Gar *see* Gar Xincun
Garabil Belentligi *122 D3 Rus.* Vozvyshennost' Karabil'. *mountain range* S Turkmenistan
Garabogaz Aylagy *122 B2 Rus.* Zaliv Kara-Bogaz-Gol. *bay* NW Turkmenistan
Garachiné *53 G5* Darién, SE Panama
Garagum *122 C3 var.* Garagumy, Qara Qum, *Eng.* Black Sand Desert, Kara Kum; *prev.* Peski Karakumy. *desert* C Turkmenistan
Garagum Canal *122 D3 var.* Kara Kum Canal, *Rus.* Karagumskiy Kanal, Karakumskiy Kanal. *canal* C Turkmenistan
Garagumy *see* Garagum
Gara Khitrino *104 D2* Shumen, NE Bulgaria
Gárassavon *see* Kaaresuvanto
Garda, Lake *96 C2 var.* Benaco, *Eng.* Lake Garda, *Ger.* Gardasee. *lake* NE Italy
Garda, Lake *see* Garda, Lago di
Gardasee *see* Garda, Lago di
Garden City *45 E5* Kansas, C USA
Garden State, The *see* New Jersey
Gardëz *123 E4 prev.* Gardiz. E Afghanistan
Gardiz *see* Gardëz
Gardner Island *see* Nikumaroro
Garegegasnjárga *see* Karigasniemi
Gargždai *106 B3* Klaipėda, W Lithuania
Garissa *73 D6* Coast, E Kenya
Garland *49 G2* Texas, SW USA
Garoe *see* Garoowe
Garonne *91 B5 anc.* Garumna. *river* S France
Garoowe *73 E5 var.* Garoe. Nugaal, N Somalia
Garoua *76 B4 var.* Garua. Nord, N Cameroon
Garrygala *see* Magtymguly
Garry Lake *37 F3 lake* Nunavut, N Canada
Garsen *73 D6* Coast, S Kenya
Garua *see* Garoua
Garumna *see* Garonne
Garwolin *98 D4* Mazowieckie, E Poland
Gar Xincun *126 A4 prev.* Gar. Xizang Zizhiqu, W China
Gary *40 B3* Indiana, N USA
Garzón *58 B4* Huila, S Colombia
Gasan-Kuli *see* Esenguly
Gascogne *91 B6 Eng.* Gascony. *cultural region* S France
Gascony *see* Gascogne
Gascoyne River *147 A5 river* Western Australia
Gaspé *39 F3* Québec, SE Canada
Gaspé, Péninsule de *39 E4 var.* Péninsule de la Gaspésie. *peninsula* Québec, SE Canada
Gaspésie, Péninsule de la *see* Gaspé, Péninsule de
Gastonia *43 E1* North Carolina, SE USA
Gastoúni *105 B6* Dytikí Elláda, S Greece
Gatchina *110 B4* Leningradskaya Oblast', NW Russian Federation
Gatineau *38 D4* Québec, SE Canada
Gatooma *see* Kadoma
Gatún, Lake *53 F4 reservoir* C Panama
Gauhāti *see* Guwāhāti
Gauja *106 D3 var.* Aa. *river* Estonia/ Latvia
Gaul/Gaule *see* France
Gauteng *see* Johannesburg, South Africa
Gāvbandī *120 D4* Hormozgān, S Iran
Gávdos *105 C8 island* SE Greece
Gavere *87 B6* Oost-Vlaanderen, NW Belgium
Gävle *85 C6 var.* Gäfle; *prev.* Gefle. Gävleborg, C Sweden
Gawler *149 B6* South Australia
Gaya *135 F3* Bihār, N India
Gaya *see* Kyjov
Gayndah *149 E5* Queensland, E Australia

Gaysin *see* Haysyn
Gaza *119 A6 Ar.* Ghazzah, *Heb.* 'Azza. NE Gaza Strip
Gaz-Achak *see* Gazojak
Gazandzhyk/Gazanjyk *see* Bereket
Gaza Strip *119 A7 Ar.* Qita Ghazzah. *disputed region* SW Asia
Gaziantep *116 D4 var.* Gazi Antep; *prev.* Aintab, Antep. Gaziantep, S Turkey
Gazi Antep *see* Gaziantep
Gazimağusa *102 D5 var.* Famagusta, *Gk.* Ammóchostos. E Cyprus
Gazli *122 D2 Rus.* Gaz-Achak. Lebap Welaýaty, C Uzbekistan
Gazojak *122 D2 Rus.* Gaz-Achak. Lebap Welaýaty, NE Turkmenistan
Gbanga *74 D5 var.* Gbarnga. N Liberia
Gbarnga *see* Gbanga
Gdańsk *98 C2 Fr.* Dantzig, *Ger.* Danzig. Pomorskie, N Poland
Gdan'skaya Bukhta/Gdańsk, Gulf of *see* Danzig, Gulf of
Gdańska, Zatoka *see* Danzig, Gulf of
Gdingen *see* Gdynia
Gdynia *98 C2 Ger.* Gdingen. Pomorskie, N Poland
Gedaref *72 C4 var.* Al Qaḍārif, El Gedaref. Gedaref, E Sudan
Gediz *116 B3* Kütahya, W Turkey
Gediz Nehri *116 A3 river* W Turkey
Geel *87 C5 var.* Gheel. Antwerpen, N Belgium
Geelong *149 C7* Victoria, SE Australia
Ge'e'mu *see* Golmud
Gefle *see* Gävle
Geilo *85 A5* Buskerud, S Norway
Gejiu *128 B6 var.* Kochiu. Yunnan, S China
Gëkdepe *see* Gökdepe
Gela *97 C7 prev.* Terranova di Sicilia. Sicilia, Italy, C Mediterranean Sea
Geldermalsen *86 C4* Gelderland, C Netherlands
Geleen *87 D6* Limburg, SE Netherlands
Gelib *see* Jilib
Gellinsor *73 E5* Galguduud, C Somalia
Gembloux *87 C6* Namur, Belgium
Gemena *77 C5* Equateur, NW Dem. Rep. Congo
Gem of the Mountains *see* Idaho
Gemona del Friuli *96 D2* Friuli-Venezia Giulia, NE Italy
Gem State *see* Idaho
Genalê Wenz *see* Juba
Genck *see* Genk
General Alvear *64 B4* Mendoza, W Argentina
General Carrera, Lago *see* Buenos Aires, Lago
General Eugenio A. Garay *64 C1 var.* Fortín General Eugenio Garay; *prev.* Yrendagué. Nueva Asunción, NW Paraguay
General José F.Uriburu *see* Zárate
General Machado *see* Camacupa
General Santos *139 F3 off.* General Santos City. Mindanao, S Philippines
General Santos City *see* General Santos
Gênes *see* Genova
Geneva *see* Genève
Geneva, Lake *95 A7 Fr.* Lac de Genève, Lac Léman, le Léman, *Ger.* Genfer See. *lake* France/Switzerland
Genève *95 A7 Eng.* Geneva, *Ger.* Genf, *It.* Ginevra. Genève, SW Switzerland
Genève, Lac de *see* Geneva, Lake
Genf *see* Genève
Genfer See *see* Geneva, Lake
Genichesk *see* Heniches'k
Genk *87 D6 var.* Genck. Limburg, NE Belgium
Gennep *86 D4* Limburg, SE Netherlands
Genoa *see* Genova
Genoa, Gulf of *96 A3 Eng.* Gulf of Genoa. *gulf* NW Italy
Genoa *102 D1 Eng.* Genoa; *anc.* Genua, *Fr.* Gênes. Liguria, NW Italy
Genovesa, Isla *60 B5 var.* Tower Island. *island* Galápagos Islands, Ecuador, E Pacific Ocean
Gent *87 B5 Eng.* Ghent, *Fr.* Gand. Oost-Vlaanderen, NW Belgium
Genua *see* Genova
Geok-Tepe *see* Gökdepe
George *78 C5* Western Cape, S South Africa
George *82 A4 river* Newfoundland and Labrador/Québec, E Canada

George, Lake *43 E3 lake* Florida, SE USA
Georgenburg *see* Jurbarkas
Georges Bank *35 D5 undersea bank* W Atlantic Ocean
George Sound *151 A7 sound* South Island, New Zealand
Georges River *148 D2 river* New South Wales, E Australia
Georgetown *59 F2 country capital* (Guyana) (Guyana) N Guyana
George Town *54 C2* Great Exuma Island, C Bahamas
George Town *54 B3 var.* Georgetown. *dependent territory capital* (Cayman Islands) Grand Cayman, SW Cayman Islands
George Town *138 B3 var.* Penang, Pinang. Pinang, Peninsular Malaysia
Georgetown *43 F2* South Carolina, SE USA
Georgetown *see* George Town
George V Land *154 C4 physical region* Antarctica
Georgia *117 F2 off.* Republic of Georgia, *Geor.* Sak'art'velo, *Rus.* Gruzinskaya SSR, Gruziya. *country* SW Asia

GEORGIA
Southwest Asia

Official name Georgia
Formation 1991 / 1991
Capital Tbilisi
Population 4.3 million / 160 people per sq mile (62 people per sq km)
Total area 26,911 sq. miles (69,700 sq. km)
Languages Georgian*, Russian, Azeri, Armenian, Mingrelian, Ossetian, Abkhazian* *(in Abkhazia)*
Religions Georgian Orthodox 74%, Muslim 10%, Russian Orthodox 10%, Armenian Apostolic Church (Orthodox) 4%, Other 2%
Ethnic mix Georgian 84%, Armenian 6%, Azeri 6%, Russian 2%, Ossetian 1%, Other 1%
Government Presidential system
Currency Lari → 100 tetri
Literacy rate 99%
Calorie consumption 2731 kilocalories

Georgia *42 D2 off.* State of Georgia, *also known as* Empire State of the South, Peach State. *state* SE USA
Georgian Bay *40 D2 lake bay* Ontario, S Canada
Georgia, Republic of *see* Georgia
Georgia, Strait of *46 A1 strait* British Columbia, W Canada
Georgi Dimitrov *see* Kostenets
Georgiu-Dezh *see* Liski
Georgiyevka *see* Korday
George von Neumayer *154 A2 German research station* Antarctica
Gera *94 C4* Thüringen, E Germany
Geráki *105 B6* Pelopónnisos, S Greece
Geraldine *151 B6* Canterbury, South Island, New Zealand
Geraldton *147 A6* Western Australia
Geral, Serra *57 D5 mountain range* S Brazil
Gerede *116 C2* Bolu, N Turkey
Gereshk *122 D5* Helmand, SW Afghanistan
Gering *44 D3* Nebraska, C USA
German East Africa *see* Tanzania
Germanicopolis *see* Çankırı
Germanium, Mare/German Ocean *see* North Sea
German Southwest Africa *see* Namibia
Germany *94 B4 off.* Federal Republic of Germany, *Bundesrepublik* Deutschland, *Ger.* Deutschland. *country* N Europe

GERMANY
Northern Europe

Official name Federal Republic of Germany
Formation 1871 / 1990
Capital Berlin
Population 82.7 million / 613 people per sq mile (237 people per sq km)
Total area 137,846 sq. miles (357,021 sq. km)
Languages German*, Turkish
Religions Protestant 34%, Roman Catholic 33%, Muslim 3.7%, Other 29.3%
Ethnic mix German 92%, Other 4%, Turkish 2%

GERMANY
(continued)

Government Parliamentary system
Currency Euro → 100 cents
Literacy rate 99%
Calorie consumption 3539 kilocalories

Germany, Federal Republic of *see* Germany
Geroliménas *105 B7* Pelopónnisos, S Greece
Gerona *see* Girona
Gerpinnes *87 C7* Hainaut, S Belgium
Gerunda *see* Girona
Gerze *116 D2* Sinop, N Turkey
Gesoriacum *see* Boulogne-sur-Mer
Gessoriacum *see* Boulogne-sur-Mer
Getafe *92 D3* Madrid, C Spain
Gevas *117 F3* Van, SE Turkey
Gevgeli *see* Gevgelija
Gevgelija *101 E6 var.* Đevđelija, Djevdjelija, *Turk.* Gevgeli. SE Macedonia
Ghaba *see* Al Ghābah
Ghana *75 E5 off.* Republic of Ghana. *country* W Africa

GHANA
West Africa

Official name Republic of Ghana
Formation 1957 / 1957
Capital Accra
Population 25.9 million / 292 people per sq mile (113 people per sq km)
Total area 92,100 sq. miles (238,540 sq. km)
Languages Twi, Fanti, Ewe, Ga, Adangme, Gurma, Dagomba (Dagbani), English*
Religions Christian 69%, Muslim 16%, Traditional beliefs 9%, Other 6%
Ethnic mix Akan 49%, Mole-Dagbani 17%, Ewe 13%, Ga and Ga-Adangbe 8%, Guan 4%
Government Presidential system
Currency Cedi → 100 pesewas
Literacy rate 72%
Calorie consumption 3003 kilocalories

Ghanzi *78 C3 var.* Khanzi. Ghanzi, W Botswana
Gharandal *119 B7* Al'Aqabah, SW Jordan
Gharbt, Jabal al *see* Liban, Jebel
Ghardaïa *70 D2* N Algeria
Gharvān *see* Gharyān
Gharyān *71 F2 var.* Gharvān. NW Libya
Ghawdex *see* Gozo
Ghaznī *123 E4 var.* Ghazni, Ghaznī. E Afghanistan
Ghazzah *see* Gaza
Gheel *see* Geel
Ghent *see* Gent
Gheorgheni *108 C4 prev.* Gheorgheni, Sîn-Miclăuş, *Ger.* Niklasmarkt, *Hu* Gyergyószentmiklós. Harghita, C Romania
Gheorghieni *see* Gheorgheni
Ghōriān *122 D4 prev.* Ghūriān. Herāt, W Afghanistan
Ghūdara *123 F3 var.* Gudara, Rus. Kudara. SE Tajikistan
Ghurdaqah *see* Al Ghurdaqah
Ghūriān *see* Ghōriān
Giamame *see* Jamaame
Giannitsá *104 B4 var.* Yiannitsá. Kentrikí Makedonía, N Greece
Gibraltar *93 G4 UK dependent territory* SW Europe
Gibraltar, Bay of *93 G5 bay* Gibraltar Spain/Mediterranean Sea Atlantic Ocean
Gibraltar, Détroit de/Gibraltar, Estrecho de *see* Gibraltar, Strait of
Gibraltar, Strait of *92 C5 Fr.* Détroit de Gibraltar, *Sp.* Estrecho de Gibraltar. *strait* Atlantic Ocean/Mediterranean Sea
Gibson Desert *147 B5 desert* Western Australia
Giedraičiai *107 C5* Utena, E Lithuania
Giessen *95 B5* Hessen, W Germany
Gifu *131 C6 var.* Gihu. Gifu, Honshū, SW Japan
Giganta, Sierra de la *50 B3 mountain range* NW Mexico
Gihu *see* Gifu
G'ijduvon *122 D2 Rus.* Gizhduvon. Buxoro Viloyati, C Uzbekistan

92 D1 var. Xixón. Asturias,
/ Spain
.iver *48 A2 river* Arizona, SW USA
rt Islands *see* Tungaru
rt River *148 C3 river* Queensland,
Australia
ebir Plateau *see* Hadabat al Jilf
.abīr
te *44 D3* Wyoming, C USA
> *see* Halmahera, Pulau
/ *47 B6* California, W USA
e, Mount *55 F1 mountain* C Saint
:ia
ma *see* Jīma
vra *see* Genève
n *147 A6* Western Australia
ır *see* Jawhar
wic *see* Ipswich
dot *58 B3* Cundinamarca,
:olombia
an *117 E2 var.* Kerasunt; *anc.*
rasus, Pharnacia. Giresun,
Turkey
nti *see* Agrigento
see Jilin
e *102 C5 Gk.* Kerýneia, Kyrenia.
:yprus
a *see* Kiruna
na *93 G2 var.* Gerona; *anc.* Gerunda.
aluña, NE Spain
orne *150 E3* Gisborne, North Island,
v Zealand
r Range *123 E3 Rus.* Gissarskiy
rebet. *mountain range* Tajikistan/
bekistan
rskiy Khrebet *see* Gissar Range
o *see* Gÿtheio
anova *96 D4* Abruzzi, C Italy
ri *see* Gyumri
giu *108 C5* Giurgiu, S Romania
72 B1 var. Al Jīzah, El Gîza, Gizeh.
:gypt
luvon *see* G'ijduvon
ko *98 D2 Ger.* Lötzen. Warmińsko-
zurskie, NE Poland
ovë *101 D5 Serb.* Đakovica.
Kosovo
ovicë *101 D5 Serb.* Đeravica.
untain* S Serbia
n *101 D5 Serb.* Gnjilane. E Kosovo
okastër *see* Gjirokastër
ikastër *101 C7 var.* Gjirokastra;
rev. Gjinokastër, Gk. Argyrokastron,
Argirocastro. Gjirokastër, S Albania
kastra *see* Gjirokastër
Haven *37 F3 var.* Uqsuqtuuq. King
illiam Island, Nunavut, NW Canada
k *85 B5* Oppland, S Norway
e Bay *39 G4* Cape Breton Island,
va Scotia, SE Canada
stone *148 E4* Queensland,
Australia
85 B5 var. Glommen. *river*
Jorway
nitsa *104 D1 prev.* Pravda,
grular. Silistra, NE Bulgaria
gow *88 C4* S Scotland, United
.ıgdom
.a Morava *see* Velika Morava
ov *111 D5* Udmurtskaya Respublika,
V Russian Federation
ritz *see* Gliwice
dale *48 B2* Arizona, SW USA
dive *44 D2* Montana, NW USA
s Falls *41 F3* New York,
E USA
am *see* Gloucester
a *100 B3 var.* Banjska Palanka.
ak-Moslavina, NE Croatia
ertind *85 A5 mountain* S Norway
ce *99 C5 Ger.* Gleiwitz. Śląskie,
oland
e *48 B2* Arizona, SW USA
gau *see* Głogów
au *see* Głogów
5w *99 B4 Ger.* Glogau, Glogow.
olnośląskie, SW Poland
ow *see* Głogów
ama *see* Gláma
amen *see* Gláma
a *see* Gláma
cester *89 D6 hist.* Caer Glou, *Lat.*
evum. C England, United Kingdom
no *99 D4 Łódź, C Poland
okoye *see* Hlybokaye
hov *see* Hlukhiv
en *see* Gløen
zno *98 C3 Ger.* Gnesen.
eilkopolskie, C Poland
see Gjilan

Gobabis *78 B3* Omaheke, E Namibia
Gobi *126 D3 desert* China/Mongolia
Gobō *131 C6* Wakayama, Honshū,
SW Japan
Godavari *124 B3 var.* Godavari.
C India
Godavari *see* Godāvari
Godhavn *see* Qeqertarsuaq
Godhra *134 C4* Gujarāt, W India
Göding *see* Hodonín
Godoy Cruz *64 B4* Mendoza,
W Argentina
Godthaab/Godthåb *see* Nuuk
Godwin Austen, Mount *see* K2
Goede Hoop, Kaap de *see* Good Hope,
Cape of
Goeie Hoop, Kaap die *see* Good Hope,
Cape of
Goeree *86 B4 island* SW Netherlands
Goes *87 B5* Zeeland, SW Netherlands
Goettingen *see* Göttingen
Gogebic Range *40 B1 hill range*
Michigan/Wisconsin, N USA
Goiânia *63 E3 prev.* Goyania. *state
capital* Goiás, C Brazil
Goiás *63 E3 off.* Estado de Goiás; *prev.*
Goiaz, Goyaz. *region* C Brazil
Goiás *63 E3 off.* Estado de Goiás; *prev.*
Goiaz, Goyaz. *state* C Brazil
Goiás, Estado de *see* Goiás
Goiaz *see* Goiás
Goidhoo Atoll *see* Horsburgh Atoll
Gojōme *130 D4* Akita, Honshū,
NW Japan
Gökçeada *104 D4 var.* Imroz Adasï, *Gk.*
Imbros. *island* NW Turkey
Gökdepe *122 C3 Rus.* Gëkdepe, Geok-
Tepe. Ahal Welaýaty, C Turkmenistan
Göksun *116 D4* Kahramanmaraş,
C Turkey
Gol *85 A5* Buskerud, S Norway
Golan Heights *119 B5 Ar.* Al Jawlān,
Heb. HaGolan. *mountain range*
SW Syria
Golaya Pristan *see* Hola Prystan'
Gołdap *98 E2 Ger.* Goldap. Warmińsko-
Mazurskie, NE Poland
Gold Coast *149 E5 cultural region*
Queensland, E Australia
Golden Bay *150 C4 bay* South Island,
New Zealand
Golden State, The *see* California
Goldingen *see* Kuldīga
Goldsboro *43 F1* North Carolina,
SE USA
Goleniów *98 B3 Ger.* Gollnow.
Zachodnio-pomorskie, NW Poland
Gollnow *see* Goleniów
Golmo *see* Golmud
Golmud *126 D4 var.* Ge'e'mu, Golmo,
Chin. Ko-erh-mu. Qinghai, C China
Golovanevsk *see* Holovanivs'k
Golub-Dobrzyń *98 C3* Kujawski-
pomorskie, C Poland
Goma *79 E6* Nord-Kivu, NE Dem.
Rep. Congo
Gombi *75 H4* Adamawa, E Nigeria
Gombroon *see* Bandar-e 'Abbās
Gomel' *see* Homyel'
Gomera *70 A3 island* Islas Canarias,
Spain, NE Atlantic Ocean
Gómez Palacio *50 D3* Durango,
C Mexico
Gonaïves *54 D3 var.* Les Gonaïves.
N Haiti
Gonâve, Île de la *54 D3 island* C Haiti
Gondar *see* Gonder
Gonder *72 C4 var.* Gondar. Āmara,
NW Ethiopia
Gondia *135 E4* Mahārāshtra, C India
Gonggar *see* Gyixong. Xizang
Zizhiqu, W China
Gongola *75 G4 river* E Nigeria
Gongtang *see* Damxung
Gonni/Gónnos *see* Gónnoi
Gónnoi *104 B4 var.* Gonnus, Gónnos;
prev. Derelí. Thessalía, C Greece
Good Hope, Cape of *78 B5 Afr.* Kaap de
Goede Hoop, Kaap die Goeie Hoop.
headland SW South Africa
Goodland *44 D4* Kansas, C USA
Goondiwindi *149 D5* Queensland,
E Australia
Goor *86 E3* Overijssel, E Netherlands
Goose Green *65 D7 var.* Prado del
Ganso. East Falkland, Falkland Islands
Goose Lake *46 B4 var.* Lago dos Gansos.
lake California/Oregon, W USA
Gopher State *see* Minnesota

Göppingen *95 B6* Baden-Württemberg,
SW Germany
Góra Kalwaria *114 D4* Mazowieckie,
C Poland
Gorakhpur *135 E3* Uttar Pradesh,
N India
Gorany *see* Harany
Goražde *100 C4* Federacija Bosne
I Hercegovine, SE Bosnia and
Herzegovina
Gorbovichi *see* Harbavichy
Goré *76 C4* Logone-Oriental, S Chad
Gorē *73 C5* Oromīya, C Ethiopia
Gore *151 B7* Southland, South Island,
New Zealand
Gorgān *120 D2 var.* Astarabad,
Astrabad, Gurgan, *prev.* Asterābād;
anc. Hyrcania. Golestán, N Iran
Gori *117 F2* C Georgia
Gorinchem *86 C4 var.* Gorkum. Zuid-
Holland, C Netherlands
Goris *117 G3* SE Armenia
Gorki *see* Horki
Gor'kiy *see* Nizhniy Novgorod
Gorkum *see* Gorinchem
Görlitz *94 D4* Sachsen, E Germany
Gorlovka *see* Horlivka
Gorna Dzhumaya *see* Blagoevgrad
Gornja Mužlja *see* Mužlja
Gornji Milanovac *100 C4* Serbia,
C Serbia
Gorodets *see* Haradzyets
Gorodishche *see* Horodyshche
Gorodnya *see* Horodnya
Gorodok *see* Haradok
Gorodok/Gorodok Yagellonski
see Horodok
Gorontalo *139 E4* Sulawesi, C Indonesia
Gorontalo, Teluk *see* Tomini, Teluk
Gorssel *86 D3* Gelderland, E Netherlands
Goryn *see* Horyn'
Gorzów Wielkopolski *98 B3 Ger.*
Landsberg, Landsberg an der Warthe.
Lubuskie, W Poland
Gosford *149 D6* New South Wales,
SE Australia
Goshogawara *130 D3 var.* Gosyogawara.
Aomori, Honshū, C Japan
Gospić *100 A3* Lika-Senj, C Croatia
Gostivar *101 D6* W FYR Macedonia
Gosyogawara *see* Goshogawara
Göteborg *85 B7 Eng.* Gothenburg.
Västra Götaland, S Sweden
Gotel Mountains *75 G5 mountain range*
E Nigeria
Gothenburg *see* Göteborg
Gotland *85 C7 island* SE Sweden
Goto-retto *131 A8 island group*
SW Japan
Gotska Sandön *85 B1 island*
SE Sweden
Gôtsu *131 B6 var.* Gôtu. Shimane,
Honshū, SW Japan
Göttingen *94 B4 var.* Goettingen.
Niedersachsen, C Germany
Gôtu *see* Gôtsu
Gouda *86 C4* Zuid-Holland,
C Netherlands
Gough Fracture Zone *67 C6 tectonic
feature* S Atlantic Ocean
Gough Island *69 B8 island* Tristan da
Cunha, S Atlantic Ocean
Gouin, Réservoir *38 D4 reservoir*
Québec, SE Canada
Goulburn *149 D6* New South Wales,
SE Australia
Gouré *75 G3* Zinder, SE Niger
Governador Valadares *63 F4* Minas
Gerais, SE Brazil
Govĭ Altayn Nuruu *127 E3 mountain
range* S Mongolia
Goyania *see* Goiânia
Goyania *see* Goiânia
Goz Beïda *76 C3* Ouaddaï, SE Chad
Gozo *102 A5 var.* Ghawdex. *island*
N Malta
Graciosa *92 A5 var.* Ilha Graciosa.
island Azores, Portugal, NE Atlantic Ocean
Graciosa, Ilha *see* Graciosa
Gradačac *100 C3* Federacija Bosne
I Hercegovine, N Bosnia and
Herzegovina

Gradaús, Serra dos *63 E3 mountain
range* C Brazil
Gradiška *see* Bosanska Gradiška
Grafton *149 E5* New South Wales,
SE Australia
Grafton *45 E1* North Dakota, N USA
Graham Land *154 A2 physical region*
Antarctica
Grajewo *98 E3* Podlaskie, NE Poland
Grampian Mountains *88 C3
mountain range* C Scotland,
United Kingdom
Gran *see* Esztergom, Hungary
Granada *52 D3* Granada, SW Nicaragua
Granada *92 D5* Andalucía, S Spain
Gran Canaria *70 A3 var.* Grand Canary.
island Islas Canarias, Spain, NE Atlantic
Ocean
Gran Chaco *64 D2 var.* Chaco. *lowland
plain* South America
Grand Bahama Island *54 B1 island*
N Bahamas
Grand Banks of Newfoundland *34 E4
undersea basin* NW Atlantic Ocean
Grand Bassa *see* Buchanan
Grand Canary *see* Gran Canaria
Grand Canyon *48 A1 canyon* Arizona,
SW USA
Grand Canyon State *see* Arizona
Grand Cayman *54 B3 island*
SW Cayman Islands
Grand Duchy of Luxembourg *see*
Luxembourg
Grande, Bahía *65 B7 bay* S Argentina
Grande Comore *see* Ngazidja
Grande de Chiloé, Isla *see* Chiloé, Isla de
Grande Prairie *37 E4* Alberta, C Canada
Grand Erg Occidental *70 D3 desert*
W Algeria
Grand Erg Oriental *71 E3 desert*
Algeria/Tunisia
Rio Grande *52 E4 var.* Río Bravo, *Sp.* Río
Bravo del Norte, Bravo del Norte. *river*
Mexico/USA
Grande Terre *55 G3 island* E West Indies
Grand Falls *39 G3* Newfoundland,
Newfoundland and Labrador, SE Canada
Grand Forks *45 E1* North Dakota,
N USA
Grandichi *see* Hrandzichy
Grand Island *45 E4* Nebraska, C USA
Grand Junction *44 C4* Colorado, C USA
Grand Paradis *see* Gran Paradiso
Grand Rapids *40 C3* Michigan, N USA
Grand Rapids *45 F1* Minnesota, N USA
Grand-Saint-Bernard, Col du *see* Great
Saint Bernard Pass
Grand-Santi *59 G3* W French Guiana
Granite State *see* New Hampshire
Gran Lago *see* Nicaragua, Lago de
Gran Malvina *see* West Falkland
Gran Paradiso *96 A2 Fr.* Grand Paradis.
mountain NW Italy
Gran Santiago *see* Santiago
Grants *88 C2* New Mexico, SW USA
Grants Pass *46 B4* Oregon, NW USA
Granville *90 B3* Manche, N France
Gratianopolis *see* Grenoble
Gratz *see* Graz
Graudenz *see* Grudziądz
Graulhet *91 C6* Tarn, S France
Grave *86 D4* Noord-Brabant,
SE Netherlands
Grayling *36 C3* Alaska, USA
Graz *95 E7 prev.* Gratz; *anc.*
Graecium. Steiermark,
SE Austria
Great Abaco *54 C1 var.* Abaco Island.
island N Bahamas
Great Alfold *see* Great Hungarian Plain
Great arası *see* Büyükağrı Dağı
Great Australian Bight *147 D7 bight*
S Australia
Great Barrier Island *150 D2 island*
New Zealand
Great Barrier Reef *148 D2 reef*
Queensland, NE Australia
Great Basin *47 C5 basin* W USA
Great Bear Lake *37 E3 Fr.* Grand Lac
de l'Ours. *lake* Northwest Territories,
NW Canada
Great Belt *85 B8 var.* Store Bælt, *Eng.*
Great Belt; *Dan.* Storebælt. *channel* Baltic
Sea/Kattegat
Great Belt *see* Storebælt
Great Britain *45 E5* Kansas, C USA
Great Britain *see* Britain

Great Dividing Range 148 D4 *mountain range* NE Australia
Greater Antilles 54 D3 *island group* West Indies
Greater Caucasus 117 G2 *mountain range* Azerbaijan/Georgia/Russian Federation Asia/Europe
Greater Sunda Islands 124 D5 *var.* Sunda Islands. *island group* Indonesia
Great Exhibition Bay 150 C1 *inlet* North Island, New Zealand
Great Exuma Island 54 C2 *island* C Bahamas
Great Falls 44 B1 Montana, NW USA
Great Grimsby *see* Grimsby
Great Hungarian Plain 99 C7 *var.* Great Alföld, Plain of Hungary, *Hung.* Alföld. *plain* SE Europe
Great Inagua 54 D2 *var.* Inagua Islands. *island* S Bahamas
Great Indian Desert *see* Thar Desert
Great Khingan Range 127 G1 *Eng.* Great Khingan Range. *mountain range* NE China
Great Khingan Range *see* Da Hinggan Ling
Great Lake *see* Tônlé Sap
Great Lakes 35 C5 *lakes* Ontario, Canada/USA
Great Lakes State *see* Michigan
Great Meteor Seamount *see* Great Meteor Tablemount
Great Meteor Tablemount 66 B3 *var.* Great Meteor Seamount. *seamount* E Atlantic Ocean
Great Nicobar 133 G3 *island* Nicobar Islands, India, NE Indian Ocean
Great Plain of China 125 E2 *plain* E China
Great Plains 45 E3 *var.* High Plains. *plains* Canada/USA
Great Rift Valley 73 C5 *var.* Rift Valley. *depression* Asia/Africa
Great Ruaha 73 C7 *river* S Tanzania
Great Saint Bernard Pass 96 A1 *Fr.* Col du Grand-Saint-Bernard, *It.* Passo del Gran San Bernardo. *pass* Italy/ Switzerland
Great Salt Lake *see* Kavīr, Dasht-e
Great Salt Lake 44 A3 *salt lake* Utah, W USA
Great Salt Lake Desert 44 A4 *plain* Utah, W USA
Great Sand Sea 71 H3 *desert* Egypt/Libya
Great Sandy Desert 146 C4 *desert* Western Australia
Great Sandy Desert *see* Ar Rub' al Khālī
Great Sandy Island *see* Fraser Island
Great Slave Lake 37 E4 *Fr.* Grand Lac des Esclaves. *lake* Northwest Territories, NW Canada
Great Socialist People's Libyan Arab Jamahiriya *see* Libya
Great Sound 42 A5 *sound* Bermuda, NW Atlantic Ocean
Great Victoria Desert 147 C5 *desert* South Australia/Western Australia
Great Wall of China 128 C4 *ancient monument* N China Asia
Great Yarmouth 89 E6 *var.* Yarmouth. E England, United Kingdom
Grebenka *see* Hrebinka
Gredos, Sierra de 92 D3 *mountain range* W Spain
Greece 105 A5 *off.* Hellenic Republic, *Gk.* Ellás; *anc.* Hellas. *country* SE Europe

GREECE
Southeast Europe

Official name Hellenic Republic
Formation 1829 / 1947
Capital Athens
Population 11.1 million / 220 people per sq mile (85 people per sq km)
Total area 50,942 sq. miles (131,940 sq. km)
Languages Greek*, Turkish, Macedonian, Albanian
Religions Orthodox Christian 98%, Muslim 1%, Other 1%
Ethnic mix Greek 98%, Other 2%
Government Parliamentary system
Currency Euro = 100 cents
Literacy rate 97%
Calorie consumption 3433 kilocalories

Greeley 44 D4 Colorado, C USA
Green Bay 40 B2 Wisconsin, N USA
Green Bay 40 B2 *lake bay* Michigan/ Wisconsin, N USA
Greeneville 43 E1 Tennessee, S USA
Greenland 82 D3 *Dan.* Grønland, *Inuit* Kalaallit Nunaat. *Danish external territory* NE North America
Greenland Sea 83 F2 *sea* Arctic Ocean
Green Mountains 41 G2 *mountain range* Vermont, NE USA
Green Mountain State *see* Vermont
Greenock 88 C4 W Scotland, United Kingdom
Green River 44 B3 Wyoming, C USA
Green River 40 C5 *river* Kentucky, C USA
Green River 44 B4 *river* Utah, W USA
Greensboro 43 F1 North Carolina, SE USA
Greenville 42 B2 Mississippi, S USA
Greenville 43 F1 North Carolina, SE USA
Greenville 43 E1 South Carolina, SE USA
Greenville 49 G2 Texas, SW USA
Greenwich 89 B8 United Kingdom
Greenwood 42 B2 Mississippi, S USA
Greenwood 43 E2 South Carolina, SE USA
Gregory Range 148 C3 *mountain range* Queensland, E Australia
Greifenberg/Greifenberg in Pommern *see* Gryfice
Greifswald 94 D2 Mecklenburg-Vorpommern, NE Germany
Grenada 42 C2 Mississippi, S USA
Grenada 55 G5 *country* SE West Indies

GRENADA
West Indies

Official name Grenada
Formation 1974 / 1974
Capital St. George's
Population 109,590 / 837 people per sq mile (322 people per sq km)
Total area 131 sq. miles (340 sq. km)
Languages English*, English Creole
Religions Roman Catholic 68%, Anglican 17%, Other 15%
Ethnic mix Black African 82%, Mulatto (mixed race) 13%, East Indian 3%, Other 2%
Government Parliamentary system
Currency East Caribbean dollar = 100 cents
Literacy rate 96%
Calorie consumption 2453 kilocalories

Grenadines, The 55 H4 *island group* Grenada/St Vincent and the Grenadines
Grenoble 91 D5 *anc.* Cularo, Gratianopolis. Isère, E France
Gresham 46 B3 Oregon, NW USA
Grevená 104 B4 Dytikí Makedonía, N Greece
Grevenmacher 87 E8 Grevenmacher, E Luxembourg
Greymouth 151 B5 West Coast, South Island, New Zealand
Grey Range 149 C5 *mountain range* New South Wales/Queensland, E Australia
Greytown *see* San Juan del Norte
Griffin 42 D2 Georgia, SE USA
Grimari 76 C4 Ouaka, C Central African Republic
Grimsby 89 E5 *prev.* Great Grimsby. E England, United Kingdom
Grobin *see* Grobiņa
Grobiņa 106 B3 *Ger.* Grobin. W Latvia
Gródek Jagielloński *see* Horodok
Grodno *see* Hrodna
Grodzisk Wielkopolski 98 B3 Wielkopolskie, C Poland
Groesbeek 86 D4 Gelderland, SE Netherlands
Grójec 98 D4 Mazowieckie, C Poland
Groningen 86 E1 Groningen, NE Netherlands
Grønland *see* Greenland
Groote Eylandt 148 B2 *island* Northern Territory, N Australia
Grootfontein 78 B3 Otjozondjupa, N Namibia
Groot Karasberge 78 B4 *mountain range* S Namibia
Gros Islet 55 F1 N Saint Lucia
Grossa, Isola *see* Dugi Otok
Grossbetschkerek *see* Zrenjanin
Grosse Morava *see* Velika Morava
Grosser Sund *see* Suur Väin

Grosseto 96 B4 Toscana, C Italy
Grossglockner 95 C7 *mountain* W Austria
Grosskanizsa *see* Nagykanizsa
Gross-Karol *see* Carei
Grosskikinda *see* Kikinda
Grossmichel *see* Michalovce
Gross-Schlatten *see* Abrud
Grosswardein *see* Oradea
Groznyy 111 B8 Chechenskaya Respublika, SW Russian Federation
Grudovo *see* Sredets
Grudziądz 98 C3 *Ger.* Graudenz. Kujawsko-pomorskie, C Poland
Grums 85 B6 Värmland, C Sweden
Grünberg/Grünberg in Schlesien *see* Zielona Góra
Grüneberg *see* Zielona Góra
Gruzinskaya SSR/Gruziya *see* Georgia
Gryazi 111 B6 Lipetskaya Oblast', W Russian Federation
Gryfice 98 B2 *Ger.* Greifenberg, Greifenberg in Pommern. Zachodniopomorskie, NW Poland
Guabito 53 E4 Bocas del Toro, NW Panama
Guadalajara 50 D4 Jalisco, C Mexico
Guadalajara 93 E3 *Ar.* Wad Al-Hajarah; *anc.* Arriaca. Castilla-La Mancha, C Spain
Guadalcanal 144 C3 *island* C Solomon Islands
Guadalquivir 92 D4 *river* W Spain
Guadalupe 50 D3 Zacatecas, C Mexico
Guadalupe Peak 48 D3 *mountain* Texas, SW USA
Guadalupe River 49 G4 *river* SW USA
Guadarrama, Sierra de 93 E2 *mountain range* C Spain
Guadeloupe 55 H3 *French overseas department* E West Indies
Guadiana 92 C4 *river* Portugal/Spain
Guadix 93 E4 Andalucía, S Spain
Guaimaca 52 C2 Francisco Morazán, C Honduras
Guajira, Península de la 58 B1 *peninsula* N Colombia
Gualaco 52 D2 Olancho, C Honduras
Gualán 52 B2 Zacapa, C Guatemala
Gualdicciolo 96 D1 NW San Marino
Gualeguaychú 64 D4 Entre Ríos, E Argentina
Guam 144 B1 *US unincorporated territory* W Pacific Ocean
Guamúchil 50 C3 Sinaloa, C Mexico
Guanabacoa 54 B2 La Habana, W Cuba
Guanajuato 51 E4 Guanajuato, C Mexico
Guanare 58 C2 Portuguesa, N Venezuela
Guanare, Río 58 D2 *river* W Venezuela
Guangdong 128 C6 *var.* Guangdong Sheng, Kuang-tung, Kwangtung, Yue. *province* S China
Guangdong Sheng *see* Guangdong
Guangju *see* Gwangju
Guangxi *see* Guangxi Zhuangzu Zizhiqu
Guangxi Zhuangzu Zizhiqu 128 C6 *var.* Guangxi, Gui, Kuang-hsi, Kwangsi, *Eng.* Kwangsi Chuang Autonomous Region. *autonomous region* S China
Guangyuan 128 B5 *var.* Kuang-yuan, Kwangyuan. Sichuan, C China
Guangzhou 128 C6 *var.* Kuang-chou, Kwangchow, *Eng.* Canton. *province capital* Guangdong, S China
Guantánamo 54 D3 Guantánamo, SE Cuba
Guantánamo, Bahía de 33 Eng. Guantanamo Bay. *US military base* SE Cuba
Guantanamo Bay *see* Guantánamo, Bahía de
Guapí 58 A4 Cauca, SW Colombia
Guaporé, Río 62 D3 *var.* Río Iténez. *river* Bolivia/Brazil
Guárico 58 D2 Veraguas, S Panama
Guasave 50 C3 Sinaloa, C Mexico
Guatemala 52 A2 *off.* Republic of Guatemala. *country* Central America

GUATEMALA
Central America

Official name Republic of Guatemala
Formation 1838 / 1838
Capital Guatemala City
Population 15.5 million / 370 people per sq mile (143 people per sq km)

GUATEMALA
(continued)

Total area 42,042 sq. miles (108,890 sq. km)
Languages Quiché, Mam, Cakchiquel, Kekchí, Spanish*
Religions Roman Catholic 65%, Protestant 33%, Other and nonreligious 2%
Ethnic mix Amerindian 60%, Mestizo 30%, Other 10%
Government Presidential system
Currency Quetzal = 100 centavos
Literacy rate 78%
Calorie consumption 2502 kilocalories

Guatemala Basin 35 B7 *undersea* E Pacific Ocean
Guatemala City *see* Ciudad de Guatemala
Guatemala, Republic of *see* Guatemala
Guaviare 56 B2 *off.* Comisaría Guaviare. *province* S Colombia
Guaviare, Comisaría *see* Guaviare
Guaviare, Río 58 D3 *river* E Colombia
Guayanas, Macizo de las *see* Guiana Highlands
Guayaquil 60 A2 *var.* Santiago de Guayaquil. Guayas, SW Ecuador
Guayaquil, Gulf of 60 A2 *var.* Gulf Guayaquil. *gulf* SW Ecuador
Guayaquil, Gulf of *see* Guayaquil, Golfo de
Guaymas 50 B2 Sonora, NW Mexico
Gubadag 122 C2 *Turkm.* Tel'man; *prev.* Tel'mansk. Dasoguz Welaýy N Turkmenistan
Guben 94 D4 *var.* Wilhelm-Pieck-Stadt Brandenburg, E Germany
Gudara *see* Ghŭdara
Gudauta 117 E1 *prev.* Gudaut'a. NW Georgia
Gudaut'a *see* Gudauta
Guéret 90 C4 Creuse, C France
Guernsey 89 D8 *British Crown Dependency* Channel Islands, NW Europe
Guerrero Negro 50 A2 Baja California Sur, W Mexico
Gui *see* Guangxi Zhuangzu Zizhiqu
Guiana *see* French Guiana
Guiana Highlands 62 D1 *var.* Macizo de las Guayanas. *mountain range* N America
Guiba *see* Juba
Guidder *see* Guider
Guider 76 B4 *var.* Guidder. Nord, N Cameroon
Guidimouni 75 G3 Zinder, S Niger
Guildford 89 D7 SE England, United Kingdom
Guilin 128 C6 *var.* Kuei-lin, Kweilin Guangxi Zhuangzu Zizhiqu, S China
Guimarães 92 B2 *var.* Guimaráes. N Portugal
Guimaráes *see* Guimarães
Guinea 74 C4 *off.* Republic of Guinea *var.* Guinée; *prev.* French Guinea, People's Revolutionary Republic of Guinea. *country* W Africa

GUINEA
West Africa

Official name Republic of Guinea
Formation 1958 / 1958
Capital Conakry
Population 11.7 million / 123 people per sq mile (48 people per sq km)
Total area 94,925 sq. miles (245,857 sq. km)
Languages Pulaar, Malinké, Soussou, French
Religions Muslim 85%, Christian 8%, Traditional beliefs 7%
Ethnic mix Peul 40%, Malinké 30%, Soussou 20%, Other 10%
Government Transitional regime
Currency Guinea franc = 100 centimes
Literacy rate 29%
Calorie consumption 2553 kilocalories

Guinea Basin 69 A5 *undersea basin* E Atlantic Ocean
Guinea-Bissau 74 B4 *off.* Republic Guinea-Bissau, *Fr.* Guinée-Bissau, *Port.* Guiné-Bissau, *prev.* Portuguese Guinea. *country* W Africa

JINEA-BISSAU
West Africa

ficial name Republic of Guinea-Bissau
rmation 1974 / 1974
pital Bissau
pulation 1.7 million / 157 people
r sq mile (60 people per sq km)
tal area 13,946 sq. miles (36,120 sq. km)
nguages Portuguese Creole, Balante,
ani, Malinké, Portuguese*
ligions Traditional beliefs 50%,
slim 40%, Christian 10%
nic mix Balante 30%, Fulani 20%,
her 16%, Mandyako 14%,
andinka 13%, Papel 7%
vernment Transitional regime
rrency CFA franc = 100 centimes
eracy rate 57%
lorie consumption 2304 kilocalories

nea-Bissau, Republic of *see*
Guinea-Bissau
nea, Gulf of *68 B4 Fr.* Golfe de
Guinée. *gulf* E Atlantic Ocean
nea, People's Revolutionary
Republic of *see* Guinea
nea, Republic of *see* Guinea
né-Bissau *see* Guinea-Bissau
née *see* Guinea
née-Bissau *see* Guinea-Bissau
née, Golfe de *see* Guinea, Gulf of
ria *59 E1* Sucre, NE Venezuela
yang *128 B6* var. Kuei-Yang,
Kuei-yang, Kueyang, Kweiyang; *prev.*
Kweichu. *province capital* Guizhou,
China
zhou *128 B6* Guangdong, SE China
arāt *134 C4* var. Gujerat. *cultural
region* W India
erat *see* Gujarāt
rānwāla *134 D2* Punjab, NE Pakistan
rāt *134 D2* Punjab, E Pakistan
barga *132 C1* Karnātaka, C India
bene *106 D3* Ger. Alt-Schwanenburg.
NE Latvia
f of Liaotung *see* Liaodong Wan
fport *42 C3* Mississippi, S USA
f, The *see* Persian Gulf
istan *see* Guliston
iston *123 E2* Rus. Gulistan. Sirdaryo
Viloyati, E Uzbekistan
kana *36 D3* Alaska, USA
lu *73 B6* N Uganda
yantsi *104 C1* Pleven, N Bulgaria
ma *see* Pishan
mbinnen *see* Gusev
mpolds *see* Humpolec
mülcine/Gümüljina *see* Komotiní
müşane *see* Gümüşhane
müşhane *117 E3* var. Gümüşane,
Gumushkhane. Gümüşhane,
NE Turkey
mushkhane *see* Gümüşhane
ney Doğu Toroslar *117 E4 mountain
range* SE Turkey
nnbjørn Fjeld *82 D4* var. Gunnbjörns
Bjerge. *mountain* C Greenland
nnbjörns Bjerge *see* Gunnbjørn Fjeld
nnedah *149 D6* New South Wales,
SE Australia
nnison *44 C5* Colorado, C USA
rbansoltan Eje *122 C2 prev.* Ýylanly,
Rus. Il'yaly. Daşoguz Welaýaty,
N Turkmenistan
sau *75 G4* Zamfara, NW Nigeria
sev *106 B4* Ger. Gumbinnen.
Kaliningradskaya Oblast', W Russian
Federation
stavus *36 D4* Alaska, USA
strow *94 C3* Mecklenburg-
Vorpommern, NE Germany
ta/Gúta *see* Kolárovo
tersloh *94 B4* Nordrhein-Westfalen,
W Germany
tta *see* Kolárovo

Guttadt *see* Dobre Miasto
Guwāhāti *135 G3 prev.* Gauhāti. Assam,
NE India
Guyana *59 F3 off.* Co-operative Republic
of Guyana; *prev.* British Guiana.
country N South America

GUYANA
South America

Official name Cooperative Republic
of Guyana
Formation 1966 / 1966
Capital Georgetown
Population 800,000 / 11 people
per sq mile (4 people per sq km)
Total area 83,000 sq. miles
(214,970 sq. km)
Languages English Creole, Hindi, Tamil,
Amerindian languages, English*
Religions Christian 57%, Hindu 28%,
Muslim 10%, Other 5%
Ethnic mix East Indian 43%, Black
African 30%, Mixed race 17%,
Amerindian 9%, Other 1%
Government Presidential system
Currency Guyanese dollar = 100 cents
Literacy rate 85%
Calorie consumption 2648 kilocalories

Guyana, Co-operative Republic of
see Guyana
Guyane *see* French Guiana
Guymon *49 E1* Oklahoma, C USA
Güzelyurt Körfezi *102 C5 Gk.* Kólpos
Mórfu, Morphou. W Cyprus
Gvardeysk *106 A4* Ger. Tapiau.
Kaliningradskaya Oblast', W Russian
Federation
Gwādar *134 A3* var. Gwadur.
Baluchistān, SW Pakistan
Gwadur *see* Gwādar
Gwalior *134 D3* Madhya Pradesh,
C India
Gwanda *78 D3* Matabeleland South,
SW Zimbabwe
Gwangju *129 E4 off.* Gwangju
Gwang-yeoksi, *prev.* Kwangju, *var.*
Guangju, Kwangchu, *Jap.* Kōshū.
SW South Korea
Gwangju Gwang-yeoksi *see* Gwangju
Gwy *see* Wye
Gyandzha *see* Gäncä
Gyangzê *126 C5* Xizang Zizhiqu,
W China
Gyaring Co *126 C5 lake* W China
Gyégu *see* Yushu
Gyergyószentmiklós *see* Gheorgheni
Gyixong *see* Gonggar
Gympie *149 E5* Queensland, E Australia
Gyomaendrőd *99 D7* Békés,
SE Hungary
Gyöngyös *99 D6* Heves, NE Hungary
Győr *99 C6* Ger. Raab, *Lat.* Arrabona.
Győr-Moson-Sopron, NW Hungary
Gýtheio *105 B6 var.* Githio; *prev.*
Ýíthion. Pelopónnisos, S Greece
Gyulafehérvár *see* Alba Iulia
Gyumri *117 F2 var.* Gümri, *Rus.*
Kumayri; *prev.* Aleksandropol',
Leninakan. W Armenia
Gyzylarbat *see* Serdar

H

Haabai *see* Ha'apai Group
Haacht *87 C6* Vlaams Brabant,
C Belgium
Haaksbergen *86 E3* Overijssel,
E Netherlands
Ha'apai Group *145 F4 var.* Haabai.
island group C Tonga
Haapsalu *106 D2* Ger. Hapsal.
Läänemaa, W Estonia
Ha'Arava *see* 'Arabah, Wādī al
Haarlem *86 C3 prev.* Harlem. Noord-
Holland, W Netherlands
Haast *151 B6* West Coast, South Island,
New Zealand
Hachijō-jima *131 D6 island* Izu-shotō,
SE Japan
Hachinohe *130 D3* Aomori, Honshū,
C Japan
Hacıqabal *117 H3 prev.* Qazimämmäd.
SE Azerbaijan
Hadâ bat al Jilf al Kabïr *72 A2 var.* Gilf
Kebir Plateau. *plateau* SW Egypt
Hadama *see* Nazrêt

Hadejia *75 G4* Jigawa, N Nigeria
Hadejia *75 G3 river* N Nigeria
Hadera *119 A6 var.* Khadera; *prev.*
Ḥadera. Haifa, N Israel
Ḥadera *see* Hadera
Hadhdhunmathi Atoll *132 A5 atoll*
S Maldives
Ha Đông *136 D3 var.* Hadong. Ha Tây,
N Vietnam
Hadong *see* Ha Đông
Hadramaut *121 C6 Eng.* Hadramaut.
mountain range S Yemen
Hadramaut *see* Ḥaḍramawt
Hadrianopolis *see* Edirne
Haerbin/Haerhpin/Ha-erh-pin *see*
Harbin
Hafnia *see* Denmark
Hafnia *see* København
Hafren *see* Severn
Hafun, Ras *see* Xaafuun, Raas
Hagåtña *144 B1 , var.* Agaña. *dependent
territory capital* (Guam) NW Guam
Hagerstown *41 E4* Maryland, NE USA
Ha Giang *136 D3* Ha Giang, N Vietnam
Hagios Evstrátios *see* Ágios Efstrátios
HaGolan *see* Golan Heights
Hagondange *90 D3* Moselle, NE France
Haguenau *90 E3* Bas-Rhin, NE France
Haibowan *see* Wuhai
Haicheng *128 D3* Liaoning, NE China
Haidarabad *see* Hyderābād
Haifa, Bay of *see* Mifrats Hefa
Haifong *see* Hai Phong
Haikou *128 C7 var.* Hai-k'ou, Hoihow,
Fr. Hoï-Hao. *province capital* Hainan,
S China
Hai-k'ou *see* Haikou
Hā'il *120 B4* Ḥā'il, NW Saudi Arabia
Hailuoto *84 D4 Swe.* Karlö. *island*
W Finland
Hainan *128 B7 var.* Hainan Sheng,
Qiong. *province* S China
Hainan Dao *128 C7 island* S China
Hainan Sheng *see* Hainan
Hainasch *see* Ainaži
Haines *36 D4* Alaska, USA
Hainichen *95 D4* Sachsen, E Germany
Hai Phong *136 D3 var.* Haifong,
Haiphong. N Vietnam
Haiphong *see* Hai Phong
Haiti *54 D3 off.* Republic of Haiti.
country C West Indies

HAITI
West Indies

Official name Republic of Haiti
Formation 1804 / 1844
Capital Port-au-Prince
Population 10.3 million / 968 people
per sq mile (374 people per sq km)
Total area 10,714 sq. miles (27,750 sq. km)
Languages French Creole*, French*
Religions Roman Catholic 55%,
Protestant 28%, Other (including
Voodoo) 16%, Nonreligious 1%
Ethnic mix Black African 95%,
Mulatto (mixed race) and European 5%
Government Presidential system
Currency Gourde = 100 centimes
Literacy rate 49%
Calorie consumption 2105 kilocalories

Haiti, Republic of *see* Haiti
Haiya *72 C3* Red Sea, NE Sudan
Hajdúhadház *99 D6* Hajdú-Bihar,
E Hungary
Hajine *see* Abū Ḥardan
Hajnówka *98 E3 Ger.* Hermhausen.
Podlaskie, NE Poland
Hakodate *130 D3* Hokkaidō, NE Japan
Hal *see* Halle
Ḥalab *118 B2 Eng.* Aleppo, *Fr.* Alep; *anc.*
Beroea. Ḥalab, NW Syria
Hala'ib Triangle *72 C3 disputed
region* NE Sudan
Ḥalāniyāt, Juzur al *121 D6 var.* Jazā'ir
Bin Ghalfān, *Eng.* Kuria Muria Islands.
island group S Oman
Halberstadt *94 C4* Sachsen-Anhalt,
C Germany
Halden *85 B6 prev.* Fredrikshald.
Østfold, S Norway
Halfmoon Bay *151 A8 var.* Oban.
Stewart Island, Southland, New Zealand
Haliacmon *see* Aliákmonas
Halifax *39 F4 province capital* Nova
Scotia, SE Canada

Halkida *see* Chalkída
Halle *87 B6 Fr.* Hal. Vlaams Brabant,
C Belgium
Halle *94 C4 var.* Halle an der Saale.
Sachsen-Anhalt, C Germany
Halle an der Saale *see* Halle
Halle-Neustadt *94 C4* Sachsen-Anhalt,
C Germany
Halley *154 B2 UK research station*
Antarctica
Hall Islands *142 B2 island group*
C Micronesia
Halls Creek *146 C3* Western Australia
Halmahera, Laut *139 F4 Eng.*
Halmahera Sea. *sea* E Indonesia
Halmahera, Pulau *139 F3 prev.* Djailolo,
Gilolo, Jailolo. *island* E Indonesia
Halmahera Sea *see* Halmahera, Laut
Halmstad *85 B7* Halland, S Sweden
Ha Long *136 E3 prev.* Hông Gai, *var.*
Hon Gai, Hongay. Quang Ninh,
N Vietnam
Hälsingborg *see* Helsingborg
Hamada *131 B6* Shimane, Honshū,
SW Japan
Hamadān *120 C3 anc.* Ecbatana.
Hamadān, W Iran
Ḥamāh *118 B3 var.* Hama; *anc.*
Epiphania, *Bibl.* Hamath. Ḥamāh,
W Syria
Hamamatsu *131 D6 var.* Hamamatu.
Shizuoka, Honshū, S Japan
Hamamatu *see* Hamamatsu
Hamar *85 B5 prev.* Storhammer.
Hedmark, S Norway
Hamath *see* Ḥamāh
Hamburg *94 B3* Hamburg, N Germany
Hamd, Wadi al *120 A4 dry watercourse*
W Saudi Arabia
Hämeenlinna *85 D5 Swe.* Tavastehus.
Etelä-Suomi, S Finland
HaMela h, Yam *see* Dead Sea
Hamersley Range *146 A4 mountain
range* Western Australia
Hamhŭng *129 E3* C North Korea
Hami *128 C3 var.* Ha-mi, *Uigh.* Kumul,
Qomul. Xinjiang Uygur Zizhiqu,
NW China
Ha-mi *see* Hami
Hamilton *42 A5 dependent territory
capital* (Bermuda) C Bermuda
Hamilton *150 D3* Waikato, North Island,
New Zealand
Hamilton *88 C4* Scotland, United
Kingdom
Hamilton *42 C2* Alabama, S USA
Hamim, Wadi al *71 G2 river* NE Libya
Hamïs Musait *see* Khamïs Mushayt
Hamm *94 B4 var.* Hamm in Westfalen.
Nordrhein-Westfalen, W Germany
Ḥammāmāt, Khalïj al *see* Hammamet,
Golfe de
Hammamet, Golfe de *102 D3 Ar.* Khalïj
al Ḥammāmāt. *gulf* NE Tunisia
Ḥammār, Hawr al *120 C3 lake* SE Iraq
Hamm in Westfalen *see* Hamm
Hampden *151 B7* Otago, South Island,
New Zealand
Hampstead *89 A7* Maryland, USA
Hamrun *102 B5* C Malta
Hämün, Daryächeh-ye *see* Şāberī,
Hāmün-e
Hamwih *see* Southampton
Hâncești *see* Hîncești
Hancewicze *see* Hantsavichy
Handan *128 C4* Hebei, E China
Hanford *47 C6* California, N USA
Hangayn Nuruu *126 D2 mountain
range* C Mongolia
Hang-chou/Hangchow *see* Hangzhou
Hangö *see* Hanko
Hangzhou *128 D5 var.* Hang-chou,
Hangchow. *province capital* Zhejiang,
SE China
Hangzhou *see* Chaniá
Hanka, Lake *see* Khanka, Lake
Hanko *85 D6 Swe.* Hangö. Etelä-Suomi,
SW Finland
Han-kou/Han-k'ou/Hankow *see* Wuhan
Hanmer Springs *151 C5* Canterbury,
South Island, New Zealand
Hannibal *45 G4* Missouri, C USA
Hannover *94 B3 Eng.* Hanover.
Niedersachsen, NW Germany
Hanöbukten *85 B7 bay* S Sweden

Ha Nôi *136 D3 Eng.* Hanoi, *Fr.* Hanoï. *country capital* (Vietnam) N Vietnam
Hanover *see* Hannover
Han Shui *127 E4 river* C China
Han-tan *see* Handan
Hantsavichy *107 B6 Pol.* Hancewicze, *Rus.* Gantsevichi. Brestskaya Voblasts', SW Belarus
Hanyang *see* Wuhan
Hanzhong *128 B5* Shaanxi, C China
Hāora *135 F4 prev.* Howrah. West Bengal, NE India
Haparanda *84 D4* Norrbotten, N Sweden
Hapsal *see* Haapsalu
Haradok *107 E5 Rus.* Gorodok. Vitsyebskaya Voblasts', N Belarus
Haradzyets *107 B6 Rus.* Gorodets. Brestskaya Voblasts', SW Belarus
Haramachi *130 D4* Fukushima, Honshū, E Japan
Harany *107 D5 Rus.* Gorany. Vitsyebskaya Voblasts', N Belarus
Harare *78 D3 prev.* Salisbury. *country capital* (Zimbabwe) Mashonaland East, NE Zimbabwe
Harbavichy *107 E6 Rus.* Gorbovichi. Mahilyowskaya Voblasts', E Belarus
Harbel *74 C5* W Liberia
Harbin *129 E2 var.* Haerbin, Ha-erh-pin, Kharbin; *prev.* Haerhpin, Pingkiang, Pinkiang. *province capital* Heilongjiang, NE China
Hardangerfjorden *85 A6 fjord* S Norway
Hardangervidda *85 A6 plateau* S Norway
Hardenberg *86 E3* Overijssel, E Netherlands
Harelbeke *87 A6 var.* Harlebeke. West-Vlaanderen, W Belgium
Harem *see* Ḥārim
Haren *86 E2* Groningen, NE Netherlands
Härer *73 D5* E Ethiopia
Hargeysa *73 D5 var.* Hargeisa. Woqooyi Galbeed, NW Somalia
Hariana *see* Haryāna
Hari, Batang *138 B4 prev.* Djambi. *river* Sumatera, W Indonesia
Ḥārim *118 B2 var.* Harem. Idlib, W Syria
Harima-nada *131 B6 sea* S Japan
Harirud *123 E4 var.* Tedzhen, *Turkm.* Tejen. *river* Afghanistan/Iran
Harlan *45 F3* Iowa, C USA
Harlebeke *see* Harelbeke
Harlem *see* Haarlem
Harlingen *86 D2 Fris.* Harns. Fryslân, N Netherlands
Harlingen *49 G5* Texas, SW USA
Harlow *89 E6* E England, United Kingdom
Harney Basin *46 B4 basin* Oregon, NW USA
Härnösand *85 C5 var.* Hernösand. Västernorrland, C Sweden
Harns *see* Harlingen
Harper *74 D5 var.* Cape Palmas. NE Liberia
Harricana *38 D3 river* Québec, SE Canada
Harris *88 B3 physical region* NW Scotland, United Kingdom
Harrisburg *41 E4 state capital* Pennsylvania, NE USA
Harrisonburg *41 E4* Virginia, NE USA
Harrison, Cape *39 F2 headland* Newfoundland and Labrador, E Canada
Harris Ridge *see* Lomonosov Ridge
Harrogate *89 D5* N England, United Kingdom
Hârşova *108 D5 prev.* Hîrşova. Constanţa, SE Romania
Harstad *84 C2* Troms, N Norway
Hartford *41 G3 state capital* Connecticut, NE USA
Hartlepool *89 D5* N England, United Kingdom
Harunabad *see* Eslāmābād-e Gharb
Har Us Gol *126 C2 lake* Hovd, W Mongolia
Har Us Nuur *126 C2 lake* NW Mongolia
Harwich *89 E6* E England, United Kingdom
Haryāna *134 D2 var.* Hariana. *cultural region* N India
Hashemite Kingdom of Jordan *see* Jordan
Hasselt *87 C6* Limburg, NE Belgium

Hassetché *see* Al Ḥasakah
Hasta Colonia/Hasta Pompeia *see* Asti
Hastings *150 E4* Hawke's Bay, North Island, New Zealand
Hastings *89 E7* SE England, United Kingdom
Hastings *45 E4* Nebraska, C USA
Haţeg *108 B4 Ger.* Wallenthal, *Hung.* Hátszeg; *prev.* Hatzeg, Hötzing. Hunedoara, SW Romania
Hátszeg *see* Haţeg
Hattem *86 D3* Gelderland, E Netherlands
Hatteras, Cape *43 G1 headland* North Carolina, SE USA
Hatteras Plain *35 D6 abyssal plain* W Atlantic Ocean
Hattiesburg *42 C3* Mississippi, S USA
Hatton Bank *see* Hatton Ridge
Hatton Ridge *80 B2 var.* Hatton Bank. *undersea ridge* N Atlantic Ocean
Hat Yai *137 C7 var.* Ban Hat Yai. Songkhla, SW Thailand
Hatzeg *see* Haţeg
Hatzfeld *see* Jimbolia
Haugesund *85 A6* Rogaland, S Norway
Haukeligrend *85 A6* Telemark, S Norway
Haukivesi *85 E5 lake* SE Finland
Hauraki Gulf *150 D2 gulf* North Island, N New Zealand
Hauroko, Lake *151 A7 lake* South Island, New Zealand
Hautes Fagnes *87 D6 Ger.* Hohes Venn. *mountain range* E Belgium
Hauts Plateaux *70 D2 plateau* Algeria/Morocco
Hauzenberg *95 D6* Bayern, SE Germany
Havana *35 D6* Illinois, N USA
Havana *see* La Habana
Havant *89 D7* S England, United Kingdom
Havelock *43 F1* North Carolina, SE USA
Havelock North *150 E4* Hawke's Bay, North Island, New Zealand
Haverfordwest *89 C6* SW Wales, United Kingdom
Havířov *99 C5* Moravskoslezský Kraj, E Czech Republic
Havre *44 C1* Montana, NW USA
Havre *see* le Havre
Havre-St-Pierre *39 F3* Québec, E Canada
Hawaii *47 A8 off.* State of Hawaii, *also known as* Aloha State, Paradise of the Pacific, *var.* Hawai'i. *state* USA, C Pacific Ocean
Hawai'i *47 B8 var.* Hawaii. *island* Hawai'ian Islands, USA, C Pacific Ocean
Hawai'ian Islands *152 D2 prev.* Sandwich Islands. *island group* Hawaii, USA
Hawaiian Ridge *152 H4 undersea ridge* N Pacific Ocean
Hawea, Lake *151 B6 lake* South Island, New Zealand
Hawera *150 D4* Taranaki, North Island, New Zealand
Hawick *88 C4* SE Scotland, United Kingdom
Hawke Bay *150 E4 bay* North Island, New Zealand
Hawkeye State *see* Iowa
Hawlêr *see* Arbil
Hawthorne *47 C6* Nevada, W USA
Hay *149 C6* New South Wales, SE Australia
HaYarden *see* Jordan
Hayastani Hanrapetut'yun *see* Armenia
Hayes *38 D3 river* Manitoba, C Canada
Hay River *37 E4* Northwest Territories, W Canada
Hays *45 E5* Kansas, C USA
Haysyn *108 D3 Rus.* Gaysin. Vinnyts'ka Oblast', C Ukraine
Hazar *122 B2 prev. Rus.* Cheleken. Balkan Welaýaty, W Turkmenistan
Heard and McDonald Islands *141 B7 Australian external territory* S Indian Ocean
Hearst *38 C4* Ontario, S Canada
Heart of Dixie *see* Alabama
Heathrow *89 A8* (London) SE England, United Kingdom
Hebei *128 C4 var.* Hebei Sheng, Hopeh, Hopei, Ji; *prev.* Chihli. *province* E China
Hebei Sheng *see* Hebei

Hebron *119 A6 var.* Al Khalīl, El Khalil, *Heb.* Hevron; *anc.* Kiriath-Arba. S West Bank
Heemskerk *86 C3* Noord-Holland, W Netherlands
Heerde *86 D3* Gelderland, E Netherlands
Heerenveen *86 D2 Fris.* It Hearrenfean. Fryslân, N Netherlands
Heerhugowaard *86 C2* Noord-Holland, NW Netherlands
Heerlen *87 D6* Limburg, SE Netherlands
Heerwegen *see* Polkowice
Hefa *119 A5 var.* Haifa, *hist.* Caiffa, Caiphas; *anc.* Sycaminum. Haifa, N Israel
Hefa, Mifraz *see* Mifrats Hefa
Hefei *128 D5 var.* Hofei, *hist.* Luchow. *province capital* Anhui, E China
Hegang *129 E2* Heilongjiang, NE China
Hei *see* Heilongjiang
Heide *94 B2* Schleswig-Holstein, N Germany
Heidelberg *95 B5* Baden-Württemberg, SW Germany
Heidenheim *see* Heidenheim an der Brenz
Heidenheim an der Brenz *95 B6 var.* Heidenheim. Baden-Württemberg, S Germany
Hei-ho *see* Nagqu
Heilbronn *95 B6* Baden-Württemberg, SW Germany
Heiligenbeil *see* Mamonovo
Heilongjiang *128 D2 var.* Hei, Heilongjiang Sheng, Hei-lung-chiang, Heilungkiang. *province* NE China
Heilong Jiang *see* Amur
Heilongjiang Sheng *see* Heilongjiang
Heiloo *86 C3* Noord-Holland, NW Netherlands
Heilsberg *see* Lidzbark Warmiński
Hei-lung-chiang/Heilungkiang *see* Heilongjiang
Heimdal *85 B5* Sør-Trøndelag, S Norway
Heinaste *see* Ainaži
Hekimhan *116 D3* Malatya, C Turkey
Helena *44 B2 state capital* Montana, NW USA
Helensville *150 D2* Auckland, North Island, New Zealand
Helgoland Bay *see* Helgoländer Bucht
Helgoländer Bucht *94 A2 var.* Helgoland Bay, Heligoland Bight. *bay* NW Germany
Heligoland Bight *see* Helgoländer Bucht
Heliopolis *see* Baalbek
Hellas *see* Greece
Hellenic Republic *see* Greece
Hellevoetsluis *86 B4* Zuid-Holland, SW Netherlands
Hellín *93 E4* Castilla-La Mancha, C Spain
Darya-ye Helmand *122 D5 var.* Rūd-e Hīrmand. *river* Afghanistan/Iran
Helmantica *see* Salamanca
Helmond *87 D5* Noord-Brabant, S Netherlands
Helsingborg *85 B7 prev.* Hälsingborg. Skåne, S Sweden
Helsingfors *see* Helsinki
Helsinki *85 D6 Swe.* Helsingfors. *country capital* (Finland) Etelä-Suomi, S Finland
Heltau *see* Cisnădie
Helvetia *see* Switzerland
Henan *128 C5 var.* Henan Sheng, Honan, Yu. *province* C China
Henderson *40 D5* Kentucky, S USA
Henderson *47 D7* Nevada, W USA
Henderson *49 H3* Texas, SW USA
Hendū Kosh *see* Hindu Kush
Hengchun *see* Hengyang
Hengduan Shan *128 A5 mountain range* SW China
Hengelo *86 E3* Overijssel, E Netherlands
Hengnan *see* Hengyang
Hengyang *128 C6 var.* Hengnan, Heng-yang; *prev.* Hengchow. Hunan, S China
Heng-yang *see* Hengyang
Henichesk *109 F4 Rus.* Genichesk. Khersons'ka Oblast', S Ukraine
Henriette, Mys *30 A3* Morbihan, NW France
Henrique de Carvalho *see* Saurimo
Henzada *see* Hinthada
Herakleion *see* Irákleio
Herāt *122 D4 var.* Herat; *anc.* Aria. Herāt, W Afghanistan
Heredia *53 E4* Heredia, C Costa Rica
Hereford *49 E2* Texas, SW USA

Herford *94 B4* Nordrhein-Westfalen, NW Germany
Héristal *see* Herstal
Herk-de-Stad *87 C6* Limburg, NE Belgium
Herlen Gol/Herlen He *see* Kerulen
Hermannstadt *see* Sibiu
Hermansverk *85 A5* Sogn Og Fjordane, S Norway
Hermhausen *see* Hajnówka
Hermiston *46 C2* Oregon, NW USA
Hermon, Mount *119 B5 Ar.* Jabal ash Shaykh. *mountain* S Syria
Hermosillo *50 B2* Sonora, NW Mexico
Hermoupolis *see* Ermoúpoli
Hernösand *see* Härnösand
Herrera del Duque *92 D3* Extremadura, W Spain
Herselt *87 C5* Antwerpen, C Belgium
Herstal *87 D6 Fr.* Héristal. Liège, E Belgium
Herzogenbusch *see* 's-Hertogenbosch
Hesse *see* Hessen
Hessen *95 B5 Eng./Fr.* Hesse. *state* C Germany
Hevron *see* Hebron
Heydebrech *see* Kędzierzyn-Koźle
Heydekrug *see* Šilutė
Heywood Islands *146 C3 island group* Western Australia
Hibbing *45 F1* Minnesota, N USA
Hibernia *see* Ireland
Hidalgo del Parral *50 C2 var.* Parral. Chihuahua, N Mexico
Hida-sanmyaku *131 C5 mountain range* Honshū, S Japan
Hierosolyma *see* Jerusalem
Hierro *70 A3 var.* Ferro. *island* Islas Canarias, Spain, NE Atlantic Ocean
High Atlas *70 C2 Eng.* High Atlas. *mountain range* C Morocco
High Atlas *see* Haut Atlas
High Plains *see* Great Plains
High Point *43 E1* North Carolina, SE USA
Hiiumaa *106 C2 Ger.* Dagden, *Swe.* Dagö. *island* W Estonia
Hikurangi *150 D2* Northland, North Island, New Zealand
Hildesheim *94 B4* Niedersachsen, N Germany
Hilla *see* Al Ḥillah
Hillaby, Mount *55 G1 mountain* N Barbados
Hill Bank *52 C1* Orange Walk, N Belize
Hillegom *86 C3* Zuid-Holland, C Netherlands
Hilo *47 B8* Hawaii, USA, C Pacific Ocean
Hilton Head Island *43 E2* South Carolina, SE USA
Hilversum *86 C3* Noord-Holland, C Netherlands
Himalaya/Himalaya Shan *see* Himalayas
Himalayas *135 E2 var.* Himalaya, *Chin.* Himalaya Shan. *mountain range* S Asia
Himeji *131 C6 var.* Himezi. Hyōgo, Honshū, SW Japan
Himezi *see* Himeji
Ḥimş *118 B4 var.* Homs; *anc.* Emesa. Ḥimş, C Syria
Hîncești *108 D4 var.* Hâncești; *prev.* Kotovsk. C Moldova
Hinchinbrook Island *148 D3* Queensland, NE Australia
Hinds *151 C6* Canterbury, South Island, New Zealand
Hindu Kush *123 F4 Per.* Hendū Kosh; *var.* Hindu Kosh. *mountain range* Afghanistan/Pakistan
Hinesville *43 E3* Georgia, SE USA
Hinnøya *84 C2* Lapp. Iinnasuolu. *island* C Norway
Hinson Bay *42 A5 bay* W Bermuda
Hinthada *136 B4* var. Henzada. Ayeyarwady, SW Myanmar (Burma)
Hios *see* Chíos
Hîrfanlı Barajı *116 C3 reservoir* C Turkey
Hirmand, Rūd-e *see* Helmand, Daryā-ye
Hirosaki *130 D3* Aomori, Honshū, C Japan
Hiroshima *131 B6 var.* Hirosima. Hiroshima, Honshū, SW Japan
Hirschberg/Hirschberg im Riesengebirge/Hirschberg in Schlesien *see* Jelenia Góra
Hirson *90 D3* Aisne, N France
Hîrşova *see* Hârşova
Hispalis *see* Sevilla

pana/Hispania *see* Spain
paniola 56 B1 *island* Dominion
Republic/Haiti
achi 131 D5 *var.* Hitati. Ibaraki,
onshū, S Japan
ati *see* Hitachi
ra 84 A4 *prev.* Hitteren. *island*
Norway
teren *see* Hitra
lmaren 85 C6 *Eng.* Lake Hjalmar.
ake C Sweden
lmar, Lake *see* Hjälmaren
rring 85 B7 Nordjylland, N Denmark
akabo Razi 136 B1 *mountain*
Myanmar (Burma)/China
byne 109 F2 *Rus.* Globino.
Poltavs'ka Oblast', NE Ukraine
khiv 109 F1 *Rus.* Glukhov. Sums'ka
Oblast', NE Ukraine
bokaye 107 D5 *Rus.* Glubokoye.
Vitsyebskaya Voblasts', N Belarus
a Bình 136 D3 Hoa Binh, N Vietnam
ang Lien Son 136 D3 *mountain range*
N Vietnam
bart 149 C8 *prev.* Hobarton, Hobart
Town. *state capital* Tasmania,
SE Australia
barton/Hobart Town *see* Hobart
bbs 49 E3 New Mexico, SW USA
bro 85 A7 Nordjylland, N Denmark
Chi Minh 137 E6 *var.* Ho Chi Minh
City; *prev.* Saigon. S Vietnam
Chi Minh City *see* Hồ Chí Minh
deida *see* Al Ḩudaydah
dmezővásárhely 99 D7 Csongrád,
SE Hungary
dna, Chott El 102 C4 *var.* Chott
el-Hodna, Ar. Shatt al-Hodna. *salt
lake* N Algeria
dna, Chott el-/Hodna, Shatt al- *see*
Hodna, Chott El
donín 99 C5 *Ger.* Göding.
Jihomoravský Kraj, SE Czech Republic
ei *see* Huy
ey *see* Huy
f 95 C5 Bayern, SE Germany
fei *see* Hefei
fu 131 B7 Yamaguchi, Honshū,
SW Japan
fuf *see* Al Hufūf
goley Islands *see* Chuuk Islands
hensalza *see* Inowrocław
henstadt *see* Zábřeh
hes Venn *see* Hautes Fagnes
he Tauern 95 C7 *mountain range*
W Austria
hhot 127 F3 *var.* Huhehot,
Huhohaote, Mong. Kukukhoto; *prev.*
Kweisui, Kwesui. Nei Mongol Zizhiqu,
N China
i An 137 E5 *prev.* Faifo. Quang Nam-
Đa Nẵng, C Vietnam
i-Hao/Hoihow *see* Haikou
kianga Harbour 150 C2 *inlet*
SE Tasman Sea
kitika 151 B5 West Coast, South
Island, New Zealand
kkaido 130 C2 *prev.* Ezo, Yeso, Yezo.
island NE Japan
la Prystan' 109 E4 *Rus.* Golaya
Pristan. Khersons'ka Oblast', S Ukraine
lbrook 48 B2 Arizona, SW USA
letown 55 G1 *prev.* Jamestown.
W Barbados
lguín 54 C2 Holguín, SE Cuba
lin Gol 127 G2 *prev.* Hulingol. Nei
Mongol Zizhiqu, N China
llabrunn 95 E6 Niederösterreich,
NE Austria
lland *see* Netherlands
llandia *see* Jayapura
lly Springs 42 C1 Mississippi, S USA
lman 37 E3 Victoria Island, Northwest
Territories, N Canada
lmsund 84 D4 Västerbotten,
N Sweden
lon 119 A6 *var.* Kholon; *prev.* Ḥolon.
Tel Aviv, C Israel
lon *see* Holon
lovanivs'k 109 E3 *Rus.* Golovanevsk.
Kirovohrads'ka Oblast', C Ukraine
lstebro 85 A7 Midtjylland,
W Denmark
lsteinborg/Holsteinsborg/
Holstenborg/Holstensborg *see*
Sisimiut
lyhead 89 C5 *Wel.* Caer Gybi.
NW Wales, United Kingdom
mbori 75 G3 Mopti, S Mali

Homs *see* Al Khums, Libya
Homs *see* Ḥimş
Homyel' 107 D7 *Rus.* Gomel'.
Homyel'skaya Voblasts', SE Belarus
Honan *see* Henan, China
Honan *see* Luoyang, China
Hondo 49 F4 Texas, SW USA
Hondo *see* Honshū
Honduras 52 C2 *off.* Republic of
Honduras. *country* Central America

HONDURAS
Central America

Official name	Republic of Honduras
Formation	1838 / 1838
Capital	Tegucigalpa
Population	8.1 million / 187 people per sq mile (72 people per sq km)
Total area	43,278 sq. miles (112,090 sq. km)
Languages	Spanish*, Garífuna (Carib), English Creole
Religions	Roman Catholic 97%, Protestant 3%
Ethnic mix	Mestizo 90%, Black 5%, Amerindian 4%, White 1%
Government	Presidential system
Currency	Lempira = 100 centavos
Literacy rate	85%
Calorie consumption	2651 kilocalories

Honduras, Golfo de *see* Honduras,
Gulf of
Honduras, Gulf of 52 C2 *Sp.* Golfo de
Honduras. *gulf* W Caribbean Sea
Honduras, Republic of *see* Honduras
Hønefoss 85 B6 Buskerud, S Norway
Honey Lake 47 B5 lake California,
W USA
Hon Gai *see* Ha Long
Hongay *see* Ha Long
Hồng Gai *see* Ha Long
Hồng Hà, Sông *see* Red River
Hong Kong 128 A1 Hong Kong, S China
Hong Kong Island 128 B2 *island*
S China Asia
Honiara 144 C3 *country capital*
(Solomon Islands) Guadalcanal,
C Solomon Islands
Honjō 130 D4 *var.* Honzyô, Yurihonjô.
Akita, Honshū, C Japan
Honolulu 47 A8 *state capital* O'ahu,
Hawaii, USA, C Pacific Ocean
Honshu 131 E5 *var.* Hondo, Honsyû.
island SW Japan
Honsyû *see* Honshū
Honte *see* Westerschelde
Hoogeveen 86 E2 Drenthe,
NE Netherlands
Hoogezand-Sappemeer 86 E2
Groningen, NE Netherlands
Hoorn 86 C2 Noord-Holland,
NW Netherlands
Hoosier State *see* Indiana
Hopa 117 E2 Artvin, NE Turkey
Hope 36 C3 British Columbia,
SW Canada
Hopedale 39 F2 Newfoundland and
Labrador, NE Canada
Hopeh/Hopei *see* Hebei
Hopkinsville 40 B5 Kentucky, S USA
Horasan 117 F3 Erzurum, NE Turkey
Horizon Deep 152 D4 *trench* W Pacific
Ocean
Horki 107 E6 *Rus.* Gorki.
Mahilyowskaya Voblasts', E Belarus
Horlivka 109 G3 *Rom.* Adâncata, *Rus.*
Gorlovka. Donets'ka Oblast', E Ukraine
Hormoz, Tangeh-ye *see* Hormuz,
Strait of
Hormuz, Strait of 120 D4 *var.* Strait of
Ormuz, *Per.* Tangeh-ye Hormoz. *strait*
Iran/Oman
Cape Horn 65 C8 *Eng.* Cape Horn.
headland S Chile
Horn, Cape *see* Hornos, Cabo de
Hornsby 148 E1 New South Wales,
SE Australia
Horodnya 109 E1 *Rus.* Gorodnya.
Chernihivs'ka Oblast', NE Ukraine
Horodok 108 B2 *Pol.* Gródek
Jagielloński, *Rus.* Gorodok, Gorodok
Yagellonski. L'vivs'ka Oblast',
NW Ukraine
Horodyshche 109 E2 *Rus.* Gorodishche.
Cherkas'ka Oblast', C Ukraine
Horoshiri-dake 130 D2 *var.* Horosiri
Dake. *mountain* Hokkaidō, N Japan

Horosiri Dake *see* Horoshiri-dake
Horsburgh Atoll 132 A4 *var.* Goidhoo
Atoll. *atoll* N Maldives
Horseshoe Bay 42 A5 *bay* W Bermuda
W Atlantic Ocean
Horseshoe Seamounts 80 A4 *seamount
range* E Atlantic Ocean
Horsham 149 B7 Victoria,
SE Australia
Horst 87 D5 Limburg, SE Netherlands
Horten 85 B6 Vestfold, S Norway
Horyn' 107 B7 *Rus.* Goryn. *river*
NW Ukraine
Hosingen 87 D7 Diekirch,
NE Luxembourg
Hospitalet *see* L'Hospitalet de Llobregat
Hotan 126 B4 *var.* Khotan, *Chin.*
Ho-t'ien. Xinjiang Uygur Zizhiqu,
NW China
Ho-t'ien *see* Hotan
Hoting 44 C4 Jämtland, C Sweden
Hot Springs 42 B1 Arkansas, C USA
Hötzing *see* Hațeg
Houayxay 136 C3 *var.* Ban Houayxay.
Bokèo, N Laos
Houghton 40 B1 Michigan, N USA
Houilles 91 B5 Yvelines, Île-de-France,
N France Europe
Houlton 41 H1 Maine, NE USA
Houma 42 B3 Louisiana, S USA
Houston 49 H4 Texas, SW USA
Hovd 126 C2 *var.* Khovd, Kobdo; *prev.*
Jirgalanta. Hovd, W Mongolia
Hove 89 E7 SE England,
United Kingdom
Hoverla, Hora 108 C3 *Rus.* Gora
Goverla. *mountain* W Ukraine
Hovsgol, Lake *see* Hövsgöl Nuur
Hövsgöl Nuur 126 D1 *var.* Lake
Hovsgol. *lake* N Mongolia
Howa, Ouadi *see* Howar, Wādi
Howar, Wadi 72 A3 *var.* Ouadi Howa.
river Chad/Sudan
Howrah *see* Hāora
Hoy 88 C2 *island* N Scotland, United
Kingdom
Hoyerswerda 94 D4 *Lus.* Wojerecy.
Sachsen, E Germany
Hpa-an 136 B4 *var.* Pa-an. Kayin State,
S Myanmar (Burma)
Hpyu *see* Phyu
Hradec Králové 99 B5 *Ger.* Königgrätz.
Královéhradecký Kraj, N Czech
Republic
Hrandzichy 107 B5 *Rus.* Grandichi.
Hrodzyenskaya Voblasts', W Belarus
Hranice 99 C5 *Ger.* Mährisch-
Weisskirchen. Olomoucký Kraj,
E Czech Republic
Hrebinka 109 E2 *Rus.* Grebenka.
Poltavs'ka Oblast', NE Ukraine
Hrodna 107 B5 *Pol.* Grodno.
Hrodzyenskaya Voblasts', W Belarus
Hrvatska *see* Croatia
Hsia-men *see* Xiamen
Hsiang-t'an *see* Xiangtan
Hsi Chiang *see* Xi Jiang
Hsing-K'ai Hu *see* Khanka, Lake
Hsi-ning/Hsining *see* Xining
Hsinking *see* Changchun
Hsin-yang *see* Xinyang
Hsu-chou *see* Xuzhou
Htawei *see* Dawei
Huacho 60 C4 Lima, W Peru
Hua Hin *see* Ban Hua Hin
Huaihua 128 C5 Hunan, S China
Huailai *see* Shacheng. Hebei,
E China
Huainan 128 D5 *var.* Huai-nan,
Hwainan. Anhui, E China
Huai-nan *see* Huainan
Huajuapan 51 F5 *var.* Huajuapan de
León. Oaxaca, SE Mexico
Huajuapan de León *see* Huajuapan
Hualapai Peak 48 A2 *mountain* Arizona,
SW USA
Huallaga, Río 60 C3 *river* N Peru
Huambo 78 B2 *Port.* Nova Lisboa.
Huambo, C Angola
Huancavelica 60 D4 Huancavelica,
SW Peru
Huancayo 60 D3 Junín, C Peru
Huang Hai *see* Yellow Sea
Huangshi 128 C5 *var.* Huang-shih,
Hwangshih. Hubei, C China
Huang-shih *see* Huangshi
Huanta 60 D4 Ayacucho, C Peru
Huánuco 60 C3 Huánuco, C Peru
Huanuni 61 F4 Oruro, W Bolivia

Huaral 60 C4 Lima, W Peru
Huarás *see* Huaraz
Huaraz 60 C3 *var.* Huarás. Ancash,
W Peru
Huarmey 60 C3 Ancash, W Peru
Huatabampo 50 C2 Sonora,
NW Mexico
Hubli 124 B3 Karnātaka, SW India
Huddersfield 89 D5 N England, United
Kingdom
Hudiksvall 85 C5 Gävleborg, C Sweden
Hudson Bay 37 G4 *bay* NE Canada
Hudson, Détroit d' *see* Hudson Strait
Hudson Strait 37 H3 *Fr.* Détroit
d'Hudson. *strait* Northwest Territories/
Québec, NE Canada
Hudur *see* Xuddur
Huê 136 E4 Th,a Thiên-Huê, C Vietnam
Huehuetenango 52 A2 Huehuetenango,
W Guatemala
Huelva 92 C4 *anc.* Onuba. Andalucía,
SW Spain
Huesca 93 F2 *anc.* Osca. Aragón,
NE Spain
Huéscar 93 E4 Andalucía, S Spain
Hughenden 148 C3 Queensland,
NE Australia
Hugo 49 G2 Oklahoma, C USA
Huhehot/Huhohaote *see* Hohhot
Huíla Plateau 78 B2 *plateau* S Angola
Huixtla 51 G5 Chiapas, SE Mexico
Hulingol *see* Holin Gol
Hull 38 D4 Québec, SE Canada
Hull *see* Kingston upon Hull
Hull Island *see* Orona
Hulst 87 B5 Zeeland, SW Netherlands
Hulun *see* Hulun Buir
Hulun Buir 127 F1 *var.* Hailar; *prev.*
Hulun. Nei Mongol Zizhiqu, N China
Hu-lun Ch'ih *see* Hulun Nur
Hulun Nur 127 F1 *var.* Hu-lun Ch'ih;
prev. Dalai Nor. *lake* NE China
Humaitá 62 D2 Amazonas, N Brazil
Humboldt River 47 C5 *river* Nevada,
W USA
Humphreys Peak 48 B1 *mountain*
Arizona, SW USA
Humpolec 99 B5 *Ger.* Gumpolds,
Humpoletz. Vysočina, C Czech
Republic
Humpoletz *see* Humpolec
Hunan 128 C6 *var.* Hunan Sheng,
Xiang. *province* S China
Hunan Sheng *see* Hunan
Hunedoara 108 B4 *Ger.* Eisenmarkt,
Hung. Vajdahunyad. Hunedoara,
SW Romania
Hünfeld 95 B5 Hessen, C Germany
Hungarian People's Republic *see*
Hungary
Hungary 99 C6 *off.* Republic of Hungary,
Ger. Ungarn, *Hung.* Magyarország, *Rom.* Ungaria, *Scr.* Madarska, *Ukr.*
Uhorshchyna; *prev.* Hungarian People's
Republic. *country* C Europe

HUNGARY
Central Europe

Official name	Republic of Hungary
Formation	1918 / 1947
Capital	Budapest
Population	10 million / 280 people per sq mile (108 people per sq km)
Total area	35,919 sq. miles (93,030 sq. km)
Languages	Hungarian* (Magyar)
Religions	Roman Catholic 52%, Calvinist 16%, Protestant 3%, Nonreligious 14%, Lutheran 3%
Ethnic mix	Magyar 90%, Roma 4%, German 3%, Serb 2%, Other 1%
Government	Parliamentary system
Currency	Forint = 100 fillér
Literacy rate	99%
Calorie consumption	2968 kilocalories

Hungary, Plain of *see* Great Hungarian
Plain
Hungary, Republic of *see* Hungary
Hunjiang *see* Baishan
Hunter Island 149 B8 *island* Tasmania,
SE Australia
Huntington 40 D4 West Virginia,
NE USA
Huntington Beach 47 B8 California,
W USA
Huntly 150 D3 Waikato, North Island,
New Zealand
Huntsville 42 D1 Alabama, S USA

Huntsville *49 G3* Texas, SW USA
Hurghada *see* Al Ghurdaqah
Huron *45 E3* South Dakota, N USA
Huron, Lake *40 D2 lake* Canada/USA
Hurukawa *see* Furukawa
Hurunui *151 C5 river* South Island, New Zealand
Húsavík *83 E4* Nordhurland Eystra, NE Iceland
Husté *see* Khust
Husum *94 B2* Schleswig-Holstein, N Germany
Huszt *see* Khust
Hutchinson *45 E5* Kansas, C USA
Hutchinson Island *43 F4 island* Florida, SE USA
Huy *87 C6 Dut.* Hoei, Hoey.
Liège, E Belgium
Huzi *see* Fuji
Hvannadalshnúkur *83 E5 volcano* S Iceland
Hvar *100 B4 It.* Lesina; *anc.* Pharus. *island* S Croatia
Hwainan *see* Huainan
Hwange *78 D3 prev.* Wankie.
Matabeleland North, W Zimbabwe
Hwang-Hae *see* Yellow Sea
Hwangshih *see* Huangshi
Hyargas Nuur *126 C2 lake* NW Mongolia
Hyderābād *134 D5 var.* Haidarabad. *state capital* Telangana, C India
Hyderābād *134 B3 var.* Haidarabad.
Sind, SE Pakistan
Hyères *91 D6* Var, SE France
Hyères, Îles d' *91 D6 island group* S France
Hypanis *see* Kuban'
Hyrcania *see* Gorgān
Hyvinge *see* Hyvinkää
Hyvinkää *85 D5 Swe.* Hyvinge. Etelä-Suomi, S Finland

I

Iader *see* Zadar
Ialomiţa *108 C5 river* SE Romania
Iaşi *108 D3 Ger.* Jassy. Iaşi, NE Romania
Ibadan *75 F5* Oyo, SW Nigeria
Ibagué *58 B3* Tolima, C Colombia
Ibar *see* Ibër
Ibarra *60 B1 var.* San Miguel de Ibarra. Imbabura, N Ecuador
Ibër *100 D4 Serb.* Ibar. *river* C Serbia
Iberia *see* Spain
Iberian Mountains *see* Ibérico, Sistema
Iberian Peninsula *80 B4 physical region* Portugal/Spain
Iberian Plain *80 B4 abyssal plain* E Atlantic Ocean
Ibérica, Cordillera *see* Ibérico, Sistema
Ibérico, Sistema *93 E2 var.* Cordillera Ibérica, *Eng.* Iberian Mountains. *mountain range* NE Spain
Ibiza *see* Eivissa
Ibo *see* Sassandra
Ica *60 D4* Ica, SW Peru
Icaria *see* Ikaría
Içá, Rio *62 C2 var.* Río Putumayo. *river* NW South America
Içá, Rio *see* Putumayo, Río
İçel *see* Mersin
Iceland *83 E4 off.* Republic of Iceland, *Dan.* Island, *Icel.* Ísland. *country* N Atlantic Ocean

ICELAND
Northwest Europe

Official name Republic of Iceland
Formation 1944 / 1944
Capital Reykjavík
Population 300,000 / 8 people per sq mile (3 people per sq km)
Total area 39,768 sq. miles (103,000 sq. km)
Languages Icelandic*
Religions Evangelical Lutheran 84%, Other (mostly Christian) 10%, Roman Catholic 3%, Nonreligious 3%
Ethnic mix Icelandic 94%, Other 5%, Danish 1%
Government Parliamentary system
Currency Icelandic króna = 100 aurar
Literacy rate 99%
Calorie consumption 3339 kilocalories

Iceland Basin *80 B1 undersea basin* N Atlantic Ocean

Icelandic Plateau *see* Iceland Plateau
Iceland Plateau *155 B6 var.* Icelandic Plateau. *undersea plateau* S Greenland Sea
Iceland, Republic of *see* Iceland
Iconium *see* Konya
Iculisma *see* Angoulême
Idabel *49 H2* Oklahoma, C USA
Idaho *46 D3 off.* State of Idaho, *also known as* Gem of the Mountains, Gem State. *state* NW USA
Idaho Falls *46 E3* Idaho, NW USA
Idensalmi *see* Iisalmi
Idfu *72 B2 var.* Edfu. SE Egypt
Idi Amin, Lac *see* Edward, Lake
Idini *74 B2* Trarza, W Mauritania
Idlib *118 B3* Idlib, NW Syria
Idre *85 B5* Dalarna, C Sweden
Iecava *106 C3* S Latvia
Ieper *87 A6 Fr.* Ypres. West-Vlaanderen, W Belgium
Ierápetra *105 D8* Kríti, Greece, E Mediterranean Sea
Ierisós *see* Ierissós
Ierissós *104 C4 var.* Ierisós. Kentrikí Makedonía, N Greece
Iferouâne *75 G2* Agadez, N Niger
Ifôghas, Adrar des *75 E2 var.* Adrar des Iforas. *mountain range* NE Mali
Iforas, Adrar des *see* Ifôghas, Adrar des
Igarka *114 D3* Krasnoyarskiy Kray, N Russian Federation
Igaunija *see* Estonia
Iglau/Iglawa/Iglawa *see* Jihlava
Iglesias *97 A5* Sardegna, Italy, C Mediterranean Sea
Igloolik *37 G2* Nunavut, N Canada
Igoumenítsa *104 A4* Ípeiros, W Greece
Iguaçu, Rio *63 E4 Sp.* Río Iguazú. *river* Argentina/Brazil
Iguaçu, Saltos do *63 E4 Sp.* Cataratas del Iguazú; *prev.* Victoria Falls. *waterfall* Argentina/Brazil
Iguala *51 E4 var.* Iguala de la Independencia. Guerrero, S Mexico
Iguala de la Independencia *see* Iguala
Iguazú, Cataratas del *see* Iguaçu, Saltos do
Iguazú, Río *see* Iguaçu, Rio
Iguid, Erg *see* Iguidi, 'Erg
Iguidi, 'Erg *70 C3 var.* Erg Iguid. *desert* Algeria/Mauritania
Ihavandhippolhu Atoll *132 A3 var.* Ihavandiffulu Atoll. *atoll* N Maldives
Ihavandiffulu Atoll *see* Ihavandhippolhu Atoll
Ihosy *79 F4* Fianarantsoa, S Madagascar
Iinnasuolu *see* Hinnøya
Iisalmi *84 E4 var.* Idensalmi. Itä-Suomi, C Finland
IJmuiden *86 C3* Noord-Holland, W Netherlands
IJssel *86 D3 var.* Yssel. *river* Netherlands
IJsselmeer *86 C2 prev.* Zuider Zee. *lake* N Netherlands
IJsselmuiden *86 D3* Overijssel, E Netherlands
IJzer *87 A6 river* W Belgium
Ikaahuk *see* Sachs Harbour
Ikaluktutiak *see* Cambridge Bay
Ikaría *105 E6 var.* Kariot, Nicaria, Nikaria; *anc.* Icaria. *island* Dodekánisa, Greece, Aegean Sea
Ikela *77 D6* Équateur, C Dem. Rep. Congo
Iki *131 A7 island* SW Japan
Ilagan *139 E1* Luzon, N Philippines
Ilave *61 E4* Puno, S Peru
Iława *98 D3 Ger.* Deutsch-Eylau. Warmińsko-Mazurskie, NE Poland
Ilebo *77 C6 prev.* Port-Francqui. Kasai-Occidental, W Dem. Rep. Congo
Île-de-France *90 C3 cultural region* N France
Ilerda *see* Lleida
Ilfracombe *89 C7* SW England, United Kingdom
Ílhavo *92 B2* Aveiro, N Portugal
Iliamna Lake *36 C3 lake* Alaska, USA
Il'ichevsk *see* Illichivs'k, Ukraine
Ilici *see* Elche
Iligan *139 E2 off.* Iligan City. Mindanao, S Philippines
Iligan City *see* Iligan
Illapel *64 B4* Coquimbo, C Chile
Illichivs'k *109 E4 Rus.* Il'ichevsk. Odes'ka Oblast', SW Ukraine
Illicis *see* Elche

Illinois *40 A4 off.* State of Illinois, *also known as* Prairie State, Sucker State. *state* C USA
Illinois River *40 B4 river* Illinois, N USA
Illurco *see* Lorca
Illuro *see* Mataró
Ilo *61 E4* Moquegua, SW Peru
Iloilo *139 E2 off.* Iloilo City. Panay Island, C Philippines
Iloilo City *see* Iloilo
Ilorin *75 F4* Kwara, W Nigeria
Ilovlya *111 B6* Volgogradskaya Oblast', SW Russian Federation
İluh *see* Batman
Il'yaly *see* Gurbansoltan Eje
Imatra *85 E5* Etelä-Suomi, SE Finland
Imbros *see* Gökçeada
Imishli *see* İmişli
İmişli *117 H3 Rus.* Imishli. C Azerbaijan
Imola *96 C3* Emilia-Romagna, N Italy
Imperatriz *63 F2* Maranhão, NE Brazil
Imperia *96 A3* Liguria, NW Italy
Impfondo *77 C5* Likouala, NE Congo
Imphal *135 H3 state capital* Manipur, NE India
Imroz Adası *see* Gökçeada
Inagua Islands *see* Little Inagua
Inagua Islands *see* Great Inagua
Inarijärvi *84 D2* Lapp. Aanaarjävri, *Swe.* Enareträsk. *lake* N Finland
İnäu *see* Ineu
Inawashiro-ko *131 D5 var.* Inawasiro Ko. *lake* Honshū, C Japan
Inawasiro Ko *see* Inawashiro-ko
İncesu *116 D3* Kayseri, Turkey Asia
Incheon *129 E4 off.* Incheon-gwang-yeoksi, *prev.* Inch'ŏn, *Jap.* Jinsen; *prev.* Chemulpo. NW South Korea
Incheon-Gwang-yeoksi *see* Incheon
Inch'ŏn *see* Incheon
Incudine, Monte *91 E7 mountain* Corse, France, C Mediterranean Sea
Indefatigable Island *see* Santa Cruz, Isla
Independence *45 F4* Missouri, C USA
Independence Fjord *83 E1 fjord* N Greenland
Independence Island *see* Malden Island
Independence Mountains *46 C4 mountain range* Nevada, W USA
India *124 B3 off.* Republic of India, *var.* Indian Union, Union of India, *Hind.* Bhārat. *country* S Asia

INDIA
South Asia

Official name Republic of India
Formation 1947 / 1947
Capital New Delhi
Population 1.25 billion / 1091 people per sq mile (421 people per sq km)
Total area 1,269,338 sq. miles (3,287,590 sq. km)
Languages Hindi*, English*, Urdu, Bengali, Marathi, Telugu, Tamil, Bihari, Gujarati, Kannada
Religions Hindu 81%, Muslim 13%, Christian 2%, Sikh 2%, Buddhist 1%, Other 1%
Ethnic mix Indo-Aryan 72%, Dravidian 25%, Mongoloid and other 3%
Government Parliamentary system
Currency Indian rupee = 100 paise
Literacy rate 63%
Calorie consumption 2459 kilocalories

India *see* Indija
Indiana *40 B4 off.* State of Indiana, *also known as* Hoosier State. *state* N USA
Indianapolis *40 C4 state capital* Indiana, N USA
Indian Church *52 C1* Orange Walk, N Belize
Indian Desert *see* Thar Desert
Indianola *45 F4* Iowa, C USA
India, Republic of *see* India
India, Union of *see* India
Indigirka *115 F2 river* NE Russian Federation
Indija *100 D3* Vojvodina, N Serbia
Indira Point *132 G3 headland* Andaman and Nicobar Islands, India, NE Indian Ocean
Indjija *see* Indija
Indomed Fracture Zone *141 B6 tectonic feature* SW Indian Ocean

Indonesia *138 B4 off.* Republic of Indonesia, *Ind.* Republik Indonesia *prev.* Dutch East Indies, Netherlands East Indies, United States of Indonesia *country* SE Asia

INDONESIA
Southeast Asia

Official name Republic of Indonesia
Formation 1949 / 1999
Capital Jakarta
Population 250 million / 360 people per sq mile (139 people per sq km)
Total area 741,096 sq. miles (1,919,440 sq. km)
Languages Javanese, Sundanese, Madurese, Bahasa Indonesian*, Dutch
Religions Sunni Muslim 86%, Christian 9%, Hindu 2%, Other 2%, Buddhist 1%
Ethnic mix Javanese 41%, Other 28%, Sundanese 15%, Coastal Malays 12%, Madurese 3%
Government Presidential system
Currency Rupiah = 100 sen
Literacy rate 93%
Calorie consumption 2713 kilocalories

Indonesian Borneo *see* Kalimantan
Indonesia, Republic of *see* Indonesia
Indonesia, Republik *see* Indonesia
Indonesia, United States of *see* Indonesia
Indore *134 D4* Madhya Pradesh, C India
Indreville *see* Châteauroux
Indus *134 C2 Chin.* Yindu He; *prev.* Yin-tu Ho. *river* S Asia
Indus Cone *see* Indus Fan
Indus Fan *112 C5 var.* Indus Cone. *undersea fan* N Arabian Sea
Indus, Mouths of the *134 B4 delta* S Pakistan
İnebolu *116 C2* Kastamonu, N Turkey
Ineu *108 A4 Hung.* Borosjenő; *prev.* Irad, W Romania
Infiernillo, Presa del *51 E4 reservoir* S Mexico
Inglewood *46 D2* California, W USA
Ingolstadt *95 C6* Bayern, S Germany
Ingulets *see* Inhulets'
Inguri *see* Enguri
Inhambane *79 E4* Inhambane, SE Mozambique
Inhulets' *109 F3 Rus.* Ingulets. Dnipropetrovs'ka Oblast', E Ukraine
I-ning *see* Yining
Inis *see* Ennis
Inis Ceithleann *see* Enniskillen
Inn *95 C6 river* C Europe
Innaanganeq *82 C1 var.* Kap York. *headland* NW Greenland
Inner Hebrides *88 B4 island group* W Scotland, United Kingdom
Inner Islands *79 H1 var.* Central Group *island group* NE Seychelles
Innisfail *148 D3* Queensland, NE Australia
Inniskilling *see* Enniskillen
Innsbruck *see* Innsbruck
Innsbruck *95 C7 var.* Innsbruck. W Austria
Inoucdjouac *see* Inukjuak
Inowraclaw *see* Inowrocław
Inowrocław *98 C3 Ger.* Hohensalza; *prev.* Inowraclaw. Kujawski-pomorskie, C Poland
I-n-Sakane, 'Erg *75 E2 desert* N Mali
I-n-Salah *70 D3 var.* In Salah. C Algeria
Insterburg *see* Chernyakhovsk
Insula *see* Lille
Inta *110 E3* Respublika Komi, NW Russian Federation
Interamna *see* Teramo
Interamna Nahars *see* Terni
International Falls *45 F1* Minnesota, N USA
Inukjuak *38 D2 var.* Inoucdjouac; *prev.* Port Harrison. Québec, NE Canada
Inuuvik *see* Inuvik
Inuvik *36 D3 var.* Inuvik. Northwest Territories, NW Canada
Invercargill *151 A7* Southland, South Island, New Zealand
Inverness *88 C3* N Scotland, United Kingdom

vestigator Ridge *141 D5 undersea*
ridge E Indian Ocean
vestigator Strait *149 B7 strait* South
  Australia
yangani *78 D3 mountain*
  NE Zimbabwe
annina *104 A4 var.* Janina, Yannina.
peiros, W Greece
a *45 F5* Kansas, C USA
nia Basin *see* Ionian Basin
nian Basin *80 D5 var.* Ionia
Basin. *undersea basin* Ionian Sea,
  C Mediterranean Sea
nian Islands *see* Iónia Nisiá
nian Sea *103 E3 Gk.* Iónio Pélagos, *It.*
  Mar Ionio. *sea* C Mediterranean Sea
nioi Nísoi *105 A5 Eng.* Ionian Islands.
region W Greece
  nio, Mar/Iónio Pélagos *see* Ionian Sea
s *105 D6 var.* Nio.
s *105 D6 var.* Nio. *island* Kykládes,
  Greece, Aegean Sea
s *105 D6 var.* Nio. *island* Kykládes,
  Greece, Aegean Sea
s *105 D6 var.* Nio. *island* Kykládes,
  Greece, Aegean Sea
lís *105 C6 prev.* Kéa. Tziá, Kykládes,
  Greece, Aegean Sea
wa *45 F3 off.* State of Iowa, *also known*
  as Hawkeye State. *state* C USA
wa City *45 G3* Iowa, C USA
wa Falls *45 G3* Iowa, C USA
el' *99 C6 var.* Ipoly, *Ger.* Eipel. *river*
  Hungary/Slovakia
iales *58 A4* Nariño, SW Colombia
oh *138 B3* Perak, Peninsular Malaysia
oly *see* Ipel'
py *76 C4* Ouaka, C Central African
  Republic
swich *149 E5* Queensland, E Australia
swich *89 E6 hist.* Gipeswic. E England,
  United Kingdom
aluit *37 H3 prev.* Frobisher Bay.
  province capital Baffin Island, Nunavut,
  NE Canada
uique *64 B1* Tarapacá, N Chile
uitos *60 C1* Loreto, N Peru
kleio *105 D7 var.* Herakleion, *Eng.*
  Candia; *prev.* Iráklion. Kríti, Greece,
  E Mediterranean Sea
áklion *see* Iráklio
an *120 C3 off.* Islamic Republic of Iran;
  prev. Persia. *country* SW Asia

RAN
Southwest Asia

Official name Islamic Republic of Iran
ormation 1502 / 1990
Capital Tehran
Population 77.4 million / 123 people
  per sq mile (47 people per sq km)
Total area 636,293 sq. miles
  (1,648,000 sq. km)
Languages Farsi*, Azeri, Luri, Gilaki,
  Mazanderani, Kurdish, Turkmen,
  Arabic, Baluchi
Religions Shi'a Muslim 89%,
  Sunni Muslim 9%, Other 2%
Ethnic mix Persian 51%, Azari 24%,
  Other 10%, Lur and Bakhtiari 8%,
  Kurdish 7%
Government Islamic theocracy
Currency Iranian rial = 100 dinars
Literacy rate 84%
Calorie consumption 3058 kilocalories

anian Plateau *120 D3 var.* Plateau of
  Iran. *plateau* N Iran
an, Islamic Republic of *see* Iran
an, Plateau of *see* Iranian Plateau
apuato *51 E4* Guanajuato, C Mexico
aq *120 B3 off.* Republic of Iraq, *Ar.*
  'Iráq. *country* SW Asia

RAQ
Southwest Asia

Official name Republic of Iraq
Formation 1932 / 1990
Capital Baghdad
Population 33.8 million / 200 people
  per sq mile (77 people per sq km)
Total area 168,753 sq. miles
  (437,072 sq. km)
Languages Arabic*, Kurdish*, Turkic
  languages, Armenian, Assyrian
Religions Shi'a Muslim 60%,
  Sunni Muslim 35%, Other (including
  Christian) 5%

IRAQ
(continued)

Ethnic mix Arab 80%, Kurdish 15%,
  Turkmen 3%, Other 2%
Government Parliamentary system
Currency New Iraqi dinar = 1000 fils
Literacy rate 79%
Calorie consumption 2489 kilocalories

'Iráq *see* Iraq
Iraq, Republic of *see* Iraq
Irbid *119 B5* Irbid, N Jordan
Irbil *see* Arbil
Ireland *89 A5 off.* Republic of Ireland, *Ir.*
  Éire. *country* NW Europe

IRELAND
Northwest Europe

Official name Ireland
Formation 1922 / 1922
Capital Dublin
Population 4.5 million / 169 people
  per sq mile (65 people per sq km)
Total area 27,135 sq. miles (70,280 sq. km)
Languages English*, Irish*
Religions Roman Catholic 87%,
  Other and nonreligious 10%, Anglican 3%
Ethnic mix Irish 99%, Other 1%
Government Parliamentary system
Currency Euro = 100 cents
Literacy rate 99%
Calorie consumption 3591 kilocalories

Ireland *80 C3 Lat.* Hibernia. *island*
  Ireland/United Kingdom
Ireland, Republic of *see* Ireland
Irian *see* New Guinea
Irian Barat *see* Papua
Irian Jaya *see* Papua
Irian, Teluk *see* Cenderawasih, Teluk
Iringa *73 C7* Iringa, C Tanzania
Iriomote-jima *130 A4 island* Sakishima-
  shotō, SW Japan
Iriona *52 D2* Colón, NE Honduras
Irish Sea *89 C5 Ir.* Muir Éireann. *sea*
  C British Isles
Irkutsk *115 E4* Irkutskaya Oblast',
  S Russian Federation
Irminger Basin *see* Reykjanes Basin
Iroise *90 A3 sea* NW France
Iron Mountain *40 B2* Michigan, N USA
Ironwood *40 B1* Michigan, N USA
Irrawaddy *136 B2 var.* Ayeyarwady.
  river W Myanmar (Burma)
Irrawaddy, Mouths of the *137 A5 delta*
  SW Myanmar (Burma)
Irtish *see* Yertis
Irtysh *see* Yertis
Irun *93 E1 Cast.* Irún. País Vasco,
  N Spain
Irún *see* Irun
Iruña *see* Pamplona
Isabela, Isla *60 A5 var.* Albemarle Island.
  island Galápagos Islands, Ecuador,
  E Pacific Ocean
Isaccea *108 D4* Tulcea, E Romania
Isachsen *37 F1* Ellef Ringnes Island,
  Nunavut, N Canada
Ísafjörður *83 E4* Vestfirðhir, NW Iceland
Isbarta *see* Isparta
Isca Damnoniorum *see* Exeter
Ise *131 C6* Mie, Honshū, SW Japan
Iseghem *see* Izegem
Isère *91 D5 river* E France
Isernia *97 D5 var.* Æsernia. Molise,
  C Italy
Ise-wan *131 C6 bay* S Japan
Isfahan *see* Eşfahān
Isha Baydhabo *see* Baydhabo
Ishigaki-jima *130 A4 island* Sakishima-
  shotō, SW Japan
Ishikari-wan *130 C2 bay* Hokkaidō,
  NE Japan
Ishim *114 C4* Tyumenskaya Oblast',
  C Russian Federation
Ishinomaki *130 D4 var.* Isinomaki.
  Miyagi, Honshū, C Japan
Ishkāshim *123 F3 var.* Ishkoshim. NE
  Afghanistan
Ishkoshim *see* Ishkāshim
Isinomaki *see* Ishinomaki
Isiro *77 E6* Orientale, NE Dem. Rep.
  Congo
Iskăr *see* Iskŭr
Iskenderun *116 D4 Eng.* Alexandretta.
  Hatay, S Turkey

İskenderun Körfezi *118 A2 Eng.* Gulf of
  Alexandretta. *gulf* S Turkey
Iskur *104 C2 var.* Iskăr. *river*
  NW Bulgaria
Yazovir Iskur *104 B2 prev.* Yazovir
  Stalin. *reservoir* W Bulgaria
Isla Cristina *92 C4* Andalucía, S Spain
Isla de León *see* San Fernando
Islāmābād *134 C1 country capital*
  (Pakistan) Federal Capital Territory
  Islāmābād, NE Pakistan
Island/Ísland *see* Iceland
Islay *88 B4 island* SW Scotland, United
  Kingdom
Isle *91 B5 river* W France
Isle of Man *89 B5 British Crown*
  Dependency NW Europe
Isles of Scilly *89 B8 island group*
  SW England, United Kingdom
Ismailia *see* Al Ismā'īliya
Ismā'iliya *see* Al Ismā'īliya
Ismid *see* İzmit
Isna *72 B2 var.* Esna. SE Egypt
Isoka *78 D1* Northern, NE Zambia
Isparta *116 B4 var.* Isbarta. Isparta,
  SW Turkey
Ispir *117 E3* Erzurum, NE Turkey
Israel *119 A7 off.* State of Israel, *var.*
  Medinat Israel, *Heb.* Yisrael, Yisra'el.
  country SW Asia

ISRAEL
Southwest Asia

Official name State of Israel
Formation 1948 / 1994
Capital Jerusalem (not internationally
  recognized)
Population 7.7 million / 981 people
  per sq mile (379 people per sq km)
Total area 8019 sq. miles (20,770 sq. km)
Languages Hebrew*, Arabic*,
  Yiddish, German, Russian, Polish,
  Romanian, Persian
Religions Jewish 76%, Muslim (mainly
  Sunni) 16%, Other 4%, Druze 2%,
  Christian 2%
Ethnic mix Jewish 76%, Arab 20%,
  Other 4%
Government Parliamentary system
Currency Shekel = 100 agorot
Literacy rate 98%
Calorie consumption 3619 kilocalories

Israel, State of *see* Israel
Issa *see* Vis
Issiq Köl *see* Issyk-Kul', Ozero
Issoire *91 C5* Puy-de-Dôme, C France
Issyk-Kul' *see* Balykchy
Issyk-Kul', Ozero *123 G2*
  var. Issiq Köl, *Kir.* Ysyk-Köl.
  lake E Kyrgyzstan
İstanbul *116 B2 Bul.* Tsarigrad, *Eng.*
  Istanbul, *prev.* Constantinople; *anc.*
  Byzantium. İstanbul, NW Turkey
İstanbul Boğazı *116 B2 var.* Bosporus
  Thracius, *Eng.* Bosphorus, Bosporus,
  Turk. Karadeniz Boğazı. *strait*
  NW Turkey
Istarska Županija *see* Istra
Istra *100 A3 off.* Istarska Županija.
  province NW Croatia
Istra *100 A3 Eng.* Istria, *Ger.* Istrien.
  cultural region NW Croatia
Istria/Istrien *see* Istra
Itabuna *63 G3* Bahia, E Brazil
Itagüí *58 B3* Antioquia, W Colombia
Itaipú, Represa de *63 E4*
  reservoir Brazil/Paraguay
Itaituba *63 E2* Pará, NE Brazil
Italia/Italiana, Republica/Italian
  Republic, The *see* Italy
Italian Somaliland *see* Somalia
Italy *96 C3 off.* The Italian Republic,
  It. Italia, Repubblica Italiana.
  country S Europe

ITALY

ITALY
(continued)

Religions Roman Catholic 85%,
  Other and nonreligious 13%, Muslim 2%
Ethnic mix Italian 94%, Other 4%,
  Sardinian 2%
Government Parliamentary system
Currency Euro = 100 cents
Literacy rate 99%
Calorie consumption 3539 kilocalories

Iténez, Río *see* Guaporé, Rio
Ithaca *41 E3* New York, NE USA
It Hearrenfean *see* Heerenveen
Itoigawa *131 C5* Niigata, Honshū,
  C Japan
Itseqqortoormiit *see* Ittoqqortoormiit
Ittoqqortoormiit *83 E3 var.*
  Itseqqortoormiit, *Dan.* Scoresbysund,
  Eng. Scoresby Sound. Tunu,
  C Greenland
Iturup, Ostrov *130 E1 island* Kuril'skiye
  Ostrova, SE Russian Federation
Itzehoe *94 B2* Schleswig-Holstein,
  N Germany
Ivalo *84 D2 Lapp.* Avveel, Avvil. Lappi,
  N Finland
Ivanava *107 B7* Pol. Janów, Janów
  Poleski, *Rus.* Ivanovo. Brestskaya
  Voblasts', SW Belarus
Ivanhoe *149 C6* New South Wales,
  SE Australia
Ivano-Frankivs'k *108 C2 Ger.* Stanislau,
  Pol. Stanisławów, *Rus.* Ivano-
  Frankovsk; *prev.* Stanislav. Ivano-
  Frankivs'ka Oblast', W Ukraine
Ivanovo *111 B5* Ivanovskaya Oblast',
  W Russian Federation
Ivanovo *see* Ivanava
Ivantsevichi/Ivatsevichi *see* Ivatsevichy
Ivatsevichy *107 B6 Pol.* Iwacewicze, *Rus.*
  Ivantsevichi, Ivatsevichi. Brestskaya
  Voblasts', SW Belarus
Ivigtut *see* Ivittuut
Ivittuut *82 B4 var.* Ivigtut. Kitaa,
  S Greenland
Iviza *see* Eivissa/Ibiza
Ivory Coast *see* Côte d'Ivoire
Ivory Coast, Republic of *see* Côte
  d'Ivoire
Ivujivik *38 D1* Québec, NE Canada
Iwacewicze *see* Ivatsevichy
Iwaki *131 D5* Fukushima, Honshū,
  SW Japan
Iwakuni *131 B7* Yamaguchi, Honshū,
  SW Japan
Iwanai *130 C2* Hokkaidō, NE Japan
Iwate *130 D3* Iwate, Honshū, N Japan
Ixtapa *51 E5* Guerrero, S Mexico
Ixtepec *51 F5* Oaxaca, SE Mexico
Iyo-nada *131 B7 sea* S Japan
Izabal, Lago de *52 B2 prev.* Golfo Dulce.
  lake E Guatemala
Izad Khvāst *120 D3* Fārs, C Iran
Izegem *87 A6 prev.* Iseghem. West-
  Vlaanderen, W Belgium
Izhevsk *111 D5 prev.* Ustinov.
  Udmurtskaya Respublika, NW Russian
  Federation
Izmail *108 D4 Rus.* Izmail. Odes'ka
  Oblast', SW Ukraine
İzmir *116 B2 var.* Ismid; *anc.* Astacus.
  Kocaeli, NW Turkey
İzmir *116 A3 prev.* Smyrna. İzmir,
  W Turkey
İznik Gölü *116 B3* lake NW Turkey
Izu-hanto *131 D6 peninsula* Honshū,
  S Japan
Izu Shichito *see* Izu-shotō
Izu-shotō *131 D6 var.* Izu Shichito.
  island group S Japan

J

Jabal ash Shifa *120 A4 desert* NW Saudi
  Arabia
Jabalpur *135 E4 prev.* Jubbulpore.
  Madhya Pradesh, C India

Jabbūl, Sabkhat al *118 B2 sabkha*
NW Syria
Jablah *118 A3 var.* Jeble, *Fr.* Djéblé. Al
Lādhiqīyah, W Syria
Jaca *93 F1* Aragón, NE Spain
Jacaltenango *52 A2* Huehuetenango,
W Guatemala
Jackson *42 B2 state capital* Mississippi,
S USA
Jackson *45 H5* Missouri, C USA
Jackson *42 C1* Tennessee, S USA
Jackson Head *151 A6 headland* South
Island, New Zealand
Jacksonville *43 E3* Florida, SE USA
Jacksonville *40 B4* Illinois, N USA
Jacksonville *43 F1* North Carolina,
SE USA
Jacksonville *49 G3* Texas, SW USA
Jacmel *54 D3 var.* Jaquemel. S Haiti
Jacob *see* Nkayi
Jacobābād *134 B3* Sind, SE Pakistan
Jadotville *see* Likasi
Jadransko More/Jadransko Morje *see*
Adriatic Sea
Jaén *60 B2* Cajamarca, N Peru
Jaén *92 D4* Andalucía, SW Spain
Jaffna *132 D3* Northern Province,
N Sri Lanka
Jagannath *see* Puri
Jagdalpur *135 E5* Chhattīsgarh, C India
Jagdaqi *127 G1* Nei Mongol Zizhiqu,
N China
Jagodina *100 D4 prev.* Svetozarevo.
Serbia, C Serbia
Jahra *see* Al Jahrā'
Jailolo *see* Halmahera, Pulau
Jaipur *134 D3 prev.* Jeypore. *state capital*
Rājasthān, N India
Jaisalmer *134 C3* Rājasthān, NW India
Jajce *100 B3* Federacija Bosne
I Hercegovine, W Bosnia and
Herzegovina
Jakarta *138 C5 prev.* Djakarta, *Dut.*
Batavia. *country capital* (Indonesia)
Jawa, C Indonesia
Jakobstad *84 D4 Fin.* Pietarsaari. Länsi-
Suomi, W Finland
Jakobstadt *see* Jēkabpils
Jalālābād *123 F4 var.* Jalalabad,
Jelalabad. Nangarhār, E Afghanistan
Jalal-Abad *see* Dzhalal-Abad, Dzhalal-
Abadskaya Oblast', Kyrgyzstan
Jalandhar *134 D2 prev.* Jullundur.
Punjab, N India
Jalapa *52 D3* Nueva Segovia,
NW Nicaragua
Jalpa *50 D4* Zacatecas, C Mexico
Jālū *71 G3 var.* Jūla. NE Libya
Jaluit Atoll *144 D2 var.* Jālwōj. *atoll*
Ralik Chain, S Marshall Islands
Jālwōj *see* Jaluit Atoll
Jamaame *73 D6 It.* Giamame; *prev.*
Margherita. Jubbada Hoose, S Somalia
Jamaica *54 A4* country W West Indies

JAMAICA
West Indies

Official name Jamaica
Formation 1962 / 1962
Capital Kingston
Population 2.8 million / 670 people
per sq mile (259 people per sq km)
Total area 4243 sq. miles (10,990 sq. km)
Languages English Creole, English*
Religions Other and nonreligious 45%,
Other Protestant 20%, Church
of God 18%, Baptist 10%, Anglican 7%
Ethnic mix Black African 91%,
Mulatto (mixed race) 7%, East Indian 1%,
European and Chinese 1%
Government Parliamentary system
Currency Jamaican dollar = 100 cents
Literacy rate 88%
Calorie consumption 2789 kilocalories

Jamaica *56 A1* island W West Indies
Jamaica Channel *54 D3 channel* Haiti/
Jamaica
Jamālpur *135 F3* Bihār, NE India
Jambi *138 B4 var.* Telanaipura; *prev.*
Djambi. Sumatera, W Indonesia
Jamdena *see* Yamdena, Pulau
James Bay *38 C3 bay* Ontario/Québec,
E Canada
James River *45 E2 river* North Dakota/
South Dakota, N USA
James River *41 E5 river* Virginia,
NE USA

Jamestown *41 E3* New York, NE USA
Jamestown *45 E2* North Dakota, N USA
Jamestown *see* Holetown
Jammu *134 D2 prev.* Jummoo. *state
capital* Jammu and Kashmir, NW India
Jammu and Kashmīr *134 D1 disputed
region* India/Pakistan
Jāmnagar *134 C4 prev.* Navanagar.
Gujarāt, W India
Jamshedpur *135 F4* Jhārkhand, NE India
Jamuna *see* Brahmaputra
Janaúba *63 F3* Minas Gerais, SE Brazil
Janesville *40 B3* Wisconsin, N USA
Janina *see* Ioánnina
Janischken *see* Joniškis
Jankovac *see* Jánoshalma
Jan Mayen *83 F4 Norwegian dependency*
N Atlantic Ocean
Jánoshalma *99 C7 SCr.* Jankovac. Bács-
Kiskun, S Hungary
Janów *see* Ivanava, Belarus
Janow/Janów *see* Jonava, Lithuania
Janów Poleski *see* Ivanava
Japan *130 C4 var.* Nippon, *Jap.* Nihon.
country E Asia

JAPAN
East Asia

Official name Japan
Formation 1590 / 1972
Capital Tokyo
Population 127 million / 874 people
per sq mile (338 people per sq km)
Total area 145,882 sq. miles
(377,835 sq. km)
Languages Japanese*, Korean, Chinese
Religions Shinto and Buddhist 76%,
Buddhist 16%, Other (including
Christian) 8%
Ethnic mix Japanese 99%, Other
(mainly Korean) 1%
Government Parliamentary system
Currency Yen = 100 sen
Literacy rate 99%
Calorie consumption 2719 kilocalories

Japan, Sea of *130 A4 var.* East Sea, *Rus.*
Yaponskoye More. *sea* NW Pacific
Ocean
Japan Trench *125 F1 trench* NW Pacific
Ocean
Japen *see* Yapen, Pulau
Japiim *62 C2 var.* Máncio Lima. Acre,
W Brazil
Japurá, Rio *62 C2 var.* Río Caquetá,
Yapurá. *river* Brazil/Colombia
Japurá, Rio *see* Caquetá, Río
Jaqué *53 G5* Darién, SE Panama
Jaquemel *see* Jacmel
Jarablos *see* Jarābulus
Jarābulus *118 C2 var.* Jarablos, Jerablus,
Fr. Djérablous. Ḩalab, N Syria
Jarbah, Jazīrat *see* Jerba, Île de
**Jardines de la Reina, Archipiélago de
los** *54 B2 island group* C Cuba
Jarid, Shaṭṭ al *see* Jerid, Chott el
Jarocin *98 C4* Wielkopolskie, C Poland
Jarosław *see* Jarosław
Jarosław *99 E5 Ger.* Jaroslau, *Rus.*
Yaroslav. Podkarpackie, SE Poland
Jarqo'rg'on *123 E3 Rus.* Dzharkurgan.
Surkhondaryo Viloyati, S Uzbekistan
Jarvis Island *145 G2 US unincorporated*
territory C Pacific Ocean
Jasło *99 D5* Podkarpackie, SE Poland
Jastrzębie-Zdrój *99 C5* Śląskie, S Poland
Jataí *63 E3* Goiás, C Brazil
Jativa *see* Xátiva
Jauf *see* Al Jawf
Jaunpiebalga *106 D3* NE Latvia
Jaunpur *135 E3* Uttar Pradesh, N India
Java *152 A3* South Dakota, N USA
Javalambre *93 E3 mountain* E Spain
Java Sea *138 D4 Ind.* Laut Jawa. *sea*
W Indonesia
Java Trench *124 D5 var.* Sunda Trench.
trench E Indian Ocean
Jawa *see* Java Sea
Jawhar *73 D6 var.* Jowhar, *It.* Giohar.
Shabeellaha Dhexe, S Somalia
Jaworów *see* Yavoriv
Jaya, Puncak *139 G4 prev.* Puntjak
Carstensz, Puntjak Sukarno. *mountain*
Papua, E Indonesia
Jayapura *139 H4 var.* Djajapura,
Dut. Hollandia; *prev.* Kotabaru,
Sukarnapura. Papua, E Indonesia
Jay Dairen *see* Dalian

Jayhawker State *see* Kansas
Jaz Murian, Hamun-e *120 E4 lake*
SE Iran
Jebba *75 F4* Kwara, W Nigeria
Jebel, Bahr el *see* White Nile
Jeble *see* Jablah
Jedda *see* Jiddah
Jędrzejów *98 D4 Ger.* Endersdorf.
Świętokrzyskie, C Poland
Jefferson City *45 G5 state capital*
Missouri, C USA
Jega *75 F4* Kebbi, NW Nigeria
Jehol *see* Chengde
Jeju-do *129 E4 Jap.* Saishū; *prev.* Cheju-
do, Quelpart. *island* S South Korea
Jeju Strait *129 E4 var.* Jeju-haehyŏp;
prev. Cheju-Strait. *strait* S South Korea
Jeju-haehyŏp *see* Jeju Strait
Jēkabpils *106 D4 Ger.* Jakobstadt.
S Latvia
Jelalabad *see* Jalālābād
Jelenia Góra *98 B4 Ger.* Hirschberg,
Hirschberg im Riesengebirge,
Hirschberg in Riesengebirge,
Hirschberg in Schlesien. Dolnośląskie,
SW Poland
Jelgava *106 C3 Ger.* Mitau. C Latvia
Jemappes *87 B6* Hainaut, S Belgium
Jember *138 D5 prev.* Djember. Jawa,
C Indonesia
Jena *94 C4* Thüringen, C Germany
Jengish Chokusu *see* Tomūr Feng
Jenin *119 A6* N West Bank
Jerablus *see* Jarābulus
Jerada *70 D2* NE Morocco
Jérémie *54 D3* SW Haiti
Jerez *see* Jerez de la Frontera, Spain
Jerez de la Frontera *92 C5 var.* Jerez;
prev. Xeres. Andalucía, SW Spain
Jerez de los Caballeros *92 C4*
Extremadura, W Spain
Jericho *119 B6 Ar.* Arīḩā, *Heb.* Yeriḩo.
E West Bank
Jerid, Chott el *71 E2 var.* Shaṭṭ al Jarīd.
salt lake SW Tunisia
Jersey *89 D8 British Crown Dependency*
Channel Islands, NW Europe
Jerusalem *103 H4 Ar.* Al Quds, Al Quds
ash Sharīf, *Heb.* Yerushalayim; *anc.*
Hierosolyma. *country capital* (Israel)
Jerusalem, NE Israel
Jesenice *95 D7 Ger.* Assling.
NW Slovenia
Jesselton *see* Kota Kinabalu
Jessore *135 G4* Khulna, W Bangladesh
Jesús María *64 C3* Córdoba,
C Argentina
Jeypore *see* Jaipur, Rājasthān, India
Jhānsi *134 D3* Uttar Pradesh, N India
Jhārkhand *135 F4 cultural region*
NE India
Jhelum *134 C2* Punjab, NE Pakistan
Ji *see* Hebei, China
Ji *see* Jilin, China
Jiangmen *128 C6* Guangdong, S China
Jiangsu *128 D4 var.* Chiang-su, Jiangsu
Sheng, Su. *province* E China
Jiangsu *see* Nanjing
Jiangsu Sheng *see* Jiangsu
Jiangxi *128 C6 var.* Chiang-hsi, Gan,
Jiangxi Sheng, Kiangsi. *province*
S China
Jiangxi Sheng *see* Jiangxi
Jiaxing *128 D5* Zhejiang, SE China
Jiayi *128 D6 prev.* Chiai; *var.* Chia-i,
Chiayi, Kiayi, Jiayi, *Jap.* Kagi.
C Taiwan
Jibuti *see* Djibouti
Jiddah *121 A5 Eng.* Jedda. Makkah,
W Saudi Arabia
Jih-k'a-tse *see* Xigazê
Jihlava *99 B5 Ger.* Iglau, *Pol.* Igława.
Vysocina, S Czech Republic
Jijili *73 D6* E Sudan
Jijel *see* Djidjelli, Djidjel, Dhexe,
C Somalia
Jilin *128 E3 var.* Chi-lin, Girin, Kirin;
prev. Yungki, Yunki. Jilin, NE China
Jilin *128 E3 var.* Chi-lin, Girin, Kirin, Ji,
Jilin Sheng, Kirin. *province* NE China
Jilong *see* Chilung; *var.*
Keelung, *Jap.* Kiirun, *Rus.* Kilon
Santissima Trinidad. N Taiwan
Jima *73 C5 var.* Jimma, *It.* Gimma.
Oromīya, C Ethiopia
Jimbolia *108 A4 Ger.* Hatzfeld, *Hung.*
Zsombolya. Timiş, W Romania
Jiménez *50 D2* Chihuahua, N Mexico
Jimma *see* Jima

Jimsar *126 C3* Xinjiang Uygur Zizhiqu,
NW China
Jin *see* Shanxi
Jin *see* Tianjin Shi
Jinan *128 C4 var.* Chinan, Chi-nan,
Tsinan. *province capital* Shandong,
E China
Jingdezhen *128 C5* Jiangxi, S China
Jinghong *128 A6 var.* Yunjinghong.
Yunnan, SW China
Jinhua *128 D5* Zhejiang, SE China
Jining *127 F3* Shandong, E China
Jinja *73 C6* S Uganda
Jinotega *52 D3* Jinotega,
NW Nicaragua
Jinotepe *52 D3* Carazo, SW Nicaragua
Jinsen *see* Incheon
Jinzhong *128 C4 var.* Yuci. Shanxi,
C China
Jinzhou *128 D3 var.* Chin-chou,
Chinchow; *prev.* Chinhsien. Liaoning,
NE China
Jirgalanta *see* Hovd
Jisr ash Shadadi *see* Ash Shadādah
Jiu *108 B5 Ger.* Schil, Schyl, *Hung.* Zsil,
Zsily. *river* S Romania
Jiujiang *128 C5* Jiangxi, S China
Jixi *129 E2* Heilongjiang, NE China
Jīzān *121 B6 var.* Qīzān. Jīzān, SW Sau
Arabia
Jizzax *123 E2 Rus.* Dzhizak. Jizzax
Viloyati, C Uzbekistan
João Belo *see* Xai-Xai
João Pessoa *63 G2 prev.* Paraíba. *state
capital* Paraíba, E Brazil
Joazeiro *see* Juazeiro
Job'urg *see* Johannesburg
Jo-ch'iang *see* Ruoqiang
Jodhpur *134 C3* Rājasthān, NW India
Joensuu *85 E5* Itä-Suomi, SE Finland
Jōetsu *131 C5 var.* Zyðetu. Niigata,
Honshū, C Japan
Jogjakarta *see* Yogyakarta
Johannesburg *78 D4 var.* Egoli,
Erautini, Gauteng, *abbrev.* Job'urg.
Gauteng, NE South Africa
Johannisburg *see* Pisz
John Day River *46 C3 river* Oregon,
NW USA
John o'Groats *88 C2* N Scotland, United
Kingdom
Johnston Atoll *143 E1 US
unincorporated territory* C Pacific
Ocean
Johor Baharu *see* Johor Bahru
Johor Bahru *138 B3 var.* Johor Baharu,
Johore Baru. Johor, Peninsular
Malaysia
Johore Baru *see* Johor Bahru
Johore Strait *138 A1 strait* Johor,
Peninsular Malaysia, Malaysia/
Singapore Asia Andaman Sea/South
China Sea
Joinvile *see* Joinville
Joinville *63 E4 var.* Joinvile. Santa
Catarina, S Brazil
Jokkmokk *84 C3 Lapp.* Dálvvadis.
Norrbotten, N Sweden
Jokyakarta *see* Yogyakarta
Joliet *40 B3* Illinois, N USA
Jonava *106 B4 var.* Janow, *Pol.* Janów.
Kaunas, C Lithuania
Jonesboro *42 B1* Arkansas, C USA
Joniškis *106 B3 Ger.* Janischken. Šiauli
N Lithuania
Jönköping *85 B7* Jönköping, S Sweden
Jonquière *39 E4* Québec, SE Canada
Joplin *45 F5* Missouri, C USA
Jordan *119 B6 off.* Hashemite Kingdom
of Jordan
Jordan *see* Al Mamlaka al Urdunīya
al Hashemīyah, Al Urdunn; *prev.*
Transjordan. *country* SW Asia

JORDAN
Southwest Asia

Official name Hashemite Kingdom
of Jordan
Formation 1946 / 1967
Capital Amman
Population 7.3 million / 213 people
per sq mile (82 people per sq km)
Total area 35,637 sq. miles (92,300 sq. km
Languages Arabic*
Religions Muslim (mainly Sunni) 92%,
Christian and Other 8%
Ethnic mix Arab 98%, Circassian 1%,
Armenian 1%
Government Monarchy

DAN
(continued)
rency Jordanian dinar = 1000 fils
racy rate 98%
orie consumption 3149 kilocalories

an 119 B5 Ar. Urdunn, Heb.
Yarden. river SW Asia
it 135 H3 Assam, NE India
75 G4 Plateau, C Nigeria
oh Bonaparte Gulf 146 D2 gulf
Australia
lateau 75 G4 plateau C Nigeria
nheimen 85 A5 mountain range
Norway
ié 118 A4 var. Junīyah. W Lebanon
e 86 D2 Fris. De Jouwer. Fryslân,
Netherlands
seno 85 E5 Etelä-Suomi, SE Finland
ar see Jawhar
rm Thurmond Reservoir see Clark
ll Lake
Aldama 50 D3 Zacatecas, C Mexico
de Fuca, Strait of 46 A1 strait
anada/USA
Fernández, Islas 57 A6 Eng.
an Fernandez Islands. island group
Chile
Fernandez Islands see Juan
rnández, Islas
eiro 63 G2 prev. Joazeiro. Bahia,
Brazil
eiro do Norte 63 G2 Ceará, E Brazil
73 B5 var. Jûbâ. country capital
outh Sudan) Central Equatoria,
South Sudan
73 D6 Amh. Genalē Wenz, It. Guiba,
om. Ganaane, Webi Jubba. river
hiopia/Somalia
a, Webi see Juba
ulpore see Jabalpur
r 93 E3 var. Jucar. river C Spain
itán 51 F5 var. Juchitán de Zaragosa.
axaca, SE Mexico
itán de Zaragoza see Juchitán
yyidat Hāmir 120 B3 Al Anbār,
raq
nburg 95 D7 Steiermark, C Austria
slavija see Serbia
lpa 52 D3 Chontales, S Nicaragua
de Fora 63 F4 Minas Gerais,
E Brazil
ountain range France/Switzerland
a 88 B4 island SW Scotland, United
ingdom
Mountains see Jura
arkas 106 B4 Ger. Georgenburg,
urburg. Tauragė, W Lithuania
urg see Jurbarkas
nala 106 C3 C Latvia
sá, Rio 62 C2 var. Río Yuruá. river
razil/Peru
aena, Rio 62 D3 river
razil
apa 52 B2 Jutiapa, S Guatemala
calpa 52 D2 Olancho, C Honduras
and 63 A7 Den. Jylland. peninsula
Denmark
avum see Salzburg
aa Morava 101 E3 Ger. Südliche
Morava. river SE Serbia
eng 78 C4 Southern, S Botswana
alan see Jutland
askylä 85 D5 Länsi-Suomi, C Finland

K

K2 126 A4 Chin. Qogir Feng, Eng. Mount
Godwin Austen. mountain China/
Pakistan
Kaafu Atoll see Male' Atoll
Kaaimanston 59 G3 Sipaliwini,
N Suriname
Kaakhka see Kaka
Kaala see Caäla
Kaamanen 84 D2 Lapp. Gámas. Lappi,
N Finland
Kaapstad see Cape Town
Kaaresuvanto 84 C3 Lapp. Gárassavon.
Lappi, N Finland
Kabale 73 B6 SW Uganda
Kabinda 77 D7 Kasai-Oriental, SE Dem.
Rep. Congo
Kabinda see Cabinda
Kābol see Kābul
Kabompo 78 C2 river W Zambia
Kābul 123 E4 prev. Kābol. country
capital (Afghanistan) Kābul,
E Afghanistan
Kabul 123 E4 var. Daryā-ye Kābul. river
Afghanistan/Pakistan
Kābul, Daryā-ye see Kabul
Kabwe 78 D2 Central, C Zambia
Kachchh, Gulf of 134 B4 var. Gulf of
Cutch, Gulf of Kutch. gulf W India
Kachchh, Rann of 134 B4 var. Rann
of Kachh, Rann of Kutch. salt marsh
India/Pakistan
Kachh, Rann of see Kachchh, Rann of
Kadan Kyun 137 B5 prev. King Island.
island Myeik Archipelago, S Myanmar
(Burma)
Kadavu 145 E4 prev. Kandavu. island
S Fiji
Kadoma 78 D3 prev. Gatooma.
Mashonaland West, C Zimbabwe
Kadugli 72 B4 Southern Kordofan,
S Sudan
Kaduna 75 G4 Kaduna, C Nigeria
Kadzhi-Say see Bokonbayevo
Kaédi 74 C3 Gorgol, S Mauritania
Kaffa see Feodosiya
Kafue 78 D2 Lusaka, SE Zambia
Kafue 78 C2 river C Zambia
Kaga Bandoro 76 C4 prev. Fort-
Crampel. Nana-Grébizi, C Central
African Republic
Kagan see Kogon
Kághet 74 D1 var. Karet. physical region
N Mauritania
Kagi see Jiayi
Kagoshima 131 B8 var. Kagosima.
Kagoshima, Kyūshū, SW Japan
Kagoshima-wan 131 A8 bay SW Japan
Kagosima see Kagoshima
Kagul see Cahul
Darya-ye Kahmard 123 E4 prev. Darya-
i-surkhab. river NE Afghanistan
Kahramanmaras 116 D4 var.
Kahraman Maraş, Maraş, Marash.
Kahramanmaraş, S Turkey
Kaiapoi 151 C6 Canterbury, South
Island, New Zealand
Kaifeng 128 C4 Henan, C China
Kai, Kepulauan 139 F4 prev. Kei Islands.
island group Maluku, SE Indonesia
Kaikohe 150 C2 Northland, North
Island, New Zealand
Kaikoura 151 C5 Canterbury, South
Island, New Zealand
Kaikoura Peninsula 151 C5 peninsula
South Island, New Zealand
Kainji Lake see Kainji Reservoir
Kainji Reservoir 75 F4 var. Kainji Lake.
reservoir W Nigeria
Kaipara Harbour 150 C2 harbour North
Island, New Zealand
Kairouan 71 E2 var. Al Qayrawān.
E Tunisia
Kaisaria see Kayseri
Kaiserslautern 95 A5 Rheinland-Pfalz,
SW Germany
Kaišiadorys 107 B5 Kaunas, S Lithuania
Kaitaia 150 C2 Northland, North Island,
New Zealand
Kajaani 84 E4 Swe. Kajana. Oulu,
C Finland
Kajan see Kayan, Sungai
Kajana see Kajaani
Kajisay see Bokonbayevo
Kaka 122 C2 Rus. Kaakhka. Ahal
Welaýaty, S Turkmenistan

Kake 36 D4 Kupreanof Island, Alaska,
USA
Kakhovka 109 F4 Khersons'ka Oblast',
S Ukraine
Kakhovs'ke Vodoskhovyshche 109 F4
Rus. Kakhovskoye Vodokhranilishche.
reservoir SE Ukraine
Kakhovskoye Vodokhranilishche see
Kakhovs'ke Vodoskhovyshche
Kākināda 132 D1 prev. Cocanada.
Andhra Pradesh, E India
Kakshaal-Too, Khrebet see
Kokshaal-Tau
Kaktovik 36 D2 Alaska, USA
Kalaallit Nunaat see Greenland
Kalahari Desert 78 B4 desert Southern
Africa
Kalaikhum see Qal'aikhum
Kalámai see Kalámata
Kalamariá 104 B4 Kentrikí Makedonía,
N Greece
Kalamás 104 A4 var. Thiamis; prev.
Thýamis. river W Greece
Kalámata 105 B6 prev. Kalámai.
Pelopónnisos, S Greece
Kalamazoo 40 C3 Michigan, N USA
Kalambaka see Kalampáka
Kálamos 105 C5 Attikí, C Greece
Kalampáka 104 B4 var. Kalambaka.
Thessalía, C Greece
Kalanchak 109 F4 Khersons'ka Oblast',
S Ukraine
Kalarash see Călăraşi
Kalasin 136 D4 var. Muang Kalasin.
E Thailand
Kālat 134 B2 var. Kelat, Khelat.
Baluchistan, SW Pakistan
Kalāt see Qalāt
Kalbarri 147 A5 Western Australia
Kalecik 116 C3 Ankara, N Turkey
Kalemie 77 E6 prev. Albertville. Katanga,
SE Dem. Rep. Congo
Kale Sultanie see Çanakkale
Kalgan see Zhangjiakou
Kalgoorlie 147 B6 Western Australia
Kalima 77 D6 Maniema, E Dem. Rep.
Congo
Kalimantan 138 D4 Eng. Indonesian
Borneo. geopolitical region Borneo,
C Indonesia
Kalinin see Tver'
Kaliningrad 106 A4 Kaliningradskaya
Oblast', W Russian Federation
Kaliningradskaya Oblast' 106 B4 var.
Kaliningrad. province and enclave
W Russian Federation
Kalinivka 109 C7 Rus. Kalinkovichi.
Homyel'skaya Voblasts', SE Belarus
Kalinkovichi see Kalinavichy
Kalisch see Kalisz
Kalispell 44 B1 Montana, NW USA
Kalisz 98 C4 Ger. Kalisch, Rus. Kalish.
prev. Calisia. Wielkopolskie, C Poland
Kalix 104 D4 Norrbotten, N Sweden
Kalixälven 84 D3 river N Sweden
Kallaste 106 E3 Tartumaa, E Estonia
Kalloní 105 D5 Lésvos, E Greece
Kalmar 85 C7 var. Calmar. Kalmar,
S Sweden
Kalmthout 87 C5 Antwerpen, N Belgium
Kálpaki 104 A4 Ípeiros, W Greece
Kalpeni Island 132 B3 island
Lakshadweep, India, N Indian Ocean
Kaltdorf see Pruszków
Kaluga 111 B5 Kaluzhskaya Oblast',
W Russian Federation
Kalush 108 C2 Ivano-
Frankivs'ka Oblast', W Ukraine
Kałusz see Kalush
Kalutara 132 D4 Western Province,
SW Sri Lanka
Kalvarija 107 B5 Kalvarija
Marijampolė, S Lithuania
Kalwaria see Kalvarija
Kálymnos 105 D6 var. Kálimnos. island
Dodekánisa, Greece, Aegean Sea
Kama 110 D4 river NW Russian
Federation
Kamarang 59 F3 W Guyana
Kambryk see Cambrai
Kamchatka 115 G3 Eng.
peninsula E Russian Federation
Kamenets-Podol'skiy see
Kam"yanets'-Podil's'kyy
Kamenka see Camenca

Kamenka Dneprovskaya see
Kam"yanka-Dniprovs'ka
Kamenskoye see Romaniv
Kamensk-Shakhtinskiy 111 B6
Rostovskaya Oblast', SW Russian
Federation
Kamina 77 D7 Katanga, S Dem. Rep.
Congo
Kamishli see Al Qāmishlī
Kamloops 37 E5 British Columbia,
SW Canada
Kammu Seamount 152 C2 guyot
N Pacific Ocean
Kampala 73 B6 country capital (Uganda)
S Uganda
Kâmpóng Cham 137 D6 prev. Kompong
Cham. Kâmpóng Cham, C Cambodia
Kâmpóng Chhnãng 137 D6 prev.
Kompong. Kâmpóng Chhnãng,
C Cambodia
Kâmpóng Saôm see Sihanoukville
Kâmpóng Spœ 137 D6 prev. Kompong
Speu. Kâmpóng Spœ, S Cambodia
Kâmpóng Thum 137 D5 prev. Trâpeăng
Phlong. Kâmpóng Thum, C Cambodia
Kâmpóng Trâbêk 137 D5 prev. Phumĭ
Kâmpóng Trâbêk, Phum Kompong
Trabek. Kâmpóng Thum, C Cambodia
Kâmpôt 137 D6 Kâmpôt, SW Cambodia
Kampuchea see Cambodia
Kampuchea, Democratic see Cambodia
Kampuchea, People's Democratic
Republic of see Cambodia
Kam"yanets'-Podil's'kyy 108 C3 Rus.
Kamenets-Podol'skiy. Khmel'nyts'ka
Oblast', W Ukraine
Kam"yanka-Dniprovs'ka 109 F3 Rus.
Kamenka-Dneprovskaya. Zaporiz'ka
Oblast', SE Ukraine
Kamyshin 111 B6 Volgogradskaya
Oblast', SW Russian Federation
Kanab 44 A1 Utah, W USA
Kanaky see New Caledonia
Kananga 77 D6 prev. Luluabourg. Kasai-
Occidental, S Dem. Rep. Congo
Kanazara see Karnātaka
Kanash 111 C5 Chuvashskaya
Respublika, W Russian Federation
Kanazawa 131 C5 Ishikawa, Honshū,
SW Japan
Kanbe 136 B4 Yangon, SW Myanmar
(Burma)
Kānchipuram 132 C2 prev.
Conjeeveram. Tamil Nādu, SE India
Kandahār 123 E5 Per. Qandahār.
S Afghanistan
Kandalaksha 110 B2 var. Kandalaksa,
Fin. Kantalahti. Murmanskaya Oblast',
NW Russian Federation
Kandangan 138 D4 Borneo, C Indonesia
Kandau see Kandava
Kandava 106 C3 Ger. Kandau. W Latvia
Kandi 75 F4 N Benin
Kandy 132 D3 Central Province,
C Sri Lanka
Kane Fracture Zone 66 B4 fracture zone
NW Atlantic Ocean
Kāne'ohe 47 A8 var. Kaneohe. O'ahu,
Hawaii, USA, C Pacific Ocean
Kaneohe see Kāne'ohe
Kanestron, Akrotírio see Palioúri,
Akrotírio
Kangān see Bandar-e Kangān
Kangaroo Island 149 A7 island South
Australia
Kangertittivaq 83 E4 Dan. Scoresby
Sund. fjord E Greenland

Kansas 45 F5 off. State of Kansas, also known as Jayhawker State, Sunflower State. state C USA
Kansas City 45 F4 Kansas, C USA
Kansas City 45 F4 Missouri, C USA
Kansas River 45 F5 river Kansas, C USA
Kansk 115 E4 Krasnoyarskiy Kray, S Russian Federation
Kansu see Gansu
Kantalahti see Kandalaksha
Kántanos 105 C7 Kríti, Greece, E Mediterranean Sea
Kantemirovka 111 B6 Voronezhskaya Oblast', W Russian Federation
Kantipur see Kathmandu
Kanton 145 F3 var. Abariringa, Canton Island; prev. Mary Island. atoll Phoenix Islands, C Kiribati
Kanye 78 C4 Southern, SE Botswana
Kaohsiung see Gaoxiong
Kaolack 74 B3 var. Kaolak. W Senegal
Kaolak see Kaolack
Kaolan see Lanzhou
Kaoma 78 C2 Western, W Zambia
Kapelle 87 B5 Zeeland, SW Netherlands
Kapellen 87 C5 Antwerpen, N Belgium
Kapka, Massif du 76 C2 mountain range E Chad
Kaplangky, Plato see Gaplañgyr Platosy
Kapoeas see Kapuas, Sungai
Kapoeta 73 C5 Eastern Equatoria, SE South Sudan
Kaposvár 99 C7 Somogy, SW Hungary
Kappeln 94 B2 Schleswig-Holstein, N Germany
Kaproncza see Koprivnica
Kapstad see Cape Town
Kapsukas see Marijampolé
Kaptsevichy 107 C7 Rus. Koptsevichi. Homyel'skaya Voblasts', SE Belarus
Kapuas, Sungai 138 C4 prev. Kapoeas. river Borneo, C Indonesia
Kapuskasing 38 C4 Ontario, S Canada
Kapyl' 107 C6 Rus. Kopyl'. Minskaya Voblasts', C Belarus
Kara-Balta 123 F2 Chuyskaya Oblast', N Kyrgyzstan
Karabil', Vozvyshennost' see Garabil Belentligi
Kara-Bogaz-Gol, Zaliv see Garabogaz Aylagy
Karabük 116 C2 Karabük, NW Turkey
Karáchi 134 B3 Sind, SE Pakistan
Karácsonkő see Piatra-Neamţ
Karadeniz see Black Sea
Karadeniz Boğazı see İstanbul Boğazı
Karaferiye see Véroia
Karaganda see Karagandy
Karagandy 114 C4 prev. Karaganda, Kaz. Qaraghandy. Karagandy, C Kazakhstan
Karaginskiy, Ostrov 115 H2 island E Russian Federation
Karagumskiy Kanal see Garagum Kanaly
Karak see Al Karak
Kara-Kala see Magtymguly
Karakax see Moyu
Karakılısse see Ağrı
Karakol 123 G2 var. Karakolka. Issyk-Kul'skaya Oblast', NE Kyrgyzstan
Karakol 123 G2 prev. Przheval'sk. Issyk-Kul'skaya Oblast', NE Kyrgyzstan
Karakolka see Karakol
Karakoram Range 134 D1 mountain range C Asia
Karaköse see Ağrı
Karakul' see Qarokül, Tajikistan
Kara Kum see Garagum
Kara Kum Canal/Karakumskiy Kanal see Garagum Kanaly
Karakumy, Peski see Garagum
Karamai see Karamay
Karaman 116 C4 Karaman, S Turkey
Karamay 126 B2 var. Karamai, Kelamayi; prev. Chin. K'o-la-ma-i. Xinjiang Uygur Zizhiqu, NW China
Karamea Bight 151 B5 gulf South Island, New Zealand
Karapelit 104 E1 Rom. Stejarul. Dobrich, NE Bulgaria
Kara-Say 123 G2 Issyk-Kul'skaya Oblast', NE Kyrgyzstan
Karasburg 78 B4 Karas, S Namibia
Kara Sea 114 D2 Eng. Kara Sea. sea Arctic Ocean
Kara Sea see Karskoye More
Kara Strait 110 E2 Eng. Kara Strait. strait N Russian Federation
Kara Strait see Karskiye Vorota, Proliv

Karatau 114 C5 Kaz. Qarataü. Zhambyl, S Kazakhstan
Karavás 105 B7 Kýthira, S Greece
Karbalā' 120 B3 var. Kerbala, Kerbela. Karbalā', S Iraq
Kardeljevo see Ploče
Kardhítsa see Kardítsa
Karditsa 105 B5 var. Kardhítsa. Thessalía, C Greece
Kärdla 106 C2 Ger. Kertel. Hiiumaa, W Estonia
Karet see Kāgheṭ
Kargı 116 C2 Çorum, N Turkey
Kargilik see Yecheng
Kariba 78 D2 Mashonaland West, N Zimbabwe
Kariba, Lake 78 D3 reservoir Zambia/Zimbabwe
Karibib 78 B3 Erongo, C Namibia
Karies see Karyés
Karigasniemi 84 D2 Lapp. Garegegasnjárga. Lappi, N Finland
Karimata, Selat 138 C4 strait W Indonesia
Karimnagar 134 D5 Telangana, C India
Karin 72 D4 Woqooyi Galbeed, N Somalia
Kariot see Ikaría
Káristos see Kárystos
Karkinits'ka Zatoka 109 E4 Rus. Karkinitskiy Zaliv. gulf S Ukraine
Karkinitskiy Zaliv see Karkinits'ka Zatoka
Karkük see Kirkük
Karleby see Kokkola
Karl-Marx-Stadt see Chemnitz
Karlo see Hailuoto
Karlovac 100 B3 Ger. Karlstadt, Hung. Károlyváros. Karlovac, C Croatia
Karlovy Vary 99 A5 Ger. Karlsbad; prev. Eng. Carlsbad. Karlovarský Kraj, W Czech Republic
Karlsbad see Karlovy Vary
Karlsburg see Alba Iulia
Karlskrona 85 C7 Blekinge, S Sweden
Karlsruhe 95 B6 var. Carlsruhe. Baden-Württemberg, SW Germany
Karlstad 85 B6 Värmland, C Sweden
Karlstadt see Karlovac
Karnāl 134 D2 Haryāna, N India
Karnātaka 132 C1 var. Kanara; prev. Maisur, Mysore. cultural region W India
Karnobat 104 D2 Burgas, E Bulgaria
Karnul see Kurnool
Karol see Carei
Károly-Fehérvár see Alba Iulia
Károlyváros see Karlovac
Karpaten see Carpathian Mountains
Kárpathos 105 E7 Kárpathos, SE Greece
Kárpathos 105 E7 It. Scarpanto; anc. Carpathos, Carpathus. island SE Greece
Karpaty see Carpathian Mountains
Karpenísi 105 B5 prev. Karpenísion. Stereá Elláda, C Greece
Karpenísion see Karpenísi
Karpílovka see Aktsyabrski
Kars 117 F2 var. Qars. Kars, NE Turkey
Karsau see Kārsava
Kārsava 106 D4 Ger. Karsau; prev. Rus. Korsovka. E Latvia
Karshi see Qarshi, Uzbekistan
Karyés 104 C4 var. Karies. Ágion Óros, N Greece
Kárystos 105 C6 var. Káristos. Évvoia, C Greece
Kasai 77 C6 var. Cassai, Kassai. river Angola/Dem. Rep. Congo
Kasaji 77 D7 Katanga, S Dem. Rep. Congo
Kasama 78 D1 Northern, N Zambia
Kasan see Koson
Kāsaragod 132 B2 Kerala, SW India
Kaschau see Košice
Kāshān 120 C3 Eṣfahān, C Iran
Kashgar see Kashi
Kashi 126 A3 Chin. Kaxgar, K'o-shih, Uigh. Kashgar. Xinjiang Uygur Zizhiqu, NW China
Kasongo 77 D6 Maniema, E Dem. Rep. Congo
Kasongo-Lunda 77 C7 Bandundu, SW Dem. Rep. Congo
Kásos 105 D7 island S Greece
Kaspiy Mangy Oypaty see Caspian Depression

Kaspiysk 111 B8 Respublika Dagestan, SW Russian Federation
Kaspiyskoye More/Kaspiy Tengizi see Caspian Sea
Kassa see Košice
Kassai see Kasai
Kassala 72 C4 Kassala, E Sudan
Kassel 94 B4 prev. Cassel. Hessen, C Germany
Kasserine 71 E2 var. Al Qaşrayn. W Tunisia
Kastamoni 116 C2 var. Castamoni, Kastamuni. Kastamonu, N Turkey
Kastamonu 116 C2 var. Kastamoni, Kastamuni. Kastamonu, N Turkey
Kastaneá see Kastaniá
Kastaniá 104 B4 prev. Kastaneá. Kentrikí Makedonía, N Greece
Kastélli see Kíssamos
Kastoría 104 B4 Dytikí Makedonía, N Greece
Kástro 105 C6 Sífnos, Kykládes, Greece, Aegean Sea
Kastsyukovichy 107 E7 Rus. Kostyukovichi. Mahilyowskaya Voblasts', E Belarus
Kastsyukowka 107 D7 Rus. Kostyukovka. Homyel'skaya Voblasts', SE Belarus

KAZAKHSTAN
Central Asia

Official name Republic of Kazakhstan
Formation 1991 / 1991
Capital Astana
Population 16.4 million / 16 people per sq mile (6 people per sq km)
Total area 1,049,150 sq. miles (2,717,300 sq. km)
Languages Kazakh*, Russian, Ukrainian, German, Uzbek, Tatar, Uighur
Religions Muslim (mainly Sunni) 47%, Orthodox Christian 44%, Other 9%
Ethnic mix Kazakh 57%, Russian 27%, Other 8%, Ukrainian 3%, German 2%
Government Presidential system
Currency Tenge = 100 tiyin
Literacy rate 99%
Calorie consumption 3107 kilocalories

Kazakhstan, Republic of see Kazakhstan
Kazakh Uplands 114 C4 Eng. Kazakh Uplands, Kirghiz Steppe. uplands C Kazakhstan
Kazakstan see Kazakhstan
Kazan' 111 C5 Respublika Tatarstan, W Russian Federation
Kazandzhik see Bereket
Kazanlůk see Kazanlŭk
Kazanlŭk 104 D2 prev. Kazanlik, Kizanlik. Kazanlŭk, C Bulgaria
Kazan-rettō see Volcano Islands
Kazatin see Kozyatyn
Kazbegi see Kazbek
Kazbek 117 F1 var. Kazbegi, Geor. Mqinvartsveri. mountain N Georgia
Käzerün 120 D4 Fārs, S Iran
Kazvin see Qazvin
Kéa 105 C6 Tziá
Kéa see Ioulís
Kéamu see Aneityum
Kearney 45 E4 Nebraska, C USA
Keban Barajı 117 E3 reservoir C Turkey
Kebkabiya 72 A4 Northern Darfur, W Sudan
Kebnekaise 84 C3 Lapp. Giebnegáisi. mountain N Sweden
Kecskemét 99 D7 Bács-Kiskun, C Hungary
Kediri 138 D5 Jawa, C Indonesia
Kędzierzyn-Kozle 99 C5 Ger. Heydebrech. Opolskie, S Poland
Keelung see Jilong
Keetmanshoop 78 B4 Karas, S Namibia
Kefallinía see Kefalloniá
Kefallonía 105 A5 var. Kefallinía. island Iónia Nisiá, Greece, C Mediterranean Sea
Kefe see Feodosiya
Kegel see Keila
Kehl 95 A6 Baden-Württemberg, SW Germany
Keijō see Seoul
Keila 106 D2 Ger. Kegel. Harjumaa, NW Estonia
Keita 75 F3 Tahoua, C Niger
Keith 149 B7 South Australia
Kék-Art 123 G2 prev. Alaykel', Alay Oshskaya Oblast', SW Kyrgyzstan
Kékes 99 D6 mountain N Hungary

ayi see Karamay
g see Klang
kiy Uzboý see Kelif Uzboý
Uzboý 122 D3 Rus. Kelifskiy
oy. salt marsh E Turkmenistan
Çayı 117 E3 river N Turkey
e 106 B4 Šiauliai, C Lithuania
76 B4 Tandjilé, SW Chad
rna 37 E5 British Columbia, Canada
46 B2 Washington, NW USA
see Kielce
ng 138 B3 var. Kluang. Johor, insular Malaysia
110 B3 Respublika Kareliya, Russian Federation
h 117 E3 Erzincan, E Turkey
man see Cukai
rovo 114 D4 prev. Shcheglovsk. nerovskaya Oblast', C Russian eration
84 D4 Lappi, NW Finland
järvi 84 D3 Swe. Kemiträsk. Lappi, inland
joki 84 D3 river NW Finland
n 123 G2 prev. Bystrovka. iyskaya Oblast', N Kyrgyzstan
ns Island see Nikumaroro
träsk see Kemijärvi
nuna 102 A5 var. Comino. island lalta
ele 84 D4 Oulu, C Finland
ten 95 B7 Bayern, S Germany
al 89 D5 NW England, United gdom
ari 139 E4 Sulawesi, C Indonesia
dy 49 G4 Texas, SW USA
na 74 C4 SE Sierra Leone
irgench see Köneürgenç
e 77 C6 Bandundu, SW Dem. . Congo
tung 136 C3 pev. Keng Tung. Shan te, E Myanmar (Burma)
Tung see Kengtung
ra 70 C2 prev. Port-Lyautey. w Morocco
ett 45 H5 Missouri, C USA
ewick 46 C2 Washington, NW USA
ra 38 A3 Ontario, S Canada
sha 40 B3 Wisconsin, N USA
au 114 B5 Yuzhnyy Kazakhstan, azakhstan
icky 40 C5 off. Commonwealth Kentucky, also known as Bluegrass te. state C USA
icky Lake 64 B5 reservoir ntucky/Tennessee, S USA
a 73 C6 off. Republic of Kenya. ntry E Africa

NYA
st Africa
cial name Republic of Kenya
nation 1963 / 1963
tal Nairobi
ulation 44.4 million / 203 people sq mile (78 people per sq km)
al area 224,961 sq. miles ,650 sq. km)
guages Swahili*, English*, Kikuyu, Kalenjin, Kamba
gions Christian 80%, Muslim 10%, itional beliefs 9%, Other 1%
nic mix Other 28%, Kikuyu 22%, ya 14%, Luo 14%, Kalenjin 11%, ba 11%
ernment Presidential system rency Kenya shilling = 100 cents racy rate 72%
rie consumption 2189 kilocalories

a, Mount see Kirinyaga
a, Republic of see Kenya
uk 45 G4 Iowa, C USA
see Tziá
ao 98 C4 Wielkopolskie, Poland
el Island see Niuatoputapu
k see Al Karak
la 132 C2 cultural region S India
sunt see Giresun
téa 105 C6 var. Keratea. Attikí, Greece
tea see Keratéa
ala/Kerbela see Karbalā'
h 109 G5 Rus. Kerch'. Avtonomna publika Krym, SE Ukraine

Kerch' see Kerch
Kerchens'ka Protska/Kerchenskiy Proliv see Kerch Strait
Kerch Strait 109 G4 var. Bosporus Cimmerius, Enikale Strait, Rus. Kerchenskiy Proliv, Ukr. Kerchens'ka Protska. strait Black Sea/Sea of Azov
Keremitlik see Lyulyakovo
Kerguelen 141 C7 island C French Southern and Antarctic Lands
Kerguelen Plateau 141 C7 undersea plateau S Indian Ocean
Kerí 105 A6 Zákynthos, Iónia Nisiá, Greece, C Mediterranean Sea
Kerikeri 150 D2 Northland, North Island, New Zealand
Kerkenah, Îles de 102 D4 var. Kerkenna Islands, Ar. Juzur Qarqannah. island group E Tunisia
Kerkenna Islands see Kerkenah, Îles de
Kerki see Atamyrat
Kérkira see Kérkyra
Kerkrade 87 D6 Limburg, SE Netherlands
Kerkuk see Kirkūk
Kérkyra 104 A4 var. Kérkira, Eng. Corfu. Kérkyra, Iónia Nisiá, Greece, C Mediterranean Sea
Kermadec Islands 152 C4 island group New Zealand, SW Pacific Ocean
Kermadec Trench 143 E4 trench SW Pacific Ocean
Kermān 120 D3 var. Kirman; anc. Carmana. Kermān, C Iran
Kermānshāh 120 C3 var. Qahremānshahr; prev. Bākhtarān. Kermānshāhān, W Iran
Kerrville 49 F4 Texas, SW USA
Kertel see Kärdla
Kerulen 127 E2 Chin. Herlen He, Mong. Herlen Gol. river China/Mongolia
Kerýneia see Girne
Kesennuma 130 D4 Miyagi, Honshū, C Japan
Keszthely 99 C7 Zala, SW Hungary
Ketchikan 36 D4 Revillagigedo Island, Alaska, USA
Kętrzyn 98 D2 Ger. Rastenburg. Warmińsko-Mazurskie, NE Poland
Kettering 89 D6 C England, United Kingdom
Kettering 40 C4 Ohio, N USA
Keupriya see Primorsko
Keurau 85 D5 Länsi-Suomi, C Finland
Keweenaw Peninsula 40 B1 peninsula Michigan, N USA
Key Largo 43 F5 Key Largo, Florida, SE USA
Keystone State see Pennsylvania
Key West 43 E5 Florida Keys, Florida, SE USA
Kezdivásárhely see Târgu Secuiesc
Khabarovsk 115 G4 Khabarovskiy Kray, SE Russian Federation
Khachmas see Xaçmaz
Khadera see Hadera
Khairpur 134 B3 Sind, SE Pakistan
Khalij as Suways 72 B2 var. Suez, Gulf of. gulf NE Egypt
Khalkidhikí see Chalkidikí
Khalkís see Chalkída
Gulf of Khambhat 134 C4 Eng. Gulf of Cambay. gulf W India
Khamis Mushayt 121 B6 var. Hamis Musait. 'Asīr, SW Saudi Arabia
Khānābād 123 E3 Kunduz, NE Afghanistan
Khān al Baghdādī see Al Baghdādī
Khandwa 134 D4 Madhya Pradesh, C India
Khanh Hung see Soc Trăng
Khaniá see Chaniá
Khanka, Lake 129 E2 var. Hsing-K'ai Hu, Lake Hanka, Chin. Xingkai Hu, Rus. Ozero Khanka. lake China/Russian Federation
Khanka, Ozero see Khanka, Lake
Khankendi see Xankändi
Khanthabouli see Savannakhét
Khanty-Mansiysk 114 C3 prev. Ostyako-Voguls'k. Khanty-Mansiyskiy Avtonomnyy Okrug-Yugra, C Russian Federation
Khān Yūnis 119 A7 var. Khān Yūnus. S Gaza Strip
Khān Yūnus see Khān Yūnis
Khanzi see Ghanzi
Kharagpur 135 F4 West Bengal, NE India

Kharbin see Harbin
Kharkiv 109 G2 Rus. Khar'kov. Kharkivs'ka Oblast', NE Ukraine
Khar'kov see Kharkiv
Kharmanli 104 D3 Khaskovo, S Bulgaria
Khartoum 72 B4 var. El Khartūm, Khartum. country capital (Sudan) Khartoum, C Sudan
Khartum see Khartoum
Khasavyurt 111 B8 Respublika Dagestan, SW Russian Federation
Khash, Dasht-e 122 D5 Eng. Khash Desert. desert SW Afghanistan
Khash Desert see Khāsh, Dasht-e
Khashm Al Qirba/Khashm al Qirbah see Khashm el Girba
Khashm el Girba 72 C4 var. Khashim Al Qirba, Khashm al Qirbah. Kassala, E Sudan
Khaskovo 104 D3 Khaskovo, S Bulgaria
Khaybar, Kowtal-e see Khyber Pass
Khaydarkan 123 F2 var. Khaydarken. Batkenskaya Oblast', SW Kyrgyzstan
Khaydarken see Khaydarkan
Khazar, Bahr-e/Khazar, Daryā-ye see Caspian Sea
Khelat see Kālat
Kherson 109 E4 Khersons'ka Oblast', S Ukraine
Kheta 115 E2 river N Russian Federation
Khíos see Chíos
Khirbet el 'Aujā et Tahtā 119 E7 E West Bank Asia
Khiva/Khiwa see Xiva
Khmel'nitskiy see Khmel 'nyts'kyy
Khmel 'nyts'kyy 108 C2 Rus. Khmel'nitskiy; prev. Proskurov. Khmel'nyts'ka Oblast', W Ukraine
Khodasy 107 E6 Rus. Khodosy. Mahilyowskaya Voblasts', E Belarus
Khodoriv 108 C2 Pol. Chodorów, Rus. Khodorov. L'vivs'ka Oblast', W Ukraine
Khodorov see Khodoriv
Khodosy see Khodasy
Khodzhent see Khujand
Khoi see Khvoy
Khojend see Khujand
Khokand see Qo'qon
Kholm see Chełm
Kholon see Holon
Khong Sedone see Muang Khôngxédôn
Khon Kaen 136 D4 var. Muang Khon Kaen. Khon Kaen, E Thailand
Khor 115 G4 Khabarovskiy Kray, SE Russian Federation
Khorat see Nakhon Ratchasima
Khorog see Khorugh
Khorugh 123 F3 Rus. Khorog. S Tajikistan
Khôst 123 F4 prev. Khowst. Khōst, E Afghanistan
Khotan see Hotan
Khouribga 70 B2 C Morocco
Khowst see Khôst
Khoy see Khvoy
Khoyniki 107 D8 Homyel'skaya Voblasts', SE Belarus
Khujand 123 E2 var. Khodzhent, Khojend, Rus. Khudzhand; prev. Leninabad, Taj. Leninobod. N Tajikistan
Khulm 123 E3 var. Kholm, Tashqurghan. Balkh, N Afghanistan
Khulna 135 G4 Khulna, SW Bangladesh
Khums see Al Khums
Khust 108 B3 var. Husht, Cz. Chust, Hung. Huszt. Zakarpats'ka Oblast', W Ukraine
Khvoy 120 C2 var. Khoi, Khoy, Āzarbāyjān-e Bākhtarī, NW Iran
Khyber Pass 134 C1 var. Kowtal-e Khaybar. pass Afghanistan/Pakistan
Kiangmai see Chiang Mai
Kiangsi see Jiangxi
Kiangsu see Jiangsu
Kiáto 105 B6 prev. Kiáton. Pelopónnisos, S Greece
Kiáton see Kiáto
Kiayi see Jiayi
Kibombo 77 D6 Maniema, E Dem. Rep. Congo

Kičevo 101 D6 SW FYR Macedonia
Kidderminster 89 D6 C England, United Kingdom
Kiel 94 B2 Schleswig-Holstein, N Germany
Kielce 98 D4 Rus. Keltsy. Świętokrzyskie, C Poland
Kieler Bucht 94 B2 bay N Germany
Kiev see Kyyiv
Kiev Reservoir 109 E1 Eng. Kiev Reservoir, Rus. Kiyevskoye Vodokhranilishche. reservoir N Ukraine
Kiev Reservoir see Kyyivs'ke Vodoskhovyshche
Kiffa 74 C3 Assaba, S Mauritania
Kigali 73 B6 country capital (Rwanda) C Rwanda
Kigoma 73 B7 Kigoma, W Tanzania
Kihnu 106 C2 var. Kihnu Saar, Ger. Kühnö. island SW Estonia
Kihnu Saar see Kihnu
Kii-suido 131 C7 strait S Japan
Kikinda 100 D3 Ger. Grosskikinda, Hung. Nagykikinda; prev. Velika Kikinda. Vojvodina, N Serbia
Kikládhes see Kykládes
Kikwit 77 C6 Bandundu, W Dem. Rep. Congo
Kilien Mountains see Qilian Shan
Kilimane see Quelimane
Kilimanjaro 69 E5 region E Tanzania
Kilimanjaro 73 C7 var. Uhuru Peak. volcano NE Tanzania
Kilingi-Nõmme 106 D3 Ger. Kurkund. Pärnumaa, SW Estonia
Kilis 116 D4 Kilis, S Turkey
Kiliya 108 D4 Rom. Chilia-Nouă. Odes'ka Oblast', SW Ukraine
Kilkenny 89 B6 Ir. Cill Chainnigh. Kilkenny, S Ireland
Kilkís 104 B3 Kentríki Makedonía, N Greece
Killarney 89 A6 Ir. Cill Airne. Kerry, SW Ireland
Killeen 49 G3 Texas, SW USA
Kilmain see Quelimane
Kilmarnock 88 C4 W Scotland, United Kingdom
Kilwa see Kilwa Kivinje
Kilwa Kivinje 73 C7 var. Kilwa. Lindi, SE Tanzania
Kimberley 78 C4 Northern Cape, C South Africa
Kimberley Plateau 146 C3 plateau Western Australia
Kimch'aek 129 E3 prev. Sŏngjin.
Kími see Kými
Kinabalu, Gunung 138 D3 mountain East Malaysia
Kindersley 37 F5 Saskatchewan, S Canada
Kindia 74 C4 Guinée-Maritime, SW Guinea
Kindley Field 42 A4 air base E Bermuda
Kindu 77 D6 prev. Kindu-Port-Empain. Maniema, C Dem. Rep. Congo
Kindu-Port-Empain see Kindu
Kineshma 111 C5 Ivanovskaya Oblast', W Russian Federation
King Abdullah Economic City 121 A5 Makkah, W Saudi Arabia
King Charles Islands see Kong Karls Land
King Christian IX Land see Kong Christian IX Land
King Frederik VI Coast see Kong Frederik VI Kyst
King Frederik VIII Land see Kong Frederik VIII Land
King Island 149 B8 island Tasmania, SE Australia
Kingissepp see Kuressaare
Kingman 48 A1 Arizona, SW USA
Kingman Reef 145 E2 US territory C Pacific Ocean
Kingsford Smith 148 E2 (Sydney) New South Wales, SE Australia
Kings Lynn 89 E6 var. Bishop's Lynn, Lynn, Lynn Regis. E England, United Kingdom
King Sound 146 B3 sound Western Australia
Kingsport 43 E1 Tennessee, S USA
Kingston 54 B5 country capital (Jamaica) E Jamaica
Kingston 38 D5 Ontario, SE Canada
Kingston 41 F3 New York, NE USA

Kingston upon Hull 89 D5 var. Hull.
  E England, United Kingdom
Kingston upon Thames 89 A8
  SE England, United Kingdom
Kingstown 55 H4 country capital
  (Saint Vincent and the Grenadines)
  Saint Vincent, Saint Vincent and the
  Grenadines
Kingstown see Dún Laoghaire
Kingsville 49 G5 Texas, SW USA
King William Island 37 F3 island
  Nunavut, N Canada
Kinneret, Yam see Tiberias, Lake
Kinrooi 87 D5 Limburg, NE Belgium
Kinshasa 77 B6 prev. Léopoldville.
  country capital (Dem. Rep. Congo)
  Kinshasa, W Dem. Rep. Congo
Kintyre 88 B4 peninsula W Scotland,
  United Kingdom
Kinyeti 73 B5 mountain S South Sudan
Kiparissía see Kyparissía
Kipili 73 B7 Rukwa, W Tanzania
Kipushi 77 D8 Katanga, SE Dem. Rep.
  Congo
Kirdzhali see Kürdzhali
Kirghizia see Kyrgyzstan
Kirghiz Range 123 F2 Rus. Kirgizskiy
  Khrebet; prev. Alexander Range.
  mountain range Kazakhstan/
  Kyrgyzstan
Kirghiz SSR see Kyrgyzstan
Kirghiz Steppe see Saryarka
Kirgizskaya SSR see Kyrgyzstan
Kirgizskiy Khrebet see Kirghiz Range
Kiriath-Arba see Hebron
Kiribati 145 F2 off. Republic of Kiribati.
  country C Pacific Ocean

KIRIBATI
Australasia & Oceania

Official name Republic of Kiribati
Formation 1979 / 1979
Capital Tarawa Atoll
Population 103,248 / 377 people
  per sq mile (145 people per sq km)
Total area 277 sq. miles (717 sq. km)
Languages English*, Kiribati
Religions Roman Catholic 55%,
  Kiribati Protestant Church 36%, Other 9%
Ethnic mix Micronesian 99%, Other 1%
Government Presidential system
Currency Australian dollar = 100 cents
Literacy rate 99%
Calorie consumption 3022 kilocalories

Kiribati, Republic of see Kiribati
Kırıkhan 116 D4 Hatay, S Turkey
Kırıkkale 116 C3 province C Turkey
Kirin see Jilin
Kirinyaga 73 C6 prev. Mount Kenya.
  volcano C Kenya
Kirishi 110 B4 var. Kirisi.
  Leningradskaya Oblast', NW Russian
  Federation
Kirisi see Kirishi
Kiritimati 145 G2 prev. Christmas
  Island. atoll Line Islands, E Kiribati
Kirkenes 84 E2 Fin. Kirkkoniemi.
  Finnmark, N Norway
Kirk-Kilissa see Kırklareli
Kirkkoniemi see Kirkenes
Kirkland Lake 38 D4 Ontario, S Canada
Kırklareli 116 A2 prev. Kirk-Kilissa.
  Kırklareli, NW Turkey
Kirkpatrick, Mount 154 B3 mountain
  Antarctica
Kirksville 45 G4 Missouri, C USA
Kirkūk 120 B3 var. Karkūk, Kerkuk. At
  Ta'mīm, N Iraq
Kirkwall 88 C2 NE Scotland, United
  Kingdom
Kirkwood 45 G4 Missouri, C USA
Kir Moab/Kir of Moab see Al Karak
Kirov 111 C5 prev. Vyatka. Kirovskaya
  Oblast', NW Russian Federation
Kirovabad see Gäncä
Kirovakan see Vanadzor
Kirovo-Chepetsk 111 D5 Kirovskaya
  Oblast', NW Russian Federation
Kirovohrad 109 E3 Rus. Kirovograd;
  prev. Kirovo, Yelizavetgrad,
  Zinov'yevsk. Kirovohrads'ka Oblast',
  C Ukraine
Kirovo/Kirovograd see Kirovohrad
Kirthar Range 134 B3 mountain range
  S Pakistan
Kiruna 84 C3 Lapp. Giron. Norrbotten,
  N Sweden

Kirun/Kirun' see Jilong
Kisalföld see Little Alföld
Kisangani 77 D5 prev. Stanleyville.
  Orientale, NE Dem. Rep. Congo
Kishinev see Chișinău
Kislovodsk 111 B7 Stavropol'skiy Kray,
  SW Russian Federation
Kismaayo 73 D6 var. Chisimayu,
  Kismayu, It. Chisimaio. Jubbada
  Hoose, S Somalia
Kismayu see Kismaayo
Kíssamos 105 C7 prev. Kastélli. Krítí,
  Greece, E Mediterranean Sea
Kissidougou 74 C4 Guinée-Forestière,
  S Guinea
Kissimmee, Lake 43 E4 lake Florida,
  SE USA
Kistna see Krishna
Kisumu 73 C6 prev. Port Florence.
  Nyanza, W Kenya
Kisvárda 99 E6 Ger. Kleinwardein.
  Szabolcs-Szatmár-Bereg, E Hungary
Kita 74 D3 Kayes, W Mali
Kitab see Kitob
Kitakyūshū 131 A7 var. Kitakyûsyû.
  Fukuoka, Kyūshū, SW Japan
Kitakyūsyû see Kitakyūshū
Kitami 130 D2 Hokkaidō, NE Japan
Kitchener 38 C5 Ontario, S Canada
Kíthnos see Kýthnos
Kitimat 36 D4 British Columbia,
  SW Canada
Kitinen 84 D3 river N Finland
Kitob 123 E3 Rus. Kitab. Qashqadaryo
  Viloyati, S Uzbekistan
Kitwe 78 D2 var. Kitwe-Nkana.
  Copperbelt, C Zambia
Kitwe-Nkana see Kitwe
Kitzbüheler Alpen 95 C7 mountain
  range W Austria
Kivalina 36 C2 Alaska, USA
Kivalo 84 D3 ridge C Finland
Kivertsi 108 C1 Pol. Kiwerce, Rus.
  Kivertsy. Volyns'ka Oblast',
  NW Ukraine
Kivertsy see Kivertsi
Kivu, Lac see Kivu, Lake
Kivu, Lake 77 E6 Fr. Lac Kivu. lake
  Rwanda/Dem. Rep. Congo
Kiwerce see Kivertsi
Kiyev see Kyyiv
Kiyevskoye Vodokhranilishche see
  Kyyivs'ke Vodoskhovyshche
Kızıl Irmak 116 C3 river C Turkey
Kizil Kum see Kyzyl Kum
Kizyl-Arvat see Serdar
Kjølen see Kölen
Kladno 99 A5 Středočeský, NW Czech
  Republic
Klagenfurt 95 D7 Slvn. Celovec.
  Kärnten, S Austria
Klaipėda 106 B3 Ger. Memel. Klaipėda,
  NW Lithuania
Klamath Falls 46 B4 Oregon, NW USA
Klamath Mountains 46 A4 mountain
  range California/Oregon, W USA
Klang 138 B3 var. Kelang; prev. Port
  Swettenham. Selangor, Peninsular
  Malaysia
Klarälven 85 B6 river Norway/Sweden
Klatovy 99 A5 Ger. Klattau. Plzeňský
  Kraj, W Czech Republic
Klattau see Klatovy
Klausenburg see Cluj-Napoca
Klazienaveen 86 E2 Drenthe,
  NE Netherlands
Kleines Ungarisches Tiefland see
  Little Alföld
Klein Karas 78 B4 Karas, S Namibia
Kleinwardein see Kisvárda
Kleisoúra 105 A5 Ípeiros, W Greece
Klerksdorp 78 D4 North-West, N South
  Africa
Klimavichy 107 E7 Rus. Klimovichi.
  Mahilyowskaya Voblasts', E Belarus
Klimovichi see Klimavichy
Klintsy 111 A5 Bryanskaya Oblast',
  W Russian Federation
Klisura 104 C2 Plovdiv, C Bulgaria
Ključ 100 B3 Federacija Bosne I
  Hercegovine, NW Bosnia and
  Herzegovina
Klobuck 98 C4 Śląskie, S Poland
Klosters 95 B7 Graubünden,
  SE Switzerland
Kluang see Keluang
Kluczbork 98 C4 Ger. Kreuzburg,
  Kreuzburg in Oberschlesien. Opolskie,
  S Poland

Klyuchevskaya Sopka, Vulkan 115 H3
  volcano E Russian Federation
Knin 100 B4 Šibenik-Knin, S Croatia
Knokke-Heist 87 A5 West-Vlaanderen,
  NW Belgium
Knoxville 42 D1 Tennessee, S USA
Knud Rasmussen Land 82 D1 physical
  region N Greenland
Kobdo see Hovd
Kōbe 131 C6 Hyōgo, Honshū,
  SW Japan
København 85 B7 Eng. Copenhagen;
  anc. Hafnia. country capital (Denmark)
  Sjælland, København, E Denmark
Kobenni 74 D3 Hodh el Gharbi,
  S Mauritania
Koblenz 95 A5 prev. Coblenz, Fr.
  Coblence; anc. Confluentes. Rheinland-
  Pfalz, W Germany
Kobrin see Kobryn
Kobryn 107 A6 Rus. Kobrin. Brestskaya
  Voblasts', SW Belarus
Kobuleti 117 F2 prev. K'obulet'i.
  W Georgia
Koçani 101 E6 NE FYR Macedonia
Kočevje 95 D8 Ger. Gottschee. S Slovenia
Koch Bihār 135 G3 West Bengal,
  NE India
Kochchi see Kochi
Kochi 132 C3 var. Cochin, Kochchi.
  Kerala SW India
Kōchi 131 B7 var. Kôti. Kōchi, Shikoku,
  SW Japan
Kochiu see Gejiu
Kodiak 36 C3 Kodiak Island, Alaska,
  USA
Kodiak Island 36 C3 island Alaska, USA
Koedoes see Kudus
Koeln see Köln
Koepang see Kupang
Ko-er-mu see Golmud
Koetai see Mahakam, Sungai
Koetaradja see Bandaaceh
Kōfu 131 D5 var. Kôhu. Yamanashi,
  Honshū, S Japan
Kogarah 148 E2 New South Wales,
  E Australia
Kogon 122 D2 Rus. Kagan. Buxoro
  Viloyati, C Uzbekistan
Kohalom see Rupea
Kohima 135 H3 state capital Nāgāland,
  E India
Koh I Noh see Büyükağrı Dağı
Kohtla-Järve 106 E2 Ida-Virumaa,
  NE Estonia
Kohu see Kōfu
Kokand see Qo'qon
Kokchetav see Kokshetau
Kokkola 84 D4 Swe. Karleby;
  prev. Swe. Gamlakarleby. Länsi-Suomi,
  W Finland
Koko 75 F4 Kebbi, W Nigeria
Kokomo 40 C4 Indiana, N USA
Koko Nor see Qinghai Hu, China
Koko Nor see Qinghai, China
Kokrines 36 C2 Alaska, USA
Kokshaal-Tau 123 G2 Rus. Khrebet
  Kakshaal-Too. mountain range China/
  Kyrgyzstan
Kokshetau 114 C4 Kaz. Kökshetaü;
  prev. Kokchetav. N Kazakhstan
Koksijde 87 A5 West-Vlaanderen,
  W Belgium
Koksoak 38 D2 river Québec, E Canada
Kokstad 78 D5 KwaZulu/Natal, E South
  Africa
Kola 139 E4 Sulawesi, C Indonesia
Ko'o-la-ma-i see Kunlun Shan
Kola Peninsula 110 C2 Eng. Kola
  Peninsula. peninsula NW Russian
  Federation
Kola Peninsula see Kol'skiy Poluostrov
Kolari 84 D3 Lappi, NW Finland
Kolárovo 99 C6 Ger. Gutta; prev.
  Guta, Hung. Gúta. Nitriansky Kraj,
  SW Slovakia
Kolberg see Kołobrzeg
Kolda 74 B3 S Senegal
Kolding 85 A7 Syddanmark, C Denmark
Kölen 84 C3 Nor. Kjølen. mountain
  range Norway/Sweden
Kolguyev, Ostrov 110 C2 island
  NW Russian Federation
Kolhāpur 132 B1 Mahārāshtra, SW India
Kolhumadulu 132 A5 var. Thaa Atoll.
  atoll S Maldives

Kolín 99 B5 Ger. Kolin. Střední Če
  C Czech Republic
Kolka 106 C2 NW Latvia
Kolkasrags 106 C2 prev. Eng. Cape
  Domesnes. headland NW Latvia
Kolkata 135 G4 prev. Calcutta. Wes
  Bengal, N India
Kollam 132 C3 var. Quilon. Kerala
  SW India
Kolmar see Colmar
Köln 94 A4 var. Koeln, Eng./Fr. Co
  prev. Cöln; anc. Colonia Agrippin
  Oppidum Ubiorum. Nordrhein-
  Westfalen, W Germany
Koło 98 C3 Wielkopolskie, C Poland
Kołobrzeg 98 B2 Ger. Kolberg.
  Zachodnio-pomorskie, NW Polar
Kolokani 74 D3 Koulikoro, W Mal
Kolomea see Kolomyya
Kolomna 111 B5 Moskovskaya Obl
  W Russian Federation
Kolomyya 108 C3 Ger. Kolomea. **
  Frankivs'ka Oblast', W Ukraine
Kolosjoki see Nikel'
Kolozsvár see Cluj-Napoca
Kolpa 100 A2 Ger. Kulpa, SCr. Kup
  river Croatia/Slovenia
Kolpino 110 B4 Leningradskaya Ob
  NW Russian Federation
Kólpos Mórfu see Güzelyurt Körfez
Kolwezi 77 D7 Katanga, S Dem. Re
  Congo
Kolyma 115 G2 river NE Russian
  Federation
Komatsu 131 C5 var. Komatu. Ish
  Honshū, SW Japan
Komatu see Komatsu
Kommunizm, Qullai see Ismoili So
  Qullai
Komoé 75 E4 var. Komoé Fleuve. r
  E Côte d'Ivoire
Komoé Fleuve see Komoé
Komotau see Chomutov
Komotiní 104 D3 var. Gümülçina, '
  Gümülcine. Anatolikí Makedonía
  Thráki, NE Greece
Kompong see Kâmpóng Chhnāng
Kompong Cham see Kâmpóng Cha
Kompong Som see Sihanoukville
Kompong Speu see Kâmpóng Spoe
Komrat see Comrat
Komsomolets, Ostrov 115 E1 islar
  Severnaya Zemlya, N Russian
  Federation
Kondolovo 104 E3 Burgas, E Bulga
Kondopoga 110 B3 Respublika Kar
  NW Russian Federation
Kondoz see Kunduz
Köneürgenç 122 C2 var. Köneürge
  Rus. Këneurgench; prev. Kunya-
  Urgench. Daşoguz Welayaty,
  N Turkmenistan
Kong Christian IX Land 82 D4 Eng
  King Christian IX Land. physical r
  SE Greenland
Kong Frederik IX Land 82 C3 phys
  SW Greenland
Kong Frederik VIII Land 83 E2 Eng.
  King Frederik VIII Land. physical
  region NE Greenland
Kong Frederik VI Kyst 82 C4 Eng.
  Frederik VI Coast. physical region
  SE Greenland
Kongola 78 C3 Caprivi, NE Namibia
Kongor 73 B5 Jonglei, E South Suda
Kong Oscar Fjord 83 E3 fjord
  E Greenland
Kongoli 76 D6 Katanga, S Dem. Rep
  Congo
Kong, Tônlé 137 E5 var. Xê Kong.
  Cambodia/Laos
Kongsberg 85 B6 Buskerud, S Norw
Kongsvinger 85 B6 Hedmark, S Nor
Königgrätz see Hradec Králové
Königshütte see Chorzów
Koninkrijk der Nederlanden see
  Netherlands
Konispol 101 C7 var. Konispoli. Vlë
  S Albania
Konispoli see Konispol
Kónitsa 104 A4 Ípeiros, W Greece

itz *see* Chojnice
jic *100 C4* Federacija Bosne I
  ercegovine, S Bosnia and
  erzegovina
osha *110 C4* Arkhangel'skaya
  blast', NW Russian Federation
otop *109 F1* Sums'ka Oblast',
  E Ukraine
stantinovka *see* Kostyantynivka
stanz *95 B7 var.* Constanz, *Eng.*
  onstance, *hist.* Kostnitz; *anc.*
  onstantia. Baden-Württemberg,
  Germany
stanza *see* Constanța
ya *116 C4 var.* Konieh, *prev.* Konia;
  nc. Iconium. Konya, C Turkey
aonik *101 D5 mountain range*
  Serbia
ar *see* Koper
er *95 D8 It.* Capodistria; *prev.* Kopar.
  W Slovenia
etdag Gershi *122 C3 mountain*
  ange Iran/Turkmenistan
etdag Gershi/Kopetdag, Khrebet
  ee Koppeh Dāgh
peh Dagh *120 D2 Rus.* Khrebet
  opetdag, *Turkm.* Köpetdag Gershi.
ountain range Iran/Turkmenistan
reinitz *see* Koprivnica
rivnica *100 B2 Ger.* Kopreinitz,
  ung. Kapronca. Koprivnica-
  riževci, N Croatia
rülü *see* Veles
tsevichi *see* Kaptsevichy
oyl' *see* Kapyl'
at *see* Nakhon Ratchasima
at Plateau *136 D4 plateau*
  Thailand
ba *135 E4* Chhattisgarh, C India
rça *see* Korçë
rçë *101 D6 var.* Korça, *Gk.* Korytsa,
  t. Corriza; *prev.* Koritsa. Korçë,
  E Albania
rcula *100 B4 It.* Curzola; *anc.* Corcyra
  Nigra. *island* S Croatia
rea Bay *127 G3 bay* China/North
  Korea
rea, Democratic People's Republic
  of *see* North Korea
rea, Republic of *see* South Korea
rea Strait *131 A7 Jap.* Chōsen-kaikyō,
  Kor. Taehan-haehyŏp. *channel* Japan/
  South Korea
rhogo *74 D4* N Côte d'Ivoire
rinthos *105 B6 anc.* Corinthus *Eng.*
  Corinth. Pelopónnisos, S Greece
ritsa *see* Korçë
riyama *131 D5* Fukushima, Honshū,
  C Japan
rla *126 C3 Chin.* K'u-erh-lo. Xinjiang
  Uygur Zizhiqu, NW China
rmend *99 B7* Vas, W Hungary
róni *105 B6* Pelopónnisos,
  S Greece
ror *144 A2* (Palau) Oreor, N Palau
rös *see* Križevci
rosten' *108 D1* Zhytomyrs'ka Oblast',
  NW Ukraine
ro Toro *76 C2* Borkou-Ennedi-
  Tibesti, N Chad
rsovka *see* Kārsava
rtrijk *87 A6 Fr.* Courtrai. West-
  Vlaanderen, W Belgium
ryak Range *115 H2 var.* Koryakskiy
  Khrebet, *Eng.* Koryak Range. *mountain*
  ange NE Russian Federation
ryak Range *see* Koryakskoy
  e Nagor'ye
ryakskiy Khrebet *see* Koryakskoye
ryazhma *110 C4* Arkhangel'skaya
  Oblast', NW Russian Federation
rytsa *see* Korçë
os *105 E6* Kos, Dodekánisa, Greece,
  Aegean Sea
s *105 E6 It.* Coo; *anc.* Cos. *island*
  Dodekánisa, Greece, Aegean Sea
-saki *131 A7 headland* Nagasaki,
  Tsushima, SW Japan
ścian *98 B4 Ger.* Kosten.
  Wielkopolskie, C Poland
ścierzyna *98 C2* Pomorskie,
  NW Poland
sciusko, Mount *see* Kosciuszko,
  Mount
sciuszko, Mount *149 C7 prev.* Mount
  Kosciusko. *mountain* New South
  Wales, SE Australia
o-shih *see* Kashi

Koshikijima-retto *131 A8 var.*
  Kosikizima Rettō. *island group*
  SW Japan
Kōshū *see* Gwangju
Košice *99 D6 Ger.* Kaschau, *Hung.* Kassa.
  Košický Kraj, E Slovakia
Kosikizima Rettō *see*
  Koshikijima-rettō
Köslin *see* Koszalin
Koson *123 E3 Rus.* Kasan. Qashqadaryo
  Viloyati, S Uzbekistan
Kosovo *101 D5 prev.* Autonomous
  Province of Kosovo and Metohija.
  country SE Europe

┌───┐
KOSOVO (not fully recognized)
  Southeast Europe

Official name Republic of Kosovo
Formation 2008 / 2008
Capital Prishtinë
Population 1.8 million / 427 people
  per sq mile (165 people per sq km)
Total area 4212 sq miles (10,908 sq km)
Languages Albanian*, Serbian*, Bosniak,
  Gorani, Roma, Turkish
Religions Muslim 92%, Roman
  Catholic 4%, Orthodox Christian 4%
Ethnic mix Albanian 92%, Serb 4%,
  Bosniak and Gorani 2%, Turkish 1%,
  Roma 1%
Government Parliamentary system
Currency Euro = 100 cents
Literacy rate 92%
Calorie consumption Not available
└───┘

Kosovo and Metohija, Autonomous
  Province of *see* Kosovo
Kosovo Polje *see* Fushë Kosovë
Kosovska Mitrovica *see* Mitrovicë
Kosrae *142 C2 prev.* Kusaie. *island*
  Caroline Islands, E Micronesia
Kossou, Lac de *74 D5 lake*
  C Côte d'Ivoire
Kostanay *114 C4 var.* Kustanay, *Kaz.*
  Qostanay. Kostanay, N Kazakhstan
Kosten *see* Kościan
Kostenets *104 C2 prev.* Georgi Dimitrov.
  Sofiya, W Bulgaria
Kostnitz *see* Konstanz
Kostroma *110 B4* Kostromskaya Oblast',
  NW Russian Federation
Kostyantynivka *109 G3 Rus.*
  Konstantinovka. Donets'ka Oblast',
  SE Ukraine
Kostyukovichi *see* Kastsyukovichy
Kostyukova *see* Kastsyukowka
Koszalin *98 B2 Ger.* Köslin.
  Zachodnio-pomorskie, NW Poland
Kota *134 D3 prev.* Kotah. Rājasthān,
  N India
Kota Baharu *see* Kota Bharu
Kota Bahru *see* Kota Bharu
Kotabaru *see* Jayapura
Kota Bharu *138 B3 var.* Kota Baharu,
  Kota Bahru. Kelantan, Peninsular
  Malaysia
Kotaboemi *see* Kotabumi
Kotabumi *138 B4 prev.* Kotaboemi.
  Sumatera, W Indonesia
Kotah *see* Kota
Kota Kinabalu *138 D3 prev.* Jesselton.
  Sabah, East Malaysia
Kotel'nyy, Ostrov *115 E2 island*
  Novosibirskiye Ostrova, N Russian
  Federation
Kotka *85 E5* Etelä-Suomi, S Finland
Kotlas *110 C4 prev.* Kotelas.
  NW Russian Federation
Kotonu *see* Cotonou
Kotor *101 C5 It.* Cattaro.
  SW Montenegro
Kotovs'k *108 D3 Rus.* Kotovsk. Odes'ka
  Oblast', SW Ukraine
Kotovsk *see* Hîncești
Kottbus *see* Cottbus
Kotto *76 D4 river* Central African
  Republic/Dem. Rep. Congo
Kotuy *115 E2 river* N Russian Federation
Koudougou *75 E4* C Burkina Faso
Koulamoutou *77 B6* Ogooué-Lolo,
  C Gabon
Koulikoro *74 D3* Koulikoro, SW Mali
Koumra *76 C4* Moyen-Chari, S Chad
Kourou *59 H3* N French Guiana
Kousséri *76 B3 prev.* Fort-Foureau.
  Extrême-Nord, NE Cameroon
Koutiala *74 D4* Sikasso, S Mali

Kouvola *85 E5* Etelä-Suomi, S Finland
Kovel' *108 C1 Pol.* Kowel. Volyns'ka
  Oblast', NW Ukraine
Kovno *see* Kaunas
Koweit *see* Kuwait
Kowel *see* Kovel'
Kowloon *128 A2* Hong Kong, S China
Kowno *see* Kaunas
Kozáni *104 B4* Dytikí Makedonía,
  N Greece
Kozara *100 B3 mountain range*
  NW Bosnia and Herzegovina
Kozarska Dubica *see* Bosanska Dubica
Kozhikode *132 C2 var.* Calicut. Kerala,
  SW India
Kozu-shima *131 D6 island* E Japan
Kozyatyn *108 D2 Rus.* Kazatin.
  Vinnyts'ka Oblast', C Ukraine
Kpalimé *75 E5 var.* Palimé. SW Togo
Krâchéh *137 D6 prev.* Kratie. Krâchéh,
  E Cambodia
Kragujevac *100 D4* Serbia, C Serbia
Krainburg *see* Kranj
Kra, Isthmus of *137 B6 isthmus*
  Malaysia/Thailand
Krakau *see* Kraków
Kraków *99 D5 Eng.* Cracow, *Ger.*
  Krakau; *anc.* Cracovia. Małopolskie,
  S Poland
Kralendijk *55 F5 dependent territory*
  capital (Bonaire) Lesser Antilles,
  S Caribbean Sea
Kraljevo *100 D4 prev.* Rankovićevo.
  Serbia, C Serbia
Kramators'k *109 G3 Rus.* Kramatorsk.
  Donets'ka Oblast', SE Ukraine
Kramatorsk *see* Kramators'k
Kramfors *85 C5* Västernorrland,
  C Sweden
Kranéa *see* Kraniá
Kraniá *104 B4 var.* Kranéa. Dytikí
  Makedonía, N Greece
Kranj *95 D7 Ger.* Krainburg. N Slovenia
Kranz *see* Zelenogradsk
Kráslava *106 D4 SE* Latvia
Krasnaye *107 C5 Rus.* Krasnoye.
  Minskaya Voblasts', C Belarus
Krasnoarmeysk *111 C6* Saratovskaya
  Oblast', W Russian Federation
Krasnodar *111 A7 prev.* Ekaterinodar,
  Yekaterinodar. Krasnodarskiy Kray,
  SW Russian Federation
Krasnodon *109 H3* Luhans'ka Oblast',
  E Ukraine
Krasnogor *see* Kallaste
Krasnogvardeyskoye *see*
  Krasnohvardiys'ke
Krasnohvardiys'ke *109 F4 Rus.*
  Krasnogvardeyskoye. Avtonomna
  Respublika Krym, S Ukraine
Krasnokamensk *115 F4* Zabaykal'skiy
  Kray, S Russian Federation
Krasnokamsk *111 D5* Permskiy Kray,
  W Russian Federation
Krasnoperekops'k *109 F4 Rus.*
  Krasnoperekopsk. Avtonomna
  Respublika Krym, S Ukraine
Krasnoperekopsk *see*
  Krasnoperekops'k
Krasnostav *see* Krasnystaw
Krasnovodsk *see* Türkmenbaşy
Krasnovodskiy Zaliv *see* Türkmenbaşy
  Aýlagy
Krasnowodsk Aylagy *see* Türkmenbaşy
  Aýlagy
Krasnoyarsk *114 D4* Krasnoyarskiy
  Kray, S Russian Federation
Krasnoye *see* Krasnaye
Krasnystaw *98 E4 Rus.* Krasnostav.
  Lubelskie, SE Poland
Krasnyy Kut *111 C6* Saratovskaya
  Oblast', W Russian Federation
Krasnyy Luch *109 H3 prev.*
  Krindachevka. Luhans'ka Oblast',
  E Ukraine
Kratie *see* Krâchéh
Krâvanh, Chuŏr Phnum *137 C6 Eng.*
  Cardamom Mountains, *Fr.* Chaîne
  des Cardamomes. *mountain range*
  W Cambodia
Krefeld *94 A4* Nordrhein-Westfalen,
  W Germany
Kreisstadt *see* Krosno Odrzańskie
Kremenchug *see* Kremenchuk
Kremenchugskoye Vodokhranilishche/
  Kremenchuts'ke Reservoir *see*
  Kremenchuts'ke Vodoskhovyshche

Kremenchuk *109 F2 Rus.* Kremenchug.
  Poltavs'ka Oblast', NE Ukraine
Kremenchuk Reservoir *109 F2*
  Eng. Kremenchuk Reservoir, *Rus.*
  Kremenchugskoye Vodokhranilishche.
  reservoir C Ukraine
Kremenets' *108 C2 Pol.* Krzemieniec,
  Rus. Kremenets. Ternopil's'ka Oblast',
  W Ukraine
Kremennaya *see* Kreminna
Kreminna *109 G2 Rus.* Kremennaya.
  Luhans'ka Oblast', E Ukraine
Kresena *see* Kresna
Kresna *104 C3 var.* Kresena.
  Blagoevgrad, SW Bulgaria
Kretikon Delagos *see* Kritikó Pélagos
Kretinga *106 B3 Ger.* Krottingen.
  Klaipėda, NW Lithuania
Kreutz *see* Cristuru Secuiesc
Kreuz *see* Križevci, Croatia
Kreuz *see* Risti, Estonia
Kreuzburg/Kreuzburg in
  Oberschlesien *see* Kluczbork
Krichëv *see* Krychaw
Krievija *see* Russian Federation
Krindachevka *see* Krasnyy Luch
Krishna *132 C1 prev.* Kistna. *river*
  C India
Krishnagiri *132 C2* Tamil Nādu,
  SE India
Kristiania *see* Oslo
Kristiansand *85 A6 var.* Christiansand.
  Vest-Agder, S Norway
Kristianstad *85 B7* Skåne, S Sweden
Kristiansund *84 A4 var.* Christiansund.
  Møre og Romsdal, S Norway
Krivoy Rog *see* Kryvyy Rih
Križevci *100 B2 Ger.* Kreuz, *Hung.*
  Körös. Varaždin, NE Croatia
Krk *100 A3 It.* Veglia; *anc.* Curieta.
  island NW Croatia
Krolevets *109 F1 Rus.* Krolevets.
  Sums'ka Oblast', NE Ukraine
Krolevets *see* Krolevets'
Królewska Huta *see* Chorzów
Kronach *95 C5* Bayern, E Germany
Kronstadt *see* Brașov
Kroonstad *78 D4* Free State, C South
  Africa
Kropotkin *111 A7* Krasnodarskiy Kray,
  SW Russian Federation
Krosno *99 D5 Ger.* Krossen.
  Podkarpackie, SE Poland
Krosno Odrzańskie *98 B3*
  Lubuskie, W Poland
Krossen *see* Krosno
Krottingen *see* Kretinga
Krško *95 E8 Ger.* Gurkfeld;
  prev. Videm-Krško. E Slovenia
Krugloye *see* Kruhlaye
Kruhlaye *107 D6 Rus.* Krugloye.
  Mahilyowskaya Voblasts', E Belarus
Krujë *101 C6 var.* Kruja, It. Croia.
  Durrës, C Albania
Krujë *see* Krujë
Krung Thep, Ao *137 C5 var.* Bight of
  Bangkok. *bay* S Thailand
Krung Thep Mahanakhon *see* Ao
  Krung Thep
Krupki *107 D6* Minskaya Voblasts',
  C Belarus
Krušné Hory *see* Erzgebirge
Kruševac *100 D4* Serbia, C Serbia
Krychaw *107 E7 Rus.* Krichëv.
  Mahilyowskaya Voblasts', E Belarus
Krym's'kyy Pivostriv *109 F5 Eng.*
  Crimea. *peninsula* S Ukraine
Kryvyy Rih *109 F3 Rus.* Krivoy Rog.
  Dnipropetrovs'ka Oblast', SE Ukraine
Krzemieniec *see* Kremenets'
Ksar al Kabir *see* Ksar-el-Kebir
Ksar al Soule *see* Er-Rachidia
Ksar-el-Kebir *70 C2 var.* Alcázar,
  Al-Kasr al-Kabir, Ksar-el-Kébir, *Ar.*
  Al-Kasr al-Kabir, *Sp.* Alcazarquivir.
  NW Morocco
Ksar-el-Kébir *see* Ksar-el-Kebir
Kuala Dungun *see* Dungun
Kuala Lumpur *138 B3 country capital*
  (Malaysia) Kuala Lumpur, Peninsular
  Malaysia

Kuala Terengganu 138 B3 *var.* Kuala Trengganu. Terengganu, Peninsular Malaysia
Kualatungkal 138 B4 Sumatera, W Indonesia
Kuang-chou *see* Guangzhou
Kuang-hsi *see* Guangxi Zhuangzu Zizhiqu
Kuang-tung *see* Guangdong
Kuang-yuan *see* Guangyuan
Kuantan 138 B3 Pahang, Peninsular Malaysia
Kuba *see* Quba
Kuban' 109 G5 *var.* Hypanis. *river* SW Russian Federation
Kubango *see* Cubango/Okavango
Kuching 138 C3 *prev.* Sarawak. Sarawak, East Malaysia
Küchnay Darwäshän 122 D5 *prev.* Küchnay Darweyshän. Helmand, S Afghanistan
Küchnay Darweyshän *see* Küchnay Darwäshän
Kuçova *see* Kuçovë
Kuçovë 101 C6 *var.* Kuçova; *prev.* Qyteti Stalin. Berat, C Albania
Kudara *see* Ghüdara
Kudus 138 C5 *prev.* Koedoes. Jawa, C Indonesia
Kuei-lin *see* Guilin
Kuei-Yang/Kuei-yang *see* Guiyang
K'u-erh-lo *see* Korla
Kueyang *see* Guiyang
Kugaaruk 37 G3 *prev.* Pelly Bay. Nunavut, N Canada
Kugluktuk 53 E3 *var.* Qurlurtuuq; *prev.* Coppermine. Nunavut, NW Canada
Kuhmo 84 E4 Oulu, E Finland
Kuhnau *see* Konin
Kühnö *see* Kihnu
Kuibyshev *see* Kuybyshevskoye Vodokhranilishche
Kuito 78 B2 Port. Silva Porto. Bié, C Angola
Kuji 130 D3 *var.* Kuzi. Iwate, Honshü, C Japan
Kukës 101 D5 *var.* Kukësi. Kukës, NE Albania
Kukësi *see* Kukës
Kukong *see* Shaoguan
Kukukhoto *see* Hohhot
Kula Kangri 135 G3 *var.* Kulhakangri. *mountain* Bhutan/China
Kuldiga 106 B3 Ger. Goldingen. W Latvia
Kuldja *see* Yining
Kulhakangri *see* Kula Kangri
Kullorsuaq 82 D2 *var.* Kuvdlorssuak. Kitaa, C Greenland
Kulm *see* Chełmno
Kulmsee *see* Chełmża
Külob 123 F3 Rus. Kulyab. SW Tajikistan
Kulpa *see* Kolpa
Kulu 116 C3 Konya, W Turkey
Kulunda Steppe *see* Ravnina Kulyndy
Kulundinskaya Ravnina *see* Ravnina Kulyndy
Kulyab *see* Külob
Kum *see* Qom
Kuma 111 B7 *river* SW Russian Federation
Kumamoto 131 A7 Kumamoto, Kyüshü, SW Japan
Kumanova *see* Kumanovo
Kumanovo 101 E5 Turk. Kumanova. N Macedonia
Kumasi 75 E5 *prev.* Coomassie. C Ghana
Kumayri *see* Gyumri
Kumba 77 A5 Sud-Ouest, W Cameroon
Kumertau 111 D6 Respublika Bashkortostan, W Russian Federation
Kumillä *see* Comilla
Kumo 75 G4 Gombe, E Nigeria
Kumon Range 136 B2 *mountain range* N Myanmar (Burma)
Kumul *see* Hami
Kunashiri *see* Kunashir, Ostrov
Kunashir, Ostrov 130 E1 *var.* Kunashiri. *island* Kuril'skiye Ostrova, SE Russian Federation
Kunda 106 E2 Lääne-Virumaa, NE Estonia
Kunduz 123 E3 *prev.* Kondoz. *province* NE Afghanistan
Kunene 69 C6 *var.* Kunene. *river* Angola/Namibia
Kunene *see* Cunene
Kungsbacka 85 B7 Halland, S Sweden

Kungur 111 D5 Permskiy Kray, NW Russian Federation
Kunlun Mountains *see* Kunlun Shan
Kunlun Shan 126 B4 Eng. Kunlun Mountains. *mountain range* NW China
Kunming 128 B6 *var.* K'un-ming; *prev.* Yunnan. *province capital* Yunnan, SW China
K'un-ming *see* Kunming
Kununurra 146 D3 Western Australia
Kunya-Urgench *see* Köneürgenç
Kuopio 85 E5 Itä-Suomi, C Finland
Kupa *see* Kolpa
Kupang 139 E5 *prev.* Koepang. Timor, C Indonesia
Kup"yans'k 109 G2 Rus. Kupyansk. Kharkivs'ka Oblast', E Ukraine
Kupyansk *see* Kup"yans'k
Kür *see* Kura
Kura 117 H3 *az.* Kür, Geor. Mtkvari, Turk. Kura Nehri. *river* SW Asia
Kura Nehri *see* Kura
Kurashiki 131 B6 *var.* Kurasiki. Okayama, Honshü, SW Japan
Kurasiki *see* Kurashiki
Kurdistan 117 F4 *cultural region* SW Asia
Kürdzhali 104 D3 *var.* Kirdzhali. Kürdzhali, S Bulgaria
Kure 131 B7 Hiroshima, Honshü, SW Japan
Küre Dağları 116 C2 *mountain range* N Turkey
Kuressaare 106 C2 Ger. Arensburg; *prev.* Kingissepp. Saaremaa, W Estonia
Kureyka 112 D2 *river* N Russian Federation
Kurgan-Tyube *see* Qürghonteppa
Kuria Muria Islands *see* Ḩalāniyāt, Juzur al
Kuril'skiye Ostrova 115 H4 Eng. Kuril Islands. *island group* SE Russian Federation
Kuril Islands *see* Kuril'skiye Ostrova
Kuril-Kamchatka Depression *see* Kuril Trench
Kuril Trench 113 F3 *var.* Kuril-Kamchatka Depression. *trench* NW Pacific Ocean
Kuril'sk 130 E1 Jap. Shana. Kuril'skiye Ostrova, Sakhalinskaya Oblats', SE Russian Federation
Ku-ring-gai 148 E1 New South Wales, E Australia
Kurisches Haff *see* Courland Lagoon
Kurkund *see* Kilingi-Nõmme
Kurnool 132 C1 *var.* Karnul. Andhra Pradesh, S India
Kursk 111 A6 Kurskaya Oblast', W Russian Federation
Kurskiy Zaliv *see* Courland Lagoon
Kuršumlija 101 D5 Serbia, S Serbia
Kurtbunar *see* Tervel
Kurtitsch/Kürtös *see* Curtici
Kuruktag 126 C3 *mountain range* NW China
Kurume 131 A7 Fukuoka, Kyüshü, SW Japan
Kurupukari 59 F3 C Guyana
Kusaie *see* Kosrae
Kushiro 130 D2 *var.* Kusiro. Hokkaidö, NE Japan
Kushka *see* Serhetabat
Kusiro *see* Kushiro
Kuskokwim Mountains 36 C3 *mountain range* Alaska, USA
Kustanay *see* Kostanay
Kustanay *see* Kostanay
Küstence/Küstendje *see* Constanţa
Kütahya 116 B3 *prev.* Kutaia. Kütahya, W Turkey
Kutai *see* Mahakam, Sungai
Kutaisi 117 F2 *prev.* var.* Kutaïsi 95 F2 W Georgia W Georgia
K'ut'aisi *see* Kutaisi
Kūt al 'Amārah *see* Al Küt
Kut al Imara *see* Al Küt
Kutaradja/Kutaraja *see* Banda Aceh
Kutch, Gulf of *see* Kachchh, Gulf of
Kutch, Rann of *see* Kachchh, Rann of
Kutina 100 B3 Sisak-Moslavina, NE Croatia
Kutno 98 C3 Łódzkie, C Poland
Kuujjuaq 39 E2 *prev.* Fort-Chimo. Québec, E Canada
Kuusamo 84 E3 Oulu, E Finland
Kuvdlorssuak *see* Kullorsuaq

Kuwait 120 C4 *off.* State of Kuwait, *var.* Dawlat al Kuwait, Koweit, Kuwete. *country* SW Asia

Kuwait *see* Al Kuwayt
Kuwait City *see* Al Kuwayt
Kuwait, Dawlat al *see* Kuwait
Kuwait, State of *see* Kuwait
Kuwajleen *see* Kwajalein Atoll
Kuwayt 120 C3 Maysän, E Iraq
Kuweit *see* Kuwait
Kuybyshev *see* Samara
Kuybyshev Reservoir 111 C5 *var.* Kuibyshev, Eng. Kuybyshev Reservoir. *reservoir* W Russian Federation
Kuybyshev Reservoir *see* Kuybyshevskoye Vodokhranilishche
Kuytun 126 B2 Xinjiang Uygur Zizhiqu, NW China
Kuzi *see* Kuji
Kuznetsk 111 B6 Penzenskaya Oblast', W Russian Federation
Kuźnica 98 E2 Białystok, NE Poland Europe
Kvaløya 84 C1 island N Norway
Kvarnbergsvattnet 84 B4 *var.* Frostviken. Jämt N Sweden
Kvarner 100 A3 *var.* Carnaro, It. Quarnero. *gulf* W Croatia
Kvitøya 83 G1 island NE Svalbard
Kwajalein Atoll 144 C1 *var.* Kwajaleen. *atoll* Ralik Chain, C Marshall Islands
Kwando *see* Cuando
Kwangchu *see* Guangzhou
Kwangchu *see* Gwangju
Kwangju *see* Gwangju
Kwango 77 C7 Port. Cuango. *river* Angola/Dem. Rep. Congo
Kwangsi/Kwangsi Chuang Autonomous Region *see* Guangxi Zhuangzu Zizhiqu
Kwangtung *see* Guangdong
Kwangyuan *see* Guangyuan
Kwanza *see* Cuanza
Kweichu *see* Guiyang
Kweilin *see* Guilin
Kweisui *see* Hohhot
Kweiyang *see* Guiyang
Kwekwe 78 D3 *prev.* Que Que. Midlands, C Zimbabwe
Kwesui *see* Hohhot
Kwidzyń 98 C2 Ger. Marienwerder. Pomorskie, N Poland
Kwigillingok 36 C3 Alaska, USA
Kwilu 77 C6 *river* W Dem. Rep. Congo
Kwito *see* Cuito
Kyabé 76 C4 Moyen-Chari, S Chad
Kyaikkami 137 B5 *prev.* Amherst. Mon State, S Myanmar (Burma)
Kyaiklat 136 B4 Ayeyarwady, SW Myanmar (Burma)
Kyaikto 136 B4 Mon State, S Myanmar (Burma)
Kyakhta 115 E5 Respublika Buryatiya, S Russian Federation
Kyaukpyu 136 A3 Rakhine State, W Myanmar (Burma)
Kyaukse 136 B3 Mandalay, C Myanmar (Burma)
Kyjov 99 C5 Ger. Gaya. Jihomoravský Kraj, SE Czech Republic
Kými 105 C5 *prev.* Kími. Évvoia, C Greece
Kými *see* Kými
Kyöngsöng *see* Seoul
Kyöto 131 C6 Kyöto, Honshü, SW Japan
Kyparissía 105 B6 *var.* Kiparissía. Pelopónnisos, S Greece
Kypros *see* Cyprus
Kyrá Panagía 105 C5 island Vóreies Sporádes, Aegean Sea

Kyrenia *see* Girne
Kyrgyz Republic *see* Kyrgyzstan
Kyrgyzstan 123 F2 *off.* Kyrgyz Republic *var.* Kirghizia; *prev.* Kirgizskaya SSR, Kirghiz SSR, Republic of Kyrgyzstan *country* C Asia

Kyrgyzstan, Republic of *see* Kyrgyzstan
Kythíra 105 C7 *var.* Kíthira, It. Cerigo, Lat. Cythera. island S Greece
Kýthnos 105 C6 Knýthnos, Kykládes, Greece, Aegean Sea
Kythnos 105 C6 *var.* Kíthnos, Therm It. Termia; anc. Cythnos. island Kykládes, Greece, Aegean Sea
Kythréa *see* Değirmenlik
Kyushu 131 B7 *var.* Kyûsyû. island SW Japan
Kyushu-Palau Ridge *see* Kyushu-Palau Ridge
Kyustendil 104 B2 anc. Pautalia. Kyustendil, W Bulgaria
Kyûsyû *see* Kyushu
Kyusyu-Palau Ridge *see* Kyushu-Palau Ridge
Kyiv 109 E2 Eng. Kiev, Rus. Kiyev. *country capital* (Ukraine) Kyyiv'ka Oblast', N Ukraine
Kyzyl 114 D4 Respublika Tyva, C Russian Federation
Kyzyl Kum 122 D2 *var.* Kizil Kum, Qizil Qum, Uzb. Qizilqum. *desert* Kazakhstan/Uzbekistan
Kyzylorda 114 B5 *var.* Kzyl-Orda, Qizil Orda, Qyzylorda; *prev.* Kzylorda, Perovsk. Kyzylorda, S Kazakhstan
Kyzylrabot *see* Qizilrabot
Kyzyl-Suu 123 G2 *prev.* Pokrovka. Issyk-Kul'skaya Oblast', NE Kyrgyzstan
Kzyl-Orda *see* Kyzylorda
Kzylorda *see* Kyzylorda

L

Laaland *see* Lolland
Laarne 87 B5 Oost-Vlaanderen, NW Belgium
La Asunción 59 E1 Nueva Esparta, NE Venezuela
Laatokka *see* Ladozhskoye, Ozero
Laâyoune 70 B3 *var.* Aaiún. *country capital* (Western Sahara) NW Western Sahara
Labe *see* Elbe
Laborca *see* Laborec
Laborec 99 E5 Hung. Laborca. *river* E Slovakia
Labrador 39 F2 *cultural region* Newfoundland and Labrador, SW Canada
Labrador Basin 34 E3 *var.* Labrador Basin. *undersea basin* NW Atlantic Ocean
Labrador Sea 82 A4 sea NW Atlantic Ocean
Labrador Sea Basin *see* Labrador Basin
Labudalin *see* Ergun
Labutta 137 A5 Ayeyarwady, SW Myanmar (Burma)
Laç 101 C6 *var.* Laci. Lezhë, C Albania
La Calera 64 B4 Valparaíso, C Chile
La Carolina 92 D4 Andalucía, S Spain

dive Islands *132 A3 Eng.* Laccadive
nds. *island group* India, N Indian
an
dive Islands/Laccadive Minicoy
Amindivi Islands, the *see*
shadweep
ba *52 D2* Atlántida, N Honduras
nás *104 B3* Kentrikí Makedonía,
reece
aux-de-Fonds *95 A7* Neuchâtel,
Switzerland
an River *149 C6 river* New South
les, SE Australia
e Laç
tat *91 D6 anc.* Citharista. Bouches-
Rhône, SE France
riga *see* Lagos
ncepción *53 E5 var.* Concepción.
iriquí, W Panama
ncepción *58 C1* Zulia,
V Venezuela
ndamine *91 C8* W Monaco
ia *41 G2* New Hampshire, NE USA
osse *40 A2* Wisconsin, N USA
uz *52 D4* Guanacaste, NW Costa
a
Ladoga *110 B3 Eng.* Lake Ladoga,
. Laatokka. *lake* NW Russian
eration
ga, Lake *see* Ladozhskoye, Ozero
smith *40 B2* Wisconsin, N USA
44 B3 Morobe, W Papua New
inea
peranza *52 C2* Intibucá,
Honduras
ette *40 C4* Indiana, N USA
ette *42 B3* Louisiana, S USA
54 A2 Pinar del Río, W Cuba
75 G4 Nassarawa, C Nigeria
che *90 B4* Sarthe, NW France
, Lac de *76 B4 lake* N Cameroon
ouat *70 D2* N Algeria
75 F5 Lagos, SW Nigeria
92 B5 anc. Lacobriga. Faro,
ortugal
s de Moreno *51 E4* Jalisco,
V Mexico
uira *70 A4* SW Western Sahara
rande *46 C3* Oregon, NW USA
uaira *66 B4* Distrito Federal,
Venezuela
nas *64 B1* Tarapacá, N Chile
nillas *61 G4* Santa Cruz, SE Bolivia
abana *54 B2 var.* Havana. *country*
pital (Cuba) Ciudad de La Habana,
Cuba
138 B4 Sumatera, W Indonesia
aye *see* 's-Gravenhage
lm *85 B7* Halland, S Sweden
re *134 D2* Punjab, NE Pakistan
95 A6 Baden-Württemberg,
Germany
85 D5 Swe. Lahtis. Etelä-Suomi,
Finland
is *see* Lahti
76 B4 prev. Behagle, De Behagle.
andjilé, S Chad
ach *see* Ljubljana
hâu *136 D3* Lai Châu, N Vietnam
see Laylā
nta *44 D5* Colorado, C USA
Charles *42 A3* Louisiana, S USA
City *43 E3* Florida, SE USA
District *89 C5 physical region*
W England, United Kingdom
Havasu City *48 A2* Arizona,
V USA
Jackson *49 H4* Texas, SW USA
land *43 E4* Florida, SE USA
side *47 C8* California, W USA
State *see* Michigan
wood *44 D4* Colorado, C USA
nau *see* Lucknow
nikós Kólpos *105 B7 gulf* S Greece
elv *84 D2 Lapp.* Leavdnja.
nmark, N Norway
on *see* Laon
ela *72 C4* Āmara, Ethiopia
bertad *52 B1* Petén, N Guatemala
gua *64 B4* Valparaíso, C Chile
92 C1 Galicia, NW Spain
pur *135 F3* Central, C Nepal
ouvière *87 B6* Hainaut, S Belgium
addalena *96 A4* Sardegna, Italy,
Mediterranean Sea
anche *see* English Channel
ar *44 D5* Colorado, C USA
armora, Punta *97 A5 mountain*
rdegna, Italy, C Mediterranean Sea

La Massana *91 A8* La Massana,
W Andorra Europe
Lambaréné *77 A6* Moyen-Ogooué,
W Gabon
Lamego *92 C2* Viseu, N Portugal
Lamesa *49 E3* Texas, SW USA
Lamezia Terme *97 D6* Calabria, SE Italy
Lamía *105 B5* Stereá Elláda, C Greece
Lamoni *45 F4* Iowa, C USA
Lampang *136 C4 var.* Muang Lampang.
Lampang, NW Thailand
Lámpeia *105 B6* Dytikí Elláda, S Greece
Lanbi Kyun *137 B6 prev.* Sullivan Island.
island Myeik Archipelago, S Myanmar
(Burma)
Lancang Jiang *see* Mekong
Lancaster *89 D5* NW England, United
Kingdom
Lancaster *47 C7* California, W USA
Lancaster *41 F4* Pennsylvania, NE USA
Lancaster Sound *37 F2 sound* Nunavut,
N Canada
Lan-chou/Lan-chow/Lanchow *see*
Lanzhou
Landao *see* Lantau Island
Landen *87 C6* Vlaams Brabant,
C Belgium
Lander *44 C3* Wyoming, C USA
Landerneau *90 A3* Finistère, NW France
Landes *91 B5 cultural region* SW France
Land of Enchantment *see* New Mexico
The Land of Opportunity *see* Arkansas
Land of Steady Habits *see* Connecticut
Land of the Midnight Sun *see* Alaska
Landsberg *see* Gorzów Wielkopolski,
Lubuskie, Poland
Landsberg an der Warthe *see* Gorzów
Wielkopolski
Land's End *89 B8 headland* SW England,
United Kingdom
Landshut *95 C6* Bayern, SE Germany
Langar *123 E2 Rus.* Lyangar. Navoiy
Viloyati, C Uzbekistan
Langfang *128 D4* Hebei, E China
Langkawi, Pulau *137 B7 island*
Peninsular Malaysia
Langres *90 D4* Haute-Marne, N France
Langsa *138 A3* Sumatera, W Indonesia
Lang Shan *127 E3 mountain range*
N China
Lang Sơn *136 D3 var.* Langson. Lang
Sơn, N Vietnam
Langson *see* Lang Sơn
Lang Suan *137 B6* Chumphon,
SW Thailand
Languedoc *91 C6 cultural region*
S France
Länkäran *117 H3 Rus.* Lenkoran'.
S Azerbaijan
Lansing *40 C3 state capital* Michigan,
N USA
Lanta, Ko *137 B7 island* S Thailand
Lantau Island *128 A2 Cant.* Tai Yue
Shan, *Chin.* Landao. *island* Hong
Kong, S China
Lan-ts'ang Chiang *see* Mekong
Lantung, Gulf of *see* Liaodong Wan
Lanzarote *70 B3 island* Islas Canarias,
Spain, NE Atlantic Ocean
Lanzhou *128 B4 var.* Lan-chou,
Lanchow, Lan-chow; *prev.* Kaolan.
province capital Gansu, C China
Lao Cai *136 D3* Lao Cai, N Vietnam
Laodicea/Laodicea ad Mare *see* Al
Lādhiqīyah
Laoet *see* Laut, Pulau
Laon *90 D3 var.* la Laon; *anc.* Laudunum.
Aisne, N France
Lao People's Democratic Republic
see Laos
La Orchila, Isla *58 D1 island*
N Venezuela
La Oroya *60 C3* Junín, C Peru
Laos *136 D4 off.* Lao People's Democratic
Republic. *country* SE Asia

LAOS
Southeast Asia

Official name Lao People's Democratic Republic
Formation 1953 / 1953
Capital Vientiane
Population 6.8 million / 76 people per sq mile (29 people per sq km)
Total area 91,428 sq. miles (236,800 sq. km)
Languages Lao*, Mon-Khmer, Yao, Vietnamese, Chinese, French

LAOS
(continued)

Religions Buddhist 65%, Other (including animist) 34%, Christian 1%
Ethnic mix Lao Loum 66%, Lao Theung 30%, Lao Soung 2%, Other 2%
Government One-party state
Currency Kip = 100 at
Literacy rate 73%
Calorie consumption 2356 kilocalories

La Palma *53 G5* Darién, SE Panama
La Palma *70 A3 island* Islas Canarias,
Spain, NE Atlantic Ocean
La Paz *61 F4 var.* La Paz de Ayacucho.
country capital (Bolivia-seat of
government) La Paz, W Bolivia
La Paz *50 B3* Baja California Sur,
NW Mexico
La Paz, Bahía de *50 B3 bay* NW Mexico
La Paz de Ayacucho *see* La Paz
La Pérouse Strait *130 D1 Jap.* Sōya-
kaikyō, *Rus.* Proliv Laperuza. *strait*
Japan/Russian Federation
Laperuza, Proliv *see* La Pérouse Strait
Lápithos *see* Lapta
Lapland *84 D3 Fin.* Lappi, *Swe.*
Lappland. *cultural region* N Europe
La Plata *64 D4* Buenos Aires,
E Argentina
La Plata *see* Sucre
La Pola *92 D1 var.* Pola de Lena.
Asturias, N Spain
Lappeenranta *85 E5 Swe.* Villmanstrand.
Etelä-Suomi, SE Finland
Lappi/Lappland *see* Lapland
Lappo *see* Lapua
Lapta *102 C5 Gk.* Lápithos. NW Cyprus
Laptev Sea *115 E2 Eng.* Laptev Sea. *sea*
Arctic Ocean
Laptev Sea *see* Laptevykh, More
Lapua *85 D5 Swe.* Lappo. Länsi-Suomi,
W Finland
Lapurdum *see* Bayonne
Łapy *98 E3* Podlaskie, NE Poland
La Quiaca *64 C2* Jujuy, N Argentina
L'Aquila *96 C4 var.* Aquila, Aquila degli
Abruzzi. Abruzzo, C Italy
Laracha *92 B1* Galicia, NW Spain
Laramie *44 C4* Wyoming, C USA
Laramie Mountains *44 C3 mountain*
range Wyoming, C USA
Laredo *93 E1* Cantabria, N Spain
Laredo *49 F5* Texas, SW USA
La Réunion *see* Réunion
Largeau *see* Faya
Largo *43 E4* Florida, SE USA
Largo, Cayo *54 B2 island* W Cuba
Lario *see* Como, Lago di
La Rioja *64 C3* La Rioja, NW Argentina
La Rioja *93 E2 autonomous community*
N Spain
Lárisa *104 B4 var.* Larissa. Thessalía,
C Greece
Larissa *see* Lárisa
Lārkāna *134 B3 var.* Larkhana. Sind,
SE Pakistan
Larkhana *see* Lārkāna
Larnaca *see* Lárnaka
Lárnaka *102 C5 var.* Larnaca, Larnax.
SE Cyprus
Larnax *see* Lárnaka
la Rochelle *90 B4 anc.* Rupella.
Charente-Maritime, W France
la Roche-sur-Yon *90 B4 prev.* Bourbon
Vendée, Napoléon-Vendée. Vendée,
NW France
La Roda *93 E3* Castilla-La Mancha,
C Spain
La Romana *55 E3* E Dominican Republic
Larvotto *91 C8* N Monaco Europe
La-sa *see* Lhasa
Las Cabezas de San Juan *92 C5*
Andalucía, S Spain
Las Cruces *48 D3* New Mexico, SW USA
La See d'Urgel *see* La Seu d'Urgell
La Serena *64 B3* Coquimbo, C Chile
La Seu d'Urgell *93 G1 prev.* La See
d'Urgel, Seo de Urgel. Cataluña,
NE Spain
La Seu d'Urgell *see* La Seu d'Urgell
la Seyne-sur-Mer *91 D6* Var, SE France
Lashio *136 B3* Shan State, E Myanmar
(Burma)
Lashkar Gāh *122 D5 var.* Lash-Kar-Gar'.
Helmand, S Afghanistan
Lash-Kar-Gar' *see* Lashkar Gāh

La Sila *97 D6 mountain range* SW Italy
La Sirena *52 D3* Región Autónoma
Atlántico Sur, E Nicaragua
Łask *98 C4* Łódzkie, C Poland
Las Lomitas *64 D2* Formosa,
N Argentina
La Solana *93 E4* Castilla-La Mancha,
C Spain
Las Palmas *70 A3 var.* Las Palmas de
Gran Canaria. Gran Canaria, Islas
Canarias, Spain, NE Atlantic Ocean
Las Palmas de Gran Canaria *see* Las
Palmas
La Spezia *96 B3* Liguria, NW Italy
Lassa *see* Lhasa
Las Tablas *53 F5* Los Santos, S Panama
Last Frontier, The *see* Alaska
Las Tunas *54 C2 var.* Victoria de las
Tunas. Las Tunas, E Cuba
La Suisse *see* Switzerland
Las Vegas *47 D7* Nevada, W USA
Latacunga *60 B1* Cotopaxi, C Ecuador
Latakia *see* Al Lādhiqīyah
la Teste *91 B5* Gironde, SW France
Latina *97 C5 prev.* Littoria. Lazio, C Italy
La Tortuga, Isla *59 E1 var.* Isla Tortuga.
island N Venezuela
La Tuque *39 E4* Québec, SE Canada
Latvia *106 C3 off.* Republic of Latvia,
Ger. Lettland, *Latv.* Latvija, Latvijas
Republika; *prev.* Latvian SSR, *Rus.*
Latviyskaya SSR. *country* NE Europe

LATVIA
Northeast Europe

Official name Republic of Latvia
Formation 1991 / 1991
Capital Riga
Population 2.1 million / 84 people per sq mile (33 people per sq km)
Total area 24,938 sq. miles (64,589 sq. km)
Languages Latvian*, Russian
Religions Other 43%, Lutheran 24%, Roman Catholic 18%, Orthodox Christian 15%
Ethnic mix Latvian 62%, Russian 27%, Other 4%, Belarussian 3%, Ukrainian 2%, Polish 2%
Government Parliamentary system
Currency Euro = 100 cents
Literacy rate 99%
Calorie consumption 3293 kilocalories

Latvian SSR/Latvija/Latvijas
Republika/Latviyskaya SSR *see* Latvia
Latvia, Republic of *see* Latvia
Laudunum *see* Laon
Laudus *see* St-Lô
Lauenburg/Lauenburg in Pommern
see Lębork
Lau Group *145 E4 island group* E Fiji
Lauis *see* Lugano
Launceston *149 C8* Tasmania,
SE Australia
La Unión *52 C2* Olancho, C Honduras
La Unión *93 F4* Murcia, SE Spain
Laurel *42 C3* Mississippi, S USA
Laurel *44 C2* Montana, NW USA
Laurentian Highlands *see* Laurentian
Mountains
Laurentian Mountains *39 E3 var.*
Laurentian Highlands, *Fr.* Les
Laurentides. *plateau* Newfoundland
and Labrador/Québec, Canada
Laurentides, Les *see* Laurentian
Mountains
Lauria *97 D6* Basilicata, S Italy
Laurinburg *43 F1* North Carolina,
SE USA
Lauru *see* Choiseul
Lausanne *95 A7 It.* Losanna. Vaud,
SW Switzerland
Laut, Pulau *138 D4 prev.* Laoet. *island*
Borneo, C Indonesia
Laval *38 D4* Québec, SE Canada
Laval *90 B3* Mayenne, NW France
La Vall d'Uixó *93 F3 var.* Vall D'Uxó.
Valenciana, E Spain
La Vega *55 E3 var.* Concepción de la
Vega. C Dominican Republic
La Vila Joíosa *see* Villajoyosa
Lávrio *105 C6 prev.* Lávrion. Attikí,
C Greece
Lávrion *see* Lávrio
Lawrence *41 G3* Massachusetts, NE USA
Lawrenceburg *42 C1* Tennessee, S USA
Lawton *49 F2* Oklahoma, C USA
La Yarada *61 E4* Tacna, SW Peru

Laylā *121 C5 var.* Laila. Ar Riyāḍ,
  C Saudi Arabia
Lazarev Sea *154 B1 sea* Antarctica
Lázaro Cárdenas *51 E5* Michoacán,
  SW Mexico
Leal *see* Lihula
Leamhcán *see* Lucan
Leamington *38 C5* Ontario, S Canada
Leavdnja *see* Lakselv
Lebak *139 E3* Mindanao, S Philippines
Lebanese Republic *see* Lebanon
Lebanon *45 G5* Missouri, C USA
Lebanon *41 G2* New Hampshire,
  NE USA
Lebanon *46 B3* Oregon, NW USA
Lebanon *118 A4 off.* Lebanese Republic,
  Ar. Al Lubnān, *Fr.* Liban. *country*
  SW Asia

LEBANON
Southwest Asia

Official name Republic of Lebanon
Formation 1941 / 1941
Capital Beirut
Population 4.8 million / 1215 people
per sq mile (469 people per sq km)
Total area 4015 sq. miles (10,400 sq. km)
Languages Arabic*, French,
Armenian, Assyrian
Religions Muslim 60%, Christian 39%,
Other 1%
Ethnic mix Arab 95%, Armenian 4%,
Other 1%
Government Parliamentary system
Currency Lebanese pound = 100 piastres
Literacy rate 90%
Calorie consumption 3181 kilocalories

Lebanon, Mount *see* Liban, Jebel
Lebap *122 D2* Lebapskiy Velayat,
  NE Turkmenistan
Lebedin *see* Lebedyn
Lebedyn *109 F2 Rus.* Lebedin. Sums'ka
  Oblast', NE Ukraine
Lębork *98 C2 var.* Lębórk, *Ger.*
  Lauenburg, Lauenburg in Pommern.
  Pomorskie, N Poland
Lebrija *92 C5* Andalucía, S Spain
Lebu *65 A5* Bío Bío, C Chile
le Cannet *91 D6* Alpes-Maritimes,
  SE France
Le Cap *see* Cap-Haïtien
Lecce *97 E6* Puglia, SE Italy
Lechainá *105 B6 var.* Lehena, Lekhainá.
  Dytikí Elláda, S Greece
Ledo Salinarius *see* Lons-le-Saunier
Leduc *37 E5* Alberta, SW Canada
Leech Lake *45 F2 lake* Minnesota, N USA
Leeds *89 D5* N England, United Kingdom
Leek *86 E2* Groningen, NE Netherlands
Leer *94 A3* Niedersachsen, NW Germany
Leeuwarden *86 D1 Fris.* Ljouwert.
  Fryslân, N Netherlands
Leeuwin, Cape *142 A5 headland*
  Western Australia
Leeward Islands *55 G3 island group*
  E West Indies
Leeward Islands *see* Sotavento, Ilhas de
Lefkáda *105 A5 prev.* Levkás.
  Lefkáda, Ioánia Nisiá, Greece,
  C Mediterranean Sea
Lefkáda *105 A5 It.* Santa Maura, *prev.*
  Levkás; *anc.* Leucas. *island* Ioánia Nisiá,
  Greece, C Mediterranean Sea
Lefká Óri *105 C7 mountain range* Kríti,
  Greece, E Mediterranean Sea
Lefkímmi *105 A5 var.* Levkímmi.
  Kérkyra, Ioánia Nisiá, Greece,
  C Mediterranean Sea
Lefkosía/Lefkoşa *see* Nicosia
Legaceaster *see* Chester
Legaspi *see* Legazpi City
Legazpi City *139 E2 var.* Legaspi.
  C Philippines
Leghorn *see* Livorno
Legnica *98 B4 Ger.* Liegnitz.
  Dolnośląskie, SW Poland
le Havre *90 B3 Eng.* Havre; *prev.* le
  Havre-de-Gràce. Seine-Maritime,
  N France
le Havre-de-Grâce *see* le Havre
Lehena *see* Lechainá
Leicester *89 D6 Lat.* Batae Coritanorum.
  C England, United Kingdom
Leiden *86 B3 prev.* Leyden; *anc.*
  Lugdunum Batavorum. Zuid-Holland,
  W Netherlands
Leie *90 D2 Fr.* Lys. *river* Belgium/France

Leinster *89 B6 Ir.* Cúige Laighean.
  cultural region E Ireland
Leipsic *see* Leipzig
Leipsoí *105 E6 island* Dodekánisa,
  Greece, Aegean Sea
Leipzig *94 C4 Pol.* Lipsk, *hist.* Leipsic;
  anc. Lipsia. Sachsen, E Germany
Leiria *92 B3 anc.* Collipo. Leiria,
  C Portugal
Leirvík *85 A6* Hordaland, S Norway
Lek *86 C4 river* SW Netherlands
Lekhainá *see* Lechainá
Lekhchevo *see* Boychinovtsi
Leksand *85 C5* Dalarna, C Sweden
Lel'chitsy *see* Lyel'chytsy
le Léman *see* Geneva, Lake
Lelystad *86 D3* Flevoland, C Netherlands
Léman, Lac *see* Geneva, Lake
le Mans *90 B3* Sarthe, NW France
Lemberg *see* L'viv
Lemesós *102 C5 var.* Limassol.
  SW Cyprus
Lemhi Range *46 D3 mountain range*
  Idaho, C USA North America
Lemnos *see* Límnos
Lemovices *see* Limoges
Lena *115 F3 river* NE Russian Federation
Lena Tablemount *141 B7 seamount*
  S Indian Ocean
Len Dao *128 C8 island* S Spratly Islands
Leninabad *see* Khujand
Leninakan *see* Gyumri
Lenine *109 G5 Rus.* Lenino. Avtonomna
  Respublika Krym, S Ukraine
Leningor *see* Leninogorsk
Leningrad *see* Sankt-Peterburg
Leningradskaya *154 B4 Russian research
  station* Antarctica
Lenino *see* Lenine, Ukraine
Leninobod *see* Khujand
Leninogorsk *see* Ridder
Leninpol' *123 F2* Talasskaya Oblast',
  NW Kyrgyzstan
Lenin-Turkmenski *see* Türkmenabat
Lenkoran' *see* Länkäran
Lenti *99 B7* Zala, SW Hungary
Lentia *see* Linz
Leoben *95 E7* Steiermark, C Austria
León *51 E4 var.* León de los Aldamas.
  Guanajuato, C Mexico
León *52 D3* León, NW Nicaragua
León *92 D1* Castilla y León, NW Spain
León de los Aldamas *see* León
Leonídi *see* Leonídio
Leonídio *105 B6 var.* Leonídi.
  Pelopónnisos, S Greece
Léopold II, Lac *see* Mai-Ndombe, Lac
Léopoldville *see* Kinshasa
Lepe *92 C4* Andalucía, S Spain
Lepel' *see* Lyepyel'
le Portel *90 C2* Pas-de-Calais, N France
Le Puglie *see* Puglia
le Puy *91 C5 prev.* le Puy-en-Velay, *hist.*
  Anicium, Podium Anicensis. Haute-
  Loire, C France
le Puy-en-Velay *see* le Puy
Léré *76 B4* Mayo-Kébbi, SW Chad
Lérida *see* Lleida
Lerma *92 D2* Castilla y León, N Spain
Leros *105 D6 island* Dodekánisa,
  Aegean Sea
Lerrnayin Gharabakh *see*
  Nagorno-Karabakh
Lerwick *88 D1* NE Scotland, United
  Kingdom
Lesbos *116 A3 anc.* Lesbos. *island*
  E Greece
Lesbos *see* Lésvos
Les Cayes *see* Cayes
Les Gonaïves *see* Gonaïves
Leshan *128 B5* Sichuan, C China
les Herbiers *90 B4* Vendée, NW France
Lesh/Leshi *see* Lezhë
Lesina *see* Hvar
Leskovac *101 E5* Serbia, SE Serbia
Lesnoy *114 C3* Sverdlovskaya Oblast',
  C Russian Federation
Lesotho *78 D4 off.* Kingdom of Lesotho;
  prev. Basutoland. *country* S Africa

LESOTHO
Southern Africa

Official name Kingdom of Lesotho
Formation 1966 / 1966
Capital Maseru
Population 2.2 million / 188 people
per sq mile (72 people per sq km)
Total area 11,720 sq. miles (30,355 sq. km)

LESOTHO
(continued)

Languages English*, Sesotho*, isiZulu
Religions Christian 90%, Traditional
beliefs 10%
Ethnic mix Sotho 99%, European
and Asian 1%
Government Parliamentary system
Currency Loti = 100 lisente;
South African rand = 100 cents
Literacy rate 76%
Calorie consumption 2595 kilocalories

Lesotho, Kingdom of *see* Lesotho
les Sables-d'Olonne *90 B4* Vendée,
  NW France
Lesser Antarctica *see* West Antarctica
Lesser Antilles *55 G4 island group*
  E West Indies
Lesser Caucasus *117 F2 Rus.* Malyy
  Kavkaz. *mountain range* SW Asia
Lesser Khingan Range *see* Xiao Hinggan
  Ling
Lesser Sunda Islands *139 E5 Eng.*
  Lesser Sunda Islands. *island group*
  C Indonesia
Lesser Sunda Islands *see* Nusa Tenggara
Leszno *98 B4 Ger.* Lissa. Wielkopolskie,
  C Poland
Lethbridge *37 E5* Alberta, SW Canada
Lethem *59 F3 S* Guyana
Leti, Kepulauan *139 F5 island group*
  E Indonesia
Letpadan *136 B4* Bago, SW Myanmar
  (Burma)
Letsôk-aw Kyun *137 B6 var.* Letsutan
  Island; *prev.* Domel Island. *island*
  Myeik Archipelago, S Myanmar
  (Burma)
Letsutan Island *see* Letsôk-aw Kyun
Lettland *see* Latvia
Lëtzebuerg *see* Luxembourg
Leucas *see* Lefkáda
Leuven *87 C6 Fr.* Louvain, *Ger.* Löwen.
  Vlaams Brabant, C Belgium
Leuze *see* Leuze-en-Hainaut
Leuze-en-Hainaut *87 B6 var.* Leuze.
  Hainaut, SW Belgium
Léva *see* Levice
Levanger *84 B4* Nord-Trøndelag,
  C Norway
Levelland *49 E2* Texas, SW USA
Leverkusen *94 A4* Nordrhein-Westfalen,
  W Germany
Levice *99 C6 Ger.* Lewentz, *Hung.* Léva,
  Lewenz. Nitriansky Kraj, SW Slovakia
Levin *150 D4* Manawatu-Wanganui,
  North Island, New Zealand
Levkás *see* Lefkáda
Levkímmi *see* Lefkímmi
Lewentz/Lewenz *see* Levice
Lewis, Isle of *88 B2 island* NW Scotland,
  United Kingdom
Lewis Range *44 B1 mountain range*
  Montana, NW USA
Lewiston *46 C2* Idaho, NW USA
Lewiston *41 G2* Maine, NE USA
Lewistown *44 C1* Montana,
  NW USA
Lexington *40 C5* Kentucky, S USA
Lexington *45 E4* Nebraska, C USA
Leyden *see* Leiden
Leyte *139 F2 island* C Philippines
Leżajsk *98 E4* Podkarpackie, SE Poland
Lezhë *101 C6 var.* Lezha; *prev.* Lesh,
  Leshi. Lezhë, NW Albania
Lhasa *126 C5 var.* La-sa, Lassa. Xizang
  Zizhiqu, W China
Lhaviyani Atoll *see* Faadhippolhu Atoll
Lhazê *126 C5 var.* Quxar. Xizang
  Zizhiqu, China E Asia
L'Hospitalet de Llobregat *93 G2 var.*
  Hospitalet. Cataluña, NE Spain
Liancourt Rocks *131 A5 Jap.* Takeshima,
  Kor. Dokdo. *island group*
  Japan/South Korea
Lianyungang *128 D4 var.* Xinpu.
  Jiangsu, E China
Liao *see* Liaoning
Liaodong Wan *127 G3 Eng.* Gulf of
  Lantung, Gulf of Liaotung. *gulf*
  NE China
Liao He *125 E1 river* NE China
Liaoning *128 D3 var.* Liao, Liaoning
  Sheng, Shengking, *hist.* Fengtien,
  Shenking. *province* NE China

Liaoning Sheng *see* Liaoning
Liaoyuan *129 E3 var.* Dongliao, S
  liao, *Jap.* Chengchiatun. Jilin, NE
Liard *see* Fort Liard
Liban *see* Lebanon
Liban, Jebel *118 B4 Ar.* Jabal al GH
  Jabal Lubnān, *Eng.* Mount Lebar
  mountain range C Lebanon
Libau *see* Liepāja
Libby *44 A1* Montana, NW USA
Liberal *45 E5* Kansas, C USA
Liberalitas Julia *see* Évora
Liberec *98 B4 Ger.* Reichenberg.
  Liberecký Kraj, N Czech Republi
Liberia *52 D4* Guanacaste, NW Co
  Rica
Liberia *74 C5 off.* Republic of Libe
  country W Africa

LIBERIA
West Africa

Official name Republic of Liberia
Formation 1847 / 1847
Capital Monrovia
Population 4.3 million / 116 people
per sq mile (45 people per sq km)
Total area 43,000 sq. miles (111,370 s
Languages Kpelle, Vai, Bassa, Kru, G
Kissi, Gola, Loma, English*
Religions Christian 40%, Traditiona
beliefs 40%, Muslim 20%
Ethnic mix Indigenous tribes (12 gr
49%, Kpellé 20%, Bassa 16%,
Gio 8%, Krou 7%
Government Presidential system
Currency Liberian dollar = 100 cent
Literacy rate 43%
Calorie consumption 2251 kilocalor

Liberia, Republic of *see* Liberia
Libian Desert *see* Libyan Desert
Libīyah, Aş Şahrā' al *see* Libyan De
Libourne *91 B5* Gironde, SW Fran
Libreville *77 A5 country capital (G
  Estuaire, NW Gabon
Libya *71 F3 prev.* Great Socialist Pe
  Libyan Arab Jamahiriya, Al Jamāl
  al 'Arabīyah al Lībīyah ash Sha'biy
  al Ishtirākīy, Libyan Arab Republi
  country N Africa

LIBYA
North Africa

Official name State of Libya
Formation 1951 / 1951
Capital Tripoli
Population 6.2 million / 9 people
per sq mile (4 people per sq km)
Total area 679,358 sq. miles
(1,759,540 sq. km)
Languages Arabic*, Tuareg
Religions Muslim (mainly Sunni) 97%
Other 3%
Ethnic mix Arab and Berber 97%,
Other 3%
Government Transitional regime
Currency Libyan dinar = 1000 dirham
Literacy rate 90%
Calorie consumption 3211 kilocalorie

Libyan Arab Republic *see* Libya
Libyan Desert *71 H4 var.* Libian De
  Ar. Aş Şahrā' al Lībīyah. *desert* N A
Libyan Plateau *103 F4 var.* Aḍ Diff
  plateau E Libya
Lichtenfels *95 C5* Bayern, SE Germa
Lichtenvoorde *86 E4* Gelderland,
  E Netherlands
Lichuan *128 C5* Hubei, C China
Lida *107 B5* Hrodzyenskaya Voblas
  W Belarus
Lidhoríkion *see* Lidoríki
Lidköping *85 B6* Västra Götaland,
  S Sweden
Lidokhorikion *see* Lidoríki
Lidoríki *105 B5 prev.* Lidhoríkion,
  Lidokhorikion. Stereá Elláda, C Gr
Lidzbark Warmiński *98 D2 Ger.*
  Heilsberg. Olsztyn, N Poland
Liechtenstein *94 D1 off.* Principality of
  Liechtenstein. *country* C Europe

LIECHTENSTEIN
Central Europe

Official name Principality of
Liechtenstein

HTENSTEIN
ntinued)
ation 1719 / 1719
al Vaduz
ation 37,000 / 597 people
] mile (231 people per sq km)
area 62 sq. miles (160 sq. km)
ages German*, Alemannish
:t, Italian
ons Roman Catholic 79%,
· 13%, Protestant 8%
: mix Liechtensteiner 66%,
· 12%, Swiss 10%, Austrian 6%,
an 3%, Italian 3%
rnment Parliamentary system
·ncy Swiss franc = 100 rappen/
·nes
·cy rate 99%
·ie consumption Not available

·enstein, Principality of *see*
·htenstein
· 87 D6 *Dut.* Luik, *Ger.* Lüttich.
·ge, E Belgium
·tz *see* Legnica
· 95 D7 Tirol, W Austria
·a 106 B3 *Ger.* Libau. W Latvia
·a *see* Lithuania
·ahof *see* Livāni
· 95 D7 Steiermark, C Austria
· 89 B6 *river* Ireland
· 144 D5 *island* Îles Loyauté, E New
·edonia
·*see* Loire
·e, Appennino 96 A2 *Eng.* Ligurian
·untains. *mountain range* NW Italy
·e, Mar *see* Ligurian Sea
·an Mountains *see* Ligure,
·ennino
·an Sea 96 A3 *Fr.* Mer
·irienne, *It.* Mar Ligure. *sea*
·enne, Mer *see* Ligurian Sea
· 47 A7 *var.* Lihue. Kaua'i,
·waii, USA
·*see* Lihu'e
·a 106 D2 *Ger.* Leal. Läänemaa,
·Estonia
·aht *see* Riga, Gulf of
· 77 D7 *prev.* Jadotville. Shaba,
·Dem. Rep. Congo
·s 85 A6 Vest-Agder, S Norway
·90 C2 *var.* l'Isle, *Dut.* Rijssel, *Flem.*
·sel, *prev.* Lisle; *anc.* Insula. Nord,
·rance
·anmoyer 85 B5 Oppland, S Norway
·røm 85 B6 Akershus, S Norway
·gwe 79 E2 *country capital* (Malawi)
·itral, W Malawi
·aeum *see* Marsala
· 60 C4 *country capital* (Peru)
·aa, W Peru
·nowa 99 D5 Małopolskie, S Poland
·ssol *see* Lemesós
·rick 89 A6 *Ir.* Luimneach. Limerick,
·Ireland
·· Vathéos *see* Sámos
·os 103 F3 *anc.* Lemnos. *island*
·reece
·ges 91 C5 *anc.* Augustoritum
·novicensium, Lemovices.
·ite-Vienne, C France
·n 53 E4 *var.* Puerto Limón. Limón,
·osta Rica
·a 52 D2 Colón, NE Honduras
·num *see* Poitiers
·usin 91 C5 *cultural region* C France
·ux 91 C6 Aude, S France
·opo 78 D3 *var.* Crocodile. *river*
·frica
·es 64 B4 Maule, C Chile
·es 51 E3 Nuevo León, NE Mexico
·es 92 D4 Andalucía, S Spain
·ln 89 D5 *anc.* Lindum, Lindum
·lonia. E England, United Kingdom
·ln 41 H2 Maine, NE USA
·ln 45 F4 *state capital* Nebraska,
·JSA
·ln Sea 34 D2 *sea* Arctic Ocean
·n 59 F3 E Guyana
·os *see* Líndos
· 73 D8 Lindi, SE Tanzania
·os 105 E7 *var.* Líndhos. Ródos,
·dekánisa, Greece, Aegean Sea
·m/Lindum Colonia *see* Lincoln
·slands 145 G3 *island group*
·Ciribati

Lingeh *see* Bandar-e Lengeh
Lingen 94 A3 *var.* Lingen an der Ems.
  Niedersachsen, NW Germany
Lingen an der Ems *see* Lingen
Lingga, Kepulauan 138 B4 *island group*
  W Indonesia
Lingling 129 C6 *prev.* Yongzhou,
  Zhishan. Hunan, S China
Linköping 85 C6 Östergötland,
  S Sweden
Linz 95 D6 *anc.* Lentia. Oberösterreich,
  N Austria
Lion, Gulf of 91 C7 *Eng.* Gulf of Lion,
  Gulf of Lions; *anc.* Sinus Gallicus. *gulf*
  S France
Lion, Gulf of/Lions, Gulf of *see* Lion,
  Golfe du
Liozno *see* Lyozna
Lipari 70 D8 *island* Isole Eolie, S Italy
Lipari Islands/Lipari, Isole *see* Eolie,
  Isole
Lipetsk 111 B5 Lipetskaya Oblast',
  W Russian Federation
Lipno 98 C3 Kujawsko-pomorskie,
  C Poland
Lipova 108 A4 *Hung.* Lippa. Arad,
  W Romania
Lipovets *see* Lypovets'
Lippa *see* Lipova
Lipsia/Lipsk *see* Leipzig
Lira 73 B6 N Uganda
Lisala 77 C5 Equateur, N Dem. Rep.
  Congo
Lisboa 92 B4 *Eng.* Lisbon; *anc.* Felicitas
  Julia, Olisipo. *country capital*
  (Portugal) Lisboa, W Portugal
Lisbon *see* Lisboa
Lisichansk *see* Lysychans'k
Lisieux 90 B3 *anc.* Noviomagus.
  Calvados, N France
Liski 111 B6 *prev.* Georgiu-Dezh.
  Voronezhskaya Oblast', W Russian
  Federation
Lisle/l'Isle *see* Lille
Lismore 149 E5 New South Wales,
  SE Australia
Lissa *see* Vis, Croatia
Lissa *see* Leszno, Poland
Lisse 86 C3 Zuid-Holland,
  W Netherlands
Litang 128 A5 *var.* Gaocheng. Sichuan,
  C China
Litani, Nahr el 119 B5 *var.* Nahr al
  Litant. *river* C Lebanon
Litant, Nahr al *see* Litani, Nahr el
Litauen *see* Lithuania
Lithgow 149 D6 New South Wales,
  SE Australia
Lithuania 106 B4 *off.* Republic of
  Lithuania, *Ger.* Litauen, *Lith.* Lietuva,
  Pol. Litwa, *Rus.* Litva; *prev.* Lithuanian
  SSR, *Rus.* Litovskaya SSR. *country*
  NE Europe

LITHUANIA
Northeast Europe
Official name Republic of Lithuania
Formation 1991 / 1991
Capital Vilnius
Population 3 million / 119 people
per sq mile (46 people per sq km)
Total area 25,174 sq. miles (65,200 sq. km)
Languages Lithuanian*, Russian
Religions Roman Catholic 77%,
Other 17%, Russian Orthodox 4%,
Protestant 1%, Old Believers 1%
Ethnic mix Lithuanian 85%, Polish 7%,
Russian 6%, Belarussian 1%, Other 1%
Government Parliamentary system
Currency Litas = 100 centu
Literacy rate 99%
Calorie consumption 3463 kilocalories

Lithuanian SSR *see* Lithuania
Lithuania, Republic of *see* Lithuania
Litóchoro 104 B4 *var.* Litohoro,
  Litókhoron. Kentrikí Makedonía,
  N Greece
Litohoro/Litókhoron *see* Litóchoro
Litovskaya SSR *see* Lithuania
Little Alföld 99 C6 *Ger.* Kleines
  Ungarisches Tiefland, *Hung.* Kisalföld,
  Slvk. Podunajská Rovina. *plain*
  Hungary/Slovakia
Little Andaman 133 F2 *island* Andaman
  Islands, India, NE Indian Ocean
Little Barrier Island 150 D2 *island*
  N New Zealand

Little Bay 93 H5 *bay* Alboran Sea,
  Mediterranean Sea
Little Cayman 54 B3 *island* E Cayman
  Islands
Little Falls 45 F2 Minnesota, N USA
Littlefield 49 E2 Texas, SW USA
Little Inagua 54 D2 *var.* Inagua Islands.
  island S Bahamas
Little Minch, The 88 B3 *strait*
  NW Scotland, United Kingdom
Little Missouri River 44 D2 *river*
  NW USA
Little Nicobar 133 G3 *island* Nicobar
  Islands, India, NE Indian Ocean
Little Rhody *see* Rhode Island
Little Rock 42 B1 *state capital* Arkansas,
  C USA
Little Saint Bernard Pass 91 D5 *Fr.* Col
  du Petit St-Bernard, *It.* Colle del Piccolo
  San Bernardo. *pass* France/Italy
Little Sound 42 A5 *bay* Bermuda,
  NW Atlantic Ocean
Littleton 44 D4 Colorado, C USA
Littoria *see* Latina
Litva/Litwa *see* Lithuania
Liu-chou/Liuchow *see* Liuzhou
Liuzhou 128 C6 *var.* Liu-chou, Liuchow.
  Guangxi Zhuangzu Zizhiqu, S China
Livanátai *see* Livanátes
Livanátes 105 B5 *prev.* Livanátai. Stereá
  Elláda, C Greece
Livāni 106 D4 *Ger.* Lievenhof. SE Latvia
Liverpool 39 F5 Nova Scotia, SE Canada
Liverpool 89 C5 NW England, United
  Kingdom
Livingston 44 B2 Montana, NW USA
Livingston 49 H3 Texas, SW USA
Livingstone 49 H3 *var.* Maramba.
  Southern, S Zambia
Livingstone Mountains 151 A7
  mountain range South Island, New
  Zealand
Livno 100 B4 Federacija Bosne
  I Hercegovine, SW Bosnia and
  Herzegovina
Livojoki 84 D4 *river* C Finland
Livonia 40 D3 Michigan, N USA
Livorno 96 B3 *Eng.* Leghorn. Toscana,
  C Italy
Lixian Jiang *see* Black River
Lixoúri 105 A5 *prev.* Lixoúrion.
  Kefallinía, Iónia Nisiá, Greece,
  C Mediterranean Sea
Lixoúrion *see* Lixoúri
Lizarra *see* Estella
Ljouwert *see* Leeuwarden
Ljubelj *see* Loibl Pass
Ljubljana 95 D7 *Ger.* Laibach, *It.*
  Lubiana; *anc.* Aemona, Emona.
  country capital (Slovenia) C Slovenia
  NE Europe
Ljungby 85 B7 Kronoberg, S Sweden
Ljusdal 85 C5 Gävleborg, C Sweden
Ljusnan 85 C5 *river* C Sweden
Llanelli 89 C6 *prev.* Llanelly. SW Wales,
  United Kingdom
Llanelly *see* Llanelli
Llanes 92 D1 Asturias, N Spain
Llanos 108 D2 *physical region* Colombia/
  Venezuela
Lleida 93 F2 *Cast.* Lérida; *anc.* Ilerda.
  Cataluña, NE Spain
Llucmajor 93 G3 Mallorca, Spain,
  W Mediterranean Sea
Loaita Island 128 C8 *island* W Spratly
  Islands
Lobatse 78 C4 *var.* Lobatsi. Kgatleng,
  SE Botswana
Lobatsi *see* Lobatse
Lobito 78 B2 Benguela, W Angola
Lob Nor *see* Lop Nur
Lobositz *see* Lovosice
Locarno 95 B8 *var.* Luggarus. Ticino,
  S Switzerland
Lochem 86 E3 Gelderland, E Netherlands
Lockport 41 F3 New York, NE USA
Lodja 77 D6 Kasai-Oriental, C Dem.
  Rep. Congo
Lodwar 73 C6 Rift Valley, NW Kenya
Łódź 98 D4 *Rus.* Lodz. Łódź, C Poland
Loei 128 C4 *var.* Loey, Muang Loei.
  C Thailand
Loey *see* Loei
Lofoten 84 B3 *var.* Lofoten Islands.
  island group C Norway

Lofoten Islands *see* Lofoten
Logan 44 B3 Utah, W USA
Logan, Mount 36 D3 *mountain* Yukon,
  W Canada
Logroño 93 E1 *anc.* Vareia, *Lat.*
  Juliobriga. La Rioja, N Spain
Loibl Pass 95 D7 *Ger.* Loiblpass, *Slvn.*
  Ljubelj. *pass* Austria/Slovenia
Loiblpass *see* Loibl Pass
Loikaw 136 B4 Kayah State, C Myanmar
  (Burma)
Loire 90 B4 *var.* Liger. *river* C France
Loja 60 B2 Loja, S Ecuador
Lokitaung 73 C5 Rift Valley, NW Kenya
Lokoja 75 G4 Kogi, C Nigeria
Loksa 106 E2 *Ger.* Loxa. Harjumaa,
  NW Estonia
Lolland 85 B8 *prev.* Laaland. *island*
  S Denmark
Lom 104 C1 *prev.* Lom-Palanka.
  Montana, NW Bulgaria
Lomami 77 D6 *river* C Dem. Rep. Congo
Lomas 60 D4 Arequipa, SW Peru
Lomas de Zamora 64 D4 Buenos Aires,
  E Argentina
Lombardia 96 B2 *Eng.* Lombardy.
  region N Italy
Lombardy *see* Lombardia
Lombok, Pulau 138 D5 *island* Nusa
  Tenggara, C Indonesia
Lomé 75 F5 *country capital* (Togo)
  S Togo
Lomela 77 D6 Kasai-Oriental, C Dem.
  Rep. Congo
Lommel 87 C5 Limburg, N Belgium
Lomond, Loch 88 B4 *lake* C Scotland,
  United Kingdom
Lomonosov Ridge 155 B3 *var.* Harris
  Ridge, *Rus.* Khrebet Homonosova.
  undersea ridge Arctic Ocean
Lomonsova, Khrebet *see* Lomonosov
  Ridge
Lom-Palanka *see* Lom
Lompoc 47 B7 California, W USA
Lom Sak 136 C4 *var.* Muang Lom Sak.
  Phetchabun, C Thailand
Łomża 98 D3 *Rus.* Lomzha. Podlaskie,
  NE Poland
Lomzha *see* Łomża
Loncoche 65 B5 Araucanía, C Chile
Londinium *see* London
London 89 A7 *anc.* Augusta, *Lat.*
  Londinium. *country capital*
  (United Kingdom) SE England,
  United Kingdom
London 38 C5 Ontario, S Canada
London 40 C5 Kentucky, S USA
Londonderry 88 B4 *var.* Derry,
  Ir. Doire. NW Northern Ireland,
  United Kingdom
Londonderry, Cape 146 C2 *cape*
  Western Australia
Londrina 63 E4 Paraná, S Brazil
Lone Star State *see* Texas
Long Bay 43 F2 *bay* W Jamaica
Long Beach 47 C8 California, W USA
Longford 89 B5 *Ir.* An Longfort.
  Longford, C Ireland
Long Island 54 D2 *island* C Bahamas
Long Island 41 G4 *island* New York,
  NE USA
Longlac 38 C3 Ontario, S Canada
Longmont 44 D4 Colorado, C USA
Longreach 148 C4 Queensland,
  E Australia
Long Strait 115 G1 *Eng.* Long Strait.
  strait NE Russian Federation
Longview 45 H5 Texas, SW USA
Longview 46 B2 Washington, NW USA
Long Xuyên 137 D6 *var.* Longxuyen. An
  Giang, S Vietnam
Longxuyen *see* Long Xuyên
Longyan 128 D6 Fujian, SE China
Longyearbyen 83 G2 *dependent
  territory capital* (Svalbard) Spitsbergen,
  W Svalbard
Lons-le-Saunier 90 D4 *anc.* Ledo
  Salinarius. Jura, E France
Lop Buri 137 C5 *var.* Loburi. Lop Buri,
  C Thailand
Lop Nor *see* Lop Nur
Lop Nur 126 C3 *var.* Lob Nor, Lop Nor,
  Lo-pu Po. *seasonal lake* NW China
Loppersum 86 E1 Groningen,
  NE Netherlands
Lo-pu Po *see* Lop Nur
Lorca 93 E4 *Ar.* Lurka; *anc.* Eliocroca,
  Lat. Illurco. Murcia, SE Spain

Lord Howe Island *142 C4 island*
E Australia
Lord Howe Rise *142 C4 undersea rise*
SW Pacific Ocean
Loreto *50 B3* Baja California Sur,
NW Mexico
Lorient *90 A3 prev.* l'Orient.

Morbihan,
NW France
l'Orient *see* Lorient
Lorn, Firth of *88 B4 inlet* W Scotland,
United Kingdom
Loro Sae *see* East Timor
Lörrach *95 A7* Baden-Württemberg,
S Germany
Lorraine *90 D3 cultural region* N France
Los Alamos *48 C1* New Mexico, SW USA
Los Amates *52 B2* Izabal, E Guatemala
Los Ángeles *65 B5* Bío Bío, C Chile
Los Ángeles *47 C7* California, W USA
Losanna *see* Lausanne
Lošinj *100 A3 Ger.* Lussin, *It.* Lussino.

island W Croatia
Loslau *see* Wodzisław Śląski
Los Mochis *50 C3* Sinaloa, C Mexico
Losonc/Losontz *see* Lučenec
Los Roques, Islas *58 D1 island group*
N Venezuela
Lot *91 B5 cultural region* S France
Lot *91 B5 river* S France
Lotagipi Swamp *73 C5 wetland* Kenya/
Sudan
Lötzen *see* Giżycko
Loualaba *see* Lualaba
Louangnamtha *136 C3 var.* Luong Nam
Tha. Louang Namtha, N Laos
Louangphabang *124 D3 var.*
Louangphrabang, Luang Prabang.
Louangphabang, N Laos
Louangphrabang *see* Louangphabang
Loubomo *see* Dolisie
Loudéac *90 A3* Côtes d'Armor,
NW France
Loudi *128 C5* Hunan, S China
Louga *74 B3* NW Senegal
Louisiade Archipelago *144 B4 island
group* SE Papua New Guinea
Louisiana *42 A2 off.* State of Louisiana,
also known as Creole State, Pelican
State. *state* S USA
Louisville *40 C5* Kentucky, S USA
Louisville Ridge *143 E4 undersea ridge*
S Pacific Ocean
Loup River *45 E4 river* Nebraska, C USA
Lourdes *91 B6* Hautes-Pyrénées, S France
Lourenço Marques *see* Maputo
Louth *89 E5* E England, United Kingdom
Loutrá *104 C4* Kentrikí Makedonía,
N Greece
Louvain *see* Leuven
Louvain-la Neuve *87 C6* Walloon
Brabant, C Belgium
Louviers *90 C3* Eure, N France
Lovech *104 C2* Lovech, N Bulgaria
Loveland *44 D4* Colorado, C USA
Lovosice *98 A4 Ger.* Lobositz. Ústecký
Kraj, NW Czech Republic
Lóvua *78 C1* Moxico, E Angola
Lowell *41 G3* Massachusetts, NE USA
Löwen *see* Leuven
Lower California *48 A4 Eng.* Lower
California. *peninsula* NW Mexico
Lower California *see* Baja California
Lower Hutt *151 D5* Wellington, North
Island, New Zealand
Lower Lough Erne *89 A5 lake*
SW Northern Ireland, United Kingdom
Lower Red Lake *45 F1 lake* Minnesota,
N USA
Lower Rhine *see* Neder Rijn
Lower Tunguska *115 E3 Eng.* Lower
Tunguska. *river* N Russian Federation
Lower Tunguska *see* Nizhnyaya
Tunguska
Lowestoft *89 E6* E England,
United Kingdom
Loxa *see* Loksa
Lo-yang *see* Luoyang
Loyauté, Îles *144 D5 island group*
S New Caledonia
Loyev *see* Loyew
Loyew *107 D8 Rus.* Loyev. Homyel'skaya
Voblasts', SE Belarus
Loznica *100 C3* Serbia, W Serbia
Lu *see* Shandong, China
Lualaba *77 D6 Fr.* Loualaba. *river*
SE Dem. Rep. Congo
Luanda *78 A1 var.* Loanda, Port. São
Paulo de Loanda. *country capital*
(Angola) Luanda, NW Angola

Luang Prabang *see* Louangphabang
Luang, Thale *137 C7 lagoon*
S Thailand
Luangua, Rio *see* Luangwa
Luangwa *73 B8 var.* Aruângua, Rio
Luangua. *river* Mozambique/Zambia
Luanshya *78 D2* Copperbelt, C Zambia
Luarca *92 C1* Asturias, N Spain
Lubaczów *99 E5 var.* Lúbaczów.
Podkarpackie, SE Poland
L'uban' *98 B4* Leningradskaya Oblast',
Russian Federation
Lubānas Ezers *see* Lubāns
Lubango *78 B2 Port.* Sá da Bandeira.
Huíla, SW Angola
Lubāns *106 D4 var.* Lubānas Ezers.
lake E Latvia
Lubao *77 D6* Kasai-Oriental, C Dem.
Rep. Congo
Lübben *94 D4* Brandenburg, E Germany
Lübbenau *94 D4* Brandenburg,
E Germany
Lubbock *49 E2* Texas, SW USA
Lübeck *94 C2* Schleswig-Holstein,
N Germany
Lubelska, Wyżyna *98 E4 plateau*
SE Poland
Lüben *see* Lubin
Lubiana *see* Ljubljana
Lubin *98 B4 Ger.* Lüben. Dolnośląskie,
SW Poland
Lublin *98 E4 Rus.* Lyublin. Lubelskie,
E Poland
Lubliniec *98 C4* Śląskie, S Poland
Lubnān, Jabal *see* Liban, Jebel
Lubny *109 F2* Poltavs'ka Oblast',
NE Ukraine
Lubsko *98 B4 Ger.* Sommerfeld.
Lubuskie, W Poland
Lubumbashi *77 E8 prev.* Élisabethville.
Shaba, SE Dem. Rep. Congo
Lubutu *77 D6* Maniema, E Dem. Rep.
Congo
Luca *see* Lucca
Lucan *89 B5 Ir.* Leamhcán. Dublin,
E Ireland
Lucanian Mountains *see* Lucano,
Appennino
Lucano, Appennino *97 D5 Eng.*
Lucanian Mountains. *mountain
range* S Italy
Lucapa *78 C1 var.* Lukapa. Lunda Norte,
NE Angola
Lucca *96 B3 anc.* Luca. Toscana, C Italy
Lucea *54 A4* W Jamaica
Lucena *139 E1 off.* Lucena City. Luzon,
N Philippines
Lucena *92 D4* Andalucía, S Spain
Lucena City *see* Lucena
Lučenec *99 D6 Ger.* Losontz, Hung.
Losonc. Banskobystrický Kraj,
C Slovakia
Lucentum *see* Alicante
Lucerna/Lucerne *see* Luzern
Luchow *see* Hefei
Łuck *see* Luts'k
Lucknow *135 E3 var.* Lakhnau. *state
capital* Uttar Pradesh, N India
Lüda *see* Dalian
Luda Kamchiya *104 D2 river* E Bulgaria
Ludasch *see* Luduş
Lüderitz *78 B4 prev.* Angra Pequena.
Karas, SW Namibia
Ludhiāna *134 D2* Punjab, N India
Ludington *40 C2* Michigan, N USA
Ludsan *see* Ludza
Luduş *108 B4 Ger.* Ludasch, Hung.
Marosludas. Mureş, C Romania
Ludvika *85 C6* Dalarna, C Sweden
Ludwigsburg *95 B6* Baden-
Württemberg, SW Germany
Ludwigsfelde *94 D3* Brandenburg,
NE Germany
Ludwigshafen *95 B5 var.* Ludwigshafen
am Rhein. Rheinland-Pfalz,
W Germany
Ludwigshafen am Rhein *see*
Ludwigshafen
Ludwigslust *94 C3* Mecklenburg-
Vorpommern, N Germany
Ludza *106 D4 Ger.* Ludsan. E Latvia
Luebo *77 C6* Kasai-Occidental,
SW Dem. Rep. Congo
Luena *78 C2 var.* Lwena, Port. Luso.
Moxico, E Angola
Lufira *77 E7 river* SE Dem. Rep. Congo
Lufkin *49 H3* Texas, SW USA
Luga *110 A4* Leningradskaya Oblast',
NW Russian Federation

Lugano *95 B8 Ger.* Lauis. Ticino,
S Switzerland
Lugansk *see* Luhans'k
Lugdunum *see* Lyon
Lugdunum Batavorum *see* Leiden
Lugenda, Rio *79 E2 river* N Mozambique
Luggarus *see* Locarno
Lugh Ganana *see* Luuq
Lugo *92 C1 anc.* Lugus Augusti. Galicia,
NW Spain
Lugoj *108 A4 Ger.* Lugosch, Hung.
Lugos. Timiş, W Romania
Lugos/Lugosch *see* Lugoj
Lugus Augusti *see* Lugo
Luguvallium/Luguvallum *see* Carlisle
Luhans'k *109 H3 Rus.* Lugansk; *prev.*
Voroshilovgrad. Luhans'ka Oblast',
E Ukraine
Luimneach *see* Limerick
Lukapa *see* Lucapa
Lukenie *77 C6 river* C Dem. Rep. Congo
Łuków *98 E4 Ger.* Bogendorf. Lubelskie,
E Poland
Lukuga *77 D7 river* SE Dem. Rep. Congo
Luleå *84 D4* Norrbotten, N Sweden
Luleälven *84 C3 river* N Sweden
Lulonga *77 C5 river* NW Dem. Rep.
Congo
Lulua *77 D7 river* S Dem. Rep. Congo
Luluabourg *see* Kananga
Lumber State *see* Maine
Lumbo *79 F2* Nampula,
NE Mozambique
Lumsden *151 A7* Southland, South
Island, New Zealand
Lund *85 B7* Skåne, S Sweden
Lüneburg *94 C3* Niedersachsen,
N Germany
Lunga, Isola *see* Dugi Otok
Lungkiang *see* Qiqihar
Lungué-Bungo *78 C2 var.*
Lungwebungu. *river* Angola/Zambia
Lungwebungu *see* Lungué-Bungo
Luninets *see* Luninyets
Luninyets *107 B7 Pol.* Łuniniec, *Rus.*
Luninets. Brestskaya Voblasts',
SW Belarus
Lunteren *86 D4* Gelderland,
C Netherlands
Luong Nam Tha *see* Louangnamtha
Luoyang *128 C4 var.* Honan, Lo-yang.
Henan, C China
Lupatia *see* Altamura
Lúrio *79 F2* Nampula, NE Mozambique
Lúrio, Rio *79 F2 river* NE Mozambique
Lurka *see* Lorca
Lusaka *78 D2 country capital* (Zambia)
Lusaka, SE Zambia
Lushnja *see* Lushnjë
Lushnjë *101 C6 var.* Lushnja. Fier,
C Albania
Luso *see* Luena
Lussin/Lussino *see* Lošinj
Lüt, Bahrat/Lut, Bahret *see* Dead Sea
Lut, Dasht-e *120 D3 var.* Kavīr-e Lūt.
desert E Iran
Lutetia/Lutetia Parisiorum *see* Paris
Lūt, Kavīr-e *see* Lut, Dasht-e
Luton *89 E6* E England, United Kingdom
Łutselk'e *37 F4 prev.* Snowdrift.
Northwest Territories, W Canada
Luts'k *108 C1 Pol.* Łuck, *Rus.* Lutsk.
Volyns'ka Oblast', NW Ukraine
Lutsk *see* Luts'k
Lüttich *see* Liège
Lutzow-Holm Bay *154 C2 var.* Lutzow-
Holm Bay. *bay* Antarctica
Lutzow-Holm Bay *see* Lützow
Holmbukta
Luuq *73 D6 It.* Lugh Ganana. Gedo,
SW Somalia
Luvua *77 D7 river* SE Dem. Rep. Congo
Luwego *73 C8 river* S Tanzania
Luxembourg *87 D8 country capital*
(Luxembourg) Luxembourg,
S Luxembourg
Luxembourg *87 D8 off.* Grand Duchy
of Luxembourg, *Fr.* Luxembourg,
Luxemburg. *country* NW Europe

LUXEMBOURG
Northwest Europe

Population 500,000 / 501 people
per sq mile (193 people per sq km
Total area 998 sq. miles (2586 sq.
Languages Luxembourgish*, Germ
French*
Religions Roman Catholic 97%,
Protestant, Orthodox Christian,
and Jewish 3%
Ethnic mix Luxembourger 62%,
Foreign residents 38%
Government Parliamentary syster
Currency Euro = 100 cents
Literacy rate 99%
Calorie consumption 3568 kilocal

Luxemburg *see* Luxembourg
Luxor *72 B2 Ar.* Al Uqsur. E Egyp
Luza *110 C4* Kirovskaya Oblast',
NW Russian Federation
Luz, Costa de la *92 C5 coastal reg*
SW Spain
Luzern *95 B7 Fr.* Lucerne, *It.* Luce
Luzern, C Switzerland
Luzon *139 E1 island* N Philippine
Luzon Strait *125 E3 strait* Philippi
Taiwan
L'viv *108 B2 Ger.* Lemberg, *Pol.*
Lwów, *Rus.* L'vov. L'vivs'ka Obl
W Ukraine
L'vov *see* L'viv
Lwena *see* Luena
Lwów *see* L'viv
Lyakhavichy *107 B6 Rus.* Lyakho
Brestskaya Voblasts', SW Belaru
Lyakhovichi *see* Lyakhavichy
Lyallpur *see* Faisalābād
Lyangar *see* Langar
Lyck *see* Ełk
Lycksele *84 C4* Västerbotten, N Sv
Lycopolis *see* Asyūţ
Lyel'chytsy *107 C7 Rus.* Lel'chitsy
Homyel'skaya Voblasts', SE Bela
Lyepyel' *107 D5 Rus.* Lepel'.
Vitsyebskaya Voblasts', N Belaru
Lyme Bay *89 C7 bay* S England, U
Kingdom
Lynchburg *41 E5* Virginia, NE US/
Lynn *see* King's Lynn
Lynn Lake *37 F4* Manitoba, C Cana
Lynn Regis *see* King's Lynn
Lyon *91 D5 Eng.* Lyons; *anc.* Lugdi
Rhône, E France
Lyons *see* Lyon
Lyozna *107 E6 Rus.* Liozno.
Vitsyebskaya Voblasts', NE Ukra
Lypovets' *108 D2 Rus.* Lipovets.
Vinnyts'ka Oblast', C Ukraine
Lys *see* Leie
Lysychans'k *109 H3 Rus.* Lisichan
Luhans'ka Oblast', E Ukraine
Lyttelton *151 C6* South Island, Ne
Zealand
Lyubotin *see* Lyubotyn
Lyubotyn *109 G2 Rus.* Lyubotin.
Kharkivs'ka Oblast', E Ukraine
Lyulyakovo *104 E2 prev.* Keremitl
Burgas, E Bulgaria
Lyusina *107 B6 Rus.* Lyusino. Bre
Voblasts', SW Belarus
Lyusino *see* Lyusina

M

Maale *see* Male'
Ma'ān *119 B7* Ma'ān, SW Jordan
Maardu *106 D2 Ger.* Maart. Harju
NW Estonia
Maart *see* Maardu
Maas *see* Meuse
Maaseik *87 D5 prev.* Maeseyck.
Limburg, NE Belgium
Maastricht *87 D6 var.* Maestricht;
Traiectum ad Mosam, Traiectum
Tungrorum. Limburg, SE Netherl
Macao *128 C6 Port.* Macau. Guang
SE China

apá 63 E1 *state capital* Amapá,
Brazil
arsca *see* Makarska
assar *see* Makassar
iu *see* Makó, Hungary
au *see* Macao
Cluer Gulf *see* Berau, Teluk
donnell Ranges 146 D4 *mountain
nge* Northern Territory, C Australia
edonia *see* Macedonia, FYR
edonia, FYR 101 D6 *off.* the Former
igoslav Republic of Macedonia, *var.*
acedonia, Mac. Makedonija, *abbrev.*
/R Macedonia, FYROM. *country*
Europe

ACEDONIA
outheast Europe

ficial name Republic of Macedonia
mation 1991 / 1991
pital Skopje
pulation 2.1 million / 212 people
sq mile (82 people per sq km)
al area 9781 sq. miles (25,333 sq. km)
guages Macedonian*, Albanian*,
kish, Romani, Serbian
gions Orthodox Christian 65%,
slim 29%, Roman Catholic 4%,
ier 2%
nic mix Macedonian 64%,
anian 25%, Turkish 4%, Roma 3%.
o 2%, Other 2%
vernment Presidential /
liamentary system
rrency Macedonian denar = 100 deni
eracy rate 98%
orie consumption 2923 kilocalories

**edonia, the Former Yugoslav
epublic of** *see* Macedonia, FYR
eió 63 G3 *state capital* Alagoas,
Brazil
hachi 60 B1 Pichincha, C Ecuador
hala 60 B2 El Oro, SW Ecuador
hanga 79 E3 Sofala, E Mozambique
hilipatnam 132 D1 *var.* Bandar
lasulipatnam. Andhra Pradesh,
India
hiques 58 C2 Zulia, NW Venezuela
ías Nguema Biyogo *see* Bioco, Isla de
ín 108 D5 Tulcea, SE Romania
kay 148 D4 Queensland,
E Australia
kay, Lake 146 C4 *salt lake* Northern
erritory/Western Australia
kenzie 37 E3 *river* Northwest
erritories, NW Canada
kenzie Bay 154 D3 *bay* Antarctica
kenzie Mountains 36 D3 *mountain
ange* Northwest Territories,
W Canada
leod, Lake 146 A4 *lake* Western
ustralia
omb 40 A4 Illinois, N USA
omer 97 A5 Sardegna, Italy,
Mediterranean Sea
on 91 D5 *anc.* Matisco, Matisco
douroum. Saône-et-Loire, C France
on 42 D2 Georgia, SE USA
on 45 G4 Missouri, C USA
quarie Ridge 154 C5 *undersea ridge*
W Pacific Ocean
uspana 51 G4 Tabasco, SE Mexico
dabá 119 B6 *var.* Mādabā, Madeba;
nc. Medeba. NW Jordan
dabá *see* Ma'dabā
dagascar 79 F3 *off.* Democratic
epublic of Madagascar, *Malg.*
adagasikara; *prev.* Malagasy Republic.
ountry W Indian Ocean

ADAGASCAR
Indian Ocean

ficial name Republic of Madagascar
rmation 1960 / 1960
pital Antananarivo
pulation 22.9 million / 102 people
sq mile (39 people per sq km)
al area 226,656 sq. miles
7,040 sq. km)
guages Malagasy*, French*, English*
igions Traditional beliefs 52%,
ristian (mainly Roman Catholic) 41%,
slim 7%
nic mix Other Malay 46%,
erina 26%, Betsimisaraka 15%,
Other 12%, Other 1%

MADAGASCAR
(continued)

Government Presidential /
parliamentary system
Currency Ariary = 5 iraimbilanja
Literacy rate 64%
Calorie consumption 2092 kilocalories

Madagascar 79 F3 *island* W Indian
Ocean
Madagascar Basin 69 E7 *undersea basin*
W Indian Ocean
Madagascar, Democratic Republic of
see Madagascar
Madagascar Plateau 69 E7 *var.*
Madagascar Ridge, Madagascar
Rise, *Rus.* Madagaskarskiy Khrebet.
undersea plateau W Indian Ocean
Madagascar Rise/Madagascar Ridge *see*
Madagascar Plateau
Madagasikara *see* Madagascar
Madagaskarskiy Khrebet *see*
Madagascar Plateau
Madang 144 B3 Madang, N Papua
New Guinea
Madaniyīn *see* Médenine
Madarska *see* Hungary
Made 86 C4 Noord-Brabant,
S Netherlands
Madeba *see* Ma'dabā
Madeira 70 A2 *var.* Ilha de Madeira.
island Madeira, Portugal, NE Atlantic
Ocean
Madeira, Ilha de *see* Madeira
Madeira Plain 66 C3 *abyssal plain*
E Atlantic Ocean
Madeira, Rio 62 D2 *var.* Río Madera.
river Bolivia/Brazil
Madeleine, Îles de la 39 F4 Eng.
Magdalen Islands. *island group*
Québec, E Canada
Madera 47 B6 California, W USA
Madera, Río *see* Madeira, Rio
Madhya Pradesh 135 E4 *prev.* Central
Provinces and Berar. *cultural region*
C India
Madinat ath Thawrah 118 C2 *var.* Ath
Thawrah. Ar Raqqah, N Syria
Madioen *see* Madiun
Madison 45 F3 South Dakota, N USA
Madison 40 B3 *state capital* Wisconsin,
N USA
Madiun 138 D5 *prev.* Madioen. Jawa,
C Indonesia
Madoera *see* Madura, Pulau
Madona 106 D4 *Ger.* Modohn. E Latvia
Madras *see* Tamil Nādu
Madras *see* Chennai
Madre de Dios, Río 61 E3 *river* Bolivia/
Peru
Madre del Sur, Sierra 51 E5 *mountain
range* S Mexico
Madre, Laguna 51 F3 *lagoon* NE Mexico
Madre, Laguna 49 G5 *lagoon* Texas,
SW USA
Madre Occidental, Sierra 50 C3 *var.*
Western Sierra Madre. *mountain range*
C Mexico
Madre Oriental, Sierra 51 E3 *var.*
Eastern Sierra Madre. *mountain range*
C Mexico
Madre, Sierra 52 B2 *var.* Sierra
de Soconusco. *mountain range*
Guatemala/Mexico
Madrid 92 D3 *country capital* (Spain)
Madrid, C Spain
Madura *see* Madurai
Madurai 132 C3 *prev.* Madura,
Mathurai. Tamil Nādu, S India
Madura, Pulau 138 D5 *prev.* Madoera.
island C Indonesia
Maebashi 131 D5 *var.* Maebasi,
Mayebashi. Gunma, Honshū, S Japan
Maebasi *see* Maebashi
Mae Nam Khong *see* Mekong
Mae Nam Nan 136 C4 *river*
NW Thailand
Mae Nam Yom 136 C4 *river* W Thailand
Maeseyck *see* Maaseik
Maestricht *see* Maastricht
Maéwo 144 D4 *prev.* Aurora. *island*
C Vanuatu
Mafia 73 D7 *island* E Tanzania
Mafraq/Muḩāfaẓat al Mafraq *see* Al
Mafraq
Magadan 115 G3 Magadanskaya Oblast',
E Russian Federation

Magallanes *see* Punta Arenas
Magallanes, Estrecho de *see* Magellan,
Strait of
Magangué 58 B2 Bolívar, N Colombia
Magdalena 61 F3 El Beni, N Bolivia
Magdalena 50 B1 Sonora, NW Mexico
Magdalena, Isla 50 B3 *island*
NW Mexico
Magdalena, Río 58 B2 *river* C Colombia
Magdalen Islands *see* Madeleine,
Îles de la
Magdeburg 94 C4 Sachsen-Anhalt,
C Germany
Magelang 138 C5 Jawa, C Indonesia
Magellan, Strait of 65 B8 *Sp.* Estrecho de
Magallanes. *strait* Argentina/Chile
Magerøy *see* Magerøya
Magerøya 84 D1 *var.* Magerøy, *Lapp.*
Máhkarávju. *island* N Norway
Maggiore, Lago *see* Maggiore, Lake
Maggiore, Lake 96 B1 *It.* Lago Maggiore.
lake Italy/Switzerland
Maglaj 100 C3 Federacija Bosne
I Hercegovine, N Bosnia and
Herzegovina
Maglie 97 E6 Puglia, SE Italy
Magna 44 B4 Utah, W USA
Magnesia *see* Manisa
Magnitogorsk 114 B4 Chelyabinskaya
Oblast', C Russian Federation
Magnolia State *see* Mississippi
Magta' Lahjar 74 C3 *var.* Magta Lahjar,
Magta' Lahjar, Magtá Lahjar. Brakna,
SW Mauritania
Magtymguly 122 C3 *prev.* Garrygala;
Rus. Kara-gala. W Turkmenistan
Magway 136 A3 *var.* Magwe. Magway,
W Myanmar (Burma)
Magyar-Becse *see* Bečej
Magyarkanizsa *see* Kanjiža
Magyarország *see* Hungary
Mahajanga 79 F2 *var.* Majunga.
Mahajanga, NW Madagascar
Mahakam, Sungai 138 D4 *var.* Koetai,
Kutai. *river* Borneo, C Indonesia
Mahalapye 78 D3 *var.* Mahalatswe.
Central, SE Botswana
Mahalatswe *see* Mahalapye
Mahān 120 D3 Kermān, E Iran
Mahānadi 135 F4 *river* E India
Mahārāshtra 134 D5 *cultural region*
W India
Mahbēs *see* El Mahbas
Mahbūbnagar 134 D5 Telangana,
C India
Mahdia 71 F2 *var.* Al Mahdīyah,
Mehdia. NE Tunisia
Mahé 79 H1 *island* Inner Islands,
NE Seychelles
Mahia Peninsula 150 E4 *peninsula*
North Island, New Zealand
Mahilyow 107 D6 *Rus.* Mogilëv.
Mahilyowskaya Voblasts', E Belarus
Máhkarávju *see* Magerøya
Mahmūd-e 'Erāqī *see* Maḩmūd-e Rāqī
Maḩmūd-e Rāqī 123 E4 *var.* Mahmūd-e
'Erāqī. Kāpīsā, NE Afghanistan
Mahón *see* Maó
Mähren *see* Moravia
Mährisch-Weisskirchen *see* Hranice
Maicao 58 C1 La Guajira, N Colombia
Mai Ceu/Mai Chio *see* Maych'ew
Maīdān Shahr 123 E4 *prev.* Meydān
Shahr. Vardak, E Afghanistan
Maidstone 89 E7 SE England, United
Kingdom
Maiduguri 75 H4 Borno, NE Nigeria
Mailand *see* Milano
Maïmana *see* Maïmanah
Maïmanah 122 D3 *var.* Meymaneh,
Maymana. Fāryāb, NW Afghanistan
Maine 41 G2 *off.* State of Maine, *also
known as* Lumber State, Pine Tree
State. *state* NE USA
Maine 90 B3 *cultural region* NW France
Maine, Gulf of 41 H3 *gulf* NE USA
Mainland 88 C2 *island* NE Scotland,
United Kingdom
Mainland 88 D1 *island* NE Scotland,
United Kingdom
Maintirano 79 F3 Mahajanga,
W Madagascar
Mainz 95 B5 *Fr.* Mayence. Rheinland-
Pfalz, SW Germany
Maio 74 A3 *var.* Mayo. *island* Ilhas de
Sotavento, SE Cape Verde
Maisur *see* Mysore, India
Maisur *see* Karnātaka, India

Maizhokunggar 126 C5 Xizang Zizhiqu,
W China
Majorca 93 G3 Eng. Majorca; *anc.*
Baleares Major. *island* Islas Baleares,
Spain, W Mediterranean Sea
Majorca *see* Mallorca
Mâjro *see* Majuro Atoll
Majunga *see* Mahajanga
Majuro Atoll 144 D2 *var.* Mâjro.
atoll Ratak Chain, SE Marshall Islands
Makale *see* Mek'elê
Makarov Basin 155 B3 *undersea basin*
Arctic Ocean
Makarska 100 B4 *It.* Macarsca.
Split-Dalmacija, SE Croatia
Makasar *see* Makassar
Makasar, Selat *see* Makassar Straits
Makassar 139 E4 *var.* Macassar,
Makasar; *prev.* Ujungpandang.
Sulawesi, C Indonesia
Makassar Straits 138 D4
Ind. Makasar Selat. *strait* C Indonesia
Makay 79 F3 *var.* Massif du Makay.
mountain range SW Madagascar
Makedonija *see* Macedonia, FYR
Makeni 74 C4 C Sierra Leone
Makeyevka *see* Makiyivka
Makhachkala 114 A4
prev. Petrovsk-Port. Respublika
Dagestan, SW Russian Federation
Makin 144 D2 *prev.* Pitt Island. *atoll*
Tungaru, W Kiribati
Makira *see* San Cristobal
Makiyivka 109 G3 *Rus.* Makeyevka;
prev. Dmitriyevsk. Donets'ka Oblast',
E Ukraine
Makkah 121 A5 Eng. Mecca. Makkah,
W Saudi Arabia
Makkovik 39 F2 Newfoundland and
Labrador, NE Canada
Makó 99 D7 *Rom.* Macău. Csongrád,
SE Hungary
Makoua 77 B5 Cuvette, C Congo
Makran Coast 120 E4 *coastal region*
SE Iran
Makrany 107 A6 *Rus.* Mokrany.
Brestskaya Voblasts', SW Belarus
Mākū 120 B2 Āzarbāyjān-e Gharbī,
NW Iran
Makurdi 75 G4 Benue, C Nigeria
Mala *see* Malaita, Solomon Islands
Malabār Coast 132 B3 *coast* SW India
Malabo 77 A5 *prev.* Santa Isabel.
country capital (Equatorial Guinea)
Isla de Bioco, NW Equatorial
Guinea
Malaca *see* Málaga
Malacca, Strait of 138 B3 *bay.* Selat
Malaka. *strait* Indonesia/Malaysia
Malacka *see* Malacky
Malacky 99 C6 *Hung.* Malacka.
Bratislavský Kraj, W Slovakia
Maladzyechna 107 C5 *Pol.* Molodeczno,
Rus. Molodechno. Minskaya Voblasts',
C Belarus
Málaga 92 D5 *anc.* Malaca. Andalucía,
S Spain
Malagarasi River 73 B7 *river*
W Tanzania Africa
Malagasy Republic *see* Madagascar
Malaita 144 C3 *var.* Mala. *island*
N Solomon Islands
Malakal 73 B5 Upper Nile, NE South
Sudan
Malakula *see* Malekula
Malang 138 D5 Jawa, C Indonesia
Malange *see* Malanje
Malanje 78 B1 *var.* Malange.
NW Angola
Mälaren 85 C6 *lake* C Sweden
Malatya 117 E4 *anc.* Melitene.
C Turkey
Malawi 79 E1 *off.* Republic of Malawi;
prev. Nyasaland, Nyasaland
Protectorate. *country* S Africa

MALAWI
Southern Africa

Official name Republic of Malawi
Formation 1964 / 1964
Capital Lilongwe
Population 16.4 million / 451 people per
sq mile (174 people per sq km)
Total area 45,745 sq. miles (118,480 sq. km)

MALAWI
(continued)

Languages Chewa, Lomwe, Yao, Ngoni, English*
Religions Protestant 55%, Roman Catholic 20%, Muslim 20%, Traditional beliefs 5%
Ethnic mix Bantu 99%, Other 1%
Government Presidential system
Currency Malawi kwacha = 100 tambala
Literacy rate 61%
Calorie consumption 2334 kilocalories

Malawi, Lake *see* Nyasa, Lake
Malawi, Republic of *see* Malawi
Malaya Viska *see* Mala Vyska
Malay Peninsula *124 D4 peninsula* Malaysia/Thailand
Malaysia *138 B3 off.* Malaysia, *var.* Federation of Malaysia; *prev.* the separate territories of Federation of Malaya, Sarawak and Sabah (North Borneo) and Singapore. *country* SE Asia

MALAYSIA
Southeast Asia

Official name Federation of Malaysia
Formation 1963 / 1965
Capital Kuala Lumpur; Putrajaya (administrative)
Population 29.7 million / 234 people per sq mile (90 people per sq km)
Total area 127,316 sq. miles (329,750 sq. km)
Languages Bahasa Malaysia*, Malay, Chinese, Tamil, English
Religions Muslim (mainly Sunni) 61%, Buddhist 19%, Christian 9%, Hindu 6%, Other 5%
Ethnic mix Malay 53%, Chinese 26%, Indigenous tribes 12%, Indian 8%, Other 1%
Government Parliamentary system
Currency Ringgit = 100 sen
Literacy rate 93%
Calorie consumption 2855 kilocalories

Malaysia, Federation of *see* Malaysia
Malbork *98 C2 Ger.* Marienburg, Marienburg in Westpreussen. Pomorskie, N Poland
Malchin *94 C3* Mecklenburg-Vorpommern, N Germany
Malden *45 H5* Missouri, C USA
Malden Island *145 G3 prev.* Independence Island. *atoll* E Kiribati
Maldives *132 A4 off.* Maldivian Dhivehi, Republic of Maldives. *country* N Indian Ocean

MALDIVES
Indian Ocean

Official name Republic of Maldives
Formation 1965 / 1965
Capital Male'
Population 394,491 / 3400 people per sq mile (1315 people per sq km)
Total area 116 sq. miles (300 sq. km)
Languages Dhivehi* (Maldivian), Sinhala, Tamil, Arabic
Religions Sunni Muslim 100%
Ethnic mix Arab–Sinhalese–Malay 100%
Government Presidential system
Currency Rufiyaa = 100 laari
Literacy rate 98%
Calorie consumption 2722 kilocalories

Maldives, Republic of *see* Maldives
Maldivian Dhivehi *see* Maldives
Male' *132 B4 Div.* Maale. *country capital* (Maldives) Male' Atoll, C Maldives
Male' Atoll *132 B4 var.* Kaafu Atoll. *atoll* C Maldives
Malekula *144 D4 var.* Malakula; *prev.* Mallicolo. *island* W Vanuatu
Malesina *105 C5* Stereá Elláda, E Greece
Malheur Lake *46 C3 lake* Oregon, NW USA
Mali *75 E3 off.* Republic of Mali, *Fr.* République du Mali; *prev.* French Sudan, Sudanese Republic. *country* W Africa

MALI
West Africa

Official name Republic of Mali
Formation 1960 / 1960
Capital Bamako
Population 15.3 million / 32 people per sq mile (13 people per sq km)
Total area 478,764 sq. miles (1,240,000 sq. km)
Languages Bambara, Fulani, Senufo, Soninke, French*
Religions Muslim (mainly Sunni) 90%, Traditional beliefs 6%, Christian 4%
Ethnic mix Bambara 52%, Other 14%, Fulani 11%, Saracolé 7%, Soninka 7%, Tuareg 5%, Mianka 4%
Government Transitional regime
Currency CFA franc = 100 centimes
Literacy rate 34%
Calorie consumption 2833 kilocalories

Malik, Wadi al *see* Milk, Wadi el
Mali Kyun *137 B5 var.* Tavoy Island. *island* Myeik Archipelago, S Myanmar (Burma)
Malin *see* Malyn
Malindi *73 D7* Coast, SE Kenya
Malines *see* Mechelen
Mali, Republic of *see* Mali
Mali, République du *see* Mali
Malkiye *see* Al Mālikiyah
Malko Tŭrnovo *104 E3* Burgas, E Bulgaria
Mallaig *88 B3* N Scotland, United Kingdom
Mallawi *72 B2 var.* Mallawi. C Egypt
Mallawi *see* Mallawī
Mallicolo *see* Malekula
Malmberget *84 C3 Lapp.* Malmivaara. Norrbotten, N Sweden
Malmédy *87 D6* Liège, E Belgium
Malmivaara *see* Malmberget
Malmö *85 B7* Skåne, S Sweden
Maloelap Atoll *144 D1 var.* Maloeļap. *atoll* E Marshall Islands
Małopolska, Wyżyna *98 D4 plateau* S Poland
Malozemel'skaya Tundra *110 D3 physical region* NW Russian Federation
Malta *106 D4* SE Latvia
Malta *44 C1* Montana, NW USA
Malta *97 C8 off.* Republic of Malta. *country* C Mediterranean Sea

MALTA
Southern Europe

Official name Republic of Malta
Formation 1964 / 1964
Capital Valletta
Population 400,000 / 3226 people per sq mile (1250 people per sq km)
Total area 122 sq. miles (316 sq. km)
Languages Maltese*, English*
Religions Roman Catholic 98%, Other and nonreligious 2%
Ethnic mix Maltese 96%, Other 4%
Government Parliamentary system
Currency Euro = 100 cents
Literacy rate 92%
Calorie consumption 3389 kilocalories

Malta *97 C8 island* Malta, C Mediterranean Sea
Malta, Canale di *see* Malta Channel
Malta Channel *97 C8 It.* Canale di Malta. *strait* Italy/Malta
Malta, Republic of *see* Malta
Maluku, Laut *see* Molucca Sea
Malung *85 B6* Dalarna, C Sweden
Malventum *see* Benevento
Malvina, Isla Gran *see* West Falkland
Malvinas, Islas *see* Falkland Islands
Malyn *108 D2 Rus.* Malin. Zhytomyrs'ka Oblast', N Ukraine
Malyy Kavkaz *see* Lesser Caucasus
Mamberamo, Sungai *139 H4 river* Papua, E Indonesia
Mambij *see* Manbij
Mamonovo *106 A4 Ger.* Heiligenbeil. Kaliningradskaya Oblast', W Russian Federation
Mamoré, Río *61 F3 river* Bolivia/Brazil
Mamou *74 C4* W Guinea
Mamoudzou *79 F2 dependent territory capital* (Mayotte) C Mayotte

Mamuno *78 C3* Ghanzi, W Botswana
Manacor *93 G3* Mallorca, Spain, W Mediterranean Sea
Manado *139 F3 prev.* Menado. Sulawesi, C Indonesia
Managua *52 D3 country capital* (Nicaragua) Managua, W Nicaragua
Managua, Lake *52 C3 var.* Xolotlán. *lake* W Nicaragua
Manakara *79 G4* Fianarantsoa, SE Madagascar
Manama *see* Al Manāmah
Mananjary *79 G3* Fianarantsoa, SE Madagascar
Manáos *see* Manaus
Manapouri, Lake *151 A7 lake* South Island, New Zealand
Manar *see* Mannar
Manas, Gora *123 E2 mountain* Kyrgyzstan/Uzbekistan
Manaus *62 D2 prev.* Manáos. *state capital* Amazonas, NW Brazil
Manavgat *116 B4* Antalya, SW Turkey
Manbij *118 C2 var.* Mambij, *Fr.* Membidj. Ḥalab, N Syria
Manchester *89 D5 Lat.* Mancunium. NW England, United Kingdom
Manchester *41 G3* New Hampshire, NE USA
Man-chou-li *see* Manzhouli
Manchurian Plain *125 E1 plain* NE China
Máncio Lima *see* Japim
Mancunium *see* Manchester
Mand *see* Mand, Rūd-e
Mandalay *136 B3* Mandalay, C Myanmar (Burma)
Mandan *45 E2* North Dakota, N USA
Mandeville *54 B5* C Jamaica
Mándra *105 C6* Attikí, C Greece
Mand, Rūd-e *120 D4 var.* Mand. *river* S Iran
Mandurah *147 A6* Western Australia
Manduria *97 E5* Puglia, SE Italy
Mandya *132 C2* Karnātaka, C India
Manfredonia *97 D5* Puglia, SE Italy
Mangai *77 C6* Bandundu, W Dem. Rep. Congo
Mangaia *145 G5 island group* S Cook Islands
Mangalia *108 D5 anc.* Callatis. Constanța, SE Romania
Mangalmé *76 C3* Guéra, SE Chad
Mangalore *132 B2* Karnātaka, W India
Manganui *see* Bloemfontein
Mango *see* Sansanné-Mango, Togo
Mangoky *79 F3 river* W Madagascar
Manhattan *45 F4* Kansas, C USA
Manicouagan, Réservoir *38 D3 lake* Québec, E Canada
Manihiki *145 F3 atoll* N Cook Islands
Manihiki Plateau *143 E3 undersea plateau* C Pacific Ocean
Maniitsoq *82 C3 var.* Manîtsoq, *Dan.* Sukkertoppen. Kitaa, S Greenland
Manila *139 E1 off.* City of Manila. *country capital* (Philippines) Luzon, N Philippines
Manila, City of *see* Manila
Manisa *116 A3 var.* Manissa, *prev.* Saruhan; *anc.* Magnesia. Manisa, W Turkey
Manissa *see* Manisa
Manitoba *37 F5 province* S Canada
Manitoba, Lake *37 F5 lake* Manitoba, S Canada
Manitoulin Island *38 C4 island* Ontario, S Canada
Manitsoq *see* Maniitsoq
Manizales *58 B3* Caldas, W Colombia
Manjimup *147 A7* Western Australia
Mankato *45 F3* Minnesota, N USA
Manlleu *93 G2* Cataluña, NE Spain
Manly *148 E1* Iowa, C USA
Mannar *132 C3 var.* Manar. Northern Province, NW Sri Lanka
Mannar, Gulf of *132 C3 gulf* India/Sri Lanka
Mannheim *95 B5* Baden-Württemberg, SW Germany
Manokwari *139 G4* New Guinea, E Indonesia
Manono *77 E7* Shaba, SE Dem. Rep. Congo
Manosque *91 D6* Alpes-de-Haute-Provence, SE France
Manra *145 F3 prev.* Sydney Island. *atoll* Phoenix Islands, C Kiribati

Mansa *78 D2 prev.* Fort Rosebery. Luapula, N Zambia
Mansel Island *37 G3 island* Nunav NE Canada
Mansfield *40 D4* Ohio, N USA
Manta *60 A2* Manabí, W Ecuador
Manteca *47 B6* California, W USA
Mantoue *see* Mantova
Mantova *96 B2 Eng.* Mantua, *Fr.* Mantoue. Lombardia, NW Italy
Mantua *see* Mantova
Manuae *145 G4 island* S Cook Islan C Cook Islands
Manukau *see* Manurewa
Manurewa *150 D3 var.* Manukau. Auckland, North Island, New Zeal
Manzanares *93 E3* Castilla-La Man C Spain
Manzanillo *54 C3* Granma, E Cuba
Manzanillo *50 D4* Colima, SW Mex
Manzhouli *127 F1 var.* Man-chou-Mongol Zizhiqu, N China
Mao *76 B3* Kanem, W Chad
Maó *93 H3 Cast.* Mahón, *Eng.* Port Mahon; *anc.* Portus Magonis. Me' Spain, W Mediterranean Sea
Maoke, Pegunungan *139 H4 Dut.* Sneeuw-gebergte, *Eng.* Snow Mountains. *mountain range* Papu E Indonesia
Maoming *128 C6* Guangdong, S Ch
Mapmaker Seamounts *125 H2 seamount range* N Pacific Ocean
Maputo *78 D4 prev.* Lourenço Mar *country capital* (Mozambique) Mar S Mozambique
Marabá *63 F2* Pará, NE Brazil
Maracaibo *58 C1* Zulia, NW Venez
Maracaibo, Gulf of *see* Venezuela, Golfo de
Maracaibo, Lake *58 C2 var.* Lake Maracaibo. *inlet* NW Venezuela
Maracaibo, Lake *see* Maracaibo, Lag
Maracay *58 D2* Aragua, N Venezue
Marada *see* Marādah
Marādah *71 G3 var.* Marada. N Lib
Maradi *75 G3* Maradi, S Niger
Maragha *see* Marāgheh
Marāgheh *120 C2 var.* Maragha. Āzarbāyjān-e Khāvarī, NW Iran
Marajó, Baía de *63 F1 bay* N Brazil
Marajó, Ilha de *63 E1 island* N Braz
Marakesh *see* Marrakech
Maramba *see* Livingstone
Maramba *see* Livingstone
Maranhão *63 F2 off.* Estado do Maranhão. *state* E Brazil
Maranhão *63 F2 off.* Estado do Maranhão. *region* E Brazil
Maranhão, Estado do *see* Maranhã
Marañón, Río *60 B2 river* N Peru
Marathon *38 C4* Ontario, S Canada
Marathónas *105 C5 prev.* Marathó Attikí, C Greece
Märäzä *see* Qobustan
Marbella *92 D5* Andalucía, S Spain
Marble Bar *146 B4* Western Austra
Marburg *see* Maribor, Slovenia
Marburg an der Lahn *94 B4 hist.* Marburg. Hessen, W Germany
March *see* Morava
Marche *96 C3 Eng.* Marches. *regio* C Italy
Marche *91 C5 cultural region* C Fran
Marche-en-Famenne *87 C7* Luxembourg, SE Belgium
Marches, Isla *60 B5 var.* Bindloe Island. *island* Galápagos Islands, Ecuador, E Pacific Ocean
Marches *see* Marche
Mar Chiquita, Laguna *64 C3 lake* C Argentina
Marcounda *see* Markounda
Mardān *134 C1* Khyber Pakhtunkh￿ N Pakistan
Mar del Plata *65 D5* Buenos Aires, E Argentina
Mardin *117 E4* Mardin, SE Turkey
Maré *144 D5 island* Iles Loyauté, E I Pacific Ocean
Marea Neagrā *see* Black Sea
Mareeba *148 D3* Queensland, NE Australia
Marek *see* Dupnitsa
Marganets *see* Marhanets'
Margarita, Isla de *59 E1 island* N Venezuela

ate 89 E7 *prev.* Mergate.
England, United Kingdom
herita *see* Jamaame
herita, Lake 73 C5 *Eng.*
e Margherita, *It.* Abbaia. *lake*
Ethiopia
herita, Lake *see* Ābaya Hāyk'
hita 108 B3 *Hung.* Margitta. Bihor,
/ Romania
itta *see* Marghita
anets' 109 F3 *Rus.* Marganets.
ipropetrovs'ka Oblast', E Ukraine
Cleofas, Isla 50 C4 *island*
Mexico
Island 149 C8 *island* Tasmania,
Australia
Madre, Isla 50 C4 *island*
Mexico
Magdalena, Isla 50 C4 *island*
Mexico
ana Trench 125 G4 *var.* Challenger
ep. *trench* W Pacific Ocean
ánské Lázně 99 A5 *Ger.* Marienbad.
lovarský Kraj, W Czech Republic
as, Islas 50 C4 *island group*
Mexico
a-Theresiopel *see* Subotica
oor 95 E7 *Ger.* Marburg.
Slovenia
za *see* Maritsa
li 73 B5 Western Equatoria,
South Sudan
e Byrd Land 154 A3 *physical region*
tarctica
e-Galante 55 G4 *var.* Ceyre to the
ribs. *island* SE Guadeloupe
enbad *see* Mariánské Lázně
enburg *see* Alūksne, Latvia
enburg *see* Malbork, Poland
enburg in Westpreussen *see*
lbork
enhausen *see* Viļaka
ental 78 B4 Hardap, SW Namibia
enwerder *see* Kwidzyń
estad 85 B6 Västra Götaland,
weden
etta 42 D2 Georgia, SE USA
ampolė 106 B4 *prev.* Kapsukas.
arijampolė, S Lithuania
ia 63 E4 São Paulo, S Brazil
a 92 B1 Galicia, NW Spain
ina Gorka *see* Mar''ina Horka
ina Horka 107 C6 *Rus.* Mar'ina
rka. Minskaya Voblasts', C Belarus
a 63 E4 Paraná, S Brazil
on 45 G3 Iowa, C USA
on 40 D4 Ohio, N USA
on, Lake 43 E2 *reservoir* South
rolina, S USA
scal Estigarribia 64 D2 Boquerón,
Paraguay
sa 104 D3 *var.* Marica, *Gk.* Évros,
rk. Meriç; *anc.* Hebrus. *river*
Europe
zburg *see* Pietermaritzburg
apol' 109 G4 *prev.* Zhdanov.
nets'ka Oblast', SE Ukraine
a 73 D6 *var.* Merca. Shabeellaha
ose, S Somalia
ham, Mount 154 B4 *mountain*
tarctica
ounda 76 C4 *var.* Marcounda.
ham, NW Central African Republic
tredwitz 95 C5 Bayern,
ermany
orough 148 D4 Queensland,
Australia
anda *see* Marmande
ande 91 B5 *anc.* Marmanda.
-et-Garonne, SW France
f Marmara 116 A2 *Eng.* Sea of
rmara. *sea* NW Turkey
ara, Sea of *see* Marmara Denizi
aris 116 A4 Muğla, SW Turkey
e 90 C3 *cultural region* N France
e 90 D3 *river* N France
76 C4 Moyen-Chari, S Chad
antsetra 79 G2 Toamasina,
Madagascar
mokotro 79 G2 *mountain*
Madagascar
ni 59 G3 *Dut.* Marowijne. *river*
nch Guiana/Suriname
sheviz *see* Topliţa
sludas *see* Ludus
súsárhely *see* Târgu Mureş
tiri 143 F4 *var.* Îlots de Bass,
rotiri. *island group* Îles Australes,
French Polynesia

Maroua 76 B3 Extrême-Nord,
N Cameroon
Marowijne *see* Maroni
Marquesas Fracture Zone 153 E3
fracture zone E Pacific Ocean
Marquette 40 B1 Michigan, N USA
Marrakech 70 C2 *var.* Marakesh, *Eng.*
Marrakesh; *prev.* Morocco. W Morocco
Marrakesh *see* Marrakech
Marrawah 149 C8 Tasmania,
SE Australia
Marree 149 B5 South Australia
Marsá al Burayqah 71 G3 *var.* Al
Burayqah. N Libya
Marsabit 73 C6 Eastern, N Kenya
Marsala 97 B7 *anc.* Lilybaeum. Sicilia,
Italy, C Mediterranean Sea
Marsberg 94 B4 Nordrhein-Westfalen,
W Germany
Marseille 91 D6 *Eng.* Marseilles;
anc. Massilia. Bouches-du-Rhône,
SE France
Marseilles *see* Marseille
Marshall 45 F2 Minnesota, N USA
Marshall 49 H2 Texas, SW USA
Marshall Islands 144 C1
off. Republic of the Marshall Islands.
country W Pacific Ocean

MARSHALL ISLANDS
Australasia & Oceania

Official name Republic of
the Marshall Islands
Formation 1986 / 1986
Capital Majuro
Population 67,000 / 996 people
per sq mile (385 people per sq km)
Total area 70 sq. miles (181 sq. km)
Languages Marshallese*, English*,
Japanese, German
Religions Protestant 90%,
Roman Catholic 8%, Other 2%
Ethnic mix Micronesian 90%, Other 10%
Government Presidential system
Currency US dollar = 100 cents
Literacy rate 91%
Calorie consumption Not available

Marshall Islands, Republic of the *see*
Marshall Islands
Marshall Seamounts 125 H3 *seamount
range* SW Pacific Ocean
Marsh Harbour 54 C1 Great Abaco,
W Bahamas
Martaban *see* Mottama
Martha's Vineyard 41 G3 *island*
Massachusetts, NE USA
Martigues 91 D6 Bouches-du-Rhône,
SE France
Martin 99 C5 *Ger.* Sankt Martin, *Hung.*
Turócszentmárton; *prev.* Turčiansky
Svätý Martin. Žilinský Kraj,
N Slovakia
Martinique 55 G4 French overseas
department E West Indies
Martinique Channel *see* Martinique
Passage
Martinique Passage 55 G4 *var.*
Dominica Channel, Martinique
Channel. *channel* Dominica/
Martinique
Marton 150 D4 Manawatu-Wanganui,
North Island, New Zealand
Martos 92 D4 Andalucía, S Spain
Marungu 77 E7 *mountain range* SE Dem.
Rep. Congo
Mary 122 D3 *prev.* Merv. Mary
Welaýaty, S Turkmenistan
Maryborough 149 D4 Queensland,
E Australia
Maryborough *see* Port Laoise
Mary Island *see* Kanton
Maryland 41 E5 *off.* State of Maryland,
also known as America in Miniature,
Cockade State, Free State, Old Line
State. *state* NE USA
Maryland *see* State of Maryland
Maryville 45 F4 Missouri, C USA
Maryville 42 D1 Tennessee, S USA
Masai Steppe 73 C7 *grassland*
NW Tanzania
Masaka 73 B6 SW Uganda
Masallı 117 H3 *Rus.* Masally.
S Azerbaijan
Masally *see* Masallı
Masasi 73 C6 Mtwara, SE Tanzania
Masawa/Massawa *see* Mits'iwa
Masaya 52 D3 Masaya, W Nicaragua

Mascarene Basin 141 B5 *undersea basin*
W Indian Ocean
Mascarene Islands 79 H4 *island group*
W Indian Ocean
Mascarene Plain 141 B5 *abyssal plain*
W Indian Ocean
Mascarene Plateau 141 B5 *undersea
plateau* W Indian Ocean
Maseru 78 D4 *country capital* (Lesotho)
W Lesotho
Mas-ḩa 119 D7 W West Bank Asia
Mashhad 120 E2 *var.* Meshed.
Khorāsān-Razavī, NE Iran
Masindi 73 B6 W Uganda
Maşīra *see* Maşīrah, Jazīrat
Masira, Gulf of 121 E5 *var.* Gulf of
Maşīra. *bay* E Oman
Masira, Gulf of *see* Maşīrah, Khalīj
Maşīrah, Jazīrat 121 E5 *var.* Maşīra.
island E Oman
Masis *see* Büyükağrı Dağı
Maskat *see* Masqaṭ
Mason City 45 F3 Iowa, C USA
Masqaṭ 121 E5 *var.* Maskat, *Eng.* Muscat.
country capital (Oman) NE Oman
Massa 96 B3 Toscana, C Italy
Massachusetts 41 G3 *off.*
Commonwealth of Massachusetts, *also
known as* Bay State, Old Bay State, Old
Colony State. *state* NE USA
Massenya 76 B3 Chari-Baguirmi,
SW Chad
Massif Central 91 C5 *plateau*
C France
Massilia *see* Marseille
Massoukou *see* Franceville
Mastanli *see* Momchilgrad
Masterton 151 D5 Wellington, North
Island, New Zealand
Masty 107 B5 *Rus.* Mosty.
Hrodzyenskaya Voblasts', W Belarus
Masuda 131 B6 Shimane, Honshū,
SW Japan
Masuku *see* Franceville
Masvingo 78 D3 *prev.* Fort Victoria,
Nyanda, Victoria. Masvingo,
SE Zimbabwe
Maşyāf 118 B3 *Fr.* Misiaf. Ḩamāh,
C Syria
Matadi 77 B6 Bas-Congo, W Dem.
Rep. Congo
Matagalpa 52 D3 Matagalpa,
C Nicaragua
Matale 132 D3 Central Province,
C Sri Lanka
Matam 74 C3 NE Senegal
Matamata 150 D3 Waikato, North
Island, New Zealand
Matamoros 50 D3 Coahuila, NE Mexico
Matamoros 51 E2 Tamaulipas, C Mexico
Matane 39 E4 Québec, SE Canada
Matanzas 54 B2 Matanzas, NW Cuba
Matara 132 D4 Southern Province,
S Sri Lanka
Mataram 138 D5 Pulau Lombok,
C Indonesia
Mataró 93 G2 *anc.* Illuro. Cataluña,
E Spain
Mataura 151 B7 Southland, South Island,
New Zealand
Mataura 151 B7 *river* South Island,
New Zealand
Mata'utu *see* Matā'utu
Matā'utu 145 E4 *var.* Mata Uta.
dependent territory capital (Wallis and
Futuna) Île Uvea, Wallis and Futuna
Matera 97 E5 Basilicata, S Italy
Mathura *see* Madurai
Matianus *see* Orūmīyeh, Daryācheh-ye
Matías Romero 51 F5 Oaxaca,
SE Mexico
Matisco/Matisco Ædourum *see* Mâcon
Mato Grosso 63 E3 *off.* Estado de Mato
Grosso; *prev.* Matto Grosso. *state*
W Brazil
Mato Grosso 63 E3 *off.* Estado de Mato
Grosso; *prev.* Matto Grosso. *region*
W Brazil
Mato Grosso do Sul 63 E4 *off.* Estado de
Mato Grosso do Sul. *state* S Brazil
Mato Grosso do Sul 63 E4 *off.* Estado de
Mato Grosso do Sul. *region* S Brazil
Mato Grosso do Sul, Estado de *see* Mato
Grosso do Sul
Mato Grosso, Estado de *see* Mato Grosso
Mato Grosso, Planalto de 56 C4 *plateau*
C Brazil
Matosinhos 92 B2 *prev.* Matozinhos.
Porto, NW Portugal

Matozinhos *see* Matosinhos
Matsue 131 B6 *var.* Matsuye, Matue.
Shimane, Honshū, SW Japan
Matsumoto 131 C5 *var.* Matumoto.
Nagano, Honshū, S Japan
Matsuyama 131 B7 *var.* Matuyama.
Ehime, Shikoku, SW Japan
Matsuye *see* Matsue
Matterhorn 95 A8 *It.* Monte Cervino.
mountain Italy/Switzerland
Matthews Ridge 59 F2 N Guyana
Matthew Town 54 D2 Great Inagua,
S Bahamas
Matto Grosso *see* Mato Grosso
Matucana 60 C4 Lima, W Peru
Matue *see* Matsue
Matumoto *see* Matsumoto
Maturín 59 E2 Monagas, NE Venezuela
Matuyama *see* Matsuyama
Mau 135 E3 *var.* Maunāth Bhanjan.
Uttar Pradesh, N India
Maui 47 B8 *island* Hawaii, USA,
C Pacific Ocean
Maun 78 C3 North-West, C Botswana
Maunāth Bhanjan *see* Mau
Mauren 94 E1 NE Liechtenstein Europe
Maurice *see* Mauritius
Mauritania 74 C2 *off.* Islamic Republic of
Mauritania, *Ar.* Mūrītāniyah. *country*
W Africa

MAURITANIA
West Africa

Official name Islamic Republic
of Mauritania
Formation 1960 / 1960
Capital Nouakchott
Population 3.9 million / 10 people
per sq mile (4 people per sq km)
Total area 397,953 sq. miles
(1,030,700 sq. km)
Languages Arabic*, Hassaniyah Arabic,
Wolof, French
Religions Sunni Muslim 100%
Ethnic mix Maure 81%, Wolof 7%,
Tukolor 5%, Other 4%, Soninka 3%
Government Presidential system
Currency Ouguiya = 5 khoums
Literacy rate 44%
Calorie consumption 2791 kilocalories

Mauritania, Islamic Republic of *see*
Mauritania
Mauritius 79 H3 *off.* Republic of
Mauritius, *Fr.* Maurice. *country*
W Indian Ocean

MAURITIUS
Indian Ocean

Official name Republic of Mauritius
Formation 1968 / 1968
Capital Port Louis
Population 1.2 million / 1671 people
per sq mile (645 people per sq km)
Total area 718 sq. miles (1860 sq. km)
Languages French Creole, Hindi, Urdu,
Tamil, Chinese, English*, French
Religions Hindu 48%, Roman Catholic
24%, Muslim 17%, Protestant 9%,
Other 2%
Ethnic mix Indo-Mauritian 68%,
Creole 27%, Sino-Mauritian 3%,
Franco-Mauritian 2%
Government Parliamentary system
Currency Mauritian rupee = 100 cents
Literacy rate 89%
Calorie consumption 3055 kilocalories

Mauritius 141 B5 *island*
W Indian Ocean
Mauritius, Republic of *see* Mauritius
Mawlamyaing *see* Mawlamyine
Mawlamyine 136 B4 *var.* Mawlamyaing,
Moulmein. Mon State, S Myanmar
(Burma)
Mawson 154 D2 Australian research
station Antarctica
Mayadin *see* Al Mayādīn
Mayaguana 54 D2 *island* SE Bahamas
Mayaguana Passage 54 D2 *passage*
SE Bahamas
Mayagüez 55 F3 W Puerto Rico
Mayamey 120 D2 Semnān, N Iran
Maya Mountains 52 B2 Sp. Montañas
Mayas. *mountain range* Belize/
Guatemala
Mayas, Montañas *see* Maya Mountains

Maych'ew 72 C4 var. Mai Chio, It. Mai Ceu. Tigray, N Ethiopia
Mayebashi see Maebashi
Mayence see Mainz
Mayfield 151 B6 Canterbury, South Island, New Zealand
Maykop 111 A7 Respublika Adygeya, SW Russian Federation
Maymana see Maïmanah
Maymyo see Pyin-Oo-Lwin
Mayo see Maio
Mayor Island 150 D3 island NE New Zealand
Mayor Pablo Lagerenza see Capitán Pablo Lagerenza
Mayotte 79 F2 French territorial collectivity E Africa
May Pen 54 B5 C Jamaica
Mayyit, Al Baḥr al see Dead Sea
Mazabuka 78 D2 Southern, S Zambia
Mazaca see Kayseri
Mazagan see El-Jadida
Mazār-e Sharīf 123 E3 var. Mazār-i Sharif. Balkh, N Afghanistan
Mazār-i Sharif see Mazār-e Sharīf
Mazatlán 50 C3 Sinaloa, C Mexico
Mažeikiai 106 B3 Telšiai, NW Lithuania
Mazirbe 106 C2 NW Latvia
Mazra'a see Al Mazra'ah
Mazury 98 D3 physical region NE Poland
Mazyr 107 C7 Rus. Mozyr'. Homyel'skaya Voblasts', SE Belarus
Mbabane 78 D4 country capital (Swaziland) NW Swaziland
Mbacké see Mbaké
Mbaïki 77 C5 var. M'Baïki. Lobaye, SW Central African Republic
M'Baïki see Mbaïki
Mbaké 74 B3 var. Mbacké. W Senegal
Mbala 78 D1 prev. Abercorn. Northern, NE Zambia
Mbale 73 C6 E Uganda
Mbandaka 77 C5 prev. Coquilhatville. Equateur, NW Dem. Rep. Congo
M'Banza Congo 78 B1 var. Mbanza Congo; prev. São Salvador, São Salvador do Congo. Dem. Rep. Congo, NW Angola
Mbanza-Ngungu 77 B6 Bas-Congo, W Dem. Rep. Congo
Mbarara 73 B6 SW Uganda
Mbé 76 B4 Nord, N Cameroon
Mbeya 77 C7 Mbeya, SW Tanzania
Mbomou/M'Bomu/Mbomu see Bomu
Mbour 74 B3 W Senegal
Mbuji-Mayi 77 D7 prev. Bakwanga. Kasai-Oriental, S Dem. Rep. Congo
McAlester 49 G2 Oklahoma, C USA
McAllen 49 G5 Texas, SW USA
McCamey 49 E3 Texas, SW USA
McComb 42 B3 Mississippi, S USA
McCook 45 E4 Nebraska, C USA
McKean Island 145 E3 island Phoenix Islands, C Kiribati
Mount McKinley 36 C3 var. Denali. mountain Alaska, USA
McKinley Park 36 C3 Alaska, USA
M'Clintock Channel 37 F2 channel Nunavut, N Canada
McMinnville 46 B3 Oregon, NW USA
McMurdo 154 B4 US research station Antarctica
McPherson 45 E5 Kansas, C USA
McPherson see Fort McPherson
Mdantsane 78 D5 Eastern Cape, SE South Africa
Mead, Lake 47 D6 reservoir Arizona/ Nevada, W USA
Mecca see Makkah
Mechelen 87 C5 Eng. Mechlin, Fr. Malines. Antwerpen, C Belgium
Mechlin see Mechelen
Mecklenburger Bucht 94 C2 bay N Germany
Mecsek 99 C7 mountain range SW Hungary
Medan 138 B3 Sumatera, E Indonesia
Medeba see Ma'dabā
Medellín 58 B3 Antioquia, NW Colombia
Médenine 71 F2 var. Madanīyīn. SE Tunisia
Medeshamstede see Peterborough
Medford 46 B4 Oregon, NW USA
Medgidia 108 D5 Constanța, SE Romania
Medgyes see Mediaș
Mediaș 108 B4 Ger. Mediasch, Hung. Medgyes. Sibiu, C Romania
Mediasch see Mediaș

Medicine Hat 37 F5 Alberta, SW Canada
Medina see Al Madīnah
Medinaceli 93 E2 Castilla y León, N Spain
Medina del Campo 92 D2 Castilla y León, N Spain
Medinat Israel see Israel
Mediolanum see Saintes, France
Mediolanum see Milano, Italy
Mediomatrica see Metz
Mediterranean Sea 102 D3 Fr. Mer Méditerranée. sea Africa/Asia/Europe
Méditerranée, Mer see Mediterranean Sea
Médoc 91 B5 cultural region SW France
Medvezh'yegorsk 110 B3 Respublika Kareliya, NW Russian Federation
Meekatharra 147 B5 Western Australia
Meemu Atoll see Mulakatholhu
Meerssen 87 D6 var. Mersen. Limburg, SE Netherlands
Meerut 134 D2 Uttar Pradesh, N India
Megáli Préspa, Límni see Prespa, Lake
Meghálaya 113 G3 cultural region NE India
Mehdia see Mahdia
Meheso see Mi'eso
Me Hka see Nmai Hka
Mehriz 120 D3 Yazd, C Iran
Mehtar Lām 123 F4 var. Mehtarlām, Meterlam, Methariam, Metharlam. Laghmān, E Afghanistan
Mehtarlām see Mehtar Lām
Meiktila 136 B3 Mandalay, C Myanmar (Burma)
Méjico see Mexico
Mejillones 64 B2 Antofagasta, N Chile
Mek'elē 72 C4 var. Makale. Tigray, N Ethiopia
Mékhé 74 B3 NW Senegal
Mekong 124 D3 var. Lan-ts'ang Chiang, Cam. Mékôngk, Chin. Lancang Jiang, Lao. Mènam Khong, Nam Khong, Th. Mae Nam Khong, Tib. Dza Chu, Vtn. Sông Tiên Giang. river SE Asia
Mékôngk see Mekong
Mekong, Mouths of the 137 E6 delta S Vietnam
Melaka 138 B3 var. Malacca. Melaka, Peninsular Malaysia
Melaka, Selat see Malacca, Strait of
Melanesia 144 D3 island group W Pacific Ocean
Melanesian Basin 142 C2 undersea basin W Pacific Ocean
Melbourne 149 C7 state capital Victoria, SE Australia
Melbourne 43 E4 Florida, SE USA
Meleda see Mljet
Melekeok 144 A2 country capital (Palau) NW Pacific Ocean
Melghir, Chott 71 E2 var. Chott Melrhir. salt lake E Algeria
Melilla 80 B5 anc. Rusaddir, Russadir. Melilla, Spain, N Africa
Melita 70 D2 enclave Spain, N Africa
Melita 37 F5 Manitoba, S Canada
Melitene see Malatya
Melitopol' 109 F4 Zaporiz'ka Oblast', SE Ukraine
Melle 87 B5 Oost-Vlaanderen, NW Belgium
Mellerud 85 B6 Västra Götaland, S Sweden
Mellieha 102 B5 E Malta
Mellizo Sur, Cerro 65 A7 mountain S Chile
Melo 64 E4 Cerro Largo, NE Uruguay
Melodunum see Melun
Melrhir, Chott see Melghir, Chott
Melsungen 94 B4 Hessen, C Germany
Melun 90 C3 anc. Melodunum. Seine-et-Marne, N France
Melville Bay/Melville Bugt see Qimusseriarsuaq
Melville Island 146 D2 island Northern Territory, N Australia
Melville Island 37 E2 island Parry Islands, Northwest Territories, NW Canada
Melville, Lake 39 F2 lake Newfoundland and Labrador, E Canada
Melville Peninsula 37 G3 peninsula Nunavut, NE Canada
Melville Sound see Viscount Melville Sound
Membidj see Manbij
Memel see Neman, NE Europe
Memel see Klaipėda, Lithuania

Memmingen 95 B6 Bayern, S Germany
Memphis 42 C1 Tennessee, S USA
Menaam see Menaldum
Menado see Manado
Ménaka 75 F3 Goa, E Mali
Menaldum 86 D1 Fris. Menaam. Fryslân, N Netherlands
Mènam Khong see Mekong
Mendaña Fracture Zone 153 F4 fracture zone E Pacific Ocean
Mende 91 C5 anc. Mimatum. Lozère, S France
Mendeleyev Ridge 155 B2 undersea ridge Arctic Ocean
Mendocino Fracture Zone 152 D2 fracture zone NE Pacific Ocean
Mendoza 64 B4 Mendoza, W Argentina
Menemen 116 A3 İzmir, W Turkey
Menengiyn Tal 127 F2 plain E Mongolia
Menongue 78 B2 var. Vila Serpa Pinto, Port. Serpa Pinto. Cuando Cubango, C Angola
Mentawai, Kepulauan 138 A4 island group W Indonesia
Meppel 86 D2 Drenthe, NE Netherlands
Meran see Merano
Merano 96 C1 Ger. Meran. Trentino-Alto Adige, N Italy
Merca see Marka
Mercedes 64 D3 Corrientes, NE Argentina
Mercedes 64 D4 Soriano, SW Uruguay
Meredith, Lake 49 E1 reservoir Texas, SW USA
Merefa 109 G2 Kharkivs'ka Oblast', E Ukraine
Mergate see Margate
Mergui see Myeik
Mergui Archipelago see Myeik Archipelago
Mérida 51 H3 Yucatán, SW Mexico
Mérida 92 C4 anc. Augusta Emerita. Extremadura, W Spain
Mérida 58 C2 Mérida, W Venezuela
Meridian 42 C2 Mississippi, S USA
Mérignac 91 B5 Gironde, SW France
Merín, Laguna see Mirim Lagoon
Merkulovichi see Myerkulavichy
Merowe 72 B3 Northern, N Sudan
Merredin 147 B6 Western Australia
Mersen see Meerssen
Mersey 89 D5 river NW England, United Kingdom
Mersin 116 C4 var. İçel. İçel, S Turkey
Mêrsrags 106 C3 NW Latvia
Meru 73 C6 Eastern, C Kenya
Merv see Mary
Merzifon 116 D2 Amasya, N Turkey
Merzig 95 A5 Saarland, SW Germany
Mesa 48 B2 Arizona, SW USA
Meseritz see Międzyrzecz
Meshed see Mashhad
Mesopotamia 57 C5 var. Mesopotamia Argentina. physical region NE Argentina
Mesopotamia Argentina see Mesopotamia
Messalo, Rio 79 E2 var. Mualo. river NE Mozambique
Messana/Messene see Messina
Messina 97 D7 var. Messana, Messene; anc. Zancle. Sicilia, Italy, C Mediterranean Sea
Messina see Musina
Messina, Strait of 97 D7 Eng. Strait of Messina. strait SW Italy
Messina, Strait of see Messina, Stretto di
Messíni 105 B6 Pelopónnisos, S Greece
Mesta see Néstos
Mestghanem see Mostaganem
Mest'ia 117 F1 prev. Mestia. Svan. Mestiya. N Georgia
Mestia see Mest'ia
Mestiya see Mest'ia
Mestre 96 C2 Veneto, NE Italy
Metairie 42 B3 Louisiana, S USA
Metán 64 C2 Salta, N Argentina
Metapán 82 B2 Santa Ana, NW El Salvador
Meta, Río 58 D3 river Colombia/ Venezuela
Meterlam see Mehtar Lām
Methariam/Metharlam see Mehtar Lām
Metis see Metz
Metković 100 B4 Dubrovnik-Neretva, SE Croatia
Métsovon 104 B4 prev. Métsovo. Ípeiros, C Greece

Métsovon see Métsovo
Metz 90 D3 anc. Divodurum Mediomatrum, Mediomatrica, Moselle, NE France
Meulaboh 138 A3 Sumatera, W Indonesia
Meuse 87 C6 Dut. Maas. river W Europe
Mexcala, Río see Balsas, Río
Mexicali 50 A1 Baja California Norte, NW Mexico
Mexicanos, Estados Unidos see Mexico
México 51 E4 var. Ciudad de México Eng. Mexico City. country capital (Mexico) México, C Mexico
Mexico 45 G4 Missouri, C USA
Mexico 50 C3 off. United Mexican States var. Méjico, México, Sp. Estados Unidos Mexicanos. country N Central America

MEXICO
North America

Official name United Mexican States
Formation 1836 / 1848
Capital Mexico City
Population 122 million / 166 people per sq mile (64 people per sq km)
Total area 761,602 sq. miles (1,972,550 sq. km)
Languages Spanish*, Nahuatl, Mayan, Zapotec, Mixtec, Otomi, Totonac, Tzotzil, Tzeltal
Religions Roman Catholic 77%, Other 14%, Protestant 6%, Nonreligious 3%
Ethnic mix Mestizo 60%, Amerindian 30%, European 9%, Other 1%
Government Presidential system
Currency Mexican peso = 100 centavos
Literacy rate 94%
Calorie consumption 3024 kilocalories

México see Mexico
Mexico City see México
México, Golfo de see Mexico, Gulf of
Mexico, Gulf of 51 F2 Sp. Golfo de México. gulf W Atlantic Ocean
Meyadine see Al Mayādīn
Meydân Shahr see Maïdân Shahr
Meymaneh see Maïmanah
Mezen' 110 D3 river NW Russian Federation
Mezőtúr 99 D7 Jász-Nagykun-Szolnok, E Hungary
Mgarr 102 A5 Gozo, N Malta
Miadziol Nowy see Myadzyel
Miahuatlán 51 F5 var. Miahuatlán de Porfirio Díaz. Oaxaca, SE Mexico
Miahuatlán de Porfirio Díaz see Miahuatlán
Miami 43 F5 Florida, SE USA
Miami 49 G1 Oklahoma, C USA
Miami Beach 43 F5 Florida, SE USA
Miäneh 120 C2 var. Miyäneh. Āžarbāyjān-e Sharqi, NW Iran
Mianyang 128 B5 Sichuan, C China
Miastko 98 C2 Ger. Rummelsburg Pommern. Pomorskie, N Poland
Mi Chai see Nong Khai
Michalovce 99 E5 Ger. Grossmichel Hung. Nagymihály. Košický Kraj E Slovakia
Michigan 40 C1 off. State of Michigan also known as Great Lakes State, State, Wolverine State. state N USA
Michigan, Lake 40 C2 lake N USA
Michurin see Tsarevo
Michurinsk 111 B5 Tambovskaya Oblast', W Russian Federation
Micoud 55 F2 SE Saint Lucia
Micronesia 144 B1 off. Federated States of Micronesia. country W Pacific

MICRONESIA
Australasia & Oceania

Official name Federated States of Micronesia
Formation 1986 / 1986
Capital Palikir (Pohnpei Island)
Population 106,104 / 392 people per sq mile (151 people per sq km)
Total area 271 sq. miles (702 sq. km)
Languages Trukese, Pohnpeian, Mortlockese, Kosraean, Yapese, English*
Religions Roman Catholic 50%, Protestant 47%, Other 3%

RONESIA
ontinued)

ic mix Chuukese 49%,
peian 24%, Other 14%, Kosraean 6%,
se 5%, Asian 2%
ernment Nonparty system
ency US dollar = 100 cents
acy rate 81%
rie consumption Not available

nesia *144 C1 island group*
Pacific Ocean
nesia, Federated States of
Micronesia
Atlantic Cordillera
Mid-Atlantic Ridge
Atlantic Ridge *66 C3 var.* Mid-
antic Cordillera, Mid-Atlantic Rise,
d-Atlantic Swell. *undersea ridge*
antic Ocean
Atlantic Rise/Mid-Atlantic Swell
Mid-Atlantic Ridge
elburg *87 B5* Zeeland,
Netherlands
elharnis *86 B4* Zuid-Holland,
Netherlands
elkerke *87 A5* West-Vlaanderen,
Belgium
le America Trench *35 B7 trench*
acific Ocean
le Andaman *133 F2 island*
daman Islands, India,
Indian Ocean
le Atlas *70 C2 Eng.* Middle Atlas.
untain range* N Morocco
le Atlas *see* Moyen Atlas
leburg Island *see* 'Eua
le Congo *see* Congo (Republic of)
lesboro *40 C5* Kentucky, S USA
lesbrough *89 D5* N England,
ited Kingdom
letown *41 F4* New Jersey, NE USA
letown *41 F3* New York, NE USA
Indian Basin *141 C5 undersea*
sin* N Indian Ocean
Indian Ridge *141 C5 var.* Central
lian Ridge. *undersea ridge*
Indian Ocean
and *38 D5* Ontario, S Canada
and *40 C3* Michigan, N USA
and *49 E3* Texas, SW USA
Pacific Mountains *152 C2 var.*
d-Pacific Seamounts.
mount range* NW Pacific Ocean
Pacific Seamounts
Mid-Pacific Mountains
ray Islands *152 D2 US territory*
acific Ocean
hów *99 D5* Małopolskie, S Poland
zyrzec Podlaski *98 E3* Lubelskie,
Poland
zyrzecz. *98 B3 Ger.* Meseritz.
buskie, W Poland
c *99 D5* Podkarpackie, SE Poland
curea-Ciuc *108 C4 Ger.*
klerburg, *Hung.* Csíkszereda.
rghita, C Romania
es del Camín *92 D1 var.* Mieres del
mino. Asturias, NW Spain
es del Camino *see* Mieres del Camín
so *73 D5 var.* Meheso, Miesso.
omiya, C Ethiopia
so *see* Mi'ēso
ats Hefa *119 A5 Eng.* Bay of Haifa;
rv. MifrazHefa. *bay* N Israel
el Asua *50 D3 var.* Miguel Auza.
catecas, C Mexico
el Auza *see* Miguel Asua
recht *86 C3* Utrecht,
Netherlands
shevichi *see* Mikashevichy
shevichy *107 C7 Pol.* Mikaszewicze,
s. Mikashevich. Brestskaya
blasts', SW Belarus
szewicze *see* Mikashevichy
aylovgrad *see* Montana
aylovka *111 B6* Volgogradskaya
blast', SW Russian Federation
nos *see* Mýkonos
e *104 C2* Lovech, N Bulgaria
n' *110 D4* Respublika Komi,
V Russian Federation
ni-sanmyaku *131 D5 mountain*
tge* Honshū, N Japan Asia
ra-jima *131 D6 island* E Japan
gro *60 B2* Guayas, SW Ecuador
n *see* Milano

Milange *79 E2* Zambézia,
NE Mozambique
Milano *96 B2 Eng.* Milan, *Ger.* Mailand;
anc. Mediolanum. Lombardia, N Italy
Milas *116 A4* Muğla, SW Turkey
Milashavichy *107 C7 Rus.* Milashevichi.
Homyel'skaya Voblasts', SE Belarus
Milashevichi *see* Milashavichy
Mildura *149 C6* Victoria, SE Australia
Mile *see* Mili Atoll
Miles *149 D5* Queensland, E Australia
Miles City *44 C2* Montana, NW USA
Milford *see* Milford Haven
Milford Haven *89 C6 prev.* Milford.
SW Wales, United Kingdom
Milford Sound *151 A6* Southland, South
Island, New Zealand
Milford Sound *151 A6 inlet* South
Island, New Zealand
Milh, Bahr al *see* Razāzah, Buhayrat ar
Mili Atoll *144 D2 var.* Mile. *atoll* Ratak
Chain, SE Marshall Islands
Mil'kovo *115 H3* Kamchatskiy Kray,
E Russian Federation
Milk River *37 E5* Alberta, SW Canada
Milk River *44 C1 river* Montana,
NW USA
Milk, Wadi el *88 B4 var.* Wadi al Malik.
river C Sudan
Milledgeville *43 E2* Georgia, SE USA
Mille Lacs Lake *45 F2 lake* Minnesota,
N USA
Millennium Island *143 F3 prev.*
Caroline Island, Thornton Island. *atoll*
Line Islands, E Kiribati
Millerovo *111 B6* Rostovskaya Oblast',
SW Russian Federation
Mílos *105 C6 island* Kykládes, Greece,
Aegean Sea
Mílos *105 C6 island* Kykládes, Greece,
Aegean Sea
Mílos *105 C7 island* Kykládes, Greece,
Aegean Sea
Milton *151 B7* Otago, South Island,
New Zealand
Milton Keynes *89 D6* SE England,
United Kingdom
Milwaukee *40 B3* Wisconsin, N USA
Mimatum *see* Mende
Min *see* Fujian
Minā' Qābūs *140 B3* NE Oman
Minas Gerais *63 F3 off.* Estado de Minas
Gerais. *state* E Brazil
Minas Gerais. *region* E Brazil
Minas Gerais, Estado de *see* Minas Gerais
Minatitlán *51 F4* Veracruz-Llave,
E Mexico
Minbu *136 A3* Magway, W Myanmar
(Burma)
Minch, The *88 B3 var.* North Minch.
strait NW Scotland, United Kingdom
Mindanao *139 E2 island* S Philippines
Mindanao Sea *see* Bohol Sea
Mindelheim *95 C6* Bayern, S Germany
Mindello *see* Mindelo
Mindelo *74 A2 var.* Mindello;
prev. Porto Grande. São Vicente,
N Cape Verde
Minden *94 B4 anc.* Minthun. Nordrhein-
Westfalen, NW Germany
Mindoro *139 E2 island* N Philippines
Mindoro Strait *139 E2 strait*
W Philippines
Mineral Wells *49 F2* Texas, SW USA
Mingächevir *117 G2 Rus.* Mingechaur,
prev. Mingechaur. C Azerbaijan
Mingáora *see* Saïdu
Mingechaur/Mingechevir *see*
Mingächevir
Mingora *see* Saïdu
Minho *92 B2 former province* N Portugal
Minho *92 B2 Sp.* Miño. *river* Portugal/
Spain
Minho, Rio *see* Miño
Minicoy Island *132 B3 island* SW India
Minius *see* Miño
Minna *75 G4* Niger, C Nigeria
Minneapolis *45 F2* Minnesota, N USA
Minnesota *45 F2 off.* State of Minnesota,
also known as Gopher State,
New England of the West,
North Star State. *state* N USA
Miño *92 B2 var.* Mino, Minius, *Port.* Rio
Minho. *see also* Minho, Rio
Minorca *93 H3 Eng.* Minorca; *anc.*
Balearis Minor. *island* Islas Baleares,
Spain, W Mediterranean Sea

Minorca *see* Menorca
Minot *45 E1* North Dakota, N USA
Minsk *107 C6 country capital* (Belarus)
Minskaya Voblasts', C Belarus
Minskaya Wzvyshsha *107 C6 mountain*
range C Belarus
Mińsk Mazowiecki *98 D3 var.* Nowo-
Minsk. Mazowieckie, C Poland
Minthun *see* Minden
Minto, Lac *38 D2 lake* Québec, C Canada
Minya *see* Al Minyā
Miraflores *50 C3* Baja California Sur,
NW Mexico
Miranda de Ebro *93 E1* La Rioja,
N Spain
Mirgorod *see* Myrhorod
Miri *138 D3* Sarawak, East Malaysia
Mirim Lagoon *63 E5 var.* Lake Mirim,
Sp. Laguna Merín. *lagoon* Brazil/
Uruguay
Mirim, Lake *see* Mirim Lagoon
Mírina *see* Mýrina
Mirjäveh *120 E4* Sīstān va Balūchestān,
SE Iran
Mirny *154 C3 Russian research station*
Antarctica
Mirnyy *115 F3* Respublika Sakha
(Yakutiya), NE Russian Federation
Mírpur Khās *134 B3* Sind, SE Pakistan
Mirtoan Sea *105 C6 Eng.* Mirtoan Sea;
anc. Myrtoum Mare. *sea* S Greece
Mírtoan Sea *see* Mirtóo Pélagos
Misiaf *see* Maşyāf
Miskito Coast *see* La Mosquitia
Miskitos, Cayos *53 E2 island group*
NE Nicaragua
Miskolc *99 D6* Borsod-Abaúj-Zemplén,
NE Hungary
Misool, Pulau *139 F4 island* Maluku,
E Indonesia
Mişrātah *71 F2 var.* Misurata.
NW Libya
Mission *49 G5* Texas, SW USA
Mississippi *42 B2 off.* State of
Mississippi, *also known as* Bayou State,
Magnolia State. *state* SE USA
Mississippi Delta *42 B4 delta* Louisiana,
S USA
Mississippi River *35 C6 river* C USA
Missoula *44 B1* Montana, NW USA
Missouri *45 F5 off.* State of Missouri,
also known as Bullion State, Show Me
State. *state* C USA
Missouri River *45 E3 river* C USA
Mistassini, Lac *38 D3 lake* Québec,
SE Canada
Mistelbach an der Zaya *95 E6*
Niederösterreich, NE Austria
Misti, Volcán *61 E4 volcano* S Peru
Misurata *see* Mişrātah
Mitau *see* Jelgava
Mitchell *151 D5* Ibaraki, Honshū, S Japan
Mitchell, Mount *43 E1 mountain* North
Carolina, SE USA
Mitchell River *148 C2 river* Queensland,
NE Australia
Mi Tho *see* My Tho
Mitilíni *see* Mytilíni
Mitilíni *105 D5* Ibaraki, Honshū, S Japan
Mitrovica *see* Mitrovicë
Mitrovica/Mitrowitz *see* Sremska
Mitrovica, Serbia
Mitrovicë *101 D5 Serb.* Mitrovica, *prev.*
Kosovska Mitrovica, Titova Mitrovica.
N Kosovo
Mits'iwa *72 C4 var.* Massawa, Massawa.
E Eritrea
Mitspe Ramon *119 A7 prev.* Mizpe
Ramon. Southern, S Israel
Mittelstadt *see* Baia Sprie
Mittelstadt *see* Baia Sprie
Mitumba Chaîne des/Mitumba Range
see Mitumba Range
Mitumba Range *77 D6 var.* Chaîne des
Mitumba, Mitumba Mountains.
range E Dem. Rep. Congo
Miueru Wantipa, Lake *78 B1 var.*
N Zambia
Miyake-jima *131 D6 island* Sakishima-
shotō, SW Japan
Miyako *130 D4 var.* Honshū, C Japan
Miyakonojō *131 B8 var.* Miyakonzyo
Miyakonzyō, Kyūshū, SW Japan
Miyakonzyō *see* Miyakonojō
Miyāneh *131 B8* Miyazaki, Kyūshū,
SW Japan
Miyazaki *131 B8* Miyazaki, Kyūshū,
SW Japan

Mizil *108 C5* Prahova, SE Romania
Miziya *104 C1* Vratsa, NW Bulgaria
Mizpe Ramon *see* Mitspe Ramon
Mjøsa *85 B6 var.* Mjøsen. *lake*
S Norway
Mjøsen *see* Mjøsa
Mladenovac *100 D4* Serbia, C Serbia
Mława *98 D3* Mazowieckie, C Poland
Mljet *101 B5* It. Meleda; *anc.* Melita.
island S Croatia
Mmabatho *78 C4* North-West, N South
Africa
Moab *44 B5* Utah, W USA
Moa Island *148 C1 island* Queensland,
NE Australia
Moanda *77 B6 var.* Mouanda. Haut-
Ogooué, SE Gabon
Moba *77 E7* Katanga,
E Dem. Rep. Congo
Mobay *see* Montego Bay
Mobaye *77 C5* Basse-Kotto, S Central
African Republic
Moberly *45 G4* Missouri, C USA
Mobile *42 C3* Alabama, S USA
Mobutu Sese Seko, Lac *see* Albert, Lake
Moçâmedes *see* Namibe
Mochudi *78 C4* Kgatleng,
SE Botswana
Mocímboa da Praia *79 F2 var.* Vila de
Mocímboa da Praia. Cabo Delgado,
N Mozambique
Môco *78 B2 var.* Morro de Môco.
mountain W Angola
Mocoa *58 A4* Putumayo, SW Colombia
Môco, Morro de *see* Môco
Mocuba *79 E3* Zambézia,
NE Mozambique
Modena *96 B3 anc.* Mutina. Emilia-
Romagna, N Italy
Modesto *47 B6* California, W USA
Modica *97 C7 anc.* Motyca. Sicilia, Italy,
C Mediterranean Sea
Modimolle *78 D4 prev.* Nylstroom.
NW South Africa
Modohn *see* Madona
Modriča *100 C3* Republika Srpska,
N Bosnia and Herzegovina
Moe *37 C7* Victoria, SE Australia
Møen *see* Møn, Denmark
Moero, Lac *see* Mweru, Lake
Moeskroen *see* Mouscron
Mogadiscio/Mogadishu *see* Muqdisho
Mogador *see* Essaouira
Mogilëv *see* Mahilyow
Mogilev-Podol'skiy *see*
Mohyliv-Podil's'kyy
Mogilno *73 C3* Kujawsko-pomorskie,
C Poland
Mohammadābād-e Rīgān *120 E4*
Kermān, SE Iran
Mohammedia *70 C2 prev.* Fédala.
NW Morocco
Mohave, Lake *47 D7 reservoir* Arizona/
Nevada, W USA
Mohawk River *41 F3 river* New York,
NE USA
Mohéli *see* Mwali
Mohns Ridge *83 F3 undersea feature*
Arctic Ocean
Moho *61 E4* Puno, SE Peru
Mohoro *73 C7* Pwani, E Tanzania
Mohyliv-Podil's'kyy *108 D3 Rus.*
Mogilev-Podol'skiy *see* Vinnyts'ka
Oblast', C Ukraine
Moi *85 A6* Rogaland, S Norway
Moili *see* Mwali
Moindou *147 A6* Province Sud, C New
Caledonia
Môisäküla *106 D3 var.* Moisekül.
S Estonia
Moissac *69 B6* Tarn-et-Garonne,
S France
Mojave Desert *47 D7 plain* California,
W USA
Mokrany *see* Makrany
Moktama *see* Mottama
Mol *107 C5* Ger. Antwerpen,
N Belgium
Mölbau *see* Madona
Moldau *see* Vltava
Moldavian SSR/Moldavskaya SSR
see Moldova
Moldo-Too, Khrebet *see* Moldo-Too,
Khrebet
Moldotau, Khrebet *see* Moldo-Too,
Khrebet
Moldo-Too, Khrebet *123 G2 prev.*
Khrebet Moldotau. *mountain range*
C Kyrgyzstan

Moldova 108 D3 off. Republic of Moldova, var. Moldavia; prev. Moldavian SSR, Rus. Moldavskaya SSR. country SE Europe

MOLDOVA
Southeast Europe

Official name Republic of Moldova
Formation 1991 / 1991
Capital Chisinau
Population 3.5 million / 269 people per sq mile (104 people per sq km)
Total area 13,067 sq. miles (33,843 sq. km)
Languages Moldovan*, Ukrainian, Russian
Religions Orthodox Christian 93%, Other 6%, Baptist 1%
Ethnic mix Moldovan 84%, Ukrainian 7%, Gagauz 5%, Russian 2%, Bulgarian 1%, Other 1%
Government Parliamentary system
Currency Moldovan leu = 100 bani
Literacy rate 99%
Calorie consumption 2837 kilocalories

Moldova Nouă 108 A4 Ger. Neumoldowa, Hung. Újmoldova. Caras-Severin, SW Romania
Moldova, Republic of see Moldova
Moldoveanul see Vârful Moldoveanu
Molfetta 97 E5 Puglia, SE Italy
Moll see Mol
Mollendo 61 E4 Arequipa, SW Peru
Mölndal 85 B7 Västra Götaland, S Sweden
Molochans'k 109 G4 Rus. Molochansk. Zaporiz'ka Oblast', SE Ukraine
Molodechno/Molodeczno see Maladzyechna
Molodezhnaya 154 C2 Russian research station Antarctica
Moloka'i 47 B8 var. Molokai. island Hawai'ian Islands, Hawaii, USA
Molokai Fracture Zone 153 E2 tectonic feature NE Pacific Ocean
Molopo 78 C4 seasonal river Botswana/ South Africa
Mólos 105 B5 Stereá Elláda, C Greece
Molotov see Severodvinsk, Arkhangel'skaya Oblast', Russian Federation
Molotov see Perm', Permskaya Oblast', Russian Federation
Moluccas 139 F4 Dut. Molukken, Eng. Moluccas; prev. Spice Islands. island group E Indonesia
Moluccas see Maluku
Molucca Sea 139 F4 Ind. Laut Maluku. sea E Indonesia
Molukken see Maluku
Mombasa 73 D7 Coast, SE Kenya
Mombetsu see Monbetsu
Momchilgrad 104 D3 prev. Mastanli. Kürdzhali, S Bulgaria
Møn 85 B8 prev. Möen. island SE Denmark
Mona, Canal de la see Mona Passage
Monaco 91 C7 var. Monaco-Ville; anc. Monoecus. country capital (Monaco) S Monaco
Monaco 91 E6 off. Principality of Monaco. country W Europe

MONACO
Southern Europe

Official name Principality of Monaco
Formation 1861 / 1861
Capital Monaco-Ville
Population 36,136 / 48,181 people per sq mile (18,531 people per sq km)
Total area 0.75 sq. miles (1.95 sq. km)
Languages French*, Italian, Monégasque, English
Religions Roman Catholic 89%, Protestant 6%, Other 5%
Ethnic mix French 47%, Other 21%, Italian 16%, Monégasque 16%
Government Monarchical / parliamentary system
Currency Euro = 100 cents
Literacy rate 99%
Calorie consumption Not available

Monaco see München
Monaco, Port de 91 C8 bay S Monaco W Mediterranean Sea
Monaco, Principality of see Monaco

Monaco-Ville see Monaco
Monahans 49 E3 Texas, SW USA
Mona, Isla 55 E3 island W Puerto Rico
Mona Passage 55 E3 Sp. Canal de la Mona. channel Dominican Republic/ Puerto Rico
Monastir see Bitola
Monbetsu 130 D2 var. Mombetsu, Monbetu. Hokkaidō, NE Japan
Monbetu see Monbetsu
Moncalieri 96 A2 Piemonte, NW Italy
Monchegorsk 110 C2 Murmanskaya Oblast', NW Russian Federation
Monclova 50 D2 Coahuila, NE Mexico
Moncton 39 F4 New Brunswick, SE Canada
Mondovì 96 A2 Piemonte, NW Italy
Monfalcone 96 D2 Friuli-Venezia Giulia, NE Italy
Monforte de Lemos 92 C1 Galicia, NW Spain
Mongo 76 C3 Guéra, C Chad
Mongolia 126 C2 Mong. Mongol Uls. country E Asia

MONGOLIA
East Asia

Official name Mongolia
Formation 1924 / 1924
Capital Ulan Bator
Population 2.8 million / 5 people per sq mile (2 people per sq km)
Total area 604,247 sq. miles (1,565,000 sq. km)
Languages Khalkha Mongolian*, Kazakh, Chinese, Russian
Religions Tibetan Buddhist 50%, Nonreligious 40%, Shamanist and Christian 6%, Muslim 4%
Ethnic mix Khalkh 95%, Kazakh 4%, Other 1%
Government Presidential / parliamentary system
Currency Tugrik (tögrög) = 100 möngö
Literacy rate 98%
Calorie consumption 2463 kilocalories

Mongolia, Plateau of 124 D1 plateau E Mongolia
Mongol Uls see Mongolia
Mongora see Saidu
Mongos, Chaine des see Bongo, Massif des
Mongu 78 C2 Western, W Zambia
Monkchester see Newcastle upon Tyne
Monkey Bay 79 E2 Southern, SE Malawi
Monkey River see Monkey River Town
Monkey River Town 52 C2 var. Monkey River. Toledo, SE Belize
Monoecus see Monaco
Mono Lake 47 C6 lake California, W USA
Monostor see Beli Manastir
Monóvar 93 F4 Cat. Monòver. Valenciana, E Spain
Monòver see Monóvar
Monroe 42 B2 Louisiana, S USA
Monrovia 74 C5 country capital (Liberia) W Liberia
Mons 87 B6 Dut. Bergen. Hainaut, S Belgium
Monselice 96 C2 Veneto, NE Italy
Montana 104 C2 prev. Ferdinand, Mikhaylovgrad. Montana, NW Bulgaria
Montana 44 B1 off. State of Montana, also known as Mountain State, Treasure State. state NW USA
Montargis 90 C4 Loiret, C France
Montauban 91 B6 Tarn-et-Garonne, S France
Montbéliard 90 D4 Doubs, E France
Mont Cenis, Col du 91 D5 pass E France
Mont-de-Marsan 91 B6 Landes, SW France
Monteagudo 61 G4 Chuquisaca, S Bolivia
Montecarlo 91 C8 Misiones, NE Argentina
Monte Caseros 64 D3 Corrientes, NE Argentina
Monte Cristi 54 D3 var. San Fernando de Monte Cristi. NW Dominican Republic
Montegiardino 96 E2 SE San Marino

Montego Bay 54 A4 var. Mobay. W Jamaica
Montélimar 91 D5 anc. Acunum Acusio, Montilium Adhemari. Drôme, E France
Montemorelos 51 E3 Nuevo León, NE Mexico
Montenegro 101 C5 Serb. Crna Gora. country SW Europe

MONTENEGRO
Southeast Europe

Official name Montenegro
Formation 2006 / 2006
Capital Podgorica
Population 600,000 / 113 people per sq mile (43 people per sq km)
Total area 5332 sq. miles (13,812 sq. km)
Languages Montenegrin*, Serbian, Albanian, Bosniak, Croatian
Religions Orthodox Christian 74%, Muslim 18%, Roman Catholic 4%, Other 4%
Ethnic mix Montenegrin 43%, Serb 32%, Other 12%, Bosniak 8%, Albanian 5%
Government Parliamentary system
Currency Euro = 100 cents
Literacy rate 98%
Calorie consumption 3568 kilocalories

Monte Patria 64 B3 Coquimbo, N Chile
Monterey 47 B6 California, W USA
Monterey see Monterrey
Monterey Bay 47 A6 bay California, W USA
Montería 58 B2 Córdoba, NW Colombia
Montero 61 G4 Santa Cruz, C Bolivia
Monterrey 51 E3 var. Monterey. Nuevo León, NE Mexico
Montes Claros 63 F3 Minas Gerais, SE Brazil
Montevideo 64 D4 country capital (Uruguay) Montevideo, S Uruguay
Montevideo 45 F2 Minnesota, N USA
Montgenèvre, Col de 91 D5 pass France/Italy
Montgomery 42 D2 state capital Alabama, S USA
Montgomery see Sāhīwāl
Monthey 95 A7 Valais, SW Switzerland
Montilium Adhemari see Montélimar
Montluçon 90 C4 Allier, C France
Montoro 92 D4 Andalucía, S Spain
Montpelier 41 G2 state capital Vermont, NE USA
Montpellier 91 C6 Hérault, S France
Montréal 39 E4 Eng. Montreal. Québec, SE Canada
Montrose 88 D3 E Scotland, United Kingdom
Montrose 44 C5 Colorado, C USA
Montserrat 55 G3 var. Emerald Isle. UK dependent territory E West Indies
Monywa 136 B3 Sagaing, C Myanmar (Burma)
Monza 96 B2 Lombardia, N Italy
Monze 78 D2 Southern, S Zambia
Monzón 93 F2 Aragón, NE Spain
Moonie 149 D5 Queensland, E Australia
Moon-Sund see Väinameri
Moora 147 A6 Western Australia
Moore 49 G1 Oklahoma, C USA
Moore, Lake 147 B6 lake Western Australia
Moorhead 45 F2 Minnesota, N USA
Moose 38 C3 Ontario, S Canada
Moosehead Lake 41 G1 lake Maine, NE USA
Moosonee 38 C3 Ontario, SE Canada
Mopti 75 E3 Mopti, C Mali
Moquegua 61 E4 Moquegua, SE Peru
Mora 85 C5 Dalarna, C Sweden
Morales 52 C2 Izabal, E Guatemala
Morant Bay 54 B5 E Jamaica
Moratalla 93 E4 Murcia, SE Spain
Morava 99 C5 var. March. river C Europe
Morava see Velika Morava, Serbia
Moravia 99 B5 Ger. Mähren. cultural region E Czech Republic
Moray Firth 88 C3 inlet N Scotland, United Kingdom
Morea see Pelopónnisos
Moreau River 44 D2 river South Dakota, N USA
Moree 149 D5 New South Wales, SE Australia

Morelia 51 E4 Michoacán, S Mexic
Morena, Sierra 92 C4 mountain re S Spain
Moreni 108 C5 Dâmbovita, S Rom:
Morgan City 42 B3 Louisiana, S U⁵
Murghāb, Daryā-ye see Murgap
Morioka 130 D4 Iwate, Honshū, C
Morlaix 90 A3 Finistère, NW Franc
Mormon State see Utah
Mornington Abyssal Plain 67 A7 abyssal plain SE Pacific Ocean
Mornington Island 148 B2 island W Australia
Morocco 70 B3 off. Kingdom of Morocco, Ar. Al Mamlakah. cou N Africa

MOROCCO
North Africa

Official name Kingdom of Morocc(
Formation 1956 / 1969
Capital Rabat
Population 33 million / 192 people per sq mile (74 people per sq km)
Total area 172,316 sq. miles (446,300 sq. km)
Languages Arabic*, Tamazight (Bert French, Spanish
Religions Muslim (mainly Sunni) 99% Other (mostly Christian) 1%
Ethnic mix Arab 70%, Berber 29%, European 1%
Government Monarchical / parliamentary system
Currency Moroccan dirham = 100 centimes
Literacy rate 67%
Calorie consumption 3334 kilocalo

Morocco see Marrakech
Morocco, Kingdom of see Morocco
Morogoro 73 C7 Morogoro, E Tan:
Moro Gulf 139 E3 gulf S Philippine
Morón 54 C2 Ciego de Ávila, C C
Mörön 126 D2 Hövsgöl, N Mongo
Morondava 79 F3 Toliara, W Madagascar
Moroni 79 F2 country capital (Corr Grande Comore, NW Comoros
Morotai, Pulau 139 F3 island Male E Indonesia
Morotiri see Marotiri
Morphou see Güzelyurt
Morrinsville 150 D3 Waikato, Nor Island, New Zealand
Morris 45 F2 Minnesota, N USA
Morris Jesup, Kap 83 E1 headlan(N Greenland
Morvan 90 D4 physical region C France
Moscow 46 C2 Idaho, NW USA
Moscow see Moskva
Mosel 95 A5 Fr. Moselle. river W F! W Europe
Mosel see Moselle
Moselle 90 D3 department NE Franc
Moselle 90 D3 Ger. Mosel. river W Europe
Moselle see Mosel
Mosgiel 151 B7 Otago, South Islan New Zealand
Moshi 73 C7 Kilimanjaro, NE Tan;
Mosjøen 84 B4 Nordland, C Norwa
Moskovskiy see Moskva
Moskva 111 B5 Eng. Moscow. country capital (Russian Federatio Gorod Moskva, W Russian Feder:
Moskva 123 E3 Rus. Moskovskiy; μ Chubek. SW Tajikistan
Moson and Magyaróvár see Mosonmagyaróvár
Mosonmagyaróvár 99 C6 Ger. Wieselburg-Ungarisch-Altenburg prev. Moson and Magyaróvár, G(Moson and Magyaróvár, Ger. Mosonmagyaróvár, E
Mosquito Coast 53 E3 var. Miskito Coast, Eng. Mosquito Coast. coas region E Nicaragua
Mosquito Coast see La Mosquitia
Mosquito Gulf 53 F4 Eng. Mosqui Gulf. gulf N Panama
Mosquitos Gulf see Mosquitos, Golf de los
Moss 85 B6 Østfold, S Norway
Mossámedes see Namibe
Mosselbaai 78 C5 var. Mosselbai, Mossel Bay. Western Cape, S Africa

elbai/Mossel Bay see Mosselbaai
endjo 77 B6 Niari, SW Congo
oró 63 G2 Rio Grande do Norte,
: Brazil
98 A4 Ger. Brüx. Ústecký Kraj,
V Czech Republic
a 102 B5 var. Musta. C Malta
aganem 70 D2 var. Mestghanem.
W Algeria
ar 100 C4 Federacija Bosne I
ercegovine, S Bosnia and Herzegovina
ty see Masty
ul see Al Mawşil
a del Cuervo 93 E3 Castilla-La
ancha, C Spain
agua, Río 52 B2 river Guatemala/
onduras
er of Presidents/Mother of States
e Virginia
ril 92 D5 Andalucía, S Spain
ru 108 B4 Gorj, SW Romania
tama 136 B4 prev. Martaban; var.
oktama. Mon State, S Myanmar
urma)
ueka 151 C5 Tasman, South Island,
ew Zealand
ul 51 H3 var. Motul de Felipe Carrillo
uerto.
tán, SE Mexico
ul de Felipe Carrillo Puerto see
otul
yca see Modica
anda see Moanda
ahoun see Black Volta
tila 77 A6 Ngounié, C Gabon
kden see Shenyang
ald Bay 37 E2 Prince Patrick Island,
orthwest Territories, N Canada
lins 90 C4 Allier, C France
lmein see Mawlamyine
andou 76 B4 Logone-Occidental,
W Chad
ng Roessei 137 D5 Bătdâmbâng,
Cambodia
n Hou see Black Volta
ntain Home 42 B1 Arkansas, C USA
ntain State see Montana
ntain State see West Virginia
nt Cook 151 B6 Canterbury, South
sland, New Zealand
nt Desert Island 41 H2 island
aine, NE USA
nt Gambier 149 B7 South Australia
nt Isa 148 B3 Queensland,
Australia
nt Magnet 147 B5 Western
ustralia
nt Pleasant 45 G4 Iowa, C USA
nt Pleasant 40 C3 Michigan, N USA
nt Vernon 40 B5 Illinois, N USA
nt Vernon 46 B1 Washington,
W USA
ardi, Dépression du 76 C2 desert
wland Chad/Sudan
ascron 87 A6 Dut. Moeskroen.
ainaut, W Belgium
ase River see Souris River
ussoro 76 B3 Kanem, W Chad
yen-Congo see Congo (Republic of)
ynoq 122 C1 Rus. Muynak.
oraqalpog'iston Respublikasi,
W Uzbekistan
yobamba 60 B2 San Martín, NW Peru
yu 126 B3 var. Karakax. Xinjiang
ygur Zizhiqu, NW China
ynkum, Peski 123 F1 Kaz.
Moyynqum. desert S Kazakhstan
ynqum see Moyynkum, Peski
zambika, Lakandranon' i see
Mozambique Channel
zambique 79 E3 off. Republic of
Mozambique; prev. People's Republic of
Mozambique; Portuguese East Africa.
ountry S Africa

OZAMBIQUE
Southern Africa

ficial name Republic of Mozambique
rmation 1975 / 1975
pital Maputo
pulation 25.8 million / 85 people
r sq mile (33 people per sq km)
tal area 309,494 sq. miles
1,590 sq. km)
nguages Makua, Xitsonga, Sena,
mwe, Portuguese*
ligions Traditional beliefs 56%,
ristian 30%, Muslim 14%

MOZAMBIQUE
(continued)

Ethnic mix Makua Lomwe 47%,
Tsonga 23%, Malawi 12%, Shona 11%,
Yao 4%, Other 3%
Government Presidential system
Currency New metical = 100 centavos
Literacy rate 51%
Calorie consumption 2267 kilocalories

Mozambique Basin see Natal Basin
Mozambique, Canal de see Mozambique
Channel
Mozambique Channel 79 E3 Fr. Canal
de Mozambique, Mal. Lakandranon' i
Mozambika. strait W Indian Ocean
Mozambique, People's Republic of see
Mozambique
Mozambique Plateau 69 D7 var.
Mozambique Rise. undersea plateau
SW Indian Ocean
Mozambique, Republic of see
Mozambique
Mozambique Rise see Mozambique
Plateau
Mozyr' see Mazyr
Mpama 77 B6 river C Congo
Mpika 78 D2 Northern, NE Zambia
Mqinvartsveri see Kazbek
Mragowo 98 D2 Ger. Sensburg.
Warmińsko-Mazurskie, NE Poland
Mthatha 78 D5 prev. Umtata. Eastern
Cape, SE South Africa
Mtkvari see Kura
Mtwara 73 D8 Mtwara, SE Tanzania
Mualo see Messalo, Rio
Muang Chiang Rai see Chiang Rai
Muang Kalasin see Kalasin
Muang Khammouan see Thakhèk
Muang Khôngxédôn 137 D5 var. Khong
Sedone. Salavan, S Laos
Muang Khon Kaen see Khon Kaen
Muang Lampang see Lampang
Muang Loei see Loei
Muang Lom Sak see Lom Sak
Muang Nakhon Sawan see Nakhon
Sawan
Muang Namo 136 C3 Oudômxai,
N Laos
Muang Nan see Nan
Muang Phalan 136 D4 var. Muang
Phalane. Savannakhét, S Laos
Muang Phalane see Muang Phalan
Muang Phayao see Phayao
Muang Phitsanulok see Phitsanulok
Muang Phrae see Phrae
Muang Roi Et see Roi Et
Muang Sakon Nakhon see Sakon
Nakhon
Muang Samut Prakan see Samut Prakan
Muang Sing 136 C3 Louang Namtha,
N Laos
Muang Ubon see Ubon Ratchathani
Muang Xaignabouri see Xaignabouli
Muar 130 B3 var. Bandar Maharani.
Johor, Peninsular Malaysia
Mucojo 79 F2 Cabo Delgado,
N Mozambique
Mudon 137 B5 Mon State, S Myanmar
(Burma)
Muenchen see München
Muenster see Münster
Mufulira 78 D2 Copperbelt, C Zambia
Mughla see Muğla
Muğla 116 A4 var. Mughla. Muğla,
SW Turkey
Mûḥ, Sabkhat al 118 C3 lake C Syria
Muhu Väin see Väinameri
Muisne 60 A1 Esmeraldas,
NW Ecuador
Mukacheve 108 B3 Hung. Munkács,
Rus. Mukachevo. Zakarpats'ka Oblast',
W Ukraine
Mukachevo see Mukacheve
Mukalla see Al Mukallā
Mukden see Shenyang
Mula 93 E4 Murcia, SE Spain
Mulakatholhu 132 B4 var. Meemu Atoll,
atoll C Maldives
Mulaku Atoll see Mulakatholhu
Muleshoe 49 E2 Texas, SW USA
Mulhacén 93 E5 var. Cerro de Mulhacén.
mountain S Spain

Mulhacén, Cerro de see Mulhacén
Mülhausen see Mulhouse
Mülheim 95 A6 var. Mulheim an
der Ruhr. Nordrhein-Westfalen,
W Germany
Mulheim an der Ruhr see Mülheim
Mulhouse 90 E4 Ger. Mülhausen. Haut-
Rhin, NE France
Müller-gerbergte see Muller,
Pegunungan
Muller, Pegunungan 138 D4 Dut.
Müller-gerbergte. mountain range
Borneo, C Indonesia
Mull, Isle of 88 B4 island W Scotland,
United Kingdom
Mulongo 77 D7 Katanga, SE Dem.
Rep. Congo
Multán 134 C2 Punjab, E Pakistan
Mumbai 134 C5 prev. Bombay. state
capital Mahārāshtra, W India
Munamägi see Suur Munamägi
Münchberg 95 C5 Bayern, E Germany
München 95 C6 var. Munich,
Eng. Munich, It. Monaco. Bayern,
SE Germany
Muncie 40 C4 Indiana, N USA
Mungbere 77 E5 Orientale, NE Dem.
Rep. Congo
Munich see München
Munkács see Mukacheve
Münster 94 A4 var. Muenster, Münster
in Westfalen. Nordrhein-Westfalen,
W Germany
Munster 89 A6 Ir. Cúige Mumhan.
cultural region S Ireland
Münster in Westfalen see Münster
Muong Xiang Ngeun 136 C4 var. Xieng
Ngeun. Louangphabang, N Laos
Muonio 84 D3 Lappi, N Finland
Muonioälv/Muoniojoki see Muonionjoki
Muonionjoki 84 D3 var. Muoniojoki,
Swe. Muonioälv. river Finland/Sweden
Muqât 118 C3 Al Mafraq, E Jordan
Muqdisho 73 D6 Eng. Mogadishu, It.
Mogadiscio. country capital (Somalia)
Banaadir, S Somalia
Mura 95 E7 SCr. Mura. river C Europe
Muradiye 117 F3 Van, E Turkey
Murapara see Murupara
Murata 96 E2 S San Marino
Murchison River 147 A5 river Western
Australia
Murcia 93 F4 Murcia, SE Spain
Murcia 93 E4 autonomous community
SE Spain
Murfreesboro 42 D1 Tennessee, S USA
Murgab see Murghob
Murgap 122 D3 Mary Welaýaty,
S Turkmenistan
Murgap 122 D3 var. Deryasy Murgap,
Murghab, Pash. Daryā-ye Morghāb,
Rus. Murgab. river Afghanistan/
Turkmenistan
Murghob 123 F3 Rus. Murgab.
SE Tajikistan
Murgon 149 E5 Queensland, E Australia
Müritänīyah see Mauritania
Müritz 94 C3 var. Müritzee.
NE Germany
Müritzee see Müritz
Murmansk 110 C2 Murmanskaya
Oblast', NW Russian Federation
Murmansk 110 C2 Murmanskaya
Oblast', NW Russian Federation
Murom 115 B5 Vladimirskaya Oblast',
W Russian Federation
Muroran 130 D3 Hokkaidō, NE Japan
Muros 92 B1 Galicia, NW Spain
Murray Fracture Zone 153 E2 fracture
zone NE Pacific Ocean
Murray Range see Murray Ridge
Murray Ridge 124 C5 var. Murray
Range. undersea ridge N Arabian Sea
Murray River 149 B6 river SE Australia
Murrumbidgee River 149 C6 river New
South Wales, SE Australia
Murska Sobota 95 E7 Ger. Olsnitz.
NE Slovenia
Murupara 150 D4 var. Murapara. Bay of
Plenty, North Island, New Zealand
Murviedro see Sagunto
Murwāra 135 E4 Madhya Pradesh,
N India
Murwillumbah 149 E5 New South
Wales, SE Australia

Murzuq, Idhan 71 F4 var. Edeyin
Murzuq. desert SW Libya
Mürzzuschlag 95 E7 Steiermark,
E Austria
Muş 117 F3 var. Mush. Muş, E Turkey
Musa, Gebel 72 C2 var. Gebel Mûsa.
mountain NE Egypt
Mûsa, Gebel see Mûsá, Jabal
Musala 104 B3 mountain W Bulgaria
Muscat see Masqaţ
Muscat and Oman see Oman
Muscatine 45 G3 Iowa, C USA
Musgrave Ranges 147 D5 mountain
range South Australia
Musina 78 D3 prev. Messina. Limpopo,
NE South Africa
Muskegon 40 C3 Michigan, N USA
Muskogean see Tallahassee
Muskogee 49 G1 Oklahoma, C USA
Musoma 73 C6 Mara, N Tanzania
Mosta see Mosta
Mustafa-Pasha see Svilengrad
Musters, Lago 65 B6 lake S Argentina
Muswellbrook 149 D6 New South
Wales, SE Australia
Mut 116 C4 İçel, S Turkey
Mu-tan-chiang see Mudanjiang
Mutare 78 E3 var. Mutari; prev. Umtali.
Manicaland, E Zimbabwe
Mutari see Mutare
Mutina see Modena
Mutsu-wan 130 D3 bay N Japan
Muttonbird Islands 151 A8 island group
SW New Zealand
Mu Us Shadi 127 E3 var. Ordos Desert;
prev. Mu Us Shamo. desert N China
Mu Us Shamo see Mu Us Shadi
Muy Muy 52 D3 Matagalpa, C Nicaragua
Muynak see Mo'ynoq
Mužlja 100 D3 Hung. Felsőmuzslya;
prev. Gornja Mužlja. Vojvodina,
N Serbia
Mwali 79 F2 var. Moili, Fr. Mohéli.
island S Comoros
Mwanza 73 B6 Mwanza, NW Tanzania
Mweka 77 C6 Kasai-Occidental, C Dem.
Rep. Congo
Mwene-Ditu 77 D7 Kasai-Oriental,
S Dem. Rep. Congo
Mweru, Lake 77 D7 var. Lac Moero. lake
Dem. Rep. Congo/Zambia
Myadel' see Myadzyel
Myadzyel 107 C5 Pol. Miadzioł Nowy,
Rus. Myadel'. Minskaya Voblasts',
N Belarus
Myanaung 136 A4 Ayeyarwady,
SW Myanmar (Burma)
Myanmar 136 A3 off. Union of Myanmar.
country SE Asia. See also Burma

MYANMAR (BURMA)
Southeast Asia

Official name Republic of the
Union of Myanmar
Formation 1948 / 1948
Capital Nay Pyi Taw
Population 53.3 million / 210 people
per sq mile (81 people per sq km)
Total area 261,969 sq. miles
(678,500 sq. km)
Languages Burmese* (Myanmar), Shan,
Karen, Rakhine, Chin, Yangbye,
Kachin, Mon
Religions Buddhist 89%, Christian 4%,
Muslim 4%, Other 3%
Ethnic mix Burman (Bamah) 68%,
Shan 9%, Karen 7%, Other 12%,
Rakhine 4%
Government Presidential system
Currency Kyat = 100 pyas
Literacy rate 92%
Calorie consumption 2528 kilocalories

Myaungmya 136 A4 Ayeyarwady,
SW Myanmar (Burma)
Myaydo see Aunglan
Myerkulavichy 107 D7 Rus.
Merkulavichy
Myingyan 136 B3 Mandalay, C Myanmar
(Burma)
Myinmu 136 B3 Sagaing, C Myanmar
(Burma)
Myitkyina 136 B2 Kachin State, N
Myanmar (Burma)

Mykolayiv *109 E4 Rus.* Nikolayev.
Mykolayivs'ka Oblast', S Ukraine
Mykonos *105 D6 var.* Míkonos. *island*
Kykládes, Greece, Aegean Sea
Myrhorod *109 F2 Rus.* Mirgorod.
Poltavs'ka Oblast', NE Ukraine
Mýrina *104 D4 var.* Mírina. Límnos,
SE Greece
Myrtle Beach *43 F2* South Carolina,
SE USA
Mýrtos *105 D8* Kríti, Greece,
E Mediterranean Sea
Myrtoum Mare *see* Mirtóo Pélagos
Myślibórz *98 B3* Zachodnio-pomorskie,
NW Poland
Mysore *132 C2 var.* Maisur. Karnātaka,
W India
Mysore *see* Karnātaka
My Tho *137 E6 var.* Mi Tho. Tiên Giang,
S Vietnam
Mytilene *see* Mytilíni
Mytilíni *105 D5 var.* Mitilíni; *anc.*
Mytilene. Lésvos, E Greece
Mzuzu *79 E2* Northern, N Malawi

N

Naberezhnyye Chelny *111 D5 prev.*
Brezhnev. Respublika Tatarstan,
W Russian Federation
Nablus *119 A6 var.* Nābulus, *Heb.*
Shekhem; *anc.* Neapolis, *Bibl.* Shechem.
N West Bank
Nābulus *see* Nablus
Nacala *79 F2* Nampula, NE Mozambique
Na-Ch'ii *see* Nagqu
Nada *see* Danzhou
Nadi *145 E4 prev.* Nandi. Viti Levu,
W Fiji
Nadur *102 A5* Gozo, N Malta
Nadvirna *108 C3 Pol.* Nadwórna, *Rus.*
Nadvornaya. Ivano-Frankivs'ka
Oblast', W Ukraine
Nadvoitsy *110 B3* Respublika Kareliya,
NW Russian Federation
Nadvornaya/Nadwórna *see* Nadvirna
Nadym *114 C3* Yamalo-Nenetskiy
Avtonomnyy Okrug, N Russian
Federation
Náfpaktos *105 B5 var.* Návpaktos.
Dytikí Elláda, C Greece
Náfplio *105 B6 prev.* Návplion.
Pelopónnisos, S Greece
Naga *139 E2 off.* Naga City; *prev.* Nueva
Caceres. Luzon, N Philippines
Naga City *see* Naga
Nagano *131 C5* Nagano, Honshū, S Japan
Nagaoka *131 D5* Niigata, Honshū,
C Japan
Nagara Pathom *see* Nakhon Pathom
Nagara Sridharmaraj *see* Nakhon Si
Thammarat
Nagara Svarga *see* Nakhon Sawan
Nagasaki *131 A7* Nagasaki, Kyūshū,
SW Japan
Nagato *131 A7* Yamaguchi, Honshū,
SW Japan
Nāgercoil *132 C3* Tamil Nādu, SE India
Nagorno-Karabakh *117 G3 var.*
Nagorno- Karabakhskaya
Avtonomnaya Oblast, *Arm.* Lerrnayin
Gharabakh, *Az.* Dağlíq Quarabağ,
Rus. Nagornyy Karabakh. *former
autonomous region* SW Azerbaijan
**Nagorno- Karabakhskaya
Avtonomnaya Oblast** *see*
Nagorno-Karabakh
Nagornyy Karabakh *see*
Nagorno-Karabakh
Nagoya *131 C6* Aichi, Honshū, SW Japan
Nāgpur *134 D4* Mahārāshtra, C India
Nagqu *126 C5 Chin.* Na-Ch'ii; *prev.* Hei-
ho. Xizang Zizhiqu, W China
Nagybánya *see* Baia Mare
Nagybecskerek *see* Zrenjanin
Nagydisznód *see* Cisnădie
Nagyenyed *see* Aiud
Nagykálló *99 E6* Szabolcs-Szatmár-
Bereg, E Hungary
Nagykanizsa *99 C7 Ger.* Grosskanizsa.
Zala, SW Hungary
Nagykároly *see* Carei
Nagykikinda *see* Kikinda
Nagykőrös *99 D7* Pest, C Hungary
Nagymihály *see* Michalovce
Nagysurány *see* Šurany
Nagyszalonta *see* Salonta
Nagyszeben *see* Sibiu

Nagyszentmiklós *see* Sânnicolau Mare
Nagyszöllős *see* Vynohradiv
Nagyszombat *see* Trnava
Nagytapolcsány *see* Topol'čany
Nagyvárad *see* Oradea
Naha *130 A3* Okinawa, Okinawa,
SW Japan
Nahariya *119 A5 prev.* Nahariyya.
Northern, N Israel
Nahariyya *see* Nahariya
Nahuel Huapí, Lago *65 B5 lake*
W Argentina
Nain *39 F2* Newfoundland and Labrador,
NE Canada
Nā'īn *120 D3* Eşfahān, C Iran
Nairobi *69 E5 country capital* (Kenya)
Nairobi Area, S Kenya
Nairobi *73 C6* Nairobi Area, S Kenya
Naissus *see* Niš
Najaf *see* An Najaf
Najima *see* Fukuoka
Najin *129 E3* NE North Korea
Najrān *121 B6 var.* Abā as Su'ūd. Najrān,
S Saudi Arabia
Nakambé *see* White Volta
Nakamura *131 B7 var.* Shimanto. Kōchi,
Shikoku, SW Japan
Nakatsugawa *131 C6 var.* Nakatugawa.
Gifu, Honshū, SW Japan
Nakatugawa *see* Nakatsugawa
Nakhichevan' *see* Naxçıvan
Nakhodka *115 G5* Primorskiy Kray,
SE Russian Federation
Nakhon Pathom *137 C5 var.* Nagara
Pathom, Nakorn Pathom. Nakhon
Pathom, W Thailand
Nakhon Ratchasima *137 C5 var.*
Khorat, Korat. Nakhon Ratchasima,
E Thailand
Nakhon Sawan *137 C5 var.* Muang
Nakhon Sawan, Nagara Svarga.
Nakhon Sawan, W Thailand
Nakhon Si Thammarat *137 C7 var.*
Nagara Sridharmaraj, Nakhon
Sithamnaraj. Nakhon Si Thammarat,
SW Thailand
Nakhon Sithamnaraj *see* Nakhon Si
Thammarat
Nakorn Pathom *see* Nakhon Pathom
Nakuru *73 C6* Rift Valley, SW Kenya
Nal'chik *111 B8* Kabardino-Balkarskaya
Respublika, SW Russian Federation
Nālūt *71 F2* NW Libya
Namakan Lake *40 A1 lake* Canada/USA
Namangan *123 F2* Namangan Viloyati,
E Uzbekistan
Nambala *78 D2* Central, C Zambia
Nam Co *126 C5 lake* W China
Nam Đinh *136 D3* Nam Ha, N Vietnam
Namib Desert *78 B3 desert* W Namibia
Namibe *78 A2 Port.* Moçâmedes,
Mossâmedes. Namibe, SW Angola
Namibia *78 B3 off.* Republic of Namibia,
var. South West Africa, *Afr.* Suidwes-
Afrika, *Ger.* Deutsch-Südwestafrika;
prev. German Southwest Africa, South-
West Africa. *country* S Africa

NAMIBIA
Southern Africa

Official name Republic of Namibia
Formation 1990 / 1994
Capital Windhoek
Population 2.3 million / 7 people
per sq mile (3 people per sq km)
Total area 318,694 sq. miles
(825,418 sq. km)
Languages Ovambo, Kavango, English*,
Bergdama, German, Afrikaans
Religions Christian 90%,
Traditional beliefs 10%
Ethnic mix Ovambo 50%, Other
tribes 22%, Kavango 9%, Damara 7%,
Herero 7%, Other 5%
Government Presidential system
Currency Namibian dollar = 100 cents;
South African rand = 100 cents
Literacy rate 76%
Calorie consumption 2086 kilocalories

Namibia, Republic of *see* Namibia
Nam Khong *see* Mekong
Namnetes *see* Nantes
Namo *see* Namu Atoll
Nampa *46 D3* Idaho, NW USA
Nampula *79 E2* Nampula,
NE Mozambique

Namsos *84 B4* Nord-Trøndelag,
C Norway
Nam Tha *136 C4 river* N Laos
Namu Atoll *144 D2 var.* Namo. *atoll*
Ralik Chain, C Marshall Islands
Namur *87 C6 Dut.* Namen. Namur,
SE Belgium
Namyit Island *128 C8 island* S Spratly
Islands
Nan *136 C4 var.* Muang Nan. Nan,
NW Thailand
Nanaimo *36 D5* Vancouver Island,
British Columbia, SW Canada
Nanchang *128 C5 var.* Nan-ch'ang,
Nanch'ang-hsien. *province capital*
Jiangxi, S China
Nan-ch'ang *see* Nanchang
Nanch'ang-hsien *see* Nanchang
Nan-ching *see* Nanjing
Nancy *90 D3* Meurthe-et-Moselle,
NE France
Nandaime *52 D3* Granada,
SW Nicaragua
Nanded *134 D5* Mahārāshtra, C India
Nandi *see* Nadi
Nándorhgy *see* Oţelu Roşu
Nandyāl *132 C1* Andhra Pradesh,
E India
Nan Hai *see* South China Sea
Naniwa *see* Ōsaka
Nanjing *128 D5 var.* Nan-ching,
Nanking; *prev.* Chianning, Chian-ning,
Kiang-ning, Jiangsu. *province capital*
Jiangsu, E China
Nanking *see* Nanjing
Nanning *128 B6 var.* Nan-ning; *prev.*
Yung-ning. Guangxi Zhuangzu
Zizhiqu, S China
Nan-ning *see* Nanning
Nanortalik *82 C5* Kitaa, S Greenland
Nanpan Jiang *136 D2 river* S China
Nanping *128 D6 var.* Nan-p'ing; *prev.*
Yenping. Fujian, SE China
Nan-p'ing *see* Nanping
Nansei Syotō Trench *see*
Ryukyu Trench
Nansen Basin *155 C4 undersea basin*
Arctic Ocean
Nansen Cordillera *see* Gakkel Ridge
Nansha Qundao *see* Spratly Islands
Nanterre *90 D1* Hauts-de-Seine,
N France
Nantes *90 B4* Bret. Naoned; *anc.*
Condivincum, Namnetes. Loire-
Atlantique, NW France
Nantucket Island *41 G3 island*
Massachusetts, NE USA
Nanumaga *145 E3 var.* Nanumanga.
atoll NW Tuvalu
Nanumanga *see* Nanumaga
Nanumea Atoll *145 E3 atoll*
NW Tuvalu
Nanyang *128 C5 var.* Nan-yang.
Henan, C China
Nan-yang *see* Nanyang
Naoned *see* Nantes
Napa *47 B6* California, W USA
Napier *150 E4* Hawke's Bay, North
Island, New Zealand
Naples *43 E5* Florida, SE USA
Naples *see* Napoli
Napo *56 A3 province* NE Ecuador
Napoléon-Vendée *see* la Roche-sur-Yon
Napoli *97 D5 Eng.* Naples; *Ger.* Neapel;
anc. Neapolis. Campania, S Italy
Napo, Río *60 C1 river* Ecuador/Peru
Naracoorte *149 B7* South Australia
Naradhivas *see* Narathiwat
Narathiwat *137 C7 var.* Naradhivas.
SW Thailand
Narbada *see* Narmada
Narbo Martius *see* Narbonne
Narbonne *91 C6 anc.* Narbo Martius.
Aude, S France
Narborough Island *see*
Fernandina, Isla
Nares Abyssal Plain *see* Nares Plain
Nares Plain *35 E6 var.* Nares Abyssal
Plain. *abyssal plain* NW Atlantic Ocean
Nares Stræde *see* Nares Strait
Nares Strait *82 D1 Dan.* Nares Stræde.
strait Canada/Greenland
Narew *98 E3 river* E Poland
Narova *see* Narva
Narovlya *see* Narowlya
Narowlya *107 C8 Rus.* Narovlya.
Homyel'skaya Voblasts', SE Belarus

Närpes *85 D5 Fin.* Närpiö. Länsi-Sue
W Finland
Närpiö *see* Närpes
Narrabri *149 D6* New South Wales,
SE Australia
Narrogin *147 B6* Western Australia
Narva *106 E2* Ida-Virumaa, NE Eston
Narva *106 E2 prev.* Narova. *river*
Estonia/Russian Federation
Narva Bay *106 E2 Est.* Narva Laht, *G*
Narva-Bucht, *Rus.* Narvskiy Zaliv.
Estonia/Russian Federation
Narva Laht *see* Narva Bay
Narva Reservoir *106 E2 Est.*
Narva Veehoidla, *Rus.* Narvskoye
Vodokhranilishche. *reservoir* Eston
Russian Federation
Narva Veehoidla *see* Narva Reservoir
Narvik *84 C3* Nordland, C Norway
Narvskiy Zaliv *see* Narva Bay
Narvskoye Vodokhranilishche *see*
Narva Reservoir
Narwa-Bucht *see* Narva Bay
Nar'yan-Mar *110 D3 prev.* Beloshche
Dzerzhinskiy. Nenetskiy Avtonomn
Okrug, NW Russian Federation
Naryn *123 G2* Narynskaya Oblast',
C Kyrgyzstan
Nassau *54 C1 country capital*
(The Bahamas) New Providence,
N Bahamas
Năsăud *108 B3 Ger.* Nussdorf, *Hung.*
Naszód. Bistriţa-Năsăud, N Roman
Nase *see* Naze
Nāshik *134 C5 prev.* Nāsik. Mahārāsh
W India
Nashua *41 G3* New Hampshire,
NE USA
Nashville *42 C1 state capital* Tenness
S USA
Näsijärvi *85 D5 lake* SW Finland
Nāsik *see* Nāshik
**Nāsir, Buhayrat/ Nāşir, Buḥayrat/
Nāşir,Buḥeiret** *see* Nasser, Lake
Nāsiri *see* Ahvāz
Nasiriya *see* An Nāşirīyah
Nasser, Lake *72 B3 var.* Buhayrat Nas
Buhayrat Nāşir, Buḥeiret Nāşir. *lak*
Egypt/Sudan
Naszód *see* Năsăud
Nata *78 C3* Central, NE Botswana
Natal *63 G2 state capital* Rio Grande d
Norte, E Brazil
Natal Basin *141 A6 var.* Mozambiqu
Basin. *undersea basin* W Indian Oce
Natanya *see* Netanya
Natchez *42 B3* Mississippi, S USA
Natchitoches *42 A2* Louisiana, S USA
Nathanya *see* Netanya
Natitingou *75 F4* NW Benin
Natsrat *see* Natzrat
Natuna Islands *see* Natuna, Kepulaua
Natuna, Kepulauan *124 D4 var.* Nat
Islands. *island group* W Indonesia
Naturaliste Plateau *141 E6 undersea
plateau* E Indian Ocean
Natzrat *119 A5 var.* Natsrat, *Ar.* En
Nazira, *Eng.* Nazareth; *prev.* Nazerat
Northern, N Israel
Naugard *see* Nowogard
Naujamiestis *106 C4* Panevėžys,
C Lithuania
Nauru *144 D2 off.* Republic of Naur
prev. Pleasant Island.
country W Pacific Ocean

NAURU
Australasia & Oceania

Official name Republic of Nauru
Formation 1968 / 1968
Capital Yaren District
Population 9434 / 1165 people
per sq mile (4444 people per sq km)
Total area 8.1 sq. miles (21 sq. km)
Languages Nauruan*, Kiribati, Chinese
Tuvaluan, English
Religions Nauruan Congregational
Church 60%, Roman Catholic 35%,
Other 5%
Ethnic mix Nauruan 93%, Chinese 5%,
European 1%, Other Pacific islanders 1%
Government Nonparty system
Currency Australian dollar = 100 cents
Literacy rate 95%
Calorie consumption Not available

Nauru, Republic of *see* Nauru
Nauta *60 C2* Loreto, N Peru

hrudak *107 C6 Pol.* Nowogródek,
, Novogrudok. Hrodzyenskaya
blasts', W Belarus
nagar *see* Jāmnagar
oolatsk *107 D5 Rus.* Novopolotsk.
syebyskaya Voblasts', N Belarus
ra *93 E2 Eng./Fr.* Navarre.
onomous community N Spain
re *see* Navarra
ssa Island *54 C3 US unincorporated
ritory* C West Indies
see Navoiy
iy *123 E2 Rus.* Navoi. Navoiy
byati, C Uzbekistan
oa *50 C2* Sonora, NW Mexico
at *see* Navolato
ato *50 C3 var.* Navolat. Sinaloa,
Mexico
aktos *see* Náfpaktos
lion *see* Náfplio
bashah *see* Nawābshāh
bshāh *134 B3 var.* Nawabashah.
d, S Pakistan
van *117 G3 Rus.* Nakhichevan'.
, Azerbaijan
s *105 D6 var.* Naxos. Náxos,
kládes, Greece, Aegean Sea
s *105 D6 island* Kykládes, Greece,
gean Sea
ro *130 D2* Hokkaidō, NE Japan
yi Taw *136 B4 country capital
yanmar (Burma)* Mandalay,
Myanmar (Burma)
'eth *see* Natzrat
a *60 D4* Ica, S Peru
a Ridge *57 A5 undersea ridge*
Pacific Ocean
ar *131 D3 var.* Nase. Kagoshima,
hami-ōshima, SW Japan
rat *see* Natzrat
li *116 A4* Aydın, SW Turkey
st *73 C5 var.* Adama, Hadama.
omīya, C Ethiopia
latando *78 B1 Port.* Salazar, Vila
azar. Cuanza Norte, NW Angola
a *76 C4* Bamingui-Bangoran,
Central African Republic
dé *77 B6* Ngounié, S Gabon
di *77 A6* Nyanga, S Gabon
ména *76 B3 var.* Ndjamena; *prev.*
t-Lamy. *country capital* (Chad)
ari-Baguirmi, W Chad
mena *see* N'Djaména
é *77 A5* Moyen-Ogooué, W Gabon
a *78 D2* Copperbelt, C Zambia
uani *see* Nzwani
h, Lough *89 B5 lake* E Northern
land, United Kingdom
Moudania *104 C4 var.* Néa
oudhania. Kentrikí Makedonía,
Greece
Moudhaniá *see* Néa Moudania
el *see* Napoli
oli *104 B4 prev.* Neápolis. Dytikí
akedonía, N Greece
oli *105 D8* Kríti, Greece,
Mediterranean Sea
oli *105 C7* Pelopónnisos, S Greece
olis *see* Neápoli, Greece
olis *see* Napoli, Italy
olis *see* Nablus, West Bank
Islands *36 A2 island group* Aleutian
ands, Alaska, USA
Zíkhni *104 C3 var.* Néa Zíkhni;
rv. Néa Zíkhna. Kentrikí Makedonía,
Greece
Zíkhna/Néa Zíkhni *see* Néa Zíkhni
j *52 B2* Quiché, W Guatemala
dag *see* Balkanabat
na, Pico da *62 C1 mountain*
W Brazil
aska *44 D4 off.* State of Nebraska,
o known as Blackwater State,
rnhusker State, Tree Planters State.
te C USA
aska City *45 F4* Nebraska, C USA
es River *49 H3 river* Texas, SW USA
chea *65 D5* Buenos Aires,
Argentina
rland *see* Netherlands
r Rijn *86 D4 Eng.* Lower Rhine.
er C Netherlands
rweert *87 D5* Limburg,
Netherlands
Be *86 E3* Gelderland, E Netherlands
pelt *87 C5* Limburg, NE Belgium
kamsk *111 D3* Respublika
shkortostan, W Russian Federation

Neftezavodsk *see* Seÿdi
Negara Brunei Darussalam *see* Brunei
Negēlē *73 D5 var.* Negelli, *It.* Neghelli.
Oromīya, C Ethiopia
Negelli *see* Negēlē
Negev *119 A7 Eng. Negev. desert* S Israel
Negev *see* HaNegev
Neghelli *see* Negēlē
Negomane *79 E2 var.* Negomano. Cabo
Delgado, N Mozambique
Negomano *see* Negomane
Negombo *132 C3* Western Province,
SW Sri Lanka
Negotin *100 E4* Serbia, E Serbia
Negra, Punto *60 A3 headland* NW Peru
Negreşti *see* Negreşti-Oaş
Negreşti-Oaş *108 B3 Hung.*
Avasfelsőfalu; *prev.* Negreşti. Satu
Mare, NE Romania
Negro, Río *65 C5 river* E Argentina
Negro, Río *62 D1 river* N South America
Negro, Río *64 D4 river* Brazil/Uruguay
Negros *139 E2 island* C Philippines
Nehbandān *120 E3* Khorāsān, E Iran
Neijiang *128 B5* Sichuan, C China
Neiva *58 B3* Huila, S Colombia
Nellore *132 D2* Andhra Pradesh, E India
Nelson *151 C5* Nelson, South Island,
New Zealand
Nelson *37 G4 river* Manitoba, C Canada
Néma *74 D3* Hodh ech Chargui,
SE Mauritania
Neman *106 B4 Ger.* Ragnit.
Kaliningradskaya Oblast', W Russian
Federation
Neman *106 A4 Bel.* Nyoman, *Ger.*
Memel, *Lith.* Nemunas, *Pol.* Niemen.
river NE Europe
Nemausus *see* Nîmes
Neméa *105 B6* Pelopónnisos, S Greece
Nemetocenna *see* Arras
Nemours *90 C3* Seine-et-Marne,
N France
Nemunas *see* Neman
Nemuro *130 E2* Hokkaidō, NE Japan
Neochóri *105 B5* Dytikí Elláda, C Greece
Nepal *135 E3 off.* Nepal. *country* S Asia

NEPAL
South Asia

Official name Nepal
Formation 1769 / 1769
Capital Kathmandu
Population 27.8 million / 526 people
per sq mile (203 people per sq km)
Total area 54,363 sq. miles
(140,800 sq. km)
Languages Nepali*, Maithili, Bhojpuri
Religions Hindu 81%, Buddhist 11%,
Muslim 4%, Other (including Christian) 4%
Ethnic mix Other 52%, Chhetri 16%,
Hill Brahman 13%, Tharu 7%, Magar 7%,
Tamang 5%
Government Transitional regime
Currency Nepalese rupee = 100 paisa
Literacy rate 57%
Calorie consumption 2580 kilocalories

Nereta *106 C4* S Latvia
Neretva *100 C4 river* Bosnia and
Herzegovina/Croatia
Neris *107 C5 Bel.* Viliya, *Pol.* Wilia; *prev.*
Pol. Wilja. *river* Belarus/Lithuania
Neris *see* Viliya
Nerva *92 C4* Andalucía, S Spain
Neryungri *115 F4* Respublika Sakha
(Yakutiya), NE Russian Federation
Neskaupstaður *83 E5* Austurland,
E Iceland
Ness, Loch *88 C3 lake* N Scotland,
United Kingdom
Nesterov *see* Zhovkva
Néstos *104 C3 Bul.* Mesta, *Turk.* Kara Su.
river Bulgaria/Greece
Nesvizh *see* Nyasvizh
Netanya *119 A6 var.* Natanya, Nathanya.
Central, C Israel
Netherlands *86 C3 off.* Kingdom of
the Netherlands, *var.* Holland, *Dut.*
Koninkrijk der Nederlanden,
Nederland. *country* NW Europe

NETHERLANDS
Northwest Europe

Official name Kingdom of the
Netherlands
Formation 1648 / 1839

NETHERLANDS
(continued)

Capital Amsterdam; The Hague
(administrative)
Population 16.8 million / 1283 people per
sq mile (495 people per sq km)
Total area 16,033 sq. miles (41,526 sq. km)
Languages Dutch*, Frisian
Religions Roman Catholic 36%,
Other 34%, Protestant 17%, Muslim 3%
Ethnic mix Dutch 82%, Other 12%,
Surinamese 2%, Turkish 2%, Moroccan 2%
Government Parliamentary system
Currency Euro = 100 cents
Literacy rate 99%
Calorie consumption 3147 kilocalories

Netherlands East Indies *see* Indonesia
Netherlands Guiana *see* Suriname
Netherlands, Kingdom of the *see*
Netherlands
Netherlands New Guinea *see* Papua
Nettilling Lake *37 G3 lake* Baffin Island,
Nunavut, N Canada
Netze *see* Noteć
Neu Amerika *see* Puławy
Neubrandenburg *94 D3* Mecklenburg-
Vorpommern, NE Germany
Neuchâtel *95 A7 Ger.* Neuenburg.
Neuchâtel, W Switzerland
Neuchâtel, Lac de *95 A7 Ger.*
Neuenburger See. *lake* W Switzerland
Neuenburg *see* Neuchâtel
Neuenburger See *see* Neuchâtel, Lac de
Neufchâteau *87 D8* Luxembourg,
SE Belgium
Neugradisk *see* Nova Gradiška
Neuhof *see* Zgierz
Neukuhren *see* Pionerskiy
Neumarkt *see* Târgu Secuiesc, Covasna,
Romania
Neumarkt *see* Târgu Mureş
Neumoldowa *see* Moldova Nouă
Neumünster *94 B2* Schleswig-Holstein,
N Germany
Neunkirchen *95 A5* Saarland,
SW Germany
Neuquén *65 B5* Neuquén, SE Argentina
Neuruppin *94 C3* Brandenburg,
NE Germany
Neusalz an der Oder *see* Nowa Sól
Neu Sandec *see* Nowy Sącz
Neusatz *see* Novi Sad
Neusiedler See *95 E6 Hung.* Fertő. *lake*
Austria/Hungary
Neusohl *see* Banská Bystrica
Neustadt *see* Baia Mare, Maramureş,
Romania
Neustadt an der Haardt *see* Neustadt an
der Weinstrasse
Neustadt an der Weinstrasse *95
B5 prev.* Neustadt an der Haardt,
hist. Niewenstat; *anc.* Nova Civitas.
Rheinland-Pfalz, SW Germany
Neustadtl *see* Novo mesto
Neustettin *see* Szczecinek
Neustrelitz *94 D3* Mecklenburg-
Vorpommern, NE Germany
Neutra *see* Nitra
Neu-Ulm *95 B6* Bayern, S Germany
Neuwied *95 A5* Rheinland-Pfalz,
W Germany
Neuzen *see* Terneuzen
Nevada *47 C5 off.* State of Nevada, *also
known as* Battle Born State, Sagebrush
State, Silver State. *state* W USA
NevadaSierra *92 D5 mountain range*
S Spain
Nevada, Sierra *47 B6 mountain range*
W USA
Nevers *90 C4 anc.* Noviodunum. Nièvre,
C France
Neves *76 E2* São Tomé, S Sao Tome and
Principe, Africa
Nevinnomyssk *111 B7* Stavropol'skiy
Kray, SW Russian Federation
Nevşehir *116 C3 var.* Nevshehr.
Nevşehir, C Turkey
Newala *73 C8* Mtwara, SE Tanzania
New Albany *40 C5* Indiana, N USA
New Amsterdam *59 G3* E Guyana
Newark *41 F4* New Jersey, NE USA
New Bedford *41 G3* Massachusetts,
NE USA
Newberg *46 B3* Oregon, NW USA
New Bern *43 F1* North Carolina,
SE USA
New Braunfels *49 G4* Texas, SW USA

Newbridge *89 B6 Ir.* An Droichead Nua.
Kildare, C Ireland
New Britain *144 B3 island* E Papua
New Guinea
New Brunswick *39 E4 Fr.* Nouveau-
Brunswick. *province* SE Canada
New Caledonia *144 D4 var.* Kanaky, Fr.
Nouvelle-Calédonie. *French overseas
territory* SW Pacific Ocean
New Caledonia *144 C5 island*
SW Pacific Ocean
New Caledonia Basin *142 C4 undersea
basin* W Pacific Ocean
Newcastle *149 D6* New South Wales,
SE Australia
Newcastle *see* Newcastle upon Tyne
Newcastle upon Tyne *88 D4 var.*
Newcastle, *hist.* Monkchester, *Lat.* Pons
Aelii. NE England, United Kingdom
New Delhi *134 D3 country capital*
(India) Delhi, N India
New England of the West *see* Minnesota
Newfoundland *39 G3 Fr.* Terre-Neuve.
island Newfoundland and Labrador,
SE Canada
Newfoundland and Labrador *39 F2 Fr.*
Terre Neuve. *province* E Canada
Newfoundland Basin *66 B3 undersea
feature* NW Atlantic Ocean
New Georgia Islands *144 C3 island
group* NW Solomon Islands
New Glasgow *39 F4* Nova Scotia,
SE Canada
New Goa *see* Panaji
New Guinea *144 A3 Dut.* Nieuw Guinea,
Ind. Irian. *island* Indonesia/Papua
New Guinea
New Hampshire *41 F2 off.* State of New
Hampshire, *also known as* Granite
State. *state* NE USA
New Haven *41 G3* Connecticut, NE USA
New Hebrides *see* Vanuatu
New Iberia *42 B3* Louisiana, S USA
New Ireland *144 C3 island* NE Papua
New Guinea
New Jersey *41 F4 off.* State of New
Jersey, *also known as* The Garden State.
state NE USA
Newman *146 B4* Western Australia
Newmarket *89 E6* E England, United
Kingdom
New Mexico *48 C2 off.* State of New
Mexico, *also known as* Land of
Enchantment, Sunshine State.
state SW USA
New Orleans *42 B3* Louisiana, S USA
New Plymouth *150 C4* Taranaki, North
Island, New Zealand
Newport *89 D7* S England, United
Kingdom
Newport *89 C7* SE Wales, United
Kingdom
Newport *40 C4* Kentucky, S USA
Newport *41 G2* Vermont, NE USA
Newport News *41 F5* Virginia, NE USA
New Providence *54 C1 island* Bahamas
Newquay *89 C7* SW England, United
Kingdom
Newry *89 B5 Ir.* An tÚir. SE Northern
Ireland, United Kingdom
New Sarum *see* Salisbury
New Siberian Islands *115 F1 Eng.*
New Siberian Islands. *island group*
N Russian Federation
New Siberian Islands *see* Novosibirskiye
Ostrova
New South Wales *149 C6 state*
SE Australia
Newton *45 G3* Iowa, C USA
Newton *45 F5* Kansas, C USA
Newtownabbey *89 B5 Ir.* Baile na
Mainistreach. E Northern Ireland,
United Kingdom
New Ulm *45 F2* Minnesota, N USA
New York *41 F4* New York, NE USA
New York *41 F3 state* NE USA
New Zealand *150 A4 country*
SW Pacific Ocean

NEW ZEALAND
Australasia & Oceania

Official name New Zealand
Formation 1947 / 1947
Capital Wellington
Population 4.5 million / 43 people
per sq mile (17 people per sq km)
Total area 103,737 sq. miles
(268,680 sq. km)

NEW ZEALAND
(continued)

Languages English*, Maori*
Religions Anglican 24%, Other 22%,
Presbyterian 18%, Nonreligious 16%,
Roman Catholic 15%, Methodist 5%
Ethnic mix European 75%, Maori 15%,
Other 7%, Samoan 3%
Government Parliamentary system
Currency New Zealand dollar = 100 cents
Literacy rate 99%
Calorie consumption 3170 kilocalories

Neyveli *132 C2* Tamil Nādu, SE India
Nezhin *see* Nizhyn
Ngangze Co *126 B5 lake* W China
Ngaoundéré *76 B4 var.* N'Gaoundéré.
  Adamaoua, N Cameroon
N'Gaoundéré *see* Ngaoundéré
Ngazidja *79 F2 Fr.* Grande Comore, *var.*
  Njazidja. *island* NW Comoros
N'Giva *78 B3 var.* Ondjiva, *Port.* Vila
  Pereira de Eça. Cunene, S Angola
Ngo *77 B6* Plateaux, SE Congo
Ngoko *77 B5 river* Cameroon/Congo
Ngourti *75 H3* Diffa, E Niger
Nguigmi *75 H3 var.* N'Guigmi. Diffa,
  SE Niger
N'Guigmi *see* Nguigmi
N'Gunza *see* Sumbe
Nguru *75 G3* Yobe, NE Nigeria
Nha Trang *137 E6* Khanh Hoa,
  S Vietnam
Niagara Falls *38 D5* Ontario, S Canada
Niagara Falls *41 E3* New York, NE USA
Niagara Falls *40 D3 waterfall* Canada/
  USA
Niamey *75 F3 country capital* (Niger)
  Niamey, SW Niger
Niangay, Lac *75 E3 lake* E Mali
Nia-Nia *77 E5* Orientale, NE Dem.
  Rep. Congo
Nias, Pulau *138 A3 island* W Indonesia
Nicaea *see* Nice
Nicaragua *52 D3 off.* Republic of
  Nicaragua. *country* Central America

NICARAGUA
Central America

Official name Republic of Nicaragua
Formation 1838 / 1838
Capital Managua
Population 6.1 million / 133 people
per sq mile (51 people per sq km)
Total area 49,998 sq. miles (129,494 sq. km)
Languages Spanish*, English Creole,
Miskito
Religions Roman Catholic 80%,
Protestant Evangelical 17%, Other 3%
Ethnic mix Mestizo 69%, White 17%,
Black 9%, Amerindian 5%
Government Presidential system
Currency Córdoba oro = 100 centavos
Literacy rate 78%
Calorie consumption 2564 kilocalories

Lake Nicaragua *52 D4 var.* Cocibolca,
  Gran Lago, *Eng.* Lake Nicaragua. *lake*
  S Nicaragua
Nicaragua, Lake *see* Nicaragua, Lago de
Nicaragua, Republic of *see* Nicaragua
Nicaria *see* Ikaría
Nice *91 D6 It.* Nizza; *anc.* Nicaea. Alpes-
  Maritimes, SE France
Nicephorium *see* Ar Raqqah
Nicholas II Land *see* Severnaya Zemlya
Nicholls Town *54 C1* Andros Island,
  NW Bahamas
Nicobar Islands *124 B4 island group*
  India, E Indian Ocean
Nicosia *102 C5 Gk.* Lefkosía, *Turk.*
  Lefkoşa. *country capital* (Cyprus)
  C Cyprus
Nicoya *52 D4* Guanacaste, W Costa Rica
Nicoya, Golfo de *52 D5 gulf* W Costa
  Rica
Nicoya, Península de *52 D4 peninsula*
  NW Costa Rica
Nida *106 A3 Ger.* Nidden. Klaipėda,
  SW Lithuania
Nidaros *see* Trondheim
Nidden *see* Nida
Nidzica *98 D3 Ger.* Niedenburg.
  Warmińsko-Mazurskie, NE Poland

Niedenburg *see* Nidzica
Niedere Tauern *99 A6 mountain range*
  C Austria
Niemen *see* Neman
Nieśwież *see* Nyasvizh
Nieuw Amsterdam *59 G3* Commewijne,
  NE Suriname
Nieuw-Bergen *86 D4* Limburg,
  SE Netherlands
Nieuwegein *86 C4* Utrecht,
  C Netherlands
Nieuw Guinea *see* New Guinea
Nieuw Nickerie *59 G3* Nickerie,
  NW Suriname
Niewenstat *see* Neustadt an der
  Weinstrasse
Niğde *116 C4* Niğde, C Turkey
Niger *75 F3 off.* Republic of Niger.
  country W Africa

NIGER
West Africa

Official name Republic of Niger
Formation 1960 / 1960
Capital Niamey
Population 17.8 million / 36 people
per sq mile (14 people per sq km)
Total area 489,188 sq. miles
(1,267,000 sq. km)
Languages Hausa, Djerma, Fulani, Tuareg,
Teda, French*
Religions Muslim 99%, Other (including
Christian) 1%
Ethnic mix Hausa 53%, Djerma and
Songhai 21%, Tuareg 11%, Fulani 7%,
Kanuri 6%, Other 2%
Government Semi-presidential system
Currency CFA franc = 100 centimes
Literacy rate 16%
Calorie consumption 2546 kilocalories

Niger *75 F4 river* W Africa
Nigeria *75 F4 off.* Federal Republic of
  Nigeria. *country* W Africa

NIGERIA
West Africa

Official name Federal Republic of Nigeria
Formation 1960 / 1961
Capital Abuja
Population 174 million / 494 people
per sq mile (191 people per sq km)
Total area 356,667 sq. miles
(923,768 sq. km)
Languages Hausa, English*, Yoruba, Ibo
Religions Muslim 50%, Christian 40%,
Traditional beliefs 10%
Ethnic mix Other 29%, Hausa 21%,
Yoruba 21%, Ibo 18%, Fulani 11%
Government Presidential system
Currency Naira = 100 kobo
Literacy rate 51%
Calorie consumption 2724 kilocalories

Nigeria, Federal Republic of *see* Nigeria
Niger, Mouths of the *75 F5 delta*
  S Nigeria
Niger, Republic of *see* Niger
Nihon *see* Japan
Niigata *131 D5* Niigata, Honshū,
  C Japan
Niihama *131 B7* Ehime, Shikoku,
  SW Japan
Ni'ihau *47 A7 var.* Niihau. *island*
  Hawaii, USA, C Pacific Ocean
Nii-jima *131 D6 island* E Japan
Nijkerk *86 D3* Gelderland,
  C Netherlands
Nijlen *87 C5* Antwerpen, N Belgium
Nijmegen *86 D4 Ger.* Nimwegen;
  anc. Noviomagus. Gelderland,
  SE Netherlands
Nikaria *see* Ikaría
Nikel' *110 C2 Finn.* Kolosjoki.
  Murmanskaya Oblast', NW Russian
  Federation
Nikiniki *139 E5* Timor, S Indonesia
Niklasmarkt *see* Gheorgheni
Nikolainkaupunki *see* Vaasa
Nikolayev *see* Mykolayiv
Nikol'sk *see* Ussuriysk
Nikol'sk-Ussuriyskiy *see* Ussuriysk
Nikopol' *109 F3* Dnipropetrovs'ka
  Oblast', SE Ukraine
Nikšić *101 C5* C Montenegro

Nikumaroro *145 E3 ; prev.* Gardner
  Island. *atoll* Phoenix Islands, C Kiribati
Nikunau *145 E3 var.* Nukunau; *prev.*
  Byron Island. *atoll* Tungaru, W Kiribati
Nile *72 B2 former province* NW Uganda
Nile *68 D3 Ar.* Nahr an Nīl. *river*
  N Africa
Nile Delta *72 B1 delta* N Egypt
Nil, Nahr an *see* Nile
Nîmes *91 C6 anc.* Nemausus, Nismes.
  Gard, S France
Nimwegen *see* Nijmegen
Nine Degree Channel *132 B3 channel*
  India/Maldives
Ninetyeast Ridge *141 D5 undersea*
  feature E Indian Ocean
Ninety Mile Beach *150 C1 beach* North
  Island, New Zealand
Ningbo *128 D5 var.* Ning-po, Yin-hsien;
  prev. Ninghsien. Zhejiang, SE China
Ning-hsia *see* Ningxia
Ninghsien *see* Ningbo
Ning-po *see* Ningbo
Ningsia/Ningsia Hui/Ningsia Hui
  Autonomous Region *see* Ningxia
Ningxia *128 B4 off.* Ningxia Huizu
  Zizhiqu, *var.* Ning-hsia, Ningsia, *Eng.*
  Ningsia Hui, Ningsia Hui Autonomous
  Region. *autonomous region* N China
Ningxia Huizu Zizhiqu *see* Ningxia
Nio *see* Íos
Niobrara River *45 E3 river* Nebraska/
  Wyoming, C USA
Nioro *74 D3 var.* Nioro du Sahel. Kayes,
  W Mali
Nioro du Sahel *see* Nioro
Niort *90 B4* Deux-Sèvres, W France
Nipigon *38 B4* Ontario, S Canada
Nipigon, Lake *38 B3 lake* Ontario,
  S Canada
Nippon *see* Japan
Niš *101 E5 Eng.* Nish, *Ger.* Nisch; *anc.*
  Naissus. Serbia, SE Serbia
Niṣab *120 B4* Al Ḩudūd ash Shamālīyah,
  N Saudi Arabia
Nisch/Nish *see* Niš
Nisibin *see* Nusaybin
Nisiros *see* Nísyros
Nisko *98 E4* Podkrapackie, SE Poland
Nismes *see* Nîmes
Nistru *see* Dniester
Nísyros *105 E7 var.* Nisiros. *island*
  Dodekánisa, Greece, Aegean Sea
Nitra *95 C6 Ger.* Neutra, *Hung.* Nyitra.
  Nitriansky Kraj, SW Slovakia
Nitra *95 C6 Ger.* Neutra, *Hung.* Nyitra.
  river W Slovakia
Niuatobutabu *see* Niuatoputapu
Niuatoputapu *145 E4 var.*
  Niuatobutabu; *prev.* Keppel Island.
  island N Tonga
Niue *145 F4 self-governing territory* in
  free association with New Zealand
  S Pacific Ocean
Niulakita *145 E3 var.* Nurakita. *atoll*
  S Tuvalu
Niutao *145 E3 atoll* NW Tuvalu
Nivernais *90 C4 cultural region*
  C France
Nizāmābād *134 D5* Telangana, C India
Nizhegorodskaya *111 C5 prev.* Nizhn'ohirs'kyy
Nizhnekamsk *111 C5* Respublika
  Tatarstan, W Russian Federation
Nizhnevartovsk *114 D3* Khanty-
  Mansiyskiy Avtonomnyy Okrug-Yugra,
  C Russian Federation
Nizhniy Novgorod *111 C5 prev.* Gor'kiy.
  Nizhegorodskaya Oblast', W Russian
  Federation
Nizhniy Odes *110 D4* Respublika Komi,
  NW Russian Federation
Nizhyn *109 E1 Rus.* Nezhin.
  Chernihivs'ka Oblast', NE Ukraine
Nizza *see* Nice
Njazidja *see* Ngazidja
Njombe *73 C8* Iringa, S Tanzania
Nkayi *77 B6 prev.* Jacob. Bouenza,
  S Congo
Nkongsamba *76 A4 var.* N'Kongsamba.
  Littoral, W Cameroon
N'Kongsamba *see* Nkongsamba
Nmai Hka *136 B2 var.* Me Hka. *river*
  N Myanmar (Burma)
Nobeoka *131 B7* Miyazaki, Kyūshū,
  SW Japan
Noboribetsu *130 D3 var.* Noboribetu.
  Hokkaidō, NE Japan
Noboribetu *see* Noboribetsu
Nogales *50 B1* Sonora, NW Mexico
Nogales *48 B3* Arizona, SW USA

Nogal Valley *see* Dooxo Nugaaleed
Noire, Rivi `ere *see* Black River
Nokia *85 D5* Länsi-Suomi,
  W Finland
Nokou *76 B3* Kanem, W Chad
Nola *77 B5* Sangha-Mbaéré, SW Central
  African Republic
Nolinsk *111 C5* Kirovskaya Oblast',
  NW Russian Federation
Nongkaya *see* Nong Khai
Nong Khai *136 C4 var.* Mi Chai,
  Nongkaya. Nong Khai, E Thailand
Nonouti *144 D2 prev.* Sydenham Island
  atoll Tungaru, W Kiribati
Noord-Beveland *86 B4 var.* North
  Beveland. *island* SW Netherlands
Noordwijk aan Zee *86 C3* Zuid-Holland,
  W Netherlands
Noordzee *see* North Sea
Nora *85 C6* Örebro, C Sweden
Norak *123 E3 Rus.* Nurek. W Tajikistan
Nord *83 F1* Avannaarsua, N Greenland
Nordaustlandet *83 G1 island*
  NE Svalbard
Norden *94 A3* Niedersachsen,
  NW Germany
Norderstedt *94 B3* Schleswig-Holstein,
  N Germany
Nordfriesische Inseln *see* North Frisian
  Islands
Nordhausen *94 C4* Thüringen,
  C Germany
Nordhorn *94 A3* Niedersachsen,
  NW Germany
Nord, Mer du *see* North Sea
Nord-Ouest, Territoires du *see*
  Northwest Territories
Nordsee/Nordsjøen/Nordsoen *see*
  North Sea
Norfolk *45 E3* Nebraska, C USA
Norfolk *41 F5* Virginia, NE USA
Norfolk Island *142 D4 Australian*
  external territory SW Pacific Ocean
Norfolk Ridge *142 D4 undersea feature*
  W Pacific Ocean
Norge *see* Norway
Norias *49 C5* Texas, SW USA
Noril'sk *114 D3* Krasnoyarskiy Kray,
  N Russian Federation
Norman *49 G1* Oklahoma, C USA
Normandes, Îles *see* Channel Islands
Normandie *90 B3 Eng.* Normandy.
  cultural region N France
Normandy *see* Normandie
Normanton *148 C3* Queensland,
  NE Australia
Norrköping *85 C6* Östergötland,
  S Sweden
Norrtälje *85 C6* Stockholm, C Sweden
Norseman *147 B6* Western Australia
Norske Havet *see* Norwegian Sea
North Albanian Alps *101 C5 Alb.*
  Bjeshkët e Namuna, *SCr.* Prokletije.
  mountain range SE Europe
Northallerton *89 D5* N England, United
  Kingdom
Northam *147 A6* Western Australia
North America *34 continent*
Northampton *89 D6* C England, United
  Kingdom
North Andaman *133 F2 island*
  Andaman Islands, India, NE Indian
  Ocean
North Australian Basin *141 E5 Fr.*
  Bassin Nord de l' Australie. *undersea*
  feature E Indian Ocean
North Bay *38 D4* Ontario, S Canada
North Beveland *see* Noord-Beveland
North Borneo *see* Sabah
North Cape *150 C1 headland* North
  Island, New Zealand
North Cape *84 D1 Eng.* North Cape.
  headland N Norway
North Cape *see* Nordkapp
North Carolina *43 E1 off.* State of North
  Carolina, *also known as* Old North
  State, Tar Heel State, Turpentine State.
  state SE USA
North Channel *40 D2 lake channel*
  Canada/USA
North Charleston *43 F2* South Carolina,
  SE USA
North Dakota *44 D2 off.* State of North
  Dakota, *also known as* Flickertail
  State, Peace Garden State, Sioux State.
  state N USA
North Devon Island *see* Devon Island
North East Frontier Agency/North
  East Frontier Agency of Assam *see*
  Arunāchal Pradesh

theast Providence Channel *54 C1*
annel N Bahamas
heim *94 B4* Niedersachsen,
Germany
thern Cook Islands *145 F4 island*
oup N Cook Islands
thern Dvina *110 C4 var.* Northern
vina. *river* NW Russian Federation
hern Ireland *88 B4 var.* The Six
ounties. *cultural region* Northern
eland, United Kingdom
thern Mariana Islands *142 B1* US
mmonwealth territory W Pacific
cean
thern Rhodesia *see* Zambia
thern Sporades *105 C5 var.* Vóreioi
orádes, *prev.* Vóríoi Sporádhes, *Eng.*
ature N Coral Sea
orthern Sporades. *island group*
Greece
thern Sporades *see* Vóreies Sporádes
thern Territory *144 A5 territory*
Australia
th European Plain *81 E3 plain*
Europe
hfield *45 F2* Minnesota, N USA
h Fiji Basin *142 D3 undersea*
ature N Coral Sea
th Frisian Islands *94 B2 var.*
ordfriesische Inseln.
Germany
h Huvadhu Atoll *132 B5 var.* Gaafu
lifu Atoll. *atoll* S Maldives
h Island *150 B2 island* N New
ealand
h Korea *129 E3 off.* Democratic
eople's Republic of Korea, *Kor.*
hosón-minjujuŭi-inmin-kanghwaguk.
ountry E Asia

ORTH KOREA
East Asia

ficial name Democratic People's
ublic of Korea
rmation 1948 / 1953
pital Pyongyang
pulation 24.9 million / 536 people
r sq mile (207 people per sq km)
tal area 46,540 sq. miles
0,540 sq. km)
nguages Korean*
ligions Atheist 100%
hnic mix Korean 100%
vernment One-party state
rrency North Korean won = 100 chon
eracy rate 99%
lorie consumption 2103 kilocalories

th Little Rock *42 B1* Arkansas,
USA
th Minch *see* Minch, The
th Mole *93 G4 harbour wall*
W Gibraltar Europe
rth Platte *45 E4* Nebraska, C USA
rth Platte River *44 D4 river* C USA
rth Pole *155 B3 pole* Arctic Ocean
th Saskatchewan *37 F5 river*
lberta/Saskatchewan, S Canada
th Sea *80 D3 Dan.* Nordsøen, *Dut.*
oordzee, *Nor.* Nordsjøen; *prev.* German
ordzee, *Fr.* Mer du Nord, *Ger.*
ordsee, *Nor.* Nordsjøen; *prev.* German
cean, *Lat.* Mare Germanicum. *sea*
W Europe
rth Siberian Lowland *115 E2 var.*
North Siberian Plain, *Eng.* North
iberian Lowland. *lowlands* N Russian
ederation
rth Siberian Lowland/North
iberian Plain *see* Severo-Sibirskaya
izmennost'
rth Star State *see* Minnesota
rth Taranaki Bight *150 C3 gulf* North
sland, New Zealand
rth Uist *88 B3 island* NW Scotland,
United Kingdom
rthwest Atlantic Mid-Ocean Canyon
4 E4 *undersea feature* N Atlantic
cean
rth West Highlands *88 C3 mountain*
range N Scotland, United Kingdom
rthwest Pacific Basin *113 G4*
undersea feature NW Pacific Ocean
rthwest Territories *37 E3 Fr.*
erritoires du Nord-Ouest. *territory*
W Canada
rthwind Plain *155 B2 undersea*
eature Arctic Ocean
rton Sound *36 C2 inlet* Alaska, USA

Norway *85 A5 off.* Kingdom of Norway,
Nor. Norge. *country* N Europe

NORWAY
Northern Europe

Official name Kingdom of Norway
Formation 1905 / 1905
Capital Oslo
Population 5 million / 42 people
per sq mile (16 people per sq km)
Total area 125,181 sq. miles (324,220 sq. km)
Languages Norwegian* (Bokmål
"book language" and Nynorsk
"new Norsk"), Sámi
Religions Evangelical Lutheran 88%,
Other and nonreligious 8%, Muslim 2%,
Pentecostal 1%, Roman Catholic 1%
Ethnic mix Norwegian 93%, Other 6%,
Sámi 1%
Government Parliamentary system
Currency Norwegian krone = 100 øre
Literacy rate 99%
Calorie consumption 3484 kilocalories

Norway, Kingdom of *see* Norway
Norwegian Basin *83 F4 undersea feature*
NW Norwegian Sea
Norwegian Sea *83 F4 var.* Norske Havet.
sea NE Atlantic Ocean
Norwich *89 E6* E England, United
Kingdom
Nösen *see* Bistrița
Noshiro *130 D4 var.* Nosiro; *prev.*
Noshirominato. Akita, Honshū,
C Japan
Noshirominato/Nosiro *see* Noshiro
Nosivka *109 E1 Rus.* Nosovka.
Chernihivs'ka Oblast', NE Ukraine
Nosop *78 C4 river* E Namibia
Nosovka *see* Nosivka
Noşratābād *120 E3* Sīstān va
Balūchestān, E Iran
Noteć *98 C3 Ger.* Netze. *river*
NW Poland
Nóties Sporádes *see* Dodekánisa
Nottingham *89 D6* C England, United
Kingdom
Nouâdhibou *74 B2 prev.* Port-Étienne.
Dakhlet Nouâdhibou, W Mauritania
Nouakchott *74 B2 country capital*
(Mauritania) Nouakchott District,
SW Mauritania
Nouméa *144 C5 dependent territory*
capital (New Caledonia) Province Sud,
S New Caledonia
Nouveau-Brunswick *see* New Brunswick
Nouvelle-Calédonie *see* New Caledonia
Nouvelle Écosse *see* Nova Scotia
Nova Civitas *see* Neustadt an der
Weinstrasse
Nova Gorica *95 D8* W Slovenia
Nova Gradiška *100 C3 Ger.* Neugradisk,
Hung. Újgradiska. Brod-Posavina,
NE Croatia
Nova Iguaçu *63 F4* Rio de Janeiro,
SE Brazil
Nova Lisboa *see* Huambo
Novara *96 B2 anc.* Novaria. Piemonte,
NW Italy
Novaria *see* Novara
Nova Scotia *39 F4 Fr.* Nouvelle Écosse.
province SE Canada
Nova Scotia *35 E5 physical region*
SE Canada
Novaya Sibir', Ostrov *115 F1 island*
Novosibirskiye Ostrova, NE Russian
Federation
Novaya Zemlya *110 D1 island group*
N Russian Federation
Novaya Zemlya Trough *see* East Novaya
Zemlya Trough
Novgorod *see* Velikiy Novgorod
Novi Grad *see* Bosanski Novi
Novi Iskŭr *104 C2* Sofiya-Grad,
W Bulgaria
Noviodunum *see* Nevers, Nièvre,
France
Noviomagus *see* Lisieux, Calvados,
France
Noviomagus *see* Nijmegen, Netherlands
Novi Pazar *101 D5 Turk.* Yenipazar.
Serbia, S Serbia
Novi Sad *100 D3 Ger.* Neusatz, *Hung.*
Újvidék. Vojvodina, N Serbia
Novoazovs'k *109 G4 Rus.* Novoazovsk.
Donets'ka Oblast', E Ukraine
Novocheboksarsk *111 C5* Chuvashskaya
Respublika, W Russian Federation

Novocherkassk *111 B7* Rostovskaya
Oblast', SW Russian Federation
Novodvinsk *110 C3* Arkhangel'skaya
Oblast', NW Russian Federation
Novograd-Volynskiy *see*
Novohrad-Volyns'kyy
Novogrudok *see* Navahrudak
Novohrad-Volyns'kyy *108 D2 Rus.*
Novograd-Volynskiy. Zhytomyrs'ka
Oblast', N Ukraine
Novokazalinsk *see* Ayteke Bi
Novokuznetsk *114 D4 prev.* Stalinsk.
Kemerovskaya Oblast', S Russian
Federation
Novolazarevskaya *154 C2 Russian
research station* Antarctica
Novo mesto *95 E8 Ger.* Rudolfswert;
prev. Ger. Neustadtl. SE Slovenia
Novomoskovsk *111 B5* Tul'skaya
Oblast', W Russian Federation
Novomoskovs'k *109 F3 Rus.*
Novomoskovsk. Dnipropetrovs'ka
Oblast', E Ukraine
Novopolotsk *see* Navapolatsk
Novoradomsk *see* Radomsko
Novo Redondo *see* Sumbe
Novorossiysk *111 A7* Krasnodarskiy
Kray, SW Russian Federation
Novoshakhtinsk *111 B6* Rostovskaya
Oblast', SW Russian Federation
Novosibirsk *114 D4* Novosibirskaya
Oblast', C Russian Federation
Novotroitsk *111 D6* Orenburgskaya
Oblast', W Russian Federation
Novotroitskoye *see* Novotroyits'ke,
Ukraine
Novotroyits'ke *109 F4 Rus.*
Novotroitskoye. Khersons'ka Oblast',
S Ukraine
Novo-Urgench *see* Urganch
Novovolyns'k *108 C1 Rus.* Novovolynsk.
Volyns'ka Oblast', NW Ukraine
Novyy Bug *see* Novyy Buh
Novyy Buh *109 E3 Rus.* Novyy Bug.
Mykolayivs'ka Oblast', S Ukraine
Novyy Dvor *see* Novy Dvor
Novyy Margilan *see* Farg'ona
Novyy Uzen' *see* Zhanaozen
Nowa Sól *98 B4 var.* Nowasól, *Ger.*
Neusalz an der Oder. Lubuskie,
W Poland
Nowasól *see* Nowa Sól
Nowogard *98 B2 var.* Nowógard, *Ger.*
Naugard. Zachodnio-pomorskie,
NW Poland
Nowogródek *see* Navahrudak
Nowo-Minsk *see* Mińsk Mazowiecki
Nowy Dwór Mazowiecki *98 D3*
Mazowieckie, C Poland
Nowy Sącz *99 D5 Ger.* Neu Sandec.
Małopolskie, S Poland
Nowy Tomyśl *98 B3 var.* Nowy Tomysl.
Wielkopolskie, C Poland
Nowy Tomysl *see* Nowy Tomyśl
Noyon *90 C3* Oise, N France
Nsanje *79 E3* Southern, S Malawi
Nsawam *75 E5* SE Ghana
Ntomba, Lac *77 C6 var.* Lac Tumba.
lake NW Dem. Rep. Congo
Nubian Desert *72 B3 desert*
NE Sudan
Nu Chiang *see* Salween
Nu'eima *119 E7* E West Bank Asia
Nueva Caceres *see* Naga
Nueva Gerona *54 B2* Isla de la Juventud,
S Cuba
Nueva Rosita *50 D2* Coahuila,
NE Mexico
Nuevitas *54 C2* Camagüey, E Cuba
Nuevo, Bajo *53 G1 island* NW Colombia
South America
Nueve Casas Grandes *50 C1* Chihuahua,
N Mexico
Nuevo, Golfo *65 C6 gulf* S Argentina
Nuevo Laredo *51 E2* Tamaulipas,
NE Mexico
Nui Atoll *145 E3 atoll* W Tuvalu
Nu Jiang *see* Salween
Nûk *see* Nuuk
Nukha *see* Şäki
Nuku'alofa *145 E5 country capital*
(Tonga) Tongatapu, S Tonga
Nukufetau Atoll *145 E3 atoll* C Tuvalu
Nukulaelae Atoll *145 E3 var.* Nukulailai.
atoll E Tuvalu
Nukulailai *see* Nukulaelae Atoll
Nukunau *see* Nikunau

Nukunonu Atoll *145 E3 island*
C Tokelau
Nukus *122 C2* Qoraqalpog'iston
Respublikasi, W Uzbekistan
Nullarbor Plain *147 C6 plateau* South
Australia/Western Australia
Nunap Isua *82 C5 var.* Uummannarsuaq,
Dan. Kap Farvel, *Eng.* Cape Farewell.
cape S Greenland
Nunavut *37 F3 territory* N Canada
Nuneaton *89 D6* C England, United
Kingdom
Nunivak Island *36 B2 island* Alaska,
USA
Nunspeet *86 D3* Gelderland,
E Netherlands
Nuoro *97 A5* Sardegna, Italy,
C Mediterranean Sea
Nuqui *58 A3* Chocó, W Colombia
Nurakita *see* Niulakita
Nurata *see* Nurota
Nurek *see* Norak
Nuremberg *see* Nürnberg
Nurmes *84 E4* Itä-Suomi, E Finland
Nürnberg *95 C5 Eng.* Nuremberg.
Bayern, S Germany
Nurota *123 E2 Rus.* Nurata. Navoiy
Viloyati, C Uzbekistan
Nusaybin *117 F4 var.* Nisibin. Manisa,
SE Turkey
Nussdorf *see* Năsăud
Nutmeg State *see* Connecticut
Nuuk *82 C4 var.* Nûk, *Dan.* Godthaab,
Godthåb. *dependent territory capital*
(Greenland) Kitaa, SW Greenland
Nyagan' *114 F3 territory* N Canada
Avtonomnyy Okrug-Yugra, N Russian
Federation
Nyainqêntanglha Shan *126 C5*
mountain range W China
Nyala *72 A4* Southern Darfur, W Sudan
Nyamapanda *78 D3* Mashonaland East,
NE Zimbabwe
Nyamtumbo *73 C8* Ruvuma, S Tanzania
Nyanda *see* Masvingo
Nyandoma *110 C4* Arkhangel'skaya
Oblast', NW Russian Federation
Nyantakara *73 B7* Kagera, NW Tanzania
Nyasa, Lake *79 E2 var.* Lake Malawi;
prev. Lago Nyassa. *lake* E Africa
Nyasaland/Nyasaland Protectorate
see Malawi
Nyassa, Lago *see* Nyasa, Lake
Nyasvizh *107 C6 Pol.* Nieśwież, *Rus.*
Nesvizh. Minskaya Voblasts', C Belarus
Nyaunglebin *136 B4* Bago, SW Myanmar
(Burma)
Nyeri *73 C6* Central, C Kenya
Nyíregyháza *99 D6* Szabolcs-Szatmár-
Bereg, NE Hungary
Nyitra *see* Nitra
Nykøbing *85 B8* Sjælland, SE Denmark
Nyköping *85 C6* Södermanland,
S Sweden
Nylstroom *see* Modimolle
Nyngan *149 D6* New South Wales,
SE Australia
Nyoman *see* Neman
Nyurba *115 F3* Respublika Sakha
(Yakutiya), NE Russian Federation
Nyzhn'ohirs'kyy *109 F4 Rus.*
Nizhnegorskiy. Avtonomna Respublika
Krym, S Ukraine
NZ *see* New Zealand
Nzega *73 C7* Tabora, C Tanzania
Nzérékoré *74 D4* SE Guinea
Nzwani *79 G1 Fr.* Anjouan, *var.*
Ndzouani. *island* SE Comoros

O

Oa'hu *47 A7 var.* Oahu. *island* Hawai'ian
Islands, Hawaii, USA
Oak Harbor *46 B1* Washington,
NW USA
Oakland *47 B6* California, W USA
Oamaru *151 B7* Otago, South Island,
New Zealand
Oaxaca *51 F5 var.* Oaxaca de Juárez; *prev.*
Antequera. Oaxaca, SE Mexico
Oaxaca de Juárez *see* Oaxaca
Ob' *112 C2 river* C Russian Federation
Obal' *107 D5 Rus.* Obol'. Vitsyebskaya
Voblasts', N Belarus
Oban *88 C4* W Scotland, United
Kingdom
Oban *see* Halfmoon Bay
Obando *see* Puerto Inírida

Obdorsk see Salekhard
Óbecse see Bečej
Obeliai *106 C4* Panevėžys, NE Lithuania
Oberhollabrunn see Tulln
Ob', Gulf of *114 D3 Eng.* Gulf of Ob.
gulf N Russian Federation
Ob', Gulf of see Obskaya Guba
Obidovichi see Abidavichy
Obihiro *130 D2* Hokkaidō, NE Japan
Obo *76 D4* Haut-Mbomou, E Central
African Republic
Obock *72 D4* E Djibouti
Obol' see Obal'
Oborniki *98 C3* Wielkolpolskie,
W Poland
Obrovo see Abrova
Ob' Tablemount *141 B7 undersea
feature* S Indian Ocean
Ocala *43 E4* Florida, SE USA
Ocaña *58 B2* Norte de Santander,
N Colombia
Ocaña *92 D3* Castilla-La Mancha,
C Spain
O Carballiño *92 C1 Cast.* Carballiño.
Galicia, NW Spain
Occidental, Cordillera *58 B2 mountain
range* W South America
Occidental, Cordillera *61 E4 mountain
range* Bolivia/Chile
Ocean Falls *36 D5* British Columbia,
SW Canada
Ocean Island see Banaba
Oceanside *47 C8* California, W USA
Ocean State see Rhode Island
Ochakiv *109 E4 Rus.* Ochakov.
Mykolayivs'ka Oblast', S Ukraine
Ochakov see Ochakiv
Ochamchira see Ochamchire
Ochamchire *117 E2 prev.* Och'amch'ire.
W Georgia
Och'amch'ire see Ochamchire
Ocho Rios *54 B4* C Jamaica
Ochrida see Ohrid
Ochrida, Lake see Ohrid, Lake
Ocotal *52 D3* Nueva Segovia,
NW Nicaragua
Ocozocuautla *51 G5* Chiapas, SE Mexico
October Revolution Island *115 E2
Eng.* October Revolution Island.
island Severnaya Zemlya, N Russian
Federation
October Revolution Island see
Oktyabr'skoy Revolyutsii, Ostrov
Ocú *53 F5* Herrera, S Panama
Ōdate *130 D3* Akita, Honshū, C Japan
Oddur see Xuddur
Ödemiş *116 A4* Izmir, SW Turkey
Ödenburg see Sopron
Odenpäh see Otepää
Odense *85 B7* Syddtjylland, C Denmark
Oder *98 B3 Cz./Pol.* Odra. *river*
C Europe
Oderhaff see Szczeciński, Zalew
Odesa *109 E4 Rus.* Odessa. Odes'ka
Oblast', SW Ukraine
Odessa *49 E3* Texas, SW USA
Odessa see Odesa
Odessus see Varna
Odienné *74 D4* NW Côte d'Ivoire
Odisha *135 F4 prev.* Orissa. *cultural
region* NE India
Ódôngk *137 D6* Kâmpóng Spoe,
S Cambodia
Odoorn *86 E2* Drenthe, NE Netherlands
Odra see Oder
Oesel see Saaremaa
Of *117 E2* Trabzon, NE Turkey
Ofanto *97 D5 river* S Italy
Offenbach *95 B5 var.* Offenbach am
Main. Hessen, W Germany
Offenbach am Main see Offenbach
Offenburg *95 B6* Baden-Württemberg,
SW Germany
Ogaadeen see Ogaden
Ogaden *73 D5 Som.* Ogaadeen. *plateau*
Ethiopia/Somalia
Ōgaki *131 C6* Gifu, Honshū, SW Japan
Ogallala *44 D4* Nebraska, C USA
Ogbomosho *75 F4 var.* Ogmoboso. Oyo,
W Nigeria
Ogden *44 B4* Utah, W USA
Ogdensburg *41 F2* New York,
NE USA
Ogmoboso see Ogbomosho
Ogulin *100 A3* Karlovac, NW Croatia
Ohio *40 C4 off.* State of Ohio, *also known
as* Buckeye State. *state* N USA
Ohio River *40 C4 river* N USA
Ohlau see Oława

Ohri see Ohrid
Ohrid *101 D6 Turk.* Ochrida, Ohri.
SW FYR Macedonia
Ohrid, Lake *101 D6 var.* Lake Ochrida,
Alb. Liqeni i Ohrit, *Mac.* Ohridsko
Ezero. *lake* Albania/FYR Macedonia
Ohridsko Ezero/Ohrit, Liqeni i see
Ohrid, Lake
Ohura *150 D3* Manawatu-Wanganui,
North Island, New Zealand
Oirschot *87 C5* Noord-Brabant,
S Netherlands
Oise *90 C3 river* N France
Oistins *55 G2* S Barbados
Ōita *131 B7* Ōita, Kyūshū, SW Japan
Ojinaga *50 D2* Chihuahua, N Mexico
Ojos del Salado, Cerro *64 B3 mountain*
W Argentina
Okaihau *150 C2* Northland, North
Island, New Zealand
Okāra *134 C2* Punjab, E Pakistan
Okavanggo see Cubango/Okavango
Okavango see Cubango
Okavango Delta *78 C3 wetland*
N Botswana
Okayama *131 B6* Okayama, Honshū,
SW Japan
Okazaki *131 C6* Aichi, Honshū, C Japan
Okeechobee, Lake *43 E4 lake* Florida,
SE USA
Okefenokee Swamp *43 E3 wetland*
Georgia, SE USA
Okhotsk *115 G3* Khabarovskiy Kray,
E Russian Federation
Okhotsk, Sea of *113 F3 sea* NW Pacific
Ocean
Okhtyrka *109 F2 Rus.* Akhtyrka.
Sums'ka Oblast', NE Ukraine
Oki-guntō see Oki-shotō
Okinawa *130 A3 island* SW Japan
Okinawa-shoto *130 A3 island group*
Nansei-shotō, SW Japan Asia
Oki-shoto *131 B6 var.* Oki-guntō. *island
group* SW Japan
Oklahoma *49 F2 off.* State of Oklahoma,
also known as The Sooner State. *state*
C USA
Oklahoma City *49 G1 state capital*
Oklahoma, C USA
Okmulgee *49 G1* Oklahoma, C USA
Oko, Wadi *72 C3 river* NE Sudan
Oktyabr'skiy *111 D6* Volgogradskaya
Oblast', SW Russian Federation
Oktyabr'skiy see Aktsyabrski
Okulovka *110 B4 var.* Okulovka.
Novgorodskaya Oblast', W Russian
Federation
Okulovka see Okulovka
Okushiri-to *130 C3 var.* Okusiri Tô.
island NE Japan
Okusiri Tô see Okushiri-tō
Oláh-Toplicza see Toplița
Öland *85 C7 island* S Sweden
Olavarría *65 D5* Buenos Aires,
E Argentina
Oława *98 C4 Ger.* Ohlau. Dolnośląskie,
SW Poland
Olbia *97 A5 prev.* Terranova Pausania.
Sardegna, Italy, C Mediterranean Sea
Old Bay State/Old Colony State see
Massachusetts
Old Dominion see Virginia
Oldebroek *86 D3* Gelderland,
E Netherlands
Oldenburg *94 B3* Niedersachsen,
NW Germany
Oldenburg *94 C2 var.* Oldenburg
in Holstein. Schleswig-Holstein,
N Germany
Oldenburg in Holstein see Oldenburg
Oldenzaal *86 E3* Overijssel,
E Netherlands
Old Harbour *54 B5* C Jamaica
Old Line State see Maryland
Old North State see North Carolina
Olëkma *115 F4 river* C Russian
Federation
Olëkminsk *115 F3* Respublika Sakha
(Yakutiya), NE Russian Federation
Oleksandrivka *109 F3 Rus.*
Aleksandrovka. Kirovohrads'ka
Oblast', C Ukraine
Oleksandriya *109 F3 Rus.* Aleksandriya.
Kirovohrads'ka Oblast', C Ukraine
Olenegorsk *110 C2* Murmanskaya
Oblast', NW Russian Federation
Olenëk *115 E3* Respublika Sakha
(Yakutiya), NE Russian Federation

Olenëk *115 E3 river* NE Russian
Federation
Oléron, Île d' *91 A5 island* W France
Olevs'k *108 D1 Rus.* Olevsk.
Zhytomyrs'ka Oblast', N Ukraine
Olevsk see Olevs'k
Ölgiy *126 C2* Bayan-Ölgiy, W Mongolia
Olhão *92 B5* Faro, S Portugal
Olifa *78 B3* Kunene, NW Namibia
Ólimbos see Ólympos
Olimpo see Fuerte Olimpo
Olisipo see Lisboa
Olita see Alytus
Oliva *93 F4* Valenciana, E Spain
Olivet *90 C4* Loiret, C France
Olmaliq *123 E2 Rus.* Almalyk. Toshkent
Viloyati, E Uzbekistan
Olmütz see Olomouc
Olomouc *99 C5 Ger.* Olmütz, *Pol.*
Ołomuniec. Olomoucký Kraj, E Czech
Republic
Ołomuniec see Olomouc
Olonets *110 B3* Respublika Kareliya,
NW Russian Federation
Olovyannaya *115 F4* Zabaykal'skiy Kray,
S Russian Federation
Olpe *94 B4* Nordrhein-Westfalen,
W Germany
Olshanka see Vil'shanka
Olsnitz see Murska Sobota
Olsztyn *98 D2 Ger.* Allenstein.
Warmińsko-Mazurskie, N Poland
Olt *108 B5 var.* Oltul, *Ger.* Alt. *river*
S Romania
Olteniţa *108 C5 prev. Eng.* Oltenitsa; *anc.*
Constantiola. Călăraşi, SE Romania
Oltenitsa see Olteniţa
Oltul see Olt
Olvera *92 D5* Andalucía, S Spain
Ol'viopol' see Pervomays'k
Olympia *46 B2 state capital* Washington,
NW USA
Olympic Mountains *46 A2 mountain
range* Washington, NW USA
Olympus, Mount *104 B4 var.* Ólimbos,
Eng. Mount Olympus. *mountain*
N Greece
Omagh *89 B5 Ir.* An Ómaigh.
W Northern Ireland, United Kingdom
Omaha *45 F4* Nebraska, C USA
Oman *121 D6 off.* Sultanate of Oman,
Ar. Salţanat 'Umān; *prev.* Muscat and
Oman. *country* SW Asia

OMAN
Southwest Asia

Official name	Sultanate of Oman
Formation	1951 / 1951
Capital	Muscat
Population	3.6 million / 44 people per sq mile (17 people per sq km)
Total area	82,031 sq. miles (212,460 sq. km)
Languages	Arabic*, Baluchi, Farsi, Hindi, Punjabi
Religions	Ibadi Muslim 75%, Other Muslim and Hindu 25%
Ethnic mix	Arab 88%, Baluchi 4%, Persian 3%, Indian and Pakistani 3%, African 2%
Government	Monarchy
Currency	Omani rial = 1000 baisa
Literacy rate	87%
Calorie consumption	Not available

Oman, Gulf of *120 E4 Ar.* Khalīj 'Umān.
gulf N Arabian Sea
Oman, Sultanate of see Oman
Omboué *77 A6* Ogooué-Maritime,
W Gabon
Omdurman *72 B4 var.* Umm Durmān.
Khartoum, C Sudan
Ometepe, Isla de *52 D4 island*
S Nicaragua
Ommen *86 E3* Overijssel, E Netherlands
Omsk *114 C4* Omskaya Oblast',
C Russian Federation
Ōmuta *131 A7* Fukuoka, Kyūshū,
SW Japan
Onda *93 F3* Valenciana, E Spain
Ondjiva see N'Giva
Öndörhaan *127 E2 var.* Undur Khan;
prev. Tsetsen Khan. Hentiy, E Mongolia
Onega *110 C3* Arkhangel'skaya Oblast',
NW Russian Federation
Onega *110 B4 river* NW Russian
Federation
Lake Onega *110 B4 Eng.* Lake Onega.
lake NW Russian Federation

Onega, Lake see Onezhskoye Ozero
Onex *95 A7* Genève, SW Switzerland
Ongole *132 D1* Andhra Pradesh, E India
Onitsha *75 G5* Anambra, S Nigeria
Onon Gol *127 E2 river* N Mongolia
Onslow *146 A4* Western Australia
Onslow Bay *43 F1 bay* North Carolina,
E USA
Ontario *38 B3 province* S Canada
Ontario, Lake *41 E3 lake* Canada/USA
Onteniente see Ontinyent
Ontinyent *93 F4 var.* Onteniente.
Valenciana, E Spain
Ontong Java Rise *125 H4 undersea
feature* W Pacific Ocean
Onuba see Huelva
Oodeypore see Udaipur
Oos-Londen see East London
Oostakker *87 B5* Oost-Vlaanderen,
NW Belgium
Oostburg *87 B5* Zeeland,
SW Netherlands
Oostende *87 A5 Eng.* Ostend,
Fr. Ostende. West-Vlaanderen,
NW Belgium
Oosterbeek *86 D4* Gelderland,
SE Netherlands
Oosterhout *86 C4* Noord-Brabant,
S Netherlands
Opatija *100 A2 It.* Abbazia. Primorje-
Gorski Kotar, NW Croatia
Opava *99 C5 Ger.* Troppau.
Moravskoslezský Kraj, E Czech
Republic
Opazova see Stara Pazova
Opelika *42 D2* Alabama, S USA
Opelousas *42 B3* Louisiana, S USA
Ophiusa see Formentera
Opmeer *86 C2* Noord-Holland,
NW Netherlands
Opochka *110 A4* Pskovskaya Oblast',
W Russian Federation
Opole *98 C4 Ger.* Oppeln. Opolskie,
S Poland
Oporto see Porto
Opotiki *150 E3* Bay of Plenty, North
Island, New Zealand
Oppeln see Opole
Oppidum Ubiorum see Köln
Oqtosh *123 E2 Rus.* Aktash. Samarqand
Viloyati, C Uzbekistan
Oradea *108 B3 prev.* Oradea Mare, *Ger.*
Grosswardein, *Hung.* Nagyvárad.
Bihor, NW Romania
Oradea Mare see Oradea
Orahovac see Rahovec
Oral see Ural'sk
Oran *70 D2 var.* Ouahran, Wahran.
NW Algeria
Orange *149 D6* New South Wales,
SE Australia
Orange *91 D6 anc.* Arausio. Vaucluse,
SE France
Orangeburg *43 E2* South Carolina,
SE USA
Orange Cone see Orange Fan
Orange Fan *69 C7 var.* Orange Cone.
undersea feature SW Indian Ocean
Orange Mouth/Orangemund see
Oranjemund
Orange River *78 B4 var.* Oranjerivier.
river S Africa
Orange Walk *52 C1* Orange Walk,
N Belize
Oranienburg *94 D3* Brandenburg,
NE Germany
Oranjemund *78 B4 var.* Orangemund;
prev. Orange Mouth. Karas,
SW Namibia
Oranjerivier see Orange River
Oranjestad *55 E5 dependent territory
capital* (Aruba) Lesser Antilles,
S Caribbean Sea
Orany see Varéna
Oraşul Stalin see Braşov
Oravicabánya see Oraviţa
Oraviţa *108 A4 Ger.* Orawitza, *Hung.*
Oravicabánya. Caraş-Severin,
SW Romania
Orawitza see Oraviţa
Orbetello *96 B4* Toscana, C Italy
Orcadas *154 A1 Argentinian research
station* South Orkney Islands,
Antarctica
Orchard Homes *44 B1* Montana,
NW USA
Ordino *91 A8* Ordino, NW Andorra
Europe
Ordos Desert see Mu Us Shadi

du 116 D2 *anc.* Cotyora. Ordu,
N Turkey
dzhonikidze 109 F3 Dnipropetrovs'ka
Oblast', E Ukraine
dzhonikidze *see* Vladikavkaz, Russian
Federation
dzhonikidze *see* Yenakiyeve, Ukraine
ealla 59 G3 E Guyana
ebro 85 C6 Örebro, C Sweden
egon 46 B3 *off.* State of Oregon, *also
known as* Beaver State, Sunset State,
Valentine State, Webfoot State. *state*
NW USA
egon City 46 B3 Oregon, NW USA
egon, State of *see* Oregon
ekhov *see* Orikhiv
el 111 B5 Orlovskaya Oblast',
W Russian Federation
em 44 B4 Utah, W USA
e Mountains 95 C5 *Cz.* Krušné Hory,
Eng. Ore Mountains. *mountain range*
Czech Republic/Germany
e Mountains *see* Erzgebirge/Krušné
Hory
enburg 111 D6 *prev.* Chkalov.
Orenburgskaya Oblast', W Russian
Federation
ense *see* Ourense
estiáda 104 D3 *prev.* Orestiás.
Anatolikí Makedonía kai Thráki,
NE Greece
estiás *see* Orestiáda
gan Peak 48 D3 *mountain* New
Mexico, SW USA
geyev *see* Orhei
hei 108 D3 *var.* Orheiu, *Rus.* Orgeyev.
N Moldova
heiu *see* Orhei
iental, Cordillera 60 D3 *mountain
range* Bolivia/Peru
iental, Cordillera 58 B3 *mountain
range* C Colombia
ihuela 93 F4 Valenciana, E Spain
ikhiv 109 G3 *Rus.* Orekhov.
Zaporiz'ka Oblast', SE Ukraine
inoco, Río 59 E2 *river* Colombia/
Venezuela
issa *see* Odisha
issaar *see* Orissaare
issaare 106 C2 *Ger.* Orissaar.
Saaremaa, W Estonia
istano 97 A5 Sardegna, Italy,
C Mediterranean Sea
ito 58 A4 Putumayo, SW Colombia
izaba, Volcán Pico de 35 C7 *var.*
Citlaltépetl. *mountain* S Mexico
kney *see* Orkney Islands
kney Islands 88 C2 *var.* Orkney,
Orkneys. *island group* N Scotland,
United Kingdom
kneys *see* Orkney Islands
lando 43 E4 Florida, SE USA
omocto 39 F4 New Brunswick,
SE Canada
rona 145 F3 *prev.* Hull Island. *atoll*
Phoenix Islands, C Kiribati
opeza *see* Cochabamba
osirá Rodhópis *see* Rhodope
Mountains
pington 89 B8 United Kingdom
schowa *see* Orsova
sha 107 E6 Vitsyebskaya Voblasts',
NE Belarus
msö *see* Vormsi
muz, Strait of *see* Hormuz, Strait of
nsköldsvik 85 C5 Västernorrland,
C Sweden
olaunum *see* Arlon
ol Dengizi *see* Aral Sea
sk 114 B4 Orenburgskaya Oblast',
W Russian Federation
şova 108 A4 *Ger.* Orschowa, *Hung.*
Orsova. Mehedinți, SW Romania
telsburg *see* Szczytno
thez 91 B6 Pyrénées-Atlantiques,
SW France
tona 96 D4 Abruzzo, C Italy
uba *see* Aruba
uro 61 F4 Oruro, W Bolivia
yokko *see* Yalu
s 86 D4 Noord-Brabant, S Netherlands

Ōsaka 131 C6 *hist.* Naniwa. Ōsaka,
Honshū, SW Japan
Ōsaki *see* Furukawa
Osa, Península de 53 E5 *peninsula*
S Costa Rica
Osborn Plateau 141 D5
undersea feature E Indian Ocean
Osca *see* Huesca
Ösel *see* Saaremaa
Osh 123 F2 Oshskaya Oblast',
SW Kyrgyzstan
Oshawa 38 D5 Ontario, SE Canada
Oshikango 78 B3 Ohangwena,
N Namibia
O-shima 131 D6 *island* S Japan
Oshkosh 40 B2 Wisconsin, N USA
Oshmyany *see* Ashmyany
Osiek *see* Osijek
Osijek 100 C3 *prev.* Osiek, Osjek, *Ger.*
Esseg, *Hung.* Eszék. Osijek-Baranja,
E Croatia
Osipenko *see* Berdyans'k
Osipovichi *see* Asipovichy
Osjek *see* Osijek
Oskaloosa 45 G4 Iowa, C USA
Oskarshamn 85 C7 Kalmar, S Sweden
Öskemen *see* Ust'-Kamenogorsk
Oskil *see* Oskol
Oskol 109 G2 *Rus.* Oskil. *river* Russian
Federation/Ukraine
Oslo 85 B6 *prev.* Christiania, Kristiania.
country capital (Norway) Oslo,
S Norway
Osmaniye 116 D4 Osmaniye, S Turkey
Osnabrück 94 A3 Niedersachsen,
NW Germany
Osogov Mountains 104 B3
var. Osogovske Planine, Osogovski
Planina, *Mac.* Osogovski Planini.
mountain range
Bulgaria/FYR Macedonia
Osogovske Planine/Osogovski Planina/
Osogovski Planini *see* Osogov
Mountains
Oşorhei *see* Târgu Mureş
Osorno 65 B5 Los Lagos, C Chile
Ossa, Serra d' 92 C4 *mountain range*
SE Portugal
Ossora 115 H2 Krasnoyarskiy Kray,
E Russian Federation
Ostee *see* Baltic Sea
Ostend/Ostende *see* Oostende
Oster 109 E1 Chernihivs'ka Oblast',
N Ukraine
Östermyra *see* Seinäjoki
Osterode/Osterode in Ostpreussen
see Ostróda
Österreich *see* Austria
Östersund 85 C5 Jämtland, C Sweden
Ostia Aterni *see* Pescara
Ostiglia 96 C2 Lombardia, N Italy
Ostrava 99 C5 Moravskoslezský Kraj,
E Czech Republic
Ostróda 98 D3 *Ger.* Osterode, Osterode
in Ostpreussen. Warmińsko-
Mazurskie, NE Poland
Ostrołęka 98 D3 *Ger.* Wiesenhof, *Rus.*
Ostrolenka. Mazowieckie, C Poland
Ostrolenka *see* Ostrołęka
Ostrov 110 A4 *Latv.* Austrava.
Pskovskaya Oblast', W Russian
Federation
Ostrovets *see* Ostrowiec Świętokrzyski
Ostrovnoy 112 C2 Murmanskaya
Oblast', NW Russian Federation
Ostrów *see* Ostrów Wielkopolski
Ostrowiec *see* Ostrowiec Świętokrzyski
Ostrowiec Świętokrzyski 98 D4
var. Ostrowiec, *Rus.* Ostrovets.
Świętokrzyskie, C Poland
Ostrów Mazowiecka 98 D3 *var.* Ostrów
Mazowiecki. Mazowieckie, NE Poland
Ostrów Mazowiecki *see* Ostrów
Mazowiecka
Ostrowo *see* Ostrów Wielkopolski
Ostrów Wielkopolski 98 C4 *var.*
Ostrów, *Ger.* Ostrowo. Wielkopolskie,
C Poland
Ostyako-Voguls'k *see* Khanty-Mansiysk
Osum *see* Osumit, Lumi i
Osumi-shoto 131 A8 *island group*
Kagoshima, Nansei-shotō, SW Japan
Asia East China Sea Pacific Ocean
Osumit, Lumi i 101 D7 *var.* Osum.
river SE Albania
Osuna 92 D4 Andalucía, S Spain
Oswego 41 F3
Otago Peninsula 151 B7 *peninsula*
South Island, New Zealand

Otaki 150 D4 Wellington, North Island,
New Zealand
Otaru 130 C2 Hokkaidō, NE Japan
Otavalo 60 B1 Imbabura, N Ecuador
Otavi 78 B3 Otjozondjupa, N Namibia
Oțelu Roşu 108 B4 *Ger.* Ferdinandsberg,
Hung. Nándorhgy. Caras-Severin,
SW Romania
Otepää 106 D3 *Ger.* Odenpäh.
Valgamaa, SE Estonia
Oti 75 E4 *river* N Togo
Otira 151 C6 West Coast, South Island,
New Zealand
Otjiwarongo 78 B3 Otjozondjupa,
N Namibia
Otorohanga 150 D3 Waikato, North
Island, New Zealand
Otranto, Canale d' *see* Otranto, Strait of
Otranto, Strait of 101 C6 *It.* Canale
d'Otranto. *strait* Albania/Italy
Otrokovice 99 C5 *Ger.* Otrokowitz.
Zlínský Kraj, E Czech Republic
Otrokowitz *see* Otrokovice
Ōtsu 131 C6 *var.* Ōtu. Shiga, Honshū,
SW Japan
Ottawa 38 D5 *country capital* (Canada)
Ontario, SE Canada
Ottawa 40 B3 Illinois, N USA
Ottawa 45 F5 Kansas, C USA
Ottawa 41 E2 *Fr.* Outaouais. *river*
Ontario/Québec, SE Canada
Ottawa Islands 38 C1 *island group*
Nunavut, C Canada
Ottignies 87 C6 Wallon Brabant,
C Belgium
Ottumwa 45 G4 Iowa, C USA
Ōtu *see* Ōtsu
Ouachita Mountains 42 A1 *mountain
range* Arkansas/Oklahoma, C USA
Ouachita River 42 B2 *river* Arkansas/
Louisiana, C USA
Ouagadougou 75 E4 *var.* Wagadugu.
country capital (Burkina Faso)
C Burkina Faso
Ouahigouya 75 E3 NW Burkina Faso
Ouahran *see* Oran
Oualâta 74 D3 *var.* Oualata. Hodh ech
Chargui, SE Mauritania
Ouanary 59 H3 E French Guiana
Ouanda Djallé 76 D4 Vakaga,
NE Central African Republic
Ouarâne 74 D2 *desert* C Mauritania
Ouargla 71 E2 *var.* Wargla. NE Algeria
Ouarzazate 70 C3 S Morocco
Oubangui 77 C5 *Fr.* Oubangui. *river*
C Africa
Oubangui *see* Ubangi
Oubangui-Chari *see* Central African
Republic
Oubangui-Chari, Territoire de l' *see*
Central African Republic
Oudjda *see* Oujda
Ouessant, Île d' 90 A3 *Eng.* Ushant.
island NW France
Ouésso 77 B5 Sangha, NW Congo
Oujda 70 D2 *Ar.* Oudjda, Ujda.
NE Morocco
Oujeft 74 C2 Adrar, C Mauritania
Oulu 84 D4 *Swe.* Uleåborg. Oulu, Finland
Oulujärvi 84 D4 *Swe.* Uleträsk. *lake*
C Finland
Oulujoki 84 D4 *Swe.* Uleälv. *river*
C Finland
Ounasjoki 84 D3 *river* N Finland
Ounianga Kébir 76 C2 Borkou-Ennedi-
Tibesti, N Chad
Oup *see* Auob
Oupeye 87 D6 Liège, E Belgium
Our 87 D6 *river* NW Europe
Ourense 92 C1 *Cast.* Orense, *Lat.*
Aurium. Galicia, NW Spain
Ourique 92 B4 Beja, S Portugal
Ours, Grand Lac de l' *see* Great Bear Lake
Ourthe 87 D7 *river* E Belgium
Ouse 89 D5 *river* N England, United
Kingdom
Outaouais *see* Ottawa
Outer Hebrides 88 B3 *var.* Western
Isles. *island group* NW Scotland,
United Kingdom
Outer Islands 79 G1 *island group*
SW Seychelles Africa W Indian Ocean
Ouvéa 144 D5 *island* Îles Loyauté,
NE New Caledonia
Ouyen 149 C6 Victoria, SE Australia
Ovalle 64 B3 Coquimbo, N Chile
Ovar 92 B2 Aveiro, N Portugal
Overflakkee 86 B4 *island*
SW Netherlands

Overijse 87 C6 Vlaams Brabant,
C Belgium
Oviedo 92 C1 *anc.* Asturias. Asturias,
NW Spain
Ovilava *see* Wels
Ovruch 108 D1 Zhytomyrs'ka Oblast',
N Ukraine
Owando 77 B5 *prev.* Fort Rousset.
Cuvette, C Congo
Owase 131 C6 Mie, Honshū, SW Japan
Owatonna 45 F3 Minnesota, N USA
Owen Fracture Zone 140 B4 *tectonic
feature* W Arabian Sea
Owen, Mount 151 C5 *mountain* South
Island, New Zealand
Owensboro 40 B5 Kentucky, S USA
Owen Stanley Range 144 B3 *mountain
range* S Papua New Guinea
Owerri 75 G5 Imo, S Nigeria
Owo 75 F5 Ondo, SW Nigeria
Owyhee River 46 C4 *river* Idaho/Oregon,
NW USA
Oxford 151 C6 Canterbury, South Island,
New Zealand
Oxford 89 D6 *Lat.* Oxonia. S England,
United Kingdom
Oxkutzcab 51 H4 Yucatán, SE Mexico
Oxnard 47 B7 California, W USA
Oxonia *see* Oxford
Oxus *see* Amu Darya
Oyama 131 C6 Tochigi, Honshū, S Japan
Oyem 77 B5 Woleu-Ntem, N Gabon
Oyo 77 B6 Cuvette, C Congo
Oyo 75 F4 Oyo, W Nigeria
Ozark 42 D3 Alabama, S USA
Ozark Plateau 45 G5 *plain* Arkansas/
Missouri, C USA
Ozarks, Lake of the 45 F5 *reservoir*
Missouri, C USA
Ozbourn Seamount 152 D4 *undersea
feature* W Pacific Ocean
Ózd 99 D6 Borsod-Abaúj-Zemplén,
NE Hungary
Ozieri 97 A5 Sardegna, Italy,
C Mediterranean Sea

P

Paamiut 82 B4 *var.* Pâmiut, *Dan.*
Frederikshåb. S Greenland
Pa-an *see* Hpa-an
Pabianice 98 C4 Łódzkie, Poland
Pabna 135 G4 Rajshahi, W Bangladesh
Pacaraima, Sierra/Pacaraim, Serra *see*
Pakaraima Mountains
Pachuca 51 E4 *var.* Pachuca de Soto.
Hidalgo, C Mexico
Pachuca de Soto *see* Pachuca
Pacific-Antarctic Ridge 154 B5
undersea feature S Pacific Ocean
Pacific Ocean 152 D3 *ocean*
Padalung *see* Phatthalung
Padang 138 B4 Sumatera, W Indonesia
Paderborn 94 B4 Nordrhein-Westfalen,
NW Germany
Padma *see* Brahmaputra
Padma *see* Ganges
Padova 96 C2 *Eng.* Padua; *anc.* Patavium.
Veneto, NE Italy
Padre Island 49 G5 *island* Texas,
SW USA
Padua *see* Padova
Paducah 40 B5 Kentucky, S USA
Paeroa 150 D3 Waikato, North Island,
New Zealand
Páfos 102 C5 *var.* Paphos. W Cyprus
Pag 100 A3 *It.* Pago. *island* Zadar,
C Croatia
Pago *see* Pag
Pago Pago 145 F4 *dependent territory
capital* (American Samoa) Tutuila,
W American Samoa
Pahiatua 150 D4 Manawatu-Wanganui,
North Island, New Zealand
Pahsien *see* Chongqing
Paide 106 D2 *Ger.* Weissenstein.
Järvamaa, N Estonia
Paihia 150 D2 Northland, North Island,
New Zealand
Päijänne 85 D5 *lake* S Finland
Paine, Cerro 65 A7 *mountain* S Chile
Painted Desert 48 B1 *desert* Arizona,
SW USA
Paisance *see* Piacenza
Paisley 88 C4 W Scotland, United
Kingdom
País Vasco 93 E1 *cultural region* N Spain

Paita *60 B3* Piura, NW Peru
Pakanbaru *see* Pekanbaru
Pakaraima Mountains *59 E3 var.* Serra
Pacaraim, Sierra Pacaraima.
mountain range N South America
Pakistan *134 A2 off.* Islamic Republic
of Pakistan, *var.* Islami Jamhuriya e
Pakistan. *country* S Asia

PAKISTAN
South Asia

Official name Islamic Republic
of Pakistan
Formation 1947 / 1971
Capital Islamabad
Population 182 million / 612 people
per sq mile (236 people per sq km)
Total area 310,401 sq. miles
(803,940 sq. km)
Languages Punjabi, Sindhi, Pashtu, Urdu*,
Baluchi, Brahui
Religions Sunni Muslim 77%, Shi'a
Muslim 20%, Hindu 2%, Christian 1%
Ethnic mix Punjabi 56%, Pathan
(Pashtun) 15%, Sindhi 14%, Mohajir 7%,
Baluchi 4%, Other 4%
Government Parliamentary system
Currency Pakistani rupee = 100 paisa
Literacy rate 55%
Calorie consumption 2428 kilocalories

Pakistan, Islamic Republic of *see*
Pakistan
Pakistan, Islami Jamhuriya e *see*
Pakistan
Paknam *see* Samut Prakan
Pakokku *136 A3* Magway, C Myanmar
(Burma)
Pak Phanang *137 C7 var.* Ban Pak
Phanang. Nakhon Si Thammarat,
SW Thailand
Pakruojis *106 C4* Šiauliai, N Lithuania
Paks *99 C7* Tolna, S Hungary
Paksé *see* Pakxé
Pakxé *137 D7 var.* Paksé. Champasak,
S Laos
Palafrugell *93 G2* Cataluña, NE Spain
Palagruža *101 B5 It.* Pelagosa. *island*
SW Croatia
Palaiá Epídavros *105 C6* Pelopónnisos,
S Greece
Palaiseau *90 D2* Essonne, N France
Palamós *93 G2* Cataluña, NE Spain
Palamuse *106 E2 Ger.* Sankt-
Bartholomäi. Jõgevamaa, E Estonia
Palanka *see* Bačka Palanka
Pālanpur *134 C4* Gujarāt, W India
Palantia *see* Palencia
Palapye *78 D3* Central, SE Botswana
Palau *144 A2 var.* Belau. *country*
W Pacific Ocean

PALAU
Australasia & Oceania

Official name Republic of Palau
Formation 1994 / 1994
Capital Melekeok
Population 21,108 / 108 people
per sq mile (42 people per sq km)
Total area 177 sq. miles (458 sq. km)
Languages Palauan*, English*, Japanese,
Angaur, Tobi, Sonsorolese
Religions Christian 66%, Modekngei 34%
Ethnic mix Palauan 74%, Filipino 16%,
Other 6%, Chinese and other Asian 4%
Government Nonparty system
Currency US dollar = 100 cents
Literacy rate 99%
Calorie consumption Not available

Palawan *139 E2 island* W Philippines
Palawan Passage *138 D2 passage*
W Philippines
Paldiski *106 D2 prev.* Baltiski, *Eng.* Baltic
Port, *Ger.* Baltischport. Harjumaa,
NW Estonia
Palembang *138 B4* Sumatera,
W Indonesia
Palencia *92 D2 anc.* Palantia, Pallantia.
Castilla y León, N Spain
Palerme *see* Palermo
Palermo *97 C7 Fr.* Palerme; *anc.*
Panhormus, Panormus. Sicilia, Italy,
C Mediterranean Sea
Pāli *134 C3* Rājasthān, N India
Palikir *144 C2 country capital*
(Micronesia) Pohnpei, E Micronesia

Palimé *see* Kpalimé
Palióúri, Akrotírio *104 C4 var.*
Akrotírio Kanestron. *headland*
N Greece
Palk Strait *132 C3 strait* India/Sri Lanka
Pallantia *see* Palencia
Palliser, Cape *151 D5 headland* North
Island, New Zealand
Palma *93 G3 var.* Palma de Mallorca.
Mallorca, Spain, W Mediterranean Sea
Palma del Río *92 D4* Andalucía,
S Spain
Palma de Mallorca *see* Palma
Palmar Sur *53 E5* Puntarenas,
SE Costa Rica
Palma Soriano *54 C3* Santiago de Cuba,
E Cuba
Palm Beach *148 E1* New South Wales,
E Australia
Palmer *154 A2 US research station*
Antarctica
Palmer Land *154 A3 physical region*
Antarctica
Palmerston *145 F4 island* S Cook Islands
Palmerston *see* Darwin
Palmerston North *150 D4*
Manawatu-Wanganui, North Island,
New Zealand
The Palmetto State *see* South Carolina
Palmi *97 D7* Calabria, SW Italy
Palmira *58 B3* Valle del Cauca,
W Colombia
Palm Springs *47 D7* California, W USA
Palmyra *see* Tudmur
Palmyra Atoll *145 G2 US privately
owned unincorporated territory*
C Pacific Ocean
Palo Alto *47 B6* California, W USA
Paloe *see* Denpasar, Bali, C Indonesia
Paloe *see* Palu
Palu *139 E4 prev.* Paloe. Sulawesi,
C Indonesia
Pamiers *91 B6* Ariège, S France
Pamir *123 F3 var.* Daryā-ye Pāmīr, *Taj.*
Dar''yoi Pomir. *river* Afghanistan/
Tajikistan
Pāmir, Daryā-ye *see* Pamir
Pamir/Pāmīr, Daryā-ye *see* Pamirs
Pamirs *123 F3 Pash.* Daryā-ye Pāmīr,
Rus. Pamir. *mountain range* C Asia
Pāmiut *see* Paamiut
Pamlico Sound *43 G1 sound* North
Carolina, SE USA
Pampa *49 E1* Texas, SW USA
Pampa Aullagas, Lago *see* Poopó, Lago
Pampas *64 C4 plain* C Argentina
Pampeluna *see* Pamplona
Pamplona *58 C2* Norte de Santander,
N Colombia
Pamplona *93 E1 Basq.* Iruña, *prev.*
Pampeluna; *anc.* Pompaelo. Navarra,
N Spain
Panaji *132 B1 var.* Pangim, Panjim, New
Goa. *state capital* Goa, W India
Panamá *53 G4 var.* Ciudad de Panama,
Eng. Panama City. *country capital*
(Panama) Panamá, C Panama
Panama *53 G5 off.* Republic of Panama.
country Central America

PANAMA
Central America

Official name Republic of Panama
Formation 1903 / 1903
Capital Panama City
Population 3.9 million / 133 people
per sq mile (51 people per sq km)
Total area 30,193 sq. miles (78,200 sq. km)
Languages English Creole,
Spanish*, Amerindian languages,
Chibchan languages
Religions Roman Catholic 84%,
Protestant 15%, Other 1%
Ethnic mix Mestizo 70%, Black 14%,
White 10%, Amerindian 6%
Government Presidential system
Currency Balboa = 100 centésimos;
US dollar
Literacy rate 94%
Calorie consumption 2644 kilocalories

Panama Basin *35 C8 undersea feature*
E Pacific Ocean
Panama Canal *53 F4 canal* E Panama
Panama City *42 D3* Florida, SE USA
Panama City *see* Panamá
Gulf of Panama *53 G5 var.* Gulf of
Panama. *gulf* S Panama

Panama, Gulf of *see* Panamá, Golfo de
Isthmus of Panama *53 G4 Eng.* Isthmus
of Panama; *prev.* Isthmus of Darien.
isthmus E Panama
Panama, Isthmus of *see* Panama,
Istmo de
Panama, Republic of *see* Panama
Panay Island *139 E2 island* C Philippines
Pančevo *100 D3 Ger.* Pantschowa, *Hung.*
Pancsova. Vojvodina, N Serbia
Pancsova *see* Pančevo
Paneas *see* Bāniyās
Panevėžys *106 C4* Panevėžys,
C Lithuania
Pangim *see* Panaji
Pangkalpinang *138 C4* Pulau Bangka,
W Indonesia
Pang-Nga *see* Phang-Nga
Panhormus *see* Palermo
Panjim *see* Panaji
Panopolis *see* Akhmīm
Pánormos *105 C7* Kríti, Greece,
E Mediterranean Sea
Panormus *see* Palermo
Pantanal *63 E3 var.* Pantanalmato-
Grossense. *swamp* SW Brazil
Pantanalmato-Grossense *see* Pantanal
Pantelleria, Isola di *91 B7 island*
SW Italy
Pantschowa *see* Pančevo
Pánuco *51 E3* Veracruz-Llave, E Mexico
Pao-chi/Paoki *see* Baoji
Paola *102 B5* E Malta
Pao-shan *see* Baoshan
Pao-t'ou/Paotow *see* Baotou
Papagayo, Golfo de *52 C4 gulf*
NW Costa Rica
Papakura *150 D3* Auckland, North
Island, New Zealand
Papantla *51 F4 var.* Papantla de Olarte.
Veracruz-Llave, E Mexico
Papantla de Olarte *see* Papantla
Papeete *145 H4 dependent territory
capital* (French Polynesia) Tahiti,
W French Polynesia
Paphos *see* Páfos
Papile *106 B3* Šiauliai, NW Lithuania
Papillion *45 F4* Nebraska, C USA
Papua *139 H4 var.* Irian Barat, West
Irian, West New Guinea, West Papua;
prev. Dutch New Guinea, Irian Jaya,
Netherlands New Guinea. *province*
E Indonesia
Papua and New Guinea, Territory of *see*
Papua New Guinea
Papua, Gulf of *144 B3 gulf* S Papua
New Guinea
Papua New Guinea *144 B3 off.*
Independent State of Papua New
Guinea; *prev.* Territory of Papua and
New Guinea. *country* NW Melanesia

PAPUA NEW GUINEA
Australasia & Oceania

Official name Independent State
of Papua New Guinea
Formation 1975 / 1975
Capital Port Moresby
Population 7.3 million / 42 people
per sq mile (16 people per sq km)
Total area 178,703 sq. miles
(462,840 sq. km)
Languages Pidgin English, Papuan,
English*, Motu, 800 (est.) native languages
Religions Protestant 60%, Roman
Catholic 37%, Other 3%
Ethnic mix Melanesian and mixed
race 100%
Government Parliamentary system
Currency Kina = 100 toea
Literacy rate 60%
Calorie consumption 2193 kilocalories

Papua New Guinea, Independent State
of *see* Papua New Guinea
Papuk *100 C3 mountain range*
NE Croatia
Pará *63 E2 off.* Estado do Pará. *state*
NE Brazil
Pará *63 E2 off.* Estado do Pará. *region*
NE Brazil
Pará *see* Belém
Paracel Islands *125 E3 disputed territory*
SE Asia
Paraćin *100 D4* Serbia, C Serbia
Paradise of the Pacific *see* Hawaii
Pará, Estado do *see* Pará
Paragua, Río *59 E3 river* SE Venezuela

Paraguay *64 C2 country*
C South America

PARAGUAY
South America

Official name Republic of Paraguay
Formation 1811 / 1938
Capital Asunción
Population 6.8 million / 44 people
per sq mile (17 people per sq km)
Total area 157,046 sq. miles
(406,750 sq. km)
Languages Guaraní, Spanish*, German
Religions Roman Catholic 90%,
Protestant (including Mennonite) 10%
Ethnic mix Mestizo 91%, Other 7%,
Amerindian 2%
Government Presidential system
Currency Guaraní = 100 céntimos
Literacy rate 94%
Calorie consumption 2698 kilocalor

Paraguay *64 D2 var.* Río Paraguay.
C South America
Paraguay, Río *see* Paraguay
Parahíba/Parahyba *see* Paraíba
Paraíba *63 G2 off.* Estado da Paraíba.
prev. Parahíba, Parahyba. *state* E B
Paraíba *63 G2 off.* Estado do Paraíba.
prev. Parahíba, Parahyba. *region*
E Brazil
Paraíba *see* João Pessoa
Paraíba, Estado da *see* Paraíba
Parakou *75 F4* C Benin
Paramaribo *59 G3 country capital*
(Suriname) Paramaribo, N Surinam
Paramushir, Ostrov *115 H3 island*
SE Russian Federation
Paraná *63 E4* Entre Ríos, E Argentin
Paraná *63 E5 off.* Estado do Paraná.
state S Brazil
Paraná *63 E5 off.* Estado do Paraná.
region S Brazil
Paraná *57 C5 var.* Alto Paraná. *river*
C South America
Paraná, Estado do *see* Paraná
Paranésti *104 C3 var.* Paranéstio.
Anatolikí Makedonía kai Thráki,
NE Greece
Paranéstio *see* Paranésti
Paraparaumu *151 D5* Wellington, N
Island, New Zealand
Parchim *94 C3* Mecklenburg-
Vorpommern, N Germany
Parczew *98 E4* Lubelskie, E Poland
Pardubice *99 B5 Ger.* Pardubitz.
Pardubický Kraj, C Czech Republic
Pardubitz *see* Pardubice
Parechcha *107 B5* Pol. Porzecze, *Ru*
Porech'ye. Hrodzyenskaya Voblast
W Belarus
Parecis, Chapada dos *62 D3 var.* Se
dos Parecis. *mountain range* W Br
Parecis, Serra dos *see* Parecis,
Chapada dos
Parenzo *see* Poreč
Parepare *139 E4* Sulawesi, C Indone
Párga *105 A5* Ípeiros, W Greece
Paria, Golfo de *see* Paria, Gulf of
Paria, Gulf of *59 E1 var.* Golfo de Pa
gulf Trinidad and Tobago/Venezue
Parika *59 F2* NE Guyana
Parikiá *105 D6* Kykládes, Greece,
Aegean Sea
Paris *90 D1 anc.* Lutetia, Lutetia
Parisiorum, Parisii. *country capit*
(France) Paris, N France
Paris *49 G2* Texas, SW USA
Parisii *see* Paris
Parkersburg *40 D4* West Virginia,
NE USA
Parkes *149 D6* New South Wales,
SE Australia
Parkhar *see* Farkhor
Parma *96 B2* Emilia-Romagna, N Ita
Parnaíba *63 F2 var.* Parnahyba. Pia
E Brazil
Pärnu *106 D2 Ger.* Pernau, *Latv.*
Pērnava; *prev. Rus.* Pernov. Pärnu
SW Estonia
Pärnu *106 D2 var.* Parnu Jõgi, *Ger.*
Pernau. *river* SW Estonia
Pärnu-Jaagupi *106 D2 Ger.* Sankt-
Jakobi. Pärnumaa, SW Estonia
Parnu Jõgi *see* Pärnu
Pärnu Laht *106 D2 Ger.* Pernauer B
bay SW Estonia

pamisus Range see Sefid Küh,
lseleh-ye
s 105 C6 island Kykládes, Greece,
egean Sea
ral 64 B4 Maule, C Chile
ral see Hidalgo del Parral
ramatta 148 D1 New South Wales,
SE Australia
ras 50 D3 var. Parras de la Fuente.
oahuila, NE Mexico
ras de la Fuente see Parras
sons 45 F5 Kansas, C USA
adena 47 C7 California, W USA
adena 49 H4 Texas, SW USA
cani 108 C3 Hung. Páskán. Iaşi,
E Romania
co 46 C2 Washington, NW USA
ewalk 94 D3 Mecklenburg-
orpommern, NE Germany
hkeni see Bolyarovo
inler 117 F3 Erzurum, NE Turkey
kán see Paşcani
ęk 98 D2 Ger. Preußisch Holland.
armiński-Mazurskie, NE Poland
ni 134 A3 Baluchistān, SW Pakistan
o de Indios 65 B6 Chubut,
Argentina
sarowitz see Požarevac
sau 95 D6 Bayern, SE Germany
so Fundo 63 E5 Rio Grande do
ul, S Brazil
tavy 107 C5 Pol. Postawy, Rus.
ostavy. Vitsyebskaya Voblasts',
W Belarus
taza, Río 60 B2 river Ecuador/Peru
to 54 A4 Nariño, SW Colombia
walys 106 C4 Panevėžys, N Lithuania
agonia 57 B7 physical region
rgentina/Chile
alung see Phatthalung
ani see Pattani
avium see Padova
ea 150 D4 Taranaki, North Island,
New Zealand
erson 41 F3 New Jersey, NE USA
hein 136 A4 var. Bassein.
ayeyarwady, SW Myanmar (Burma)
mos 105 D6 island Dodekánisa,
Greece, Aegean Sea
na 135 F3 var. Azimabad. state capital
ihār, N India
nos 117 H3 Ağrı, E Turkey
os, Lagoa dos 63 E5 lagoon
Brazil
ra 105 B5 Eng. Patras; prev. Pátrai.
Dytikí Elláda, S Greece
rai/Patras see Pátra
tani 137 C7 var. Patani. Pattani,
W Thailand
taya 137 C5 Chon Buri, S Thailand
uca, Río 52 D2 river E Honduras
) B6 Pyrénées-Atlantiques,
W France
alatuk 37 E3 Northwest Territories,
NW Canada
angde 136 B4 Bago, C Myanmar
(Burma)
atalia see Kyustendil
via 96 B2 anc. Ticinum. Lombardia,
N Italy
ilosta 106 B3 W Latvia
olikeni 104 D2 Veliko Tŭrnovo,
N Bulgaria
lodar 114 C4 Pavlodar,
NE Kazakhstan
lograd see Pavlohrad
lohrad 109 G3 Rus. Pavlograd.
Dnipropetrovs'ka Oblast', E Ukraine
wai, Pulau 138 A2 island
SW Singapore Asia
vn, Nam 136 B3 river C Myanmar
(Burma)
x Augusta see Badajoz
x Julia see Beja
xoi 105 A5 island Iónia Nisiá, Greece,
C Mediterranean Sea
vo Obispo see Chetumal
ysandú 64 D4 Paysandú, W Uruguay
zar 117 E2 Rize, NE Turkey
zardzhik 104 C3 prev. Tatar
Pazardzhik. Pazardzhik, SW Bulgaria
ace Garden State see North Dakota
ach State see Georgia
arl Islands 53 G5 Eng. Pearl Islands.
sland see Panama
arl Islands see Perlas, Archipiélago
de las
rl Lagoon 53 E3 Eng. Pearl Lagoon.
agoon E Nicaragua

Pearl Lagoon see Perlas, Laguna de
Pearl River 42 B3 river Louisiana/
Mississippi, S USA
Pearsall 49 F4 Texas, SW USA
Peawanuk 38 C2 Ontario, S Canada
Peć see Pejë
Pechora 110 D3 Respublika Komi,
NW Russian Federation
Pechora 110 D3 river NW Russian
Federation
Pechora Sea 110 D2 Eng. Pechora Sea.
sea NW Russian Federation
Pechora Sea see Pechorskoye More
Pecos 49 E3 Texas, SW USA
Pecos River 49 E3 river New Mexico/
Texas, SW USA
Pécs 99 C7 Ger. Fünfkirchen, Lat.
Sopianae. Baranya, SW Hungary
Pedra Lume 74 A3 Sal, NE Cape Verde
Pedro Cays 54 C3 island group Greater
Antilles, S Jamaica North America
N Caribbean Sea W Atlantic Ocean
Pedro Juan Caballero 64 D2 Amambay,
E Paraguay
Peer 87 D5 Limburg, NE Belgium
Pegasus Bay 151 C6 bay South Island,
New Zealand
Pegu see Bago
Pehuajó 64 C4 Buenos Aires, E Argentina
Pei-ching see Beijing/Beijing Shi
Peine 94 B3 Niedersachsen, C Germany
Pei-p'ing see Beijing/Beijing Shi
Peipsi Järv/Peipus-See see Peipus, Lake
Peipus, Lake 106 E3 Est. Peipsi Järv, Ger.
Peipus-See, Rus. Chudskoye Ozero.
lake Estonia/Russian Federation
Peiraiás 105 C6 prev. Piraiévs, Eng.
Piraeus. Attikí, C Greece
Pejë 101 D5 Serb. Peć. W Kosovo
Pëk see Phônsaven
Pekalongan 138 C4 Jawa, C Indonesia
Pekanbaru 138 B3 var. Pakanbaru.
Sumatera, W Indonesia
Pekin 40 B4 Illinois, N USA
Peking see Beijing/Beijing Shi
Pelagie 97 B8 island group SW Italy
Pelagosa see Palagruža
Pelican State see Louisiana
Pelly Bay see Kugaaruk
Pélmonostor see Beli Manastir
Peloponnese 105 B6 var. Morea, Eng.
Peloponnese; anc. Peloponnesus.
peninsula S Greece
Peloponnese/Peloponnesus see
Pelopónnisos
Pematangsiantar 138 B3 Sumatera,
W Indonesia
Pemba 79 F2 prev. Port Amelia,
Porto Amélia. Cabo Delgado,
NE Mozambique
Pemba 73 D7 island E Tanzania
Pembroke 38 D4 Ontario, SE Canada
Penang see Pinang, Pulau, Peninsular
Malaysia
Penang see George Town
Penas, Golfo de 65 A7 gulf S Chile
Penderma see Bandırma
Pendleton 46 C3 Oregon, NW USA
Pend Oreille, Lake 46 D2 lake Idaho,
NW USA
Peneius see Pineiós
Peng-pu see Bengbu
Penibético, Sistema see Béticos, Sistemas
Peniche 92 B3 Leiria, W Portugal
Peninsular State see Florida
Penninae, Alpes/Pennine, Alpi see
Pennine Alps
Pennine Alps 95 A8 Fr. Alpes Pennines,
It. Alpi Pennine, Lat. Alpes Penninae.
mountain range Italy/Switzerland
Pennine Chain see Pennines
Pennines 89 D5 var. Pennine Chain.
mountain range N England, United
Kingdom
Pennines, Alpes see Pennine Alps
Pennsylvania 41 E4 off. Commonwealth
of Pennsylvania, also known as
Keystone State. state NE USA
Penobscot River 41 G2 river Maine,
NE USA
Penong 149 A6 South Australia
Penonomé 53 F5 Coclé, C Panama
Penrhyn 145 G3 atoll N Cook Islands
Penrhyn Basin 143 F3 undersea feature
C Pacific Ocean
Penrith 148 D1 New South Wales,
SE Australia
Penrith 89 D5 NW England, United
Kingdom

Pensacola 42 C3 Florida, SE USA
Pentecost 144 D4 Fr. Pentecôte. island
C Vanuatu
Pentecôte see Pentecost
Penza 111 C6 Penzenskaya Oblast',
W Russian Federation
Penzance 89 C7 SW England, United
Kingdom
Peoria 40 B4 Illinois, N USA
Perchtoldsdorf 95 E6 Niederösterreich,
NE Austria
Percival Lakes 146 C4 lakes Western
Australia
Perdido, Monte 93 F1 mountain
NE Spain
Perece Vela Basin see West Mariana
Basin
Pereira 58 B3 Risaralda, W Colombia
Peremyshl see Przemyśl
Pergamino 64 C4 Buenos Aires,
E Argentina
Périgueux 91 C5 anc. Vesuna.
Dordogne, SW France
Perito Moreno 65 B6 Santa Cruz,
S Argentina
Perleberg 94 C3 Brandenburg,
N Germany
Perlepe see Prilep
Perm' 114 C3 prev. Molotov. Permskiy
Kray, NW Russian Federation
Pernambuco 63 G2 off. Estado de
Pernambuco. region E Brazil
Pernambuco 63 G2 off. Estado de
Pernambuco. state E Brazil
Pernambuco see Recife
Pernambuco Abyssal Plain see
Pernambuco Plain
Pernambuco, Estado de see
Pernambuco
Pernambuco Plain 67 C5 var.
Pernambuco Abyssal Plain. undersea
feature E Atlantic Ocean
Pernau see Pärnu
Pernauer Bucht see Pärnu Laht
Përnava see Pärnu
Pernik 104 B2 prev. Dimitrovo. Pernik,
W Bulgaria
Pernov see Pärnu
Perote 51 F4 Veracruz-Llave, E Mexico
Pérouse see Perugia
Perovsk see Kyzylorda
Perpignan 91 C6 Pyrénées-Orientales,
S France
Perryton 49 F1 Texas, SW USA
Perryville 45 H5 Missouri, C USA
Persia see Iran
Persian Gulf 120 C4 var. Gulf, The,
Ar. Khalīj al 'Arabī, Per. Khalīj-e Fars.
gulf SW Asia
Perth 147 A6 state capital Western
Australia
Perth 88 C4 C Scotland, United Kingdom
Perth Basin 141 E6 undersea feature
SE Indian Ocean
Peru 60 C3 off. Republic of Peru. country
W South America

PERU
South America

Official name Republic of Peru
Formation 1824 / 1941
Capital Lima
Population 30.4 million / 62 people
per sq mile (24 people per sq km)
Total area 496,223 sq. miles
(1,285,200 sq. km)
Languages Spanish*, Quechua*, Aymara
Religions Roman Catholic 81%, Other 19%
Ethnic mix Amerindian 45%,
Mestizo 37%, White 15%, Other 3%
Government Presidential system
Currency New sol = 100 céntimos
Literacy rate 94%
Calorie consumption 2624 kilocalories

Peru see Beru
Peru Basin 67 A5 undersea feature
E Pacific Ocean
Peru-Chile Trench 56 A4 undersea
feature E Pacific Ocean
Perugia 96 C4 Fr. Pérouse; anc. Perusia.
Umbria, C Italy
Perugia, Lake of see Trasimeno, Lago
Peru, Republic of see Peru
Perusia see Perugia
Péruwelz 87 B6 Hainaut, SW Belgium
Pervomays'k 109 E3 prev. Ol'viopol'.
Mykolayivs'ka Oblast', S Ukraine

Pervyy Kuril'skiy Proliv 115 H3 strait
E Russian Federation
Pesaro 96 C3 anc. Pisaurum. Marche,
C Italy
Pescara 96 D4 anc. Aternum, Ostia
Aterni. Abruzzo, C Italy
Peshāwar 134 C1 Khyber Pakhtunkhwa,
N Pakistan
Peshkopi 101 C6 var. Peshkopia,
Peshkopija. Dibër, NE Albania
Peshkopia/Peshkopija see Peshkopi
Pessac 91 B5 Gironde, SW France
Petach-Tikva see Petah Tikva
Petah Tikva 119 A6 var. Petach-Tikva,
Petah Tiqwa, Petah Tikva; prev. Petaḥ
Tiqwa. Tel Aviv, C Israel
Petaḥ Tiqwa see Petah Tikva
Petakh Tikva/Petah Tiqwa see Petah
Tikva
Pétange 87 D8 Luxembourg,
SW Luxembourg
Petchaburi see Phetchaburi
Peterborough 149 B6 South Australia
Peterborough 38 D5 Ontario, SE Canada
Peterborough 89 E6 prev.
Medeshamstede. E England, United
Kingdom
Peterhead 88 D3 NE Scotland, United
Kingdom
Peter I Øy 154 A3 Norwegian
dependency Antarctica
Petermann Bjerg 83 E3 mountain
C Greenland
Petersburg 41 E5 Virginia, NE USA
Peters Mine 59 F3 var. Peter's Mine.
N Guyana
Petit St-Bernard, Col du see Little Saint
Bernard Pass
Peto 51 H4 Yucatán, SE Mexico
Petoskey 40 C2 Michigan, N USA
Petra see Wādī Mūsá
Petrich 104 C3 Blagoevgrad, SW Bulgaria
Petrikau see Piotrków Trybunalski
Petrikov see Pyetrykaw
Petrinja 100 B3 Sisak-Moslavina,
C Croatia
Petroaleksandrovsk see To'rtko'l
Petrodvorets 110 A4 Fin. Pietarhovi.
Leningradskaya Oblast', NW Russian
Federation
Petrograd see Sankt-Peterburg
Petrokov see Piotrków Trybunalski
Petropavl see Petropavlovsk
Petropavlovsk 114 C4 Kaz. Petropavl.
Severnyy Kazakhstan, N Kazakhstan
Petropavlovsk-Kamchatskiy 115
H3 Kamchatskiy Kray, E Russian
Federation
Petroşani 108 B4 var. Petroşeni,
Ger. Petroschen, Hung. Petrozsény.
Hunedoara, W Romania
Petroschen/Petroşeni see Petroşani
Petroskoi see Petrozavodsk
Petrovgrad see Zrenjanin
Petrovsk-Port see Makhachkala
Petrozavodsk 114 B2 Fin. Petroskoi.
Respublika Kareliya, NW Russian
Federation
Petrozsény see Petroşani
Pettau see Ptuj
Pevek 115 G1 Chukotskiy Avtonomnyy
Okrug, NE Russian Federation
Pezinok 99 C6 Ger. Bösing, Hung. Bazin.
Bratislavský Kraj, W Slovakia
Pforzheim 95 B6 Baden-Württemberg,
SW Germany
Pfungstadt 95 B5 Hessen, W Germany
Phangan, Ko 137 C6 island SW Thailand
Phang-Nga 137 B7 var. Pang-Nga,
Phangnga. Phangnga, SW Thailand
Phangnga see Phang-Nga
Phan Rang/Phanrang see Phan Rang-
Thap Cham
Phan Rang-Thap Cham 137 E6 var.
Phanrang, Phan Rang, Phan Rang
Thap Cham. Ninh Thuận, S Vietnam
Phan Thiết 137 E6 Bình Thuận,
S Vietnam
Pharnacia see Giresun
Pharus see Hvar
Phatthalung 137 C7 var. Padalung,
Patalung. Phatthalung, SW Thailand
Phayao 136 C4 var. Muang Phayao.
Phayao, NW Thailand
Phenix City 42 D2 Alabama, S USA
Phet Buri see Phetchaburi
Phetchaburi 137 C5 var. Bejraburi,
Petchaburi, Phet Buri. Phetchaburi,
SW Thailand

Philadelphia 41 F4 Pennsylvania, NE USA
Philadelphia see 'Amman
Philippine Basin 125 F3 undersea feature W Pacific Ocean
Philippine Islands 139 E1 island group W Pacific Ocean
Philippines 139 E1 off. Republic of the Philippines. country SE Asia

PHILIPPINES
Southeast Asia

Official name Republic of the Philippines
Formation 1946 / 1946
Capital Manila
Population 98.4 million / 855 people per sq mile (330 people per sq km)
Total area 115,830 sq. miles (300,000 sq. km)
Languages Filipino*, English*, Tagalog, Cebuano, Ilocano, Hiligaynon, many other local languages
Religions Roman Catholic 81%, Protestant 9%, Muslim 5%, Other (including Buddhist) 5%
Ethnic mix Other 34%, Tagalog 28%, Cebuano 13%, Ilocano 9%, Hiligaynon 8%, Bisaya 8%
Government Presidential system
Currency Philippine peso = 100 centavos
Literacy rate 95%
Calorie consumption 2608 kilocalories

Philippine Sea 125 F3 sea W Pacific Ocean
Philippines, Republic of the see Philippines
Philippine Trench 142 A1 undersea feature W Philippine Sea
Philippopolis see Plovdiv
Phitsanulok 136 C4 var. Bisnulok, Muang Phitsanulok, Pitsanulok. Phitsanulok, C Thailand
Phlórina see Flórina
Phnom Penh see Phnum Penh
Phnum Penh 137 D6 var. Phnom Penh. country capital (Cambodia) Phnum Penh, S Cambodia
Phoenix 48 B2 state capital Arizona, SW USA
Phoenix Islands 145 E3 island group C Kiribati
Phôngsali 136 C3 var. Phong Saly. Phôngsali, N Laos
Phong Saly see Phôngsali
Phônsaven 136 D4 var. Xieng Khouang; prev. Pèk, Xiangkhoang. Xiangkhoang, N Laos
Phou Louang see Annamite Mountains
Phrae 136 C4 var. Muang Phrae, Prae. Phrae, NW Thailand
Phra Nakhon Si Ayutthaya see Ayutthaya
Phra Thong, Ko 137 B6 island SW Thailand
Phuket 137 B7 var. Bhuket, Puket, Mal. Ujung Salang; prev. Junkseylon, Salang. Phuket, SW Thailand
Phuket, Ko 137 B7 island SW Thailand
Phumĭ Kâmpóng Trâbêk see Kâmpóng Trâbêk
Phumĭ Kompong Trabek see Kâmpóng Trâbêk
Phum Samrong see Sâmraông
Phu Vinh see Tra Vinh
Phyu 136 B4 var. Hpyu, Pyu. Bago, C Myanmar (Burma)
Piacenza 96 B2 Fr. Paisance; anc. Placentia. Emilia-Romagna, N Italy
Piatra-Neamţ 108 C4 Hung. Karácsonkő. Neamţ, NE Romania
Piauhy see Piauí
Piauí 63 F2 off. Estado do Piauí; prev. Piauhy. state E Brazil
Piauí 63 F2 off. Estado do Piauí; prev. Piauhy. region E Brazil
Piauí, Estado do see Piauí
Picardie 90 C3 Eng. Picardy. cultural region N France
Picardy see Picardie
Piccolo San Bernardo, Colle di see Little Saint Bernard Pass
Pichilemu 64 B4 Libertador, C Chile
Pico 92 A5 var. Ilha do Pico. island Azores, Portugal, NE Atlantic Ocean
Pico, Ilha do see Pico

Picos 63 F2 Piauí, E Brazil
Picton 151 C5 Marlborough, South Island, New Zealand
Piedmont see Piemonte
Piedras Negras 51 E2 var. Ciudad Porfirio Díaz. Coahuila, NE Mexico
Pielavesi 84 D4 lake C Finland
Pielinen 84 E4 var. Pielisjärvi. lake E Finland
Pielisjärvi see Pielinen
Piemonte 96 A2 Eng. Piedmont. region NW Italy
Pierre 45 E3 state capital South Dakota, N USA
Piešt'any 99 C6 Ger. Pistyan, Hung. Pöstyén. Trnavský Kraj, W Slovakia
Pietarhovi see Petrodvorets
Pietari see Sankt-Peterburg
Pietarsaari see Jakobstad
Pietermaritzburg 78 C5 var. Maritzburg. KwaZulu/Natal, E South Africa
Pietersburg see Polokwane
Bay of Pigs 54 B2 Eng. Bay of Pigs. bay SE Cuba
Pigs, Bay of see Cochinos, Bahía de
Pihkva Järv see Pskov, Lake
Pijijiapán 51 G5 Chiapas, SE Mexico
Pikes Peak 44 C5 mountain Colorado, C USA
Pikeville 40 D5 Kentucky, S USA
Pikinni see Bikini Atoll
Piła 98 B3 Ger. Schneidemühl. Wielkopolskie, C Poland
Pilar 64 D3 var. Villa del Pilar. Ñeembucú, S Paraguay
Pilcomayo, Río 57 C5 river C South America
Pilos see Pýlos
Pilsen see Plzeň
Pilzno see Plzeň
Pinang see Pinang, Pulau, Peninsular Malaysia
Pinang see George Town
Pinang, Pulau 138 B3 var. Penang, Pinang; prev. Prince of Wales Island. island Peninsular Malaysia
Pinar del Río 54 A2 Pinar del Río, W Cuba
Píndhos/Píndhos Óros see Píndos
Píndos 104 A4 var. Píndhos Óros, Eng. Pindus Mountains; prev. Píndhos. mountain range C Greece
Pindus Mountains see Píndos
Pine Bluff 42 B2 Arkansas, C USA
Pine Creek 146 D2 Northern Territory, N Australia
Pinega 110 C3 river NW Russian Federation
Pineioś 104 B4 var. Piniós; anc. Peneius. river C Greece
Pineland 49 H3 Texas, SW USA
Pines, Akrotírio 104 C4 var. Akrotírio Pínnes. headland N Greece
Pines, The Isle of see Juventud, Isla de la
Pine Tree State see Maine
Pingdingshan 128 C4 Henan, C China
Pingkiang see Harbin
Ping, Mae Nam 136 B4 river W Thailand
Piniós see Pineioś
Pinkiang see Harbin
Pínnes, Akrotírio see Pínes, Akrotírio
Pinos, Isla de see Juventud, Isla de la
Pinotepa Nacional 51 F5 var. Santiago Pinotepa Nacional. Oaxaca, SE Mexico
Pinsk 107 B7 Pol. Pińsk. Brestskaya Voblasts', SW Belarus
Pinta, Isla 60 A5 var. Abingdon. island Galápagos Islands, Ecuador, E Pacific Ocean
Piombino 96 B3 Toscana, C Italy
Pioneer Mountains 46 D3 mountain range Montana, N USA North America
Pionerskiy 106 A4 Ger. Neukuhren. Kaliningradskaya Oblast', W Russian Federation
Piotrków Trybunalski 98 D4 Ger. Petrikau, Rus. Petrokov. Łódzkie, C Poland
Piraeus/Piraiévs see Peiraías
Pírgos see Pýrgos
Pirineos see Pyrenees
Piripiri 63 F2 Piauí, E Brazil
Pirna 94 D4 Sachsen, E Germany
Pirot 101 E5 Serbia, SE Serbia
Piryatin see Pyryatyn
Pisa 96 B3 var. Pisae. Toscana, C Italy
Pisae see Pisa

Pisaurum see Pesaro
Pisco 60 D4 Ica, SW Peru
Písek 99 A5 Budějovický Kraj, S Czech Republic
Pishan 126 A3 var. Guma. Xinjiang Uygur Zizhiqu, NW China
Pishpek see Bishkek
Pistoia 96 B3 anc. Pistoria, Pistoriæ. Toscana, C Italy
Pistoria/Pistoriæ see Pistoia
Pistyan see Piešt'any
Pisz 98 D3 Ger. Johannisburg. Warmińsko-Mazurskie, NE Poland
Pita 74 C4 NW Guinea
Pitalito 58 B4 Huila, S Colombia
Pitcairn Group of Islands see Pitcairn, Henderson, Ducie & Oeno Islands
Pitcairn Island 143 G4 island S Pitcairn Group of Islands
Pitcairn, Henderson, Ducie & Oeno Islands 143 G4 var. Pitcairn Group of Islands. UK dependent territory C Pacific Ocean
Piteå 84 D4 Norrbotten, N Sweden
Pitești 108 B5 Argeș, S Romania
Pitsanulok see Phitsanulok
Pitt Island see Makin
Pittsburg 45 F5 Kansas, C USA
Pittsburgh 41 E4 Pennsylvania, NE USA
Pittsfield 41 F3 Massachusetts, NE USA
Piura 60 B2 Piura, NW Peru
Pivdennyy Buh 109 E3 Rus. Yuzhnyy Bug. river S Ukraine
Placentia see Piacenza
Placetas 54 B2 Villa Clara, C Cuba
Plainview 49 E2 Texas, SW USA
Pláka 105 C7 Kykládes, Greece, Aegean Sea
Planeta Rica 58 B2 Córdoba, NW Colombia
Planken 94 E1 C Liechtenstein Europe
Plano 49 G2 Texas, SW USA
Plasencia 92 C3 Extremadura, W Spain
Plate, River 64 D4 var. River Plate. estuary Argentina/Uruguay
Plate, River see Plata, Río de la
Platinum 36 C3 Alaska, USA
Plattensee see Balaton
Platte River 45 E4 river Nebraska, C USA
Plattsburgh 41 F2 New York, NE USA
Plauen 94 C4 var. Plauen im Vogtland. Sachsen, E Germany
Plauen im Vogtland see Plauen
Plavinas 106 D4 Ger. Stockmannshof. S Latvia
Plây Cu see Plei Ku
Pleasant Island see Nauru
Plei Ku 137 E5 prev. Plây Cu. Gia Lai, C Vietnam
Plenty, Bay of 150 E3 bay North Island, New Zealand
Plérin 90 A3 Côtes d'Armor, NW France
Plesetsk 110 C3 Arkhangel'skaya Oblast', NW Russian Federation
Pleshchenitsy see Plyeshchanitsy
Pleskau see Pskov
Pleskauer See see Pskov, Lake
Pleskava see Pskov
Pleszew 98 C4 Wielkopolskie, C Poland
Pleven 104 C2 prev. Plevna. Pleven, N Bulgaria
Plevlja/Plevlje see Pljevlja
Plevna see Pleven
Pljevlja 101 C4 prev. Plevlja, Plevlje. N Montenegro
Plocce see Ploče
Ploče 100 B4 It. Plocce; prev. Kardeljevo. Dubrovnik-Neretva, SE Croatia
Płock 98 D3 Ger. Plozk. Mazowieckie, C Poland
Plöcken Pass 95 C7 Ger. Plöckenpass, It. Passo di Monte Croce Carnico. pass SW Austria
Plöckenpass see Plöcken Pass
Ploești see Ploiești
Ploiești 108 C5 prev. Ploești. Prahova, SE Romania
Plomári 105 D5 prev. Plomárion. Lésvos, E Greece
Plomárion see Plomári
Plońsk 98 D3 Mazowieckie, C Poland
Plovdiv 104 C3 prev. Eumolpias; anc. Evmolpia, Philippopolis, Lat. Trimontium. Plovdiv, C Bulgaria
Plozk see Płock
Plungė 106 B3 Telšiai, W Lithuania
Plyeshchanitsy 107 D5 Rus. Pleshchenitsy. Minskaya Voblasts', N Belarus

Plymouth 89 C7 SW England, United Kingdom
Plzeň 99 A5 Ger. Pilsen, Pol. Pilzno. Plzeňský Kraj, W Czech Republic
Po 80 D4 river N Italy
Pobedy, Pik see Tomür Feng
Po, Bocche del see Po, Foci del
Pocahontas 42 B1 Arkansas, C USA
Pocatello 46 E4 Idaho, NW USA
Pochinok 111 A5 Smolenskaya Oblast', W Russian Federation
Pocking 95 D6 Bayern, SE Germany
Poděbrady 99 B5 Ger. Podiebrad. Středočeský Kraj, C Czech Republic
Podgorica 101 C5 prev. Titograd. country capital (Montenegro) S Montenegro
Podiebrad see Poděbrady
Podil's'ka Vysochina 108 D3 plateau W Ukraine
Podium Anicensis see le Puy
Podol'sk 111 B5 Moskovskaya Oblast', W Russian Federation
Podravska Slatina see Slatina
Podujevě 101 D5 Serb. Podujevo. N Kosovo
Podujevo see Podujevě
Podunajská Rovina see Little Alföld
Poetovio see Ptuj
Pogradec 101 D6 var. Pogradeci. K SE Albania
Pogradeci see Pogradec
Pohjanlahti see Bothnia, Gulf of
Pohnpei 144 C2 prev. Ponape Asce Island. island E Micronesia
Poitiers see Poitiers
Poinsett, Cape 154 D4 headland Antarctica
Point de Galle see Galle
Pointe-à-Pitre 55 G3 Grande Terre C Guadeloupe
Pointe-Noire 77 B6 Kouilou, S Congo
Point Lay 36 C2 Alaska, USA
Poitiers 90 B4 prev. Poictiers; anc. Limonum. Vienne, W France
Poitou 90 B4 cultural region W France
Pokharā 135 E3 Western, C Nepal
Pokrovka see Kyzyl-Suu
Pokrovs'ke 109 G3 Rus. Pokrovskoe. Dnipropetrovs'ka Oblast', E Ukraine
Pokrovskoye see Pokrovs'ke
Pola see Pula
Pola de Lena see La Pola
Poland 98 B4 off. Republic of Poland var. Polish Republic, Pol. Polska, Rzeczpospolita Polska; prev. Pol. Rzeczpospolita Ludowa, The Polish People's Republic. country C Europe

POLAND
Northern Europe

Official name Republic of Poland
Formation 1918 / 1945
Capital Warsaw
Population 38.2 million / 325 people per sq mile (126 people per sq km)
Total area 120,728 sq. miles (312,685 sq. km)
Languages Polish*
Religions Roman Catholic 93%, Other nonreligious 5%, Orthodox Christian 2%
Ethnic mix Polish 98%, Other 2%
Government Parliamentary system
Currency Zloty = 100 groszy
Literacy rate 99%
Calorie consumption 3485 kilocalories

Poland, Republic of see Poland
Polatlı 116 C3 Ankara, C Turkey
Polatsk 107 D5 Rus. Polotsk. Vitsyebskaya Voblasts', N Belarus
Pol-e Khomrī see Pul-e Khumrī
Poli see Pólis
Polikastro/Polikastron see Polýkastro
Polikrayshte see Dolna Oryakhovitsa
Pólis 102 C5 var. Poli. W Cyprus
Polish People's Republic, The see Poland
Polish Republic see Poland
Polkowice 98 B4 Ger. Heerwegen. Dolnośląskie, W Poland
Pollença 93 G3 Mallorca, Spain, W Mediterranean Sea
Pologi see Polohy
Polohy 109 G3 Rus. Pologi. Zaporiz'ka Oblast', SE Ukraine
Polokwane 78 D4 prev. Pietersburg. Limpopo, NE South Africa

olonne *108 D2 Rus.* Polonnoye.
Khmel'nyts'ka Oblast', NW Ukraine
olonnoye *see* Polonne
olotsk *see* Polatsk
**olska/Polska, Rzeczpospolita/Polska
Rzeczpospolita Ludowa** *see* Poland
olski Trümbesh *104 D2 prev.* Polsko
Kosovo. Ruse, N Bulgaria
olsko Kosovo *see* Polski Trümbesh
oltava *109 F2* Poltavs'ka Oblast',
NE Ukraine
oltoratsk *see* Aşgabat
olva *106 E3 Ger.* Pölwe. Põlvamaa,
SE Estonia
olwe *see* Põlva
olyarnyy *110 C2* Murmanskaya Oblast',
NW Russian Federation
olykastro *104 B3 var.* Polikastro; *prev.*
Polikastron. Kentrikí Makedonía,
N Greece
olynesia *143 F4 island group* C Pacific
Ocean
omeranian Bay *94 D2 Ger.*
Pommersche Bucht, *Pol.* Zatoka
Pomorska. *bay* Germany/Poland
omir, Dar"yoi *see* Pamir/Pāmīr,
Daryā-ye
ommersche Bucht *see* Pomeranian Bay
omorska, Zatoka *see* Pomeranian Bay
omorskiy Proliv *110 D2 strait*
NW Russian Federation
o, Mouth of the *96 C2 var.* Bocche del
Po. *river* NE Italy
ompaelo *see* Pamplona
ompano Beach *43 F5* Florida, SE USA
onape Ascension Island *see* Pohnpei
onca City *49 G1* Oklahoma, C USA
once *55 F3* C Puerto Rico
ondicherry *132 C2 var.* Puducchheri,
Fr. Pondichéry. Pondicherry, SE India
onferrada *92 C1* Castilla y León,
NW Spain
oniatowa *98 E4* Lubelskie, E Poland
ons Aelii *see* Newcastle upon Tyne
ons Vetus *see* Pontevedra
onta Delgada *92 B5* São Miguel,
Azores, Portugal, NE Atlantic Ocean
onta Grossa *42 E4* Paraná, S Brazil
ontarlier *90 D4* Doubs, E France
onteareas *92 B2* Galicia, NW Spain
onte da Barca *92 B2* Viana do Castelo,
N Portugal
ontevedra *92 B1 anc.* Pons Vetus.
Galicia, NW Spain
ontiac *40 D3* Michigan, N USA
ontianak *138 C4* Borneo, C Indonesia
ontisarae *see* Pontoise
ontivy *90 A3* Morbihan, NW France
ontoise *90 C3 anc.* Briva Isarae, Cergy-
Pontoise, Pontisarae. Val-d'Oise,
N France
onziane Island *97 C5 island* C Italy
oole *89 D7* S England, United Kingdom
oona *see* Pune
oopó, Lago *61 F4 var.* Lago Pampa
Aullagas. *lake* W Bolivia
opayán *58 B4* Cauca, SW Colombia
operinge *87 A6* West-Vlaanderen,
W Belgium
oplar Bluff *45 G5* Missouri, C USA
opocatépetl *51 E4 volcano* S Mexico
opper *see* Poprad
oprad *99 D5 Ger.* Deutschendorf,
Hung. Poprád. Prešovský Kraj,
E Slovakia
oprád *99 D5 Ger.* Popper, *Hung.*
Poprád. *river* Poland/Slovakia
orbandar *134 B4* Gujarāt, W India
orcupine Plain *80 B3 undersea feature*
E Atlantic Ocean
ordenone *96 C2 anc.* Portenau. Friuli-
Venezia Giulia, NE Italy
oreč *100 A2* It. Parenzo. Istra,
NW Croatia
orech'ye *see* Parechcha
ori *85 D5 Swe.* Björneborg. Länsi-
Suomi, SW Finland
orirua *151 D5* Wellington, North
Island, New Zealand
orkhov *110 A4* Pskovskaya Oblast',
W Russian Federation
orlamar *59 E1* Nueva Esparta,
NE Venezuela
óros *105 C6* Póros, S Greece
óros *105 A5* Kefallinía, Iónia Nisiá,
Greece, C Mediterranean Sea
orsangenfjorden *84 D2 Lapp.*
Porsáńguvuotna. *fjord* N Norway
orsáńguvuotna *see* Porsangenfjorden

Porsgrunn *85 B6* Telemark, S Norway
Portachuelo *61 G4* Santa Cruz, C Bolivia
Portadown *89 B5 Ir.* Port An Dúnáin.
S Northern Ireland, United Kingdom
Portalegre *92 C3 anc.* Ammaia, Amoea.
Portalegre, E Portugal
Port Alexander *36 D4* Baranof Island,
Alaska, USA
Port Alfred *78 D5* Eastern Cape, S South
Africa
Port Amelia *see* Pemba
Port An Dúnáin *see* Portadown
Port Angeles *46 B1* Washington,
NW USA
Port Antonio *54 B5* NE Jamaica
Port Arthur *49 H4* Texas, SW USA
Port Augusta *149 B6* South Australia
Port-au-Prince *54 D3 country capital*
(Haiti) C Haiti
Port Blair *133 F2* Andaman and Nicobar
Islands, SE India
Port Charlotte *43 E4* Florida, SE USA
Port Darwin *see* Darwin
Port d'Envalira *91 B8* E Andorra Europe
Port Douglas *148 D3* Queensland,
NE Australia
Port Elizabeth *78 C5* Eastern Cape,
S South Africa
Portenau *see* Pordenone
Porterville *47 C7* California, W USA
Port-Étienne *see* Nouâdhibou
Port Florence *see* Kisumu
Port-Francqui *see* Ilebo
Port-Gentil *77 A6* Ogooué-Maritime,
W Gabon
Port Harcourt *75 G5* Rivers, S Nigeria
Port Hardy *36 D5* Vancouver Island,
British Columbia, SW Canada
Port Harrison *see* Inukjuak
Port Hedland *146 B4* Western Australia
Port Huron *40 D3* Michigan, N USA
Portimão *92 B4 var.* Vila Nova de
Portimão. Faro, S Portugal
Port Jackson *148 E1 harbour* New South
Wales, E Australia
Portland *149 B7* Victoria, SE Australia
Portland *41 G2* Maine, NE USA
Portland *46 B3* Oregon, NW USA
Portland *49 G4* Texas, SW USA
Portland Bight *54 B5 bay* S Jamaica
Portlaoighise *see* Port Laoise
Port Laoise *89 B6 var.* Portlaoise,
Portlaoighise; *prev.* Maryborough.
C Ireland
Portlaoise *see* Port Laoise
Port Lavaca *49 G4* Texas, SW USA
Port Lincoln *149 A6* South Australia
Port Louis *79 H3 country capital*
(Mauritius) NW Mauritius
Port-Lyautey *see* Kénitra
Port Macquarie *149 E6* New South
Wales, SE Australia
Port Mahon *see* Maó
Portmore *54 B5* C Jamaica
Port Moresby *144 B3 country capital*
(Papua New Guinea) Central/National
Capital District,
SW Papua New Guinea
Port Natal *see* Durban
Porto *92 B2 Eng.* Oporto; *anc.* Portus
Cale. Porto, NW Portugal
Porto Alegre *63 E5 var.* Pôrto Alegre.
state capital Rio Grande do Sul, S Brazil
Porto Alegre *76 E2* São Tomé,
S Sao Tome and Principe, Africa
Porto Alegre *76 E2* São Tomé,
S Sao Tome and Principe, Africa
Porto Alexandre *see* Tombua
Porto Bello *see* Portobelo
Portobelo *53 G4 var.* Porto Bello, Puerto
Bello. Colón, N Panama
Port O'Connor *49 G4* Texas, SW USA
Porto Edda *see* Sarandë
Portoferraio *96 B4* Toscana, C Italy
Port-of-Spain *55 H5 country capital*
(Trinidad and Tobago) Trinidad,
Trinidad and Tobago
Porto Grande *see* Mindelo
Portogruaro *96 C2* Veneto, NE Italy
Porto-Novo *75 F5 country capital*
(Benin) S Benin
Porto Rico *see* Puerto Rico
Porto Santo *70 A2 var.* Ilha do Porto
Santo. *island* Madeira, Portugal,
NE Atlantic Ocean
Porto Santo, Ilha do *see* Porto Santo
Port Torres *97 A5* Sardegna, Italy,
C Mediterranean Sea

Porto Velho *62 D2 var.* Velho. *state
capital* Rondônia, W Brazil
Portoviejo *60 A2 var.* Puertoviejo.
Manabí, W Ecuador
Port Pirie *149 B6* South Australia
Port Rex *see* East London
Port Said *see* Bûr Sa'îd
Portsmouth *89 D7* S England, United
Kingdom
Portsmouth *41 G3* New Hampshire,
NE USA
Portsmouth *40 D4* Ohio, N USA
Portsmouth *41 F5* Virginia, NE USA
Port Stanley *see* Stanley
Port Sudan *72 C3* Red Sea, NE Sudan
Port Swettenham *see* Klang/Pelabuhan
Klang
Port Talbot *89 C7* S Wales, United
Kingdom
Portugal *92 B3 off.* Portuguese Republic.
country SW Europe

PORTUGAL
Southwest Europe

Official name	The Portuguese Republic
Formation	1139 / 1640
Capital	Lisbon
Population	10.6 million / 299 people per sq mile (115 people per sq km)
Total area	35,672 sq. miles (92,391 sq. km)
Languages	Portuguese*
Religions	Roman Catholic 92%, Protestant 4%, Nonreligious 3%, Other 1%
Ethnic mix	Portuguese 98%, African and other 2%
Government	Parliamentary system
Currency	Euro = 100 cents
Literacy rate	94%
Calorie consumption	3456 kilocalories

Portuguese East Africa
see Mozambique
Portuguese Guinea *see* Guinea-Bissau
Portuguese Republic *see* Portugal
Portuguese Timor *see* East Timor
Portuguese West Africa *see* Angola
Portus Cale *see* Porto
Portus Magnus *see* Almería
Portus Magonis *see* Maó
Port-Vila *144 D4 var.* Vila. *country
capital* (Vanuatu) Éfaté, C Vanuatu
Porvenir *65 B8* Magallanes, S Chile
Porvoo *85 E6 Swe.* Borgå. Etelä-Suomi,
S Finland
Porzecze *see* Parechcha
Posadas *64 D3* Misiones, NE Argentina
Poschega *see* Požega
Posen *see* Poznań
Posnania *see* Poznań
Postavy/Postawy *see* Pastavy
Posterholt *87 D5* Limburg,
SE Netherlands
Postojna *95 D8 Ger.* Adelsberg, *It.*
Postumia. SW Slovenia
Postumia *see* Postojna
Pöstyén *see* Piešt'any
Potamós *105 C7* Antikýthira, S Greece
Potentia *see* Potenza
Potenza *97 D5 anc.* Potentia. Basilicata,
S Italy
Poti *117 F2 prev.* P'ot'i. W Georgia
P'ot'i *see* Poti
Potiskum *75 G4* Yobe, NE Nigeria
Potomac River *41 E5 river* NE USA
Potosí *61 F4* Potosí, S Bolivia
Potsdam *94 D3* Brandenburg,
NE Germany
Potwar Plateau *134 C2 plateau*
NE Pakistan
Poŭthĭsăt *137 D6 prev.* Pursat.
NW Cambodia
Po, Valle del *see* Po Valley
Po Valley *96 B2* Italian *valley*
N Italy
Považská Bystrica *99 C5 Ger.*
Waagbistritz, *Hung.* Vágbeszterce.
Trenčiansky Kraj, W Slovakia
Poverty Bay *150 E4 inlet* North Island,
New Zealand
Póvoa de Varzim *92 B2* Porto,
NW Portugal
Powder River *44 D2 river* Montana/
Wyoming, NW USA
Powell *44 C2* Wyoming, C USA
Powell, Lake *44 B5 lake* Utah, W USA
Požarevac *100 D4 Ger.* Passarowitz.
Serbia, NE Serbia

Poza Rica *51 F4 var.* Poza Rica de
Hidalgo. Veracruz-Llave, E Mexico
Poza Rica de Hidalgo *see* Poza Rica
Požega *100 D4 prev.* Slavonska Požega,
Ger. Poschega, *Hung.* Pozsega. Požega-
Slavonija, NE Croatia
Požega *100 D4* Serbia
Poznań *98 C3 Ger.* Posen, Posnania.
Wielkopolskie, C Poland
Pozoblanco *92 D4* Andalucía, S Spain
Pozsega *see* Požega
Pozsony *see* Bratislava
Pozzallo *97 C8* Sicilia, Italy,
C Mediterranean Sea
Prachatice *99 A5 Ger.* Prachatitz.
Jihočeský Kraj, S Czech Republic
Prachatitz *see* Prachatice
Prado del Ganso *see* Goose Green
Prae *see* Phrae
Prag/Praga/Prague *see* Praha
Praha *99 A5 Eng.* Prague, *Ger.* Prag, *Pol.*
Praga. *country capital* (Czech Republic)
Středočeský Kraj, NW Czech Republic
Praia *74 A3 country capital* (Cape Verde)
Santiago, S Cape Verde
Prairie State *see* Illinois
Prathet Thai *see* Thailand
Prato *96 B3* Toscana, C Italy
Pratt *45 E5* Kansas, C USA
Prattville *42 D2* Alabama, S USA
Pravda *see* Glavinitsa
Pravia *92 C1* Asturias, N Spain
Preny *see* Prienai
Prenzlau *94 D3* Brandenburg,
NE Germany
Prerau *see* Přerov
Přerov *99 C5 Ger.* Prerau. Olomoucký
Kraj, E Czech Republic
Preschau *see* Prešov
Prescott *48 B2* Arizona, SW USA
Preševo *101 D5* Serbia, SE Serbia
Presidente Epitácio *63 E4* São Paulo,
S Brazil
Presidente Stroessner *see* Ciudad
del Este
Prešov *99 D5 var.* Preschau, *Ger.*
Eperies, *Hung.* Eperjes. Prešovský Kraj,
E Slovakia
Prespa, Lake *101 D6 Alb.* Liqen i
Prespës, *Gk.* Límni Megáli Préspa,
Limni Prespa, *Mac.* Prespansko
Ezero, *Serb.* Prespansko Jezero. *lake*
SE Europe
**Prespa, Limni/Prespansko Ezero/
Prespansko Jezero/Prespës, Liqen i**
see Prespa, Lake
Presque Isle *41 H1* Maine, NE USA
Pressburg *see* Bratislava
Preston *89 D5* NW England, United
Kingdom
Prestwick *88 C4* W Scotland, United
Kingdom
Pretoria *78 D4 country capital*
(South Africa) (admin. capital)
NE South Africa
Preussisch Eylau *see* Bagrationovsk
Preußisch Holland *see* Pasłęk
Preussisch-Stargard *see* Starogard
Gdański
Préveza *105 A5* Ípeiros, W Greece
Pribilof Islands *36 A3 island group*
Alaska, USA
Priboj *100 C4* Serbia, W Serbia
Price *44 B4* Utah, W USA
Prichard *42 C3* Alabama, S USA
Priekulé *106 B3 Ger.* Prökuls. Klaipéda,
W Lithuania
Prienai *106 B4 Pol.* Preny. Kaunas,
S Lithuania
Prieska *78 C4* Northern Cape, C South
Africa
Prijedor *100 B3* Republika Srpska,
NW Bosnia and Herzegovina
Prijepolje *100 D4* Serbia, W Serbia
Prikaspiyskaya Nizmennost' *see*
Caspian Depression
Prilep *101 D6 Turk.* Perlepe. S FYR
Macedonia
Primorsk *106 A4 Ger.* Fischhausen.
Kaliningradskaya Oblast', W Russian
Federation
Primorsk/Primorskoye *see* Prymors'k
Prince Albert *37 F5* Saskatchewan,
S Canada
Prince Edward Island *39 F4 Fr.* Île-du
Prince-Édouard. *province* SE Canada

Prince Edward Islands *69 E8 island group* S South Africa
Prince George *37 E5* British Columbia, SW Canada
Prince of Wales Island *148 B1 island* Queensland, E Australia
Prince of Wales Island *37 F2 island* Queen Elizabeth Islands, Nunavut, NW Canada
Prince of Wales Island *see* Pinang, Pulau
Prince Patrick Island *37 E2 island* Parry Islands, Northwest Territories, NW Canada
Prince Rupert *36 D4* British Columbia, SW Canada
Prince's Island *see* Príncipe
Princess Charlotte Bay *148 C2 bay* Queensland, NE Australia
Princess Elizabeth Land *154 C3 physical region* Antarctica
Príncipe *77 A5 var.* Príncipe Island, *Eng.* Prince's Island. *island* N Sao Tome and Principe
Príncipe Island *see* Príncipe
Prinzapolka *53 E3* Región Autónoma Atlántico Norte, NE Nicaragua
Pripet *107 C7 Bel.* Prypyats', *Ukr.* Pryp"yat'. *river* Belarus/Ukraine
Pripet Marshes *107 B7 wetland* Belarus/Ukraine
Prishtinë *101 D5 var.* Pristina, *Serb.* Priština. C Kosovo
Priština/Pristina *see* Prishtinë
Privas *91 D5* Ardèche, E France
Prizren *101 D5* S Kosovo
Probolinggo *138 D5* Jawa, C Indonesia
Probstberg *see* Wyszków
Progreso *51 H3* Yucatán, SE Mexico
Prokhladnyy *111 B8* Kabardino-Balkarskaya Respublika, SW Russian Federation
Prokletije *see* North Albanian Alps
Prōkuls *see* Priekulė
Prokuplje *101 D5* Serbia, SE Serbia
Prome *see* Pyay
Promyshlennyy *110 E3* Respublika Komi, NW Russian Federation
Prościejów *see* Prostějov
Proskurov *see* Khmel 'nyts'kyy
Prossnitz *see* Prostějov
Prostějov *99 C5 Ger.* Prossnitz, *Pol.* Prościejów. Olomoucký Kraj, E Czech Republic
Provence *91 D6 cultural region* SE France
Providence *41 G3 state capital* Rhode Island, NE USA
Providence *see* Fort Providence
Providencia, Isla de *53 F3 island* NW Colombia, Caribbean Sea
Providenıya *155 B1* Chukotskiy Avtonomnyy Okrug, NE Russian Federation
Provo *44 B4* Utah, W USA
Prudhoe Bay *36 D2* Alaska, USA
Prusa *see* Bursa
Pruszków *98 D3 Ger.* Kaltdorf. Mazowieckie, C Poland
Prut *108 D4 Ger.* Pruth. *river* E Europe
Pruzana *see* Pruzhany
Pruzhany *107 B6 Pol.* Prużana. Brestskaya Voblasts', SW Belarus
Prychornomor'ska Nyzovyna *see* Black Sea Lowland
Prydniprovs'ka Nyzovyna/ Prydnyaprowskaya Nizina *see* Dnieper Lowland
Prydz Bay *154 D3 bay* Antarctica
Pryluky *109 E2 Rus.* Priluki. Chernihivs'ka Oblast', NE Ukraine
Prymors'k *109 G4 Rus.* Primorsk; *prev.* Primorskoye. Zaporiz'ka Oblast', SE Ukraine
Pryp"yat'/Prypyats' *see* Pripet
Przemyśl *99 E5 Rus.* Peremyshl. Podkarpackie, C Poland
Przheval'sk *see* Karakol
Psará *105 D5 island* E Greece
Psel *109 F2 Rus.* Psël. *river* Russian Federation/Ukraine
Psël *see* Psel
Pskov *114 B2 Ger.* Pleskau, *Latv.* Pleskava. Pskovskaya Oblast', W Russian Federation
Pskov, Lake *116 E3 Est.* Pihkva Järv, *Ger.* Pleskauer See, *Rus.* Pskovskoye Ozero. *lake* Estonia/Russian Federation
Pskovskoye Ozero *see* Pskov, Lake
Ptich' *see* Ptsich

Ptsich *107 C7 Rus.* Ptich'. Homyel'skaya Voblasts', SE Belarus
Ptsich *107 C7 Rus.* Ptich'. *river* SE Belarus
Ptuj *95 E7 Ger.* Pettau; *anc.* Poetovio. NE Slovenia
Pucallpa *60 C3* Ucayali, C Peru
Puck *98 C2* Pomorskie, N Poland
Pudasjärvi *84 D4* Oulu, C Finland
Puebla *51 F4 var.* Puebla de Zaragoza. Puebla, S Mexico
Puebla de Zaragoza *see* Puebla
Pueblo *44 D5* Colorado, C USA
Puerto Acosta *61 E4* La Paz, W Bolivia
Puerto Aisén *65 B6* Aisén, S Chile
Puerto Ángel *51 F5* Oaxaca, SE Mexico
Puerto Argentino *see* Stanley
Puerto Ayacucho *58 D3* Amazonas, SW Venezuela
Puerto Baquerizo Moreno *60 B5 var.* Baquerizo Moreno. Galápagos Islands, Ecuador, E Pacific Ocean
Puerto Barrios *52 C2* Izabal, E Guatemala
Puerto Bello *see* Portobelo
Puerto Berrío *58 B2* Antioquia, C Colombia
Puerto Cabello *58 D1* Carabobo, N Venezuela
Puerto Cabezas *53 E2 var.* Bilwi. Región Autónoma Atlántico Norte, NE Nicaragua
Puerto Carreño *58 D3* Vichada, E Colombia
Puerto Cortés *52 C2* Cortés, NW Honduras
Puerto Cumarebo *58 C1* Falcón, N Venezuela
Puerto Deseado *65 C7* Santa Cruz, SE Argentina
Puerto Escondido *51 F5* Oaxaca, SE Mexico
Puerto Francisco de Orellana *60 B1 var.* Coca. NE Ecuador
Puerto Gallegos *see* Río Gallegos
Puerto Inírida *58 D3 var.* Obando. Guainía, E Colombia
Puerto La Cruz *59 E1* Anzoátegui, NE Venezuela
Puerto Lempira *53 E2* Gracias a Dios, E Honduras
Puerto Limón *see* Limón
Puertollano *92 D4* Castilla-La Mancha, C Spain
Puerto López *58 C1* La Guajira, N Colombia
Puerto Maldonado *61 E3* Madre de Dios, E Peru
Puerto México *see* Coatzacoalcos
Puerto Montt *65 B5* Los Lagos, C Chile
Puerto Natales *65 B7* Magallanes, S Chile
Puerto Obaldía *53 H5* Kuna Yala, NE Panama
Puerto Plata *55 E3 var.* San Felipe de Puerto Plata. N Dominican Republic
Puerto Presidente Stroessner *see* Ciudad del Este
Puerto Princesa *139 E2 off.* Puerto Princesa City. Palawan, W Philippines
Puerto Princesa City *see* Puerto Princesa
Puerto Príncipe *see* Camagüey
Puerto Rico *55 F3 off.* Commonwealth of Puerto Rico; *prev.* Porto Rico. *US commonwealth territory* C West Indies
Puerto Rico *56 B1 island* C West Indies
Puerto Rico, Commonwealth of *see* Puerto Rico
Puerto Rico Trench *56 B1 trench* NE Caribbean Sea
Puerto San José *see* San José
Puerto San Julián *65 B7 var.* San Julián. Santa Cruz, SE Argentina
Puerto Suárez *61 H4* Santa Cruz, E Bolivia
Puerto Vallarta *50 D4* Jalisco, SW Mexico
Puerto Varas *65 B5* Los Lagos, C Chile
Puerto Viejo *53 E4* Heredia, NE Costa Rica
Puertoviejo *see* Portoviejo
Puget Sound *46 B1 sound* Washington, NW USA
Puglia *97 E5 var.* Le Puglie, *Eng.* Apulia. *region* SE Italy
Pukaki, Lake *151 B6 lake* South Island, New Zealand

Pukekohe *150 D3* Auckland, North Island, New Zealand
Puket *see* Phuket
Pukhavichy *107 C6 Rus.* Pukhovichi. Minskaya Voblasts', C Belarus
Pukhovichi *see* Pukhavichy
Pula *100 A3 It.* Pola; *prev.* Pulj. Istra, NW Croatia
Pulaski *40 D5* Virginia, NE USA
Puławy *98 D4 Ger.* Neu Amerika. Lubelskie, E Poland
Pul-e Khumrī *123 E4 prev.* Pol-e Khomrī, *var.* Pul-i-Khumri. Baghlān, NE Afghanistan
Pul-i-Khumri *see* Pul-e Khumrī
Pulj *see* Pula
Pullman *46 C2* Washington, NW USA
Pułtusk *98 D3* Mazowieckie, C Poland
Puná, Isla *60 A2 island* SW Ecuador
Pune *134 C5 prev.* Poona. Mahārāshtra, W India
Punjab *134 C2 prev.* West Punjab, Western Punjab. *province* E Pakistan
Puno *61 E4* Puno, SE Peru
Punta Alta *65 C5* Buenos Aires, E Argentina
Punta Arenas *65 B8 prev.* Magallanes. Magallanes, S Chile
Punta Gorda *52 C2* Toledo, SE Belize
Punta Gorda *53 E4* Región Autónoma Atlántico Sur, SE Nicaragua
Puntarenas *52 D4* Puntarenas, W Costa Rica
Punto Fijo *58 C1* Falcón, N Venezuela
Pupuya, Nevado *61 E4 mountain* W Bolivia
Puri *135 F5 var.* Jagannath. Odisha, E India
Puriramya *see* Buriram
Purmerend *86 C3* Noord-Holland, C Netherlands
Pursat *see* Poŭthĭsăt, Poŭthĭsăt, W Cambodia
Purus, Río *62 C2 var.* Río Purús. *river* Brazil/Peru
Pusan *see* Busan
Pushkino *see* Biläsuvar
Püspökladány *99 D6* Hajdú-Bihar, E Hungary
Putorana, Gory/Putorana Mountains *see* Putorana, Plato
Putorana Mountains *115 E3 var.* Gory Putorana, *Eng.* Putorana Mountains. *mountain range* N Russian Federation
Putrajaya *138 B3 administrative capital* (Malaysia) Kuala Lumpur, Peninsular Malaysia
Puttalam *132 C3* North Western Province, W Sri Lanka
Puttgarden *94 C2* Schleswig-Holstein, N Germany
Putumayo, Río *58 B5 var.* Içá, Rio. *river* NW South America
Putumayo, Río *see* Içá, Rio
Puurmani *106 D2 Est.* Talkhof. Jõgevamaa, E Estonia
Pyatigorsk *111 B7* Stavropol'skiy Kray, SW Russian Federation
Pyatikhatki *see* P"yatykhatky
P"yatykhatky *109 F3 Rus.* Pyatikhatki. Dnipropetrovs'ka Oblast', E Ukraine
Pyay *136 B4 var.* Prome, Pye. Bago, C Myanmar (Burma)
Pye *see* Pyay
Pyetrykaw *107 C7 Rus.* Petrikov. Homyel'skaya Voblasts', SE Belarus
Pyin-Oo-Lwin *136 B3 var.* Maymyo. Mandalay, C Myanmar (Burma)
Pylos *105 B6 var.* Pílos. Pelopónnisos, S Greece
P'yŏngyang *129 E3 var.* P'yŏngyang-si, *Eng.* Pyongyang. *country capital* (North Korea) SW North Korea
P'yŏngyang-si *see* P'yŏngyang
Pyramid Lake *47 C5 lake* Nevada, W USA
Pyrenaei Montes *see* Pyrenees
Pyrenees *102 B2 Fr.* Pyrénées, *Sp.* Pirineos; *anc.* Pyrenaei Montes. *mountain range* SW Europe
Pýrgos *105 B6 var.* Pírgos. Dytikí Elláda, S Greece
Pyritz *see* Pyrzyce
Pyryatyn *109 E2 Rus.* Piryatin. Poltavs'ka Oblast', NE Ukraine
Pyrzyce *98 B3 Ger.* Pyritz. Zachodnio-pomorskie, NW Poland

Pyu *see* Phyu
Pyuntaza *136 B4* Bago, SW Myanmar (Burma)

Q

Qā' al Jafr *119 C7 lake* S Jordan
Qaanaaq *82 D1 var.* Qânâq, *Dan.* Th Avannaarsua, N Greenland
Qabātiya *119 E6* N West Bank Asia
Qābis *see* Gabès
Qābis, Khalīj *see* Gabès, Golfe de
Qacentina *see* Constantine
Qafṣah *see* Gafsa
Qagan Us *see* Dulan
Qahremānshahr *see* Kermānshāh
Qaidam Pendi *126 C4 basin* C China
Qal'aikhum *123 F3 Rus.* Kalaikhum. S Tajikistan
Qalāt *123 E5 Per.* Kalāt. Zābul, S Afghanistan
Qal'at Bīshah *121 B5* 'Asīr, SW Saudi Arabia
Qalqiliya *see* Qalqilya
Qalqilya *119 D6 var.* Qalqiliya. Centr W West Bank Asia
Qamanittuaq *see* Baker Lake
Qamdo *126 D5* Xizang Zizhiqu, W China
Qamishly *see* Al Qāmishlī
Qânâq *see* Qaanaaq
Qaqortoq *82 C4 Dan.* Julianehåb. Kitaa, S Greenland
Qaraghandy/Qaraghandy Oblysy *see* Karagandy
Qara Qum *see* Garagum
Qarataū *see* Karatau, Zhambyl, Kazakhstan
Qarkilik *see* Ruoqiang
Qarokŭl *123 F3 Rus.* Karakul'. E Tajikistan
Qarqannah, Juzur *see* Kerkenah, Îles de
Qars *see* Kars
Qarshi *123 E3 Rus.* Karshi; *prev.* Bek-Budi. Qashqadaryo Viloyati, S Uzbekistan
Qasigianguit *see* Qasigiannguit
Qasigiannguit *82 C3 var.* Qasigianguit *Dan.* Christianshåb. Kitaa, C Greenland
Qaṣr al Farāfirah *72 B2 var.* Qasr Farāfra. W Egypt
Qasr Farāfra *see* Qaṣr al Farāfirah
Qaṭanā *119 B5 var.* Katana. Rif Dimashq, S Syria
Qatar *120 C4 off.* State of Qatar, *Ar.*

QATAR
Southwest Asia

Official name State of Qatar
Formation 1971 / 1971
Capital Doha
Population 2.2 million / 518 people per sq mile (200 people per sq km)
Total area 4416 sq. miles (11,437 sq. km)
Languages Arabic*
Religions Muslim (mainly Sunni) 95%, Other 5%
Ethnic mix Qatari 20%, Indian 20%, Other Arab 20%, Nepalese 13%, Filipino 10%, Other 10%, Pakistani 7%
Government Monarchy
Currency Qatar riyal = 100 dirhams
Literacy rate 97%
Calorie consumption Not available

Dawlat Qatar. *country* SW Asia
Qatar, State of *see* Qatar
Qattara Depression *72 A1 var.* Munkhafad al Qattārah *var.* Monkha el Qattara, *Eng.* Qattara Depression. *desert* NW Egypt
Qattārah, Munkhafad al *see* Qattara Depression
Qattâra, Monkhafad el *see* Qattara Depression
Qausuittuq *see* Resolute
Qazaqstan/Qazaqstan Respublikasy *see* Kazakhstan
Qazimämmäd *see* Hacıqabal
Qazris *see* Cáceres
Qazvīn *120 C2 var.* Kazvin. Qazvīn, N Iran
Qena *see* Qinā
Qeqertarssuaq *see* Qeqertarsuaq

ertarsuaq *82 C3 var.* Qeqertarssuaq,
an. Godhavn. Kitaa, S Greenland
ertarsuaq *82 C3 island* W Greenland
ertarsuup Tunua *82 C3 Dan.* Disko
ugt. *inlet* W Greenland
veh *see* Qorveh
hm *120 D4 var.* Jazīreh-ye Qeshm,
eshm Island. *island* S Iran
hm Island/Qeshm, Jazīreh-ye
ee Qeshm
an Shan *126 D3 var.* Kilien
Iountains. *mountain range*
l China
usseriarsuaq *82 C2 Dan.*
Ielville Bugt, *Eng.* Melville Bay. *bay*
W Greenland
ā *72 B2 var.* Qena; *anc.* Caene,
Iaenepolis. E Egypt
g *see* Qinghai
gdao *128 D4 var.* Ching-Tao,
h'ing-tao, Tsingtao, Tsintao, *Ger.*
'singtau. Shandong, E China
ghai *126 C4 var.* Chinghai, Koko
Ior, Qing, Qinghai Sheng, Tsinghai.
rovince C China
ghai Hu *126 D4 var.* Ch'ing Hai,
'sing Hai, *Mong.* Koko Nor. *lake*
C China
ghai Sheng *see* Qinghai
huangdao *128 D3* Hebei, E China
zhou *128 B6* Guangxi Zhuangzu
izhiqu, S China
ng *see* Hainan
jhar *128 D2 var.* Ch'i-ch'i-ha-
rh, Tsitsihar; *prev.* Lungkiang.
leilongjiang, NE China
a *126 B4* Xinjiang Uygur Zizhiqu,
NW China
a Ghazzah *see* Gaza Strip
ai *126 C3* Xinjiang Uygur Zizhiqu,
NW China
ān *see* Jīzān
il Orda *see* Kyzylorda
il Qum/Qïzïlqum *see* Kyzyl Kum
ilrabot *123 G3 Rus.* Kyzylrabot.
SE Tajikistan
bustan *117 H2 prev.* Märäzä.
E Azerbaijan
gir Feng *see* K2
m *120 C3 var.* Kum, Qum. Qom,
N Iran
molangma Feng *see* Everest, Mount
mul *see* Hami
'qon *123 F2 var.* Khokand,
Rus. Kokand. Farg'ona Viloyati,
E Uzbekistan
rveh *120 C3 var.* Qerveh, Qurveh.
Kordestān, W Iran
stanay/Qostanay Oblysy *see*
Kostanay
ubaiyāt *118 B4 var.* Al Qubayyāt.
N Lebanon
ussantina *see* Constantine
ang Ngai *137 E5 var.* Quangngai,
Quang Nghia. Quang Ngai, C Vietnam
angngai *see* Quang Ngai
ang Nghia *see* Quang Ngai
an Long *see* Ca Mau
anzhou *128 D6 var.* Ch'uan-chou,
Tsinkiang; *prev.* Chin-chiang. Fujian,
SE China
anzhou *128 C6* Guangxi Zhuangzu
Zizhiqu, S China
'Appelle *37 F5 river* Saskatchewan,
S Canada
arles, Pegunungan *139 E4 mountain
range* Sulawesi, C Indonesia
arnero *see* Kvarner
artu Sant' Elena *97 A6* Sardegna,
Italy, C Mediterranean Sea
ba *117 H2 Rus.* Kuba. N Azerbaijan
bba *see* Ba'qūbah
ébec *39 E4 var.* Quebec. *province*
capital* Québec, SE Canada
ébec *38 D3 var.* Quebec. *province*
SE Canada
een Charlotte Islands *36 C5 Fr.* Îles
de la Reine-Charlotte. *island group*
British Columbia, SW Canada
een Charlotte Sound *36 C5 sea area*
British Columbia, W Canada
een Elizabeth Islands *33 E1 Fr.* Îles
de la Reine-Élisabeth. *island group*
Nunavut, N Canada
eensland *148 B4 state* N Australia
eenstown *151 B7* Otago, South Island,
New Zealand
eenstown *78 D5* Eastern Cape,
S South Africa

Quelimane *79 E3 var.* Kilimane,
Kilmain, Quilimane. Zambézia,
NE Mozambique
Quelpart *see* Jeju-do
Quepos *53 E4* Puntarenas, S Costa Rica
Que Que *see* Kwekwe
Quera *see* Chur
Querétaro *51 E4* Querétaro de Arteaga,
C Mexico
Quesada *53 E4 var.* Ciudad Quesada,
San Carlos. Alajuela, N Costa Rica
Quetta *134 B2* Baluchistān,
SW Pakistan
Quetzalcoalco *see* Coatzacoalcos
Quezaltenango *52 A2 var.*
Quetzaltenango. Quezaltenango,
W Guatemala
Quibdó *58 A3* Chocó, W Colombia
Quilimane *see* Quelimane
Quillabamba *60 D3* Cusco, C Peru
Quilon *see* Kollam
Quimper *90 A3 anc.* Quimper Corentin.
Finistère, NW France
Quimper Corentin *see* Quimper
Quimperlé *90 A3* Finistère, NW France
Quincy *40 A4* Illinois, N USA
Qui Nhon/Quinhon *see* Quy Nhơn
Quissico *79 E4* Inhambane,
S Mozambique
Quito *60 B1 country capital* (Ecuador)
Pichincha, N Ecuador
Qulyndy Zhazyghy *see* Ravnina Kulyndy
Qum *see* Qom
Qurein *see* Al Kuwayt
Qūrghonteppa *123 E3 Rus.* Kurgan-
Tyube. SW Tajikistan
Qurlurtuuq *see* Kugluktuk
Qurveh *see* Qorveh
Quşayr *see* Al Quşayr
Quxar *see* Lhazê
Quy Nhơn *137 E5 var.* Quinhon, Qui
Nhon. Binh Dinh, C Vietnam
Qyteti Stalin *see* Kuçovë
Qyzylorda *see* Kyzylorda

R

Raab *100 B1 Hung.* Rába. *river* Austria/
Hungary
Raab *see* Rába
Raab *see* Győr
Raahe *84 D4 Swe.* Brahestad. Oulu,
W Finland
Raalte *86 D3* Overijssel, E Netherlands
Raamsdonksveer *86 C4* Noord-Brabant,
S Netherlands
Raasiku *106 D2* Ger. Rasik. Harjumaa,
NW Estonia
Rába *99 B7 Ger.* Raab. *river* Austria/
Hungary
Rába *see* Raab
Rabat *70 C2 var.* al Dar al Baida. *country
capital* (Morocco) NW Morocco
Rabat *102 B5* W Malta
Rabat *see* Victoria
Rabbah Ammon/Rabbath Ammon
see 'Ammon
Rabinal *52 B2* Baja Verapaz,
C Guatemala
Rabka *99 D5* Małopolskie, S Poland
Rábnița *see* Ribnița
Rabyanah Ramlat *71 G4 var.* Rebiana
Sand Sea, Şaḩrā' Rabyānah. *desert*
SE Libya
Rabyanah, Şaḩrā' *see* Rabyanah, Ramlat
Rácari *see* Durankulak
Race, Cape *39 H3 headland*
Newfoundland, Newfoundland and
Labrador, E Canada
Rach Gia *137 D6* Kiên Giang,
S Vietnam
Racine *40 B3* Wisconsin, N USA
Rácz-Becse *see* Bečej
Rădăuți *108 C4* Ger. Radautz, *Hung.*
Rádóci. Suceava, N Romania
Rádóci *see* Rădăuți
Rádeyilíkóe *see* Fort Good Hope
Radom *98 D4* Mazowieckie, C Poland
Radomyshl' *108 D2* Zhytomyrs'ka
Oblast', N Ukraine
Radoviš *101 E6 prev.* Radovište.
E Macedonia
Radovište *see* Radoviš
Radviliškis *106 B4* Lithuania

Radzyń Podlaski *98 E4* Lubelskie,
E Poland
Rae-Edzo *see* Edzo
Raetihi *150 D4* Manawatu-Wanganui,
North Island, New Zealand
Rafa *see* Rafah
Rafaela *64 C3* Santa Fe, E Argentina
Rafah *119 A7 var.* Rafa, Rafaḥ, *Heb.*
Rafiaḥ, Raphiah. SW Gaza Strip
Rafḥah *120 B4* Al Ḥudūd ash
Shamālīyah, N Saudi Arabia
Rafiah *see* Rafah
Raga *73 A5* Western Bahr el Ghazal,
W South Sudan
Ragged Island Range *54 C2 island group*
S Bahamas
Ragnit *see* Neman
Ragusa *97 C7* Sicilia, Italy,
C Mediterranean Sea
Ragusa *see* Dubrovnik
Rahachow *107 D7 Rus.* Rogachëv.
Homyel'skaya Voblasts', SE Belarus
Rahaeng *see* Tak
Rahaţ, Ḩarrat *121 B5 lava flow* W Saudi
Arabia
Raḥīmyār Khān *134 C3* Punjab,
SE Pakistan
Rahovec *101 D5 Serb.* Orahovac.
W Kosovo
Raiatea *145 G4 island* Îles Sous le Vent,
W French Polynesia
Rāichūr *132 C1* Karnātaka, C India
Raidestos *see* Tekirdağ
Rainier, Mount *34 A4 volcano*
Washington, NW USA
Rainy Lake *38 A4 lake* Canada/USA
Raippaluoto *see* Replot
Rairaha *see* Sale
Rairakhol *see* Rairangpur
Raisen *see* Raisin
Raisin *see* Raisen
Raithby *see* Rajgangpur
Raiwala Bazar *see* Rajgangpur

Ra's al 'Ayn *118 D1 var.* Ras al'Ain. Al
Ḩasakah, N Syria
Ra's an Naqb *119 B7* Ma'ān, S Jordan
Raseiniai *106 B4* Kaunas, C Lithuania
Rasht *120 C2 var.* Resht. Gīlān, NW Iran
Rasik *see* Raasiku
Rîşnov *108 C4 prev.* Rîşno, Rozsnyó,
Hung. Barcarozsnyó. Brașov,
C Romania
Rastenburg *see* Kętrzyn
Ratak Chain *144 D1 island group* Ratak
Chain, E Marshall Islands
Ratān *85 C5* Jämtland, C Sweden
Rat Buri *see* Ratchaburi
Ratchaburi *137 C5 var.* Rat Buri.
Ratchaburi, W Thailand
Ratisbon/Ratisbona/Ratisbonne *see*
Regensburg
Rat Islands *36 A2 island group* Aleutian
Islands, Alaska, USA
Ratlām *134 D4 prev.* Rutlam. Madhya
Pradesh, C India
Ratnapura *132 D4* Sabaragamuwa
Province, S Sri Lanka
Raton *48 D1* New Mexico, SW USA
Rättvik *85 C5* Dalarna, C Sweden
Raudhatain *see* Ar Rawḑatayn
Raufarhöfn *83 E4* Norðurland Eystra,
NE Iceland
Raukawa *see* Cook Strait
Raukumara Range *150 E3 mountain
range* North Island, New Zealand
Rāulakela *see* Rāurkela
Rauma *85 D5 Swe.* Raumo. Länsi-
Suomi, SW Finland
Raumo *see* Rauma
Rāurkela *135 F4 var.* Rāulakela,
Rourkela. Odisha, E India
Ravenna *96 C3* Emilia-Romagna,
N Italy
Ravi *134 C2 river* India/Pakistan
Ravnina Kulyndy *114 C4 prev.* Kulunda
Steppe, *Kaz.* Qulyndy Zhazyghy, *Rus.*
Kulundinskaya Ravnina. *grassland*
Kazakhstan/Russian Federation
Rāwalpindi *134 C1* Punjab, NE Pakistan
Rawa Mazowiecka *98 D4* Łódzkie,
C Poland
Rawicz *98 C4 Ger.* Rawitsch.
Wielkopolskie, C Poland
Rawitsch *see* Rawicz
Rawlins *44 C3* Wyoming, C USA
Rawson *65 C6* Chubut, SE Argentina
Rayak *118 B4 var.* Rayaq, Riyāq.
E Lebanon
Rayaq *see* Rayak
Rayong *137 C5* Rayong, S Thailand
Raz, Pointe du *90 A3 headland* NW France
Razāzah, Buhayrat ar *120 B3 var.* Baḥr
al Milḥ. *lake* C Iraq
Razdolnoye *see* Rozdol'ne
Razgrad *104 D2* Razgrad, N Bulgaria
Razim, Lacul *108 E1 prev.* Razelm.
lagoon NW Black Sea
Reading *89 D7* S England, United
Kingdom
Reading *41 F4* Pennsylvania, NE USA
Realicó *64 C4* La Pampa, C Argentina
Reàng Kesei *137 D5* Bătdâmbâng,
W Cambodia
Greater Antarctica *see* East Antarctica
Rebecca, Lake *147 C6 lake* Western
Australia
Rebiana Sand Sea *see* Rabyānah, Ramlat
Rebun-to *130 C2 island* NE Japan
Rechitsa *see* Rechytsa
Rechytsa *107 D7 Rus.* Rechitsa.
Brestskaya Voblasts', SW Belarus
Recife *65 G2 prev.* Pernambuco. *state
capital* Pernambuco, E Brazil
Recklinghausen *94 A4* Nordrhein-
Westfalen, W Germany
Recogne *87 C7* Luxembourg, SE Belgium
Reconquista *64 D3* Santa Fe,
C Argentina
Red Deer *37 E5* Alberta, SW Canada
Redding *47 B5* Ille-et-Vilaine, NW France
Red River *52 C1 var.* Yuan, *Chin.* Yuan
Jiang, *Vtn.* Sông Hông Hà. *river* China/
Vietnam
Red River *35 C6 river* S USA
Red River *49 G3 river* Louisiana, S USA
Red Sea *72 C3 var.* Sinus Arabicus. *sea*
Africa/Asia
Red Wing *45 G2* Minnesota, N USA
Reefton *151 C5* West Coast, South
Island, New Zealand

Reese River *47 C5 river* Nevada, W USA
Refahiye *117 E3* Erzincan, C Turkey
Regensburg *95 C6 Eng.* Ratisbon, *Fr.*
  Ratisbonne, *hist.* Ratisbona; *anc.* Castra
  Regina, Reginum. Bayern, SE Germany
Regenstauf *95 C6* Bayern, SE Germany
Rēgestān *122 D5 prev.* Rīgestān, *var.*
  Registan. *desert region* S Afghanistan
Reggane *70 D3* C Algeria
Reggio Calabria *see* Reggio di Calabria
Reggio di Calabria *97 D7 var.* Reggio
  Calabria, *Gk.* Rhegion; *anc.* Regium,
  Rhegium. Calabria, SW Italy
Reggio Emilia *see* Reggio nell'Emilia
Reggio nell'Emilia *96 B2 var.* Reggio
  Emilia, *abbrev.* Reggio; *anc.* Regium
  Lepidum. Emilia-Romagna, N Italy
Reghin *108 C4 Ger.* Sächsisch-Reen,
  Hung. Szászrégen; *prev.* Reghinul
  Săsesc, *Ger.* Sächsisch-Regen. Mureş,
  C Romania
Reghinul Săsesc *see* Reghin
Regina *37 F5 province capital*
  Saskatchewan, S Canada
Reginum *see* Regensburg
Registan *see* Rēgestān
Regium *see* Reggio di Calabria
Regium Lepidum *see* Reggio nell'Emilia
Rehoboth *78 B3* Hardap, C Namibia
Rehovot *119 A6 ; prev.* Reḥovot. Central,
  C Israel
Reḥovot *see* Rehovot
Reichenau *see* Bogatynia, Poland
Reichenberg *see* Liberec
Reid *147 D6* Western Australia
Reikjavik *see* Reykjavík
Ré, Île de *90 A4 island* W France
Reims *90 D3 Eng.* Rheims; *anc.*
  Durocortorum, Remi. Marne,
  N France
Reindeer Lake *37 F4 lake* Manitoba/
  Saskatchewan, C Canada
Reine-Charlotte, Îles de la *see* Queen
  Charlotte Islands
Reine-Élisabeth, Îles de la *see* Queen
  Elizabeth Islands
Reinga, Cape *150 C1 headland* North
  Island, New Zealand
Reinosa *92 D1* Cantabria, N Spain
Reka *see* Rijeka
Rekhovot *see* Rehovot
Reliance *37 F4* Northwest Territories,
  C Canada
Remi *see* Reims
Rendina *see* Rentína
Rendsburg *94 B2* Schleswig-Holstein,
  N Germany
Rengat *138 B4* Sumatera, W Indonesia
Reni *108 D4* Odes'ka Oblast',
  SW Ukraine
Rennell *144 C4 var.* Mu Ngawa. *island*
  S Solomon Islands
Rennes *90 B3 Bret.* Roazon; *anc.*
  Condate. Ille-et-Vilaine, NW France
Reno *47 C5* Nevada, W USA
Renqiu *128 C4* Hebei, E China
Rentina *105 B5 var.* Rendina. Thessalía,
  C Greece
Reps *see* Rupea
Repulse Bay *37 G3* Northwest
  Terretories, N Canada
Reschitza *see* Reşiţa
Resht *see* Rasht
Resicabánya *see* Reşiţa
Resistencia *64 D3* Chaco, NE Argentina
Reşiţa *108 A4 Ger.* Reschitza, *Hung.*
  Resicabánya. Caraş-Severin,
  W Romania
Resolute *37 F2 Inuit* Qausuittuq.
  Cornwallis Island, Nunavut, N Canada
Resolution Island *39 E1 island* Nunavut,
  NE Canada
Resolution Island *151 A7 island*
  SW New Zealand
Réunion *79 H4 off.* La Réunion. *French
  overseas department* W Indian Ocean
Réunion *141 B5 island* W Indian Ocean
Reus *93 F2* Cataluña, E Spain
Reutlingen *95 B6* Baden-Württemberg,
  S Germany
Reuver *87 D5* Limburg, SE Netherlands
Reval/Revel *see* Tallinn
Revillagigedo Island *50 B5 island*
  Alexander Archipelago, Alaska, USA
Rexburg *46 E3* Idaho, NW USA
Reyes, Isla del *53 G5 island* Archipiélago
  de las Perlas, SE Panama

Reykjanes Basin *82 C5 var.* Irminger
  Basin. *undersea basin* N Atlantic Ocean
Reykjanes Ridge *80 A1 undersea ridge*
  N Atlantic Ocean
Reykjavík *83 E5 var.* Reikjavik.
  country capital (Iceland)
  Höfudhborgarsvaedhi, W Iceland
Reynosa *51 E2* Tamaulipas, C Mexico
Rezā'īyeh, Daryācheh-ye *see* Orūmīyeh,
  Daryācheh-ye
Rezé *90 A4* Loire-Atlantique,
  NW France
Rēzekne *106 D4 Ger.* Rositten; *prev. Rus.*
  Rezhitsa. SE Latvia
Rezhitsa *see* Rēzekne
Rezovo *104 E3 Turk.* Rezve. Burgas,
  E Bulgaria
Rezve *see* Rezovo
Rhaedestus *see* Tekirdağ
Rhegion/Rhegium *see* Reggio di Calabria
Rheims *see* Reims
Rhein *see* Rhine
Rheine *94 A3 var.* Rheine in Westfalen.
  Nordrhein-Westfalen, NW Germany
Rheine in Westfalen *see* Rheine
Rheinisches Schiefergebirge *95 A5 var.*
  Rhine State Uplands, *Eng.* Rhenish
  Slate Mountains. *mountain range*
  W Germany
Rhenish Slate Mountains *see* Rheinisches
  Schiefergebirge
Rhin *see* Rhine
Rhine *80 D4 Dut.* Rijn, *Fr.* Rhin, *Ger.*
  Rhein. *river* W Europe
Rhinelander *40 B2* Wisconsin, N USA
Rhine State Uplands *see* Rheinisches
  Schiefergebirge
Rho *96 B2* Lombardia, N Italy
Rhode Island *41 G3 off.* State of Rhode
  Island and Providence Plantations, *also
  known as* Little Rhody, Ocean State.
  state NE USA
Rhodes *105 E7 var.* Ródhos, *Eng.*
  Rhodes, *It.* Rodi; *anc.* Rhodos. *island*
  Dodekánisa, Greece, Aegean Sea
Rhodes *see* Ródos
Rhodesia *see* Zimbabwe
Rhodope Mountains *104 C3 var.*
  Rodhópi Óri, *Bul.* Rhodope Planina,
  Rodopi, *Gk.* Orosirá Rodhópis, *Turk.*
  Dospad Dagh. *mountain range*
  Bulgaria/Greece
Rhodope Planina *see* Rhodope
  Mountains
Rhône *80 C4 river* France/Switzerland
Rhum *88 B3 var.* Rum. *island*
  W Scotland, United Kingdom
Ribble *89 D5 river* NW England, United
  Kingdom
Ribeira *see* Santa Uxía de Ribeira
Ribeirão Preto *63 F4* São Paulo,
  S Brazil
Riberalta *61 F2* El Beni, N Bolivia
Ribniţa *108 D3 var.* Rābniţa, *Rus.*
  Rybnitsa. NE Moldova
Rice Lake *40 A2* Wisconsin, N USA
Richard Toll *74 B3* N Senegal
Richfield *44 B4* Utah, W USA
Richland *46 C2* Washington, NW USA
Richmond *151 C5* Tasman, South Island,
  New Zealand
Richmond *40 C5* Kentucky, S USA
Richmond *41 E5 state capital* Virginia,
  NE USA
Richmond Range *151 C5 mountain
  range* South Island, New Zealand
Ricobayo, Embalse de *92 C2 reservoir*
  NW Spain
Ricomagus *see* Riom
Ridder *114 D4 Kaz.* Leningor; *prev.*
  Leninogorsk. Vostochnyy Kazakhstan,
  E Kazakhstan
Ridgecrest *47 C7* California, W USA
Ried *see* Ried im Innkreis
Ried im Innkreis *95 D6 var.* Ried.
  Oberösterreich, NW Austria
Riemst *87 D6* Limburg, NE Belgium
Riesa *94 D4* Sachsen, E Germany
Rift Valley *see* Great Rift Valley
Riga *106 C3 Eng.* Riga. *country capital*
  (Latvia) C Latvia
Rīgaer Bucht *see* Riga, Gulf of
Riga, Gulf of *106 C3 Est.* Liivi Laht, *Ger.*
  Rīgaer Bucht, *Latv.* Rīgas Jūras Līcis,
  Rus. Rizhskiy Zaliv; *prev. Est.* Riia Laht.
  gulf Estonia/Latvia
Rīgas Jūras Līcis *see* Riga, Gulf of
Rīgestān *see* Rēgestān
Riia Laht *see* Riga, Gulf of

Riihimäki *85 D5* Etelä-Suomi, S Finland
Rijeka *100 A2 Ger.* Sankt Veit am Flaum,
  It. Fiume, *Slvn.* Reka; *anc.* Tarsatica.
  Primorje-Gorski Kotar, NW Croatia
Rijn *see* Rhine
Rijssel *see* Lille
Rijssen *86 E3* Overijssel, E Netherlands
Rimah, Wadi ar *120 B4 var.* Wādī ar
  Rummah. *dry watercourse* C Saudi
  Arabia
Rimini *96 C3 anc.* Ariminum. Emilia-
  Romagna, N Italy
Rîmnicu-Sărat *see* Râmnicu Sărat
Rîmnicu Vîlcea *see* Râmnicu Vâlcea
Rimouski *39 E4* Québec, SE Canada
Ringebu *85 B5* Oppland, S Norway
Ringen *see* Rõngu
Ringkøbing Fjord *85 A7 fjord*
  W Denmark
Ringvassøya *84 C2 Lapp.* Ráneš. *island*
  N Norway
Rio *see* Rio de Janeiro
Riobamba *60 B1* Chimborazo,
  C Ecuador
Rio Branco *56 B3 state capital* Acre,
  W Brazil
Rio Branco, Território de *see* Roraima
Río Bravo *51 E2* Tamaulipas, C Mexico
Río Cuarto *64 C4* Córdoba, C Argentina
Rio de Janeiro *63 F4 var.* Rio. *state
  capital* Rio de Janeiro, SE Brazil
Río Gallegos *65 B7 var.* Gallegos, Puerto
  Gallegos. Santa Cruz, S Argentina
Rio Grande *63 E5 var.* São Pedro do
  Rio Grande do Sul. Rio Grande do
  Sul, S Brazil
Rio Grande *50 D3* Zacatecas, C Mexico
Rio Grande do Norte *63 G2 off.* Estado
  do Rio Grande do Norte. *state* E Brazil
Rio Grande do Norte *63 G2 off.* Estado
  do Rio Grande do Norte. *region* E Brazil
Rio Grande do Norte, Estado do *see* Rio
  Grande do Norte
Rio Grande do Sul *63 E5 off.* Estado do
  Rio Grande do Sul. *state* S Brazil
Rio Grande do Sul *63 E5 off.* Estado do
  Rio Grande do Sul. *region* S Brazil
Rio Grande do Sul, Estado do *see* Rio
  Grande do Sul
Rio Grande Plateau *see* Rio Grande Rise
Rio Grande Rise *57 E6 var.* Rio Grande
  Plateau. *undersea plateau*
  SW Atlantic Ocean
Ríohacha *58 B1* La Guajira, N Colombia
Río Lagartos *51 H3* Yucatán, SE Mexico
Riom *91 C5 anc.* Ricomagus. Puy-de-
  Dôme, C France
Río Verde *51 E4 var.* Rioverde. San Luis
  Potosí, C Mexico
Rioverde *see* Río Verde
Ripoll *93 G2* Cataluña, NE Spain
Rishiri-tō *130 C2 var.* Rishiri Tô. *island*
  NE Japan
Rishiri Tô *see* Rishiri-tō
Rîşno *see* Râşnov
Risti *106 D2 Ger.* Kreuz. Läänemaa,
  W Estonia
Rivas *52 D4* Rivas, SW Nicaragua
Rivera *64 D3* Rivera, NE Uruguay
River Falls *40 A2* Wisconsin, N USA
Riverside *47 C7* California, W USA
Riverton *151 A7* Southland, South
  Island, New Zealand
Riverton *44 C3* Wyoming, C USA
Rivière-du-Loup *39 E4* Québec,
  SE Canada
Rivne *108 C2 Pol.* Równe, *Rus.* Rovno.
  Rivnens'ka Oblast', NW Ukraine
Rivoli *96 A2* Piemonte, NW Italy
Riyadh/Riyād, Minţaqat ar *see* Ar Riyāḍ
Riyāq *see* Rayak
Rize *117 E2* Rize, NE Turkey
Rizhao *128 D4* Shandong, E China
Rizhskiy Zaliv *see* Riga, Gulf of
Rkiz *74 C3* Trarza, W Mauritania
Road Town *55 F3 dependent territory
  capital* (British Virgin Islands) Tortola,
  C British Virgin Islands
Roanne *91 C5 anc.* Rodunma. Loire,
  E France
Roanoke *41 E5* Virginia, NE USA
Roanoke River *43 F1 river* North
  Carolina/Virginia, SE USA
Roatán *52 C2 var.* Coxen Hole, Coxin
  Hole. Islas de la Bahía, N Honduras
Roat Kampuchea *see* Cambodia
Roazon *see* Rennes
Robbie Ridge *143 E3 undersea ridge*
  W Pacific Ocean

Robert Williams *see* Caála
Robinson Range *147 B5 mountain ra[..]*
  Western Australia
Robson, Mount *37 E5 mountain* Briti[..]
  Columbia, SW Canada
Robstown *49 G4* Texas, SW USA
Roca Partida, Isla *50 B5 island*
  W Mexico
Rocas, Atol das *63 G2 island* E Brazil
Rochefort *87 C7* Namur, SE Belgium
Rochefort *90 B4 var.* Rochefort sur M[e]
  Charente-Maritime, W France
Rochefort sur Mer *see* Rochefort
Rochester *45 G3* Minnesota, N USA
Rochester *41 G2* New Hampshire,
  NE USA
Rocheuses, Montagnes/Rockies *see*
  Rocky Mountains
Rockall Bank *80 B2 undersea bank*
  N Atlantic Ocean
Rockall Trough *80 B2 trough* N Atlan[..]
  Ocean
Rockdale *148 E2* Texas, SW USA
Rockford *40 B3* Illinois, N USA
Rockhampton *148 D4* Queensland,
  E Australia
Rock Hill *43 E1* South Carolina, SE U[S]
Rockingham *147 A6* Western Australi[a]
Rock Island *40 B3* Illinois, N USA
Rock Sound *54 C1* Eleuthera Island,
  C Bahamas
Rock Springs *44 C3* Wyoming, C USA
Rockstone *59 F3* C Guyana
Rocky Mount *43 F1* North Carolina,
  SE USA
Rocky Mountains *34 B4 var.* Rockies,
  Fr. Montagnes Rocheuses. *mountain*
  range Canada/USA
Roden *86 E2* Drenthe, NE Netherlands
Rodez *91 C5 anc.* Segodunum. Aveyro[n]
  S France
Rodhópi Óri *see* Rhodope Mountains
Ródhos/Rodi *see* Ródos
Rodopi *see* Rhodope Mountains
Rodosto *see* Tekirdağ
Rodunma *see* Roanne
Roermond *87 D5* Limburg,
  SE Netherlands
Roeselare *87 A6 Fr.* Roulers; *prev.*
  Rousselaere. West-Vlaanderen,
  W Belgium
Rogachëv *see* Rahachow
Rogatica *100 C4* Republika Srpska,
  SE Bosnia and Herzegovina
Rogers *42 A1* Arkansas, C USA
Roger Simpson Island *see* Abemama
Roi Ed *see* Roi Et
Roi Et *137 D5 var.* Muang Roi Et, Roi
  Ed. Roi Et, E Thailand
Roja *106 C2* NW Latvia
Rokiškis *106 C4* Panevėžys,
  NE Lithuania
Rokycany *99 A5 Ger.* Rokytzan.
  Plzeňský Kraj, W Czech Republic
Rokytzan *see* Rokycany
Rólas, Ilha das *76 E2 island* S Sao Tom[e]
  and Principe, Africa, E Atlantic Ocean
Rolla *45 G5* Missouri, C USA
Røm *see* Rømø
Roma *96 C4 Eng.* Rome. *country capito[l]*
  (Italy) Lazio, C Italy
Roman *124 D5* Queensland, E Australia
Roman *104 C2* Vratsa, NW Bulgaria
Roman *108 C4 Hung.* Románvásár.
  Neamţ, NE Romania
Romania *108 B4 Bul.* Rumŭniya, *Ger.*
  Rumänien, *Hung.* Románia, *Rom.*
  România, *SCr.* Rumunjska, *Ukr.*
  Rumuniya. *prev.* Republica Socialistă
  România, Roumania, Rumania,
  Socialist Republic of Romania, *prev.*
  Rom. România. *country* SE Europe

ROMANIA
Southeast Europe

Official name Romania	
Formation 1878 / 1947	
Capital Bucharest	
Population 21.7 million / 244 people per sq mile (94 people per sq km)	
Total area 91,699 sq. miles (237,500 sq. km)	
Languages Romanian*, Hungarian (Magyar), Romani, German	
Religions Romanian Orthodox 87%, Protestant 5%, Roman Catholic 5%, Greek Orthodox 1%, Greek Catholic (Uniate) 1%, Other 1%	

OMANIA
(continued)

nic mix Romanian 89%, Magyar 7%,
ma 2%, Other 2%
overnment Presidential system
irrency New Romanian leu = 100 bani
eracy rate 99%
alorie consumption 3363 kilocalories

mânia, Republica Socialistă *see*
Romania
mania, Socialist Republic of *see*
Romania
maniv 109 F3 *Rus.* Dneprodzerzhinsk,
prev. Dniprodzerzhyns'k, *prev.*
Kamenskoye. Dnipropetrovs'ka
Oblast', E Ukraine
mánvásár *see* Roman
ne *see* Roma
minia *see* Romania
nny 109 F2 Sums'ka Oblast',
NE Ukraine
no 85 A7 *Ger.* Röm. *island*
SW Denmark
ncador, Serra do 56 D4 *mountain
range* C Brazil
nda 92 D5 Andalucía, S Spain
ndônia 62 D3 *off.* Estado de
Rondônia; *prev.* Território de
Rondônia. *state* W Brazil
ndônia 62 D3 *off.* Estado de
Rondônia; *prev.* Território de
Rondônia. *region* W Brazil
ndônia, Estado de *see* Rondônia
ndônia, Território de *see* Rondônia
ndonópolis 63 E3 Mato Grosso,
W Brazil
ngelap Atoll 144 D1
var. Rönlap. *atoll* Ralik Chain,
NW Marshall Islands
nlap *see* Rongelap Atoll
nne 85 B8 Bornholm, E Denmark
nne Ice Shelf 154 B4 *ice shelf*
Antarctica
osendaal 87 C5 Noord-Brabant,
S Netherlands
oosevelt Island 154 B4 *island*
Antarctica
raima 62 D1 *off.* Estado de Roraima;
prev. Território de Rio Branco,
Território de Roraima. *state* N Brazil
raima 62 D1 *off.* Estado de Roraima;
prev. Território de Rio Branco,
Território de Roraima. *region* N Brazil
raima, Estado de *see* Roraima
raima, Mount 59 E3 *mountain*
N South America
raima, Território de *see* Roraima
ros 85 B5 Sør-Trøndelag, S Norway
ss 151 B6 West Coast, South Island,
New Zealand
sa, Lake 54 D2 *lake* Great Inagua,
S Bahamas
sario 64 D4 Santa Fe, C Argentina
sario 64 D2 San Pedro, C Paraguay
sario *see* Rosarito
sarito 50 A1 *var.* Rosario. Baja
California Norte, NW Mexico
scianum *see* Rossano
scommon 40 C2 Michigan, N USA
seau 55 G4 *prev.* Charlotte
Town. *country capital* (Dominica)
SW Dominica
seburg 46 B4 Oregon, NW USA
senau *see* Rožňava
enberg 49 G4 Texas, SW USA
enberg *see* Ružomberok, Slovakia
engarten 94 B3 Niedersachsen,
N Germany
senheim 95 C6 Bayern, S Germany
sia 93 H5 W Gibraltar Europe
sia Bay 93 H5 *bay* SW Gibraltar
Europe W Mediterranean Sea
Atlantic Ocean
siori de Vede 108 B5 Teleorman,
S Romania
sitten *see* Rēzekne
slavl' 111 A5 Smolenskaya Oblast',
W Russian Federation
smalen 86 C4 Noord-Brabant,
Netherlands
ssano 97 E6 *anc.* Roscianum.
Calabria, SW Italy
ss Ice Shelf 154 B4 *ice shelf* Antarctica

Rossiyskaya Federatsiya *see* Russian
Federation
Rosso 74 B3 Trarza, SW Mauritania
Rossosh' 111 B6 Voronezhskaya Oblast',
W Russian Federation
Ross Sea 154 B4 *sea* Antarctica
Rostak *see* Ar Rustāq
Rostock 94 C2 Mecklenburg-
Vorpommern, NE Germany
Rostov *see* Rostov-na-Donu
Rostov-na-Donu 111 B7 *var.* Rostov,
Eng. Rostov-on-Don. Rostovskaya
Oblast', SW Russian Federation
Rostov-on-Don *see* Rostov-na-Donu
Roswell 48 D2 New Mexico,
SW USA
Rota 144 B1 *island* S Northern Mariana
Islands
Rotcher Island *see* Tamana
Rothera 154 A2 UK *research station*
Antarctica
Rotomagus *see* Rouen
Rotorua 150 D3 Bay of Plenty, North
Island, New Zealand
Rotorua, Lake 150 D3 *lake* North Island,
New Zealand
Rotterdam 86 C4 Zuid-Holland,
SW Netherlands
Rottweil 95 B6 Baden-Württemberg,
S Germany
Rotuma 145 E4 *island* NW Fiji Oceania
S Pacific Ocean
Roubaix 90 C2 Nord, N France
Rouen 90 C3 *anc.* Rotomagus. Seine-
Maritime, N France
Roulers *see* Roeselare
Roumania *see* Romania
Round Rock 49 G3 Texas, SW USA
Rourkela *see* Rāurkela
Rousselaere *see* Roeselare
Roussillon 91 C6 *cultural region*
S France
Rouyn-Noranda 38 D4 Québec,
SE Canada
Rovaniemi 84 D3 Lappi, N Finland
Rovigno *see* Rovinj
Rovigo 96 C2 Veneto, NE Italy
Rovinj 100 A3 *It.* Rovigno. Istra,
NW Croatia
Rovno *see* Rivne
Rovuma, Rio 79 F2 *var.* Ruvuma. *river*
Mozambique/Tanzania
Rovuma, Rio *see* Ruvuma
Równe *see* Rivne
Roxas City 139 E2 Panay Island,
C Philippines
Royale, Isle 40 B1 *island* Michigan,
N USA
Royan 91 B5 Charente-Maritime,
W France
Rozdol'ne 109 F4 *Rus.* Razdolnoye.
Avtonomna Respublika Krym,
S Ukraine
Rožňava 99 D6 *Ger.* Rosenau, *Hung.*
Rozsnyó. Košický Kraj, E Slovakia
Rózsahegy *see* Ružomberok
Rozsnyó *see* Rāşnov, Romania
Rozsnyó *see* Rožňava, Slovakia
Ruanda *see* Rwanda
Ruapehu, Mount 150 D4 *volcano* North
Island, New Zealand
Ruapuke Island 151 B8 *island*
SW New Zealand
Ruatoria 150 E3 Gisborne, North Island,
New Zealand
Ruawai 150 D2 Northland, North Island,
New Zealand
Rubezhnoye *see* Rubizhne
Rubizhne 109 H3 *Rus.* Rubezhnoye.
Luhans'ka Oblast', E Ukraine
Ruby Mountains 47 D5 *mountain range*
Nevada, W USA
Rucava 106 B3 SW Latvia
Rudensk *see* Rudzyensk
Rūdiškės 107 B5 Vilnius, S Lithuania
Rudnik *see* Dolni Chiflik
Rudny *see* Rudnyy
Rudnyy 114 C4 *var.* Rudny. Kostanay,
N Kazakhstan
Rudolf, Lake *see* Turkana, Lake
Rudolfswert *see* Novo mesto
Rudzyensk 107 C6 *Rus.* Rudensk.
Minskaya Voblasts', C Belarus
Rufiji 73 C7 *river* E Tanzania
Rufino 64 C4 Santa Fe, C Argentina
Rugāji 106 D4 E Latvia
Rügen 94 D2 *headland* NE Germany
Ruggell 94 E1 N Liechtenstein
Europe

Ruhja *see* Rūjiena
Ruhnu 106 C2 *var.* Ruhnu Saar, *Swe.*
Runö. *island* SW Estonia
Ruhnu Saar *see* Ruhnu
Rujen *see* Rūjiena
Rūjiena 106 D3 *Est.* Ruhja, *Ger.* Rujen.
N Latvia
Rukwa, Lake 73 B7 *lake* SE Tanzania
Rum *see* Rhum
Ruma 100 D3 Vojvodina, N Serbia
Rumadiya *see* Ar Ramādī
Rumania/Rumänien *see* Romania
Rumbek 73 B5 El Buhayrat, C South
Sudan
Rum Cay 54 D2 *island* C Bahamas
Rumia 98 C2 Pomorskie, N Poland
Rummah, Wādī ar *see* Rimah, Wādī ar
Rummelsburg in Pommern *see* Miastko
Rumuniya/Rumŭniya/Rumunjska
see Romania
Runanga 151 B5 West Coast, South
Island, New Zealand
Rundu 78 C3 *var.* Runtu. Okavango,
NE Namibia
Runö *see* Ruhnu
Runtu *see* Rundu
Ruoqiang 126 C3 *var.* Jo-ch'iang, *Uigh.*
Charkhlik, Charkhliq, Qarkilik.
Xinjiang Uygur Zizhiqu, NW China
Rupea 108 C4 *Ger.* Reps, *Hung.*
Kőhalom; *prev.* Cohalm. Braşov,
C Romania
Rupel 87 B5 *river* N Belgium
Rupella *see* la Rochelle
Rupert, Rivière de 38 D3 *river* Québec,
C Canada
Rusaddir *see* Melilla
Ruschuk/Rusçuk *see* Ruse
Ruse 104 D1 *var.* Ruschuk, Rustchuk,
Turk. Rusçuk. Ruse, N Bulgaria
Russadir *see* Melilla
Russellville 42 A1 Arkansas, C USA
Russia *see* Russian Federation
Russian America *see* Alaska
Russian Federation 112 D2 *off.* Russian
Federation, *var.* Russia, *Latv.* Krievija,
Rus. Rossiyskaya Federatsiya. *country*
Asia/Europe

RUSSIAN FEDERATION
Europe / Asia

Official name Russian Federation
Formation 1480 / 1991
Capital Moscow
Population 143 million / 22 people
per sq mile (8 people per sq km)
Total area 6,592,735 sq. miles
(17,075,200 sq. km)
Languages Russian*, Tatar, Ukrainian,
Chavash, various other national languages
Religions Orthodox Christian 75%,
Muslim 14%, other 11%
Ethnic mix Russian 80%, Other 12%,
Tatar 4%, Ukrainian 2%, Bashkir 1%,
Chavash 1%
Government Presidential /
parliamentary system
Currency Russian rouble = 100 kopeks
Literacy rate 99%
Calorie consumption 3358 kilocalories

Russian Federation *see* Russian
Federation
Rustaq *see* Ar Rustāq
Rustavi 117 G2 *prev.* Rust'avi.
SE Georgia
Rust'avi *see* Rustavi
Rustchuk *see* Ruse
Ruston 42 B2 Louisiana, S USA
Rutanzige, Lake *see* Edward, Lake
Rutba *see* Ar Ruţbah
Rutlam *see* Ratlām
Rutland 41 F2 Vermont, NE USA
Rutog 126 A4 *var.* Rutög, Rutok. Xizang
Zizhiqu, W China
Rutok *see* Rutog
Ruvuma 69 E5 *var.* Rio Rovuma. *river*
Mozambique/Tanzania
Ruvuma *see* Rovuma, Rio
Ruwenzori 77 E5 *mountain range* Dem.
Rep. Congo/Uganda
Ruzhany 107 B6 Brestskaya Voblasts',
SW Belarus
Ružomberok 99 C5 *Ger.* Rosenberg,
Hung. Rózsahegy. Žilinský Kraj,
N Slovakia
Rwanda 73 B6 *off.* Rwandese Republic;
prev. Ruanda. *country* C Africa

RWANDA
Central Africa

Official name Republic of Rwanda
Formation 1962 / 1962
Capital Kigali
Population 11.8 million / 1225 people
per sq mile (473 people per sq km)
Total area 10,169 sq. miles (26,338 sq. km)
Languages Kinyarwanda*, French*,
Kiswahili, English*
Religions Christian 94%, Muslim 5%,
Traditional beliefs 1%
Ethnic mix Hutu 85%, Tutsi 14%,
Other (including Twa) 1%
Government Presidential system
Currency Rwanda franc = 100 centimes
Literacy rate 66%
Calorie consumption 2148 kilocalories

Rwandese Republic *see* Rwanda
Ryazan' 111 B5 Ryazanskaya Oblast',
W Russian Federation
Rybach'ye *see* Balykchy
Rybinsk 110 B4 *prev.* Andropov.
Yaroslavskaya Oblast', W Russian
Federation
Rybnik 99 C5 Śląskie, S Poland
Rybnitsa *see* Rîbniţa
Ryde 148 E1 United Kingdom
Ryki 98 D4 Lubelskie, E Poland
Rykovo *see* Yenakiyeve
Rypin 98 C3 Kujawsko-pomorskie,
C Poland
Ryssel *see* Lille
Rysy 99 C5 *mountain* S Poland
Ryukyu Islands 130 A2 *Eng.* Ryukyu
Islands. *island group* SW Japan
Ryukyu Islands 125 E3 *Eng.* Ryukyu
Islands. *island group* SW Japan
Ryukyu Islands *see* Nansei-shotō
Ryukyu Islands *see* Nansei-shotō
Ryukyu Trench 125 F3 *var.* Nansei Syotō
Trench. *trench* S East China Sea
Rzeszów 99 E5 Podkarpackie, SE Poland
Rzhev 110 B4 Tverskaya Oblast',
W Russian Federation

S

Saale 94 C4 *river* C Germany
Saalfeld 95 C5 *var.* Saalfeld an der Saale.
Thüringen, C Germany
Saalfeld an der Saale *see* Saalfeld
Saarbrücken 95 A6 *Fr.* Sarrebruck.
Saarland, SW Germany
Sääre 106 C2 *var.* Sjar. Saaremaa,
W Estonia
Saare *see* Saaremaa
Saaremaa 106 C2 *Ger.* Oesel, Ösel; *prev.*
Saare. *island* W Estonia
Saariselkä 84 D2 *Lapp.* Suoločielgi.
Lappi, N Finland
Sab' Ābār 118 C4 *var.* Sab'a Biyar, Sa'b
Bi'ār. Ḥimş, C Syria
Sab'a Biyar *see* Sab' Ābār
Šabac 100 D3 Serbia, W Serbia
Sabadell 93 G2 Cataluña, E Spain
Sabah 138 D3 *prev.* British North
Borneo, North Borneo. *state* East
Malaysia
Sabanalarga 58 B1 Atlántico,
N Colombia
Sabaneta 58 C1 Falcón, N Venezuela
Sabaria *see* Szombathely
Sab'atayn, Ramlat as 121 C6 *desert*
C Yemen
Sabaya 61 F4 Oruro, S Bolivia
Sa'b Bi'ār *see* Sab' Ābār
Saberi, Hamun-e 122 C5 *var.*
Daryācheh-ye Hāmun, Daryācheh-ye
Sīstān. *lake* Afghanistan/Iran
Sabhā 71 F3 C Libya
Sabi *see* Save
Sabinas 51 E2 Coahuila, NE Mexico
Sabinas Hidalgo 51 E2 Nuevo León,
NE Mexico
Sabine River 49 H3 *river* Louisiana/
Texas, SW USA
Sabkha *see* As Sabkhah
Sable, Cape 43 E5 *headland* Florida,
SE USA
Sable Island 39 G4 *island* Nova Scotia,
SE Canada
Şabyā 121 B6 Jīzān, SW Saudi Arabia
Sabzawar *see* Sabzevār

Sabzevār *120 D2 var.* Sabzawar. Khorāsān-Razavī, NE Iran
Sachsen *94 D4 Eng.* Saxony, *Fr.* Saxe. *state* E Germany
Sachs Harbour *37 E2 var.* Ikaahuk. Banks Island, Northwest Territories, N Canada
Sächsisch-Reen/Sächsisch-Regen *see* Reghin
Sacramento *47 B5 state capital* California, W USA
Sacramento Mountains *48 D2 mountain range* New Mexico, SW USA
Sacramento River *47 B5 river* California, W USA
Sacramento Valley *47 B5 valley* California, W USA
Sá da Bandeira *see* Lubango
Şa'dah *121 B6* NW Yemen
Sado *see* Sadoga-shima
Sadoga-shima *131 C5 var.* Sado. *island* C Japan
Saena Julia *see* Siena
Safad *see* Tsefat
Şafāqis *see* Sfax
Şafāshahr *120 D3 var.* Deh Bīd. Fārs, C Iran
Safed *see* Tsefat
Säffle *85 B6* Värmland, C Sweden
Safford *48 C3* Arizona, SW USA
Safi *70 B2* W Morocco
Sagaing *136 B3* Sagaing, C Myanmar (Burma)
Sagami-nada *131 D6 inlet* SW Japan
Sagan *see* Żagań
Sāgar *134 D4 prev.* Saugor. Madhya Pradesh, C India
Sagarmāthā *see* Everest, Mount
Sagebrush State *see* Nevada
Saghez *see* Saqqez
Saginaw *40 C3* Michigan, N USA
Saginaw Bay *40 D2 lake bay* Michigan, N USA
Sagua la Grande *54 B2* Villa Clara, C Cuba
Sagunto *93 F3 Cat.* Sagunt, *Ar.* Murviedro; *anc.* Saguntum. Valenciana, E Spain
Sagunt/Saguntum *see* Sagunto
Sahara *68 B3 desert* Libya/Algeria
Şaharā al Gharbīya *72 B2 Eng.* Western Desert, *desert* Egypt
Saharan Atlas *70 D2 var.* Saharan Atlas. *mountain range* Algeria/Morocco
Saharan Atlas *see* Atlas Saharien
Sahel *74 D3 physical region* C Africa
Sāhilīyah, Jibāl as *118 B3 mountain range* NW Syria
Sāhīwāl *134 C2 prev.* Montgomery. Punjab, E Pakistan
Saïda *119 A5 var.* Şaydā, Sayida; *anc.* Sidon. W Lebanon
Sa'īdābād *see* Sīrjān
Saidpur *135 G3 var.* Syedpur. Rajshahi, NW Bangladesh
Saïdu *134 C1 var.* Mingora, Mongora; *prev.* Mingāora. Khyber Pakhtunkhwa, N Pakistan
Saigon *see* Hồ Chi Minh
Saimaa *85 E5 lake* SE Finland
St Albans *89 E6 anc.* Verulamium. E England, United Kingdom
Saint Albans *40 D5* West Virginia, NE USA
St Andrews *88 C4* E Scotland, United Kingdom
Saint Anna Trough *see* Svyataya Anna Trough
St. Ann's Bay *54 B4* C Jamaica
St. Anthony *39 G3* Newfoundland and Labrador, SE Canada
Saint Augustine *43 E3* Florida, SE USA
St Austell *89 C7* SW England, United Kingdom
St.Botolph's Town *see* Boston
St-Brieuc *90 A3* Côtes d'Armor, NW France
St. Catharines *38 D5* Ontario, S Canada
St-Chamond *91 D5* Loire, E France
Saint Christopher and Nevis, Federation of *see* Saint Kitts and Nevis
Saint Christopher-Nevis *see* Saint Kitts and Nevis

Saint Cloud *45 F2* Minnesota, N USA
Saint Croix *55 F3 island* S Virgin Islands (US)
Saint Croix River *40 A2 river* Minnesota/Wisconsin, N USA
St David's Island *42 B5 island* E Bermuda
St-Denis *79 G4 dependent territory capital* (Réunion) NW Réunion
St-Dié *90 E4* Vosges, NE France
St-Egrève *91 D5* Isère, E France
Sainte Marie, Cap *see* Vohimena, Tanjona
Saintes *91 B5 anc.* Mediolanum. Charente-Maritime, W France
St-Étienne *91 D5* Loire, E France
St-Flour *91 C5* Cantal, C France
St-Gall/Saint Gall/St. Gallen *see* Sankt Gallen
St-Gaudens *91 B6* Haute-Garonne, S France
Saint George *149 D5* Queensland, E Australia
St George *42 B4* N Bermuda
St George *44 A5* Utah, W USA
St. George's *55 G5 country capital* (Grenada) SW Grenada
St-Georges *39 E4* Québec, SE Canada
St-Georges *59 H3* E French Guiana
Saint George's Channel *89 B6 channel* Ireland/Wales, United Kingdom
St George's Island *42 B4 island* E Bermuda
Saint Helena *69 B6 UK dependent territory* C Atlantic Ocean
St Helier *89 D8 British Crown Dependency capital* (Jersey) S Jersey, Channel Islands
St.Iago de la Vega *see* Spanish Town
Saint Ignace *40 C2* Michigan, N USA
St-Jean, Lac *39 E4 lake* Québec, SE Canada
Saint Joe River *46 D2 river* Idaho, NW USA North America
St. John *39 F4* New Brunswick, SE Canada
Saint-John *see* Saint John
Saint John *41 H1 Fr.* Saint-John. *river* Canada/USA
St John's *55 G3 country capital* (Antigua and Barbuda) Antigua, Antigua and Barbuda
St. John's *39 H3 province capital* Newfoundland and Labrador, E Canada
Saint Joseph *45 F4* Missouri, C USA
St Julian's *see* San Ġiljan
St Kilda *88 A3 island* NW Scotland, United Kingdom
Saint Kitts and Nevis *55 F3 off.* Federation of Saint Christopher and Nevis, *var.* Saint Christopher-Nevis. *country* E West Indies

---

SAINT KITTS & NEVIS
West Indies

Official name Federation of Saint Christopher and Nevis
Formation 1983 / 1983
Capital Basseterre
Population 51,134 / 368 people per sq mile (142 people per sq km)
Total area 101 sq. miles (261 sq km)
Languages English*, English Creole
Religions Anglican 33%, Methodist 29%, Other 22%, Moravian 9%, Roman Catholic 7%
Ethnic mix Black 95%, Mixed race 3%, White 1%, Other and Amerindian 1%
Government Parliamentary system
Currency Eastern Caribbean dollar = 100 cents
Literacy rate 98%
Calorie consumption 2507 kilocalories

St-Laurent *see* St-Laurent-du-Maroni
St-Laurent-du-Maroni *59 H3 var.* St-Laurent. NW French Guiana
Saint-Laurent, Fleuve *see* St. Lawrence
St. Lawrence *39 E4 Fr.* Fleuve St-Laurent. *river* Canada/USA
St. Lawrence, Gulf of *39 F3 gulf* NW Atlantic Ocean
Saint Lawrence Island *36 B2 island* Alaska, USA
St-Lô *90 B3 anc.* Briovera, Laudus. Manche, N France

St-Louis *90 E4* Haut-Rhin, NE France
Saint Louis *74 B3* NW Senegal
Saint Louis *45 G4* Missouri, C USA
Saint Lucia *55 E1 country* SE West Indies

SAINT LUCIA
West Indies

Official name Saint Lucia
Formation 1979 / 1979
Capital Castries
Population 162,781 / 690 people per sq mile (267 people per sq km)
Total area 239 sq. miles (620 sq. km)
Languages English*, French Creole
Religions Roman Catholic 90%, Other 10%
Ethnic mix Black 83%, Mulatto (mixed race) 13%, Asian 3%, Other 1%
Government Parliamentary system
Currency Eastern Caribbean dollar = 100 cents
Literacy rate 95%
Calorie consumption 2629 kilocalories

Saint Lucia Channel *55 H4 channel* Martinique/Saint Lucia
St-Malo *90 B3* Ille-et-Vilaine, NW France
St-Malo, Golfe de *90 A3 gulf* NW France
Saint Martin *see* Sint Maarten
St.Matthew's Island *see* Zadetkyi Kyun
St.Matthias Group *144 B3 island group* NE Papua New Guinea
St. Moritz *95 B7 Ger.* Sankt Moritz, *Rmsch.* San Murezzan. Graubünden, SE Switzerland
St-Nazaire *90 A4* Loire-Atlantique, NW France
Saint Nicholas *see* São Nicolau
Saint-Nicolas *see* Sint-Niklaas
St-Omer *90 C2* Pas-de-Calais, N France
Saint Paul *45 F2 state capital* Minnesota, N USA
St-Paul, Île *141 C6 var.* St.Paul Island. *island* Île St-Paul, NE French Southern and Antarctic Lands, Antarctica, Indian Ocean
St.Paul Island *see* St-Paul, Île
St Peter Port *89 D8 British Crown Dependency capital* (Guernsey) C Guernsey, Channel Islands
Saint Petersburg *43 E4* Florida, SE USA
Saint Petersburg *see* Sankt-Peterburg
St-Pierre and Miquelon *39 G4 Fr.* Îles St-Pierre et Miquelon. *French territorial collectivity* NE North America
St-Quentin *90 C3* Aisne, N France
Saint Thomas *see* São Tomé, Sao Tome and Principe
Saint Thomas *see* Charlotte Amalie, Virgin Islands (US)
Saint Ubes *see* Setúbal
Saint Vincent *55 G4 island* N Saint Vincent and the Grenadines
Saint Vincent *see* São Vicente
Saint Vincent and the Grenadines *55 H4 country* SE West Indies

---

SAINT VINCENT & THE GRENADINES
West Indies

Official name Saint Vincent and the Grenadines
Formation 1979 / 1979
Capital Kingstown
Population 103,220 / 788 people per sq mile (304 people per sq km)
Total area 150 sq. miles (389 sq km)
Languages English*, English Creole
Religions Anglican 47%, Methodist 28%, Roman Catholic 13%, Other 12%
Ethnic mix Black 66%, Mulatto (mixed race) 19%, Other 12%, Carib 2%, Asian 1%
Government Parliamentary system
Currency Eastern Caribbean dollar = 100 cents
Literacy rate 88%
Calorie consumption 2960 kilocalories

Saint Vincent, Cape *see* São Vicente, Cabo de
Saint Vincent Passage *55 H4 passage* Saint Lucia/Saint Vincent and the Grenadines

---

Saint Yves *see* Setúbal
Saipan *142 B1 island/country capital* (Northern Mariana Islands) S Northern Mariana Islands
Saishū *see* Jeju-do
Sajama, Nevado *61 F4 mountain* W Bolivia
Sajószentpéter *99 D6* Borsod-Abaúj-Zemplén, NE Hungary
Sakākah *120 B4* Al Jawf, NW Saudi Arabia
Sakakawea, Lake *44 D1 reservoir* North Dakota, N USA
Sak'art'velo *see* Georgia
Sakata *130 D4* Yamagata, Honshū, C Japan
Sakhalin *115 G4 var.* Sakhalin. *island* SE Russian Federation
Sakhalin *see* Sakhalin, Ostrov
Sakhon Nakhon *see* Sakon Nakhon
Şäki *117 G2 Rus.* Sheki; *prev.* Nukha. NW Azerbaijan
Saki *see* Saky
Sakishima-shoto *130 A3 var.* Sakisima Syotō. *island group* SW Japan
Sakisima Syotō *see* Sakishima-shotō
Sakiz *see* Chíos
Sakiz-Adasi *see* Chíos
Sakon Nakhon *136 D4 var.* Muang Sakon Nakhon, Sakhon Nakhon. Sakon Nakhon, E Thailand
Saky *109 F5 Rus.* Saki. Avtonomna Respublika Krym, S Ukraine
Sal *74 A3 island* Ilhas de Barlavento, NE Cape Verde
Sala *85 C6* Västmanland, C Sweden
Salacgrīva *106 C3 Est.* Salatsi. N Latvia
Sala Consilina *97 D5* Campania, S Italy
Salado, Río *62 D5 river* E Argentina
Salado, Río *64 C3 river* C Argentina
Şalālah *121 D6* SW Oman
Salamá *52 B2* Baja Verapaz, C Guatemala
Salamanca *64 B4* Coquimbo, C Chile
Salamanca *92 D2 anc.* Helmantica, Salmantica. Castilla y León, NW Spain
Salamīyah *118 B3 var.* As Salamiyah. Ḥamāh, W Syria
Salang *see* Phuket
Salantai *106 B3* Klaipėda, NW Lithuania
Salatsi *see* Salacgrīva
Salavan *137 D5 var.* Saravan, Saravane. Salavan, S Laos
Salavat *111 D6* Respublika Bashkortostan, W Russian Federation
Sala y Gomez *153 F4 island* Chile, E Pacific Ocean
Sala y Gomez Fracture Zone *see* Sala y Gomez Ridge
Sala y Gomez Ridge *153 G4 var.* Sala y Gomez Fracture Zone. *fracture zone* SE Pacific Ocean
Salazar *see* N'Dalatando
Šalčininkai *107 C5* Vilnius, SE Lithuania
Salduba *see* Zaragoza
Saldus *106 B3 Ger.* Frauenburg. W Latvia
Sale *149 C7* Victoria, SE Australia
Salé *70 C2* NW Morocco
Salekhard *114 D3 prev.* Obdorsk. Yamalo-Nenetskiy Avtonomnyy Okrug, N Russian Federation
Salem *132 C2* Tamil Nādu, SE India
Salem *46 B3 state capital* Oregon, NW USA
Salerno *97 D5 anc.* Salernum. Campania, S Italy
Salerno, Gulf of *97 C5 Eng.* Gulf of Salerno. *gulf* S Italy
Salerno, Gulf of *see* Salerno, Golfo di
Salernum *see* Salerno
Salihorsk *107 C7 Rus.* Soligorsk. Minskaya Voblasts', S Belarus
Salima *79 E2* Central, C Malawi
Salina *45 E5* Kansas, C USA
Salina Cruz *51 F5* Oaxaca, SE Mexico
Salinas *60 A2* Guayas, W Ecuador
Salinas *47 B6* California, W USA
Salisbury *89 D7 var.* New Sarum. S England, United Kingdom
Salisbury *see* Harare
Sállan *see* Sørøya
Salliq *see* Coral Harbour
Sallyana *see* Salyān
Salmantica *see* Salamanca
Salmon River *46 D3 river* Idaho, NW USA

10on River Mountains 46 D3
 ountain range Idaho, NW USA
   85 D6 Länsi-Suomi, SW Finland
n-de-Provence 91 D6 Bouches-du-
 1ône, SE France
nica/Salonika *see* Thessaloníki
nta 108 A3 Hung. Nagyszalonta.
 .hor, W Romania
1k 111 B7 Rostovskaya Oblast',
 W Russian Federation
see As Salţ
a 64 C2 Salta, NW Argentina
ash 89 C7 SW England, United
 ingdom
illo 51 E3 Coahuila, NE Mexico
Lake City 44 B4 *state capital* Utah,
 W USA
o 64 D4 Salto, N Uruguay
on Sea 47 D8 *lake* California,
 W USA
ador 63 G3 *prev.* São Salvador. *state*
 apital Bahia, E Brazil
veen 124 C2 *Bur.* Thanlwin, *Chin.* Nu
 hiang, Nu Jiang. *river* SE Asia
än 135 E3 *var.* Sallyana. Mid
 Vestern, W Nepal
burg 95 D6 *anc.* Juvavum. Salzburg,
 Austria
gitter 94 C4 *prev.* Watenstedt-
 alzgitter. Niedersachsen, C Germany
wedel 94 C3 Sachsen-Anhalt,
 f Germany
ac *see* Bosanski Šamac
akhixai *see* Attapu
alayuca 50 C1 Chihuahua, N Mexico
ar 139 F2 *island* C Philippines
ara 114 B3 *prev.* Kuybyshev.
 amarskaya Oblast', W Russian
 ederation
narang *see* Semarang
narinda 138 D4 Borneo, C Indonesia
narkand *see* Samarqand
narkandsli/Samarkandskoye *see*
 'emirtau
narobriva *see* Amiens
narqand 123 E2 *Rus.* Samarkand.
 amarqand Viloyati, C Uzbekistan
nawa *see* As Samāwah
nbalpur 135 F4 Odisha, E India
nbava 79 G2 Antsiranana,
 NE Madagascar
nbir 108 B2 *Rus.* Sambor. L'vivs'ka
 Oblast', NW Ukraine
nbor *see* Sambir
nbre 90 D2 *river* Belgium/France
nfya 78 D2 Luapula, N Zambia
ninatal 94 E2 *valley* Austria/
 .iechtenstein Europe
nän *see* Semnän
n Neua *see* Xam Nua
noa 145 E4 *off.* Independent State
 of Samoa. *var.* Sāmoa; *prev.* Western
 Samoa. *country* W Polynesia

AMOA
Australasia & Oceania

Jfficial name Independent State
 Samoa
irmation 1962 / 1962
apital Apia
opulation 200,000 / 183 people
 per sq mile (71 people per sq km)
atal area 1104 sq. miles (2860 sq. km)
nguages Samoan*, English*
eligions Christian 99%, Other 1%
thnic mix Polynesian 91%,
 ronesian 7%, Other 2%
overnment Parliamentary system
urrency Tala = 100 sene
:eracy rate 99%
alorie consumption 2872 kilocalories

noa *see* Samoa
noa Basin 143 E3 *undersea basin*
 N Pacific Ocean
noa, Independent State of *see* Samoa
nobor 100 A2 Zagreb, N Croatia
nos 105 E6 *prev.* Limín Vathéos.
nos 105 D6 *island* Dodekánisa,
 Aegean Sea
nothrace *see* Samothráki
nothráki 104 D4 Samothráki,
 NE Greece
nothraki 104 C4 *anc.* Samothrace.
 sland NE Greece

Sampit 138 C4 Borneo, C Indonesia
Sâmraông 137 D5 *prev.* Phumĭ
 Sâmraông, Phum Samrong. Siĕmréab,
 NW Cambodia
Samsun 116 D2 *anc.* Amisus. Samsun,
 N Turkey
Samt'redia 117 F2 *prev.* Samtredia.
 W Georgia
Samtredia *see* Samt'redia
Samui, Ko 137 C6 *island* SW Thailand
Samut Prakan 137 C5 *var.* Muang Samut
 Prakan, Paknam. Samut Prakan,
 C Thailand
San 74 D3 Ségou, C Mali
San 99 E5 *river* SE Poland
Şan'ā' 121 B6 *Eng.* Sana. *country capital*
 (Yemen) W Yemen
Sana 100 B3 *river* NW Bosnia and
 Herzegovina
Sanae 154 B2 *South African research*
 station Antarctica
Sanaga 77 B5 *river* C Cameroon
San Ambrosio, Isla 57 A5 *Eng.* San
 Ambrosio Island. *island* W Chile
San Ambrosio Island *see* San Ambrosio,
 Isla
Sanandaj 120 C3 *prev.* Sinneh.
 Kordestān, W Iran
San Andrés, Isla de 53 F3 *island*
 NW Colombia, Caribbean Sea
San Andrés Tuxtla 51 F4 *var.* Tuxtla.
 Veracruz-Llave, E Mexico
San Angelo 49 F3 Texas, SW USA
San Antonio 52 B2 Toledo, S Belize
San Antonio 64 B4 Valparaíso, C Chile
San Antonio 49 F4 Texas, SW USA
San Antonio Oeste 65 C5 Río Negro,
 E Argentina
San Antonio River 49 G4 *river* Texas,
 SW USA
Sanata Uxía de Ribeira 92 B1 *var.*
 Ribeira. Galicia, NW Spain
Sanāw 121 C6 *var.* Sanaw. NE Yemen
San Benedicto, Isla 50 B4 *island*
 W Mexico
San Benito 52 B1 Petén, N Guatemala
San Benito 49 G5 Texas, SW USA
San Bernardino 47 C7 California,
 W USA
San Blas 50 C3 Sinaloa, C Mexico
San Blas, Cape 42 D3 *headland* Florida,
 SE USA
San Blas, Cordillera de 53 G4 *mountain*
 range NE Panama
San Carlos 52 D4 Río San Juan,
 S Nicaragua
San Carlos 48 B2 Arizona, SW USA
San Carlos 52 D4 Quesada, Costa Rica
San Carlos de Ancud *see* Ancud
San Carlos de Bariloche 65 B5 Río
 Negro, W Argentina
San Carlos del Zulia 58 C2 Zulia,
 W Venezuela
San Clemente Island 47 B8 *island*
 Channel Islands, California, W USA
San Cristóbal 58 C2 Táchira,
 W Venezuela
San Cristobal 144 C4 *var.* Makira. *island*
 SE Solomon Islands
San Cristóbal *see* San Cristóbal de
 Las Casas
San Cristóbal de Las Casas 51 G5 *var.*
 San Cristóbal. Chiapas, SE Mexico
San Cristóbal, Isla 60 B5 *var.* Chatham
 Island. *island* Galápagos Islands,
 Ecuador, E Pacific Ocean
Sancti Spíritus 54 B2 Sancti Spíritus,
 C Cuba
Sandakan 138 D3 Sabah, East Malaysia
Sandalwood Island *see* Sumba, Pulau
Sandanski 104 C3 *prev.* Sveti Vrach.
 Blagoevgrad, SW Bulgaria
Sanday 88 D2 *island* NE Scotland, United
 Kingdom
Sanders 48 C2 Arizona, SW USA
Sand Hills 44 D3 *mountain range*
 Nebraska, C USA
San Diego 47 C8 California, W USA
Sandnes 85 A6 Rogaland, S Norway
Sandomierz 98 D4 *Rus.* Sandomir.
 Świętokrzyskie, C Poland
Sandomir *see* Sandomierz
Sandoway *see* Thandwe
Sandpoint 46 C1 Idaho, NW USA
Sand Springs 49 G1 Oklahoma, C USA
Sandusky 40 D3 Ohio, N USA
Sandvika 85 A6 Akershus, S Norway

Sandviken 85 C6 Gävleborg, C Sweden
Sandwich Island *see* Efate
Sandwich Islands *see* Hawai'ian Islands
Sandy Bay 93 H5 Saskatchewan,
 C Canada
Sandy City 44 B4 Utah, W USA
Sandy Lake 38 B3 *lake* Ontario,
 C Canada
San Esteban 52 D2 Olancho, C Honduras
San Eugenio/San Eugenio del Cuareim
 see Artigas
San Felipe 58 D1 Yaracuy,
 NW Venezuela
San Felipe de Puerto Plata *see* Puerto
 Plata
San Felix, Isla 57 A5 *Eng.* San Felix
 Island. *island* W Chile
San Felix Island *see* San Félix, Isla
San Fernando 92 C5 *prev.* Isla de León.
 Andalucía, S Spain
San Fernando 55 H5 Trinidad, Trinidad
 and Tobago
San Fernando 46 D1 California, W USA
San Fernando 58 D2 *var.* San Fernando
 de Apure. Apure, C Venezuela
San Fernando de Apure *see* San
 Fernando
San Fernando del Valle de Catamarca
 64 C3 *var.* Catamarca. Catamarca,
 NW Argentina
San Fernando de Monte Cristi *see*
 Monte Cristi
San Francisco 47 B6 California, W USA
San Francisco del Oro 50 C2
 Chihuahua, N Mexico
San Francisco de Macorís 55 E3
 C Dominican Republic
San Francisco *see* Tacuarembó
San Gabriel 60 B1 Carchi, N Ecuador
San Gabriel Mountains 46 E1 *mountain*
 range California, USA
Sangihe, Kepulauan *see* Sangir,
 Kepulauan
San Giljan 102 B5 *var.* St Julian's.
 N Malta
Sangir, Kepulauan 139 F3 *var.*
 Kepulauan Sangihe. *island group*
 N Indonesia
Sāngli 132 B1 Mahārāshtra, W India
Sangmélima 77 B5 Sud, S Cameroon
Sangre de Cristo Mountains 48 D1
 mountain range Colorado/New Mexico,
 C USA
San Ignacio 52 B1 *prev.* Cayo, El Cayo.
 Cayo, W Belize
San Ignacio 61 F3 El Beni, N Bolivia
San Ignacio 50 B2 Baja California Sur,
 NW Mexico
San Joaquin Valley 47 B7 *valley*
 California, W USA
San Jorge, Gulf of 65 C6 *var.* Gulf of San
 Jorge. *gulf* S Argentina
San Jorge, Gulf of *see* San Jorge, Golfo
San José 53 E4 *country capital*
 (Costa Rica) San José, C Costa Rica
San José 61 G3 *var.* San José de
 Chiquitos. Santa Cruz, E Bolivia
San José 52 B3 *var.* Puerto San José.
 Escuintla, S Guatemala
San Jose 47 B6 California, W USA
San José *see* San José del Guaviare,
 Colombia
San José de Chiquitos *see* San José
San José de Cúcuta *see* Cúcuta
San José del Guaviare 58 C4 *var.* San
 José. Guaviare, S Colombia
San Juan 64 B4 San Juan, W Argentina
San Juan 55 F3 *dependent territory*
 capital (Puerto Rico) NE Puerto Rico
San Juan *see* San Juan de los Morros
San Juan Bautista 64 D3 Misiones,
 S Paraguay
San Juan Bautista *see* Villahermosa
San Juan Bautista Tuxtepec *see* Tuxtepec
San Juan de Alicante *see* Sant Joan
 d'Alacant
San Juan del Norte 53 E4 *var.* Greytown.
 Río San Juan, SE Nicaragua
San Juan de los Morros 58 D2 *var.* San
 Juan. Guárico, N Venezuela
San Juanito, Isla 50 C4 *island* C Mexico
San Juan Mountains 48 D1 *mountain*
 range Colorado, C USA
San Juan, Río 53 E4 *river* Costa Rica/
 Nicaragua
San Juan River 48 C1 *river* Colorado/
 Utah, W USA

San Julián *see* Puerto San Julián
Sankt-Bartholomäi *see* Palamuse
Sankt Gallen 95 B7 *var.* St. Gallen, *Eng.*
 Saint Gall, *Fr.* St-Gall. Sankt Gallen,
 NE Switzerland
Sankt-Georgen *see* Sfântu Gheorghe
Sankt-Jakobi *see* Pärnu-Jaagupi,
 Pärnumaa, Estonia
Sankt Martin *see* Martin
Sankt Moritz *see* St. Moritz
Sankt-Peterburg 110 B4 *prev.*
 Leningrad, Petrograd, *Eng.*
 Saint Petersburg, *Fin.* Pietari.
 Leningradskaya Oblast', NW Russian
 Federation
Sankt Pölten 95 E6 Niederösterreich,
 N Austria
Sankt Veit am Flaum *see* Rijeka
Sankuru 77 D6 *river* C Dem. Rep. Congo
Şanlıurfa 117 E4 *prev.* Sanli Urfa, Urfa;
 anc. Edessa. şanlıurfa, S Turkey
Sanli Urfa *see* Şanlıurfa
San Lorenzo 61 G5 Tarija, S Bolivia
San Lorenzo 60 A1 Esmeraldas,
 N Ecuador
San Lorenzo, Isla 60 C4 *island* W Peru
Sanlúcar de Barrameda 92 C5
 Andalucía, S Spain
San Luis 64 C4 San Luis, C Argentina
San Luis 52 B2 Petén, NE Guatemala
San Luis *see* San Luis Río Colorado
San Luis Obispo 47 B7 California,
 W USA
San Luis Potosí 51 E3 San Luis Potosí,
 C Mexico
San Luis Río Colorado 50 A1 *var.*
 San Luis Río Colorado. Sonora,
 NW Mexico
San Luis Río Colorado *see* San Luis Río
 Colorado
San Marcos 52 A2 San Marcos,
 W Guatemala
San Marcos 49 G4 Texas, SW USA
San Marcos de Arica *see* Arica
San Marino 96 E1 *country capital* (San
 Marino) C San Marino
San Marino 96 D1 *off.* Republic of San
 Marino. *country* S Europe

SAN MARINO
Southern Europe

Official name Republic of San Marino
Formation 1631 / 1631
Capital San Marino
Population 32,448 / 1352 people
 per sq mile (532 people per sq km)
Total area 23.6 sq. miles (61 sq. km)
Languages Italian*
Religions Roman Catholic 93%,
 Other and nonreligious 7%
Ethnic mix Sammarinese 88%,
 Italian 10%, Other 2%
Government Parliamentary system
Currency Euro = 100 cents
Literacy rate 99%
Calorie consumption Not available

San Marino, Republic of *see* San Marino
San Martín 154 A2 *Argentinian research*
 station Antarctica
San Mateo 59 E2 Anzoátegui,
 NE Venezuela
San Matías 61 H3 Santa Cruz, E Bolivia
San Matías, Gulf of 65 C5 *var.* Gulf of
 San Matías. *gulf* E Argentina
San Matías, Gulf of *see* San Matías, Golfo
Sanmenxia 128 C4 *var.* Shan Xian.
 Henan, C China
Sânmiclăuş Mare *see* Sânnicolau Mare
San Miguel 52 C3 San Miguel,
 SE El Salvador
San Miguel 50 D2 Coahuila,
 N Mexico
San Miguel de Ibarra *see* Ibarra
San Miguel de Tucumán 64 C3 *var.*
 Tucumán. Tucumán, N Argentina
San Miguelito 53 G4 Panamá, C Panama
San Miguel, Río 61 G3 *river* E Bolivia
San Murezzan *see* St. Moritz
Sannär *see* Sennar
Sânnicolaul-Mare *see* Sânnicolau Mare
Sânnicolau Mare 108 A4 *var.*
 Sânnicolaul-Mare, *Hung.*
 Nagyszentmiklós; *prev.* Sânmiclăuş
 Mare, Sinnicolau Mare. Timiş,
 W Romania

Sanok 99 E5 Podkarpackie, SE Poland
San Pablo 61 F5 Potosí, S Bolivia
San Pedro 52 C1 Corozal, NE Belize
San-Pédro 74 D5 S Côte d'Ivoire
San Pedro 50 D3 var. San Pedro de las
Colonias. Coahuila, NE Mexico
San Pedro de la Cueva 50 C2 Sonora,
NW Mexico
San Pedro de las Colonias see San Pedro
San Pedro de Lloc 60 B3 La Libertad,
NW Peru
San Pedro Mártir, Sierra 50 A1
mountain range NW Mexico
San Pedro Sula 52 C2 Cortés,
NW Honduras
San Rafael 64 B4 Mendoza, W Argentina
San Rafael Mountains 47 C7 mountain
range California, W USA
San Ramón de la Nueva Orán 64 C2
Salta, N Argentina
San Remo 96 A3 Liguria, NW Italy
San Salvador 52 B3 country capital
(El Salvador) San Salvador,
SW El Salvador
San Salvador 54 D2 prev. Watlings
Island. island E Bahamas
San Salvador de Jujuy 64 C2 var. Jujuy.
Jujuy, N Argentina
San Salvador, Isla 60 A4 island Ecuador
Sansanné-Mango 75 E4 var. Mango.
N Togo
San Sebastián 93 E1 País Vasco, N Spain
see also Donostia
Sansepolcro 96 C3 Toscana, C Italy
San Severo 97 D5 Puglia, SE Italy
Santa Ana 61 F3 El Beni, N Bolivia
Santa Ana 52 B3 Santa Ana,
NW El Salvador
Santa Ana 46 D2 California, W USA
Sarnia de Coro see Coro
Santa Ana Mountains 46 E2 mountain
range California, W USA
Santa Barbara 50 C2 Chihuahua,
N Mexico
Santa Barbara 47 C7 California, W USA
Santa Cataliña de Armada 92 B1 var.
Santa Comba. Galicia, NW Spain
Santa Catalina Island 47 B8 island
Channel Islands, California, W USA
Santa Catarina 63 E5 off. Estado de
Santa Catarina. region S Brazil
Santa Catarina 63 E5 off. Estado de
Santa Catarina. state S Brazil
Santa Catarina, Estado de see Santa
Catarina
Santa Clara 54 B2 Villa Clara, C Cuba
Santa Clarita 46 D1 California, USA
Santa Comba see Santa Cataliña de
Armada
Santa Cruz 76 E2 São Tomé, S Sao Tome
and Principe, Africa
Santa Cruz 47 B6 California, W USA
Santa Cruz 61 G4 department E Bolivia
Santa Cruz Barillas see Barillas
Santa Cruz del Quiché 52 B2 Quiché,
W Guatemala
Santa Cruz de Tenerife 70 A3
Tenerife, Islas Canarias, Spain,
NE Atlantic Ocean
Santa Cruz, Isla 60 B5 var. Indefatigable
Island, Isla Chávez. island Galápagos
Islands, Ecuador, E Pacific Ocean
Santa Cruz Islands 144 D3 island group
E Solomon Islands
Santa Cruz, Río 65 B7 river S Argentina
Santa Elena 52 B1 Cayo, W Belize
Santa Fe 64 C4 Santa Fe, C Argentina
Santa Fe 48 D1 state capital New Mexico,
SW USA
Santa Fe see Bogotá
Santa Fe de Bogotá see Bogotá
Santa Genoveva 50 B3 mountain
NW Mexico
Santa Isabel 144 C3 var. Bughotu. island
S Solomon Islands
Santa Isabel see Malabo
Santa Lucia Range 47 B7 mountain
range California, W USA
Santa Margarita, Isla 50 B3 island
NW Mexico
Santa Maria 63 E5 Rio Grande do Sul,
S Brazil
Santa Maria 47 B7 California, USA
Santa Maria 92 A5 island Azores,
Portugal, NE Atlantic Ocean
Santa Maria del Buen Aire
see Buenos Aires

Santa María, Isla 60 A5 var. Isla
Floreana, Charles Island. island
Galápagos Islands, Ecuador, E Pacific
Ocean
Santa Marta 58 B1 Magdalena,
N Colombia
Santa Maura see Lefkáda
Santa Monica 46 D1 California, W USA
Santana 76 E2 São Tomé, C Sao Tome
and Príncipe, Africa
Santander 92 D1 Cantabria, N Spain
Santarém 63 E2 Pará, N Brazil
Santarém 92 B3 anc. Scalabis. Santarém,
W Portugal
Santa Rosa 64 C4 La Pampa,
C Argentina
Santa Rosa see Santa Rosa de Copán
Santa Rosa de Copán 52 C2 var. Santa
Rosa. Copán, W Honduras
Santa Rosa Island 47 B8 island
California, W USA
Santa Uxía de Ribeira 92 B1
var. Ribeira. NW Spain
Sant Carles de la Ràpita see Sant Carles
de la Ràpita
Sant Carles de la Ràpita 93 F3 var. Sant
Carles de la Rápida. Cataluña,
NE Spain
Santiago 64 B4 var. Gran Santiago.
country capital (Chile) Santiago,
C Chile
Santiago 55 E3 var. Santiago de los
Caballeros. N Dominican Republic
Santiago 53 F5 Veraguas, S Panama
Santiago 74 A3 var. São Tiago. island
Ilhas de Sotavento, S Cape Verde
Santiago see Santiago de Compostela
Santiago see Santiago de Cuba, Cuba
Santiago de Compostela 92 B1 var.
Santiago, Eng. Compostella; anc.
Campus Stellae. Galicia, NW Spain
Santiago de Cuba 54 C3 var. Santiago.
Santiago de Cuba, E Cuba
Santiago de Guayaquil see Guayaquil
Santiago del Estero 64 C3 Santiago del
Estero, C Argentina
Santiago de los Caballeros see Santiago,
Dominican Republic
Santiago de los Caballeros see Ciudad de
Guatemala, Guatemala
Santiago Pinoteca Nacional see
Pinoteca Nacional
Santiago, Río 60 B2 river N Peru
Santi Quaranta see Sarandë
Santíssima Trinidad see Jilong
Sant Joan d'Alacant 93 F4 Cast. San
Juan de Alicante. País Valenciano,
E Spain
Sant Julià de Lòria 91 A8 Sant Julià de
Lòria, SW Andorra Europe
Santo see Espiritu Santo
Santo Antão 74 A2 island Ilhas de
Barlavento, N Cape Verde
Santo António 76 E1 Príncipe,
N Sao Tome and Principe, Africa
Santo Domingo 55 D3 prev. Ciudad
Trujillo. country capital (Dominican
Republic) SE Dominican Republic
Santo Domingo de los Colorados 60 B1
Pichincha, NW Ecuador
Santo Domingo Tehuantepec see
Tehuantepec
Santorini see Thíra
Santorini 105 D7 island Kykládes,
Greece, Aegean Sea
San Tomé de Guayana see Ciudad
Guayana
Santos 63 F4 São Paulo, S Brazil
Santos Plateau 57 D5 undersea plateau
SW Atlantic Ocean
Santo Tomé 64 D3 Corrientes,
NE Argentina
Santo Tomé de Guayana see Ciudad
Guayana
San Valentín, Cerro 65 A6 mountain
S Chile
San Vicente 52 C3 San Vicente,
C El Salvador
São Francisco, Rio 63 F3 river E Brazil
São Hill 73 C7 Iringa, S Tanzania
São João da Madeira 92 B2 Aveiro,
N Portugal
São Jorge 92 A5 island Azores, Portugal,
NE Atlantic Ocean
São Luís 63 F2 state capital Maranhão,
NE Brazil
São Mandol see São Manuel, Rio

São Manuel, Rio 63 E3 var. São Mandol,
Teles Pirés. river C Brazil
São Marcos, Baía de 63 F1 bay
N Brazil
São Miguel 92 A5 island Azores,
Portugal, NE Atlantic Ocean
Saona, Isla 55 E3 island SE Dominican
Republic
Saône 91 D5 river E France
São Nicolau 74 A3 Eng. Saint Nicholas.
island Ilhas de Barlavento, N Cape
Verde
São Paulo 63 E4 state capital São Paulo,
S Brazil
São Paulo 63 E4 off. Estado de São
Paulo. region S Brazil
São Paulo 63 E4 off. Estado de São
Paulo. state S Brazil
São Paulo de Loanda see Luanda
São Paulo, Estado de see São Paulo
São Pedro do Rio Grande do Sul see
Rio Grande
São Roque, Cabo de 63 G2 headland
E Brazil
São Salvador see Salvador, Brazil
São Salvador/São Salvador do Congo
see M'Banza Congo, Angola
São Tiago see Santiago
São Tomé 77 A5 country capital
(Sao Tome and Príncipe)
São Tomé, S Sao Tome and Principe
São Tomé 76 E2 Eng. Saint Thomas.
island S Sao Tome and Principe
Sao Tome and Principe 76 D1 off.
Democratic Republic of Sao Tome and
Príncipe. country E Atlantic Ocean

SÃO TOMÉ & PRÍNCIPE
West Africa

Official name The Democratic Republic
of São Tomé and Príncipe
Formation 1975 / 1975
Capital São Tomé
Population 200,000 / 539 people
per sq mile (208 people per sq km)
Total area 386 sq. miles (1001 sq. km)
Languages Portuguese Creole,
Portuguese*
Religions Roman Catholic 84%, Other 16%
Ethnic mix Black 90%, Portuguese
and Creole 10%
Government Presidential system
Currency Dobra = 100 céntimos
Literacy rate 70%
Calorie consumption 2676 kilocalories

**Sao Tome and Principe, Democratic
Republic of** see Sao Tome and Principe
São Tomé, Pico de 76 D2 mountain
São Tomé, C Sao Tome and Principe,
Africa
São Vicente 74 A3 Eng. Saint Vincent.
island Ilhas de Barlavento, N Cape
Verde
São Vicente, Cabo de 92 B5 Eng.
Cape Saint Vincent, Port. Cabode São
Vicente. cape S Portugal
São Vicente, Cabo de see São Vicente,
Cabo de
Sápai see Sápes
Sapele 75 F5 Delta, S Nigeria
Sápes 104 D3 var. Sápai. Anatolikí
Makedonía kai Thráki, NE Greece
Sapir 119 B7 prev. Sappir. Southern,
S Israel
Sa Pobla 93 G3 Mallorca, Spain,
W Mediterranean Sea
Sappir see Sapir
Sapporo 130 D2 Hokkaidō, NE Japan
Sapri 97 D6 Campania, S Italy
Sapulpa 49 G1 Oklahoma, C USA
Saqqez 120 C2 var. Saghez, Sakiz,
Saqqiz. Kordestán, NW Iran
Saqqiz see Saqqez
Sara Buri 137 C5 var. Saraburi. Saraburi,
C Thailand
Saraburi see Sara Buri
Saragossa see Zaragoza
Saragt see Sarahs
Saraguro 60 B2 Loja, S Ecuador
Sarahs 122 D3 var. Saragt, Rus. Serakhs.
Ahal Welaýaty, S Turkmenistan
Sarajevo 100 C4 country capital
(Bosnia and Herzegovina) Federacija
Bosne I Hercegovine, SE Bosnia and
Herzegovina

Sarakhs 120 E2 Khorásán-Razaví, E Iran
Saraktash 111 D6 Orenburgskaya
Oblast', W Russian Federation
Saran' 114 C4 Kaz. Saran. Karagandy,
C Kazakhstan
Saranda see Sarandë
Sarandë 101 C7 var. Saranda, It. Porto
Edda; prev. Santi Quaranta. Vlorë,
S Albania
Saransk 111 C5 Respublika Mordoviya,
W Russian Federation
Sarasota 43 E4 Florida, SE USA
Saratov 114 B3 Saratovskaya Oblast',
W Russian Federation
Saravan/Saravane see Salavan
Sarawak 138 D3 state East Malaysia
Sarawak see Kuching
Sarcelles 90 D1 Val-d'Oise, Île-de-
France, N France Europe
Sardinia 97 A5 Eng. Sardinia. island
Italy, C Mediterranean Sea
Sardinia see Sardegna
Sarera, Teluk see Cenderawasih, Teluk
Sargasso Sea 66 B4 sea W Atlantic Ocean
Sargodha 134 C2 Punjab, NE Pakistan
Sarh 76 C4 prev. Fort-Archambault.
Moyen-Chari, S Chad
Sārī 120 D2 var. Sari, Sári. Mázandarán,
N Iran
Saría 105 E7 island SE Greece
Sarıkamış 117 F3 Kars, NE Turkey
Sarikol Range 123 G3 Rus. Sarykol'skiy
Khrebet. mountain range China/
Tajikistan
Sark 89 D8 Fr. Sercq. island Channel
Islands
Şarkışla 116 D3 Sivas, C Turkey
Sarmiento 65 B6 Chubut, S Argentina
Sarnia 38 C5 Ontario, S Canada
Sarny 108 C1 Rivnens'ka Oblast',
NW Ukraine
Sarochyna 107 D5 Rus. Sorochino.
Vitsyebskaya Voblasts', N Belarus
Sarov 111 C5 prev. Sarova. Respublika
Mordoviya, SW Russian Federation
Sarova see Sarov
Sarpsborg 85 B6 Østfold, S Norway
Sarrebruck see Saarbrücken
Sartène 91 E7 Corse, France,
C Mediterranean Sea
Sarthe 90 B4 river N France
Sárti 104 C4 Kentrikí Makedonía,
N Greece
Saruhan see Manisa
Sarykol'skiy Khrebet see Sarikol Range
Sary-Tash 123 F2 Oshskaya Oblast',
SW Kyrgyzstan
Saryyesik-Atyrau, Peski 123 G1 desert
E Kazakhstan
Sasebo 131 A7 Nagasaki, Kyūshū,
SW Japan
Saskatchewan 37 F5 province
SW Canada
Saskatchewan 37 F5 river Manitoba/
Saskatchewan, C Canada
Saskatoon 37 F5 Saskatchewan, S Canada
Sasovo 111 B5 Ryazanskaya Oblast',
W Russian Federation
Sassandra 74 D5 S Côte d'Ivoire
Sassandra 74 D5 var. Ibo, Sassandra
Fleuve. river S Côte d'Ivoire
Sassandra Fleuve see Sassandra
Sassari 97 A5 Sardegna, Italy,
C Mediterranean Sea
Sassenheim 86 C3 Zuid-Holland,
W Netherlands
Sassnitz 94 D2 Mecklenburg-
Vorpommern, NE Germany
Sathmar see Satu Mare
Sátoraljaújhely 99 D6 Borsod-Abaúj-
Zemplén, NE Hungary
Satpura Range 134 D4 mountain range
C India
Satsuma-Sendai 131 A8 Kagoshima,
Kyūshū, SW Japan
Satsunan-shoto 130 A3 island group
Nansei-shotō, SW Japan Asia
Sattanen 84 D3 Lappi, NE Finland
Satu Mare 108 B3 Ger. Sathmar,
Hung. Szatmárrnémeti. Satu Mare,
NW Romania
Sau see Sava
Saudi Arabia 121 B5 off. Kingdom
of Saudi Arabia, Al 'Arabīyah as
Su'ūdīyah, Ar. Al Mamlakah al
'Arabīyah as Su'ūdīyah. country
SW Asia

SAUDI ARABIA
Southwest Asia

Official name Kingdom of Saudi Arabia
Formation 1932 / 1932
Capital Riyadh
Population 28.8 million / 35 people
per sq mile (14 people per sq km)
Total area 756,981 sq. miles
(1,960,582 sq. km)
Languages Arabic*
Religions Sunni Muslim 85%,
Shi'a Muslim 15%
Ethnic mix Arab 72%, Foreign residents
(mostly south and southeast Asian) 20%,
Afro-Asian 8%
Government Monarchy
Currency Saudi riyal = 100 halalat
Literacy rate 94%
Calorie consumption 3122 kilocalories

Saudi Arabia, Kingdom of *see* Saudi
Arabia
Sauer *see* Sûre
Saugor *see* Sāgar
Saulkrasti *106 C3* C Latvia
Sault Sainte Marie *38 C4* Michigan,
N USA
Sault Ste. Marie *40 C1* Michigan,
N USA
Sault Ste. Marie *38 C4* Ontario,
S Canada
Saumur *78 B4* Maine-et-Loire,
NW France
Saurimo *78 C1 Port.* Henrique de
Carvalho, Vila Henrique de Carvalho.
Lunda Sul, NE Angola
Sava *107 E6* Mahilyowskaya Voblasts',
E Belarus
Savá *52 D2* Colón, N Honduras
Savai'i *145 E4 island* NW Samoa
Savannah *43 E2* Georgia, SE USA
Savannah River *43 E2 river* Georgia/
South Carolina, SE USA
Savannakhét *136 D4 var.* Khanthabouli.
Savannakhét, S Laos
Savanna-La-Mar *54 A5* W Jamaica
Savaria *see* Szombathely
Save *100 B3 Eng.* Save, *Ger.* Sau, *Hung.*
Száva. *river* SE Europe
Save *see* Sava
Save, Rio *79 E3 var.* Sabi. *river*
Mozambique/Zimbabwe
Saverne *90 E3 var.* Zabern; *anc.* Tres
Tabernae. Bas-Rhin, NE France
Savigliano *96 A2* Piemonte, NW Italy
Savissivik *see* Savissivik
Saviñski *see* Savinskiy
Savinskiy *110 C3 var.* Savinski.
Arkhangel'skaya Oblast', NW Russian
Federation
Savissivik *82 D1 var.* Savigsivik.
Avannaarsua, N Greenland
Savoie *91 D5 cultural region* E France
Savona *96 A2* Liguria, NW Italy
Savu Sea *139 E5 Ind.* Laut Sawu. *sea*
S Indonesia
Sawakin *see* Suakin
Sawdiri *see* Sodiri
Sawhāj *72 B2 var.* Sawhâj, *var.* Sohâg,
Suliag. C Egypt
Sawhāj *see* Sawhāj
Sawqirah *121 D6 var.* Suqrah. S Oman
Sawu, Laut *see* Savu Sea
Saxe *see* Sachsen
Saxony *see* Sachsen
Sayaboury *see* Xaignabouli
Sayanskiy Khrebet *112 D3 mountain
range* S Russian Federation
Sayat *122 D3 Rus.* Sayat. Lebap
Welaýaty, E Turkmenistan
Sayaxché *52 B2* Petén, N Guatemala
Saydā/Sayida *see* Saïda
Sayhūt *121 D6* E Yemen
Saynshand *127 E2* Dornogovĭ,
SE Mongolia
Sayre *41 E3* Pennsylvania, NE USA
Say'ūn *121 C6 var.* Saywūn. C Yemen
Saywūn *see* Say'ūn
Scalabis *see* Santarém
Scandinavia *66 D2 geophysical region*
NW Europe
Scarborough *89 D5* N England, United
Kingdom
Scarpanto *see* Kárpathos
Scebeli *see* Shebeli
Schaan *94 E1* W Liechtenstein Europe

Schaerbeek *87 C6* Brussels, C Belgium
Schaffhausen *95 B7 Fr.* Schaffhouse.
Schaffhausen, N Switzerland
Schaffhouse *see* Schaffhausen
Schagen *86 C2* Noord-Holland,
NW Netherlands
Schaulen *see* Šiauliai
Schebschi Mountains *see* Shebshi
Mountains
Scheessel *94 B3* Niedersachsen,
NW Germany
Schefferville *39 E2* Québec, E Canada
Schelde *see* Scheldt
Scheldt *87 B5 Dut.* Schelde, *Fr.* Escaut.
river W Europe
Schell Creek Range *47 D5 mountain
range* Nevada, W USA
Schenectady *41 F3* New York, NE USA
Schertz *49 G4* Texas, SW USA
Schiermonnikoog *86 D1* Fris.
Skiermûntseach. *island*
Waddeneilanden, N Netherlands
Schijndel *86 D4* Noord-Brabant,
S Netherlands
Schil *see* Jiu
Schiltigheim *90 E3* Bas-Rhin, NE France
Schivelbein *see* Świdwin
Schleswig *94 B2* Schleswig-Holstein,
N Germany
Schleswig-Holstein *94 B2 state*
N Germany
Schlettstadt *see* Sélestat
Schlochau *see* Człuchów
Schneekoppe *see* Sněžka
Schneidemühl *see* Piła
Schoden *see* Skuodas
Schönebeck *94 C4* Sachsen-Anhalt,
C Germany
Schönlanke *see* Trzcianka
Schooten *see* Schoten
Schoten *87 C5 var.* Schooten.
Antwerpen, N Belgium
Schouwen *86 B4 island* SW Netherlands
Schwabenalb *see* Schwäbische Alb
Schwäbische Alb *95 B6 var.*
Schwabenalb, *Eng.* Swabian Jura.
mountain range S Germany
Schwandorf *95 C5* Bayern, SE Germany
Schwaz *95 C7* Tirol, W Austria
Schweidnitz *see* Świdnica
Schweinfurt *95 B5* Bayern, SE Germany
Schweiz *see* Switzerland
Schwerin *94 C3* Mecklenburg-
Vorpommern, N Germany
Schwertberg *see* Świecie
Schwiebus *see* Świebodzin
Schwyz *95 B7 var.* Schwiz. Schwyz,
C Switzerland
Schyl *see* Jiu
Scio *see* Chíos
Scoresby Sound/Scoresbysund *see*
Ittoqqortoormiit
Scoresby Sund *see* Ittoqqortoormiit
Scotia Sea *57 C8 sea* SW Atlantic Ocean
Scotland *88 C3 cultural region* Scotland,
UK
Scott Base *154 B4* NZ research station
Antarctica
Scott Island *154 B5 island* Antarctica
Scottsbluff *44 D3* Nebraska, C USA
Scottsboro *42 D1* Alabama, S USA
Scottsdale *48 B2* Arizona, SW USA
Scranton *41 F3* Pennsylvania, NE USA
Scrobesbyrig' *see* Shrewsbury
Scupi *see* Skopje
Scutari *see* Shkodër
Scutari, Lake *101 C5 Alb.* Liqeni i
Shkodrës, *SCr.* Skadarsko Jezero. *lake*
Albania/Montenegro
Scyros *see* Skýros
Searcy *42 B1* Arkansas, C USA
Seattle *46 B2* Washington, NW USA
Sébaco *52 D3* Matagalpa, W Nicaragua
Sebaste/Sebastia *see* Sivas
Sebastián Vizcaíno, Bahía *50 A2 bay*
NW Mexico
Sebastopol *see* Sevastopol'
Sebenico *see* Šibenik
Sechura, Bahía de *60 A3 bay* NW Peru
Secunderābād *134 D5 var.*
Sikandarabad. Telangana, C India
Sedan *90 D3* Ardennes, N France
Seddon *151 D5* Marlborough, South
Island, New Zealand
Seddonville *151 C5* West Coast, South
Island, New Zealand
Sédhiou *74 B3* SW Senegal

Sedlez *see* Siedlce
Sedona *48 B2* Arizona, SW USA
Sedunum *see* Sion
Seeland *see* Sjælland
Seenu Atoll *see* Addu Atoll
Seesen *94 B4* Niedersachsen, C Germany
Segestica *see* Sisak
Segezha *110 B3* Respublika Kareliya,
NW Russian Federation
Seghedin *see* Szeged
Segna *see* Senj
Segodunum *see* Rodez
Ségou *74 D3 var.* Segu. Ségou, C Mali
Segovia *92 D2* Castilla y León, C Spain
Segoviao Wangki *see* Coco, Río
Segu *see* Ségou
Séguédine *75 H2* Agadez, NE Niger
Seguin *49 G4* Texas, SW USA
Segura *93 E4 river* S Spain
Seinäjoki *85 D5 Swe.* Östermyra. Länsi-
Suomi, W Finland
Seine *90 D1 river* N France
Seine, Baie de la *90 B3 bay* N France
Sejong City *128 E4 administrative
capital* (South Korea) C South Korea
Sekondi *see* Sekondi-Takoradi
Sekondi-Takoradi *75 E5 var.* Sekondi.
S Ghana
Selânik *see* Thessaloníki
Selenga *127 E1 Mong.* Selenge Mörön.
river Mongolia/Russian Federation
Selenge Mörön *see* Selenga
Sélestat *90 E4 Ger.* Schlettstadt. Bas-
Rhin, NE France
Seleucia *see* Silifke
Selfoss *83 E5* Sudhurland, SW Iceland
Sélibabi *74 C3 var.* Sélibaby. Guidimaka,
S Mauritania
Sélibaby *see* Sélibabi
Selma *47 C6* California, W USA
Selseln-ye Sefid Kuh *122 D4 Eng.*
Paropamisus Range. *mountain range*
W Afghanistan
Selway River *46 D2 river* Idaho,
NW USA North America
Selwyn Range *148 B3 mountain range*
Queensland, C Australia
Selzaete *see* Zelzate
Semarang *138 C5 var.* Samarang. Jawa,
C Indonesia
Sembé *77 B5* Sangha, NW Congo
Semendria *see* Smederevo
Semey *114 D4 prev.* Semipalatinsk.
Vostochnyy Kazakhstan, E Kazakhstan
Semezhevo *see* Syemyezhava
Seminole *49 E3* Texas, SW USA
Seminole, Lake *42 D3 reservoir* Florida/
Georgia, SE USA
Semipalatinsk *see* Semey
Semnān *120 D3 var.* Samnān. Semnān,
N Iran
Semois *87 C8 river* SE Belgium
Sendai *130 D4* Miyagi, Honshū,
C Japan
Sendai-wan *130 D4 bay* E Japan
Senec *99 C6 Ger.* Wartberg, *Hung.* Szenc;
prev. Szempcz. Bratislavský Kraj,
W Slovakia
Senegal *74 B3 off.* Republic of Senegal,
Fr. Sénégal. *country* W Africa

SENEGAL
West Africa

Official name Republic of Senegal
Formation 1960 / 1960
Capital Dakar
Population 14.1 million / 190 people
per sq mile (73 people per sq km)
Total area 75,749 sq. miles (196,190 sq. km)
Languages Wolof, Pulaar, Serer, Diola,
Mandinka, Malinké, Soninké, French*
Religions Sunni Muslim 95%, Christian
(mainly Roman Catholic) 4%, Traditional
beliefs 1%
Ethnic mix Wolof 43%, Serer 15%,
Peul 14%, Other 14%, Toucouleur 9%,
Diola 5%
Government Presidential system
Currency CFA franc = 100 centimes
Literacy rate 52%
Calorie consumption 2426 kilocalories

Senegal *74 C3 Fr.* Sénégal. *river* W Africa
Senegal, Republic of *see* Senegal
Senftenberg *94 D4* Brandenburg,
E Germany

Senia *see* Senj
Senica *99 C6 Ger.* Senitz, *Hung.* Szenice.
Trnavský Kraj, W Slovakia
Seniça *see* Sjenica
Senitz *see* Senica
Senj *100 A3 Ger.* Zengg, *It.* Segna; *anc.*
Senia. Lika-Senj, NW Croatia
Senja *84 C2 prev.* Senjen. *island*
N Norway
Senjen *see* Senja
Sejong City *128 E4 country capital*
(South Korea) (administrative)
C South Korea
Senkaku-shoto *130 A3 island group*
SW Japan
Senlis *90 C3* Oise, N France
Sennar *72 C4 var.* Sannâr. Sinnar,
C Sudan
Senones *see* Sens
Sens *90 C3 anc.* Agendicum, Senones.
Yonne, C France
Sensburg *see* Mrągowo
Sên, Stœng *137 D5 river* C Cambodia
Senta *100 D3 Hung.* Zenta. Vojvodina,
N Serbia
Seo de Urgel *see* La Seu d'Urgell
Seoul *129 E4 off.* Seoul Teukbyeolsi,
prev. Sŏul, *Jap.* Keijō; *prev.* Kyŏngsŏng.
country capital (South Korea)
NW South Korea
Seoul Teukbyeolsi *see* Seoul
Şepşi-Sângeorz/Sepsiszentgyörgy *see*
Sfântu Gheorghe
Sept-Îles *39 E3* Québec,
SE Canada
Seraing *87 D6* Liège, E Belgium
Serakhs *see* Sarahs
Seram, Laut *139 F4 Eng.* Ceram Sea. *sea*
E Indonesia
Pulau Seram *139 F4 var.* Serang,
Eng. Ceram. *island* Maluku,
E Indonesia
Serang *138 C5* Jawa, C Indonesia
Serang *see* Seram, Pulau
Serasan, Selat *138 C3 strait* Indonesia/
Malaysia
Serbia *100 D4 off.* Federal Republic
of Serbia; *prev.* Yugoslavia, *SCr.*
Jugoslavija. *country* SE Europe

SERBIA
Southeast Europe

Official name Republic of Serbia
Formation 2006 / 2008
Capital Belgrade
Population 9.5 million / 318 people
per sq mile (128 people per sq km)
Total area 29,905 sq. miles (77,453 sq. km)
Languages Serbian*, Hungarian (Magyar)
Religions Orthodox Christian 85%,
Roman Catholic 6%, Other 6%,
Muslim 3%
Ethnic mix Serb 83%, Other 10%,
Magyar 4%, Bosniak 2%, Roma 1%
Government Parliamentary system
Currency Dinar = 100 para
Literacy rate 98%
Calorie consumption 2823 kilocalories

Serbia, Federal Republic of *see* Serbia
Sercq *see* Sark
Serdar *122 C2 prev.* Rus. Gyzyrlabat,
Kizyl-Arvat. Balkan Welaýaty,
W Turkmenistan
Serdica *see* Sofiya
Serdobol' *see* Sortavala
Serenje *78 D2* Central, E Zambia
Seres *see* Sérres
Seret/Sereth *see* Siret
Serhetabat *122 D4 prev. Rus.*
Gushgy, Kushka. Mary Welaýaty,
S Turkmenistan
Sérifos *105 C6 anc.* Seriphos. *island*
Kykládes, Greece, Aegean Sea
Seriphos *see* Sérifos
Serov *114 C3* Sverdlovskaya Oblast',
C Russian Federation
Serowe *78 D3* Central, SE Botswana
Serpa Pinto *see* Menongue
Serpent's Mouth, The *59 F2 Sp.* Boca
de la Serpiente. *strait* Trinidad and
Tobago/Venezuela
Serpiente, Boca de la *see* Serpent's
Mouth, The
Serpukhov *111 B5* Moskovskaya Oblast',
W Russian Federation

Sérrai see Sérres
Serrana, Cayo de 53 F2 island group
  NW Colombia South America
Serranilla, Cayo de 53 F2 island
  group NW Colombia South America
  Caribbean Sea
Serravalle 96 E1 N San Marino
Sérres 104 C3 var. Seres; prev. Sérrai.
  Kentrikí Makedonía, NE Greece
Sesdlets see Siedlce
Sesto San Giovanni 96 B2 Lombardia,
  N Italy
Sesvete 100 B2 Zagreb, N Croatia
Setabis see Xàtiva
Sète 91 C6 prev. Cette. Hérault, S France
Setesdal 85 A6 valley S Norway
Sétif 71 E2 var. Stif. N Algeria
Setté Cama 77 A6 Ogooué-Maritime,
  SW Gabon
Setúbal 92 B4 Eng. Saint Ubes, Saint
  Yves. Setúbal, W Portugal
Setúbal, Baía de 92 B4 bay W Portugal
Seul, Lac 38 B3 lake Ontario, S Canada
Sevan 117 G2 C Armenia
Sevan, Lake 117 G3 Eng. Lake Sevan,
  Rus. Ozero Sevan. lake E Armenia
Sevan, Lake/Sevan, Ozero
  see Sevana Lich
Sevastopol' 109 F5 Eng. Sebastopol.
  Avtonomna Respublika Krym, Ukraine
Severn 38 B2 river Ontario, S Canada
Severn 89 D6 Wel. Hafren. river
  England/Wales, United Kingdom
Severnaya Zemlya 115 E2 var. Nicholas
  II Land. island group N Russian
  Federation
Severnyy 110 E3 Respublika Komi,
  NW Russian Federation
Severodonetsk see Syeverodonets'k
Severodvinsk 110 C3 prev. Molotov,
  Sudostroy. Arkhangel'skaya Oblast',
  NW Russian Federation
Severomorsk 110 C2 Murmanskaya
  Oblast', NW Russian Federation
Seversk 114 D4 Tomskaya Oblast',
  C Russian Federation
Sevier Lake 44 A4 lake Utah, W USA
Sevilla 92 C4 Eng. Seville; anc. Hispalis.
  Andalucía, SW Spain
Seville see Sevilla
Sevlievo 104 D2 Gabrovo, N Bulgaria
Sevluš/Sevlyush see Vynohradiv
Seward's Folly see Alaska
Seychelles 79 G1 off. Republic of
  Seychelles. country W Indian Ocean

SEYCHELLES
Indian Ocean

Official name Republic of Seychelles
Formation 1976 / 1976
Capital Victoria
Population 90,846 /874 people
  per sq mile (336 people per sq km)
Total area 176 sq. miles (455 sq. km)
Languages French Creole*,
  English*, French*
Religions Roman Catholic 82%,
  Anglican 6%, Other (including Muslim)
  6%, Other Christian 3%, Hindu 2%,
  Seventh-day Adventist 1%
Ethnic mix Creole 89%, Indian 5%,
  Other 4%, Chinese 2%
Government Presidential system
Currency Seychelles rupee = 100 cents
Literacy rate 92%
Calorie consumption 2426 kilocalories

Seychelles, Republic of see Seychelles
Seyðisfjörður 83 E5 Austurland,
  E Iceland
Seydi 122 D2 Rus. Seydi; prev.
  Neftezavodsk. Lebap Welaýaty,
  E Turkmenistan
Seyhan see Adana
Sfákia see Chóra Sfakíon
Sfântu Gheorghe 108 C4 Ger. Sankt-
  Georgen, Hung. Sepsiszentgyörgy;
  prev. Şepsi-Sângeorz, Sfîntu Gheorghe.
  Covasna, C Romania
Sfax 71 F2 Ar. Şafāqis. E Tunisia
Sfîntu Gheorghe see Sfântu Gheorghe
's-Gravenhage 86 B4 var. Den Haag,
  Eng. The Hague, Fr. La Haye.
  country capital (Netherlands)
  (seat of government)
  Zuid-Holland, W Netherlands

's-Gravenzande 86 B4 Zuid-Holland,
  W Netherlands
Shaan/Shaanxi Sheng see Shaanxi
Shaanxi 128 B5 var. Shaan, Shaanxi
  Sheng, Shan-hsi, Shenshi, Shensi.
  province C China
Shabani see Zvishavane
Shabeelle, Webi see Shebeli
Shache 126 A3 var. Yarkant. Xinjiang
  Uygur Zizhiqu, NW China
Shacheng see Huailai
Shackleton Ice Shelf 154 D3 ice shelf
  Antarctica
Shaddādī see Ash Shadādah
Shāhābād see Eslāmābād-e Gharb
Sha Hi see Orūmīyeh, Daryācheh-ye
Shahjahanabad see Delhi
Shahr-e Kord 120 C3 var. Shahr Kord.
  Chahār Maḩall va Bakhtīārī, C Iran
Shahr Kord see Shahr-e Kord
Shāhrūd 120 D2 prev. Emāmrūd,
  Emāmshahr. Semnān, N Iran
Shalkar 114 B4 var. Chelkar.
  Aktyubinsk, W Kazakhstan
Shām, Bādiyat ash see Syrian Desert
Shana see Kuril'sk
Shandi see Shendi
Shandong 128 D4 var. Lu, Shandong
  Sheng, Shantung. province E China
Shandong Sheng see Shandong
Shanghai 128 D5 var. Shang-hai.
  Shanghai Shi, E China
Shangrao 128 D5 Jiangxi, S China
Shan-hsi see Shaanxi, China
Shan-hsi see Shanxi, China
Shannon 89 A6 Ir. An tSionainn. river
  W Ireland
Shan Plateau 136 B3 plateau E Myanmar
  (Burma)
Shansi see Shanxi
Shantar Islands see Shantarskiye Ostrova
Shantarskiye Ostrova 115 G3 Eng.
  Shantar Islands. island group E Russian
  Federation
Shantou 128 D6 var. Shan-t'ou, Swatow.
  Guangdong, S China
Shan-t'ou see Shantou
Shantung see Shandong
Shanxi 128 C4 var. Jin, Shan-hsi,
  Shansi, Shanxi Sheng. province
  C China
Shan Xian see Sanmenxia
Shanxi Sheng see Shanxi
Shaoguan 128 C6 var. Shao-kuan,
  Cant. Kukong; prev. Ch'u-chiang.
  Guangdong, S China
Shao-kuan see Shaoguan
Shaqrā' 120 B4 Ar Riyāḍ,
  C Saudi Arabia
Shaqrā see Shuqrah
Shar 114 D5 var. Charsk. Vostochnyy
  Kazakhstan, E Kazakhstan
Shari 130 D2 Hokkaidō, NE Japan
Shari see Chari
Sharjah see Ash Shāriqah
Shark Bay 147 A5 bay
  Western Australia
Sharqi, Al Jabal ash/Sharqi, Jebel esh
  see Anti-Lebanon
Shashe 78 D3 var. Shashi. river
  Botswana/Zimbabwe
Shashi see Shashe
Shatskiy Rise 125 G1 undersea rise
  N Pacific Ocean
Shawnee 49 G1 Oklahoma, C USA
Shaykh, Jabal ash see Hermon, Mount
Shchadryn 107 D7 Rus. Shchedrin.
  Homyel'skaya Voblasts', SE Belarus
Shchedrin see Shchadryn
Shcheglovsk see Kemerovo
Shchëkino 111 B5 Tul'skaya Oblast',
  W Russian Federation
Shchors 109 E1 Chernihivs'ka Oblast',
  N Ukraine
Shchuchin see Shchuchyn
Shchuchinsk 114 C4 prev. Shchuchye.
  Akmola, N Kazakhstan
Shchuchye see Shchuchinsk
Shchuchyn 107 B5 Pol. Szczuczyn
  Nowogródzki, Rus. Shchuchin.
  Hrodzyenskaya Voblasts', W Belarus
Shebekino 111 A6 Belgorodskaya
  Oblast', W Russian Federation
Shebelē Wenz, Wabē see Shebeli
Shebeli 73 D5 Amh. Wabē Shebelē Wenz,
  It. Scebeli, Som. Webi Shabeelle. river
  Ethiopia/Somalia

Sheberghān see Shibirghān
Sheboygan 40 B2 Wisconsin, N USA
Shebshi Mountains 76 A4 var.
  Schebschi Mountains. mountain range
  E Nigeria
Shechem see Nablus
Shedadi see Ash Shadādah
Sheffield 89 D5 N England, United
  Kingdom
Shekhem see Nablus
Sheki see Şäki
Shelby 44 B1 Montana, NW USA
Sheldon 45 F3 Iowa, C USA
Shelekhov Gulf 115 G2 Eng. Shelekhov
  Gulf. gulf E Russian Federation
Shelekhov Gulf see Shelikhova, Zaliv
Shendi 72 C4 var. Shandī. River Nile,
  NE Sudan
Shengking see Liaoning
Shenking see Liaoning
Shenshi/Shensi see Shaanxi
Shenyang 128 D3 Chin. Shen-yang, Eng.
  Moukden, Mukden; prev. Fengtien.
  province capital Liaoning, NE China
Shen-yang see Shenyang
Shepetivka 108 D2 Rus. Shepetovka.
  Khmel'nyts'ka Oblast', NW Ukraine
Shepetovka see Shepetivka
Shepparton 149 C7 Victoria,
  SE Australia
Sherbrooke 39 E4 Québec, SE Canada
Shereik 72 C3 River Nile, N Sudan
Sheridan 44 C2 Wyoming, C USA
Sherman 49 G2 Texas, SW USA
's-Hertogenbosch 86 C4 Fr. Bois-le-
  Duc, Ger. Herzogenbusch. Noord-
  Brabant, S Netherlands
Shetland Islands 88 D1 island group
  NE Scotland, United Kingdom
Shevchenko see Aktau
Shibarghan see Shibirghān
Shiberghan/Shiberghān see Shibirghān
Shibirghān 123 E3 var. Sheberghan,
  Shiberghan, Shiberghān. Jowzjān,
  N Afghanistan
Shibetsu 130 D2 var. Sibetu. Hokkaidō,
  NE Japan
Shibh Jazirat Sinā' see Sinai
Shibirghān 123 E3 var. Shibarghān,
  Shiberghan, Shiberghān; prev.
  Sheberghān. Jowzjān, N Afghanistan
Shibushi-wan 131 B8 bay SW Japan
Shigatse see Xigazê
Shih-chia-chuang/Shihmen see
  Shijiazhuang
Shihezi 126 C2 Xinjiang Uygur Zizhiqu,
  NW China
Shiichi see Shyichy
Shijiazhuang 128 C4 var. Shih-chia-
  chuang; prev. Shihmen. province
  capital Hebei, E China
Shikārpur 134 B3 Sind, S Pakistan
Shikoku 131 C7 var. Sikoku. island
  SW Japan
Shikoku Basin 125 F2 var. Sikoku Basin.
  undersea basin N Philippine Sea
Shikotan, Ostrov 130 E2 Jap. Shikotan-
  tō. island NE Russian Federation
Shikotan-tō see Shikotan, Ostrov
Shilabo 73 D5 Sumalē, E Ethiopia
Shiliguri 135 F3 prev. Siliguri. West
  Bengal, NE India
Shilka 115 F4 river S Russian Federation
Shillong 135 G3 state capital Meghālaya,
  NE India
Shimanto see Nakamura
Shimbir Berris see Shimbiris
Shimbiris 72 E4 var. Shimbir Berris.
  mountain N Somalia
Shimoga 132 C2 Karnātaka, W India
Shimonoseki 131 A7 var. Simonoseki,
  hist. Akamagaseki, Bakan. Yamaguchi,
  Honshū, SW Japan
Shinano-gawa 131 C5 var. Sinano Gawa.
  river Honshū, S Japan
Shindand 122 D4 prev. Shīndand. Herāt,
  W Afghanistan
Shīndand see Shindand
Shingū 131 C6 var. Singū. Wakayama,
  Honshū, SW Japan
Shinjō 130 D4 var. Sinzyō. Yamagata,
  Honshū, C Japan
Shinyanga 73 C7 Shinyanga,
  NW Tanzania
Shiprock 48 C1 New Mexico, SW USA
Shīrāz 120 D4 var. Shīrāz. Fārs, S Iran
Shishchitsy see Shyshchytsy

Shivpuri 134 D3 Madhya Pradesh,
  C India
Shizugawa 130 D4 Miyagi, Honshū,
  NE Japan
Shizuoka 131 D6 var. Sizuoka. Shizuoka,
  Honshū, S Japan
Shklov see Shklow
Shklow 107 D6 Rus. Shklov.
  Mahilyowskaya Voblasts', E Belarus
Shkodër 101 C5 var. Shkodra, It. Scutari,
  SCr. Skadar. Shkodër, NW Albania
Shkodra see Shkodër
Shkodrës, Liqeni i see Scutari, Lake
Shkumbini, Lumi i 101 C6 var.
  Shkumbi, Shkumbin. river C Albania
Shkumbi/Shkumbin see Shkumbinit,
  Lumi i
Sholāpur see Solāpur
Shostka 109 F1 Sums'ka Oblast',
  NE Ukraine
Show Low 48 B2 Arizona, SW USA
Show Me State see Missouri
Shpola 109 E3 Cherkas'ka Oblast',
  N Ukraine
Shqipëria/Shqipërisë, Republika e
  see Albania
Shreveport 42 A2 Louisiana, S USA
Shrewsbury 89 D6 hist. Scrobesbyrig'.
  W England, United Kingdom
Shu 114 C5 Kaz. Shū. Zhambyl,
  SE Kazakhstan
Shuang-liao see Liaoyuan
Shū, Kazakhstan see Shu
Shumagin Islands 36 B3 island group
  Alaska, USA
Shumen 104 D2 Shumen, NE Bulgaria
Shumilina 107 E5 Rus. Shumilino.
  Vitsyebskaya Voblasts', NE Belarus
Shumilino see Shumilina
Shunsen see Chuncheon
Shuqrah 121 B7 var. Shaqrā. SW Yemen
Shwebo 136 B3 Sagaing, C Myanmar
  (Burma)
Shyichy 107 C7 Rus. Shiichi.
  Homyel'skaya Voblasts', SE Belarus
Shymkent 114 B5 prev. Chimkent.
  Yuzhnyy Kazakhstan, S Kazakhstan
Shyshchytsy 107 C6 Rus. Shishchitsy.
  Minskaya Voblasts', C Belarus
Siam see Thailand
Siam, Gulf of see Thailand, Gulf of
Sian see Xi'an
Siang see Brahmaputra
Siangtan see Xiangtan
Šiauliai 106 B4 Ger. Schaulen. Šiauliai,
  N Lithuania
Siazan' see Siyäzän
Sibay 111 D6 Respublika Bashkortostan,
  W Russian Federation
Šibenik 100 B4 It. Sebenico. Šibenik-
  Knin, S Croatia
Siberia 115 E3 var. Siberia. physical
  region NE Russian Federation
Siberia see Sibir'
Siberoet see Siberut, Pulau
Siberut, Pulau 138 A4 prev. Siberoet.
  island Kepulauan Mentawai,
  W Indonesia
Sibi 134 B2 Baluchistān, SW Pakistan
Sibir' 115 E3 var. Siberia. physical region
  NE Russian Federation
Sibiti 77 B6 Lékoumou, S Congo
Sibiu 108 B4 Ger. Hermannstadt, Hung.
  Nagyszeben. Sibiu, C Romania
Sibolga 138 B3 Sumatera, W Indonesia
Sibu 138 D3 Sarawak, East Malaysia
Sibut 76 C4 prev. Fort-Sibut. Kémo,
  S Central African Republic
Sibuyan Sea 139 E2 sea
  W Pacific Ocean
Sichon 137 C6 var. Ban Sichon, Si Chon.
  Nakhon Si Thammarat, SW Thailand
Si Chon see Sichon
Sichuan 128 B5 var. Chuan, Sichuan
  Sheng, Ssu-ch'uan, Szechuan,
  Szechwan. province C China
Sichuan Pendi 128 B5 basin C China
Sichuan Sheng see Sichuan
Sicilian Channel see Sicily, Strait of
Sicily 97 C7 Eng. Sicily; anc. Trinacria.
  island Italy, C Mediterranean Sea
Sicily, Strait of 97 B7 var. Sicilian
  Channel. strait C Mediterranean Sea
Sicuani 61 E4 Cusco, S Peru
Sidári 104 A4 Kérkyra, Iónia Nisiá,
  Greece, C Mediterranean Sea
Sidas 138 C4 Borneo, C Indonesia

derno 97 D7 Calabria, SW Italy
dhirókastron see Sidirókastro
di Barrāni 72 A1 NW Egypt
di Bel Abbès 70 D2 var. Sidi bel Abbès,
  Sidi-Bel-Abbès. NW Algeria
dirókastro 104 C3 prev.
  Sidhirókastron. Kentrikí Makedonía,
  NE Greece
dley, Mount 154 B4 mountain
  Antarctica
dney 44 D1 Montana, NW USA
dney 44 B4 Nebraska, C USA
dney 40 C4 Ohio, N USA
don see Saïda
dra see Surt
dra/Sidra, Gulf of see Surt, Khalij,
  N Libya
debenbürgen see Transylvania
edlce 98 E3 Ger. Sedlez, Rus. Sesdlets.
  Mazowieckie, C Poland
egen 94 B4 Nordrhein-Westfalen,
  W Germany
emiatycze 98 E3 Podlaskie, NE Poland
aena 96 B3 Fr. Sienne; anc. Saena Julia.
  Toscana, C Italy
aenne see Siena
aeradz 98 C4 Sieradz, C Poland
aerpc 98 D3 Mazowieckie, C Poland
.erra Leone 74 C4 off. Republic of Sierra
  Leone. country W Africa

SIERRA LEONE
West Africa

Official name Republic of Sierra Leone
Formation 1961 / 1961
Capital Freetown
Population 6.1 million / 221 people
per sq mile (85 people per sq km)
Total area 27,698 sq. miles (71,740 sq. km)
Languages Mende, Temne, Krio, English*
Religions Muslim 60%, Christian 30%,
Traditional beliefs 10%
Ethnic mix Mende 35%, Temne 32%,
Other 21%, Limba 8%, Kuranko 4%
Government Presidential system
Currency Leone = 100 cents
Literacy rate 44%
Calorie consumption 2333 kilocalories

ierra Leone Basin 66 C4 undersea basin
  E Atlantic Ocean
ierra Leone, Republic of see Sierra
  Leone
ierra Leone Ridge see Sierra Leone Rise
ierra Leone Rise 66 C4 var. Sierra
  Leone Ridge, Sierra Leone Schwelle.
  undersea rise E Atlantic Ocean
ierra Leone Schwelle see Sierra
  Leone Rise
ierra Vista 48 B3 Arizona, SW USA
ífnos 105 C6 anc. Siphnos. island
  Kykládes, Greece, Aegean Sea
igli 138 A3 Sumatera, W Indonesia
iglufjörður 83 E4 Nordhurland Vestra,
  N Iceland
ignal Peak 48 A2 mountain Arizona,
  SW USA
ignan see Xi'an
igny 154 A2 UK research station South
  Orkney Islands, Antarctica
iguatepeque 52 C2 Comayagua,
  W Honduras
iguiri 74 D4 NE Guinea
ihanoukville 137 D6 var. Kâmpông
  Saôm; prev. Kompong Som. Kâmpóng
  Saôm, SW Cambodia
iilinjärvi 84 E4 Itä-Suomi, C Finland
iirt 117 F4 var. Sert; anc. Tigranocerta.
  Siirt, SE Turkey
ikandarabad see Secunderábåd
ikasso 74 D4 Sikasso, S Mali
ikeston 45 H5 Missouri, C USA
ikhote-Alin', Khrebet 115 G4
  mountain range SE Russian Federation
iking see Xi'an
iklós 99 C7 Baranya, SW Hungary
ikoku see Shikoku
ikoku Basin see Shikoku Basin
ilalé 106 B4 Tauragė, W Lithuania
ilchar 135 G3 Assam, NE India
ilesia 98 B4 physical region SW Poland
ilifke 116 C4 anc. Seleucia. Içel,
  S Turkey
iliguri see Shiliguri
iling Co 126 C5 lake W China
ilinhot see Xilinhot

Silistra 104 E1 var. Silistria; anc.
  Durostorum. Silistra, NE Bulgaria
Silistria see Silistra
Sillamäe 106 E2 Ger. Sillamäggi. Ida-
  Virumaa, NE Estonia
Sillamäggi see Sillamäe
Sillein see Žilina
Šilutė 106 B4 Ger. Heydekrug. Klaipėda,
  W Lithuania
Silvan 117 E4 Diyarbakır, SE Turkey
Silva Porto see Kuito
Silver State see Colorado
Silver State see Nevada
Simanichy 107 C7 Rus. Simonichi.
  Homyel'skaya Voblasts', SE Belarus
Simav 116 B3 Kütahya, W Turkey
Simav Çayı 116 A3 river NW Turkey
Simbirsk see Ul'yanovsk
Simeto 97 C7 river Sicilia, Italy,
  C Mediterranean Sea
Simeulue, Pulau 138 A3 island
  NW Indonesia
Simferopol' 109 F5 Avtonomna
  Respublika Krym, S Ukraine
Simitla 104 C3 Blagoevgrad, SW Bulgaria
Şimlăul Silvaniei/Şimleul Silvaniei see
  Şimleu Silvaniei
Şimleu Silvaniei 108 B3 Hung.
  Szilágysomlyó; prev. Şimlăul Silvaniei,
  Şimleul Silvaniei. Sălaj, NW Romania
Simonichi see Simanichy
Simonoseki see Shimonoseki
Simpelveld 87 D6 Limburg,
  SE Netherlands
Simplon Pass 95 B8 pass S Switzerland
Simpson see Fort Simpson
Simpson Desert 148 B4 desert Northern
  Territory/South Australia
Sinai 72 C2 var. Sinai Peninsula, Ar.
  Shibh Jazīrat Sīnā', Sīnā. physical
  region NE Egypt
Sinaia 108 C4 Prahova, SE Romania
Sinano Gawa see Shinano-gawa
Sinā/Sinai Peninsula see Sinai
Sincelejo 58 B2 Sucre, NW Colombia
Sind see Sindh
Sindelfingen 95 B6 Baden-Württemberg,
  SW Germany
Sindh 134 B3 prev. Sind. province
  SE Pakistan
Sindi 106 D2 Ger. Zintenhof. Pärnumaa,
  SW Estonia
Sines 92 B4 Setúbal, S Portugal
Singan see Xi'an
Singapore 138 B3 country capital
  (Singapore) (Singapore) S Singapore
Singapore 138 A1 off. Republic of
  Singapore. country SE Asia

SINGAPORE
Southeast Asia

Official name Republic of Singapore
Formation 1965 / 1965
Capital Singapore
Population 5.4 million / 22,881 people
per sq mile (8852 people per sq km)
Total area 250 sq. miles (648 sq. km)
Languages Mandarin*, Malay*,
Tamil*, English*
Religions Buddhist 55%, Taoist 22%,
Muslim 16%, Hindu, Christian, and Sikh 7%
Ethnic mix Chinese 74%, Malay 14%,
Indian 9%, Other 3%
Government Parliamentary system
Currency Singapore dollar = 100 cents
Literacy rate 93%
Calorie consumption Not available

Singapore, Republic of see Singapore
Singen 95 B6 Baden-Württemberg,
  S Germany
Singida 73 C7 Singida, C Tanzania
Singkang 139 E4 Sulawesi, C Indonesia
Singkawang 138 C3 Borneo, C Indonesia
Singora see Songkhla
Singū see Shingū
Sining see Xining
Siniscola 97 A5 Sardegna, Italy,
  C Mediterranean Sea
Sinj 100 B4 Split-Dalmacija, SE Croatia
Sinkiang/Sinkiang Uighur
  Autonomous Region see Xinjiang
  Uygur Zizhiqu
Sinnamarie see Sinnamary
Sinnamary 59 H3 var. Sinnamarie.
  N French Guiana

Sinneh see Sanandaj
Sînnicolau Mare see Sânnicolau Mare
Sinoe, Lacul see Sinoie, Lacul
Sinoie, Lacul 108 D5 prev. Lacul Sinoe.
  lagoon SE Romania
Sinop 116 D2 anc. Sinope. Sinop, Turkey
Sinope see Sinop
Sinsheim 95 B6 Baden-Württemberg,
  SW Germany
Sint Maarten 55 G3 Eng. Saint Martin.
  island Lesser Antilles
Sint-Michielsgestel 86 C4 Noord-
  Brabant, S Netherlands
Sin-Miclăuş see Gheorgheni
Sint-Niklaas 87 B5 Fr. Saint-Nicolas.
  Oost-Vlaanderen, N Belgium
Sint-Pieters-Leeuw 87 B6 Vlaams
  Brabant, C Belgium
Sintra 92 B3 prev. Cintra. Lisboa,
  W Portugal
Sinujiif 73 E5 Nugaal, NE Somalia
Sinus Aelaniticus see Aqaba, Gulf of
Sinus Gallicus see Lion, Golfe du
Sinyang see Xinyang
Sinzyô see Shinjō
Sion 95 A7 Ger. Sitten; anc. Sedunum.
  Valais, SW Switzerland
Sioux City 45 F3 Iowa, C USA
Sioux Falls 45 F3 South Dakota, N USA
Sioux State see North Dakota
Siphnos see Sífnos
Siping 128 D3 var. Ssu-p'ing, Szeping;
  prev. Ssu-p'ing-chieh. Jilin, NE China
Siple, Mount 154 A4 mountain Siple
  Island, Antarctica
Siquirres 53 E4 Limón, E Costa Rica
Siracusa 97 D7 Eng. Syracuse. Sicilia,
  Italy, C Mediterranean Sea
Sir Edward Pellew Group 148 B2 island
  group Northern Territory, NE Australia
Siret 108 C3 var. Siretul, Ger. Sereth, Rus.
  Seret. river Romania/Ukraine
Siretul see Siret
Siria see Syria
Sirikit Reservoir 136 C4 lake N Thailand
Sīrjān 120 D4 prev. Sa'īdābād. Kermān,
  S Iran
Sirna see Sýrna
Şırnak 117 F4 Şırnak, SE Turkey
Síros see Sýros
Sirte see Surt
Sirte, Gulf of 71 F2 Eng. Gulf of Sidra,
  Gulf of Sirti, Sidra. gulf N Libya
Sirti, Gulf of see Surt, Khalij
Sisak 100 B3 var. Siscia, Ger. Sissek,
  Hung. Sziszek; anc. Segestica. Sisak-
  Moslavina, C Croatia
Siscia see Sisak
Sisimiut 82 C3 var. Holsteinborg,
  Holsteinsborg, Holstenborg,
  Holstensborg. Kitaa, S Greenland
Sissek see Sisak
Sīstān, Daryācheh-ye see Şāberī,
  Hāmūn-e
Sitaş Cristuru see Cristuru Secuiesc
Siteía 105 D8 var. Sitía. Kríti, Greece,
  E Mediterranean Sea
Sitges 93 G2 Cataluña, NE Spain
Sitía see Siteía
Sittang see Sittoung
Sittard 87 D5 Limburg, SE Netherlands
Sitten see Sion
Sittoung 136 B4 var. Sittang. river
  S Myanmar (Burma)
Sittwe 136 A3 var. Akyab. Rakhine State,
  W Myanmar (Burma)
Siuna 52 D3 Región Autónoma Atlántico
  Norte, NE Nicaragua
Siut see Asyūt
Sivas 116 D3 anc. Sebastia, Sebaste.
  Sivas, C Turkey
Siverek 117 F4 Şanlıurfa, S Turkey
Siwa see Sīwah
Sīwah 72 A2 var. Siwa. NW Egypt
Six Counties, The see Northern Ireland
Six-Fours-les-Plages 91 D6 Var,
  SE France
Siyäzän 117 H2 Rus. Siazan'.
  NE Azerbaijan
Sjar see Sääre
Sjenica 101 D5 Turk. Seniça. Serbia,
  SW Serbia
Skadar see Shkodër
Skadarsko Jezero see Scutari, Lake
Skagerak see Skagerrak
Skagerrak 85 A6 var. Skagerak. channel
  N Europe

Skagit River 46 B1 river Washington,
  NW USA
Skalka 84 C3 Lapp. Skalkká. lake
  N Sweden
Skalkká see Skalka
Skarżysko-Kamienna 98 D4
  Świętokrzyskie, C Poland
Skaudvilė 106 B4 Tauragė, SW Lithuania
Skegness 89 E6 E England, United
  Kingdom
Skellefteå 84 D4 Västerbotten, N Sweden
Skellefteälven 84 C4 river N Sweden
Ski 85 B6 Akershus, S Norway
Skiathos 105 C5 Skíathos, Greece, Aegean Sea
Skídal' 107 B5 Rus. Skidel'.
  Hrodzyenskaya Voblasts', W Belarus
Skidel' see Skidal'
Skiermûntseach see Schiermonnikoog
Skierniewice 98 D3 Łódzkie,
  C Poland
Skiftet 106 C1 strait Finland Atlantic
  Ocean Baltic Sea Gulf of Bothnia/Gulf
  of Finland
Skíros see Skýros
Skópelos 105 C5 Skópelos, Vóreies
  Sporádes, Greece, Aegean Sea
Skopje 101 D6 var. Üsküb, Turk. Üsküp;
  prev. Skoplje; anc. Scupi. country
  capital (Macedonia) N FYR Macedonia
Skoplje see Skopje
Skovorodino 115 F4 Amurskaya Oblast',
  SE Russian Federation
Skudnesfjorden 85 A6 fjord S Norway
Skuodas 106 B3 Ger. Schoden, Pol.
  Szkudy. Klaipėda, NW Lithuania
Skye, Isle of 88 B3 island NW Scotland,
  United Kingdom
Skylge see Terschelling
Skýros 105 C5 var. Skíros. Skńyros,
  Vóreies Sporádes, Greece, Aegean Sea
Skýros 105 C5 var. Skyros; anc. Scyros.
  island Vóreies Sporádes, Greece,
  Aegean Sea
Slagelse 85 B7 Sjælland, E Denmark
Slatina 100 C3 Hung. Szlatina; prev.
  Podravska Slatina. Virovitica-
  Podravina, NE Croatia
Slatina 108 B5 Olt, S Romania
Slavgorod see Slawharad
Slavonski Brod 100 C3 Ger. Brod, Hung.
  Bród; prev. Brod, Brod na Savi. Brod-
  Posavina, NE Croatia
Slavuta 108 C2 Khmel'nyts'ka Oblast',
  NW Ukraine
Slavyansk see Slov"yans'k
Slawharad 107 E7 Rus. Slavgorod.
  Mahilyowskaya Voblasts', E Belarus
Sławno 98 C2 Zachodnio-pomorskie,
  NW Poland
Slēmānī see As Sulaymānīyah
Sliema 102 B5 N Malta
Sligo 89 A5 Ir. Sligeach. Sligo,
  NW Ireland
Sliven 104 D2 var. Slivno. Sliven,
  C Bulgaria
Slivnitsa 104 B2 Sofiya, W Bulgaria
Slivno see Sliven
Slobozia 108 C5 Ialomiţa, SE Romania
Slonim 107 B6 Pol. Słonim.
  Hrodzyenskaya Voblasts', W Belarus
Słonim see Slonim
Slovakia 99 C6 off. Slovak Republika,
  Ger. Slowakei, Hung. Szlovákia, Slvk.
  Slovensko. country C Europe

SLOVAKIA
Central Europe

Official name Slovak Republic
Formation 1993 / 1993
Capital Bratislava
Population 5.5 million / 290 people
per sq mile (112 people per sq km)
Total area 18,859 sq. miles (48,845 sq. km)
Languages Slovak*, Hungarian
(Magyar), Czech
Religions Roman Catholic 69%,
Nonreligious 13%, Other 13%,
Greek Catholic (Uniate) 4%,
Orthodox Christian 1%
Ethnic mix Slovak 86%, Magyar 10%,
Roma 2%, Czech 1%, Other 1%
Government Parliamentary system
Currency Euro = 100 cents
Literacy rate 99%
Calorie consumption 2902 kilocalories

Slovak Ore Mountains *see* Slovenské rudohorie
Slovenia *95 D8 off.* Republic of Slovenia, *Ger.* Slowenien, *Slvn.* Slovenija. *country* SE Europe

SLOVENIA
Central Europe

Official name Republic of Slovenia
Formation 1991 / 1991
Capital Ljubljana
Population 2 million / 269 people per sq mile (104 people per sq km)
Total area 7820 sq. miles (20,253 sq. km)
Languages Slovenian*
Religions Roman Catholic 58%, Other 28%, Atheist 10%, Orthodox Christian 2%, Muslim 2%
Ethnic mix Slovene 83%, Other 12%, Serb 2%, Croat 2%, Bosniak 1%
Government Parliamentary system
Currency Euro = 100 cents
Literacy rate 99%
Calorie consumption 3173 kilocalories

Slovenia, Republic of *see* Slovenia
Slovenija *see* Slovenia
Slovenská Republika *see* Slovakia
Slovenské rudohorie *99 D6 Eng.* Slovak Ore Mountains, *Ger.* Slowakisches Erzgebirge, *Ungarisches Erzgebirge. mountain range* C Slovakia
Slovensko *see* Slovakia
Slov"yans'k *109 G3 Rus.* Slavyansk. Donets'ka Oblast', E Ukraine
Slowakei *see* Slovakia
Slowakisches Erzgebirge *see* Slovenské rudohorie
Slowenien *see* Slovenia
Slubice *98 B3 Ger.* Frankfurt. Lubuskie, W Poland
Sluch *108 D1 river* NW Ukraine
Słupsk *98 C2 Ger.* Stolp. Pomorskie, N Poland
Slutsk *107 C6* Minskaya Voblasts', S Belarus
Smallwood Reservoir *39 F2 lake* Newfoundland and Labrador, S Canada
Smara *70 B3 var.* Es Semara. N Western Sahara
Smarhon' *107 C5 Pol.* Smorgonie, *Rus.* Smorgon'. Hrodzyenskaya Voblasts', W Belarus
Smederevo *100 D4 Ger.* Semendria. Serbia, N Serbia
Smederevska Palanka *100 D4* Serbia, C Serbia
Smela *see* Smila
Smila *109 E2 Rus.* Smela. Cherkas'ka Oblast', C Ukraine
Smilten *see* Smiltene
Smiltene *106 D3 Ger.* Smilten. N Latvia
Smøla *84 A4 island* W Norway
Smolensk *111 A5* Smolenskaya Oblast', W Russian Federation
Smorgon'/Smorgonie *see* Smarhon'
Smyrna *see* İzmir
Snake *34 B4 river* Yukon, NW Canada
Snake River *46 C3 river* NW USA
Snake River Plain *46 D4 plain* Idaho, NW USA
Sneek *86 D2 Fris.* Snits. Fryslân, N Netherlands
Sneeuw-gebergte *see* Maoke, Pegunungan
Snežka *98 B4 Ger.* Schneekoppe, *Pol.* Śnieżka. *mountain* N Czech Republic/Poland
Śniardwy, Jezioro *98 D2 Ger.* Spirdingsee. *lake* NE Poland
Sniečkus *see* Visaginas
Śnieżka *see* Snežka
Snina *99 E5 Hung.* Szinna. Prešovský Kraj, E Slovakia
Snits *see* Sneek
Snowdonia *89 C6 mountain range* NW Wales, United Kingdom
Snowdrift *see* Łutselk'e
Snow Mountains *see* Maoke, Pegunungan
Snyder *49 F3* Texas, SW USA
Sobradinho, Barragem de *see* Sobradinho, Represa de
Sobradinho, Represa de *63 F2 var.* Barragem de Sobradinho. *reservoir* E Brazil

Sochi *111 A7* Krasnodarskiy Kray, SW Russian Federation
Société, Îles de la/Society Islands *see* Société, Archipel de la
Society Islands *145 G4 var.* Archipel de Tahiti, Îles de la Société, *Eng.* Society Islands. *island group* W French Polynesia
Soconusco, Sierra de *see* Madre, Sierra
Socorro *48 D2* New Mexico, SW USA
Socorro, Isla *50 B5 island* W Mexico
Socotra *121 C7 var.* Sokotra, *Eng.* Socotra. *island* SE Yemen
Socotra *see* Suquṭrā
Soc Trăng *137 D6 var.* Khanh Hung. Soc Trăng, S Vietnam
Socuéllamos *93 E3* Castilla-La Mancha, C Spain
Sodankylä *84 D3* Lappi, N Finland
Sodari *see* Sodiri
Söderhamn *85 C5* Gävleborg, C Sweden
Södertälje *85 C6* Stockholm, C Sweden
Sodiri *72 B4 var.* Sawdirī, Sodari. Northern Kordofan, C Sudan
Soekaboemi *see* Sukabumi
Soemba *see* Sumba, Pulau
Soengaipenoeh *see* Sungaipenuh
Soerabaja *see* Surabaya
Soerakarta *see* Surakarta
Sofia *see* Sofiya
Sofiya *104 C2 var.* Sophia, *Eng.* Sofia, *Lat.* Serdica. *country capital* (Bulgaria) Sofiya-Grad, W Bulgaria
Sogamoso *58 B3* Boyacá, C Colombia
Sognefjorden *85 A5 fjord* NE North Sea
Sohâg *see* Sawhāj
Sohar *see* Şuḥār
Sohm Plain *66 B3 abyssal plain* NW Atlantic Ocean
Sohrau *see* Żory
Sokal' *108 C2 Rus.* Sokal. L'vivs'ka Oblast', NW Ukraine
Söke *116 A4* Aydın, SW Turkey
Sokhumi *117 E1 Rus.* Sukhumi. NW Georgia
Sokodé *75 F4* C Togo
Sokol *110 C4* Vologodskaya Oblast', NW Russian Federation
Sokółka *98 E3* Podlaskie, NE Poland
Sokolov *99 A5 Ger.* Falkenau an der Eger; *prev.* Falknov nad Ohří. Karlovarský Kraj, W Czech Republic
Sokone *74 B3* W Senegal
Sokoto *75 F3* Sokoto, NW Nigeria
Sokoto *75 F4 river* NW Nigeria
Sokotra *see* Suquṭrā
Solāpur *124 B3 var.* Sholāpur. Mahārāshtra, W India
Solca *108 C3 Ger.* Solka. Suceava, N Romania
Sol, Costa del *92 D5 coastal region* S Spain
Soldeu *91 B7* NE Andorra Europe
Solec Kujawski *98 C3* Kujawsko-pomorskie, C Poland
Soledad *58 B1* Atlántico, N Colombia
Isla Soledad *see* East Falkland
Soligorsk *see* Salihorsk
Solikamsk *114 C3* Permskiy Kray, NW Russian Federation
Sol'-Iletsk *111 D6* Orenburgskaya Oblast', W Russian Federation
Solingen *94 A4* Nordrhein-Westfalen, W Germany
Solka *see* Solca
Sollentuna *85 C6* Stockholm, C Sweden
Solo *see* Surakarta
Solok *138 B4* Sumatera, W Indonesia
Solomon Islands *144 C3 prev.* British Solomon Islands Protectorate. *country* W Solomon Islands N Melanesia W Pacific Ocean

SOLOMON ISLANDS
Australasia & Oceania

Official name Solomon Islands
Formation 1978 / 1978
Capital Honiara
Population 600,000 / 56 people per sq mile (21 people per sq km)
Total area 10,985 sq. miles (28,450 sq. km)
Languages English*, Pidgin English, Melanesian Pidgin, around 120 native languages
Religions Church of Melanesia 34%,

SOLOMON ISLANDS
(continued)

Roman Catholic 19%, South Seas Evangelical Church 17%, Methodist 11%, Seventh-day Adventist 10%, Other 9%
Ethnic mix Melanesian 93%, Polynesian 4%, Micronesian 2%, Other 1%
Government Parliamentary system
Currency Solomon Islands dollar = 100 cents
Literacy rate 77%
Calorie consumption 2473 kilocalories

Solomon Islands *144 C3 island group* Papua New Guinea/Solomon Islands
Solomon Sea *144 B3 sea* W Pacific Ocean
Soltau *94 B3* Niedersachsen, NW Germany
Sol'tsy *110 A4* Novgorodskaya Oblast', W Russian Federation
Solun *see* Thessaloníki
Solwezi *78 D2* North Western, NW Zambia
Sōma *130 D4* Fukushima, Honshū, C Japan
Somalia *73 D5 off.* Somali Democratic Republic, *Som.* Jamuuriyada Demuqraadiga Soomaaliyeed, Soomaaliya; *prev.* Italian Somaliland, Somaliland Protectorate. *country* E Africa

SOMALIA
East Africa

Official name Federal Republic of Somalia
Formation 1960 / 1960
Capital Mogadishu
Population 10.5 million / 43 people per sq mile (17 people per sq km)
Total area 246,199 sq. miles (637,657 sq. km)
Languages Somali*, Arabic*, English, Italian
Religions Sunni Muslim 99%, Christian 1%
Ethnic mix Somali 85%, Other 15%
Government Nonparty system
Currency Somali shilin = 100 senti
Literacy rate 24%
Calorie consumption 1696 kilocalories

Somali Basin *69 E5 undersea basin* W Indian Ocean
Somali Democratic Republic *see* Somalia
Somaliland *73 D5 disputed territory* N Somalia
Somaliland Protectorate *see* Somalia
Sombor *100 C3 Hung.* Zombor. Vojvodina, NW Serbia
Someren *87 D5* Noord-Brabant, SE Netherlands
Somerset *42 A5 var.* Somerset Village. W Bermuda
Somerset *40 C5* Kentucky, S USA
Somerset Island *42 A5 island* W Bermuda
Somerset Island *37 F2 island* Queen Elizabeth Islands, Nunavut, NW Canada
Somerset Village *see* Somerset
Somers Islands *see* Bermuda
Somerton *48 A2* Arizona, SW USA
Someş *108 B3 river* Hungary/Romania Europe
Somme *90 C2 river* N France
Sommerfeld *see* Lubsko
Somotillo *52 C3* Chinandega, NW Nicaragua
Somoto *52 D3* Madriz, NW Nicaragua
Songea *73 D4* Ruvuma, S Tanzania
Sŏngjin *see* Kimch'aek
Songkhla *137 C7 var.* Songkla, *Mal.* Singora. Songkhla, SW Thailand
Songkla *see* Songkhla
Sonoran Desert *48 A3 var.* Desierto de Altar. *desert* Mexico/USA
Sonsonate *52 B3* Sonsonate, W El Salvador
Soochow *see* Suzhou
Soomaaliya/Soomaaliyeed, Jamuuriyada Demuqraadiga *see* Somalia
Soome Laht *see* Finland, Gulf of
Sop Hao *136 D3* Houaphan, N Laos

Sophia *see* Sofiya
Sopianae *see* Pécs
Sopot *98 C2 Ger.* Zoppot. Pomorskie, N Poland
Sopron *99 B6 Ger.* Ödenburg. Győr-Moson-Sopron, NW Hungary
Sorau/Sorau in der Niederlausitz *see* Żary
Sorgues *91 D6* Vaucluse, SE France
Sorgun *116 B3* Yozgat, C Turkey
Soria *93 E2* Castilla y León, N Spain
Soroca *108 D3 Rus.* Soroki. N Moldova
Sorochino *see* Sarochyna
Soroki *see* Soroca
Sorong *139 F4* Papua, E Indonesia
Sørøy *see* Sørøya
Sørøya *84 C2 var.* Sørøy, *Lapp.* Sállan. *island* N Norway
Sortavala *110 B3 prev.* Serdobol'. Respublika Kareliya, NW Russian Federation
Sotavento, Ilhas de *74 A3 var.* Leeward Islands. *island group* S Cape Verde
Sotkamo *84 E4* Oulu, C Finland
Souanké *77 B5* Sangha, NW Congo
Soueida *see* As Suwaydā'
Soufli *104 D3 prev.* Souflion. Anatolikí Makedonía kai Thráki, NE Greece
Souflion *see* Soufli
Soufrière *55 F2* W Saint Lucia
Soukhné *see* As Sukhnah
Sŏul *see* Seoul
Soûr *119 A5 var.* Şūr; *anc.* Tyre. SW Lebanon
Souris River *45 E1 var.* Mouse River. *river* Canada/USA
Soúrpi *105 B5* Thessalía, C Greece
Sousse *71 F2 var.* Sūsah. NE Tunisia
South Africa *78 C4 off.* Republic of South Africa, *Afr.* Suid-Afrika. *country* S Africa

SOUTH AFRICA
Southern Africa

Official name Republic of South Africa
Formation 1934 / 1994
Capital Pretoria; Cape Town; Bloemfontein
Population 52.8 million / 112 people per sq mile (43 people per sq km)
Total area 471,008 sq. miles (1,219,912 sq. km)
Languages English*, isiZulu*, isiXhosa*, Afrikaans*, Sepedi*, Setswana*, Sesotho*, Xitsonga*, siSwati*, Tshivenda*, isiNdebele*
Religions Christian 68%, Traditional beliefs and animist 29%, Muslim 2%, Hindu 1%
Ethnic mix Black 89%, White 9%, Asian 2%
Government Presidential system
Currency Rand = 100 cents
Literacy rate 94%
Calorie consumption 3007 kilocalories

South Africa, Republic of *see* South Africa
South America *56 continent*
Southampton *89 D7 hist.* Hamwih, *Lat.* Clausentum. S England, United Kingdom
Southampton Island *37 G3 island* Nunavut, NE Canada
South Andaman *133 F2 island* Andaman Islands, India, NE Indian Ocean
South Australia *149 A5 state* S Australia
South Australian Basin *142 B5 undersea basin* SW Indian Ocean
South Bend *40 C3* Indiana, N USA
South Beveland *see* Zuid-Beveland
South Bruny Island *149 C8 island* Tasmania, SE Australia
South Carolina *43 E2 off.* State of South Carolina, *also known as* The Palmetto State. *state* SE USA
South Carpathians *see* Carpaţii Meridionali
South China Basin *125 E4 undersea basin* SE South China Sea
South China Sea *125 E4 Chin.* Nan Hai, *Ind.* Laut Cina Selatan, *Vtn.* Biên Đông. *sea* SE Asia
South Dakota *44 D2 off.* State of South Dakota, *also known as* The Coyote State, Sunshine State. *state* N USA

theast Indian Ridge 141 D7
ndersea ridge Indian Ocean/Pacific
Ocean
theast Pacific Basin 153 E5 *var.*
elling Hausen Mulde. *undersea basin*
E Pacific Ocean
th East Point 149 C7 *headland*
Victoria, S Australia
thend-on-Sea 89 E6 E England,
South Island, New Zealand
thern Alps 151 B6 *mountain range*
South Island, New Zealand
thern Cook Islands 145 F4 *island*
roup S Cook Islands
thern Cross 147 B6 Western
Australia
thern Indian Lake 37 F4 *lake*
Manitoba, C Canada
thern Ocean 67 B7 *ocean* Atlantic
Ocean/Indian Ocean/Pacific Ocean
thern Uplands 88 C4 *mountain*
range S Scotland, United Kingdom
th Fiji Basin 142 D4 *undersea basin*
S Pacific Ocean
th Geomagnetic Pole 154 B3 *pole*
Antarctica
th Georgia 57 D8 *island* South
Georgia and the South Sandwich
Islands, SW Atlantic Ocean
th Goulburn Island 146 E2 *island*
Northern Territory, N Australia
th Huvadhu Atoll 132 A5 *atoll*
S Maldives
th Indian Basin 141 D7 *undersea*
basin Indian Ocean/Pacific Ocean
th Island 151 C6 *island* S New
Zealand
th Korea 129 E4 *off.* Republic of
Korea, *Kor.* Taehan Min'guk. *country*
E Asia

OUTH KOREA
East Asia

fficial name Republic of Korea
rmation 1948 / 1953
apital Seoul; Sejong City
pulation 49.3 million / 1293 people
per sq mile (499 people per sq km)
tal area 38,023 sq. miles (98,480 sq. km)
anguages Korean*
ligions Mahayana Buddhist 47%,
otestant 38%, Roman Catholic 11%,
onfucianist 3%, Other 1%
hnic mix Korean 100%
overnment Presidential system
urrency South Korean won = 100 chon
teracy rate 99%
alorie consumption 3329 kilocalories

OUTH SUDAN
East Africa

fficial name Republic of South Sudan
rmation 2011 / 2011
apital Juba
pulation 11.3 million / 45 people
per sq mile (18 people per sq km)
tal area 248,777 sq. miles
44,329 sq. km)
anguages Arabic, Dinka, Nuer,
ande, Bari, Shilluk,
rituko, English*

SOUTH SUDAN
(continued)

Religions Over 50% Christian beliefs
Ethnic mix Dinka 40%, Nuer 15%,
Bari 10%, Shilluk/Anwak 10%,
Azande 10%, Arab 10%, Other 5%
Government Transitional regime
Currency South Sudan pound = 100
piastres
Literacy rate 37%
Calorie consumption Not available

South Taranaki Bight 150 C4 *bight*
SE Tasman Sea
South Tasmania Plateau *see* Tasman
Plateau
South Uist 88 B3 *island* NW Scotland,
United Kingdom
South-West Africa/South West Africa
see Namibia
South West Cape 151 A8 *headland*
Stewart Island, New Zealand
Southwest Indian Ocean Ridge *see*
Southwest Indian Ridge
Southwest Indian Ridge 141 B6 *var.*
Southwest Indian Ocean Ridge.
undersea ridge SW Indian Ocean
Southwest Pacific Basin 143 E4 *var.*
South Pacific Basin. *undersea basin*
SE Pacific Ocean
Sovereign Base Area 102 C5 *uk military*
installation S Cyprus
Soweto 78 D4 Gauteng,
NE South Africa
Sōya-kaikyō *see* La Pérouse Strait
Spain 92 D3 *off.* Kingdom of Spain, *Sp.*
España; *anc.* Hispania, Iberia, *Lat.*
Hispana. *country* SW Europe

SPAIN
Southwest Europe

Official name Kingdom of Spain
Formation 1492 / 1713
Capital Madrid
Population 46.9 million / 243 people per
sq mile (94 people per sq km)
Total area 194,896 sq. miles
(2504,782 sq. km)
Languages Spanish*, Catalan*,
Galician*, Basque*
Religions Roman Catholic 96%, Other 4%
Ethnic mix Castilian Spanish 72%,
Catalan 17%, Galician 6%, Basque 2%,
Other 2%, Roma 1%
Government Parliamentary system
Currency Euro = 100 cents
Literacy rate 98%
Calorie consumption 3183 kilocalories

Spain, Kingdom of *see* Spain
Spalato *see* Split
Spanish Town 54 B5 *hist.* St.Iago de la
Vega. C Jamaica
Sparks 47 C5 Nevada, W USA
Sparta *see* Spárti
Spartanburg 43 E1 South Carolina,
SE USA
Spárti 105 B6 *Eng.* Sparta. Pelopónnisos,
S Greece
Spearfish 44 D2 South Dakota,
N USA
Speightstown 55 G1 NW Barbados
Spencer 45 F3 Iowa, C USA
Spencer Gulf 149 B6 *gulf* South Australia
Spey 88 C3 *river* NE Scotland, United
Kingdom
Spice Islands *see* Maluku
Spiess Seamount 67 C7 *seamount*
S Atlantic Ocean
Spijkenisse 86 B4 Zuid-Holland,
SW Netherlands
Spili 105 C8 Kríti, Greece,
E Mediterranean Sea
Spin Böldak 123 E5 *prev.* Spin Buldak.
Kandahār, S Afghanistan
Spin Buldak *see* Spin Böldak
Spirdingsee *see* Śniardwy, Jezioro
Spitsbergen 83 F2 *island* NW Svalbard
Split 100 B4 *It.* Spalato. Split-Dalmacija,
S Croatia
Spoği 106 D4 SE Latvia
Spokane 46 C2 Washington, NW USA
Spratly Islands 138 B2 *Chin.* Nansha
Qundao. *disputed territory* SE Asia
Spree 94 D4 *river* E Germany

Springfield 40 B4 *state capital* Illinois,
N USA
Springfield 41 G3 Massachusetts,
NE USA
Springfield 45 G5 Missouri, C USA
Springfield 40 C4 Ohio, N USA
Springfield 46 B3 Oregon, NW USA
Spring Garden 59 F2 NE Guyana
Spring Hill 43 E4 Florida, SE USA
Springs Junction 151 C5 West Coast,
South Island, New Zealand
Springsure 148 D4 Queensland,
E Australia
Sprottau *see* Szprotawa
Spruce Knob 41 E4 *mountain* West
Virginia, NE USA
Srbinje *see* Foča
Srbobran 100 D3 *var.* Bácsszenttamás,
Hung. Szenttamás. Vojvodina,
N Serbia
Srebrenica 100 C4 Republika Srpska,
E Bosnia and Herzegovina
Sredets 104 D2 *prev.* Syulemeshlii. Stara
Zagora, C Bulgaria
Sredets 104 E2 *prev.* Grudovo. Burgas,
E Bulgaria
Sremska Mitrovica 100 C3 *prev.*
Mitrovica, *Ger.* Mitrowitz. Vojvodina,
NW Serbia
Srepok, Sông *see* Srêpôk, Tônle
Srêpôk, Tônle 137 E5 *var.* Sông Srepok.
river Cambodia/Vietnam
Sri Aman 138 C3 Sarawak, East Malaysia
Sri Jayewardenapura Kotte 132 D3
administrative capital (Sri Lanka)
var. Sri Jayawardanapura. Western
Province, W Sri Lanka
Srīkākulam 135 F5 Andhra Pradesh,
E India
Sri Lanka 132 D3 *off.* Democratic
Socialist Republic of Sri Lanka; *prev.*
Ceylon. *country* S Asia

SRI LANKA
South Asia

Official name Democratic Socialist
Republic of Sri Lanka
Formation 1948 / 1948
Capital Colombo;
Sri Jayewardenapura Kotte
Population 21.3 million / 852 people
per sq mile (329 people per sq km)
Total area 25,332 sq. miles (65,610 sq. km)
Languages Sinhala*, Tamil*,
Sinhala-Tamil, English
Religions Buddhist 69%, Hindu 15%,
Muslim 8%, Christian 8%
Ethnic mix Sinhalese 74%, Tamil 18%,
Moor 7%, Other 1%
Government Presidential /
parliamentary system
Currency Sri Lanka rupee = 100 cents
Literacy rate 91%
Calorie consumption 2488 kilocalories

Sri Lanka, Democratic Socialist
Republic of *see* Sri Lanka
Srinagarind Reservoir 137 C5 *lake*
W Thailand
Srpska, Republika 100 B3 *republic*
Bosnia and Herzegovina
Ssu-ch'uan *see* Sichuan
Ssu-p'ing/Ssu-p'ing-chieh *see* Siping
Stabroek 87 B5 Antwerpen,
N Belgium
Stade 94 B3 Niedersachsen,
NW Germany
Stadskanaal 86 E2 Groningen,
NE Netherlands
Stafford 89 D6 C England, United
Kingdom
Staicele 106 D3 N Latvia
Staierdorf-Anina *see* Anina
Stájerlakanina *see* Anina
Stakhanov 109 H3 Luhans'ka Oblast',
E Ukraine
Stalin *see* Varna
Stalinabad *see* Dushanbe
Stalingrad *see* Volgograd
Stalino *see* Donets'k
Stalinobod *see* Dushanbe
Stalinsk *see* Novokuznetsk
Stalinski Zaliv *see* Varnenski Zaliv
Stalin, Yazovir *see* Iskŭr, Yazovir
Stalowa Wola 98 E4 Podkarpackie,
SE Poland

Stamford 41 F3 Connecticut, NE USA
Stampalia *see* Astypálaia
Stanislau *see* Ivano-Frankivs'k
Stanislav *see* Ivano-Frankivs'k
Stanisławów *see* Ivano-Frankivs'k
Stanke Dimitrov *see* Dupnitsa
Stanley 65 D7 *var.* Port Stanley, Puerto
Argentino. *dependent territory capital*
(Falkland Islands) East Falkland,
Falkland Islands
Stanleyville *see* Kisangani
Stann Creek *see* Dangriga
Stanovoy Khrebet 113 E3 *mountain*
range SE Russian Federation
Stanthorpe 149 D5 Queensland,
E Australia
Staphorst 86 D2 Overijssel,
E Netherlands
Starachowice 98 D4 Świętokrzyskie,
C Poland
Stara Kanjiža *see* Kanjiža
Stara Pazova 100 D3 *Ger.* Altpasua,
Hung. Ópazova. Vojvodina, N Serbia
Stara Planina *see* Balkan Mountains
Stara Zagora 104 D2 *Lat.* Augusta
Trajana. Stara Zagora, C Bulgaria
Starbuck Island 145 G5 *prev.* Volunteer
Island. *island* E Kiribati
Stargard in Pommern *see* Stargard
Szczeciński
Stargard Szczeciński 98 B3 *Ger.*
Stargard in Pommern. Zachodnio-
pomorskie, NW Poland
Stari Bečej *see* Bečej
Starobel'sk *see* Starobil's'k
Starobil's'k 109 H2 *Rus.* Starobel'sk.
Luhans'ka Oblast', E Ukraine
Starobin 107 C7 *var.* Starobyn.
Minskaya Voblasts', S Belarus
Starobyn *see* Starobin
Starogard Gdański 98 C2 *Ger.*
Preussisch-Stargard. Pomorskie,
N Poland
Starokonstantinov *see* Starokostyantyniv
Starokostyantyniv 108 D2 *Rus.*
Starokonstantinov. Khmel'nyts'ka
Oblast', NW Ukraine
Starominskaya 111 A7 Krasnodarskiy
Kray, SW Russian Federation
Staryya Darohi 107 C6 *Rus.* Staryye
Dorogi. Minskaya Voblasts', S Belarus
Staryye Dorogi *see* Staryya Darohi
Staryy Oskol 111 B6 Belgorodskaya
Oblast', W Russian Federation
State College 41 E4 Pennsylvania,
NE USA
Staten Island *see* Estados, Isla de los
Statesboro 43 E2 Georgia, SE USA
States, The *see* United States of America
Staunton 41 E5 Virginia, NE USA
Stavanger 85 A6 Rogaland, S Norway
Stavers Island *see* Vostok Island
Stavropol' 111 B7 *prev.* Voroshilovsk.
Stavropol'skiy Kray, SW Russian
Federation
Stavropol' *see* Tol'yatti
Steamboat Springs 44 C4 Colorado,
C USA
Steenwijk 86 D2 Overijssel,
N Netherlands
Steier *see* Steyr
Steierdorf/Steierdorf-Anina *see* Anina
Steinamanger *see* Szombathely
Steinkjer 84 B4 Nord-Trøndelag,
C Norway
Stejarul *see* Karapelit
Stendal 94 C3 Sachsen-Anhalt,
C Germany
Stepanakert *see* Xankändi
Stephenville 49 F3 Texas, SW USA
Sterling 44 D4 Colorado, C USA
Sterling 40 B3 Illinois, N USA
Sterlitamak 114 B3 Respublika
Bashkortostan, W Russian Federation
Stettin *see* Szczecin
Stettiner Haff *see* Szczeciński, Zalew
Stevenage 89 E6 E England, United
Kingdom
Stevens Point 40 B2 Wisconsin, N USA
Stewart Island 151 A8 *island* S New
Zealand
Steyerlak-Anina *see* Anina
Steyr 95 D6 *var.* Steier. Oberösterreich,
N Austria
St. Helena Bay 78 B5 *bay* SW South
Africa
Stif *see* Sétif

Stillwater *49 G1* Oklahoma, C USA
Štip *101 E6* E FYR Macedonia
Stirling *88 C4* C Scotland, United Kingdom
Stjørdalshalsen *84 B4* Nord-Trøndelag, C Norway
St-Maur-des-Fossés *90 E2* Val-de-Marne, Île-de-France, N France Europe
Stockach *95 B6* Baden-Württemberg, S Germany
Stockholm *85 C6 country capital* (Sweden) Stockholm, C Sweden
Stockmannshof *see* Pļaviņas
Stockton *47 B6* California, W USA
Stockton Plateau *49 E4 plain* Texas, SW USA
Stŏeng Trêng *137 D5 prev.* Stung Treng. Stŏeng Trêng, NE Cambodia
Stoke *see* Stoke-on-Trent
Stoke-on-Trent *89 D6 var.* Stoke. C England, United Kingdom
Stolbce *see* Stowbtsy
Stolbtsy *see* Stowbtsy
Stolp *see* Słupsk
Stolpmünde *see* Ustka
Stómio *104 B4* Thessalía, C Greece
Store Bælt *see* Storebælt
Storebælt *see* Storebælt
Støren *85 B5* Sør-Trøndelag, S Norway
Storfjorden *83 G2 fjord* S Norway
Storhammer *see* Hamar
Stornoway *88 B2* NW Scotland, United Kingdom
Storsjön *85 B5 lake* C Sweden
Storuman *84 C4* Västerbotten, N Sweden
Storuman *84 C4 lake* N Sweden
Stowbtsy *107 C6 Pol.* Stolbce, *Rus.* Stolbtsy. Minskaya Voblasts', C Belarus
Strabane *89 B5 Ir.* An Srath Bán. N Northern Ireland, United Kingdom
Strakonice *99 A5 Ger.* Strakonitz. Jihočeský Kraj, S Czech Republic
Strakonitz *see* Strakonice
Stralsund *94 D2* Mecklenburg-Vorpommern, NE Germany
Stranraer *89 C5* S Scotland, United Kingdom
Strasbourg *90 E3 Ger.* Strassburg; *anc.* Argentoratum. Bas-Rhin, NE France
Strǎşeni *108 D3 var.* Strasheny. C Moldova
Strasheny *see* Strǎşeni
Strassburg *see* Strasbourg, France
Strassburg *see* Aiud, Romania
Stratford *150 D4* Taranaki, North Island, New Zealand
Strathfield *148 E2* New South Wales, E Australia
Straubing *95 C6* Bayern, SE Germany
Strehaia *108 B5* Mehedinți, SW Romania
Strelka *114 D4* Krasnoyarskiy Kray, C Russian Federation
Strigonium *see* Esztergom
Strofilia *see* Strofyliá
Strofyliá *105 C5 var.* Strofilia. Évvoia, C Greece
Stromboli *97 D6 island* Isole Eolie, S Italy
Stromeferry *88 C3* N Scotland, United Kingdom
Strömstad *85 B6* Västra Götaland, S Sweden
Strömsund *84 C4* Jämtland, C Sweden
Struga *101 D6* SW FYR Macedonia
Struma *see* Strymónas
Strumica *101 E6* E FYR Macedonia
Strumyani *104 C3* Blagoevgrad, SW Bulgaria
Strymónas *104 C3 Bul.* Struma. *river* Bulgaria/Greece
Stryy *108 B2* L'vivs'ka Oblast', NW Ukraine
Studholme *151 B6* Canterbury, South Island, New Zealand
Stuhlweissenberg *see* Székesfehérvár
Stung Treng *see* Stŏeng Trêng
Sturgis *44 D3* South Dakota, N USA
Stuttgart *95 B6* Baden-Württemberg, SW Germany
Stykkishólmur *83 E4* Vesturland, W Iceland
Styr *108 C1 Rus.* Styr'. *river* Belarus/Ukraine
Su *see* Jiangsu
Suakin *72 C3 var.* Sawakin. Red Sea, NE Sudan

Subačius *106 C4* Panevėžys, NE Lithuania
Subaykhān *118 E3* Dayr az Zawr, E Syria
Subotica *100 D2 Ger.* Maria-Theresiopel, *Hung.* Szabadka. Vojvodina, N Serbia
Suceava *108 C3 Ger.* Suczawa, *Hung.* Szucsava. Suceava, NE Romania
Su-chou *see* Suzhou
Suchow *see* Xuzhou, Jiangsu, China
Suchow *see* Suzhou, Jiangsu, China
Sucker State *see* Illinois
Sucre *61 F4 hist.* Chuquisaca, La Plata. *country capital* (Bolivia-legal capital) Chuquisaca, S Bolivia
Suczawa *see* Suceava
Sudan *72 A4 off.* Republic of Sudan, *Ar.* Jumhuriyat as-Sudan; *prev.* Anglo-Egyptian Sudan. *country* N Africa

SUDAN
East Africa

Official name Republic of the Sudan
Formation 1956 / 2011
Capital Khartoum
Population 38 million / 53 people per sq mile (20 people per sq km)
Total area 718,722 sq. miles (1,861,481 sq. km)
Languages Arabic*, Nubian, Beja, Fur
Religions Nearly the whole population is Muslim (mainly Sunni)
Ethnic mix Arab 60%, Other 18%, Nubian 10%, Beja 8%, Fur 3%, Zaghawa 1%
Government Presidential system
Currency New Sudanese pound = 100 piastres
Literacy rate 73%
Calorie consumption 2346 kilocalories

Sudanese Republic *see* Mali
Sudan, Jumhuriyat as- *see* Sudan
Sudan, Republic of *see* Sudan
Sudbury *38 C4* Ontario, S Canada
Sudd *71 B5 swamp region* C South Sudan
Sudeten *98 B4 var.* Sudetes, Sudetic Mountains, *Cz./Pol.* Sudety. *mountain range* Czech Republic/Poland
Sudetes/Sudetic Mountains/Sudety *see* Sudeten
Südkarpaten *see* Carpații Meridionalii
Südliche Morava *see* Južna Morava
Sudong, Pulau *138 A2 island* SW Singapore Asia
Sudostroy *see* Severodvinsk
Sue *73 B5 river* W South Sudan
Sueca *93 F3* Valenciana, E Spain
Sue Wood Bay *42 B5 bay* W Bermuda North America W Atlantic Ocean
Suez *72 B1 Ar.* As Suways, El Suweis. NE Egypt
Suez Canal *72 B1 Ar.* Qanāt as Suways. *canal* NE Egypt
Suez, Gulf of *see* Khalij as Suways
Suğla Gölü *116 C4 lake* SW Turkey
Şuḩār *121 D5 var.* Sohar. N Oman
Sühbaatar *127 E1* Selenge, N Mongolia
Suhl *95 C5* Thüringen, C Germany
Suicheng *see* Suixi
Suid-Afrika *see* South Africa
Suixi *128 C6 var.* Suicheng. Guangdong, S China
Sujāwal *134 B3* Sind, SE Pakistan
Sukabumi *138 C5 prev.* Soekaboemi. Jawa, C Indonesia
Sukagawa *131 D5* Fukushima, Honshū, C Japan
Sukarnapura *see* Jayapura
Sukarno, Puntjak *see* Jaya, Puncak
Sukhne *see* As Sukhnah
Sukhona *110 C4 var.* Tot'ma. *river* NW Russian Federation
Sukhumi *see* Sokhumi
Sukkertoppen *see* Maniitsoq
Sukkur *134 B3* Sind, SE Pakistan
Sukumo *131 B7* Kōchi, Shikoku, SW Japan
Sulaimaniya *see* As Sulaymānīyah
Sulaiman Range *134 C2 mountain range* C Pakistan
Sula, Kepulauan *139 E4 island group* C Indonesia
Sulawesi, Laut *see* Celebes Sea

Sulechów *98 B3 Ger.* Züllichau. Lubuskie, W Poland
Suliag *see* Sawhāj
Sullana *60 B2* Piura, NW Peru
Sullivan Island *see* Lanbi Kyun
Sulphur Springs *49 G2* Texas, SW USA
Sultānābād *see* Arāk
Sulu Archipelago *139 E3 island group* SW Philippines
Sülüktü *see* Sulyukta
Sulu, Laut *see* Sulu Sea
Sulu Sea *139 E2 var.* Laut Sulu. *sea* SW Philippines
Sulyukta *123 E2 Kir.* Sülüktü. Batkenskaya Oblast', SW Kyrgyzstan
Sumatra *137 B8 Eng.* Sumatra. *island* W Indonesia
Sumatra *see* Sumatera
Šumava *see* Bohemian Forest
Sumba, Pulau *139 E5 Eng.* Sandalwood Island; *prev.* Soemba. *island* Nusa Tenggara, C Indonesia
Sumba, Selat *139 E5 strait* Nusa Tenggara, S Indonesia
Sumbawanga *73 B7* Rukwa, W Tanzania
Sumbe *78 B2 var.* N'Gunza, *Port.* Novo Redondo. Cuanza Sul, W Angola
Sumeih *73 B5* Southern Darfur, S Sudan
Sumgait *see* Sumqayıt, Azerbaijan
Summer Lake *46 B4 lake* Oregon, NW USA
Summit *93 H5* Alaska, USA
Sumqayıt *117 H2 Rus.* Sumgait. E Azerbaijan
Sumy *109 F2* Sums'ka Oblast', NE Ukraine
Sunbury *149 C7* Victoria, SE Australia
Sunda Islands *see* Greater Sunda Islands
Sunda, Selat *138 B5 strait* Jawa/Sumatera, SW Indonesia
Sunda Trench *see* Java Trench
Sunderland *88 D4 var.* Wearmouth. NE England, United Kingdom
Sundsvall *85 C5* Västernorrland, C Sweden
Sunflower State *see* Kansas
Sungaipenuh *138 B4 prev.* Soengaipenoeh. Sumatera, W Indonesia
Sunnyvale *47 A6* California, W USA
Sunset State *see* Oregon
Sunshine State *see* Florida
Sunshine State *see* New Mexico
Sunshine State *see* South Dakota
Suntar *115 F3* Respublika Sakha (Yakutiya), NE Russian Federation
Sunyani *75 E5* W Ghana
Suoločielgi *see* Saariselkä
Suomenlahti *see* Finland, Gulf of
Suomen Tasavalta/Suomi *see* Finland
Suomussalmi *84 E4* Oulu, E Finland
Suŏng *137 D6* Kâmpóng Cham, C Cambodia
Suoyarvi *110 B3* Respublika Kareliya, NW Russian Federation
Supe *60 C3* Lima, W Peru
Supérieur, Lac *see* Superior, Lake
Superior, Lake *40 B1 Fr.* Lac Supérieur. *lake* Canada/USA
Suqrah *see* Şawqirah
Şūr *121 E5* NE Oman
Şūr *see* Soûr
Surabaja *see* Surabaya
Surabaya *138 D5 prev.* Surabaja, Soerabaja. Jawa, C Indonesia
Surakarta *138 C5 Eng.* Solo; *prev.* Soerakarta. Jawa, S Indonesia
Šurany *99 C6 Hung.* Nagysurány. Nitriansky Kraj, SW Slovakia
Sūrat *134 C4* Gujarāt, W India
Suratdhani *see* Surat Thani
Surat Thani *131 C6 var.* Suratdhani. Surat Thani, SW Thailand
Surazh *107 E5* Vitsyebskaya Voblasts', NE Belarus
Surdulica *101 E5* Serbia, SE Serbia
Sûre *87 D7 var.* Sauer. *river* W Europe
Surendranagar *134 C4* Gujarāt, W India
Surfers Paradise *149 E5* Queensland, E Australia
Surgut *114 D3* Khanty-Mansiyskiy Avtonomnyy Okrug-Yugra, C Russian Federation
Surin *137 D5* Surin, E Thailand

Surinam *see* Suriname
Suriname *59 G3 off.* Republic of Suriname, *var.* Surinam; *prev.* Dutch Guiana, Netherlands Guiana. *country* N South America

SURINAME
South America

Official name Republic of Suriname
Formation 1975 / 1975
Capital Paramaribo
Population 500,000 / 8 people per sq mile (3 people per sq km)
Total area 63,039 sq. miles (163,270 sq. km)
Languages Sranan (Creole), Dutch*, Javanese, Sarnami Hindi, Saramaccan, Chinese, Carib
Religions Hindu 27%, Protestant 25%, Roman Catholic 23%, Muslim 20%, Traditional beliefs 5%
Ethnic mix East Indian 27%, Creole 18%, Black 15%, Javanese 15%, Mixed race 13%, Other 6%, Amerindian 4%, Chinese 2%
Government Presidential / parliamentary system
Currency Suriname dollar = 100 cents
Literacy rate 95%
Calorie consumption 2727 kilocalories

Suriname, Republic of *see* Suriname
Süriya/Süriyah, Al-Jumhūrīyah al-'Arabiyah as- *see* Syria
Surkhab, Darya-i- *see* Kahmard, Daryā-ye
Surkhob *123 F3 river* C Tajikistan
Surt *71 G2 var.* Sidra, Sirte. N Libya
Surtsey *83 E5 island* S Iceland
Suruga-wan *131 D6 bay* SE Japan
Susa *96 A2* Piemonte, NE Italy
Sūsah *see* Sousse
Susanville *47 B5* California, W USA
Susitna *36 C3* Alaska, USA
Susteren *87 D5* Limburg, SE Netherlands
Susuman *115 G3* Magadanskaya Oblast', E Russian Federation
Sutlej *134 C2 river* India/Pakistan
Suur Munamägi *106 D3 var.* Munamägi, *Ger.* Eier-Berg. *mountain* SE Estonia
Suur Väin *106 C2 Ger.* Grosser Sund. *strait* W Estonia
Suva *145 E4 country capital* (Fiji) Viti Levu, W Fiji
Suvalkai/Suvalki *see* Suwałki
Suvorovo *104 E2 prev.* Vetrino. Varna, E Bulgaria
Suwałki *98 E2 Lith.* Suvalkai, *Rus.* Suvalki. Podlaskie, NE Poland
Şuwār *see* Aş Şuwār
Suways, Qanāt as *see* Suez Canal
Suweida *see* As Suwaydā'
Suzhou *128 D5 var.* Soochow, Su-chou, Suchow; *prev.* Wuhsien. Jiangsu, E China
Svalbard *83 E1 Norwegian dependency* Arctic Ocean
Svartisen *84 C3 glacier* C Norway
Svay Riĕng *137 D6* Svay Riĕng, S Cambodia
Sveg *85 B5* Jämtland, C Sweden
Svenstavik *85 C5* Jämtland, C Sweden
Sverdlovsk *see* Yekaterinburg
Sverige *see* Sweden
Sveti Vrach *see* Sandanski
Svetlogorsk *see* Svyetlahorsk
Svetlograd *111 B7* Stavropol'skiy Kray, SW Russian Federation
Svetlovodsk *see* Svitlovods'k
Svetozarevo *see* Jagodina
Svilengrad *104 D3 prev.* Mustafa-Pasha. Khaskovo, S Bulgaria
Svitlovods'k *109 F3 Rus.* Svetlovodsk. Kirovohrads'ka Oblast', C Ukraine
Svizzera *see* Switzerland
Svobodnyy *115 G4* Amurskaya Oblast', SE Russian Federation
Svyataya Anna Trough *155 C4 var.* Saint Anna Trough. *trough* N Kara Sea
Svyetlahorsk *107 D7 Rus.* Svetlogorsk. Homyel'skaya Voblasts', SE Belarus
Swabian Jura *see* Schwäbische Alb
Swakopmund *78 B3* Erongo, W Namibia
Swan Islands *53 E1 island group* NE Honduras North America
Swansea *89 C7 Wel.* Abertawe. S Wales, United Kingdom
Swarzędz *98 C3* Poznań, W Poland

atow see Shantou
aziland 78 D4 off. Kingdom of
Swaziland. country S Africa

WAZILAND
Southern Africa

fficial name Kingdom of Swaziland
ormation 1968 / 1968
apital Mbabane
opulation 1.2 million / 181 people
er sq mile (70 people per sq km)
otal area 6704 sq. miles (17,363 sq. km)
anguages English*, siSwati*,
iZulu, Xitsonga
eligions Traditional beliefs 40%, Other
0%, Roman Catholic 20%, Muslim 10%
thnic mix Swazi 97%, Other 3%
overnment Monarchy
urrency Lilangeni = 100 cents
teracy rate 83%
alorie consumption 2275 kilocalories

aziland, Kingdom of see Swaziland
eden 84 B4 off. Kingdom of Sweden,
Swe. Sverige. country N Europe

WEDEN
Northern Europe

fficial name Kingdom of Sweden
ormation 1523 / 1921
apital Stockholm
opulation 9.6 million / 60 people
er sq mile (23 people per sq km)
otal area 173,731 sq. miles
449,964 sq. km)
anguages Swedish*, Finnish, Sámi
eligions Evangelical Lutheran 75%,
Other 13%, Muslim 5%, Other
rotestant 5%, Roman Catholic 2%
thnic mix Swedish 86%, Foreign-born
r first-generation immigrant 12%,
innish and Sámi 2%
overnment Parliamentary system
urrency Swedish krona = 100 öre
teracy rate 99%
alorie consumption 3160 kilocalories

reden, Kingdom of see Sweden
reetwater 49 F3 Texas, SW USA
ridnica 98 B4 Ger. Schweidnitz.
Walbrzych, SW Poland
idwin 98 B2 Ger. Schivelbein.
Zachodnio-pomorskie, NW Poland
iebodzice 98 B4 Ger. Freiburg in
Schlesien, Swiebodzice. Walbrzych,
SW Poland
iebodzin 98 B3 Ger. Schwiebus.
Lubuskie, W Poland
iecie 98 C3 Ger. Schwertberg.
Kujawsko-pomorskie, C Poland
indon 89 D7 S England, United
Kingdom
inemünde see Świnoujście
inoujście 98 B2 Ger. Swinemünde.
Zachodnio-pomorskie, NW Poland
iss Confederation see Switzerland
itzerland 95 A7 off. Swiss
Confederation, Fr. La Suisse, Ger.
Schweiz, It. Svizzera; anc. Helvetia.
country C Europe

WITZERLAND
Central Europe

Official name Swiss Confederation
ormation 1291 / 1857
Capital Bern
opulation 8.1 million / 528 people
er sq mile (204 people per sq km)
otal area 15,942 sq. miles (41,290 sq. km)
anguages German*, Swiss-German,
rench*, Italian*, Romansch
eligions Roman Catholic 42%,
rotestant 35%, Other and
onreligious 19%, Muslim 4%
thnic mix German 64%, French 20%,
Other 9.5%, Italian 6%, Romansch
.5%Government Parliamentary system
urrency Swiss franc = 100
appen/centimes
iteracy rate 99%
alorie consumption 3487 kilocalories

caminum see Hefa
denham Island see Nonouti

Sydney 148 D1 state capital New South
Wales, SE Australia
Sydney 39 G4 Cape Breton Island, Nova
Scotia, SE Canada
Sydney Island see Manra
Syedpur see Saidpur
Syemyezhava 107 C6 Rus. Semezhevo.
Minskaya Voblasts', C Belarus
Syene see Aswān
Syeverodonets'k 109 H3 Rus.
Severodonetsk. Luhans'ka Oblast',
E Ukraine
Syktyvkar 110 D4 prev. Ust'-Sysol'sk.
Respublika Komi, NW Russian
Federation
Sylhet 135 G3 Sylhet, NE Bangladesh
Synel'nykove 109 G3 Dnipropetrovs'ka
Oblast', E Ukraine
Syowa 154 C2 Japanese research station
Antarctica
Syracuse 41 E3 New York, NE USA
Syracuse see Siracusa
Syrdar'ya 114 B4 Sirdaryo Viloyati,
E Uzbekistan
Syria 118 B3 off. Syrian Arab Republic,
var. Siria, Syrie, Ar. Al-Jumhūrīyah
al-'Arabīyah as-Sūrīyah, Sūrīya.
country SW Asia

SYRIA
Southwest Asia

Official name Syrian Arab Republic
Formation 1941 / 1967
Capital Damascus
Population 21.9 million / 308 people
per sq mile (119 people per sq km)
Total area 71,498 sq. miles (184,180 sq. km)
Languages Arabic*, French, Kurdish,
Armenian, Circassian, Turkic languages,
Assyrian, Aramaic
Religions Sunni Muslim 74%, Alawi 12%,
Christian 10%, Druze 3%, Other 1%
Ethnic mix Arab 90%, Kurdish 9%,
Armenian, Turkmen, and Circassian 1%
Government Presidential system
Currency Syrian pound = 100 piastres
Literacy rate 85%
Calorie consumption 3106 kilocalories

Syrian Arab Republic see Syria
Syrian Desert 119 D5 Ar. Al Hamad,
Bādiyat ash Shām. desert SW Asia
Syrie see Syria
Sýrna 105 E7 var. Sirna. island Kykládes,
Greece, Aegean Sea
Sýros 105 C6 var. Síros. island Kykládes,
Greece, Aegean Sea
Syulemeshlii see Sredets
Syvash, Zaliv see Syvash, Zatoka
Syvash, Zatoka 109 F4 Rus. Zaliv Syvash.
inlet S Ukraine
Syzran' 111 C6 Samarskaya Oblast',
W Russian Federation
Szabadka see Subotica
Szamotuły 98 B3 Poznań, W Poland
Szászrégen see Reghin
Szatmárrnémeti see Satu Mare
Száva see Sava
Szczecin 98 B3 Eng./Ger. Stettin.
Zachodnio-pomorskie, NW Poland
Szczecinek 98 B2 Ger. Neustettin.
Zachodnio-pomorskie, NW Poland
Szczeciński, Zalew 98 A2 var. Stettiner
Haff, Ger. Oderhaff. bay Germany/
Poland
Szczuczyn Nowogródzki see
Shchuchyn
Szczytno 98 D3 Ger. Ortelsburg.
Warmińsko-Mazurskie,
NE Poland
Szechuan/Szechwan see Sichuan
Szeged 99 D7 Ger. Szegedin, Rom.
Seghedin. Csongrád, SE Hungary
Szegedin see Szeged
Székelykeresztúr see Cristuru Secuiesc
Székesfehérvár 99 C6 Ger.
Stuhlweissenberg; anc. Alba Regia.
Fejér, W Hungary
Szeklerburg see Miercurea-Ciuc
Szekler Neumarkt see Târgu Secuiesc
Szekszárd 99 C7 Tolna, S Hungary
Szempcz/Szenc see Senec
Szenice see Senica
Szenttamás see Srbobran
Szeping see Siping
Szilágysomlyó see Șimleu Silvaniei

Szinna see Snina
Sziszek see Sisak
Szitás-Keresztúr see Cristuru Secuiesc
Szkudy see Skuodas
Szlatina see Slatina
Szlovákia see Slovakia
Szolnok 99 D6 Jász-Nagykun-Szolnok,
C Hungary
Szombathely 99 B6 Ger. Steinamanger;
anc. Sabaria, Savaria. Vas,
W Hungary
Szprotawa 98 B4 Ger. Sprottau.
Lubuskie, W Poland
Sztálinváros see Dunaújváros
Szucsava see Suceava

T

Tabariya, Bahrat see Tiberias, Lake
Table Rock Lake 49 G1 reservoir
Arkansas/Missouri, C USA
Tábor 99 B5 Jihočeský Kraj, S Czech
Republic
Tabora 73 B7 Tabora, W Tanzania
Tabriz 120 C2 var. Tebriz; anc. Tauris.
Āzarbāyjān-e Sharqi, NW Iran
Tabuaeran 145 G2 prev. Fanning Island.
atoll Line Islands, E Kiribati
Tabūk 120 A4 Tabūk, NW Saudi Arabia
Täby 85 C6 Stockholm, C Sweden
Tachau see Tachov
Tachov 99 A5 Ger. Tachau. Plveňský
Kraj, W Czech Republic
Tacloban 139 F2 off. Tacloban City.
Leyte, C Philippines
Tacloban City see Tacloban
Tacna 61 E4 Tacna, SE Peru
Tacoma 46 B2 Washington, NW USA
Tacuarembó 64 D4 prev. San Fructuoso.
Tacuarembó, C Uruguay
Tademait, Plateau du 70 D3 plateau
C Algeria
Tadmor/Tadmur see Tudmur
Tādpatri 132 C2 Andhra Pradesh,
E India
Tadzhikistan see Tajikistan
Taegu see Daegu
Taehan-haehyŏp see Korea Strait
Taehan Min'guk see South Korea
Taejŏn see Daejeon
Tafassâsset, Ténéré du 75 G2 desert
N Niger
Tafila/Ţafilah, Muḩāfaẕat aṭ see Aṭ
Ţafilah
Taganrog 111 A7 Rostovskaya Oblast',
SW Russian Federation
Taganrog, Gulf of 109 G4 Rus.
Taganrogskiy Zaliv, Ukr. Taharoz'ka
Zatoka. gulf Russian Federation/
Ukraine
Taganrogskiy Zaliv see Taganrog,
Gulf of
Taguatinga 63 F3 Tocantins,
C Brazil
Tagus 92 C3 Port. Rio Tejo, Sp. Río Tajo.
river Portugal/Spain
Tagus Plain 80 A4 abyssal plain
E Atlantic Ocean
Taharoz'ka Zatoka see Taganrog,
Gulf of
Tahat 71 E4 mountain SE Algeria
Tahiti 145 H4 island Îles du Vent,
W French Polynesia
Tahiti, Archipel de see Société, Archipel
de la
Tahlequah 49 G1 Oklahoma,
C USA
Tahoe, Lake 47 B5 lake California/
Nevada, W USA
Tahoua 75 F3 Tahoua, W Niger
Taibei 128 D6 var. T'aipei; Jap. Taihoku;
prev. Daihoku. country capital
(Taiwan) N Taiwan
Taichū see Taizhong
T'aichung see Taizhong
Taiden see Daejeon
Taieri 151 B7 river South Island, New
Zealand
Taihape 150 D4 Manawatu-Wanganui,
North Island, New Zealand
Taihoku see Taibei
Taikyū see Daegu
Tailem Bend 149 B7 South Australia
T'ainan see Tainan
Tainan 128 D6 prev. Dainan, T'ainan.
S Taiwan

Taipei see Taibei
Taiping 138 B3 Perak, Peninsular
Malaysia ·
Taiwan 128 D6 off. Republic of China,
var. Formosa, Formo'sa. country
E Asia

TAIWAN
East Asia

Official name Republic of China (ROC)
Formation 1949 / 1949
Capital Taipei
Population 23.3 million / 1871 people
per sq mile (722 people per sq km)
Total area 13,892 sq. miles (35,980 sq. km)
Languages Amoy Chinese, Mandarin
Chinese*, Hakka Chinese
Religions Buddhist, Confucianist, and
Taoist 93%, Christian 5%, Other 2%
Ethnic mix Han (pre-20th-century
migration) 84%, Han (20th-century
migration) 14%, Aboriginal 2%
Government Presidential system
Currency Taiwan dollar = 100 cents
Literacy rate 98%
Calorie consumption 2959 kilocalories

Taiwan see Taizhong
T'aiwan Haihsia/Taiwan Haixia see
Taiwan Strait
Taiwan Strait 128 D6 var. Formosa
Strait, Chin. T'aiwan Haihsia, Taiwan
Haixia. strait China/Taiwan
Taiyuan 128 C4 var. T'ai-yuan, T'ai-
yüan; prev. Yangku. province capital
Shanxi, C China
T'ai-yuan/T'ai-yüan see Taiyuan
Taizhong 128 D6 Jap. Taichū; prev.
T'aichung, Taiwan. C Taiwan
Ta'izz 121 B7 SW Yemen
Tajikistan 123 E3 off. Republic of
Tajikistan, Rus. Tadzhikistan, Taj.
Jumhurii Tojikiston; prev. Tajik S.S.R.
country C Asia

TAJIKISTAN
Central Asia

Official name Republic of Tajikistan
Formation 1991 / 1991
Capital Dushanbe
Population 8.2 million / 148 people
per sq mile (57 people per sq km)
Total area 55,251 sq. miles (143,100 sq. km)
Languages Tajik*, Uzbek, Russian
Religions Sunni Muslim 95%,
Shi'a Muslim 3%, Other 2%
Ethnic mix Tajik 80%, Uzbek 15%,
Other 3%, Russian 1%, Kyrgyz 1%
Government Presidential system
Currency Somoni = 100 diram
Literacy rate 99%
Calorie consumption 2101 kilocalories

Tajikistan, Republic of see Tajikistan
Tajik S.S.R see Tajikistan
Tajo, Río see Tagus
Tak 136 C4 var. Rahaeng. Tak,
W Thailand
Takao see Gaoxiong
Takaoka 131 C5 Toyama, Honshū,
SW Japan
Takapuna 150 D2 Auckland, North
Island, New Zealand
Takeshima see Liancourt Rocks
Takhiatash see Taxiatosh
Takhtakupyr see Taxtako'pir
Takikawa 130 D2 Hokkaidō,
NE Japan
Takla Makan Desert 126 B3 Eng. Takla
Makan Desert. desert NW China
Takla Makan Desert see Taklimakan
Shamo
Takow see Gaoxiong
Takutea 145 G4 island S Cook Islands
Talabriga see Aveiro, Portugal
Talabriga see Talavera de la Reina, Spain
Talachyn 107 D6 Rus. Tolochin.
Vitsyebskaya Voblasts', NE Belarus
Talamanca, Cordillera de 53 E5
mountain range S Costa Rica
Talara 60 B2 Piura, NW Peru
Talas 123 F2 Talasskaya Oblast',
NW Kyrgyzstan
Talaud, Kepulauan 139 F3 island group
E Indonesia

Talavera de la Reina *92 D3 anc.*
Caesarobriga, Talabriga. Castilla-La
Mancha, C Spain
Talca *64 B4* Maule, C Chile
Talcahuano *65 B5* Bío Bío, C Chile
Taldykorgan *114 C5 Kaz.* Taldyqorghan;
prev. Taldy-Kurgan. Taldykorgan,
SE Kazakhstan
Taldy-Kurgan/Taldyqorghan *see*
Taldykorgan
Ta-lien *see* Dalian
Taliq-an *see* Tāloqān
Tal'ka *107 C6* Minskaya Voblasts',
C Belarus
Talkhof *see* Puurmani
Tallahassee *42 D3 prev.* Muskogean.
state capital Florida, SE USA
Tall al Abyaḍ *see* At Tall al Abyaḍ
Tallin *see* Tallinn
Tallinn *106 D2 Ger.* Reval, *Rus.* Tallin;
prev. Revel. *country capital* (Estonia)
Harjumaa, NW Estonia
Tall Kalakh *118 B4 var.* Tell Kalakh.
Ḥimṣ, C Syria
Tallulah *42 B2* Louisiana,
S USA
Talnakh *114 D3* Krasnoyarskiy Kray,
N Russian Federation
Tal'ne *109 E3 Rus.* Tal'noye. Cherkas'ka
Oblast', C Ukraine
Tal'noye *see* Tal'ne
Taloga *49 F1* Oklahoma, C USA
Tāloqān *123 E3 var.* Taliq-an. Takhār,
NE Afghanistan
Talsen *see* Talsi
Talsi *106 C3 Ger.* Talsen.
NW Latvia
Taltal *64 B2* Antofagasta, N Chile
Talvik *84 D2* Finnmark, N Norway
Tamabo, Banjaran *138 D3 mountain
range* East Malaysia
Tamale *75 E4* C Ghana
Tamana *145 E3 prev.* Rotcher Island.
atoll Tungaru, W Kiribati
Tamanrasset *71 E4 var.* Tamenghest.
S Algeria
Tamar *89 C7 river* SW England, United
Kingdom
Tamar *see* Tudmur
Tamatave *see* Toamasina
Tamazunchale *51 E4* San Luis Potosí,
C Mexico
Tambacounda *74 C3* SE Senegal
Tambov *111 B6* Tambovskaya Oblast',
W Russian Federation
Tambura *73 B5* Western Equatoria,
SW South Sudan
Tamchaket *see* Tâmchekkeṭ
Tâmchekkeṭ *74 C3 var.* Tamchaket.
Hodh el Gharbi, S Mauritania
Tamenghest *see* Tamanrasset
Tamil Nādu *132 C3 prev.* Madras.
cultural region SE India
Tam Ky *137 E5* Quang Nam-f̣a Nâng,
C Vietnam
Tammerfors *see* Tampere
Tampa *43 E4* Florida, SE USA
Tampa Bay *43 E4 bay* Florida,
SE USA
Tampere *85 D5 Swe.* Tammerfors.
Länsi-Suomi, W Finland
Tampico *51 E3* Tamaulipas,
C Mexico
Tamworth *149 D6* New South Wales,
SE Australia
Tanabe *131 C7* Wakayama, Honshū,
SW Japan
Tana Bru *84 D2* Finnmark, N Norway
Tanais *see* Don
Lake Tana *72 C4 var.* Lake Tana. *lake*
NW Ethiopia
Tana, Lake *see* T'ana Hāyk'
Tanami Desert *146 D3 desert* Northern
Territory, N Australia
Tananarive *see* Antananarivo
Ţăndărei *108 D5* Ialomiţa, SE Romania
Tandil *65 D5* Buenos Aires, E Argentina
Tandjoengkarang *see* Bandar Lampung
Tanega-shima *131 B8 island* Nansei-
shotō, SW Japan
Tanen Taunggyi *see* Tane Range
Tane Range *136 B4* Bur. Tanen
Taunggyi. *mountain range*
W Thailand
Tanezrouft *70 C4 desert* Algeria/Mali
Ţanf, Jabal aţ *118 D4 mountain* SE Syria
Tanga *73 C7* Tanga, E Tanzania

Tanganyika and Zanzibar *see* Tanzania
Tanganyika, Lake *73 B7 lake*
E Africa
Tanger *70 C2 var.* Tangiers, Tangier, *Fr./
Ger.* Tangerk, *Sp.* Tánger; *anc.* Tingis.
NW Morocco
Tangerk *see* Tanger
Tanggula Shan *126 C4 mountain*
W China
Tangier *see* Tanger
Tangiers *see* Tanger
Tangra Yumco *126 B5 var.* Tangro Tso.
lake W China
Tangro Tso *see* Tangra Yumco
Tangshan *128 D3 var.* T'ang-shan.
Hebei, E China
T'ang-shan *see* Tangshan
Tanimbar, Kepulauan *139 F5 island
group* Maluku, E Indonesia
Tanintharyi *137 B6 prev.* Tenasserim.
S Myanmar (Burma)
**Tanjungkarang/Tanjungkarang-
Telukbetung** *see* Bandar Lampung
Tanna *144 D4 island* S Vanuatu
Tannenhof *see* Krynica
Tan-Tan *70 B3* SW Morocco
Tanzania *73 C7 off.* United Republic of
Tanzania, *Swa.* Jamhuri ya Muungano
wa Tanzania; *prev.* German East Africa,
Tanganyika and Zanzibar. *country*
E Africa

TANZANIA
East Africa

Official name United Republic
of Tanzania
Formation 1964 / 1964
Capital Dodoma
Population 49.3 million / 144 people
per sq mile (56 people per sq km)
Total area 364,898 sq. miles
(945,087 sq. km)
Languages Kiswahili*, Sukuma, Chagga,
Nyamwezi, Hehe, Makonde, Yao,
Sandawe, English*
Religions Christian 63%, Muslim 35%,
Other 2%
Ethnic mix Native African (over 120
tribes) 99%, European, Asian, and Arab 1%
Government Presidential system
Currency Tanzanian shilling = 100 cents
Literacy rate 68%
Calorie consumption 2167 kilocalories

Tanzania, Jamhuri ya Muungano wa
see Tanzania
Tanzania, United Republic of *see*
Tanzania
Taoudenit *see* Taoudenni
Taoudenni *75 E2 var.* Taoudenit.
Tombouctou, N Mali
Tapa *106 E2 Ger.* Taps. Lääne-Virumaa,
NE Estonia
Tapachula *51 G5* Chiapas, SE Mexico
Tapaiu *see* Gvardeysk
Tapajós, Rio *63 E2 var.* Tapajóz. *river*
NW Brazil
Tapajóz *see* Tapajós, Rio
Taps *see* Tapa
Ţarābulus *71 F2 var.* Ţarābulus al Gharb,
Eng. Tripoli. *country capital* (Libya)
NW Libya
Ţarābulus al Gharb *see* Ţarābulus
Ţarābulus/Ţarābulus ash Shām *see*
Tripoli
Taraclia *108 D4 Rus.* Tarakilya.
S Moldova
Tarakilya *see* Taraclia
Mount Taranaki *150 C4 var.* Egmont.
volcano North Island, New Zealand
Tarancón *93 E3* Castilla-La Mancha,
C Spain
Taranto *97 E5 var.* Tarentum. Puglia,
SE Italy
Taranto, Gulf of *97 E6 Eng.* Gulf of
Taranto. *gulf* S Italy
Taranto, Gulf of *see* Taranto, Golfo di
Tarapoto *60 C2* San Martín, N Peru
Tarare *91 D5* Rhône, E France
Tarascon *91 D6* Bouches-du-Rhône,
SE France
Tarawa *144 D2 atoll* Tungaru, W Kiribati
Taraz *114 C5 prev.* Aulie Ata, Auliye-
Ata, Dzhambul, Zhambyl. Zhambyl,
S Kazakhstan

Tarazona *93 E2* Aragón, NE Spain
Tarbes *91 B6 anc.* Bigorra. Hautes-
Pyrénées, S France
Tarcoola *149 A6* South Australia
Taree *149 D6* New South Wales,
SE Australia
Tarentum *see* Taranto
Târgovişte *108 C5 prev.* Tîrgovişte.
Dâmboviţa, S Romania
Targu Jiu *108 B4 prev.* Tîrgu Jiu. Gorj,
W Romania
Târgul-Neamţ *see* Târgu-Neamţ
Târgul-Săcuiesc *see* Târgu Secuiesc
Târgu Mureş *108 B4 prev.* Oşorhei,
Tîrgu Mures, *Ger.* Neumarkt, *Hung.*
Marosvásárhely. Mureş, C Romania
Târgu-Neamţ *108 C3 var.* Târgul-
Neamţ; *prev.* Tîrgu-Neamţ. Neamţ,
NE Romania
Târgu Ocna *108 C4 Hung.* Aknavásár;
prev. Tîrgu Ocna. Bacău, E Romania
Târgu Secuiesc *108 C4 Ger.*
Neumarkt, Szekler Neumarkt, *Hung.*
Kezdivásárhely; *prev.* Chezdi-Oşorhei,
Târgul-Săcuiesc, Tîrgu Secuiesc.
Covasna, E Romania
Tar Heel State *see* North Carolina
Tarija *61 G5* Tarija, S Bolivia
Tarim *121 C6* C Yemen
Tarim Basin *124 C2 Eng.* Tarim Basin.
basin NW China
Tarim Basin *see* Tarim Pendi
Tarim He *126 B3 river* NW China
Tarma *60 C3* Junín, C Peru
Tarn *91 C6 cultural region* S France
Tarn *91 C6 river* S France
Tarnobrzeg *98 D4* Podkarpackie,
SE Poland
Tarnopol *see* Ternopil'
Tarnów *99 D5* Małopolskie, S Poland
Tarraco *see* Tarragona
Tarragona *93 G2 anc.* Tarraco.
Cataluña, E Spain
Tarrasa *see* Terrassa
Tàrrega *93 F2 var.* Tarrega. Cataluña,
NE Spain
Tarsatica *see* Rijeka
Tarsus *116 C4* İçel, S Turkey
Tartous/Tartouss *see* Ţarţūs
Tartu *106 D3 Ger.* Dorpat; *prev. Rus.*
Yurev, Yury'ev. Tartumaa,
SE Estonia
Ţarţūs *118 A3 off.* Muḩāfaẓat Ţarţūs,
var. Tartous, Tartus. *governorate*
W Syria
Ţarţūs, Muḩāfaẓat *see* Ţarţūs
Ta Ru Tao, Ko *137 B7 island* S Thailand
Tarvisio *96 D2* Friuli-Venezia Giulia,
NE Italy
Tarvisium *see* Treviso
Tashauz *see* Daşoguz
Tashi Chho Dzong *see* Thimphu
Tashkent *see* Toshkent
Tash-Kömür *see* Tash-Kumyr
Tash-Kumyr *123 F2 Kir.* Tash-
Kömür. Dzhalal-Abadskaya Oblast',
W Kyrgyzstan
Tashqurghan *see* Khulm
Tasikmalaja *see* Tasikmalaya
Tasikmalaya *138 C5 prev.* Tasikmalaja.
Jawa, C Indonesia
Tasman Basin *142 C5 var.* East
Australian Basin. *undersea basin*
S Tasman Sea
Tasman Bay *151 C5 inlet* South Island,
New Zealand
Tasmania *149 B8 prev.* Van Diemen's
Land. *state* SE Australia
Tasmania *152 B4 island* SE Australia
Tasman Plateau *142 C5 var.* South
Tasmania Plateau. *undersea plateau*
SW Tasman Sea
Tasman Sea *142 C5 sea* SW Pacific
Ocean
Tassili-n-Ajjer *71 E4 plateau*
E Algeria
Tatabánya *99 C6* Komárom-Esztergom,
NW Hungary
Tatar Pazardzhik *see* Pazardzhik
Tathlith *121 B5 'Asīr,* S Saudi Arabia
Tatra Mountains *99 D5 Ger.* Tatra,
Hung. Tátra, *Pol./Slvk.* Tatry.
mountain range Poland/Slovakia
Tatra/Tátra *see* Tatra Mountains
Tatry *see* Tatra Mountains
Ta-t'ung/Tatung *see* Datong

Tatvan *117 F3* Bitlis, SE Turkey
Ta'ū *145 F4 var.* Tau. *island* Manua'a
Islands, E American Samoa
Taukum, Peski *123 G1 desert*
SE Kazakhstan
Taumarunui *150 D4* Manawatu-
Wanganui, North Island, New Zealand
Taungdwingyi *136 B3* Magway,
C Myanmar (Burma)
Taunggyi *136 B3* Shan State, C Myanmar
(Burma)
Taungoo *136 B4* Bago, C Myanmar
(Burma)
Taunton *89 C7* SW England, United
Kingdom
Taupo *150 D3* Waikato, North Island,
New Zealand
Taupo, Lake *150 D3 lake* North Island,
New Zealand
Tauragė *106 B4 Ger.* Tauroggen.
Tauragė, SW Lithuania
Tauranga *150 D3* Bay of Plenty, North
Island, New Zealand
Tauris *see* Tabrīz
Tauroggen *see* Tauragė
Taurus Mountains *116 C4 Eng.* Tau
Mountains. *mountain range* S Turkey
Taurus Mountains *see* Toros Dağları
Tavas *116 B4* Denizli, SW Turkey
Tavastehus *see* Hämeenlinna
Tavira *92 C5* Faro, S Portugal
Tavoy *see* Dawei
Tavoy Island *see* Mali Kyun
Ta Waewae Bay *151 A7 bay* South
Island, New Zealand
Tawakoni, Lake *49 G2 reservoir* Texas,
SW USA
Tawau *138 D3* Sabah, East Malaysia
Ţawkar *see* Tokar
Tawzar *see* Tozeur
Taxco *51 E4 var.* Taxco de Alarcón.
Guerrero, S Mexico
Taxco de Alarcón *see* Taxco
Taxiatosh *122 C2 Rus.* Takhiatash.
Qoraqalpog'iston Respublikasi,
W Uzbekistan
Taxtako'pir *122 D1 Rus.* Takhtakupyr.
Qoraqalpog'iston Respublikasi,
NW Uzbekistan
Tay *88 C3 river* C Scotland, United
Kingdom
Taylor *49 G3* Texas, SW USA
Taymā' *120 A4* Tabūk, NW Saudi Arabia
Taymyr, Ozero *115 E2 lake* N Russian
Federation
Taymyr, Poluostrov *115 E2 peninsula*
N Russian Federation
Taz *114 D3 river* N Russian Federation
Tbilisi *117 G2 var.* T'bilisi, *Eng.* Tiflis.
country capital (Georgia) SE Georgia
T'bilisi *see* Tbilisi
Tchad *see* Chad
Tchad, Lac *see* Chad, Lake
Tchien *see* Zwedru
Tchongking *see* Chongqing
Tczew *98 C2 Ger.* Dirschau. Pomorskie,
N Poland
Te Anau *151 A7* Southland, South Island,
New Zealand
Te Anau, Lake *151 A7 lake* South Island,
New Zealand
Teapa *51 G4* Tabasco, SE Mexico
Teate *see* Chieti
Tebingtinggi *138 B3* Sumatera,
N Indonesia
Tebriz *see* Tabrīz
Techirghiol *108 D5* Constanţa,
SE Romania
Tecomán *50 D4* Colima, SW Mexico
Tecpan *51 E5 var.* Tecpan de Galeana.
Guerrero, S Mexico
Tecpan de Galeana *see* Tecpan
Tecuci *108 C4* Galaţi, E Romania
Tedzhen *see* Harīrūd/Tejen
Tedzhen *see* Tejen
Tees *89 D5 river* N England, United
Kingdom
Tefé *62 D2* Amazonas, N Brazil
Tegal *138 C4* Jawa, C Indonesia
Tegelen *87 D5* Limburg,
SE Netherlands
Tegucigalpa *52 C3 country capital*
(Honduras) Francisco Morazán,
SW Honduras
Teheran *see* Tehrān
Tehrān *120 C3 var.* Teheran. *country
capital* (Iran) Tehrān, N Iran

huacán *51 F4* Puebla, S Mexico
huantepec *51 F5* var. Santo Domingo
Tehuantepec. Oaxaca, SE Mexico
ulf of Tehuantepec *51 F5* var. Gulf of
Tehuantepec. *gulf* S Mexico
huantepec, Gulf of *see* Tehuantepec,
Golfo de
thmus of Tehuantepec *51 F5* var.
Isthmus of Tehuantepec. *isthmus*
SE Mexico
huantepec, Isthmus of *see*
Tehuantepec, Istmo de
jen *122 C3 Rus.* Tedzhen. Ahal
Welaýaty, S Turkmenistan
jen *see* Harīrūd
jo, Rio *see* Tagus
Kao *150 C1* Northland, North Island,
New Zealand
kax *51 H4* var. Tekax de Álvaro
Obregón. Yucatán, SE Mexico
kax de Álvaro Obregón *see* Tekax
keli *114 C5* Almaty, SE Kazakhstan
kirdağ *116 A2 It.* Rodosto; *anc.*
Bisanthe, Raidestos, Rhaedestus.
Tekirdağ, NW Turkey
Kuiti *150 D3* Waikato, North Island,
New Zealand
ela *52 C2* Atlántida, NW Honduras
elanaipura *see* Jambi
elangana *134 D5 cultural region*
SE India
el Aviv-Jaffa *see* Tel Aviv-Yafo
el Aviv-Yafo *119 A6* var. Tel Aviv-Jaffa.
Tel Aviv, C Israel
les Pirés *see* São Manuel, Rio
elish *104 C2 prev.* Azizie. Pleven,
N Bulgaria
ell Abiad/Tell Abyad *see* At Tall al
Abyaḍ
ell Kalakh *see* Tall Kalakh
ell Shedadi *see* Ash Shadādah
el'man/Tel'mansk *see* Gubadag
eloekbetoeng *see* Bandar Lampung
elo Martius *see* Toulon
elschen *see* Telšiai
elšiai *106 B3 Ger.* Telschen. Telšiai,
NW Lithuania
elukbetung *see* Bandar Lampung
emerin *100 D3* Vojvodina, N Serbia
emeschburg/Temeschwar *see*
Timişoara
emesvár/Temeswar *see* Timişoara
emirtau *114 C4 prev.* Samarkandski,
Samarkandskoye. Karagandy,
C Kazakhstan
empio Pausania *97 A5* Sardegna, Italy,
C Mediterranean Sea
emple *49 G3* Texas, SW USA
emuco *65 B5* Araucanía, C Chile
emuka *151 B6* Canterbury, South
Island, New Zealand
enasserim *see* Tanintharyi
énenkou *74 D3* Mopti, C Mali
énéré *75 G3 physical region* C Niger
enerife *70 A3 island* Islas Canarias,
Spain, NE Atlantic Ocean
engger Shamo *127 E3 desert*
N China
engréla *74 D4 var.* Tingréla. N Côte
d'Ivoire
enkodogo *75 E4* S Burkina Faso
ennant Creek *148 A3* Northern
Territory, C Australia
ennessee *42 C1 off.* State of Tennessee,
also known as The Volunteer State.
state SE USA
ennessee River *42 C1 river* S USA
enos *see* Tínos
epelena *see* Tepelenë
epelenë *101 C7 var.* Tepelena, *It.*
Tepeleni. Gjirokastër, S Albania
epic *50 D4* Nayarit, C Mexico
eplice *98 A4 Ger.* Teplitz; *prev.* Teplice-
Šanov, Teplitz-Schönau. Ústecký Kraj,
NW Czech Republic
eplice-Šanov/Teplitz/Teplitz-Schönau
see Teplice
equila *50 D4* Jalisco, SW Mexico
eraina *145 E2* Washington Island.
atoll Line Islands, E Kiribati
eramo *96 C4 anc.* Interamna. Abruzzi,
C Italy
ercan *117 E3* Erzincan, NE Turkey
erceira *92 A5 var.* Ilha Terceira. *island*
Azores, Portugal, NE Atlantic Ocean
erceira, Ilha *see* Terceira

Terekhovka *see* Tsyerakhowka
Teresina *63 F2 var.* Therezina. *state
capital* Piauí, NE Brazil
Termez *see* Termiz
Termia *see* Kýthnos
Términos, Laguna de *51 G4 lagoon*
SE Mexico
Termiz *123 E3 Rus.* Termez.
Surkhondaryo Viloyati, S Uzbekistan
Termoli *96 D4* Molise, C Italy
Terneuzen *87 B5 var.* Neuzen. Zeeland,
SW Netherlands
Terni *96 C4 anc.* Interamna Nahars.
Umbria, C Italy
Ternopil' *108 C2 Pol.* Tarnopol, *Rus.*
Ternopol'. Ternopil's'ka Oblast',
W Ukraine
Ternopol' *see* Ternopil'
Terracina *97 C5* Lazio, C Italy
Terranova di Sicilia *see* Gela
Terranova Pausania *see* Olbia
Terrassa *93 G2 Cast.* Tarrasa. Cataluña,
E Spain
Terre Adélie *154 C4 physical region*
Antarctica
Terre Haute *40 B4* Indiana, N USA
Terre Neuve *see* Newfoundland and
Labrador
Terschelling *86 C1 Fris.* Skylge. *island*
Waddeneilanden, N Netherlands
Teruel *93 F3 anc.* Turba. Aragón,
E Spain
Tervel *104 E1 prev.* Kurtbunar, *Rom.*
Curtbunar. Dobrich, NE Bulgaria
Tervueren *see* Tervuren
Tervuren *87 C6 var.* Tervueren. Vlaams
Brabant, C Belgium
Teseney *72 C4 var.* Tessenei. W Eritrea
Tessalit *75 E2* Kidal, NE Mali
Tessaoua *75 G3* Maradi, S Niger
Tessenderlo *87 C5* Limburg,
NE Belgium
Tessenei *see* Teseney
Testigos, Islas los *59 E1 island group*
N Venezuela
Tete *79 E2* Tete, NW Mozambique
Teterow *94 C3* Mecklenburg-
Vorpommern, NE Germany
Tétouan *70 C2 var.* Tetouan, Tetuán.
N Morocco
Tetovo *101 D5* Razgrad, N Bulgaria
Tetschen *see* Děčín
Tetuán *see* Tétouan
Teverya *see* Tverya
Te Waewae Bay *151 A7 bay* South
Island, New Zealand
Texarkana *42 A2* Arkansas, C USA
Texarkana *49 H2* Texas, SW USA
Texas *49 F3 off.* State of Texas, *also
known as* Lone Star State. *state* S USA
Texas City *49 H4* Texas, SW USA
Texel *86 C2 island* Waddeneilanden,
NW Netherlands
Texoma, Lake *49 G2 reservoir*
Oklahoma/Texas, C USA
Teziutlán *51 F4* Puebla, S Mexico
Thaa Atoll *see* Kolhumadulu
Thai, Ao *see* Thailand, Gulf of
Thai Binh *136 D3* Thai Binh, N Vietnam
Thailand *137 C5 off.* Kingdom of
Thailand, *Th.* Prathet Thai; *prev.* Siam.
country SE Asia

THAILAND
Southeast Asia

Official name Kingdom of Thailand
Formation 1238 / 1907
Capital Bangkok
Population 61 million / 340 people
per sq mile (131 people per sq km)
Total area 198,455 sq. miles
(514,000 sq. km)
Languages Thai*, Chinese, Malay, Khmer,
Mon, Karen, Miao
Religions Buddhist 95%, Muslim 4%,
Other (including Christian) 1%
Ethnic mix Thai 83%, Chinese 12%,
Malay 3%, Khmer and Other 2%
Government Transitional regime
Currency Baht = 100 satang
Literacy rate 96%
Calorie consumption 2757 kilocalories

Thailand, Gulf of *137 C6 var.* Gulf of
Siam, *Th.* Ao Thai, *Vtn.* Vinh Thai Lan.
gulf SE Asia

Thailand, Kingdom of *see* Thailand
Thai Lan, Vinh *see* Thailand, Gulf of
Thai Nguyên *136 D3* Bắc Thái,
N Vietnam
Thakhèk *136 D4 var.* Muang
Khammouan. Khammouan, C Laos
Thamarīd *see* Thamarīt
Thamarīt *121 D6 var.* Thamarīd,
Thumrayt. SW Oman
Thames *150 D3* Waikato, North Island,
New Zealand
Thames *89 B8 river* S England, United
Kingdom
Thandwe *136 A4 var.* Sandoway.
Rakhine State, W Myanmar (Burma)
Thanh Hoa *136 D3* Thanh Hoa,
N Vietnam
Thanintari Taungdan *see* Bilauktaung
Range
Thanlwin *see* Salween
Thar Desert *134 C3 var.* Great Indian
Desert, Indian Desert. *desert* India/
Pakistan
Tharthar, Buhayrat ath *120 B3 lake*
C Iraq
Thásos *104 C4* Thásos, E Greece
Thásos *104 C4 island* E Greece
Thaton *136 B4* Mon State, S Myanmar
(Burma)
Thayetmyo *136 A4* Magway, C Myanmar
(Burma)
The Crane *55 H2 var.* Crane.
S Barbados
The Dalles *46 B3* Oregon, NW USA
The Flatts Village *see* Flatts Village
The Hague *see* 's-Gravenhage
Theodosia *see* Feodosiya
The Pas *37 F5* Manitoba, C Canada
Therezina *see* Teresina
Thérma *105 D6* Ikaría, Dodekánisa,
Greece, Aegean Sea
Thermaic Gulf *104 B4 Eng.* Thermaic
Gulf; *anc.* Thermaicus Sinus. *gulf*
N Greece
Thermaic Gulf/Thermaicus Sinus *see*
Thermaïkós Kólpos
Thermía *see* Kýthnos
Thérmo *105 B5* Dytikí Elláda,
C Greece
The Rock *93 H4* New South Wales,
SE Australia
The Sooner State *see* Oklahoma
Thessaloníki *104 C3 Eng.* Salonica,
Salonika, *SCr.* Solun, *Turk.* Selânik.
Kentrikí Makedonía, N Greece
The Valley *55 G3 dependent territory
capital* (Anguilla) E Anguilla
The Village *59 G1* Oklahoma,
C USA
The Volunteer State *see* Tennessee
Thiamis *see* Kalamás
Thian Shan *see* Tien Shan
Thibet *see* Xizang Zizhiqu
Thief River Falls *45 F1* Minnesota,
N USA
Thienen *see* Tienen
Thiers *91 C5* Puy-de-Dôme, C France
Thiès *73 B3* W Senegal
Thikombia *see* Cikobia
Thimbu *see* Thimphu
Thimphu *135 G3 var.* Thimbu; *prev.*
Tashi Chho Dzong. *country capital*
(Bhutan) W Bhutan
Thionville *90 D3 Ger.* Diedenhofen.
Moselle, NE France
Thíra *105 D7 var.* Santoríni. Kykládes,
Greece, Aegean Sea
Thiruvananthapuram *132 C3 var.*
Trivandrum, Tiruvantapuram. *state
capital* Kerala, SW India
Thitu Island *128 C8 island* NW Spratly
Islands
Tholen *86 B4 island* SW Netherlands
Thomasville *42 D3* Georgia,
SE USA
Thompson *37 F4* Manitoba, C Canada
Thonon-les-Bains *91 D5* Haute-Savoie,
E France
Thorenburg *see* Turda
Thorlákshöfn *83 E5* Sudhurland,
SW Iceland
Thorn *see* Toruń
Thornton Island *see* Millennium Island
Thorshavn *see* Tórshavn
Thospitis *see* Van Gölü
Thouars *90 B4* Deux-Sèvres,
W France

Thoune *see* Thun
Thracian Sea *104 D4 Gk.* Thrakikó
Pélagos; *anc.* Thracium Mare. *sea*
Greece/Turkey
Thracium Mare/Thrakikó Pélagos *see*
Thracian Sea
Three Gorges Reservoir *129 C5*
reservoir C China
Three Kings Islands *150 C1 island
group* N New Zealand
Thrissur *132 C3 var.* Trichūr. Kerala,
SW India
Thuin *87 B7* Hainaut, S Belgium
Thule *see* Qaanaaq
Thumrayt *see* Thamarīt
Thun *95 A7 Fr.* Thoune. Bern,
W Switzerland
Thunder Bay *38 B4* Ontario, S Canada
Thuner See *95 A7 lake* C Switzerland
Thung Song *137 C7 var.* Cha Mai.
Nakhon Si Thammarat, SW Thailand
Thurso *88 C2* N Scotland, United
Kingdom
Thýamis *see* Kalamás
Tianjin *128 D4 var.* Tientsin. Tianjin
Shi, E China
Tianjin *see* Tianjin Shi
Tianjin Shi *128 D4 var.* Jin, Tianjin,
T'ien-ching, Tientsin. *municipality*
E China
Tian Shan *see* Tien Shan
Tianshui *128 B4* Gansu, C China
Tiba *see* Chiba
Tiber *96 C4 Eng.* Tiber. *river* C Italy
Tiber *see* Tevere, Italy
Tiber *see* Tívoli, Italy
Tiberias *see* Tverya
Tiberias, Lake *119 B5 var.* Chinnereth,
Sea of Bahr Tabariya, Sea of Galilee, *Ar.*
Bahrat Tabariya, *Heb.* Yam Kinneret.
lake N Israel
Tibesti *76 C2 var.* Tibesti Massif, *Ar.*
Tibïstï. *mountain range* N Africa
Tibesti Massif *see* Tibesti
Tibet *see* Xizang Zizhiqu
Tibetan Autonomous Region *see* Xizang
Zizhiqu
Tibet, Plateau of *126 B4 var.* Xizang
Gaoyuan, *Eng.* Plateau of Tibet. *plateau*
W China
Tibet, Plateau of *see* Qingzang Gaoyuan
Tibïstï *see* Tibesti
Tibni *see* At Tibnī
Tiburón, Isla *50 B2 var.* Isla del Tiburón.
island NW Mexico
Tiburón, Isla del *see* Tiburón, Isla
Tichau *see* Tychy
Tichît *74 D2 var.* Tichitt. Tagant,
C Mauritania
Tichitt *see* Tichît
Ticinum *see* Pavia
Ticul *51 H3* Yucatán, SE Mexico
Tidjikdja *see* Tidjikja
Tidjikja *74 C2 var.* Tidjikdja; *prev.* Fort-
Cappolani. Tagant, C Mauritania
T'ien-ching *see* Tianjin Shi
Tienen *87 C6 var.* Thienen, *Fr.*
Tirlemont. Vlaams Brabant,
C Belgium
Tiên Giang, Sông *see* Mekong
Tien Shan *126 B3 Chin.* Thian Shan,
Tian Shan, Tien Shan, *Rus.* Tiyan'-
Shan'. *mountain range* C Asia
Tientsin *see* Tianjin
Tierp *85 C6* Uppsala, C Sweden
Tierra del Fuego *65 B8 island* Argentina/
Chile
Tiflis *see* Tbilisi
Tifton *42 D3* Georgia, SE USA
Tifu *139 F4* Pulau Buru, E Indonesia
Tighina *108 D4 Rus.* Bendery; *prev.*
Bender. E Moldova
Tigranocerta *see* Siirt
Tigris *120 B2 Ar.* Dijlah, *Turk.* Dicle.
river Iraq/Turkey
Tiguentourine *71 E3* E Algeria
Tijuana *50 A1* Baja California Norte,
NW Mexico
Tikhoretsk *111 A7* Krasnodarskiy Kray,
SW Russian Federation
Tikhvin *110 B4* Leningradskaya Oblast',
NW Russian Federation
Tiki Basin *143 G3 undersea basin*
S Pacific Ocean
Tikinsso *52 C4 river* C Guinea
Tikirarjuaq *see* Whale Cove

Tiksi 115 F2 Respublika Sakha (Yakutiya), NE Russian Federation
Tilburg 86 C4 Noord-Brabant, S Netherlands
Tilimsen see Tlemcen
Tilio Martius see Toulon
Tillabéri 75 F3 var. Tillabéry. Tillabéri, W Niger
Tillabéry see Tillabéri
Tílos 105 E7 island Dodekánisa, Greece, Aegean Sea
Timan Ridge 110 D3 Eng. Timan Ridge. ridge NW Russian Federation
Timan Ridge see Timanskiy Kryazh
Timaru 151 B6 Canterbury, South Island, New Zealand
Timbaki/Timbákion see Tympáki
Timbedgha 74 D3 var. Timbédra. Hodh ech Chargui, SE Mauritania
Timbédra see Timbedgha
Timbuktu see Tombouctou
Timiş 108 A4 county SW Romania
Timişoara 108 A4 Ger. Temeschwar, Temeswar, Hung. Temesvár;   - prev. Temeschburg. Timiş, W Romania
Timmins 38 C4 Ontario, S Canada
Timor 125 F5 island Nusa Tenggara, C Indonesia
Timor Sea 125 F5 sea E Indian Ocean
Timor Timur see East Timor
Timor Trench see Timor Trough
Timor Trough 125 F5 var. Timor Trench. trough NE Timor Sea
Timrå 85 C5 Västernorrland, C Sweden
Tindouf 70 C3 W Algeria
Tineo 92 C1 Asturias, N Spain
Tingis see Tanger
Tingo María 60 C3 Huánuco, C Peru
Tingréla see Tengréla
Tinhosa Grande 76 E2 island N Sao Tome and Principe, Africa, E Atlantic Ocean
Tinhosa Pequena 76 E1 island N Sao Tome and Principe, Africa, E Atlantic Ocean
Tinian 144 B1 island S Northern Mariana Islands
Tínos 105 D6 Tínos, Kykládes, Greece, Aegean Sea
Tínos 105 D6 anc. Tenos. island Kykládes, Greece, Aegean Sea
Tip 101 E6 Papua, E Indonesia
Tipitapa 52 D3 Managua, W Nicaragua
Tip Top Mountain 38 C4 mountain Ontario, S Canada
Tirana see Tiranë
Tiranë 101 C6 var. Tirana. country capital (Albania) Tiranë, C Albania
Tiraspol 108 D4 Rus. Tiraspol'. E Moldova
Tiraspol' see Tiraspol
Tiree 88 B3 island W Scotland, United Kingdom
Tîrgoviște see Târgoviște
Tîrgu Jiu see Targu Jiu
Tîrgu Mureș see Târgu Mureș
Tîrgu-Neamț see Târgu-Neamț
Tîrgu Ocna see Târgu Ocna
Tîrgu Secuiesc see Târgu Secuiesc
Tîrlemont see Tienen
Tírnavos see Týrnavos
Tírnovo see Veliko Túrnovo
Tirol 95 C7 off. Land Tirol, var. Tyrol, It. Tirolo. state W Austria
Tirol, Land see Tirol
Tirolo see Tirol
Tirreno, Mare see Tyrrhenian Sea
Tiruchchirāppalli 132 C3 prev. Trichinopoly. Tamil Nādu, SE India
Tiruppattūr 132 C2 Tamil Nādu, SE India
Tiruvantapuram see Thiruvanathapuram
Tisa 103 F1 Ger. Theiss, Rom./Slvn./ SCr. Tisa, Rus. Tissa, Ukr. Tysa. river SE Europe
Tisa see Tisza
Tiszakécske 99 D7 Bács-Kiskun, C Hungary
Titano, Monte 96 E1 mountain C San Marino
Titicaca, Lake 61 E4 lake Bolivia/ Peru
Titograd see Podgorica

Titose see Chitose
Titova Mitrovica see Mitrovicë
Titovo Užice see Užice
Titu 108 C5 Dâmbovița, S Romania
Titule 77 D5 Orientale, N Dem. Rep. Congo
Tiverton 89 C7 SW England, United Kingdom
Tívoli 96 C4 anc. Tiber. Lazio, C Italy
Tizimín 51 H3 Yucatán, SE Mexico
Tizi Ouzou 71 E1 var. Tizi-Ouzou. N Algeria
Tizi-Ouzou see Tizi Ouzou
Tiznit 70 B3 SW Morocco
Tjilatjap see Cilacap
Tjirebon see Cirebon
Tlaquepaque 50 D4 Jalisco, C Mexico
Tlascala see Tlaxcala
Tlaxcala 51 F4 var. Tlascala, Tlaxcala de Xicohténcatl. Tlaxcala, C Mexico
Tlaxcala de Xicohténcatl see Tlaxcala
Tlemcen 70 D2 var. Tilimsen, Tlemsen. NW Algeria
Tlemsen see Tlemcen
Toamasina 79 G3 prev./Fr. Tamatave. Toamasina, E Madagascar
Toba, Danau 138 B3 lake Sumatera, W Indonesia
Tobago 55 H5 island NE Trinidad and Tobago
Toba Kakar Range 134 B2 mountain range NW Pakistan
Tobol see Tobyl
Tobol'sk 114 C3 Tyumenskaya Oblast', C Russian Federation
Tobruch/Tobruk see Ţubruq
Tobyl 114 C4 prev. Tobol. river Kazakhstan/Russian Federation
Tocantins 63 E3 off. Estado do Tocantins. state C Brazil
Tocantins 63 E3 off. Estado do Tocantins. region C Brazil
Tocantins, Estado do see Tocantins
Tocantins, Rio 63 F2 river N Brazil
Tocoa 52 D2 Colón, N Honduras
Tocopilla 64 B2 Antofagasta, N Chile
Todi 96 C4 Umbria, C Italy
Todos os Santos, Baía de 63 G3 bay E Brazil
Toetoes Bay 151 B8 bay South Island, New Zealand
Tofua 145 E4 island Ha'apai Group, C Tonga
Togo 75 E4 off. Togolese Republic; prev. French Togoland. country W Africa

TOGO
West Africa

Official name The Togolese Republic
Formation 1960 / 1960
Capital Lomé
Population 6.2 million / 295 people per sq mile (114 people per sq km)
Total area 21,924 sq. miles (56,785 sq. km)
Languages Ewe, Kabye, Gurma, French*
Religions Christian 47%, Traditional beliefs 33%, Muslim 14%, Other 6%
Ethnic mix Ewe 46%, Other African 41%, Kabye 12%, European 1%
Government Presidential system
Currency CFA franc = 100 centimes
Literacy rate 60%
Calorie consumption 2366 kilocalories

Togolese Republic see Togo
Tojikiston, Jumhurii see Tajikistan
Tokanui 151 B7 Southland, South Island, New Zealand
Tokar 72 C3 var. Ţawkar. Red Sea, NE Sudan
Tokat 116 D3 Tokat, N Turkey
Tokelau 145 E3 NZ overseas territory W Polynesia
Tôketerebes see Trebišov
Tokio see Tôkyô
Tokmak 109 G4 var. Velykyy Tokmak. Zaporiz'ka Oblast', SE Ukraine
Tokmak see Tomok
Tokoroa 150 D3 Waikato, North Island, New Zealand
Tokounou 74 C4 C Guinea
Tokushima 131 C6 var. Tokusima. Tokushima, Shikoku, SW Japan
Tôkyô 130 A1 var. Tokio. country capital (Japan) Tôkyô, Honshû, S Japan

Tôkyô-wan 130 A2 bay S Japan
Tolbukhin see Dobrich
Toledo 92 D3 anc. Toletum. Castilla-La Mancha, C Spain
Toledo 40 D3 Ohio, N USA
Toledo Bend Reservoir 49 G3 reservoir Louisiana/Texas, SW USA
Toletum see Toledo
Toliara 79 F4 var. Toliary; prev. Tuléar. Toliara, SW Madagascar
Toliary see Toliara
Tolmein see Tolmin
Tolmin 95 D7 Ger. Tolmein, It. Tolmino. W Slovenia
Tolmino see Tolmin
Tolna 99 C7 Ger. Tolnau. Tolna, S Hungary
Tolnau see Tolna
Tolochin see Talachyn
Tolosa 93 E1 País Vasco, N Spain
Tolosa see Toulouse
Toluca 51 E4 var. Toluca de Lerdo. México, S Mexico
Toluca de Lerdo see Toluca
Tol'yatti 111 C6 prev. Stavropol'. Samarskaya Oblast', W Russian Federation
Tomah 40 B2 Wisconsin, N USA
Tomakomai 130 D2 Hokkaidô, NE Japan
Tomar 92 B3 Santarém, W Portugal
Tomaschow see Tomaszów Mazowiecki
Tomaschow see Tomaszów Lubelski
Tomaszów see Tomaszów Mazowiecki
Tomaszów Lubelski 98 E4 Ger. Tomaschow. Lubelskie, E Poland
Tomaszów Mazowiecka see Tomaszów Mazowiecki
Tomaszów Mazowiecki 98 D4 var. Tomaszów Mazowiecka; prev. Tomaszów, Ger. Tomaschow. Łódzkie, C Poland
Tombigbee River 42 C3 river Alabama/ Mississippi, S USA
Tombouctou 75 E3 Eng. Timbuktu. Tombouctou, N Mali
Tombua 78 A2 Port. Porto Alexandre. Namibe, SW Angola
Tomelloso 93 E3 Castilla-La Mancha, C Spain
Tomini, Gulf of see Teluk Tomini
Tomini, Teluk 139 E4 Eng. Gulf of Tomini; prev. Teluk Gorontalo. bay Sulawesi, C Indonesia
Tomok 123 G2 prev. Tokmak. Chuyskaya Oblast', N Kyrgyzstan
Tomsk 114 D4 Tomskaya Oblast', C Russian Federation
Tomür Feng 126 B3 pre. Pik Pobedy; Kyrg. Jengish Chokusu. mountain China/Kyrgyzstan
Toncontín 52 C3 prev. Tegucigalpa. country capital (Honduras) Francisco Morazán, SW Honduras
Tonezh see Tonyezh
Tonga 145 E4 off. Kingdom of Tonga, var. Friendly Islands. country SW Pacific Ocean

TONGA
Australasia & Oceania

Official name Kingdom of Tonga
Formation 1970 / 1970
Capital Nuku'alofa
Population 106,322 / 382 people per sq mile (148 people per sq km)
Total area 289 sq. miles (748 sq. km)
Languages English*, Tongan*
Religions Free Wesleyan 41%, Other 17%, Roman Catholic 16%, Church of Jesus Christ of Latter-day Saints 14%, Free Church of Tonga 12%
Ethnic mix Tongan 98%, Other 2%
Government Monarchy
Currency Pa'anga (Tongan dollar) = 100 seniti
Literacy rate 99%
Calorie consumption Not available

Tonga, Kingdom of see Tonga
Tongatapu 145 E5 island Tongatapu Group, S Tonga
Tongatapu Group 145 E5 island group S Tonga

Tonga Trench 143 E3 trench S Pacific Ocean
Tongchuan 128 C4 Shaanxi, C China
Tongeren 87 D6 Fr. Tongres. Limburg, NE Belgium
Tongking, Gulf of see Tonkin, Gulf of
Tongliao 127 G2 Nei Mongol Zizhiqu, N China
Tongres see Tongeren
Tongshan see Xuzhou, Jiangsu, China
Tongtian He 126 C4 var. Zhi Qu. river C China
Tonj 73 B5 Warap, W South Sudan
Tonkin, Gulf of 128 B7 var. Tongking Gulf of, Chin. Beibu Wan, Vtn. Vinh Bắc Bô. gulf China/Vietnam
Tônlé Sap 137 D5 Eng. Great Lake. lake W Cambodia
Tonopah 47 C6 Nevada, W USA
Tonyezh 107 C7 Rus. Tonezh. Homyel'skaya Voblasts', SE Belarus
Tooele 44 B4 Utah, W USA
Toowoomba 149 E5 Queensland, E Australia
Topeka 45 F4 state capital Kansas, C USA
Toplicza see Toplița
Toplița 108 C3 Ger. Töplitz, Hung. Maroshevíz; prev. Toplița Română, Hung. Oláh-Toplicza, Toplicza. Harghita, C Romania
Toplița Română/Töplitz see Toplița
Topol'čany 99 C6 Hung. Nagytapolcsány. Nitrianský Kraj, W Slovakia
Topolovgrad 104 D3 prev. Kavaklı. Khaskovo, S Bulgaria
Topolya see Bačka Topola
Top Springs Roadhouse 148 E3 Northern Territory, N Australia
Torda see Turda
Torez 109 H3 Donets'ka Oblast', SE Ukraine
Torgau 94 D4 Sachsen, E Germany
Torhout 87 A5 West-Vlaanderen, W Belgium
Torino 96 A2 Eng. Turin. Piemonte, NW Italy
Tornacum see Tournai
Torneå see Tornio
Torneträsk 84 C3 lake N Sweden
Tornio 84 D4 Swe. Torneå. Lappi, NW Finland
Tornionjoki 84 D3 river Finland/Sweden
Tornionjoki 84 D3 river Finland/Sweden
Toro 92 D2 Castilla y León, N Spain
Toronto 38 D5 province capital Ontario, S Canada
Torquay 89 C7 SW England, United Kingdom
Torrance 46 D2 California, W USA
Torre, Alto da 92 B3 mountain C Portugal
Torre del Greco 97 D5 Campania, S Italy
Torrejón de Ardoz 93 E3 Madrid, C Spain
Torrelavega 92 D1 Cantabria, N Spain
Torrens, Lake 149 A6 salt lake South Australia
Torrent 93 F3 Cas. Torrente, var. Torrent de l'Horta. Valenciana, E Spain
Torrent de l'Horta/Torrente see Torrent
Torreón 50 D3 Coahuila, NE Mexico
Torres Strait 148 C1 strait Australia/ Papua New Guinea
Torres Vedras 92 B3 Lisboa, C Portugal
Torrington 44 D3 Wyoming, C USA
Tórshavn 83 F5 Dan. Thorshavn. Dependent territory capital Faroe Islands
To'rtkŭl 122 D7 var. Türtkül, Rus. Turtkul'; prev. Petroaleksandrovsk. Qoraqalpog'iston Respublikasi, W Uzbekistan
Tortoise Islands see Galápagos Islands
Tortosa 93 F2 anc. Dertosa. Cataluña, E Spain
Tortue, Montagne 59 H3 mountain range C French Guiana
Tortuga, Isla see La Tortuga, Isla

ruń 98 C3 *Ger.* Thorn. Toruń, Kujawsko-pomorskie, C Poland

rva 106 D3 *Ger.* Tõrwa. Valgamaa, S Estonia

rwa *see* Tõrva

rzhok 110 B4 Tverskaya Oblast', W Russian Federation

sa-wan 131 B7 *bay* SW Japan

scana 96 B3 *Eng.* Tuscany. *region* C Italy

scano, Archipelago 96 B4 *Eng.* Tuscan Archipelago. *island group* C Italy

shkent 123 E2 *Eng./Rus.* Tashkent. *country capital* (Uzbekistan) Toshkent Viloyati, E Uzbekistan

tana 93 E4 Murcia, SE Spain

t'ma *see* Sukhona

ttori 131 B6 Tottori, Honshū, SW Japan

uájil 74 C2 Tiris Zemmour, N Mauritania

uggourt 71 E2 NE Algeria

ukoto 74 C3 Kayes, W Mali

ul 90 D3 Meurthe-et-Moselle, NE France

ulon 91 D6 *anc.* Telo Martius, Tilio Martius. Var, SE France

ulouse 91 B6 *anc.* Tolosa. Haute-Garonne, S France

ungoo *see* Taungoo

uraine 90 B4 *cultural region* C France

urane *see* Ða Nãng

urcoing 90 C2 Nord, N France

urnai 87 A6 *var.* Tournay, *Dut.* Doornik; *anc.* Tornacum. Hainaut, SW Belgium

urnay *see* Tournai

urs 90 B4 *anc.* Caesarodunum, Turoni. Indre-et-Loire, C France

varkovskiy 111 B5 Tul'skaya Oblast', W Russian Federation

wer Island *see* Genovesa, Isla

wnsville 148 D3 Queensland, NE Australia

woeti Meer *see* Towuti, Danau

wraghoudi 122 D4 Herãt, NW Afghanistan

wson 41 F4 Maryland, NE USA

wuti, Danau 139 E4 *Dut.* Towoeti Meer. *lake* Sulawesi, C Indonesia

yama 131 C5 Toyama, Honshū, SW Japan

yama-wan 131 B5 *bay* W Japan

yohara *see* Yuzhno-Sakhalinsk

yota 131 C6 Aichi, Honshū, SW Japan

yzeur 71 E2 *var.* Tawzar. W Tunisia

ablous *see* Tripoli

abzon 117 E2 *Eng.* Trebizond; *anc.* Trapezus, *Tib.* Tarabosan. NE Turkey

aiectum ad Mosam/Traiectum Tungorum *see* Maastricht

aiskirchen 95 E6 Niederösterreich, NE Austria

rajani Portus *see* Civitavecchia

rajectum ad Rhenum *see* Utrecht

rakai 107 C5 *Ger.* Traken, *Pol.* Troki. Vilnius, SE Lithuania

raken *see* Trakai

ralee 89 A6 *Ir.* Trá Lí. SW Ireland

rá Lí *see* Tralee

ralles Aydin *see* Aydın

rang 137 C7 Trang, S Thailand

ransantarctic Mountains 154 B3 *mountain range* Antarctica

ransilvania *see* Transylvania

ransilvaniei, Alpi *see* Carpații Meridionalii

ransjordan *see* Jordan

ransnistria 108 D3 *cultural region* NE Moldova

ranssylvanische Alpen/ Transylvanian Alps *see* Carpații Meridionalii

ransylvania 108 B4 *Eng.* Ardeal, Transilvania, *Ger.* Siebenbürgen, *Hung.* Erdély. *cultural region* NW Romania

ransylvanian Alps 108 B4 *var.* Alpi Transilvaniei, Carpații Sudici, *Eng.* South Carpathians, Transylvanian Alps, *Ger.* Südkarpaten, Transsylvanische Alpen, *Hung.* Déli-Kárpátok, Erdélyi-Havasok. *mountain range* C Romania

rapani 97 B7 *anc.* Drepanum. Sicilia, Italy, C Mediterranean Sea

Trãpeãng Vẽng *see* Kâmpóng Thum

Trapezus *see* Trabzon

Traralgon 149 C7 Victoria, SE Australia

Trasimenischersee *see* Trasimeno, Lago

Trasimeno, Lago 96 C4 *Eng.* Lake of Perugia, *Ger.* Trasimenischersee. *lake* C Italy

Traù *see* Trogir

Traverse City 40 C2 Michigan, N USA

Tra Vinh 137 D6 *var.* Phu Vinh. Tra Vinh, S Vietnam

Travis, Lake 49 F3 *reservoir* Texas, SW USA

Travnik 100 C4 Federacija Bosne I Hercegovine, C Bosnia and Herzegovina

Trbovlje 95 E7 *Ger.* Trifail. C Slovenia

Treasure State *see* Montana

Třebíč 99 B5 *Ger.* Trebitsch. Vysočina, C Czech Republic

Trebinje 101 C5 Republika Srpska, S Bosnia and Herzegovina

Trebišov 99 D6 *Hung.* Tőketerebes. Košický Kraj, E Slovakia

Trebitsch *see* Třebíč

Trebnitz *see* Trzebnica

Tree Planters State *see* Nebraska

Trélazé 90 B4 Maine-et-Loire, NW France

Trelew 65 C6 Chubut, SE Argentina

Tremelo 87 C5 Vlaams Brabant, C Belgium

Trenčín 99 C5 *Ger.* Trentschin, *Hung.* Trencsén. Trenčiansky Kraj, W Slovakia

Trencsén *see* Trenčín

Trengganu, Kuala *see* Kuala Terengganu

Trenque Lauquen 64 C4 Buenos Aires, E Argentina

Trent *see* Trento

Trento 96 C2 *Eng.* Trent, *Ger.* Trient; *anc.* Tridentum. Trentino-Alto Adige, N Italy

Trenton 41 F4 *state capital* New Jersey, NE USA

Trentschin *see* Trenčín

Tres Arroyos 65 D5 Buenos Aires, E Argentina

Treskavica 100 C4 *mountain range* SE Bosnia and Herzegovina

Tres Tabernae *see* Saverne

Treves/Trèves *see* Trier

Treviso 96 C2 *anc.* Tarvisium. Veneto, NE Italy

Trichinopoly *see* Tiruchchirāppalli

Trichūr *see* Thrissur

Tridentum/Trient *see* Trento

Trier 95 A5 *Eng.* Treves, *Fr.* Trèves; *anc.* Augusta Treverorum. Rheinland-Pfalz, SW Germany

Triesen 94 E2 SW Liechtenstein Europe

Triesenberg 94 E2 SW Liechtenstein

Trieste 96 D2 *Slvn.* Trst. Friuli-Venezia Giulia, NE Italy

Trifail *see* Trbovlje

Tríkala 104 B4 *prev.* Trikkala. Thessalía, C Greece

Trikkala *see* Tríkala

Trimontium *see* Plovdiv

Trinacria *see* Sicilia

Trincomalee 132 D3 *var.* Trinkomali. Eastern Province, NE Sri Lanka

Trindade, Ilha da 67 C5 *island* Brazil, A' Atlantic Ocean

Trinidad 61 F3 El Beni, N Bolivia

Trinidad 60 A4 Flores, S Uruguay

Trinidad 44 D5 Colorado, C USA

Trinidad 55 H5 *island* C Trinidad and Tobago

Trinidad and Tobago 55 H5 *off.* Republic of Trinidad and Tobago. *country* SE West Indies

TRINIDAD & TOBAGO
West Indies

Official name Republic of Trinidad and Tobago
Formation 1962 / 1962
Capital Port-of-Spain
Population 1.3 million / 656 people per sq mile (254 people per sq km)
Total area 1980 sq. miles (5128 sq. km)
Languages English Creole, English*, Hindi, French, Spanish

TRINIDAD & TOBAGO
(continued)

Religions Roman Catholic 26%, Hindu 23%, Other and nonreligious 23%, Anglican 8%, Baptist 7%, Pentecostal 7%, Muslim 6%
Ethnic mix East Indian 40%, Black 38%, Mixed race 20%, White and Chinese 1% Other 1%
Government Parliamentary system
Currency Trinidad and Tobago dollar = 100 cents
Literacy rate 99%
Calorie consumption 2889 kilocalories

Trinidad and Tobago, Republic of *see* Trinidad and Tobago

Trinité, Montagnes de la 59 H3 *mountain range* C French Guiana

Trinity River 49 G3 *river* Texas, SW USA

Trinkomali *see* Trincomalee

Trípoli 105 B6 *prev.* Trípolis. Pelopónnisos, S Greece

Tripoli 118 B4 *var.* Tarãbulus, Țarãbulus ash Shãm, Trãblous; *anc.* Tripolis. N Lebanon

Tripoli *see* Țarãbulus

Tripolis *see* Trípoli, Greece

Tripolis *see* Tripoli, Lebanon

Tristan da Cunha 69 B7 *UK dependent territory* SE Atlantic Ocean

Triton Island 128 B7 *island* S Paracel Islands

Trivandrum *see* Thiruvananthapuram

Trnava 99 C6 *Ger.* Tyrnau, *Hung.* Nagyszombat. Trnavský Kraj, W Slovakia

Trnovo *see* Veliko Tŭrnovo

Trogir 100 B4 *It.* Traù. Split-Dalmacija, S Croatia

Troglav 100 B4 *mountain* Bosnia and Herzegovina/Croatia

Trois-Rivières 39 E4 Québec, SE Canada

Troki *see* Trakai

Trollhättan 85 B6 Västra Götaland, S Sweden

Tromsø 84 C2 *Fin.* Tromssa. Troms, N Norway

Tromssa *see* Tromsø

Trondheim 84 B4 *Ger.* Drontheim; *prev.* Nidaros, Trondhjem. Sør-Trøndelag, S Norway

Trondheimsfjorden 84 B4 *fjord* S Norway

Trondhjem *see* Trondheim

Troódos 102 C5 *var.* Troodos Mountains. *mountain range* C Cyprus

Troodos Mountains *see* Troódos

Troppau *see* Opava

Troy 42 D3 Alabama, S USA

Troy 41 F3 New York, NE USA

Troyan 104 C2 Lovech, N Bulgaria

Troyes 90 D3 *anc.* Augustobona Tricassium. Aube, N France

Trst *see* Trieste

Trstenik 100 E4 Serbia, C Serbia

Trucial States *see* United Arab Emirates

Trujillo 118 C3 *var.* Tadmur, Tamar, *Gk.* Palmyra, *Bibl.* Tadmor. Ḥimṣ, C Syria

Trujillo 60 B3 La Libertad, NW Peru

Trujillo 92 C3 Extremadura, W Spain

Truk Islands *see* Chuuk Islands

Trŭn 104 B2 Pernik, W Bulgaria

Truro 39 F4 Nova Scotia, SE Canada

Truro 89 C7 SW England, United Kingdom

Trzcianka 98 B3 *Ger.* Schönlanke. Pila, Wielkopolskie, C Poland

Trzebnica 98 C4 *Ger.* Trebnitz. Dolnoślaskie, SW Poland

Tsalka 117 F5 *var.* Chalkidike, C Aegean Sea

Tsamkong *see* Zhanjiang

Tsangpo *see* Brahmaputra

Tsarevo 104 E2 *prev.* Michurin. Burgas, E Bulgaria

Tsarigrad *see* İstanbul

Tsaritsyn *see* Volgograd

Tschakathurn *see* Čakovec

Tschaslau *see* Čáslav

Tschenstochau *see* Częstochowa

Tsefat 119 B5 *var.* Safed, *Ar.* Safad; *prev.* Zefat. Northern, N Israel

Tselinograd *see* Astana

Tsetsen Khan *see* Öndörhaan

Tsetserleg 126 D2 Arhangay, C Mongolia

Tshela 77 B6 Bas-Congo, W Dem. Rep. Congo

Tshikapa 77 C7 Kasai-Occidental, SW Dem. Rep. Congo

Tshuapa 77 D6 *river* C Dem. Rep. Congo

Tsinan *see* Jinan

Tsing Hai *see* Qinghai Hu, China

Tsinghai *see* Qinghai, China

Tsingtao/Tsingtau *see* Qingdao

Tsinkiang *see* Quanzhou

Tsintao *see* Qingdao

Tsitsihar *see* Qiqihar

Tsu 131 C6 *var.* Tu. Mie, Honshū, SW Japan

Tsugaru-kaikyo 130 C3 *strait* N Japan

Tsumeb 78 B3 Oshikoto, N Namibia

Tsuruga 131 C6 *var.* Turuga. Fukui, Honshū, SW Japan

Tsuruoka 130 D4 *var.* Turuoka. Yamagata, Honshū, C Japan

Tsushima 131 A7 *var.* Tsushima-tō, Tusima. *island group* SW Japan

Tsushima-tō *see* Tsushima

Tsyerakhowka 107 D8 *Rus.* Terekhovka. Homyel'skaya Voblasts', SE Belarus

Tsyurupinsk *see* Tsyurupyns'k

Tsyurupyns'k 109 E4 *Rus.* Tsyurupinsk. Khersons'ka Oblast', S Ukraine

Tu *see* Tsu

Tuamotu, Archipel des *see* Tuamotu, Îles

Tuamotu Fracture Zone 143 H3 *fracture zone* E Pacific Ocean

Tuamotu Islands 145 H4 *var.* Archipel des Tuamotu, Dangerous Archipelago, Tuamotu Islands. *island group* N French Polynesia

Tuamotu Islands *see* Tuamotu, Îles

Tuapi 53 E2 Región Autónoma Atlántico Norte, NE Nicaragua

Tuapse 111 A7 Krasnodarskiy Kray, SW Russian Federation

Tuba City 48 B1 Arizona, SW USA

Tubbergen 86 E3 Overijssel, E Netherlands

Tubeke *see* Tubize

Tubize 87 B6 *Dut.* Tubeke. Walloon Brabant, C Belgium

Tubmanburg 74 C5 NW Liberia

Tubruq 71 H2 *Eng.* Tobruk, *It.* Tobruch. NE Libya

Tubuai, Îles/Tubuai Islands *see* Australes, Îles

Tucker's Town 42 B5 E Bermuda

Tuckum *see* Tukums

Tucson 48 B3 Arizona, SW USA

Tucumán *see* San Miguel de Tucumán

Tucumcari 49 E2 New Mexico, S W USA

Tucupita 59 E2 Delta Amacuro, NE Venezuela

Tucuruí, Represa de 63 F2 *reservoir* NE Brazil

Tudela 93 E2 *Basq.* Tutera; *anc.* Tutela. Navarra, N Spain

Tudmur 118 C3 *var.* Tadmur, Tamar, *Gk.* Palmyra, *Bibl.* Tadmor. Ḥimṣ, C Syria

Tuguegarao 139 E1 Luzon, N Philippines

Tuktoyaktuk 37 E3 Northwest Territories, NW Canada

Tukums 106 C3 *Ger.* Tuckum. W Latvia

Tula 111 B5 Tul'skaya Oblast', W Russian Federation

Tulancingo 51 E4 Hidalgo, C Mexico

Tulare Lake Bed 47 C7 *salt flat* California, W USA

Tulcán 60 B1 Carchi, N Ecuador

Tulcea 108 D5 Tulcea, E Romania

Tul'chin *see* Tul'chyn

Tul'chyn 108 D3 *Rus.* Tul'chin. Vinnyts'ka Oblast', C Ukraine

Tuléar *see* Toliara

Tulia 49 E2 Texas, SW USA

Tūlkarm 119 D7 West Bank, Israel

Tulle 91 C5 *anc.* Tutela. Corrèze, C France

Tulln 95 E6 *var.* Oberhollabrunn. Niederösterreich, NE Austria

Tully 148 D3 Queensland, NE Australia

Tulsa 49 G1 Oklahoma, C USA

Tuluá 58 B3 Valle del Cauca, W Colombia

Tulun *115 E4* Irkutskaya Oblast', S Russian Federation
Tumaco *58 A4* Nariño, SW Colombia
Tumba, Lac *see* Ntomba, Lac
Tumbes *60 A2* Tumbes, NW Peru
Tumkūr *132 C2* Karnātaka, W India
Tumuc-Humac Mountains *63 E1 var.* Serra Tumucumaque. *mountain range* N South America
Tumucumaque, Serra *see* Tumuc-Humac Mountains
Tunca Nehri *see* Tundzha
Tunduru *73 C8* Ruvuma, S Tanzania
Tundzha *104 D3 Turk.* Tunca Nehri. *river* Bulgaria/Turkey
Tungabhadra Reservoir *132 C2 lake* S India
Tungaru *145 E2 prev.* Gilbert Islands. *island group* W Kiribati
T'ung-shan *see* Xuzhou
Tungsten *36 D4* Northwest Territories, W Canada
Tung-t'ing Hu *see* Dongting Hu
Tunis *71 E1 var.* Tūnis. *country capital* (Tunisia) NE Tunisia
Tunis, Golfe de *102 D3 Ar.* Khalīj Tūnis. *gulf* NE Tunisia
Tunisia *71 F2 off.* The Tunisian Republic, *Ar.* Al Jumhūrīyah at Tūnisīyah, *Fr.* République Tunisienne. *country* N Africa

---

TUNISIA
North Africa

Official name The Tunisian Republic
Formation 1956 / 1956
Capital Tunis
Population 11 million / 183 people per sq mile (71 people per sq km)
Total area 63,169 sq. miles (163,610 sq. km)
Languages Arabic*, French
Religions Muslim (mainly Sunni) 98%, Christian 1%, Jewish 1%
Ethnic mix Arab and Berber 98%, Jewish 1%, European 1%
Government Transitional regime
Currency Tunisian dinar = 1000 millimes
Literacy rate 80%
Calorie consumption 3362 kilocalories

Tunisian Republic *see* Tunisia
Tunisienne, République *see* Tunisia
Tūnisīyah, Al Jumhūrīyah at *see* Tunisia
Tūnis, Khalīj *see* Tunis, Golfe de
Tunja *58 B3* Boyacá, C Colombia
Tuong Buong *see* Tương Đương
Tương Đương *136 D4 var.* Tuong Buong. Nghê An, N Vietnam
Tüp *see* Tyup
Tupelo *42 C2* Mississippi, S USA
Tupiza *61 G5* Potosí, S Bolivia
Turabah *121 B5* Makkah, W Saudi Arabia
Turangi *150 D4* Waikato, North Island, New Zealand
Turan Lowland *122 C2 var.* Turan Plain, *Kaz.* Turan Oypaty, *Rus.* Turanskaya Nizmennost', *Turk.* Turan Pesligi, *Uzb.* Turan Pasttekisligi. *plain* C Asia
Turan Oypaty/Turan Pesligi/Turan Plain/Turanskaya Nizmennost' *see* Turan Lowland
Turan Pasttekisligi *see* Turan Lowland
Ṭurayf *120 A3* Al Ḥudūd ash Shamālīyah, NW Saudi Arabia
Turba *see* Teruel
Turbat *134 A3* Baluchistān, SW Pakistan
Turčiansky Svätý Martin *see* Martin
Turda *108 B4 Ger.* Thorenburg, *Hung.* Torda. Cluj, NW Romania
Turek *98 C3* Wielkopolskie, C Poland
Turfan *see* Turpan
Turin *see* Torino
Turkana, Lake *73 C6 var.* Lake Rudolf. *lake* N Kenya
Turkestan *see* Turkistan
Turkey *116 B3 off.* Republic of Turkey, *Turk.* Türkiye Cumhuriyeti. *country* SW Asia

---

TURKEY
Asia / Europe

Official name Republic of Turkey
Formation 1923 / 1939

TURKEY
(continued)

Capital Ankara
Population 74.9 million / 252 people per sq mile (97 people per sq km)
Total area 301,382 sq. miles (780,580 sq. km)
Languages Turkish*, Kurdish, Arabic, Circassian, Armenian, Greek, Georgian, Ladino
Religions Muslim (mainly Sunni) 99%, Other 1%
Ethnic mix Turkish 70%, Kurdish 20%, Other 8%, Arab 2%
Government Parliamentary system
Currency new Turkish lira = 100 kurus
Literacy rate 95%
Calorie consumption 3680 kilocalories

Turkey, Republic of *see* Turkey
Turkish Republic of Northern Cyprus *102 D5 disputed territory* Cyprus
Turkistan *114 B5 prev.* Turkestan. Yuzhnyy Kazakhstan, S Kazakhstan
Türkiye Cumhuriyeti *see* Turkey
Türkmenabat *122 D3 prev. Rus.* Chardzhev, Chardzhou, Chardzhui, Lenin-Turkmenski, *Turkm.* Chärjew. Lebap Welaýaty, E Turkmenistan
Türkmen Aylagy *122 B2 Rus.* Turkmenskiy Zaliv. *lake gulf* W Turkmenistan
Turkmenbashi *see* Türkmenbaşy
Türkmenbaşy *122 B2 Rus.* Turkmenbashi; *prev.* Krasnovodsk. Balkan Welaýaty, W Turkmenistan
Türkmenbaşy Aylagy *122 A2 prev. Rus.* Krasnovodskiy Zaliv, *Turkm.* Krasnowodsk Aylagy. *lake Gulf* W Turkmenistan
Turkmenistan *122 B2 ; prev.* Turkmenskaya Soviet Socialist Republic. *country* C Asia

---

TURKMENISTAN
Central Asia

Official name Turkmenistan
Formation 1991 / 1991
Capital Ashgabat
Population 5.1 million / 27 people per sq mile (10 people per sq km)
Total area 188,455 sq. miles (488,100 sq. km)
Languages Turkmen*, Uzbek, Russian, Kazakh, Tatar
Religions Sunni Muslim 89%, Orthodox Christian 9%, Other 2%
Ethnic mix Turkmen 85%, Other 6%, Uzbek 5%, Russian 4%
Government Presidential system
Currency Manat = 100 tenge
Literacy rate 99%
Calorie consumption 2883 kilocalories

Turkmenskaya Soviet Socialist Republic *see* Turkmenistan
Turkmenskiy Zaliv *see* Türkmen Aylagy
Turks and Caicos Islands *55 E2 UK dependent territory* N West Indies
Turku *85 D6 Swe.* Åbo. Länsi-Suomi, SW Finland
Turlock *47 B6* California, W USA
Turnagain, Cape *150 D4 headland* North Island, New Zealand
Turnau *see* Turnov
Turnhout *87 C5* Antwerpen, N Belgium
Turnov *98 B4 Ger.* Turnau. Liberecký Kraj, N Czech Republic
Turnovo *see* Veliko Tŭrnovo
Turnu Măgurele *108 B5 var.* Turnu-Măgurele. Teleorman, S Romania
Turnu Severin *see* Drobeta-Turnu Severin
Turócszentmárton *see* Martin
Turoni *see* Tours
Turpan *126 C3 var.* Turfan. Xinjiang Uygur Zizhiqu, NW China
Turpan Depression *see* Turpan Pendi
Turpan Pendi *126 C3 Eng.* Turpan Depression. *depression* NW China
Turpentine State *see* North Carolina
Turuga *see* Tsuruga

Turuoka *see* Tsuruoka
Tuscaloosa *42 C2* Alabama, S USA
Tuscan Archipelago *see* Toscano, Archipelago
Tuscany *see* Toscana
Tusima *see* Tsushima
Tutela *see* Tulle, France
Tutela *see* Tudela, Spain
Tutera *see* Tudela
Tuticorin *132 C3* Tamil Nādu, SE India
Tutrakan *104 D1* Silistra, NE Bulgaria
Tutuila *145 F4 island* W American Samoa
Tuvalu *145 E3 prev.* Ellice Islands. *country* SW Pacific Ocean

---

TUVALU
Australasia & Oceania

Official name Tuvalu
Formation 1978 / 1978
Capital Funafuti Atoll
Population 10,698 / 1070 people per sq mile (411 people per sq km)
Total area 10 sq. miles (26 sq. km)
Languages Tuvaluan, Kiribati, English*
Religions Church of Tuvalu 97%, Bahá'í 1%, Seventh-day Adventist 1%, Other 1%
Ethnic mix Polynesian 96%, Micronesian 4%
Government Nonparty system
Currency = 100 cents; Tuvaluan dollar = 100 cents
Literacy rate 98%
Calorie consumption Not available

Tuwayq, Jabal *121 C5 mountain range* C Saudi Arabia
Tuxpan *50 D4* Jalisco, C Mexico
Tuxpan *50 D4* Nayarit, C Mexico
Tuxpán *51 F4 var.* Tuxpán de Rodríguez Cano. Veracruz-Llave, E Mexico
Tuxpán de Rodríguez Cano *see* Tuxpán
Tuxtepec *51 F4 var.* San Juan Bautista Tuxtepec. Oaxaca, S Mexico
Tuxtla *51 G5 var.* Tuxtla Gutiérrez. Chiapas, SE Mexico
Tuxtla *see* San Andrés Tuxtla
Tuxtla Gutiérrez *see* Tuxtla
Tuy Hoa *137 E5* Phu Yên, S Vietnam
Tuz, Lake *116 C3 lake* C Turkey
Tver' *110 B4 prev.* Kalinin. Tverskaya Oblast', W Russian Federation
Tverya *119 B5 var.* Tiberias; *prev.* Teverya. Northern, N Israel
Twin Falls *46 D4* Idaho, NW USA
Tyan'-Shan' *see* Tien Shan
Tychy *99 D5 Ger.* Tichau. Śląskie, S Poland
Tyler *49 G3* Texas, SW USA
Tylos *see* Bahrain
Tympáki *105 C8 var.* Timbaki; *prev.* Timbákion. Kríti, Greece, E Mediterranean Sea
Tynda *115 F4* Amurskaya Oblast', SE Russian Federation
Tyne *88 D4 river* N England, United Kingdom
Tyôsi *see* Chôshi
Tyras *see* Dniester
Tyre *see* Soûr
Tyrnau *see* Trnava
Týrnavos *104 B4 var.* Tírnavos. Thessalía, C Greece
Tyrol *see* Tirol
Tyros *see* Bahrain
Tyrrhenian Sea *97 B6 It.* Mare Tirreno. *sea* N Mediterranean Sea
Tyumen' *114 C3* Tyumenskaya Oblast', C Russian Federation
Tyup *123 G2 Kir.* Tüp. Issyk-Kul'skaya Oblast', NE Kyrgyzstan
Tywyn *89 C6* W Wales, United Kingdom
Tzekung *see* Zigong
Tziá *105 C6 prev.* Kéa, Kéos; *anc.* Ceos. *island* Kykládes, Greece, Aegean Sea

---

U

Uaco Cungo *78 B1* C Angola
UAE *see* United Arab Emirates
Uanle Uen *see* Wanlaweyn
Uaupés, Rio *see* Vaupés, Río

Ubangi-Shari *see* Central African Republic
Ube *131 B7* Yamaguchi, Honshū, SW Japan
Ubeda *93 E4* Andalucía, S Spain
Uberaba *63 F4* Minas Gerais, SE Brazil
Uberlândia *63 F4* Minas Gerais, S E Brazil
Ubol Rajadhani/Ubol Ratchathani *see* Ubon Ratchathani
Ubon Ratchathani *137 D5 var.* Muang Ubon, Ubol Rajadhani, Ubol Ratchathani, Ubon Ratchathani. Ubon Ratchathani, E Thailand
Ubrique *92 D5* Andalucía, S Spain
Ubsu-Nur, Ozero *see* Uvs Nuur
Ucayali, Río *60 D3 river* C Peru
Uchiura-wan *130 D3 bay* NW Pacific Ocean
Uchkuduk *see* Uchquduq
Uchquduq *122 D2 Rus.* Uchkuduk. Navoiy Viloyati, N Uzbekistan
Uchtagan Gumy/Uchtagan, Peski *see* Uçtagan Gumy
Uçtagan Gumy *122 C2 var.* Uchtaga Gumy, *Rus.* Peski Uchtagan. *desert* NW Turkmenistan
Udaipur *134 C3 prev.* Oodeypore. Rājasthān, N India
Uddevalla *85 B6* Västra Götaland, S Sweden
Udine *96 D2 anc.* Utina. Friuli-Venezia Giulia, NE Italy
Udintsev Fracture Zone *154 A5 tectonic feature* S Pacific Ocean
Udipi *see* Udupi
Udon Ratchathani *see* Ubon Ratchathani
Udon Thani *136 C4 var.* Ban Mak Khaeng, Udorndhani. Udon Thani, N Thailand
Udorndhani *see* Udon Thani
Udupi *132 B2 var.* Udipi. Karnātaka, SW India
Uele *77 D5 var.* Welle. *river* NE Dem. Rep. Congo
Uelzen *94 C3* Niedersachsen, N Germany
Ufa *111 D6* Respublika Bashkortostan, W Russian Federation
Ugâle *106 C2* NW Latvia
Uganda *73 B6 off.* Republic of Uganda. *country* E Africa

UGANDA
East Africa

Official name Republic of Uganda
Formation 1962 / 1962
Capital Kampala
Population 37.6 million / 488 people per sq mile (188 people per sq km)
Total area 91,135 sq. miles (236,040 sq. km)
Languages Luganda, Nkole, Chiga, Lango, Acholi, Teso, Lugbara, English*
Religions Christian 85%, Muslim (mainly Sunni) 12%, Other 3%
Ethnic mix Other 50%, Baganda 17%, Banyakole 10%, Basoga 9%, Iteso 7%, Bakiga 7%
Government Presidential system
Currency Uganda shilling = 100 cents
Literacy rate 73%
Calorie consumption 2279 kilocalories

Uganda, Republic of *see* Uganda
Uhorshchyna *see* Hungary
Uhuru Peak *see* Kilimanjaro
Uíge *78 B1* Port. Carmona, Vila Marechal Carmona. Uíge, NW Angola
Uinta Mountains *44 B4 mountain ra* Utah, W USA
Uitenhage *78 C5* Eastern Cape, S Sou Africa
Uithoorn *86 C3* Noord-Holland, C Netherlands
Ujda *see* Oujda
Ujelang Atoll *144 C1 var.* Wujlān. *a* Ralik Chain, W Marshall Islands
Ujgradiska *see* Nova Gradiška
Ujmoldova *see* Moldova Nouă
Ujpand, Pulau *see* Makassar
Ujung Salang *see* Phuket
Ujvidék *see* Novi Sad
UK *see* United Kingdom

ta *114 C3* Respublika Komi,
W Russian Federation
ah *47 B5* California, W USA
nergë *106 C4* Pol. Wilkomierz.
ilnius, C Lithuania
aina *see* Ukraine
aine *108 C2 off.* Ukraine, *Rus.*
kraina, *Ukr.* Ukrayina; *prev.*
krainian Soviet Socialist Republic,
krainskay S.S.R. *country* SE Europe

RAINE
Eastern Europe

ficial name Ukraine
mation 1991 / 1991
pital Kiev
pulation 45.2 million / 194 people
sq mile (75 people per sq km)
al area 233,089 sq. miles
3,700 sq. km)
guages Ukrainian*, Russian, Tatar
igions Christian (mainly Orthodox)
%, Other 5%
nic mix Ukrainian 78%, Russian 17%,
her 5%
vernment Presidential system
rrency Hryvna = 100 kopiykas
eracy rate 99%
lorie consumption 3142 kilocalories

aine *see* Ukraine
ainian Soviet Socialist Republic
e Ukraine
ainskay S.S.R/Ukrayina *see*
kraine
anbaatar *127 E2 Eng.* Ulan Bator;
rev. Urga. *country capital* (Mongolia)
öv, C Mongolia
ngom *126 C2* Uvs, NW Mongolia
n Bator *see* Ulaanbaatar
nhad *see* Chifeng
n-Ude *115 E4 prev.* Verkhneudinsk.
espublika Buryatiya, S Russian
ederation
ilv *see* Oulujoki
räsk *see* Oulujärvi
: *86 E4* Gelderland, E Netherlands
apool *88 C3* N Scotland, United
ingdom
n *95 B6* Baden-Württemberg,
Germany
an *129 E4 Jap.* Urusan. SE South
orea
er *89 B5 province* Northern Ireland,
nited Kingdom/Ireland
ngur Hu *126 B2* lake NW China
ru *147 D5 var.* Ayers Rock. *monolith*
orthern Territory, C Australia
anovka *109 E3* Kirovohrads'ka
blast', C Ukraine
anovsk *111 C5 prev.* Simbirsk.
l'yanovskaya Oblast', W Russian
ederation
án *51 H3* Yucatán, SE Mexico
an' *109 E3 Rus.* Uman. Cherkas'ka
blast', C Ukraine
an *see* Uman'
anak/Uumanaq *see* Uummannaq
ān, Khalij *see* Oman, Gulf of
ān, Saltanat *see* Oman
brian-Machigiano, Appennino *see*
Jmbro-Marchigiano, Appennino
bro-Marchigiano, Appennino
6 *C3 Eng.* Umbrian-Machigian
ountains. *mountain range* C Italy
eä *84 C4* Västerbotten, N Sweden
eälven *84 C4 river* N Sweden
iat *36 D2* Alaska, USA
m Buru *72 A4* Western Darfur,
/ Sudan
m Durmān *see* Omdurman
m Ruwaba *72 C4 var.* Umm
uwābah, Um Ruwāba. Northern
ordofan, C Sudan
m Ruwābah *see* Umm Ruwaba
nak Island *36 A3 island* Aleutian
lands, Alaska, USA
Ruwāba *see* Umm Ruwaba
tali *see* Mutare
tata *see* Mthatha
a *100 B3 river* Bosnia and
erzegovina/Croatia
c *100 B3 river* W Bosnia and
erzegovina
laska Island *36 A3 island* Aleutian
lands, Alaska, USA

'Unayzah *120 B4 var.* Anaiza. Al Qaşīm,
C Saudi Arabia
Unci *see* Almería
Uncía *61 F4* Potosí, C Bolivia
Uncompahgre Peak *44 B5 mountain*
Colorado, C USA
Undur Khan *see* Öndörhaan
Ungaria *see* Hungary
Ungarisches Erzgebirge *see* Slovenské
rudohorie
Ungarn *see* Hungary
Ungava Bay *39 E1 bay* Québec,
E Canada
Ungava Peninsula *see* Ungava,
Péninsule d'
Ungava, Péninsule d' *38 D1 Eng.*
Ungava Peninsula. *peninsula* Québec,
SE Canada
Ungeny *see* Ungheni
Ungheni *108 D3 Rus.* Ungeny.
W Moldova
Unguja *see* Zanzibar
Üngüz Angyrsyndaky Garagum *122 C2*
Rus. Zaunguzskiye Garagumy. *desert*
N Turkmenistan
Ungvár *see* Uzhhorod
Unimak Island *36 B3 island* Aleutian
Islands, Alaska, USA
Union *43 E1* South Carolina, SE USA
Union City *42 C1* Tennessee, S USA
Union of Myanmar *see* Burma
United Arab Emirates *121 C5 Ar.* Al
Imārāt al 'Arabīyah al Muttaḩidah,
abbrev. UAE; *prev.* Trucial States.
country SW Asia

UNITED ARAB EMIRATES
Southwest Asia

Official name United Arab Emirates
Formation 1971 / 1972
Capital Abu Dhabi
Population 9.3 million / 288 people
per sq mile (111 people per sq km)
Total area 32,000 sq. miles (82,880 sq. km)
Languages Arabic*, Farsi, Indian and
Pakistani languages, English
Religions Muslim (mainly Sunni) 96%,
Christian, Hindu, and other 4%
Ethnic mix Asian 60%, Emirian 25%,
Other Arab 12%, European 3%
Government Monarchy
Currency UAE dirham = 100 fils
Literacy rate 90%
Calorie consumption 3215 kilocalories

United Arab Republic *see* Egypt
United Kingdom *89 B5*
off. United Kingdom of Great Britain
and Northern Ireland, *abbrev.* UK.
country NW Europe

UNITED KINGDOM
Northwest Europe

Official name United Kingdom of Great
Britain and Northern Ireland
Formation 1707 / 1922
Capital London
Population 63.1 million / 676 people
per sq mile (261 people per sq km)
Total area 94,525 sq. miles
(244,820 sq. km)
Languages English*, Welsh* *(in Wales)*,
Scottish Gaelic, Irish Gaelic
Religions Anglican 45%, Other and
nonreligious 36%, Roman Catholic 9%,
Presbyterian 4%, Muslim 3%,
Methodist 2%, Hindu 1%
Ethnic mix English 80%, Scottish 9%,
West Indian, Asian, and other 5%,
Northern Irish 3%, Welsh 3%
Government Parliamentary system
Currency Pound sterling = 100 pence
Literacy rate 99%
Calorie consumption 3414 kilocalories

**United Kingdom of Great Britain and
Northern Ireland** *see*
United Kingdom
United Mexican States *see* Mexico
United Provinces *see* Uttar Pradesh
United States of America *35 B5*
off. United States of America,
var. America, The States,
abbrev. U.S., USA.
country North America

UNITED STATES
North America

Official name United States of America
Formation 1776 / 1959
Capital Washington D.C.
Population 320 million / 88 people
per sq mile (34 people per sq km)
Total area 3,717,792 sq. miles
(9,626,091 sq. km)
Languages English, Spanish, Chinese,
French, German, Tagalog, Vietnamese,
Italian, Korean, Russian, Polish
Religions Protestant 52%, Roman
Catholic 25%, Other and nonreligious
20%, Jewish 2%, Muslim 1%
Ethnic mix White 60%, Hispanic 17%,
Black American/African 14%,
Asian 6%, Other 3%
Government Presidential system
Currency US dollar = 100 cents
Literacy rate 99%
Calorie consumption 3639 kilocalories

United States of America *see* United
States of America
Unst *88 D1 island* NE Scotland, United
Kingdom
Ünye *116 D2* Ordu, W Turkey
Upala *52 D4* Alajuela, NW Costa Rica
Upata *59 E2* Bolívar, E Venezuela
Upemba, Lac *77 D7 lake* SE Dem.
Rep. Congo
Upernavik *82 C2 var.* Upernivik. Kitaa,
C Greenland
Upernivik *see* Upernavik
Upington *78 C4* Northern Cape,
W South Africa
'Upolu *145 F4 island* SE Samoa
Upper Klamath Lake *46 A4 lake*
Oregon, NW USA
Upper Lough Erne *89 A5 lake*
SW Northern Ireland, United
Kingdom
Upper Red Lake *45 F1 lake* Minnesota,
N USA
Upper Volta *see* Burkina Faso
Uppsala *85 C6* Uppsala, C Sweden
Uqsuqtuuq *see* Gjoa Haven
Ural *114 B4 Kaz.* Zayyq. *river*
Kazakhstan/Russian Federation
Ural Mountains *114 C3 var.* Ural'skiy
Khrebet, *Eng.* Ural Mountains.
mountain range Kazakhstan/Russian
Federation
Ural Mountains *see* Ural'skiye Gory
Ural'sk *114 B3 Kaz.* Oral. Zapadnyy
Kazakhstan, NW Kazakhstan
Ural'skiy Khrebet *see* Ural'skiye Gory
Uraricoera *62 D1* Roraima, N Brazil
Ura-Tyube *see* Ürooteppa
Urbandale *45 F3* Iowa, C USA
Urdunn *see* Jordan
Uren' *111 C5* Nizhegorodskaya Oblast',
W Russian Federation
Urga *see* Ulaanbaatar
Urganch *122 D2 Rus.* Urgench; *prev.*
Novo-Urgench. Xorazm Viloyati,
W Uzbekistan
Urgench *see* Urganch
Urgut *123 E3* Samarqand Viloyati,
C Uzbekistan
Urmia, Lake *121 C2 var.* Matianus, Sha
Hi, Urumi Yeh, *Eng.* Lake Urmia; *prev.*
Daryācheh-ye Rezā'īyeh. *lake*
NW Iran
Urmia, Lake *see* Orūmīyeh, Daryācheh-ye
Uroševac *see* Ferizaj
Ürooteppa *123 E2 Rus.* Ura-Tyube.
NW Tajikistan
Uruapan *51 E4 var.* Uruapan del
Progreso. Michoacán, SW Mexico
Uruapan del Progreso *see* Uruapan
Uruguai, Rio *see* Uruguay
Uruguay *64 D4 off.* Oriental Republic
of Uruguay; *prev.* La Banda Oriental.
country E South America

URUGUAY
South America

Official name Eastern Republic
of Uruguay
Formation 1828 / 1828
Capital Montevideo
Population 3.4 million / 50 people
per sq mile (19 people per sq km)

URUGUAY
(continued)

Total area 68,039 sq. miles (176,220 sq. km)
Languages Spanish*
Religions Roman Catholic 66%, Other
and nonreligious 30%,
Jewish 2%, Protestant 2%
Ethnic mix White 90%, Mestizo 6%,
Black 4%
Government Presidential system
Currency Uruguayan peso
= 100 centésimos
Literacy rate 98%
Calorie consumption 2939 kilocalories

Uruguay *64 D3 var.* Rio Uruguai, Río
Uruguay. *river* E South America
Uruguay, Oriental Republic of *see*
Uruguay
Uruguay, Rio *see* Uruguay
Urumchi *see* Ürümqi
Urumi Yeh *see* Orūmīyeh, Daryācheh-ye
Ürümqi *126 C3 var.* Tihwa, Urumchi,
Urumqi, Urumtsi, Wu-lu-k'o-mu-shi,
Wu-lu-mu-ch'i; *prev.* Ti-hua. Xinjiang
Uygur Zizhiqu, NW China
Urumtsi *see* Ürümqi
Urundi *see* Burundi
Urup, Ostrov *115 H4 island* Kuril'skiye
Ostrova, SE Russian Federation
Urusan *see* Ulsan
Urziceni *108 C5* Ialomiţa, SE Romania
Usa *110 E3 river* NW Russian
Federation
Uşak *116 B3 prev.* Ushak. Uşak,
W Turkey
Ushak *see* Uşak
Ushant *see* Ouessant, Île d'
Ushuaia *65 B8* Tierra del Fuego,
S Argentina
Usinsk *110 E3* Respublika Komi,
NW Russian Federation
Üsküb/Üsküp *see* Skopje
Usmas Ezers *106 B3 lake* NW Latvia
Usol'ye-Sibirskoye *115 E4* Irkutskaya
Oblast', C Russian Federation
Ussel *91 C5* Corrèze, C France
Ussuriysk *115 G5 prev.* Nikol'sk,
Nikol'sk-Ussuriyskiy, Voroshilov.
Primorskiy Kray, SE Russian
Federation
Ustica *97 B6 island* S Italy
Ust'-Ilimsk *115 E4* Irkutskaya Oblast',
C Russian Federation
Ústí nad Labem *98 A4 Ger.* Aussig.
Ústecký Kraj, NW Czech Republic
Ustinov *see* Izhevsk
Ustka *98 C2 Ger.* Stolpmünde.
Pomorskie, N Poland
Ust'-Kamchatsk *115 H2* Kamchatskiy
Kray, E Russian Federation
Ust'-Kamenogorsk *114 D5 Kaz.*
Öskemen. Vostochnyy Kazakhstan,
E Kazakhstan
Ust'-Kut *115 E4* Irkutskaya Oblast',
C Russian Federation
Ust'-Olenëk *115 E3* Respublika Sakha
(Yakutiya), NE Russian Federation
Ustrzyki Dolne *99 E5* Podkarpackie,
SE Poland
Ust'-Sysol'sk *see* Syktyvkar
Ustyurt Plateau *122 B1 var.* Ust
Urt, Uzb. Ustyurt Platosi. *plateau*
Kazakhstan/Uzbekistan
Ustyurt Platosi *see* Ustyurt Plateau
Usulután *52 C3* Usulután,
SE El Salvador
Usumacinta, Rio *52 B1 river*
Guatemala/Mexico
Usumbura *see* Bujumbura
U.S./USA *see* United States of America
Utah *44 B4 off.* State of Utah, *also known
as* Beehive State, Mormon State. *state*
W USA
Utah Lake *44 B4 lake* Utah,
W USA
Utena *106 C4* Utena, E Lithuania
Utica *41 F3* New York, NE USA
Utina *see* Udine
Utrecht *86 C4 Lat.* Trajectum ad
Rhenum. Utrecht, C Netherlands
Utsunomiya *131 D5 var.* Utunomiya.
Tochigi, Honshū, S Japan
Uttarakhand *135 E2 cultural region*
N India

249

Uttar Pradesh *135 E3 prev.* United Provinces, United Provinces of Agra and Oudh. *cultural region* N India
Utunomiya *see* Utsunomiya
Uulu *106 D2* Pärnumaa, SW Estonia
Uummannaq *82 C3 var.* Umanak, Umanaq. Kitaa, C Greenland
Uummannarsuaq *see* Nunap Isua
Uvalde *49 F4* Texas, SW USA
Uvaravichy *107 D7 Rus.* Uvarovichi. Homyel'skaya Voblasts', SE Belarus
Uvarovichi *see* Uvaravichy
Uvea, Île *145 E4 island* N Wallis and Futuna
Uvs Nuur *126 C1 var.* Ozero Ubsu-Nur. *lake* Mongolia/Russian Federation
'Uwaynāt, Jabal al *88 A3 var.* Jebel Uweinat. *mountain* Libya/Sudan
Uweinat, Jebel *see* 'Uwaynāt, Jabal al
Uyo *75 G5* Akwa Ibom, S Nigeria
Uyuni *61 F5* Potosí, W Bolivia
Uzbekistan *122 D2 off.* Republic of Uzbekistan. *country* C Asia

---

UZBEKISTAN
Central Asia

Official name Republic of Uzbekistan
Formation 1991 / 1991
Capital Tashkent
Population 9 million / 167 people per sq mile (65 people per sq km)
Total area 172,741 sq. miles (447,400 sq. km)
Languages Uzbek*, Russian, Tajik, Kazakh
Religions Sunni Muslim 88%, Orthodox Christian 9%, Other 3%
Ethnic mix Uzbek 80%, Russian 6%, Other 6%, Tajik 5%, Kazakh 3%
Government Presidential system
Currency Som = 100 tiyin
Literacy rate 99%
Calorie consumption 2675 kilocalories

Uzbekistan, Republic of *see* Uzbekistan
Uzhgorod *see* Uzhhorod
Uzhhorod *108 B2 Rus.* Uzhgorod; *prev.* Ungvár. Zakarpats'ka Oblast', W Ukraine
Užice *100 D4 prev.* Titovo Užice. Serbia, W Serbia

V

Vaal *78 D4 river* C South Africa
Vaals *87 D6* Limburg, SE Netherlands
Vaasa *85 D5 Swe.* Vasa; *prev.* Nikolainkaupunki. Länsi-Suomi, W Finland
Vaassen *86 D3* Gelderland, E Netherlands
Vác *99 C6 Ger.* Waitzen. Pest, N Hungary
Vadodara *134 C4 prev.* Baroda. Gujarāt, W India
Vaduz *94 E2 country capital* (Liechtenstein) W Liechtenstein
Vág *see* Váh
Vágbeszterce *see* Považská Bystrica
Váh *99 C5 Ger.* Waag, *Hung.* Vág. *river* W Slovakia
Váhtjer *see* Gällivare
Väinameri *106 C2 prev.* Muhu Väin, Ger. Moon-Sund. *sea* E Baltic Sea
Vajdahunyad *see* Hunedoara
Valachia *see* Wallachia
Valday *110 B4* Novgorodskaya Oblast', W Russian Federation
Valdecañas, Embalse de *92 D3 reservoir* W Spain
Valdepeñas *93 E4* Castilla-La Mancha, C Spain
Valdez *36 C3* Alaska, USA
Valdia *see* Weldiya
Valdivia *65 B5* Los Ríos, C Chile
Val-d'Or *38 D4* Québec, SE Canada
Valdosta *43 E3* Georgia, SE USA
Valence *91 D5 anc.* Valentia, Valentia Julia, Ventia. Drôme, E France
Valencia *93 F3* Valenciana, E Spain
Valencia *46 D1* California, USA
Valencia *58 D1* Carabobo, N Venezuela
Valencia, Gulf of *93 F3 var.* Gulf of Valencia. *gulf* E Spain
Valencia, Gulf of *see* Valencia, Golfo de

Valenciana *93 F3 var.* Valencia, *Cat.* València; *anc.* Valentia. *autonomous community* NE Spain
Valencia/València *see* Valenciana
Valenciennes *90 D2* Nord, N France
Valentia *see* Valence, France
Valentia *see* Valenciana
Valentia Julia *see* Valence
Valentine State *see* Oregon
Valera *58 C2* Trujillo, NW Venezuela
Valetta *see* Valletta
Valga *106 D3 Ger.* Walk, *Latv.* Valka. Valgamaa, S Estonia
Valira *91 A8 river* Andorra/Spain
Valjevo *100 C4* Serbia, W Serbia
Valjok *see* Válljohka
Valka *106 D3 Ger.* Walk. N Latvia
Valka *see* Valga
Valkenswaard *87 D5* Noord-Brabant, S Netherlands
Valladolid *51 H3* Yucatán, SE Mexico
Valladolid *92 D2* Castilla y León, NW Spain
Vall D'Uxó *see* La Vall d'Uixó
Valle De La Pascua *58 D2* Guárico, N Venezuela
Valledupar *58 B1* Cesar, N Colombia
Vallejo *47 B6* California, W USA
Vallenar *64 B3* Atacama, N Chile
Valletta *97 C8 prev.* Valetta. *country capital* (Malta) E Malta
Valley City *45 E2* North Dakota, N USA
Válljohka *84 D2 var.* Valjok. Finnmark, N Norway
Valls *93 G2* Cataluña, NE Spain
Valmiera *106 D3 Est.* Volmari, *Ger.* Wolmar. N Latvia
Valona *see* Vlorë
Valozhyn *107 C5 Pol.* Wołożyn, *Rus.* Volozhin. Minskaya Voblasts', C Belarus
Valparaíso *64 B4* Valparaíso, C Chile
Valparaíso *40 C3* Indiana, N USA
Valverde del Camino *92 C4* Andalucía, S Spain
Van *117 F3* Van, E Turkey
Vanadzor *117 F2 prev.* Kirovakan. N Armenia
Vancouver *36 D5* British Columbia, SW Canada
Vancouver *46 B3* Washington, NW USA
Vancouver Island *36 D5 island* British Columbia, SW Canada
Vanda *see* Vantaa
Van Diemen Gulf *146 D2 gulf* Northern Territory, N Australia
Van Diemen's Land *see* Tasmania
Vaner, Lake *see* Vänern
Vänern *85 B6 Eng.* Lake Vaner; *prev.* Lake Vener. *lake* S Sweden
Vangaindrano *79 G4* Fianarantsoa, SE Madagascar
Van Horn *48 D3* Texas, SW USA
Lake Van *117 F3 Eng.* Lake Van; *anc.* Thospitis. *salt lake* E Turkey
Van, Lake *see* Van Gölü
Vannes *90 A3 anc.* Dariorigum. Morbihan, NW France
Vantaa *85 D6 Swe.* Vanda. Etelä-Suomi, S Finland
Vanua Levu *145 E4 island* N Fiji
Vanuatu *144 C4 off.* Republic of Vanuatu; *prev.* New Hebrides. *country* SW Pacific Ocean

---

VANUATU
Australasia & Oceania

Official name Republic of Vanuatu
Formation 1980 / 1980
Capital Port Vila
Population 300,000 / 64 people per sq mile (25 people per sq km)
Total area 4710 sq. miles (12,200 sq. km)
Languages Bislama* (Melanesian pidgin), English*, French*, other indigenous languages
Religions Presbyterian 37%, Other 19%, Anglican 15%, Roman Catholic 15%, Traditional beliefs 8%, Seventh-day Adventist 6%
Ethnic mix ni-Vanuatu 94%, European 4%, Other 2%
Government Parliamentary system
Currency Vatu = 100 centimes
Literacy rate 83%
Calorie consumption 2820 kilocalories

Vanuatu, Republic of *see* Vanuatu
Van Wert *40 C4* Ohio, N USA
Vapincum *see* Gap
Varakļāni *106 D4* C Latvia
Vārānasi *135 E3 prev.* Banaras, Benares, *hist.* Kasi. Uttar Pradesh, N India
Varangerfjorden *84 E2 Lapp.* Várjjatvuotna. *fjord* N Norway
Varangerhalvøya *84 D2 Lapp.* Várnjárga. *peninsula* N Norway
Varannó *see* Vranov nad Topľ ou
Varasd *see* Varaždin
Varaždin *100 B2 Ger.* Warasdin, *Hung.* Varasd. Varaždin, N Croatia
Varberg *85 B7* Halland, S Sweden
Vardar *101 E6 Gk.* Axiós. *river* FYR Macedonia/Greece
Varde *85 A7* Syddtjylland, W Denmark
Vareia *see* Logroño
Varèna *107 B5 Pol.* Orany. Alytus, S Lithuania
Varese *96 B2* Lombardia, N Italy
Vârful Moldoveanu *108 B4 var.* Moldoveanul; *prev.* Vîrful Moldoveanu. *mountain* C Romania
Várjjatvuotna *see* Varangerfjorden
Varkaus *85 E5* Itä-Suomi, C Finland
Varna *104 E2 prev.* Stalin; *anc.* Odessus. Varna, E Bulgaria
Varnenski Zaliv *104 E2 prev.* Stalinski Zaliv. *bay* E Bulgaria
Várnjárga *see* Varangerhalvøya
Varshava *see* Warszawa
Vasa *see* Vaasa
Vasiliki *105 A5* Lefkáda, Iónia Nisiá, Greece, C Mediterranean Sea
Vasilishki *107 B5 Pol.* Wasiliszki. Hrodzyenskaya Voblasts', W Belarus
Vasil'kov *see* Vasyl'kiv
Vaslui *108 D4 var.* Vaslui, C Romania
Västerås *85 C6* Västmanland, C Sweden
Vasyl'kiv *109 E2 var.* Vasil'kov. Kyyivs'ka Oblast', N Ukraine
Vaté *see* Efate
Vatican City *97 A7 off.* Vatican City. *country* S Europe

VATICAN CITY
Southern Europe

Official name State of the Vatican City
Formation 1929 / 1929
Capital Vatican City
Population 839 / 4935 people per sq mile (1907 people per sq km)
Total area 0.17 sq. miles (0.44 sq. km)
Languages Italian*, Latin*
Religions Roman Catholic 100%
Ethnic mix Most resident lay persons are Italian
Government Papal state
Currency Euro = 100 cents
Literacy rate 99%
Calorie consumption Not available

Vatican City *see* Vatican City
Vatnajökull *83 E5 glacier* SE Iceland
Vatter, Lake *see* Vättern
Vättern *85 B6 Eng.* Lake Vatter; *prev.* Lake Vetter. *lake* S Sweden
Vaughn *48 D2* New Mexico, SW USA
Vaupés, Río *58 C4 var.* Río Uaupés. *river* Brazil/Colombia
Vava'u Group *145 E4 island group* N Tonga
Vavuniya *132 D3* Northern Province, N Sri Lanka
Vawkavysk *107 B6 Pol.* Wołkowysk, *Rus.* Volkovysk. Hrodzyenskaya Voblasts', W Belarus
Växjö *85 C7 var.* Vexiö. Kronoberg, S Sweden
Vaygach, Ostrov *110 E2 island* NW Russian Federation
Veendam *86 E2* Groningen, NE Netherlands
Veenendaal *86 D4* Utrecht, C Netherlands
Vega *84 B4 island* C Norway
Veglia *see* Krk
Veisiejai *107 B5* Alytus, S Lithuania
Vejer de la Frontera *92 C5* Andalucía, S Spain

Veldhoven *87 D5* Noord-Brabant, S Netherlands
Velebit *100 A3 mountain range* C Croatia
Velenje *95 E7 Ger.* Wöllan. N Slovenia
Veles *101 E6 Turk.* Köprülü. C FYR Macedonia
Velho *see* Porto Velho
Velika Kikinda *see* Kikinda
Velika Morava *100 D4 var.* Glavn'a Morava, Morava, *Ger.* Grosse Morava. *river* C Serbia
Velikaya *113 G2 river* NE Russian Federation
Veliki Bečkerek *see* Zrenjanin
Velikiye Luki *110 A4* Pskovskaya Oblast', W Russian Federation
Velikiy Novgorod *110 B4 prev.* Novgorod. Novgorodskaya Oblast', W Russian Federation
Veliko Tûrnovo *104 D2 prev.* Tirnovo, Trnovo, Tûrnovo. Veliko Tûrnovo, N Bulgaria
Velingrad *104 C3* Pazardzhik, C Bulgaria
Vel'ký Krtíš *99 D6* Banskobystrický Kraj, C Slovakia
Vellore *132 C2* Tamil Nādu, SE India
Velobriga *see* Viana do Castelo
Velsen *see* Velsen-Noord
Velsen-Noord *86 C3 var.* Velsen. Noord-Holland, W Netherlands
Vel'sk *110 C4 var.* Velsk. Arkhangel'skaya Oblast', NW Russia
Velvendós *see* Velvéntos
Velvéntos *104 B4 var.* Velvendós. C Greece
Velykyy Tokmak *see* Tokmak
Vendôme *90 C4* Loir-et-Cher, C France
Venedig *see* Venezia
Vener, Lake *see* Vänern
Venetia *see* Venezia
Venezia *96 C2 Eng.* Venice, *Fr.* Venise Ger. Venedig; *anc.* Venetia. Veneto, NE Italy
Venezia, Golfo di *see* Venice, Gulf of
Venezuela *58 D2 off.* Republic of Venezuela; *prev.* Estados Unidos de Venezuela, United States of Venezu *country* N South America

---

VENEZUELA
South America

Official name Bolivarian Republic of Venezuela
Formation 1830 / 1830
Capital Caracas
Population 30.4 million / 89 people per sq mile (34 people per sq km)
Total area 352,143 sq. miles (912,050 sq. km)
Languages Spanish*, Amerindian languages
Religions Roman Catholic 96%, Protestant 2%, Other 2%
Ethnic mix Mestizo 69%, White 20%, Black 9%, Amerindian 2%
Government Presidential system
Currency Bolívar fuerte = 100 céntimo
Literacy rate 96%
Calorie consumption 2880 kilocalorie

Venezuela, Estados Unidos de *see* Venezuela
Venezuela, Gulf of *58 C1 Eng.* Gulf o Maracaibo, Gulf of Venezuela. *gulf* NW Venezuela
Venezuela, Gulf of *see* Venezuela, Golfo de
Venezuelan Basin *56 B1 undersea ba* E Caribbean Sea
Venezuela, Republic of *see* Venezuela
Venezuela, United States of *see* Venezuela
Venice *42 C4* Louisiana, S USA
Venice *see* Venezia
Venice, Gulf of *96 C2 It.* Golfo di Venezia, *Slvn.* Beneški Zaliv. *gulf* N Adriatic Sea
Venise *see* Venezia
Venlo *87 D5 prev.* Venloo. Limburg, SE Netherlands

loo *see* Venlo
ata *106 B3 Ger.* Windau. *river* Latvia/
Lithuania
ata Belgarum *see* Winchester
atia *see* Valence
atimiglia *96 A3* Liguria, NW Italy
atspils *106 B2 Ger.* Windau.
NW Latvia
ra *64 D3* Santa Fe, C Argentina
racruz *51 F4 var.* Veracruz Llave.
racruz Llave *see* Veracruz
veracruz-Llave, E Mexico
racruz Llave *see* Veracruz
rcellae *see* Vercelli
rcelli *96 A2 anc.* Vercellae. Piemonte,
NW Italy
rdal *see* Verdalsøra
rdalsøra *84 B4 var.* Verdal. Nord-
Trøndelag, C Norway
rde, Cabo *see* Cape Verde
rde, Costa *92 D1 coastal region*
N Spain
rden *94 B3* Niedersachsen,
NW Germany
ria *see* Véroia
rkhnedvinsk *see* Vyerkhnyadzvinsk
rkhneudinsk *see* Ulan-Ude
rkhoyanskiy Khrebet *115 F3*
mountain range NE Russian Federation
rmillion *45 F3* South Dakota, N USA
rmont *41 F2 off.* State of Vermont,
also known as Green Mountain State.
state NE USA
rnal *44 B4* Utah, W USA
rnon *49 F2* Texas, SW USA
rőcze *see* Virovitica
roia *104 B4 var.* Veria, Vérroia,
Turk. Karaferiye. Kentrikí Makedonía,
N Greece
rona *96 C2* Veneto, NE Italy
rroia *see* Véroia
rsailles *90 D1* Yvelines, N France
rsecz *see* Vršac
rulamium *see* St Albans
rviers *87 D6* Liège, E Belgium
sdre *87 D6 river* E Belgium
NE Bulgaria
sontio *see* Besançon
soul *90 D4 anc.* Vesulum, Vesulum.
Haute-Saône, E France
sterålen *84 B2 island group*
N Norway
stfjorden *84 C3 fjord* C Norway
stmannaeyjar *83 E5* Sudhurland,
S Iceland
sulium/Vesulum *see* Vesoul
suna *see* Périgueux
suvius *97 D5 Eng.* Vesuvius. *volcano*
S Italy
suvius *see* Vesuvio
szprém *99 C7 Ger.* Veszprim.
Veszprém, W Hungary
szprim *see* Veszprém
trino *see* Suvorovo
trino *see* Vyetryna
tter, Lake *see* Vättern
urne *87 A5 var.* Furnes. West-
Vlaanderen, W Belgium
xiö *see* Växjö
acha *61 F4* La Paz, W Bolivia
ana de Castelo *see* Viana do Castelo
ana do Castelo *92 B2 var.* Viana do
Castelo; *anc.* Velobriga. Viana do
Castelo, NW Portugal
anen *86 C4* Utrecht, C Netherlands
angchan *136 C3 var.* Vieng Pou
Kha. Louang Namtha, N Laos
areggio *96 B3* Toscana, C Italy
country capital (Laos) C Laos
angchan *136 C3 var.* Vieng Pou
Kha. Louang Namtha, N Laos
borg *85 A7* Midtjylland,
N Denmark
c *93 G2 var.* Vich; *anc.* Ausa, Vicus
Ausonensis. Cataluña, NE Spain
centia *see* Vicenza
cenza *96 C2 anc.* Vicentia. Veneto,
NE Italy
ch *see* Vic
chy *91 C5* Allier, C France
cksburg *42 B2* Mississippi, S USA
ctoria *79 H1 country capital*
(Seychelles) Mahé, SW Seychelles
ctoria *36 D5 province capital*
Vancouver Island, British Columbia,
SW Canada
ctoria *102 A5 var.* Rabat. Gozo,
NW Malta

Victoria *49 G4* Texas, SW USA
Victoria *149 C7 state* SE Australia
Victoria *see* Masvingo, Zimbabwe
Victoria Bank *see* Vitória Seamount
Victoria de Durango *see* Durango
Victoria de las Tunas *see* Las Tunas
Victoria Falls *78 C3* Matabeleland
North, W Zimbabwe
Victoria Falls *78 C2 waterfall* Zambia/
Zimbabwe
Victoria Falls *see* Iguaçu, Saltos do
Victoria Island *37 F3 island* Northwest
Territories/Nunavut, NW Canada
Victoria, Lake *73 B6 var.* Victoria
Nyanza. *lake* E Africa
Victoria Land *154 C4 physical region*
Antarctica
Victoria Nyanza *see* Victoria, Lake
Victoria River *146 D3 river* Northern
Territory, N Australia
Victorville *47 C7* California, W USA
Vicus Ausonensis *see* Vic
Vicus Elbii *see* Viterbo
Vidalia *43 E2* Georgia, SE USA
Videm-Krško *see* Krško
Viďeň *see* Wien
Vidin *104 B1 anc.* Bononia. Vidin,
NW Bulgaria
Vidzy *107 C5* Vitsyebskaya Voblasts',
NW Belarus
Viedma *65 C5* Río Negro,
E Argentina
Vieja, Sierra *48 D3 mountain range*
Texas, SW USA
Vieng Pou Kha *see* Viangphoukha
Vienna *see* Wien, Austria
Vienna *see* Vienne, France
Vienne *91 D5 anc.* Vienna. Isère,
E France
Vienne *90 B4 river* W France
Vientiane *see* Viangchan
Vientos, Paso de los *see* Windward
Passage
Vierzon *90 C4* Cher, C France
Viesite *106 C4 Ger.* Eckengraf. S Latvia
Vietnam *136 D4 off.* Socialist Republic of
Vietnam, *Vtn.* Công Hoa Xa Hôi Chu
Nghia Viêt Nam. *country* SE Asia

VIETNAM
Southeast Asia

Official name Socialist Republic
of Vietnam
Formation 1976 / 1976
Capital Hanoi
Population 91.7 million / 730 people
per sq mile (282 people per sq km)
Total area 127,243 sq. miles
(329,560 sq. km)
Languages Vietnamese*, Chinese, Thai,
Khmer, Muong, Nung, Miao, Yao, Jarai
Religions Other 74%, Buddhist 14%,
Roman Catholic 7%, Cao Dai 3%,
Protestant 2%
Ethnic mix Vietnamese 86%, Other 8%,
Muong 2%, Tay 2%, Thai 2%
Government One-party state
Currency Dong = 10 hao = 100 xu
Literacy rate 94%
Calorie consumption 2703 kilocalories

Vietnam, Socialist Republic of *see*
Vietnam
Vietri *see* Viêt Tri
Viêt Tri *136 D3 var.* Vietri. Vinh Phu,
N Vietnam
Vieux Fort *55 F2* S Saint Lucia
Vigo *92 B2* Galicia, NW Spain
Viipuri *see* Vyborg
Vijayawāda *132 D1 prev.* Bezwada.
Andhra Pradesh, SE India
Vila *see* Port-Vila
Vila Artur de Paiva *see* Cubango
Vila da Ponte *see* Cubango
Vila de João Belo *see* Xai-Xai
Vila de Mocímboa da Praia *see*
Mocímboa da Praia
Vila do Conde *92 B2* Porto,
NW Portugal
Vila do Zumbo *78 D2 prev.*
Vila do Zumbu, Zumbo. Tete,
NW Mozambique
Vila do Zumbu *see* Vila do Zumbo
Vilafranca del Penedès *93 G2 var.*
Villafranca del Panadés. Cataluña,
NE Spain

Vila General Machado *see* Camacupa
Vila Henrique de Carvalho *see*
Saurimo
Viļaka *106 D4 Ger.* Marienhausen.
NE Latvia
Vilalba *92 C1* Galicia, NW Spain
Vila Marechal Carmona *see* Uíge
Vila Nova de Gaia *92 B2* Porto,
NW Portugal
Vila Nova de Portimão *see* Portimão
Vila Pereira de Eça *see* N'Giva
Vila Real *92 C2 var.* Vila Rial. Vila Real,
N Portugal
Vila Rial *see* Vila Real
Vila Robert Williams *see* Caála
Vila Salazar *see* N'Dalatando
Vileyka *see* Vilyeyka
Vilhelmina *84 C4* Västerbotten,
N Sweden
Vilhena *62 D3* Rondônia, W Brazil
Vília *105 C5* Attikí, C Greece
Viliya *107 C5 Lith.* Neris. *river*
W Belarus
Viliya *see* Neris
Viljandi *106 D2 Ger.* Fellin.
Viljandimaa, S Estonia
Vilkaviškis *106 B4* Pol. Wyłkowyszki.
Marijampolė, SW Lithuania
Villa Acuña *50 D2 var.* Ciudad Acuña.
Coahuila, NE Mexico
Villa Bella *61 F2* El Beni, N Bolivia
Villacarrillo *93 E4* Andalucía,
S Spain
Villa Cecilia *see* Ciudad Madero
Villach *95 D7 Slvn.* Beljak. Kärnten,
S Austria
Villacidro *97 A5* Sardegna, Italy,
C Mediterranean Sea
Villa Concepción *see* Concepción
Villa del Pilar *see* Pilar
Villafranca de los Barros *92 C4*
Extremadura, W Spain
Villafranca del Panadés *see* Vilafranca
del Penedès
Villahermosa *51 G4 prev.* San Juan
Bautista. Tabasco, SE Mexico
Villajoyosa *93 F4 Cat.* La Vila Joíosa.
Valenciana, E Spain
Villa María *64 C4* Córdoba,
C Argentina
Villa Martín *61 F5* Potosí, SW Bolivia
Villa Mercedes *64 C4* San Juan,
Argentina
Villanueva *50 D3* Zacatecas, C Mexico
Villanueva de la Serena *92 C3*
Extremadura, W Spain
Villanueva de los Infantes *93 E4*
Castilla-La Mancha, C Spain
Villarrica *64 D2* Guairá, SE Paraguay
Villavicencio *58 B3* Meta, C Colombia
Villaviciosa *92 D1* Asturias,
N Spain
Villazón *61 G5* Potosí, S Bolivia
Villena *93 F4* Valenciana, E Spain
Villeurbanne *91 D5* Rhône,
E France
Villingen-Schwenningen *95 B6* Baden-
Württemberg, S Germany
Villmanstrand *see* Lappeenranta
Vilna *see* Vilnius
Vilnius *107 C5 Pol.* Wilno, *Ger.* Wilna;
prev. Rus. Vilna. *country capital*
(Lithuania) Vilnius, SE Lithuania
Vil'shanka *109 E3 Rus.* Olshanka.
Kirovohrads'ka Oblast', C Ukraine
Vilvoorde *87 C6 Fr.* Vilvorde. Vlaams
Brabant, C Belgium
Vilvorde *see* Vilvoorde
Vilyeyka *107 C5 Pol.* Wilejka, *Rus.*
Vileyka. Minskaya Voblasts',
NW Belarus
Vilyuy *115 F3 river* NE Russian
Federation
Viña del Mar *64 B4* Valparaíso,
C Chile
Vinarós *93 F3* Valenciana, E Spain
Vincennes *40 B4* Indiana, N USA
Vindhya Mountains *see* Vindhya Range
Vindhya Range *134 D4 var.* Vindhya
Mountains. *mountain range* N India
Vindobona *see* Wien
Vineland *41 F4* New Jersey, NE USA
Vinh *136 D4* Nghê An, N Vietnam
Vinh Loi *see* Bac Liêu
Vinishte *104 C2* Montana, NW Bulgaria
Vinita *49 G1* Oklahoma, C USA

Vinkovci *100 C3 Ger.* Winkowitz, *Hung.*
Vinkovcze. Vukovar-Srijem, E Croatia
Vinkovcze *see* Vinkovci
Vinnitsa *see* Vinnytsya
Vinnytsya *108 D2 Rus.* Vinnitsa.
Vinnyts'ka Oblast', C Ukraine
Vinogradov *see* Vynohradiv
Vinson Massif *154 A3 mountain*
Antarctica
Viranşehir *117 E4* Şanlıurfa,
SE Turkey
Vîrful Moldoveanu *see* Vârful
Moldoveanu
Virginia *45 G1* Minnesota, N USA
Virginia *41 E5 off.* Commonwealth
of Virginia, *also known as* Mother
of Presidents, Mother of States, Old
Dominion. *state* NE USA
Virginia Beach *41 F5* Virginia,
NE USA
Virgin Islands *see* British Virgin Islands
Virgin Islands (US) *55 F3 var.* Virgin
Islands of the United States; *prev.*
Danish West Indies. *US unincorporated
territory* E West Indies
Virgin Islands of the United States *see*
Virgin Islands (US)
Viröchey *137 E5* Rôtânôkiri,
NE Cambodia
Virovitica *100 C2 Ger.* Virovititz,
Hung. Verőcze; *prev. Ger.* Werowitz.
Virovitica-Podravina, NE Croatia
Virovititz *see* Virovitica
Virton *87 D8* Luxembourg,
SE Belgium
Virtsu *106 D2 Ger.* Werder. Läänemaa,
W Estonia
Vis *100 B4 It.* Lissa; *anc.* Issa. *island*
S Croatia
Vis *see* Fiți
Visaginas *106 C4 prev.* Sniečkus. Utena,
E Lithuania
Visākhapatnam *135 E5 var.*
Vishakhapatnam. Andhra Pradesh,
SE India
Visalia *47 C6* California, W USA
Visby *85 C7 Ger.* Wisby. Gotland,
SE Sweden
Viscount Melville Sound *37 F2 prev.*
Melville Sound. *sound* Northwest
Territories, N Canada
Visé *87 D6* Liège, E Belgium
Viseu *92 C2 var.* Vizeu. Viseu,
N Portugal
Vishakhapatnam *see* Visākhapatnam
Vislinskiy Zaliv *see* Vistula Lagoon
Visoko *100 C4* Federacija Bosne I
Hercegovine, C Bosnia and
Herzegovina
Visttasjohka *84 D3 river* N Sweden
Vistula *98 C2 Eng.* Vistula, *Ger.*
Weichsel. *river* C Poland
Vistula *see* Wisła
Vistula Lagoon *98 C2 Ger.* Frisches Haff,
Pol. Zalew Wiślany, *Rus.* Vislinskiy
Zaliv. *lagoon* Poland/Russian
Federation
Vitebsk *see* Vitsyebsk
Viterbo *96 C4 anc.* Vicus Elbii. Lazio,
C Italy
Viti *see* Fiji
Viti Levu *145 E4 island* W Fiji
Vitim *115 F4 river* C Russian Federation
Vitória *63 F4 state capital* Espírito Santo,
SE Brazil
Vitória Bank *see* Vitória Seamount
Vitória da Conquista *63 F3* Bahia,
E Brazil
Vitoria-Gasteiz *93 E1 var.* Vitoria, *Eng.*
Vittoria. País Vasco, N Spain
Vitória Seamount *67 B5 var.* Victoria
Bank, Vitória Bank. *seamount*
C Atlantic Ocean
Vitré *90 B3* Ille-et-Vilaine, NW France
Vitsyebsk *107 E5 Rus.* Vitebsk.
Vitsyebskaya Voblasts', NE Belarus
Vittoria *97 C7* Sicilia, Italy,
C Mediterranean Sea
Vittoria *see* Vitoria-Gasteiz
Vizcaya, Golfo de *see* Biscay, Bay of
Vizianagaram *135 E5 var.* Vizianagram.
Andhra Pradesh, E India
Vizianagram *see* Vizianagaram
Vjosës, Lumi i *101 C7 var.* Vijosa,
Vijosë, *Gk.* Aóos.
river Albania/Greece

Vlaanderen 87 A6 Eng. Flanders, Fr. Flandre. cultural region Belgium/France
Vlaardingen 86 B4 Zuid-Holland, SW Netherlands
Vladikavkaz 111 B8 prev. Dzaudzhikau, Ordzhonikidze. Respublika Severnaya Osetiya, SW Russian Federation
Vladimir 111 B5 Vladimirskaya Oblast', W Russian Federation
Vladimirovka see Yuzhno-Sakhalinsk
Vladimir-Volynskiy see Volodymyr-Volyns'kyy
Vladivostok 115 G5 Primorskiy Kray, SE Russian Federation
Vlagtwedde 86 E2 Groningen, NE Netherlands
Vlasotince 101 E5 Serbia, SE Serbia
Vlieland 86 C1 Fris. Flylân. island Waddeneilanden, N Netherlands
Vlijmen 86 C4 Noord-Brabant, S Netherlands
Vlissingen 87 B5 Eng. Flushing, Fr. Flessingue. Zeeland, SW Netherlands
Vlodava see Włodawa
Vlonë/Vlora see Vlorë
Vlorë 101 C7 prev. Vlonë, It. Valona, Vlora. Vlorë, SW Albania
Vlotslavsk see Włocławek
Vöcklabruck 95 D6 Oberösterreich, NW Austria
Vogelkop see Doberai, Jazirah
Vohimena, Tanjona 79 F4 Fr. Cap Sainte Marie. headland S Madagascar
Voiron 91 D5 Isère, E France
Vojvodina 100 D3 Ger. Wojwodina. Vojvodina, N Serbia
Volga 111 B7 river NW Russian Federation
Volga Uplands 81 G3 var. Volga Uplands. mountain range W Russian Federation
Volga Uplands see Privolzhskaya Vozvyshennost'
Volgodonsk 111 B7 Rostovskaya Oblast', SW Russian Federation
Volgograd 111 B7 prev. Stalingrad, Tsaritsyn. Volgogradskaya Oblast', SW Russian Federation
Volkhov 110 B4 Leningradskaya Oblast', NW Russian Federation
Volkovysk see Vawkavysk
Volmari see Valmiera
Volnovakha 109 G3 Donets'ka Oblast', SE Ukraine
Volodymyr-Volyns'kyy 108 C1 Pol. Włodzimierz, Rus. Vladimir-Volynskiy. Volyns'ka Oblast', NW Ukraine
Vologda 110 B4 Vologodskaya Oblast', W Russian Federation
Vólos 105 B5 Thessalía, C Greece
Volozhin see Valozhyn
Vol'sk 111 C6 Saratovskaya Oblast', W Russian Federation
Volta 75 E5 river SE Ghana
Volta Blanche see White Volta
Volta, Lake 75 E5 reservoir SE Ghana
Volta Noire see Black Volta
Volturno 97 D5 river S Italy
Volunteer Island see Starbuck Island
Volzhskiy 111 B6 Volgogradskaya Oblast', SW Russian Federation
Vónnu 106 E3 Ger. Wendau. Tartumaa, SE Estonia

Vosges 90 E4 mountain range NE France
Vostochnyy Sayan 115 E4 Mong. Dzüün Soyonï Nuruu, Eng. Eastern Sayans. mountain range Mongolia/Russian Federation
Vostock Island see Vostok Island
Vostok 154 C3 Russian research station Antarctica
Vostok Island 145 G3 var. Vostock Island; prev. Stavers Island. island Line Islands, SE Kiribati
Voznesens'k 109 E3 Rus. Voznesensk. Mykolayivs'ka Oblast', S Ukraine
Vranje 101 E5 Serbia, SE Serbia
Vranov see Vranov nad Topl'ou
Vranov nad Topl'ou 99 D5 var. Vranov, Hung. Varannó. Prešovský Kraj, E Slovakia
Vratsa 104 C2 Vratsa, NW Bulgaria
Vrbas 100 C3 Vojvodina, NW Serbia
Vrbas 100 C3 river N Bosnia and Herzegovina
Vršac 100 E3 Ger. Werschetz, Hung. Versecz. Vojvodina, NE Serbia
Vsetín 99 C5 Ger. Wsetin. Zlínský Kraj, E Czech Republic
Vučitrn see Vushtrri
Vukovar 100 C3 Hung. Vukóvár. Vukovar-Srijem, E Croatia
Vulcano, Isola 97 C7 island Isole Eolie, S Italy
Vung Tau 137 E6 prev. Fr. Cape Saint Jacques, Cap Saint-Jacques. Ba Ria-Vung Tau, S Vietnam
Vusan see Busan
Vushtrri 101 D5 Serb. Vučitrn. K Kosovo
Vyatka 111 C5 river NW Russian Federation
Vyatka see Kirov
Vyborg 110 B3 Fin. Viipuri. Leningradskaya Oblast', NW Russian Federation
Vyerkhnyadzvinsk 107 D5 Rus. Verkhnedvinsk. Vitsyebskaya Voblasts', N Belarus
Vyetryna 107 D5 Rus. Vetrino. Vitsyebskaya Voblasts', N Belarus
Vynohradiv 108 B3 Cz. Sevluš, Hung. Nagyszöllős, Rus. Vinogradov; prev. Sevlyush. Zakarpats'ka Oblast', W Ukraine

W

Wa 75 E4 NW Ghana
Waag see Váh
Waagbistritz see Považská Bystrica
Waal 86 C4 river S Netherlands
Wabash 40 C4 Indiana, N USA
Wabash River 40 B5 river N USA
Waco 49 G3 Texas, SW USA
Wad Al-Hajarah see Guadalajara
Waddān 71 F3 NW Libya
Waddenzee 86 C1 var. Wadden Zee. sea SE North Sea
Wadden Zee see Waddenzee
Waddington, Mount 36 D5 mountain British Columbia, SW Canada
Wādī as Sīr 119 B6 var. Wadi es Sir. 'Ammān, NW Jordan
Wadi es Sir see Wādī as Sīr
Wadi Halfa 72 B3 var. Wādī Ḥalfā'. Northern, N Sudan
Wādī Mūsā 119 B7 var. Petra. Ma'ān, S Jordan
Wad Medani see Wad Medani
Wad Medani 72 C4 var. Wad Madanī. Gezira, C Sudan
Waflia 139 F4 Pulau Buru, E Indonesia
Wagadugu see Ouagadougou
Wagga Wagga 149 C7 New South Wales, SE Australia
Wagin 147 B7 Western Australia
Wāh 134 C1 Punjab, NE Pakistan
Wahai 139 F4 Pulau Seram, E Indonesia
Wahaybah, Ramlat Al see Wahībah, Ramlat Āl
Wahiawā 47 A8 var. Wahiawa. O'ahu, Hawaii, USA, C Pacific Ocean
Wahībah, Ramlat Ahl see Wahībah, Ramlat Āl
Wahībah Sands 121 E5 var. Ramlat Ahl Wahībah, Ramlat Al Wahaybah, Eng. Wahībah Sands. desert N Oman

Wahībah Sands see Wahībah, Ramlat Āl
Wahpeton 45 F2 North Dakota, N USA
Wahran see Oran
Waiau 151 A7 river South Island, New Zealand
Waigeo, Pulau 139 G4 island Maluku, E Indonesia
Waikaremoana, Lake 150 E4 lake North Island, New Zealand
Wailuku 47 B8 Maui, Hawaii, USA, C Pacific Ocean
Waimate 151 B6 Canterbury, South Island, New Zealand
Waiouru 150 D4 Manawatu-Wanganui, North Island, New Zealand
Waipara 151 C6 Canterbury, South Island, New Zealand
Waipawa 150 E4 Hawke's Bay, North Island, New Zealand
Waipukurau 150 D4 Hawke's Bay, North Island, New Zealand
Wairau 151 C5 river South Island, New Zealand
Wairoa 150 E4 Hawke's Bay, North Island, New Zealand
Wairoa 150 D2 river North Island, New Zealand
Waitaki 151 B6 river South Island, New Zealand
Waitara 150 D4 Taranaki, North Island, New Zealand
Waitzen see Vác
Waiuku 150 D3 Auckland, North Island, New Zealand
Wakasa-wan 131 C6 bay C Japan
Wakatipu, Lake 151 A7 lake South Island, New Zealand
Wakayama 131 C6 Wakayama, Honshū, SW Japan
Wake Island 152 C2 US unincorporated territory NW Pacific Ocean
Wake Island 142 D1 atoll NW Pacific Ocean
Wakkanai 130 C1 Hokkaidō, NE Japan
Walachei/Walachia see Wallachia
Wałbrzych 98 B4 Ger. Waldenburg, Waldenburg in Schlesien. Dolnośląskie, SW Poland
Walcourt 87 C7 Namur, S Belgium
Walcz 98 B3 Ger. Deutsch Krone. Zachodnio-pomorskie, NW Poland
Waldenburg/Waldenburg in Schlesien see Wałbrzych
Waldia see Weldiya
Wales 36 C2 Alaska, USA
Wales 89 C6 Wel. Cymru. cultural region Wales, United Kingdom
Walgett 149 D5 New South Wales, SE Australia
Walk see Valga, Estonia
Walk see Valka, Latvia
Walker Lake 47 C5 lake Nevada, W USA
Wallachia 108 B5 var. Walachia, Ger. Walachei, Rom. Valachia. cultural region S Romania
Walla Walla 46 C2 Washington, NW USA
Wallenthal see Hateg
Wallis and Futuna 145 E4 Fr. Territoire de Wallis et Futuna. French overseas territory C Pacific Ocean
Wallis et Futuna, Territoire de see Wallis and Futuna
Walnut Ridge 42 B1 Arkansas, C USA
Waltenberg see Zalău
Walthamstow 89 B7 Waltham Forest, SE England, United Kingdom
Walvisbaai see Walvis Bay
Walvis Bay 78 A4 Afr. Walvisbaai. Erongo, NW Namibia
Walvish Ridge see Walvis Ridge
Walvis Ridge 69 B7 var. Walvish Ridge. undersea ridge E Atlantic Ocean
Wan see Anhui
Wanaka 151 B6 Otago, South Island, New Zealand
Wanaka, Lake 151 A6 lake South Island, New Zealand
Wanchuan see Zhangjiakou
Wandel Sea 83 E1 sea Arctic Ocean
Wandsworth 89 A8 Wandsworth, SE England, United Kingdom
Wanganui 150 D4 Manawatu-Wanganui, North Island, New Zealand

Wangaratta 149 C7 Victoria, SE Australia
Wankie see Hwange
Wanki, Río see Coco, Río
Wanlaweyn 73 D6 var. Wanle Weyn, It. Uanle Uen. Shabeellaha Hoose, SW Somalia
Wanle Weyn see Wanlaweyn
Wanxian see Wanzhou
Wanzhou 128 B5 var. Wanxian. Chongqing Shi, C China
Warangal 135 E5 Telangana, C India
Warasdin see Varaždin
Warburg 94 B4 Nordrhein-Westfalen, W Germany
Ware 37 E4 British Columbia, W Canada
Waremme 87 C6 Liège, E Belgium
Waren 94 C3 Mecklenburg-Vorpommern, NE Germany
Wargla see Ouargla
Warkworth 150 D2 Auckland, North Island, New Zealand
Warnemünde 94 C2 Mecklenburg-Vorpommern, NE Germany
Warner 49 G1 Oklahoma, C USA
Warnes 61 G4 Santa Cruz, C Bolivia
Warrego River 149 C5 seasonal river New South Wales/Queensland, E Australia
Warren 40 D3 Michigan, N USA
Warren 40 D3 Ohio, N USA
Warren 41 E3 Pennsylvania, NE USA
Warri 75 F5 Delta, S Nigeria
Warrnambool 149 B7 Victoria, SE Australia
Warsaw/Warschau see Warszawa
Warszawa 98 D3 Eng. Warsaw, Ger. Warschau, Rus. Varshava. country capital (Poland) Mazowieckie, C Poland
Warta 98 B3 Ger. Warthe. river W Poland
Wartberg see Senec
Warthe see Warta
Warwick 149 E5 Queensland, E Australia
Washington 44 A2 NE England, United Kingdom
Washington D.C. 41 E4 country capital (USA) District of Columbia, NE USA
Washington Island see Teraina
Washington, Mount 41 G2 mountain New Hampshire, NE USA
Wash, The 89 E6 inlet E England, United Kingdom
Wasiliszki see Vasilishki
Waspam 53 E2 var. Waspán. Región Autónoma Atlántico Norte, NE Nicaragua
Waspán see Waspam
Watampone 139 E4 var. Bone. Sulawesi, C Indonesia
Watenstedt-Salzgitter see Salzgitter
Waterbury 41 F3 Connecticut, NE USA
Waterford 89 B6 Ir. Port Láirge. Waterford, S Ireland
Waterloo 45 G3 Iowa, C USA
Watertown 41 F2 New York, NE USA
Watertown 45 F2 South Dakota, N USA
Waterville 41 G2 Maine, NE USA
Watford 89 A7 E England, United Kingdom
Watlings Island see San Salvador
Watsa 77 E5 Orientale, NE Dem. Rep. Congo
Watts Bar Lake 42 D1 reservoir Tennessee, S USA
Wau 73 B5 var. Wāw. Western Bahr el Ghazal, S South Sudan
Waukegan 40 B3 Illinois, N USA
Waukesha 40 B3 Wisconsin, N USA
Wausau 40 B2 Wisconsin, N USA
Waverly 45 G3 Iowa, C USA
Wavre 87 C6 Walloon Brabant, C Belgium
Wāw see Wau
Wawa 38 C4 Ontario, S Canada
Waycross 43 E3 Georgia, SE USA
Wearmouth see Sunderland
Webfoot State see Oregon
Webster City 45 F3 Iowa, C USA
Weddell Plain 154 A2 abyssal plain SW Atlantic Ocean

dell Sea 154 A2
 SW Atlantic Ocean
ner 94 A3 Niedersachsen,
W Germany
rt 87 D5 Limburg, SE Netherlands
sp 86 C3 Noord-Holland,
Netherlands
orzewo 98 D2 Ger. Angerburg.
 armińsko-Mazurskie, NE Poland
hsel see Wisła
nar 94 C4 Thüringen, C Germany
senburg see Alba Iulia, Romania
ssenburg 94 C4 Thüringen, C Germany
senburg in Bayern 95 C6 Bayern,
 Germany
ssenstein see Paide
skirchen see Bela Crkva
swampach 87 D7 Diekirch,
 Luxembourg
herowo 98 C2 Pomorskie,
W Poland
chman Hall 55 G1 C Barbados
diya 72 C4 var. Waldia, It. Valdia.
mara, N Ethiopia
kom 78 C4 Free State,
South Africa
e see Uele
esley Islands 148 B2 island group
 ueensland, N Australia
ington 151 D5 country capital
 New Zealand) Wellington, North
land, New Zealand
ington 45 F5 Kansas, C USA
ington see Wellington, Isla
ington, Isla 65 A7 var. Wellington.
 and S Chile
s 46 D4 Nevada, W USA
sford 150 D2 Auckland, North
 Island, New Zealand
s, Lake 147 C5 lake Western
 ustralia
s 95 C6 anc. Ovilava. Oberösterreich,
 Austria
ibley 89 A8 Alberta, W Canada
mel 87 B6 Vlaams Brabant,
 Belgium
atchee 46 B2 Washington,
W USA
chi 75 E4 W Ghana
-chou/Wenchow see Wenzhou
dau see Võnnu
zhou 128 D5 var. Wen-chou,
 enchow. Zhejiang, SE China
da 78 C4 Kgalagadi, S Botswana
der see Virtsu
enöv see Voranava
kendam 86 C4 Noord-Brabant,
Netherlands
owitz see Virovitica
ro see Võru
schetz see Vršac
enberg see Rakvere
er 94 B3 river NW Germany
sel Islands 148 B1 island group
 orthern Territory, N Australia
t Antarctica 154 A3 prev. Lesser
 ntarctica. physical region Antarctica
t Australian Basin see Wharton
asin
t Bank 119 A6 disputed region
W Asia
t Bend 40 B3 Wisconsin, N USA
t Bengal 135 F4 cultural region
E India
t Cape 151 A7 headland South
 land, New Zealand
t Des Moines 45 F3 Iowa, C USA
terland 94 B2 Schleswig-Holstein,
 Germany
tern Australia 146 B4 state
 / Australia
tern Bug see Bug
tern Carpathians 99 E7 mountain
 ange W Romania Europe
tern Desert 72 B2 var. Sahara el
 harbiya, Eng. Western Desert. desert
 Egypt
tern Desert see Ṣaḥrā' al Gharbīya
tern Dvina 85 E7 Bel. Dzvina, Ger.
 Düna. Latv. Daugava, Rus. Zapadnaya
 vina. river W Europe
tern Ghats 134 C5 mountain range
 / India
tern Isles see Outer Hebrides
tern Punjab see Punjab
tern Sahara 70 B3 disputed territory
 Africa

Western Samoa see Samoa
Western Sayans 114 D4 Eng. Western
 Sayans. mountain range S Russian
 Federation
Western Sayans see Zapadnyy Sayan
Western Scheldt 87 B5 Eng. Western
 Scheldt; prev. Honte. inlet S
 North Sea
Western Scheldt see Westerschelde
Western Sierra Madre see Madre
 Occidental, Sierra
West Falkland 65 C7 var. Gran Malvina,
 Isla Gran Malvina. island W Falkland
 Islands
West Fargo 45 F2 North Dakota,
 N USA
West Frisian Islands 86 C1 Eng.
 West Frisian Islands. island group
 N Netherlands
West Frisian Islands see
 Waddeneilanden
West Irian see Papua
Westliche Morava see Zapadna Morava
West Mariana Basin 142 B1 var. Perece
 Vela Basin. undersea feature W Pacific
 Ocean
West Memphis 42 B1 Arkansas,
 C USA
West New Guinea see Papua
Weston-super-Mare 89 D7 SW England,
 United Kingdom
West Palm Beach 43 F4 Florida,
 SE USA
West Papua see Papua
Westport 151 C5 West Coast, South
 Island, New Zealand
West Punjab see Punjab
West River see Xi Jiang
West Siberian Plain 114 C3 Eng.
 West Siberian Plain. plain C Russian
 Federation
West Siberian Plain see Zapadno-
 Sibirskaya Ravnina
West Virginia 40 D4 off. State of West
 Virginia, also known as Mountain State.
 state NE USA
Wetar, Pulau 139 F5 island Kepulauan
 Damar, E Indonesia
Wetzlar 95 B5 Hessen, W Germany
Wevok 36 C2 var. Wewuk. Alaska,
 USA
Wewuk see Wevok
Wexford 89 B6 Ir. Loch Garman.
 SE Ireland
Weyburn 37 F5 Saskatchewan, S Canada
Weymouth 89 D7 S England, United
 Kingdom
Wezep 86 D3 Gelderland,
 E Netherlands
Whakatane 150 E3 Bay of Plenty,
 North Island, New Zealand
Whale Cove 37 G3 var. Tikirarjuaq.
 Nunavut, C Canada
Whangarei 150 D2 Northland, North
 Island, New Zealand
Wharton Basin 141 D5 var. West
 Australian Basin. undersea feature
 E Indian Ocean
Whataroa 151 B6 West Coast, South
 Island, New Zealand
Wheatland 44 D3 Wyoming, C USA
Wheeler Peak 48 D1 mountain New
 Mexico, SW USA
Wheeling 40 D4 West Virginia,
 NE USA
Whitby 89 D5 N England, United
 Kingdom
Whitefish 44 B1 Montana, NW USA
Whitehaven 89 C5 NW England,
 United Kingdom
Whitehorse 36 D4 territory capital
 Yukon, W Canada
White Nile 72 B4 Ar. Al Baḥr al Abyaḍ,
 An Nīl al Abyaḍ, Baḥr el Jebel. river
 NE South Sudan
White Nile 72 B4 var. Bahr el Jebel.
 river S Sudan
White River 44 D3 river South Dakota,
 N USA
White Sea 110 C3 Eng. White Sea. sea
 NW Russian Federation
White Sea see Beloye More
White Volta 75 E4 var. Nakambé, Fr.
 Volta Blanche. river Burkina Faso/
 Ghana
Whitianga 150 D2 Waikato, North
 Island, New Zealand

Whitney, Mount 47 C6 mountain
 California, W USA
Whitsunday Group 148 D3 island group
 Queensland, E Australia
Whyalla 149 B6 South Australia
Wichita 45 F5 Kansas, C USA
Wichita Falls 49 F2 Texas, SW USA
Wichita River 49 F2 river Texas,
 SW USA
Wickenburg 48 B2 Arizona, SW USA
Wicklow 89 B6 Ir. Cill Mhantáin. county
 E Ireland
Wicklow Mountains 89 B6 Ir. Sléibhte
 Chill Mhantáin. mountain range
 E Ireland
Wieliczka 99 D5 Małopolskie, S Poland
Wieluń 98 C4 Sieradz, C Poland
Wien 95 E6 Eng. Vienna, Hung.
 Bécs, Slvk. Viedeň, Slvn. Dunaj; anc.
 Vindobona. country capital (Austria)
 Wien, NE Austria
Wiener Neustadt 95 E6
 Niederösterreich, E Austria
Wierden 86 E3 Overijssel,
 E Netherlands
Wiesbaden 95 B5 Hessen,
 W Germany
Wieselburg und Ungarisch-Altenburg/
 Wieselburg-Ungarisch-Altenburg
 see Mosonmagyaróvár
Wiesenhof see Ostrołęka
Wight, Isle of 89 D7 island United
 Kingdom
Wigorna Ceastre see Worcester
Wijchen 86 D4 Gelderland,
 SE Netherlands
Wijk bij Duurstede 86 D4 Utrecht,
 C Netherlands
Wilcannia 149 C6 New South Wales,
 SE Australia
Wilejka see Vilyeyka
Wilhelm, Mount 144 B3 mountain
 C Papua New Guinea
Wilhelm-Pieck-Stadt see Guben
Wilhelmshaven 94 B3 Niedersachsen,
 NW Germany
Wilia/Wilja see Neris
Wilkes Barre 41 F3 Pennsylvania,
 NE USA
Wilkes Land 154 C4 physical region
 Antarctica
Wiłkomierz see Ukmergė
Willard 48 D2 New Mexico,
 SW USA
Willcox 48 C3 Arizona, SW USA
Willebroek 87 B5 Antwerpen,
 C Belgium
Willemstad 55 E5 dependent territory
 capital (Curaçao) Lesser Antilles, S
 Caribbean Sea
Williston 44 D1 North Dakota,
 N USA
Wilmington 41 F4 Delaware,
 NE USA
Wilmington 43 F2 North Carolina,
 SE USA
Wilmington 40 C4 Ohio, N USA
Wilna/Wilno see Vilnius
Wilrijk 87 C5 Antwerpen, N Belgium
Winchester 89 D7 hist. Wintanceaster,
 Lat. Venta Belgarum. S England,
 United Kingdom
Winchester 41 E4 Virginia, NE USA
Windau see Ventspils, Latvia
Windau see Venta, Latvia/Lithuania
Windhoek 78 B3 Ger. Windhuk.
 country capital (Namibia) Khomas,
 C Namibia
Windhuk see Windhoek
Windorah 148 C4 Queensland,
 C Australia
Windsor 38 C5 Ontario, S Canada
Windsor 89 D7 S England, United
 Kingdom
Windsor 41 G3 Connecticut,
 NE USA
Windward Islands 55 H4 island group
 E West Indies
Windward Islands see Barlavento,
 Ilhas de, Cape Verde
Windward Passage 54 D3 Sp. Paso de
 los Vientos. channel Cuba/Haiti
Winisk 38 C2 river Ontario, C Canada
Winkowitz see Vinkovci
Winnebago, Lake 40 B2 lake
 Wisconsin, N USA
Winnemucca 47 C5 Nevada, W USA

Winnipeg 37 G5 province capital
 Manitoba, S Canada
Winnipeg, Lake 37 G5 lake Manitoba,
 C Canada
Winnipegosis, Lake 38 A3 lake
 Manitoba, C Canada
Winona 45 G3 Minnesota, N USA
Winschoten 86 E2 Groningen,
 NE Netherlands
Winsen 94 B3 Niedersachsen,
 N Germany
Winston Salem 43 E1 North Carolina,
 SE USA
Winsum 86 D1 Groningen,
 NE Netherlands
Wintanceaster see Winchester
Winterswijk 86 E4 Gelderland,
 E Netherlands
Winterthur 95 B7 Zürich,
 NE Switzerland
Winton 148 C4 Queensland,
 E Australia
Winton 151 A7 Southland,
 South Island, New Zealand
Wisby see Visby
Wisconsin 40 A2 off. State of
 Wisconsin, also known as
 Badger State. state N USA
Wisconsin Rapids 40 B2 Wisconsin,
 N USA
Wisconsin River 40 B3 river
 Wisconsin, N USA
Wiślany, Zalew see Vistula Lagoon
Wismar 94 C2 Mecklenburg-
 Vorpommern, N Germany
Wittenberge 94 C3 Brandenburg,
 N Germany
Wittlich 95 A5 Rheinland-Pfalz,
 SW Germany
Wittstock 94 C3 Brandenburg,
 NE Germany
W. J. van Blommesteinmeer 59 G3
 reservoir E Suriname
Władysławowo 98 C2 Pomorskie,
 N Poland
Włocławek 98 C3 Ger./Rus. Vlotslavsk.
 Kujawsko-pomorskie, C Poland
Włodawa 98 E4 Rus. Vlodava. Lubelskie,
 SE Poland
Włodzimierz see Volodymyr-Volyns'kyy
Wlotzkasbaken 78 B3 Erongo,
 W Namibia
Wodonga 149 C7 Victoria,
 SE Australia
Wodzisław Śląski 99 C5 Ger. Loslau.
 Śląskie, S Poland
Wojerecy see Hoyerswerda
Wōjjä see Wotje Atoll
Wojwodina see Vojvodina
Woking 89 D7 SE England, United
 Kingdom
Wolf, Isla 60 A4 island Galápagos
 Islands, W Ecuador, South America
Wolfsberg 95 D7 Kärnten,
 SE Austria
Wolfsburg 94 C3 Niedersachsen,
 N Germany
Wolgast 94 D2 Mecklenburg-
 Vorpommern, NE Germany
Wołkowysk see Vawkavysk
Wöllan see Velenje
Wollaston Lake 37 F4 Saskatchewan,
 C Canada
Wollongong 149 D6 New South Wales,
 SE Australia
Wolmar see Valmiera
Wołożyn see Valozhyn
Wolvega 86 D2 Fris. Wolvegea. Fryslân,
 N Netherlands
Wolvegea see Wolvega
Wolverhampton 89 D6 C England,
 United Kingdom
Wolverine State see Michigan
Wŏnsan 107 E3 SE North Korea
Woodburn 46 B3 Oregon,
 NW USA
Woodland 47 B5 California,
 W USA
Woodruff 40 B2 Wisconsin, N USA
Woods, Lake of the 38 A3 Canada/USA
Woodville 150 D4 Manawatu-
 Wanganui, North Island,
 New Zealand
Woodward 49 F1 Oklahoma, C USA
Worcester 78 C5 Western Cape,
 SW South Africa

Worcester 89 D6 hist. Wigorna Ceaster. W England, United Kingdom
Worcester 41 G3 Massachusetts, NE USA
Workington 89 C5 NW England, United Kingdom
Worland 44 C3 Wyoming, C USA
Wormatia see Worms
Worms 95 B5 anc. Augusta Vangionum, Borbetomagus, Wormatia. Rheinland-Pfalz, SW Germany
Worms see Vormsi
Worthington 45 F3 Minnesota, N USA
Wotje Atoll 144 D1 var. Wōjjä. atoll Ratak Chain, E Marshall Islands
Woudrichem 86 C4 Noord-Brabant, S Netherlands
Wrangel Island 115 F1 Eng. Wrangel Island. island NE Russian Federation
Wrangel Island see Vrangelya, Ostrov
Wrangel Plain 155 B2 undersea feature Arctic Ocean
Wrocław 98 C4 Eng./Ger. Breslau. Dolnośląskie, SW Poland
Września 98 C3 Wielkopolskie, C Poland
Wsetin see Vsetín
Wuchang see Wuhan
Wuday'ah 121 C6 spring/well S Saudi Arabia
Wuhai 127 E3 var. Haibowan. Nei Mongol Zizhiqu, N China
Wuhan 128 C5 var. Han-kou, Han-k'ou, Hanyang, Wuchang, Wu-han; prev. Hankow. province capital Hubei, C China
Wu-han see Wuhan
Wuhsien see Suzhou
Wuhsi/Wu-hsi see Wuxi
Wuhu 128 D5 var. Wu-na-mu. Anhui, E China
Wujlän see Ujelang Atoll
Wukari 75 G4 Taraba, E Nigeria
Wuliang Shan 128 A6 mountain range SW China
Wu-lu-k'o-mu-shi/Wu-lu-mu-ch'i see Ürümqi
Wu-na-mu see Wuhu
Wuppertal 94 A4 prev. Barmen-Elberfeld. Nordrhein-Westfalen, W Germany
Würzburg 95 B5 Bayern, SW Germany
Wusih see Wuxi
Wuxi 128 D5 var. Wuhsi, Wu-hsi, Wusih. Jiangsu, E China
Wuyi Shan 125 E3 mountain range SE China
Wye 89 C6 Wel. Gwy. river England/ Wales, United Kingdom
Wyłkowyszki see Vilkaviškis
Wyndham 146 D3 Western Australia
Wyoming 40 C3 Michigan, N USA
Wyoming 44 B3 off. State of Wyoming, also known as Equality State. state C USA
Wyszków 98 D3 Ger. Probstberg. Mazowieckie, NE Poland

X

Xaafuun, Raas 72 E4 var. Ras Hafun. cape NE Somalia
Xaçmaz 117 H2 Rus. Khachmas. N Azerbaijan
Xaignabouli 136 C4 prev. Muang Xaignabouri, Fr. Sayaboury. Xaignabouli, N Laos
Xai-Xai 79 E4 prev. João Belo, Vila de João Belo. Gaza, S Mozambique
Xalapa 51 F4 Veracruz-Llave, Mexico
Xam Nua 136 D3 var. Sam Neua. Houaphan, N Laos
Xankändi 117 G3 Rus. Khankendi; prev. Stepanakert. SW Azerbaijan
Xánthi 104 C3 Anatolikí Makedonía kai Thráki, NE Greece
Xàtiva 93 F3 Cas. Xátiva; anc. Setabis, var. Jativa. Valenciana, E Spain
Xauen see Chefchaouen
Xäzär Dänizi see Caspian Sea
Xeres see Jerez de la Frontera
Xiaguan see Dali
Xiamen 128 D6 var. Hsia-men; prev. Amoy. Fujian, SE China
Xi'an 128 C4 var. Changan, Sian, Signan, Siking, Singan, Xian. province capital Shaanxi, C China

Xiang see Hunan
Xiangkhoang see Phônsaven
Xiangtan 128 C5 var. Hsiang-t'an, Siangtan. Hunan, S China
Xiao Hinggan Ling 128 D2 Eng. Lesser Khingan Range. mountain range NE China
Xichang 128 B5 Sichuan, C China
Xieng Khouang see Phônsaven
Xieng Ngeun see Muong Xiang Ngeun
Xigazê 126 C5 var. Jih-k'a-tse, Shigatse, Xigaze. Xizang Zizhiqu, W China
Xi Jiang 124 D3 var. Hsi Chiang, Eng. West River. river S China
Xilinhot 127 F2 var. Silinhot. Nei Mongol Zizhiqu, N China
Xilokastro see Xylókastro
Xin see Xinjiang Uygur Zizhiqu
Xingkai Hu see Khanka, Lake
Xingu, Rio 63 E2 river C Brazil
Xingxingxia 126 D3 Xinjiang Uygur Zizhiqu, NW China
Xining 127 E4 var. Hsining, Hsi-ning, Sining. province capital Qinghai, C China
Xinjiang see Xinjiang Uygur Zizhiqu
Xinjiang Uygur Zizhiqu 126 B3 var. Sinkiang, Sinkiang Uighur Autonomous Region, Xin, Xinjiang. autonomous region NW China
Xinpu see Lianyungang
Xinxiang 128 C4 Henan, C China
Xinyang 128 C5 var. Hsin-yang, Sinyang. Henan, C China
Xinzo de Limia 92 C2 Galicia, NW Spain
Xiqing Shan 124 D2 mountain range C China
Xiva 122 D2 Rus. Khiva, Khiwa. Xorazm Viloyati, W Uzbekistan
Xixón see Gijón
Xizang see Xizang Zizhiqu
Xizang Gaoyuan see Qingzang Gaoyuan
Xizang Zizhiqu 126 B4 var. Thibet, Tibetan Autonomous Region, Xizang, Eng. Tibet. autonomous region W China
Xolotlán see Managua, Lago de
Xucheng see Xuwen
Xuddur 73 D5 var. Hudur, It. Oddur. Bakool, SW Somalia
Xuwen 128 C7 var. Xucheng. Guangdong, S China
Xuzhou 128 D4 var. Hsu-chou, Suchow, Tongshan; prev. T'ung-shan. Jiangsu, E China
Xylókastro 105 B5 var. Xilokastro. Pelopónnisos, S Greece

Y

Ya'an 128 B5 var. Yaan. Sichuan, C China
Yabēlo 73 C5 Oromīya, C Ethiopia
Yablis 53 E2 Región Autónoma Atlántico Norte, NE Nicaragua
Yablonovyy Khrebet 115 F4 mountain range S Russian Federation
Yabrai Shan 127 E3 mountain range NE China
Yafran 71 F2 NW Libya
Yaghan Basin 67 B7 undersea feature SE Pacific Ocean
Yagotin see Yahotyn
Yahotyn 109 E2 Rus. Yagotin. Kyyivs'ka Oblast', N Ukraine
Yahualica 50 D4 Jalisco, SW Mexico
Yakima 46 B2 Washington, NW USA
Yakima River 46 B2 river Washington, NW USA
Yakoruda 104 C3 Blagoevgrad, SW Bulgaria
Yaku-shima 131 B8 island Nansei-shotō, SW Japan
Yakutat 36 D4 Alaska, USA
Yakutsk 115 F3 Respublika Sakha (Yakutiya), NE Russian Federation
Yala 137 C7 Yala, SW Thailand
Yalizava 107 D6 Rus. Yelizovo. Mahilyowskaya Voblasts', E Belarus
Yalong Jiang 128 A5 river C China
Yalova 116 B3 Yalova, NW Turkey
Yalpug, Ozero see Yalpuh, Ozero
Yalpuh, Ozero 108 D4 Rus. Ozero Yalpug. lake SW Ukraine
Yalta 109 F5 Avtonomna Respublika Krym, S Ukraine

Yalu 125 E2 Chin. Yalu Jiang, Jap. Oryokko, Kor. Amnok-kang. river China/North Korea
Yalu Jiang see Yalu
Yamaguchi 131 B7 var. Yamaguti. Yamaguchi, Honshū, SW Japan
Yamal, Poluostrov 114 D2 peninsula N Russian Federation
Yamaniyah, Al Jumhūrīyah al see Yemen
Yambio 73 B5 var. Yambiyo. Western Equatoria, S South Sudan
Yambiyo see Yambio
Yambol 104 D2 Turk. Yanboli. Yambol, E Bulgaria
Yamdena, Pulau 139 G5 prev. Jamdena. island Kepulauan Tanimbar, E Indonesia
Yamoussoukro 74 D5 country capital (Côte d'Ivoire) C Côte d'Ivoire
Yamuna 134 D3 prev. Jumna. river N India
Yana 115 F2 river NE Russian Federation
Yanboli see Yambol
Yanbu' al Baḥr 121 A5 Al Madīnah, W Saudi Arabia
Yangambi 77 D5 Orientale, N Dem. Rep. Congo
Yangchow see Yangzhou
Yangiyo'l 123 E2 Rus. Yangiyul'. Toshkent Viloyati, E Uzbekistan
Yangiyul' see Yangiyo'l
Yangku see Taiyuan
Yangon 136 B4 Eng. Rangoon. Yangon, S Myan mar (Burma)
Yangtze 128 B4 var. Yangtze Kiang, Eng. Yangtze. river C China
Yangtze see Chang Jiang
Yangtze Kiang see Chang Jiang
Yangzhou 128 D5 var. Yangchow. Jiangsu, E China
Yankton 45 E3 South Dakota, N USA
Yannina see Ioánnina
Yanskiy Zaliv 113 F2 bay N Russian Federation
Yantai 128 D4 var. Yan-t'ai; prev. Chefoo, Chih-fu. Shandong, E China
Yaoundé 77 B5 var. Yaunde. country capital (Cameroon) Centre, S Cameroon
Yap 144 A1 island Caroline Islands, W Micronesia
Yapanskoye More East Sea/ Japan, Sea of
Yapen, Pulau 139 G4 prev. Japen. island E Indonesia
Yap Trench 142 B2 var. Yap Trough. undersea feature SE Philippine Sea
Yap Trough see Yap Trench
Yapurá, Rio see Caquetá, Río, Brazil/ Colombia
Yapurá see Japurá, Rio, Brazil/ Colombia
Yaqui, Río 50 C2 river NW Mexico
Yaransk 111 C5 Kirovskaya Oblast', NW Russian Federation
Yarega 110 D4 Respublika Komi, NW Russian Federation
Yarkant see Shache
Yarlung Zangbo Jiang see Brahmaputra
Yarmouth 39 F5 Nova Scotia, SE Canada
Yarmouth see Great Yarmouth
Yaroslav see Jarosław
Yaroslavl' 110 B4 Yaroslavskaya Oblast', W Russian Federation
Yarumal 58 B2 Antioquia, NW Colombia
Yasyel'da 107 B7 river Brestskaya Voblasts', SW Belarus Europe
Yatsushiro 131 A7 var. Yatusiro. Kumamoto, Kyūshū, SW Japan
Yatusiro see Yatsushiro
Yaunde see Yaoundé
Yavari see Javari, Rio
Rio Yavari 62 C2 var. Yavarí. river Brazil/Peru
Yaviza 53 H5 Darién, SE Panama
Yavoriv 108 B2 Pol. Jaworów, Rus. Yavorov. L'vivs'ka Oblast', NW Ukraine
Yavorov see Yavoriv
Yazd 120 D3 var. Yezd. Yazd, C Iran

Yazoo City 42 B2 Mississippi, S USA
Yding Skovhoj 85 A7 hill C Denmark
Ýdra 105 C6 var. Ídhra. island Ýdra, S Greece
Ye 137 B5 Mon State, S Myanmar (Burma)
Yecheng 126 A3 var. Kargilik. Xinjiang Uygur Zizhiqu, NW China
Yefremov 111 B5 Tul'skaya Oblast', W Russian Federation
Yekaterinburg 114 C3 prev. Sverdlovsk. Sverdlovskaya Oblast', C Russian Federation
Yekaterinodar see Krasnodar
Yekaterinoslav see Dnipropetrovs'k
Yelets 111 B5 Lipetskaya Oblast', W Russian Federation
Yelisavetpol see Gäncä
Yelizavetgrad see Kirovohrad
Yelizovo see Yalizava
Yell 88 D1 island NE Scotland, United Kingdom
Yellowhammer State see Alabama
Yellowknife 37 E4 territory capital Northwest Territories, W Canada
Yellow River 128 C4 var. Yellow River. river C China
Yellow River see Huang He
Yellow Sea 128 D4 Chin. Huang Hai, Kor. Hwang-Hae. sea E Asia
Yellowstone River 44 C2 river Montana Wyoming, NW USA
Yel'sk 107 C7 Homyel'skaya Voblasts', SE Belarus
Yelwa 75 F4 Kebbi, W Nigeria
Yemen 121 C7 off. Republic of Yemen, Ar. Al Jumhūrīyah al Yamanīyah, al Yaman. country SW Asia

Yemen, Republic of see Yemen
Yemva 110 D4 prev. Zheleznodorozhnyy. Respublika Komi, NW Russian Federation
Yenakiyeve 109 G3 Rus. Yenakiyevo prev. Ordzhonikidze, Rykovo. Donets'ka Oblast', E Ukraine
Yenakiyevo see Yenakiyeve
Yenangyaung 136 A3 Magway, W Myanmar (Burma)
Yendi 75 E4 NE Ghana
Yengisar 126 A3 Xinjiang Uygur Zizhiqu, NW China
Yenierenköy 102 D4 var. Yialousa, Gk. Agialoúsa. NE Cyprus
Yenipazar see Novi Pazar
Yenisey 114 D3 river Mongolia/Russia Federation
Yenping see Nanping
Yeovil 89 D7 SW England, United Kingdom
Yeppoon 148 D4 Queensland, E Australia
Yerevan 117 F3 Eng. Erivan. country capital (Armenia) C Armenia
Yeriho see Jericho
Yertis 114 C4 var. Irtish, Kaz. Ertis; prev. Irtysh. river C Asia
Yerushalayim see Jerusalem
Yesil' 114 C4 Kaz. Esil. river Kazakhstan/Russian Federation
Yeso see Hokkaidō
Yeu, Île d' 90 A4 island NW France
Yevlakh see Yevlax
Yevlax 117 G2 Rus. Yevlakh. C Azerbaijan

patoriya 109 F5 Avtonomna
espublika Krym, S Ukraine
a 109 H4 river SW Russian
deration
rishche see Yezyaryshcha
a see Hokkaidō
aryshcha 107 E5 Rus. Yezershche.
E Belarus
tsyebskaya Voblasts',
ousa see Yenierenköy
nitsá see Giannitsá
ang 128 C5 Hubei, C China
zeli 116 D3 Sivas, N Turkey
huan 128 B4 var. Yinch'uan,
n-ch'uan, Yinchwan.
rovince capital Ningxia,
China
hwan see Yinchuan
tu He see Indus
hsien see Ningbo
ng 126 B2 var. I-ning, Uigh. Gulja,
uldja. Xinjiang Uygur Zizhiqu,
W China
tu Ho see Indus
ion see Gýtheio
yakarta 138 C5 prev. Djokjakarta,
gjakarta, Jokyakarta. Jawa,
Indonesia
ohama 131 D5 Aomori, Honshū,
Japan
ohama 130 A2 Kanagawa, Honshū,
Japan
ote 130 D4 Akita, Honshū,
Japan
a 75 H4 Adamawa, E Nigeria
ago 131 B6 Tottori, Honshū,
N Japan
g'an 128 D6 var. Yongan. Fujian,
S China
gzhou see Lingling
kers 41 F3 New York,
E USA
ne 90 C4 river C France
al 58 C3 var. El Yopal. Casanare,
Colombia
k 89 D5 anc. Eboracum,
ouracum. N England,
nited Kingdom
k 45 E4 Nebraska, C USA
k, Cape 148 C1 headland
ueensland, NE Australia
k, Kap see Innaanganeq
kton 37 F5 Saskatchewan,
Canada
o 52 C2 Yoro, C Honduras
hkar-Ola 111 C5 Respublika Mariy
I, W Russian Federation
onbulag see Altay
angstown 40 D4 Ohio, N USA
th, Isle of 54 A2 var. Isla de Pinos,
g. Isle of Youth; prev. The Isle of the
ines. island W Cuba
th, Isle of see Juventud, Isla de la
es see Ieper
a 46 B4 California, W USA
ndagüé see
eneral Eugenio A. Garay
el see IJssel
k-Köl see Issyk-Kul', Ozero
k-Köl see Balykchy
n see Henan
n see Red River
n Jiang see Red River
a City 47 B5 California, W USA
atán, Canal de see
ucatan Channel
atan Channel 51 H3 Sp. Canal de
ucatán. channel Cuba/Mexico
atan Peninsula 35 C7 Eng. Yucatan
eninsula, NE Australia
atan Peninsula see Yucatán,
enínsula de
i see Jinzhong
see Guangdong
Shan, Tai see Lantau Island
yang 128 C5 Hunan, S China
goslavia see Serbia
havichy 107 D5 Rus. Yukhovichi.
itsyebskaya Voblasts', N Belarus
khovichi see Yukhavichy
on 36 D3 prev. Yukon Territory,
r. Territoire du Yukon. territory
NW Canada
kon River 36 C2 river
Canada/USA

Yukon, Territoire du see Yukon
Yukon Territory see Yukon
Yulin 128 C6 Guangxi Zhuangzu
Zizhiqu, S China
Yuma 48 A2 Arizona, SW USA
Yumendong 128 A3 prev. Laojunmiao.
Gansu, N China
Yun see Yunnan
Yungki see Jilin
Yung-ning see Nanning
Yunjinghong see Jinghong
Yunki see Jilin
Yunnan 128 A6 var. Yun, Yunnan
Sheng, Yünnan, Yun-nan. province
SW China
Yunnan see Kunming
Yunnan Sheng see Yunnan
Yünnan/Yun-nan see Yunnan
Yurev see Tartu
Yurihonjō see Honjō
Yuruá, Río see Juruá, Rio
Yury'ev see Tartu
Yushu 128 D4 var. Gyêgu. Qinghai,
C China
Yuty 64 D3 Caazapá, S Paraguay
Yuzhno-Sakhalinsk 115 H4 Jap.
Toyohara; prev. Vladimirovka. Ostrov
Sakhalin, Sakhalinskaya Oblast',
SE Russian Federation
Yuzhnyy Bug see Pivdennyy Buh
Yuzhou see Chongqing
Ýylanly see Gurbansoltan Eje

Z

Zaandam see Zaanstad
Zaanstad 86 C3 prev. Zaandam. Noord-
Holland, C Netherlands
Zabaykal'sk 115 F5 Zabaykal'skiy Kray,
S Russian Federation
Zabern see Saverne
Zabid 121 B7 W Yemen
Žabinka see Zhabinka
Ząbkowice see Ząbkowice Śląskie
Ząbkowice Śląskie 98 B4 var.
Ząbkowice, Ger. Frankenstein,
Frankenstein in Schlesien.
Dolnośląskie, SW Poland
Zábřeh 99 C5 Ger. Hohenstadt.
Olomoucký Kraj, E Czech Republic
Zacapa 52 B2 Zacapa, E Guatemala
Zacatecas 50 D3 Zacatecas, C Mexico
Zacatepec 51 E4 Morelos, S Mexico
Zacháro 105 B6 var. Zaharo, Zakháro.
Dytikí Elláda, S Greece
Zadar 100 A3 It. Zara; anc. Iader.
Zadar, SW Croatia
Zadetkyi Kyun 137 B6 var. St.Matthew's
Island. island Myeik Archipelago,
S Myanmar (Burma)
Zafra 92 C4 Extremadura, W Spain
Żagań 98 B4 var. Żagań, Żegań, Ger.
Sagan. Lubuskie, W Poland
Zagazig see Az Zaqāziq
Zágráb see Zagreb
Zagreb 100 B2 Ger. Agram, Hung.
Zágráb. country capital (Croatia)
Zagreb, N Croatia
Zagros Mountains 120 C3 Eng. Zagros
Mountains. mountain range W Iran
Zagros Mountains see Zāgros, Kūhhā-ye
Zaharo see Zacháro
Zähedän 120 E4 var. Zahidan; prev.
Duzdab. Sīstān va Balūchestān,
SE Iran
Zahidan see Zähedän
Zaḥlah see Zaḥlé
Zahlé 118 B4 var. Zaḥlah. C Lebanon
Záhony 99 E6 Szabolcs-Szatmár-Bereg,
NE Hungary
Zaire see Congo (river)
Zaire see Congo (Democratic Republic of)
Zaječar 101 E4 Serbia, E Serbia
Zakataly see Zaqatala
Zakháro see Zacháro
Zakhidnyy Buh/Zakhodni Buh see Bug
Zākhō see Zaxo
Zākhū see Zaxo
Zakopane 99 D5 Małopolskie, S Poland
Zákynthos 105 A6 var. Zákinthos, It.
Zante. island Iónia Nísoí, Greece,
C Mediterranean Sea
Zalaegerszeg 99 B7 Zala, W Hungary
Zalău 108 B3 Ger. Waltenberg, Hung.
Zilah; prev. Ger. Sillenmarkt. Sălaj,
NW Romania

Žalim 121 B5 Makkah, W Saudi Arabia
Zambesi/Zambeze see Zambezi
Zambezi 78 C2 North Western,
W Zambia
Zambezi 78 D2 var. Zambesi, Port.
Zambeze. river S Africa
Zambia 78 C2 off. Republic of Zambia;
prev. Northern Rhodesia. country
S Africa

ZAMBIA
Southern Africa

Official name Republic of Zambia
Formation 1964 / 1964
Capital Lusaka
Population 14.5 million / 51 people
per sq mile (20 people per sq km)
Total area 290,584 sq. miles
(752,614 sq. km)
Languages Bemba, Tonga, Nyanja, Lozi,
Lala-Bisa, Nsenga, English*
Religions Christian 63%, Traditionalbeliefs
36%, Muslim and Hindu 1%
Ethnic mix Bemba 34%, Other
African 26%, Tonga 16%, Nyanja 14%,
Lozi 9%, European 1%
Government Presidential system
Currency New Zambian kwacha
= 100 ngwee
Literacy rate 61%
Calorie consumption 1937 kilocalories

Zambia, Republic of see Zambia
Zamboanga 139 E3 off. Zamboanga City.
Mindanao, S Philippines
Zamboanga City see Zamboanga
Zambrów 98 E3 Łomża, E Poland
Zamora 92 D2 Castilla y León,
NW Spain
Zamora de Hidalgo 50 D4 Michoacán,
SW Mexico
Zamość 98 E4 Rus. Zamoste. Lubelskie,
E Poland
Zamoste see Zamość
Zancle see Messina
Zanda 126 A4 Xizang Zizhiqu, W China
Zanesville 40 D4 Ohio, N USA
Zanjān 120 C2 var. Zenjan, Zinjan.
Zanjān, NW Iran
Zante see Zákynthos
Zanthus 147 C6 Western Australia,
S Australia Oceania
Zanzibar 73 D7 Zanzibar, E Tanzania
Zanzibar 73 C7 Swa. Unguja. island
E Tanzania
Zaozhuang 128 D4 Shandong,
E China
Zapadna Morava 100 D4 Ger. Westliche
Morava. river C Serbia
Zapadnaya Dvina 110 A4 Tverskaya
Oblast', W Russian Federation
Zapadnaya Dvina see Western Dvina
Zapadnyy Bug see Bug
Zapala 65 B5 Neuquén, W Argentina
Zapiola Ridge 67 B6 undersea feature
SW Atlantic Ocean
Zapolyarnyy 110 C2 Murmanskaya
Oblast', NW Russian Federation
Zaporizhzhya 109 F3 Rus. Zaporozh'ye;
prev. Aleksandrovsk. Zaporiz'ka
Oblast', SE Ukraine
Zaporozh'ye see Zaporizhzhya
Zapotiltic 50 D4 Jalisco, SW Mexico
Zaqatala 117 G2 Rus. Zakataly.
NW Azerbaijan
Zara 116 D3 Sivas, C Turkey
Zara see Zadar
Zarafshan see Zarafshon
Zarafshon 122 D2 Rus. Zarafshan.
Navoiy Viloyati, N Uzbekistan
Zaragoza 93 F2 Eng. Saragossa; anc.
Caesaraugusta, Salduba. Aragón,
NE Spain
Zarand 120 D3 Kermān, C Iran
Zaranj 122 D5 Nīmrōz, SW Afghanistan
Zarasai 106 C4 Utena, E Lithuania
Zárate 64 D4 prev. General José
F.Uriburu. Buenos Aires, E Argentina
Zarautz 93 E1 var. Zarauz. País Vasco,
N Spain
Zarauz see Zarautz
Zaraza 59 E2 Guárico, N Venezuela
Zarghūn Shahr 123 E4 var. Katawaz.
Paktīkā, SE Afghanistan
Zaria 75 G4 Kaduna, C Nigeria

Zarós 105 D8 Kríti, Greece,
E Mediterranean Sea
Zarqa see Az Zarqā'
Žary 98 B4 Ger. Sorau, Sorau in der
Niederlausitz. Lubuskie, W Poland
Zaunguzskiye Garagumy see Üngüz
Angyrsyndaky Garagum
Zavertse see Zawiercie
Zavet 104 D1 Razgrad, NE Bulgaria
Zavidovići 100 C3 Federacija Bosne
I Hercegovine, N Bosnia and
Herzegovina
Zawia see Az Zāwiyah
Zawiercie 98 D4 Rus. Zavertse. Śląskie,
S Poland
Zawilah 71 F3 var. Zuwaylah, It. Zueila.
C Libya
Zaxo 120 B2 var. Zākhū, var. Zākhō.
Dahūk, N Iraq
Zaysan Köl see Zaysan, Ozero
Zaysan, Ozero 114 D5 Kaz. Zaysan Köl.
lake E Kazakhstan
Zayyq see Zhayyk
Zbarazh 108 C2 Ternopil's'ka Oblast',
W Ukraine
Zduńska Wola 98 C4 Sieradz, C Poland
Zealand 85 B8 Eng. Zealand, Ger.
Seeland. island E Denmark
Zealand see Sjælland
Zeebrugge 87 A5 West-Vlaanderen,
NW Belgium
Zeewolde 86 D3 Flevoland,
C Netherlands
Zefat see Tsefat
Zegań see Żagań
Zeiden see Codlea
Zeist 86 C4 Utrecht, C Netherlands
Zele 87 B5 Oost-Vlaanderen,
NW Belgium
Zelenoborskiy 110 B2 Murmanskaya
Oblast', NW Russian Federation
Zelenograd 111 B5 Moskovskaya
Oblast', W Russian Federation
Zelenogradsk 106 A4 Ger. Cranz,
Kranz. Kaliningradskaya Oblast',
W Russian Federation
Zelle see Celle
Zel'va 107 B6 Pol. Zelwa. Hrodzyenskaya
Voblasts', W Belarus
Zelwa see Zel'va
Zelzate 87 B5 var. Selzaete. Oost-
Vlaanderen, NW Belgium
Žemaičių Aukštumas 106 B4 physical
region W Lithuania
Zemst 87 C5 Vlaams Brabant,
C Belgium
Zemun 100 D3 Serbia, N Serbia
Zengg see Senj
Zenica 100 C4 Federacija Bosne I
Hercegovine, C Bosnia and
Herzegovina
Zenta see Senta
Zeravshan 123 E3 Taj./Uzb. Zarafshon.
river Tajikistan/Uzbekistan
Zevenaar 86 D4 Gelderland,
SE Netherlands
Zevenbergen 86 C4 Noord-Brabant,
S Netherlands
Zeya 113 E3 river SE Russian Federation
Zgerzh see Zgierz
Zgierz 98 C4 Ger. Neuhof, Rus. Zgerzh.
Łódź, C Poland
Zgorzelec 98 B4 Ger. Görlitz.
Dolnośląskie, SW Poland
Zhabinka 107 A6 Pol. Żabinka.
Brestskaya Voblasts', SW Belarus
Zhambyl see Taraz
Zhanaozen 114 A4 Kaz. Zhangaözen;
prev. Novyy Uzen'. Mangistau,
W Kazakhstan
Zhangaözen see Zhanaozen
Zhangaqazaly see Ayteke Bi
Zhang-chia-k'ou see Zhangjiakou
Zhangdian see Zibo
Zhangjiakou 128 C3 var. Changkiakow,
Zhang-chia-k'ou, Eng. Kalgan; prev.
Wanchuan. Hebei, E China
Zhangzhou 128 D6 Fujian, SE China
Zhanjiang 128 C7 var. Chanchiang,
Chan-chiang, Cant. Tsamkong, Fr.
Fort-Bayard. Guangdong, S China
Zhaoqing 128 C6 Guangdong, S China
Zhayyk see Ural
Zhdanov see Mariupol'
Zhe see Zhejiang
Zhejiang 128 D5 var. Che-chiang,
Chekiang, Zhe, Zhejiang Sheng.
province SE China
Zhejiang Sheng see Zhejiang

Zheleznodorozhnyy *106 A4*
Kaliningradskaya Oblast', W Russian
Federation
Zheleznodorozhnyy *see* Yemva
Zheleznogorsk *111 A5* Kurskaya Oblast',
W Russian Federation
Zhëltyye Vody *see* Zhovti Vody
Zhengzhou *128 C4* var. Ch'eng-chou,
Chengchow; *prev.* Chenghsien.
province capital Henan, C China
Zhezkazgan *114 C4 Kaz.* Zhezqazghan;
prev. Dzhezkazgan. Karagandy,
C Kazakhstan
Zhezqazghan *see* Zhezkazgan
Zhidachov *see* Zhydachiv
Zhishan *see* Lingling
Zhitkovichi *see* Zhytkavichy
Zhitomir *see* Zhytomyr
Zhlobin *107 D7* Homyel'skaya Voblasts',
SE Belarus
Zhmerinka *see* Zhmerynka
Zhmerynka *108 D2 Rus.* Zhmerinka.
Vinnyts'ka Oblast', C Ukraine
Zhodino *see* Zhodzina
Zhodzina *107 D6 Rus.* Zhodino.
Minskaya Voblasts', C Belarus
Zholkev/Zholkva *see* Zhovkva
Zhonghua Renmin Gongheguo *see*
China
Zhosaly *114 B4 prev.* Dzhusaly.
Kzylorda, SW Kazakhstan
Zhovkva *108 B2 Pol.* Żółkiew, *Rus.*
Zholkev, Zholkva; *prev.* Nesterov.
L'vivs'ka Oblast', NW Ukraine
Zhovti Vody *109 F3 Rus.* Zhëltyye Vody.
Dnipropetrovs'ka Oblast', E Ukraine
Zhovtneve *109 E4 Rus.* Zhovtnevoye.
Mykolayivs'ka Oblast', S Ukraine
Zhovtnevoye *see* Zhovtneve
Zhi Qu *see* Tongtian He
Zhydachiv *108 B2 Pol.* Żydaczów,
Rus. Zhidachov. L'vivs'ka Oblast',
NW Ukraine
Zhytkavichy *107 C7 Rus.* Zhitkovichi.
Homyel'skaya Voblasts',
SE Belarus
Zhytomyr *108 D2 Rus.* Zhitomir.
Zhytomyrs'ka Oblast', NW Ukraine
Zibo *128 D4 var.* Zhangdian. Shandong,
E China
Zichenau *see* Ciechanów

Zielona Góra *98 B4 Ger.* Grünberg,
Grünberg in Schlesien, Grüneberg.
Lubuskie, W Poland
Zierikzee *86 B4* Zeeland,
SW Netherlands
Zigong *128 B5 var.* Tzekung. Sichuan,
C China
Ziguinchor *74 B3* SW Senegal
Zilah *see* Zalău
Žilina *99 C5 Ger.* Sillein, *Hung.* Zsolna.
Žilinský Kraj, N Slovakia
Zillenmarkt *see* Zalău
Zimbabwe *78 D3 off.* Republic of
Zimbabwe; *prev.* Rhodesia. *country*
S Africa

ZIMBABWE
Southern Africa

Official name Republic of Zimbabwe
Formation 1980 / 1980
Capital Harare
Population 14.1 million / 94 people
per sq mile (36 people per sq km)
Total area 150,803 sq. miles
(390,580 sq. km)
Languages Shona, isiNdebele, English*
Religions Syncretic 50%, Christian 25%,
Traditional beliefs 24%,
Other (including Muslim) 1%
Ethnic mix Shona 71%, Ndebele 16%,
Other African 11%, White 1%, Asian 1%
Government Presidential system
Currency Zimbabwe dollar suspended in
2009; nine other currencies are
legal tender
Literacy rate 84%
Calorie consumption 2210 kilocalories

Zimbabwe, Republic of *see* Zimbabwe
Zimnicea *108 C5* Teleorman, S Romania
Zimovniki *111 B7* Rostovskaya Oblast',
SW Russian Federation
Zinder *75 G3* Zinder, S Niger
Zinov'yevsk *see* Kirovohrad
Zintenhof *see* Sindi
Zipaquirá *58 B3* Cundinamarca,
C Colombia
Zittau *94 D4* Sachsen, E Germany
Zlatni Pyasŭtsi *104 E2* Dobrich,
NE Bulgaria

Zlín *99 C5 prev.* Gottwaldov. Zlínský
Kraj, E Czech Republic
Złoczów *see* Zolochiv
Złotów *98 C3* Wielkopolskie, C Poland
Znamenka *see* Znam"yanka
Znam"yanka *109 F3 Rus.* Znamenka.
Kirovohrads'ka Oblast', C Ukraine
Żnin *98 C3* Kujawsko-pomorskie,
C Poland
Zoetermeer *86 C4* Zuid-Holland,
W Netherlands
Zólkiew *see* Zhovkva
Zolochev *see* Zolochiv
Zolochiv *109 G2 Rus.* Zolochev.
Kharkivs'ka Oblast', E Ukraine
Zolochiv *108 C2 Pol.* Złoczów, *var.*
Zolochev. L'vivs'ka Oblast',
W Ukraine
Zolote *109 H3 Rus.* Zolotoye. Luhans'ka
Oblast', E Ukraine
Zolotonosha *109 E2* Cherkas'ka Oblast',
C Ukraine
Zolotoye *see* Zolote
Zólyom *see* Zvolen
Zomba *79 E2* Southern, S Malawi
Zombor *see* Sombor
Zongo *77 C5* Equateur, N Dem. Rep.
Congo
Zonguldak *116 C2* Zonguldak,
NW Turkey
Zonhoven *87 D6* Limburg, NE Belgium
Zoppot *see* Sopot
Żory *99 C5 var.* Zory, *Ger.* Sohrau.
Śląskie, S Poland
Zouar *76 C2* Borkou-Ennedi-Tibesti,
N Chad
Zouérat *74 C2 var.* Zouérate, Zouïrât.
Tiris Zemmour, N Mauritania
Zouérate *see* Zouérat
Zouïrât *see* Zouérat
Zrenjanin *100 D3 prev.* Petrovgrad,
Veliki Bečkerek, *Ger.* Grossbetschkerek,
Hung. Nagybecskerek. Vojvodina,
N Serbia
Zsil/Zsily *see* Jiu
Zsolna *see* Žilina
Zsombolya *see* Jimbolia
Zsupanya *see* Županja
Zubov Seamount *67 D5 undersea
feature* E Atlantic Ocean
Zueila *see* Zawīlah

Zug *95 B7 Fr.* Zoug. Zug, C Switzerla[nd]
Zugspitze *95 C7 mountain* S Germa[ny]
Zuid-Beveland *87 B5 var.* South
Beveland. *island* SW Netherlands
Zuider Zee *see* IJsselmeer
Zuidhorn *86 E1* Groningen,
NE Netherlands
Zuidlaren *86 E2* Drenthe,
NE Netherlands
Zula *72 C4* E Eritrea
Züllichau *see* Sulechów
Zumbo *see* Vila do Zumbo
Zundert *87 C5* Noord-Brabant,
S Netherlands
Zunyi *128 B5* Guizhou, S China
Županja *100 C3 Hung.* Zsupanya.
Vukovar-Srijem, E Croatia
Zürich *95 B7 Eng./Fr.* Zurich, *It.* Zu[rigo]
Zürich, N Switzerland
Zurich, Lake *95 B7 Eng.* Lake Zurich
lake NE Switzerland
Zurich, Lake *see* Zürichsee
Zurigo *see* Zürich
Zutphen *86 D3* Gelderland,
E Netherlands
Zuwārah *71 F2* NW Libya
Zuwaylah *see* Zawilah
Zuyevka *111 D5* Kirovskaya Oblast',
NW Russian Federation
Zvenigorodka *see* Zvenyhorodka
Zvenyhorodka *109 E2 Rus.*
Zvenigorodka. Cherkas'ka Oblast',
C Ukraine
Zvishavane *78 D3 prev.* Shabani.
Matabeleland South, S Zimbabwe
Zvolen *99 C6 Ger.* Altsohl, *Hung.*
Zólyom. Banskobystrický Kraj,
C Slovakia
Zvornik *100 C4* E Bosnia and
Herzegovina
Zwedru *74 D5 var.* Tchien. E Liberia
Zwettl *95 E6* Wien, NE Austria
Zwevegem *87 A6* West-Vlaanderen,
W Belgium
Zwickau *95 C5* Sachsen, E Germany
Zwolle *86 D3* Overijssel, E Netherla[nds]
Żydaczów *see* Zhydachiv
Zyőetu *see* Jōetsu
Żyrardów *98 D3* Mazowieckie, C Pol[and]
Zyryanovsk *114 D3* Vostochnyy
Kazakhstan, E Kazakhstan

Key to map pages

North & West Asia 112-113

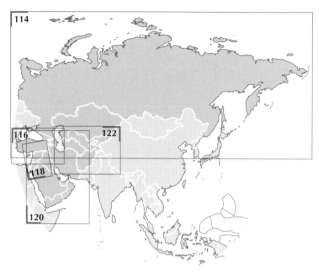

114

116

122

118

120

South & East Asia 124-125

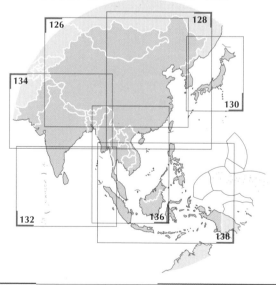

126

128

134

130

132

136

138